# ENVIRONMENTAL SCIENCE

SECOND EDITION

# ENVIRONMENTAL SCIENCE
## The Way the World Works

**Bernard J. Nebel**  *Department of Biology, Catonsville Community College*

*with editorial assistance by* **Edward J. Kormondy**, *University of Hawaii*

*with writing assistance by* **Nancy K. Minkott**, *free-lance journalist, Bethesda, Md.*

Prentice-Hall, Inc., Englewood Cliffs, New Jersey 07632

**Library of Congress Cataloging-in-Publication Data**

Nebel, Bernard J.
  Environmental science.

  Bibliography: p.
  Includes index.
  1. Ecology.  2. Human ecology.  3. Pollution—Environmental aspects.  I. Title.
QH541.N39  1987      574.5      86-91496
ISBN 0-13-283037-X

Dedicated to my daughter, Tamra and my son, Christopher.

Editorial/production supervision: Fay Ahuja
Cover design and interior design: Lee Cohen
Photo research: Tobi Zausner
Photo editor: Lorinda Morris
Manufacturing buyer: Barbara Kittle
Page layout: Peggy Finnerty and Steven Frim

**Environmental Science**, Second Edition
Bernard J. Nebel

Cover Photo: Four By Five

Part Opener Photos
PART I: NASA
    II: Erika Stone/Peter Arnold, Inc.
   III: Bjorn Bolstad/Peter Arnold, Inc.
   IV: Judy Aronson/Peter Arnold, Inc.
    V: Hans Pfletschinger/Peter Arnold, Inc.
   VI: Malcolm S. Kirk/Peter Arnold, Inc.

Printed in the United States of America

10  9  8  7  6  5  4  3

ISBN 0-13-283037-X  01

Prentice-Hall International (UK) Limited, *London*
Prentice-Hall of Australia Pty. *Sydney*
Prentice-Hall Canada Inc., *Toronto*
Prentice-Hall Hispanoamericana, S.A., *Mexico*
Prentice-Hall of India Private *New Delhi*
Prentice-Hall of Japan, Inc., *Tokyo*
Prentice-Hall of Southeast Asia Pty. Ltd., *Singapore*
Editora Prentice-Hall do Brasil, Ltda., *Rio de Janeiro*

# CONTENTS IN BRIEF

# CONTENTS

## 8

## Water, the Water Cycle, and Water Management    216

# PART IV
# POLLUTION    253

## 9

## Water Pollution due to Sediments    257

## 10

## Water Pollution due to Sewage    268

## 11

## Water Pollution due to Eutrophication    292

## 12

## Groundwater Pollution and Toxic Wastes    308

13

## Air Pollution and Its Control  327

14

## Acid Precipitation and Factors Affecting Climate  361

15

## Risks and Economics of Pollution  392

## PART V
## PESTS AND PEST CONTROL  399

16

## The Pesticide Treadmill  402

## 22

### Nuclear Power, Coal, and Synthetic Fuels — 534

## 23

### Solar Energy, Other "Renewable" Energy Sources, and Conservation — 564

## 24

### Land Use — 600

# PREFACE

The theme and objective of the second edition of *Environmental Science: The Way the World Works* is to: (1) develop an understanding of the principles underlying the support and maintenance of all ecosystems; (2) show how our management (or lack of management) of the human ecosystem, in many cases, is counter to these principles, and how environmental problems, if not tragedies, occur as a result; (3) show how solutions to environmental problems lie in the direction of recognizing ecological principles and managing our human systems accordingly.

While the first edition of *Environmental Science: The Way the World Works* was very successful, the second edition has been entirely rewritten and to some extent reorganized, as well as updated, to achieve the objectives even more fully. The systematically organized, easy-to-read style suitable for the typical nonscience, freshman student has been maintained. Additionally, the "bulky" chapters of the first edition have been divided into two or more separate chapters. Then related chapters are grouped into six parts, each with an introductory overview. This has been done to increase the focus on individual topics and issues, and to enable more flexibility in teaching. An introductory chapter explaining how scientific facts, principles, and theories are derived, and their value and limitations in solving problems has been added. Finally, *Environmental Science: The Way the World Works* has been given an attractive, four-color design with many new photographs and line drawings.

Special effort has been made to present each topic in the context of our developing understanding of the issue. Our recognition of the problem, actions that have been taken, and their relative success or failure is given as a basis for discussion of present and future courses of action. The importance of political activism in attaining desired ends is stressed.

## A SYNOPSIS

**Part I** *(Chapters 1 to 4)* develops the concepts of what ecosystems are, how they function, how they maintain (or lose) stability, and how they evolve. Greater than usual attention is given to providing a thorough understanding of nutrient recycling and energy flow (Chapter 2), but this is done without resorting to the "heavy" chemistry which caused problems for some students using the first edition. However, a presentation of the basic chemistry is still included in an appendix. Likewise, evolution is given a more thorough treatment (Chapter 4). In each of these chapters there is a concluding section, "Implications for Humans," which sets the stage for the more thorough discussion of environmental issues which comes in the following Parts.

**Part II** *(Chapters 5 to 6)* addresses the issues of human populations and economic development, stressing the importance of quality as opposed to quantity of life. The fact that sound ecological management is increasingly necessary to maintain (much more to improve) the standard of living of growing populations is stressed.

**Part III** *(Chapters 7 to 8)* describes the all-important resources of soil and water, and shows how the natural processes which maintain these resources are being upset or undercut by various human activities. It emphasizes how prudent water and soil management is central to sustaining all natural ecosystems as well as agricultural systems.

**Part IV** (*Chapters 9 to 15*) provides a comprehensive understanding of sources, effects, and control procedures for the major kinds of pollutants affecting water and air. Sediments, sewage, excess nutrients (eutrophication), toxic chemicals, important air pollutants including those of indoor air, and acid precipitation are each given full chapter treatments. Then, Chapter 15 provides a discussion regarding weighing the relative costs of pollution control against the risks of not controlling pollution.

**Part V** (*Chapters 16 to 17*) addresses the self-defeating nature and pollution hazards of traditional chemical pest control procedures, describes various natural control methods potentially available, and stresses the importance of ecological pest management.

**Part VI** (*Chapters 18 to 24*) addresses mineral, biological, energy, and land resources. Regarding biological resources (Chapter 19), the values of natural biota in providing a genetic bank essential for maintaining agriculture, and in performing the natural services of maintaining air and water quality, as well as scientific, aesthetic, and commercial values, are stressed as reasons for requiring their conservation and preservation. Three chapters on energy (21 to 23) provide a comprehensive understanding concerning the enigma of oil supplies, of nuclear power and its pros and cons, and of various solar and other renewable alternatives and conservation. How we choose to use land and associated lifestyles is shown (Chapter 24) to be a basic factor affecting all other resources including air and water.

## PEDAGOGICAL AIDS

*Organization.* Each chapter of the second edition of *Environmental Science: The Way the World Works* is rigorously organized and outlined according to a hierarchy of learning objectives, which facilitates students' gaining a clear, comprehensive understanding of each topic.

*Concept Frameworks.* Each chapter is preceded by a *Concept Framework* which consists of the outline of the chapter and parallel study questions page-keyed to the text. The concept frameworks may be used as: (1) a convenient preview of the chapter; (2) a learning guide; (3) a reference to quickly locate particular sections; (4) as a summary and review.

*Vocabulary Words.* New terms appear in bold-face type where they are first introduced and defined, and later in italics to emphasize the term's use in context. Then, all such words are entered in the Glossary.

*Illustrations and Photographs.* Nearly 500 photographs and line drawings, many in four-color, enhance the attractiveness of the text. Moreover, all illustrations have been prepared and photographs selected especially to help convey particular concepts and increase understanding. They are all keyed to specific points in the text.

*Appendixes.* A list of environmental organizations; the metric system and equivalent English units, energy units and equivalents; and basic chemical concepts are included.

## SUPPLEMENTARY MATERIALS

The text is accompanied by a Student Study Guide and an Instructor's Manual with test questions keyed to specific sections of the outline (see Concept Frameworks) for each chapter.

## ACKNOWLEDGEMENTS

In bringing the second edition of this text to fruition my many thanks and much credit must go to:

Nancy Minkoff, my writing assistant who helped convert my rough drafts into a finished manuscript.

Edward Kormondy, for playing an ongoing role as a reviewer and editorial consultant.

Peter S. Dixon, University of California, Irvine; David Pimentel, Cornell University; Fred Racle, Michigan State University; and Ronald Ward, Grand Valley State Colleges, Allendale, Michigan, for laboriously reviewing the entire manuscript and making extensive comments and corrections which I found extremely helpful. Likewise to Richard Andren, Montgomery County Community College, Blue Bell, Pennsylvania, for reviewing a large portion.

The exceptionally dedicated and capable people at Prentice-Hall, particularly Fay Ahuja for her tremendous amount of work as Production Editor; Elaine Luthy for an exceptionally fine job of copy editing; Lee Cohen for designing; and Robert Sickles, Biology Editor, for his constant support and encouragement.

My colleagues at Catonsville Community College, particularly Carol Daihl, Steve Simon, Gary Kaiser, David Jeffrey, Barbara Carr, and David Hargrove, for their interest and for providing support in many ways over the years.

My daughter, Tamra; my son, Christopher; Mary Kintner, Olive Blumenfeld, and Udo Essien for helping out at critical times with typing, proofreading, and other tasks.

All those who helped out with the first edi-

tion, for without the first there would be no second: Jean Nebel, Emily (McNamara) Bookholtz, David Hunley, Mary Beyer, Terri Leonnig, Lynn Carr, Kathy (Yaw) Zegwitz, Joan Truby, Byron Daudelin, Kai A. Nebel, Jack Anderson, Joseph Newcomer, and Bruce Welch.

Countless people who have been most helpful in providing information, reports, and photographs. I apologize for not listing them individually.

All my students for continually providing the incentive and the testing ground for this work.

*Bernard J. Nebel*

# Introduction
# WHAT IS SCIENCE?
# Why Study
# Environmental Science?

## CONCEPT FRAMEWORK

| Outline | | Study Questions |
|---|---|---|

1. What is meant by *objective understanding?* What is its benefit? Relate science and the aim of scientists to objective understanding.

2. State why the scientific method is important and describe steps of the scientific method.

3. All scientific information is based on what? What requirements are demanded of scientific observations? What are accepted as scientific facts?

4. What is a *hypothesis?* How is it tested?

5. What is the purpose of scientific experimentation? What is a *controlled experiment?* Why is it important to have a control? Why is it important to test only one fact at a time?

6. What is a scientific *theory*? Contrast scientific facts with theories. Can theories be tested? How?

7. What is a *natural law*? What is the predictive value of natural laws?

8. What are the three main functions of instruments in science?

9. Contrast *science* and *technology*. Describe how they are mutually supporting.

10. Give two reasons for continuing controversy over scientific views.

11. Can science make value judgments? Can science aid in making value judgments? How? Can scientific theories and principles be ignored? What are the consequences?

12. Define *ecology*. Define *environmental science*. Describe the importance of understanding and applying environmental science. Why does it depend on more than just scientists?

Acid rain, toxic wastes, air pollution, water pollution, endangered species, shortages of resources, nuclear wastes, energy problems, groundwater depletion and contamination, soil erosion, overpopulation—these are some of the many environmental problems we all face. In modern society we have learned to look to scientists for solutions; but many of these problems are perceived as the direct or indirect result of progress in science and technology. Therefore, is science the "hero" or really the "culprit"? Science is commonly seen as a mysterious maze of complex instruments, experiments, and theories, which fascinate some, alienate others, and confuse many. As students about to begin the study of environmental science, you should understand what science really is, what science can do, and, even more importantly, what it cannot do. Through a better understanding of science in general and environmental science in particular, we can indeed solve our environmental problems and build a more secure future.

## WHAT IS SCIENCE?

Solutions to some problems may be found through an empirical or trial-and-error approach. However, an accurate and objective understanding of the problem generally enables more effective problem solving. For example, little headway was made in curing infectious diseases when it was thought that they were caused by evil spirits, sins, God's wrath, or emotional feelings. In contrast, the discovery of bacteria and other microbes and the recognition in the late 1800s of their role in causing disease led to the development of effective defenses and cures now used in modern medicine and public health. Likewise, until recently, we believed that mental illness was "all in the head." Today, the recognition of real chemical imbalances permits more effective treatment of certain illnesses. As we learn more about how cells function, we can expect to find more effective treatments for cancer and other diseases.

In brief, the aim of scientists is to develop an accurate, objective picture of what the world is and how it works in the sense of causes and effects—a picture that is free of the distortions that may result from emotional biases, prejudices, errors, or deception. This picture, so far as it has been developed, is the subject matter of science, generally categorized into such areas as chemistry, physics, and biology.

## The Scientific Method

Just as important as the subject matter, however, is the **scientific method,** the method of acquiring objective information. How can we be sure that any piece of information is factual? The scientific method begins with observations of things and events, and progresses through the formulation of cause-and-effect relationships called **hypotheses.** In turn, hypotheses are tested by making further observations or through controlled experiments. Finally a theory is developed, which also undergoes testing, and gradually an overall objective understanding emerges. We shall look at each of these steps in more detail.

### OBSERVATIONS AND FACTS

All scientific information is based on **observations.** However, observations in science are restricted to impressions gained through one or more of the five basic senses in their normal state: that is, seeing, hearing, touching, smelling, and tasting. Only observations made through these senses are accepted because it is felt that only these senses provide us with objective information. For example, contrast your relative confidence in what you see in real life versus what you see in a dream. Indeed, it is well recognized that even with real-life observations there may be distortions due to illusions, errors, or prejudices. Therefore a further demand is made.

All observations must be subject to confirmation and verification. That is, other investigators must be able to repeat the observation, perhaps using different techniques and tests, and confirm, independently, that the original observer is correct. Observations that do not stand the test of confirmation and verification are dismissed. For example, some claim to have seen spaceships with alien creatures, but this information is not generally accepted as factual because such observations have not been repeatable or confirmable. On the other hand, observations that do stand the test of confirmation and verification become accepted as scientific fact (Fig. 0-1).

All Scientific Information Is Based on Careful, Thorough Observations Using the Basic Five Senses.

Further, Observations Must Be Verifiable by Others before They Are Accepted as Factual.

**FIGURE 0-1**
Scientific observations and facts.

This process of observation leads to a wealth of factual information concerning the physical characteristics of all sorts of things, both living and nonliving. It also gives us a wealth of information concerning how things act: objects heavier than air fall when dropped; water boils and apparently disappears when heated; plants grown on poor soil do not thrive.

### HYPOTHESES AND THEIR TESTING

After making an observation, the almost automatic mental response is to find some explanation or cause for the observed event. When the real cause is unknown, the first explanation may be no more than a tentative guess; it is called a *hypothesis.* Of course, the hypothesis may be correct or incorrect. The key point is that hypotheses are tested to see if they are logically consistent with other observations.

A specific thought process used in testing the consistency of a hypothesis is the *"if . . . then . . ."* phrase. One reasons that, if the hypothesis is true, then such and such should logically follow. Finding the "then" to be true provides support for the correctness of the hypothesis. If a hy-

pothesis is inconsistent with other observations it is discarded as false; hypotheses that remain logically consistent with all other observations are gradually accepted as factual. For example, *if* water becomes vapor when heated, *then* we should be able to observe the recondensation of vapor back to water. This is consistent with observations. On the other hand, a hypothesis that water actually vanishes into nothing on boiling is inconsistent with our general observation that things do not just vanish—they go somewhere.

### CONTROLLED EXPERIMENTS

Some hypotheses may be tested by making further direct observations, but the testing of others may involve performing experiments. However, the results of experiments may be readily misinterpreted and give rise to capricious information unless they are rigorously **controlled.** Effectively this means that all the factors involved in the experiment are controlled in such a way that the factor being tested provides the only difference between an **experimental** or **test group** and a **control group** that is used for comparison.

To illustrate, the observation that plants do not grow well in some soils gives rise to the hypothesis that plants require certain nutrients from soil. Analysis of soils that support good growth reveals that these soils contain nitrogen, phosphorus, potassium, and other chemicals. We can then ask if these elements are required. Specifically, for example, is nitrogen required? The thought process is, *if* nitrogen is required, *then* plants will not grow in soil lacking nitrogen. The controlled experiment is as follows: investigators develop a nutrient medium in which they can control all of the chemicals. Quartz sand irrigated with a water solution containing the respective amounts of various soluble compounds found in good soil is such a medium. Now, the growth of plants in the complete medium, the *control group,* can be compared with the growth of plants in a medium that is the same in every respect except for the lack of a nitrogen compound. This group of plants is the *experimental* or *test group.* Importantly, all other conditions such as light, temperature, and water must also be the same for the two groups.

A key feature of the controlled experiment is that it must enable a direct comparison between the experimental group, which receives the test treatment, and the control group, in which all fac-

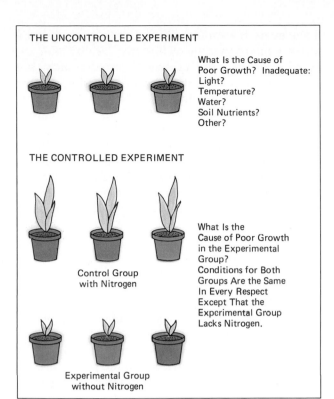

THE UNCONTROLLED EXPERIMENT

What Is the Cause of
Poor Growth? Inadequate:
Light?
Temperature?
Water?
Soil Nutrients?
Other?

THE CONTROLLED EXPERIMENT

Control Group
with Nitrogen

What Is the
Cause of Poor Growth
in the Experimental
Group?
Conditions for Both
Groups Are the Same
In Every Respect
Except That the
Experimental Group
Lacks Nitrogen.

Experimental Group
without Nitrogen

**FIGURE 0-2**
The uncontrolled experiment cannot be interpreted because
any number of factors—singly or in combination—may be
responsible for the lack of growth observed. The controlled
experiment demonstrates that lack of nitrogen is responsible
for the lack of growth, because this is the only factor that
differs between the two groups.

tors are exactly the same except for the one item
being tested. Finding that plants in the minus-ni-
trogen medium fail to grow while those in the
complete medium grow well allows the experi-
menter to conclude that nitrogen is required since
it is the only variable in the experiment (Fig 0-2).

Do you see the importance of the control?
Could the same conclusion be reached without
the presence of the control? No, because potentially a number of other unrecognized factors
could also be responsible for the failure of the
plants to grow. Only the growth of the control
plants provides concrete evidence that such factors are not present in the experiment. Similarly,
do you see the importance of having just one variable? If two or more variables (differences) exist,
the observed effect could be due to any one or
some combination of the variables and the inves-
tigator has no way of distinguishing what specifically caused the final outcome.

Another important consideration is the size
of the group. Inevitably, variation exists among
individuals. Confusing random differences with
effects caused by the treatment is a likely pitfall if
there are only one or two individuals in the experimental and/or control groups. Thus, the
groups must be large enough to permit individual
differences to be distinguished from differences
resulting from treatment.

Finally, the demand for confirmation and
verification applies. Other investigators must be
able to conduct the same experiment and obtain
consistent observations before the conclusion is
accepted as factual.

Indeed there are many "myths," such as the
one that plants do better when talked to or when
soft music is played. These myths apparently derive from casual observations without adequate
controls or replicates (Fig 0-3). At least they have
not stood the test of verification in controlled experiments where sound is the only variable. On
the other hand, a plant's requirement for nitrogen
and various other elements has been confirmed.

**FIGURE 0-3**
Untested hypotheses can lead to "myths."

Importantly, however, in the course of developing and conducting controlled experiments, investigators frequently make additional observations that lead to further questions, hypotheses, and experimental testing. For example, testing of various plant species for their nitrogen requirements led to the finding that certain plants, mainly those belonging to the legume (pea, bean) family, do not require nitrogen compounds in the soil medium. Pursuing this observation has revealed that the roots of legumes have nodules that house bacteria that convert nitrogen gas from the air into nitrogen compounds the plant can utilize, a process known as fixing nitrogen.

## FORMULATION OF THEORIES

Our basic questions concerning the world are generally very broad: What is life? What makes plants grow? Unfortunately, individual observations and experiments can only answer extremely specific questions. However, innumerable specific observations and experiments provide the pieces that fit together into a larger picture, a **theory.** Constructing a theory is a process similar to a detective finding clues (observations) and then fitting them together into a picture (a theory) of "who dunnit." A theory, then, is a conceptual formulation that provides a logical explanation or framework for all the facts (objective information gained through observations and experiments). Importantly, a theory is not a fact itself because it does not lend itself to direct observation. It remains a logical construct that provides a rational explanation or interpretation regarding causes and phenomena underlying certain observations. Nevertheless, theories may be tested and confirmed or rejected.

Theories are tested in much the same way that hypotheses are tested. The formulation of a theory will suggest further aspects that should be observable directly or through experimentation—"*if . . . then. . . .*" When observations are made that are contrary to the theory, the theory is modified to incorporate the new findings in a logical way, or it is rejected in favor of another theory that does provide a rational explanation. In either case, a theory is gradually developed which is consistent with all observations, experiments, and empirical evidence and which can be used to reliably predict outcomes in the sense of "*if . . . then. . . .*" When theories reach this state, we

have every reason to believe that they represent a correct interpretation of objective reality. For example, we have never seen atoms as such, but innumerable observations and experiments are coherently explainable by the idea that all gases, liquids, and solids consist of various combinations of a few more than 100 kinds of atoms. Thus, we have formulated and accepted the atomic theory of matter. Further, we use this theory to predict characteristics of materials we have not yet made and outcomes of experiments not yet performed.

Similarly, we have not observed evolution as such, but the theory of evolution provides a logically consistent framework for innumerable observations and experiments.

## PRINCIPLES AND NATURAL LAWS

In the course of conducting experiments, certain phenomena may stand out as precisely predictable with no known exceptions. We may define such phenomena as **basic principles** or **natural laws.** For example, innumerable tests reveal a uniform attractive force that pulls everything to the surface of the earth and also holds planets and satellites in orbit. We define this phenomenon as the law of gravity. Similarly, laws of thermodynamics (energy) are derived from unvarying experience regarding energy transfers. Another example lies in the observation that in chemical reactions atoms are only rearranged; they are not created, destroyed, or changed. This gives rise to the law of conservation of matter.

Once defined, natural laws have tremendous predictive value. Any attempt to violate a natural law can be predicted to fail with reasonable certainty since there are no known exceptions.

## The Role of Instruments in Science

The use of complex instruments often gives science an aura of mystery. Yet most scientific instruments simply extend and quantify our powers of observation. For example, microscopes and telescopes extend our ability to see smaller and more distant objects. Various radiation detectors enable us to see (or hear) X-rays and radio waves, which are wavelengths shorter and longer, respectively, than those of visible light. Importantly, all instruments used in science are themselves subjected to tests and verification until we are confident that they are giving us a real representation of objective phenomena as opposed to creating their own images or illusions.

Instruments as simple as a meter stick or as complex as a radio telescope also enable us to quantify our observations; that is, they enable us to measure exact quantities. For example, we may feel cold but a thermometer enables us to determine exactly how cold it is. We may feel an electric shock, but an appropriate meter enables us to say exactly what the voltage is. Comparisons and verification of different observations and events would be impossible if it were not for this quantification.

Instruments additionally help us achieve conditions and perform manipulations required for controlled experiments.

## Science and Technology

Basic or pure science is motivated by the desire to know and understand. **Technology** is motivated by the desire to solve a particular problem or achieve a specific result. Technology may muddle forward based on the trial-and-error approach and gradually build on experience. Before there was any scientific understanding of theories or principles, many early, highly successful machines were developed in this way. But many failures also occurred. The trial-and-error approach is exceedingly wasteful in terms of time, money, and occasionally human lives.

Once principles or natural laws are revealed and theories are formulated, they can be efficiently applied to technological development to achieve desired results. For instance, it is conceivable that we could have sent someone to the moon using the trial-and-error process, but at what cost? By strictly adhering to the theories and principles involved, we were able to do it on the first shot.

Thus, there is no question that technological development is aided by theories formulated through basic science. But basic science also benefits from technology in two primary ways. First, technology serves in the development of new instruments that may extend our ability to observe and test hypotheses. Second, technology is the "proving ground" for new theories and principles.

As the saying goes, "if one progresses logically from a theory and reaches an illogical conclusion, something is wrong with the theory." The fact that a technological innovation works demonstrates most of all that our understanding of the theories and principles on which it is based is correct. For example, the fact that we were able to send men to the moon and back underscores that theories concerning planetary motions and gravitational forces are correct. On the other hand, if an attempt does not produce the theoretical result, it indicates that either our understanding is amiss or we are not adhering to the principles or theory. Thus, science and technology are mutually supporting.

In conclusion, the scientific method enables us to gain an understanding of how the world works. In turn, using technology and working within the framework of this understanding, we can achieve desired objectives. Importantly, science does not and cannot allow us magically to achieve objectives. We do not, for example, "wish" people to the moon and back. But, by working within the framework of the theories and natural laws involved, we are able to achieve this result.

## Unresolved Questions and Controversy

Obviously, developing objective, scientific understanding is an ongoing process. Much study and/or experimentation is required before all alternative hypotheses are adequately tested and satisfactory theories are developed. Progress is particularly slow when the questions involve events such as climatic changes or dieoffs of forests, which do not lend themselves readily to controlled experiments. In the meantime, controversy may be rife as alternative hypotheses and theories are proposed and questions remain unresolved. However, the controversy often does not involve scientific understanding so much as it involves value judgments.

## SCIENCE AND VALUE JUDGMENTS

Our endeavors in science can produce an understanding enabling us to achieve various objectives. But science does not and cannot tell us what objectives to strive for. Value judgments remain the responsibility of human individuals and society. For example, science provides us with an understanding of atoms; however, whether or not we use this understanding to build nuclear weapons is a human decision based on values and judgments quite apart from science. Yet, understanding gained through science may aid us in making value decisions by helping us to see the consequences of certain actions. The *"if . . . then . . ."* formula fits in here. For example, *if* we have a full-scale nuclear conflict, *then* virtually all life on earth will be destroyed. But again, it should be emphasized that whether or not the nuclear trigger is pulled will be a human, not a scientific, judgment.

Further, science itself does not demand that we adhere to principles or theories. We can willfully or carelessly ignore certain principles and theories of construction, for instance. Science can only predict that the outcome of such behavior will be the collapse of the building.

This predictive ability of science should not be confused with "fortune telling." The future is not predestined fate that slowly unwinds like a string from a ball. But, with a correct understanding of the pertinent theories and principles, we can predict the results that various actions will bring. Then we can choose an objective and, by systematically working according to the constraints of principles and theories, we can achieve that objective. In short, the future will be what we make it (Fig. 0-4).

## ENVIRONMENTAL SCIENCE AND ITS APPLICATION

Scientists making observations and developing theories and principles in different areas produce the various disciplines of science such as biology, chemistry, and physics. As scientists make observations in new areas, new fields of science are created. In the mid-1800s, scientists began to study the broader aspects of how plants and animals interact with each other and their **environment**. This area of biology became known as **ecology.** In the 1960s it became widely recognized that ecological principles and theories did not apply just to wild plants and animals in natural environments. They are applicable to humans and apply on a global scale. This offshoot of ecology, the study of ecological principles and their application to the human situation, has become known as *environmental science.*

It should be clear that if our human system, involving agriculture, industry, and so on, is to

**FIGURE 0-4**
Understanding scientific principles and theories allows one to work toward goals in a meaningful way.

function satisfactorily over the long run, it must be operated in accordance with ecological theories and principles. Here we find cause for serious concern. We are still, for the most part, operating our human system according to biases and emotional feelings that bear little relationship to the real, objective world. Such nonobjective views are expressed by statements such as, "We can dominate and totally control nature"; "Growth is progress and progress is good"; "Resources are unlimited or if we exhaust one source, we can always find another"; "Nature can absorb and take care of all wastes"; "Humans are separate from nature and not subject to nature's laws"; "Science and technology can solve all of our problems despite our careless actions."

This kind of thinking has brought us problems such as pollution, loss of soil productivity, depletion of water and other resources, energy problems, and overpopulation, to name just a few. Whether or not these problems are solved depends, first, on our understanding of ecological principles and theories and, second, on our ability to manage our human ecological system accordingly. Significantly, scientists are not the only people who will determine the end result. Scientists can describe the theories and principles and make predictions regarding the outcomes of various courses of action. But the actual decision of which course to follow rests with the understanding, values, and judgments of society as a whole. Thus, each one of us is a participant, whether we recognize it or not, in the ecological management or mismanagement of the earth. It is necessary for us to understand the ecological principles and the environmental implications of various actions if we are to make rational value decisions.

Part 1 of this text is devoted to developing an understanding of ecological theory and principles and recognizing how they apply to the human situation. The following parts address specific environmental and resource problems and how they may be resolved if ecological principles are recognized and adhered to.

# WHAT ECOSYSTEMS ARE AND HOW THEY WORK

In 1968 astronauts returned with photographs of the earth taken from the moon. These photographs made it clear as never before that the earth is just a sphere suspended in the void of space. It is like a self-contained spaceship on an everlasting journey. There is no home base to which to return for repairs, more provisions, or disposal of wastes; there is just the continuous radiation from the sun. Indeed, the term "Spaceship Earth" was coined by futurist Buckminster Fuller as a result of this new perspective on our planet.

Who is at the controls of Spaceship Earth?

Unfortunately, no one! But Spaceship Earth is equipped with an amazing array of self-providing mechanisms. Enormously diverse plant and animal species interact in ways such that each obtains its needs from and provides for the support of others. Air and water are constantly repurified and recycled. Then there are self-regulating mechanisms as well, which tend to keep all the systems in balance with each other.

But now problems are arising. In particular, the human species is multiplying out of all proportion to others. This is placing greater and greater demands on all systems and, at the same time, it is undercutting their productivity through pollution and overexploitation. The natural regulatory mechanisms are being upset. It is clear that such behavior aboard Spaceship Earth cannot be sustained without catastrophic consequences. Nor can we afford the happy-go-lucky luxury of trial-and-error learning when the fate of the whole world is at stake. We must gain an understanding of how Spaceship Earth works and then we must learn to conduct our activities within this context.

Here in Part I our objective is to provide a general framework of understanding concerning the way our spaceship works. This understanding is gained through a study of natural ecosystems: what they are, how they function, how they are regulated, and how they develop and change. In keeping with the scientfic method, we shall approach each area by describing the basic observations that have been made and showing how these observations have led to the formation of operating theories and principles. Finally, the understanding of these theories and principles will enable us to see more clearly where current trends are headed and how certain human activities must be modified if modern society is to be sustained.

# WHAT IS AN ECOSYSTEM?

---

## CONCEPT FRAMEWORK

| Outline | | Study Questions |
|---|---|---|
| **I. DEFINITION AND EXAMPLES OF ECOSYSTEMS** | **15** | 1. Define *ecosystem* and name and characterize five major ecosystems found in North America. Are these ecosystems separate and distinct? Define *species* and *biota*. Distinguish between *ecosystems, biomes,* and the *biosphere.* |
| **II. WHY DO DIFFERENT REGIONS SUPPORT DIFFERENT ECOSYSTEMS?** | **20** | |
| **A. Abiotic Factors** | **20** | 2. Define and give examples of *abiotic factors.* Explain how different regions may vary in terms of abiotic factors, particularly temperature and precipitation. |
| 1. Optimums, Ranges of Tolerance, and Limiting Factors | **21** | 3. For any abiotic factor, what is meant by an organism's *optimum, range of tolerance,* and *limits of tolerance?* |

13

4. What is meant by the *law of limiting factors?* Describe its practical application. Describe how rainfall and temperature are primarily responsible for the distribution of major biomes. What is the effect of other abiotic factors? Distinguish between *climate* and *microclimate.*

5. Define and give examples of *biotic factors.*

6. List the three major components of *biotic structure.*

7. What organisms are *producers?* Describe the role of producers in terms of the chemical process of photosynthesis, changes between organic and inorganic states of material, and sources of energy. Define and distinguish between *autotrophs* and *heterotrophs* on the basis of energy source.

8. What organisms are *consumers?* Distinguish between various categories of consumers, namely, *primary* and *secondary consumers, herbivores, carnivores, parasites,* and *detritus feeders.*

9. Define *detritus.* What is the ecological role of decomposers? Distinguish between *decomposers* and *detritus feeders.*

10. Describe and distinguish among *food chains, food webs,* and *trophic levels.* Define *biomass* and explain how it changes at different trophic levels.

11. Name and describe other kinds of biotic relationships.

12. Summarize the concept of an ecosystem.

13. Give some examples of *physical barriers.* What is the likely consequence when a species invades another ecosystem?

14. Discuss how the concept of an ecosystem applies to humans. List ways in which the human ecological system is upsetting natural ecosystems. What are the implications?

The concept of an ecosystem is central to the entire study of environmental science. Our objective in this chapter is to understand what ecosystems are and why different ecosystems are found in different areas of the world.

## DEFINITION AND EXAMPLES OF ECOSYSTEMS

An **ecosystem** may be defined as a grouping of plants, animals and microbes interacting with each other and their environment. Furthermore, the interrelationships are such that the entire grouping may perpetuate itself, perhaps indefinitely. This definition is a very condensed description of what is observed in nature and can be best understood by considering some examples with which you are probably already familiar. A quick photographic tour of the United States shows, in the East, deciduous (leaf-shedding) forests that turn brilliant colors in the fall before the leaves drop; prairies or grasslands in the Central States; deserts with distinctive cacti in the Southwest; and evergreen, coniferous forests in the northern and western Mountain States (Fig. 1-1).

(a)

**FIGURE 1-1**
(a) Deciduous forest biome. One of many types, this is a mixed oak-hickory stand in West Virginia. (Photo by William Bierley.) (b) Grassland biome. Western Nebraska. (Earth Scenes/Lynn M. Stone.)

(b)

**FIGURE 1-1** (cont.)
(c) Desert biome. Saguaro
cactus and creosote bush in the
Sonoran Desert in Big Bend
National Park in southwestern
Texas. (Photo by Tamra Nebel.)

(d)

(d) Boreal forest biome. Western red cedar and Alaska cedar,
Alaska. (Earth Scenes/Breck P. Kent.)

We know that particular animals are associated with each of these plant communites: deer, squirrels, and woodchucks in the eastern deciduous forests; bison in the prairies (before they were killed off by early settlers); and moose in coniferous forests. Letting our thoughts range on, we may think of African savannah with fascinating grazing animals such as giraffes, zebras, and various antelopes; tropical forests in equatorial regions with monkeys and parrots; or caribou and reindeer in the arctic tundra (Fig. 1-2).

Looking at these examples more closely, we note that each consists, first, of a distinctive **plant community.** The term *plant community* refers to all the species of plants growing together in an area. A **species** is defined as a population of plants, animals, or microbes of which all the members do or potentially can interbreed to produce viable offspring. The plant community is generally characterized by a certain dominant species, various species of deciduous trees in the forests of the eastern United States, for example. But many other plant species may also be found. Second, each distinctive plant community supports a more or less distinctive array of animal species. Again, there is a tendency to focus on large mammals, but actually smaller animals, such as mice, birds,

**FIGURE 1-2**
(a) Savannas. Savannas are primarily grasslands with scattered deciduous trees. They are mostly located between tropical rain forests and deserts. Rainfall is between 80 and 160 cm per year but long dry periods and periodic fires preclude most tree species. This photo is Masai Mara, Kenya, Africa. (Animals Animals/Zig Leszczynski.)

(b)

(b) Tundra biome, Alaska. The northward extent of boreal forest is limited by permafrost, subsoil which remains permanently frozen. Permafrost prevents the growth of trees, but the surface layer of soil which thaws each summer supports an abundant growth of grasses and other low vegetation. Caribou are the dominant grazing animal of the North American Tundra. (Animals Animals/David C. Fritts.)

various species of insects, earthworms, and other such organisms, are far more abundant in terms of both numbers and **biomass,** the total combined weight of all individuals. Third, and still less conspicuous, is an even larger biomass of **microbes,** mainly bacteria, and fungi, which are present in the soil and feed mainly on dead plant and animal material.

We refer to the total array of plants, animals, and microbes as the **biota.** Finally, we are aware that these biota or groupings of organisms existed in different regions long before humans did and,

if they were not disturbed by humans, they would continue to exist, perhaps indefinitely. To summarize, each of the examples represents a grouping of plants, animals, and microbes interacting with each other and their environment. Thus, each is an ecosystem. The study of ecosystems and the interactions that occur among organisms and between organisms and the environment is the science of **ecology;** scientists who conduct these studies are **ecologists.**

An ecosystem may be very extensive, covering millions of square kilometers (1 km$^2$ = 0.4

mi²), or it may be as small as a pond or a woods. The significant point is that it can be defined, at least for study purposes, as a more or less specific grouping of plants and animals interacting with each other and their environment. Very large terrestrial ecosystems such as the ones mentioned above are referred to as **biomes.** The distribution

of the major biomes of North America is shown in Figure 1-3, and a world distribution is shown in Figure 1-4. Similarly, oceans may be divided into separate ecosystems such as coral reefs, continental shelves, and deep oceans. These major biomes contain a variety of related ecosystems within them.

Whether large or small, ecosystems usually do not have distinct boundaries, but one ecosystem grades into the next through a transitional region that shares many of the species and environmental characteristics of the two adjacent systems

**FIGURE 1-3**
Major biomes of the United States and Canada. (Adapted from Amos Turk, Jonathan Turk, Janet Wittes, and Robert Wittes, *Environmental Science*, 2nd ed., p. 58. Philadelphia: W. B. Saunders Company, 1978.)

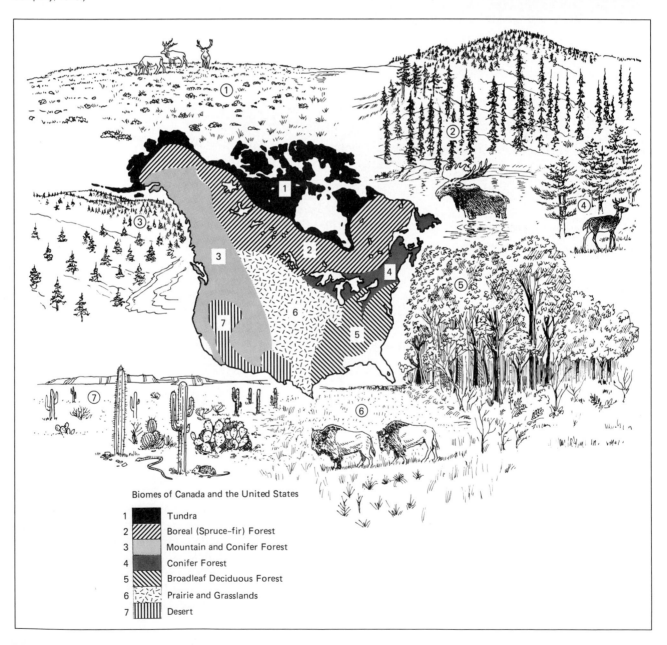

Biomes of Canada and the United States

| | | |
|---|---|---|
| 1 | ■ | Tundra |
| 2 | | Boreal (Spruce–fir) Forest |
| 3 | | Mountain and Conifer Forest |
| 4 | | Conifer Forest |
| 5 | | Broadleaf Deciduous Forest |
| 6 | | Prairie and Grasslands |
| 7 | | Desert |

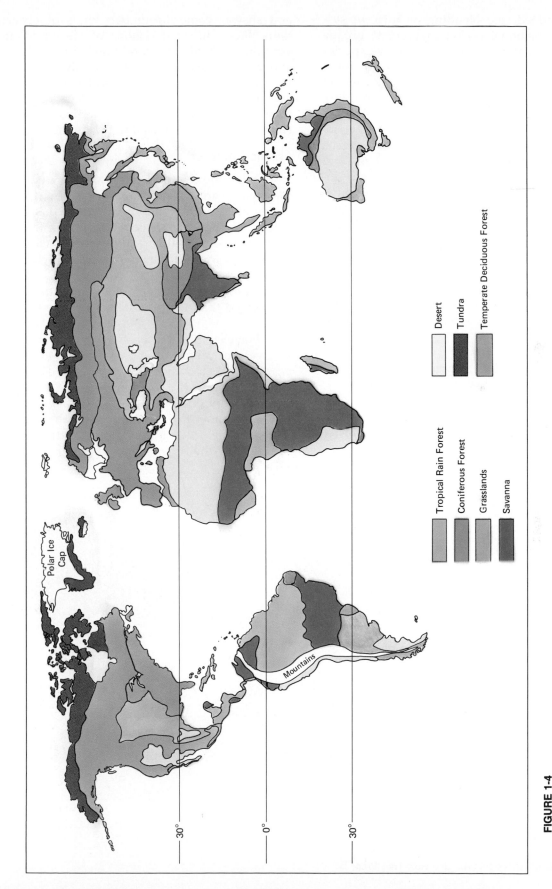

**FIGURE 1-4**
**The vegetative biomes, world view. Frequently these major biomes are divided into a number of subtypes.** (Reproduced by permission from D. K. Northington and J. R. Goodin, *The Botanical World*, St. Louis: Times Mirror/Mosby College Publishing, 1984.)

Tropical Rain Forest

Coniferous Forest

Grasslands

Savanna

Desert

Tundra

Temperate Deciduous Forest

Polar Ice Cap

Mountains

30°

0°

30°

(Fig. 1-5). Indeed, the blending of adjacent systems in a transitional region may create unique environments that have many distinctive species as well as those that are common to the adjoining ecosystems (Fig. 1-6). Thus, the transition region may be studied as an ecosystem in its own right.

The blending of one ecosystem into another should emphasize that ecosystems are not isolated. Many kinds of plants and animals may be found in two or more different ecosystems and some species, such as migratory birds, may inhabit different ecosystems at different times of the year. Further, what happens in one ecosystem definitely affects another. For example, water draining from the land into a lake may carry soil sediments and nutrients that profoundly affect the lake's ecosystem.

In the final analysis, all the organisms of the earth, including humans, interact with each other in an overall global system that we refer to as the **biosphere.**

## WHY DO DIFFERENT REGIONS SUPPORT DIFFERENT ECOSYSTEMS?

Why different ecosystems occur in different regions and why they remain more or less restricted to these areas is an intriguing question. The general answer comes from two kinds of observations. First, different regions of the world have very different climatic conditions. Second, plants and animals are usually specifically adapted to particular conditions. It is logical that plants and animals are limited to regions or locations where their **adaptations** correspond to the prevailing conditions, but let's examine this in more detail.

### Abiotic Factors

All of the chemical-physical factors of the environment are called **abiotic** factors (from *a*, "without," and *bio*, "life"). The most conspicuous abiotic factors are precipitation (rainfall plus equivalent snowfall) and temperature; we are familiar with the fact that these vary widely in different parts of the world, but the variation may be even more than we commonly recognize. More is involved than total rainfall or average temperature.

For example, the eastern United States has an average rainfall of about 100 cm (40 in.) per year which, for the most part, is distributed equally through the 12 months of the year. This creates a very different environmental effect from the same amount of rainfall coming almost exclusively during 6 months of the year, the wet season, leaving the other half of the year as the dry

**FIGURE 1-5**
Ecosystems are not isolated from one another. One ecosystem blends into the next through a transitional region that contains many species common to the two adjacent systems.

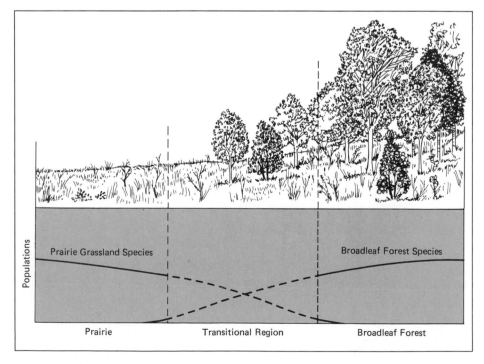

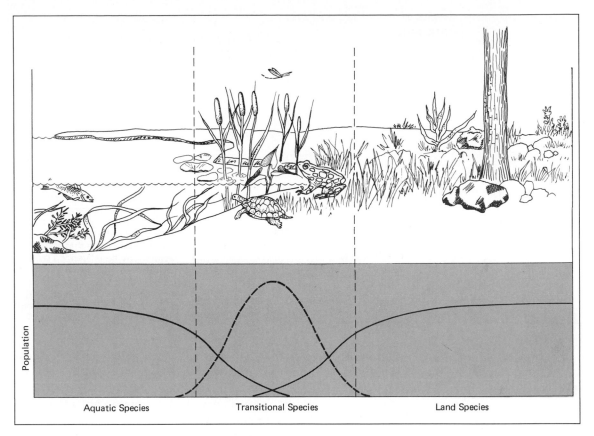

**FIGURE 1-6**
The transitional region may create a unique habitat that harbors specialized species. Thus, the transitional region itself may form a unique ecosystem.

Within the figure:

Population

Aquatic Species    Transitional Species    Land Species

season. Likewise, an average temperature of 20° C (68° F) in an area where temperature does not drop below freezing is very different from the same average resulting from a hot summer and a cold winter. In fact, the cold temperature extreme—no freezing temperatures, slight freezing, or several weeks of hard freezing temperatures—proves to be more significant biologically than the average temperature. Further, different amounts and distributions of precipitation may be combined with different temperature patterns; thus, there are numerous combinations involving just these two factors.

Yet, many other abiotic factors may also be involved. These include type and depth of soil, availability of essential nutrients, wind, fire, salinity (saltiness), light, day length, terrain, and pH (the measure of acidity or alkalinity of soil or water). To illustrate, consider terrain: in the Northern Hemisphere, north-facing slopes generally have cooler temperatures than south-facing

slopes. Or consider soil type: a sandy soil, because it does not hold water well, gives the same effect as less rainfall. Or consider wind: by increasing evaporation, it may also give the effect of relatively drier conditions. However, these and other factors may exert a critical effect in their own right.

In summary, we can see that the abiotic factors, all of which are continually present to some degree, interact with each other to create a matrix of an infinite number of different environmental conditions (Fig. 1-7).

## OPTIMUMS, RANGES OF TOLERANCE, AND LIMITING FACTORS

We shall now turn our attention to the ways different species "fit" different environmental conditions. We shall focus on plants because they illustrate the principles most clearly.

From field observations (observations of things as they exist in nature as opposed to laboratory experiments), one may guess that different species of plants vary widely in their tolerance to (ability to withstand) different abiotic factors. This

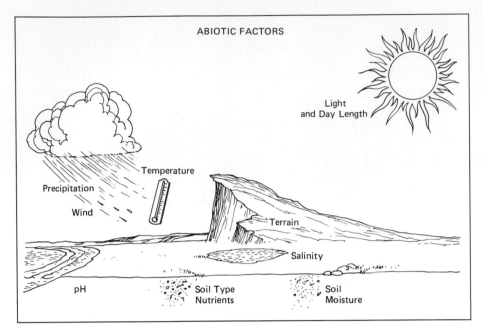

ABIOTIC FACTORS

Light and Day Length

Temperature

Precipitation

Wind

Terrain

Salinity

pH

Soil Type Nutrients

Soil Moisture

**FIGURE 1-7**
Abiotic factors. Abiotic factors include all the chemical-physical factors of the environment, the most significant of which are indicated. Observe that abiotic factors may be influenced by one another. Also note that abiotic factors may vary widely throughout the day and/or year; the extremes are frequently more significant than averages.

The temperature at which maximum growth rate is found is called the **optimum** temperature. The total range of temperature that will sustain growth is called the **range of tolerance** (for temperature). The temperatures below and above which the plants will not grow at all are called the **limits of tolerance.**

Similar experimentation has been done to test most other abiotic factors. For each factor tested, results follow the same general pattern. There is an optimum, which gives maximum hypothesis has been tested and verified and is seen even more clearly through the following kinds of experiments.

Plants are grown in a series of chambers in which all the abiotic factors can be controlled; thus, a single factor being tested can be varied in a systematic way while all other factors are kept constant. For example, light, soil, water, and so on are kept constant, but the temperature is varied from one chamber to the next. Note that this kind of controlled experiment is necessary to distinguish the effect of temperature from other factors. The results show that, as temperature is raised from a low point, the plants grow better and better until they reach some maximum rate of growth. If temperature is raised still further, however, the plants become increasingly stressed; they do not grow as well, suffer injury, and die. This is pictured graphically in Figure 1-8.

growth, a range of tolerance above and below the optimum that will sustain less vigorous growth, and finally, limits above and below which the plants cannot survive. Of course, not every species has been tested for all factors; however, the consistency of such observations leads us to conclude that this is a fundamental biological principle. We can generalize then that each species has an optimum, a range of tolerance, and a limit of tolerance with respect to every factor.

In addition to the principle of optimums, this line of experimentation demonstrates that species may differ markedly with respect to the point at which the optimum and limits of tolerance occur. For instance, what may be too little water for one species may be the optimum for a second and may result in the death of a third. Some plants cannot tolerate any freezing temperatures (i.e., any exposure to 0°C [32°F] or less is fatal). Others can tolerate slight but not intense freezing, and some actually require several weeks of freezing temperatures in order to complete their life cycles. The same can be said for other factors. But, while optimums and limits of tolerance may differ for different species, there may be considerable overlap in their ranges of tolerance.

Thus, controlled experiments support the hypothesis that species differ in their adaptation to various abiotic factors. The geographic distribution of a plant species may be determined by

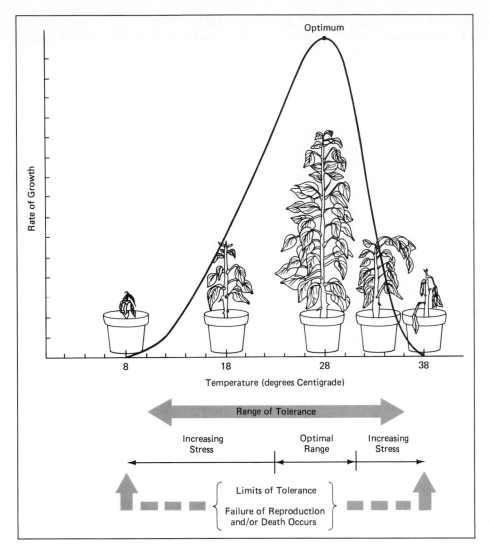

**FIGURE 1-8**
For every factor influencing growth, reproduction, and survival, there is an optimum, a range of tolerance, and then limits beyond which death occurs. Optimums and limits of tolerance differ for different species.

the degree to which its requirements are matched by the abiotic factors present. A species may flourish where it finds conditions optimal. It may survive but do poorly in conditions that differ from its optimum. But it will not survive where any abiotic factor is beyond its limit of tolerance for that factor.

### LAW OF LIMITING FACTORS

It is important to emphasize that either *too much* or *too little* of any single abiotic factor may limit or prevent growth despite all other factors being kept at or near their optimums. This observation is generalized as the **law of limiting factors,** also known as **Liebig's law of miminums.**

Justus von Liebig formulated it in 1840 in connection with his observations regarding nutrients, but it has been generalized now to include any and all abiotic factors. The single factor that is found to be limiting growth (or other response) of an organism is called the **limiting factor.** Importantly, the limiting factor may be a problem of too much of something as well as too little. For example, plants may be killed by overwatering or overfertilizing, a common "accident" among amateur gardeners, as well as by underwatering or underfertilizing.

The reason a species in one ecosystem does not penetrate indefinitely into an adjacent ecosystem is often because it confronts one or more abiotic factors in the adjacent system which are limiting. However, biological factors such as predation, disease, parasites, and competition from other species may also be limiting factors. These biological factors will be discussed at greater length in the next section on biotic factors. For now, let us consider some examples of limiting abiotic factors.

With respect to plants, the abiotic factor that is most commonly limiting in natural terrestrial ecosystems is water. Water is the main factor responsible for the separation of major biomes into forests, grasslands, and deserts. This occurs as follows: The optimal amount of rainfall for many tree species is about 150 cm (60 in.) of rainfall per year; they reach their limit of tolerance at about 75 cm (30 in.) per year. Grasses have a much lower limit of tolerance for water, about 25 cm (10 in.) per year, but there are many species of cacti and other specialized plants that can survive with as little as 5 to 10 cm (2 to 4 in.) per year. Consequently, the natural ecosystems of regions with more than 100 cm (40 in.) of rainfall per year are typically forests. Regions with 25 to 75 cm (10 to 30 in.) of rainfall are typically grasslands, and regions with less than 25 cm (10 in.) are only sparsely vegetated with species such as cacti, sagebrush, and tumbleweed. Such areas we recognize as deserts. In intermediate ranges of rainfall, as one might expect, forests grade into grasslands and grasslands grade into deserts.

Temperature also plays a role in limiting major plant communities. However, except for extreme cold, which gives rise to tundra or permanent ice, the effect of temperature is largely superimposed on that of rainfall. That is, 100 cm or more of rainfall will support the growth of a forest, but temperature will determine the kind of forest. Spruces and firs cope best with the very severe winters and short growing seasons found in northern regions and/or at high elevations. Deciduous trees, which drop leaves and go into a period of dormancy, also are splendid at coping with freezing winter temperatures, but they require a longer growing season. Thus, deciduous tree species predominate in more temperate latitudes where rainfall is adequate. Finally, in tropical forests, broadleaf evergreen trees predominate because these species, which cannot tolerate freezing temperatures, are most successful where there is a continuous growing season. Likewise, a hot desert has different species than a cold desert,

but areas receiving less than 25 cm or so of precipitation will be deserts with only a few drought-tolerant species in either case.

Temperature also exerts some influence by its effect on the evaporation of water; water evaporates faster at higher temperatures. Consequently, the transitions from deserts to grasslands and from grasslands to forests are found at higher precipitation levels in hot regions and at lower precipitation levels in cold regions. These interactions between temperatures and rainfall are illustrated in Figure 1-9.

In the far north only the surface layer of soil thaws each summer; below a few inches the soil remains permanently frozen (permafrost). This factor limits the northern extension of spruce-fir conifer forests but still permits the growth of the small, hardy plants that occupy the tundra. Of course, still colder temperatures limit the tundra vegetation and produce the polar ice caps.

In conclusion, the distribution of the plant species that characterize the major biomes of the earth are determined largely by the abiotic factors of rainfall and temperature. Other abiotic factors, however, are frequently the limiting factors that cause variation within the major biome. For example, within the deciduous forests of the eastern United States, oaks and hickories generally predominate on rocky, poor, well-drained soils; beeches and maples are found on the richer soils. In other words, within the deciduous forest biome, soil type is frequently the factor that determines the distribution of certain tree species. Likewise, relative abundance or lack of certain nutrients in the soil may determine the distribution of various species within grasslands.

In certain cases an abiotic factor other than rainfall or temperature may be the primary limiting factor. For example, the strip of land adjacent to the coast frequently receives a salty spray from the ocean, a factor that relatively few plants can tolerate. Consequently this strip is frequently occupied by a unique community of salt-tolerant plants (Fig. 1-10). Another example is a rocky outcrop with little or no soil. Such an area may have a rich community of mosses and lichens similar to a tundra, but here the limiting factor is the lack of soil. Salt concentration is commonly the factor limiting the distribution of aquatic plants and animals between fresh and salt waters. Available light is the factor that delimits amounts and kind

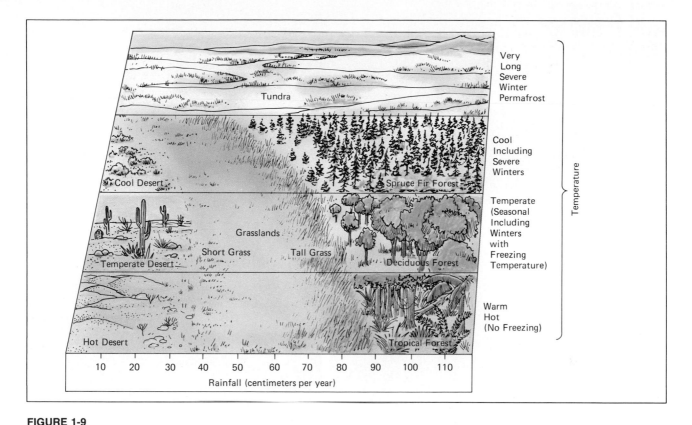

**FIGURE 1-9**
Abiotic factors and major biomes. Moisture is generally the overriding factor determining the type of biome that may be supported in a region. Given adequate moisture, an area will generally support a forest. Temperature, however, determines the kind of forest. The situation is similar for grasslands and deserts. At cooler temperatures there is a shift toward less precipitation because lower temperatures reduce evaporative water loss. Temperature becomes the overriding factor only when it is severe enough to sustain permafrost.

of vegetation beneath trees in a forest. There is almost no vegetation under a dense evergreen forest because of lack of light. In a deciduous forest there are ground species that take advantage of the lack of cover in the early spring; other species take advantage of the light in late fall after the leaves drop. Fire, which will be discussed further in Chapter 3, is also a very significant factor that limits some species but not others.

Especially within a transition area, a secondary abiotic factor may be crucial. For example, consider an area with about 25 cm (10 in.) of rainfall, which is the borderline amount between desert and grassland. In such an area, a soil with good water-holding capacity will support grass whereas a sandy soil with little ability to hold water will support only desert species.

Frequently, ecologists speak in terms of **microclimates.** The prevailing precipitation and temperature pattern of the region creates the overall climate that determines the major biome. However, any number of factors may come into play and cause the actual conditions on or near the ground to be markedly different. The particular conditions from the ground up to a height of 2 m (about 6 ft) comprise what is called the microclimate. Thus in considering an organism's interrelationship with its environment one must consider more than the overall climatic factors of the region; one must also consider the microclimate of its particular location. Again, we should emphasize the fact that all abiotic factors tend to interact with each other to create the resulting environment.

## Biotic Factors

An ecosystem always involves more than a single species of plant interacting with abiotic factors. The plant community is invariably comprised of a number of species which may compete with each other, but which may also provide mu-

**FIGURE 1-10**
Specific abiotic factors may determine the type of vegetation.
Along the Pacific coast in central California, usual vegetation is
precluded by salt spray, but certain succulent species which
are salt-tolerant thrive. (Photos by author.)

tual support. The plant community almost invariably supports a host of various animals, fungi, bacteria, and other microbes. Thus, each species does not just interact with abiotic factors; it is also constantly interacting with other species, as well, gaining food, shelter, or other benefits from some, competing with others, and being eaten by still others. All interactions with other species are classified as **biotic factors.** Importantly, some of the biotic factors affecting any species are positive, while others are negative, and some may be neutral. Major interrelationships among species will become clear as we study the structure of ecosystems.

## BIOTIC STRUCTURE OF ECOSYSTEMS

Structure refers to parts and the way they fit together to make the whole. Thus, the biotic structure of ecosystems refers to categories of organisms, the parts, and the way they interact together within the entire system. Despite the diversity of ecosystems, rain forests to deserts, ecologists find that they all have a structural similarity. That is, they all have the same basic categories of organisms that interact together in the same ways. The major categories of organisms are called **producers, consumers,** and **decomposers.**

**Producers.** Producers include all plants that carry on photosynthesis, a process we shall describe in a moment. Most plants that carry on photosynthesis are easily identified by their having leaves, needles, and/or other parts that are green. They range in diversity from microscopic single-celled algae through medium-sized plants such as grass and cacti to giant trees. Of course, each ecosystem has its distinctive plant producers but the most important feature is that they contain the pigment **chlorophyll.** Chlorophyll is responsible for the green color but, most importantly, *chlorophyll absorbs light energy which, in turn, enables the plant to carry on photosynthesis.* **Photosynthesis** is the process whereby plants use light energy, absorbed by chlorophyll, to convert carbon dioxide from the air and water into sugar. Oxygen gas is formed as a byproduct and is released into the atmosphere (Fig. 1-11). In addition, producers are able to manufacture all of the complex chemicals of their bodies from the sugar and a few additional simple chemical nutrients absorbed from the soil or water.

We speak of the simple chemicals that make up air, water, and minerals of rock and soil as **inorganic,** whereas the complex chemicals such as protein, fats, and carbohydrates that make up tissues of plants and animals are **organic** (Fig. 1-12). In short, plants that carry on photosynthesis are able to produce all their complex *organic* chemicals from simple *inorganic* chemicals present in the environment by using light as an energy source. In so doing, some of the *energy* from light, as well as the chemical elements from air, water, and soil, is also incorporated into the organic compounds.

This is extremely significant because we find that all other organisms in the ecosystem, including humans, must feed on complex organic material for their source of energy as well as nutrients. Thus, you can see the significance of producers. They *produce* the food for all the rest of the organisms in the system. Indeed, this distinction allows us to divide all living things into two categories: **autotrophs** (*auto,* "self," *and troph,* "feeding") and **heterotrophs** (*hetero,* "other," *and troph,* "feeding"). Producers are autotrophs; all other organisms are heterotrophs. Importantly, not all plants are producers; all fungi (mushrooms, molds, and other such organisms) and a few higher plants, such as Indian pipe (Fig. 1-13), do not have chlorophyll and they do not carry on photosynthesis. Like animals, they feed on preexistent organic matter. Like animals, these plants are heterotrophs.

**Consumers.** Consumers are primarily animals. Again, the diversity is tremendous, ranging in size from microscopic protozoans to blue whales and elephants weighing many tons, and they include such diverse groups as worms, fish and shellfish, insects and related organisms, reptiles, mammals, and birds. The range of species of animals present in an ecosystem depends on the presence of suitable food sources and other requirements such as underbrush for shelter or places for nesting and/or rearing young, as well as suitable abiotic factors. Many animals have extremely specific requirements in terms of what they can eat and the kind of habitat (e.g., forest, brush, open field) they need to survive. Hence, consumers may be divided into many subcategories according to how and on what they feed.

Animals, whether as large as elephants or as small as tiny mites, that feed directly on producers, are called **primary consumers.** Animals that

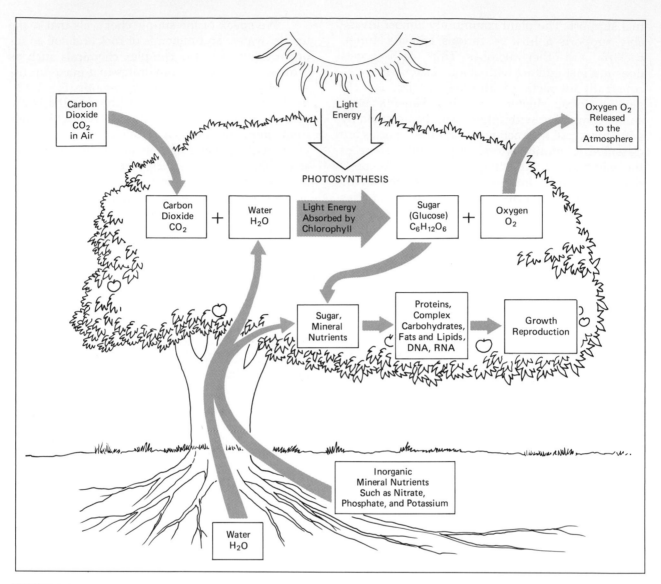

**FIGURE 1-11**

Producers. Green plants, which contain chlorophyll, can absorb light energy and use it to produce glucose from carbon dioxide and water, releasing oxygen as a byproduct. The glucose, along with a few additional mineral nutrients from the soil, is used in the production of all plant tissues leading to growth.

feed on primary consumers are called **secondary consumers.** For example, a rabbit that feeds upon carrots is a primary consumer; a fox that feeds upon rabbits is a secondary consumer. There may also be third, fourth, or even higher levels of consumers. Certain animals may occupy more than one position on the consumer scale. For instance, humans are primary consumers when they eat vegetables, secondary consumers when they eat beef, and third-level consumers when they eat fish that

feed on other organisms that in turn feed on algae.

Primary consumers, those animals that eat plant material, are also called **herbivores.** Secondary and higher orders of consumers are called **carnivores.** Those that feed on both plants and animals are called **omnivores.**

In a relationship in which one animal attacks, kills, and feeds on another, the animal that attacks and kills is called the **predator;** the animal that is killed is called the **prey.** Together, the two animals are said to have a **predator-prey** relationship. For example, foxes and rabbits have a predator-prey relationship, the fox being the predator, the rabbit being the prey.

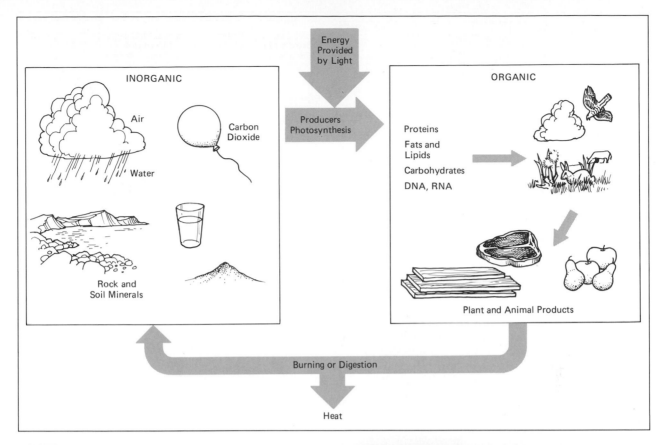

**FIGURE 1-12**
*Organic* and *inorganic*. Simple states of matter found in air, water, and rock and soil minerals are said to be *inorganic*. The complex states found in plant and animal tissues are said to be *organic*. The change from inorganic to organic is enabled by producers using light energy in the process of photosynthesis. Organic materials will break down to inorganic materials again through burning or digestion, releasing energy. Chemically, organic compounds involve carbon-carbon and carbon-hydrogen bonds not found in inorganic materials.

**FIGURE 1-13**
Indian pipe, a flowering plant that is not a producer. It does not carry on photosynthesis but derives its energy from other organic matter as do animals.
(Photo by author.)

**Parasites** form another important category of consumers. Rather than devouring their food as such, parasites become intimately associated with their "prey" and feed on it over an extended period of time, typically without killing it (at least not immediately) but usually doing harm to it. The plant or animal that is fed upon is called the **host;** thus we speak of a **host-parasite** relationship. These feeding relationships are summarized in Figure 1-14.

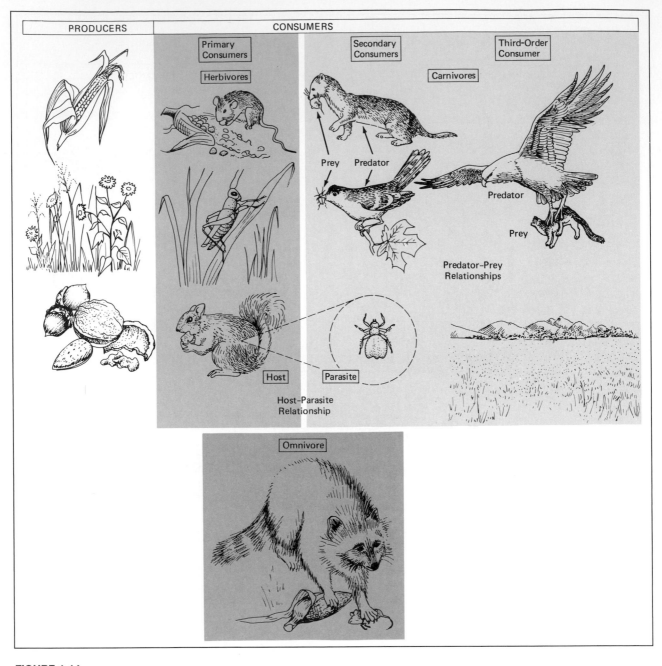

| PRODUCERS | CONSUMERS | | |
|---|---|---|---|
| | Primary Consumers | Secondary Consumers | Third-Order Consumer |
| | Herbivores | Carnivores | |

Prey    Predator

Predator

Prey

Predator–Prey Relationships

Host    Parasite

Host–Parasite Relationship

Omnivore

**FIGURE 1-14**
The most common feeding relationships among organisms.

There are many kinds of host-parasite relationships. Some parasites, such as tapeworms (Fig. 1-15a) live inside their host and are called **endoparasites.** Other parasites, such as ticks, lice, or lampreys (Fig. 1-15b) attach themselves to the outside of their host and are called **ectoparasites.** As these examples illustrate, a wide variety of organisms may be parasitic. Virtually every major group of living things has at least some members that are parasitic. Bacteria and other microorganisms that cause disease of plants or animals are highly specialized endoparasites. Many serious plant diseases and some animal diseases (such as athlete's foot) are caused by parasitic fungi. Even some plants such as dodder (Fig. 1-15c) are parasitic on other plants. Ecologically speaking, parasitic plants must actually be considered consumers.

(a)

(b)

(c)

**FIGURE 1-15**
Diversity of parasites. Nearly every major biological group of organisms has at least some members that are parasitic on others. Shown here are (a) intestinal tapeworm, an endoparasite (USDA photo); (b) lamprey, an ectoparasite (U.S. Fish and Wildlife Service photo); (c) dodder, a plant parasite (Photo by Steve Simon).

The above categories refer to consumers that generally feed on other *living* organisms. Herbivores feed on living plant parts; carnivores capture living prey, and parasites feed only on living hosts. However, much of the vegetation in an ecosystem dies quite apart from being consumed. The natural leaf fall in forests and the dieback of grasses in unfavorable seasons are examples. Likewise, fecal wastes of animals and dead animal bodies represent unconsumed organic matter. This dead plant and animal material is called **detritus.** In turn, there are many organisms that are specialized to feed on detritus and we refer to this category of consumers as **detritus feeders.** Examples of detritus feeders include vultures, earthworms, millipedes, crayfish, termites, ants, wood beetles, and so forth. As with regular consumers, one can identify *primary* detritus feeders, those that feed directly on detritus, *secondary* detritus feeders, and so on.

**Decomposers.**     Finally, much of the detritus in an ecosystem, particularly dead leaves and wood, is apparently not eaten by detritus feeders, but rots, decays, or decomposes. Actually, this rotting or decomposition is caused by the feeding activity of certain organisms called **decomposers.** Decomposers, or at leat their feeding parts, are microscopic, so all we normally see is the decomposition rather than the organism.

Decomposers consist of two classes of organisms: **fungi** and **bacteria.** Fungi include such organisms as molds, mushrooms, shelf fungi, coral fungi, and puffballs (Fig. 1-16). The part we recognize as the mushroom, shelf, or puffball is just the fruiting body, or reproductive structure, and is only a small portion of the whole organism. Beneath is an extensive network of microscopic rootlike filaments, called **mycelia,** that penetrate through the dead leaves, wood, or other detritus. The mycelia secrete digestive enzymes that break down the detritus material into simpler organic nutrients that can be absorbed into the fungal cells (Fig. 1-17). While mushrooms occasionally appear to be growing on inorganic soil, their mycelia are actually feeding upon organic material in the soil. Bacteria, which are microscopic, single-celled organisms, receive their nourishment in the

**FIGURE 1-16**

Decomposers. Thousands of species of fungi, a few of the major types of which are shown here, are *decomposers*. They are plantlike organisms, but they feed on dead organic matter, much like animals. The result of their feeding is observed as the rotting, decay, or decomposition of the dead organic material, such as wood or dead leaves. Many species of bacteria are also decomposers. (Photos by author.)

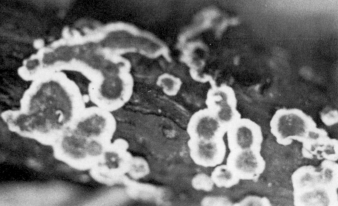

**FIGURE 1-16** (*cont.*)

same way as fungi. Much of the detritus in an ecosystem would not be consumed and would simply accumulate to the detriment of the system if it were not for these organisms.

Fungi and bacteria produce reproductive spores in tremendous abundance and their micro-scopic size allows them to be carried easily by air currents, assuring that they are present virtually everywhere in the environment. Therefore, their growth and hence the rotting or decay of organic matter occurs wherever suitable conditions of temperature and moisture prevail unless specific measures are taken to prevent it. Consider for example, what happens to food materials that are not preserved in some way.

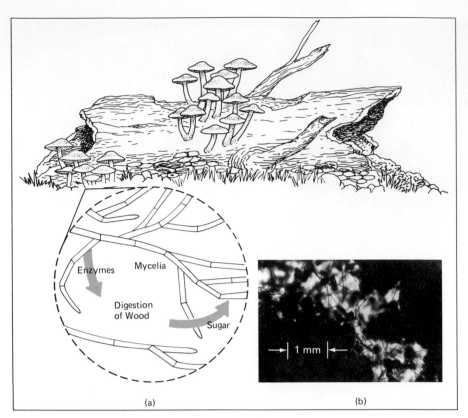

Within the figure: Enzymes · Mycelia · Digestion of Wood · Sugar · 1 mm · (a) · (b)

You may note that decomposers are highly specialized detritus feeders, which is *animallike in function.* Yet fungi, at least, are *plantlike in appearance* and they reproduce by spores as do many other plants. Should they be considered plants or animals? Actually, they are so distinctive unto themselves that taxomonists have decided that fungi and bacteria belong neither to the plant kingdom nor the animal kingdom; thus two additional kingdoms have been designated, one for bacteria and the other for fungi.

In passing it is significant to note that most species of bacteria are harmless decomposers feeding only on dead organic matter. Only a relatively few species are parasites that cause disease; then, some species are photosynthetic and, hence, are producers ecologically speaking.

Decomposers are not the end in themselves. Bacteria and fungi are an important food for other organisms such as protozoans, mites, insects, and worms living in the soil or water (Fig. 1-18). Many of the fungi, such as button mushrooms, are also considered a great delicacy by people. Thus, there are links between the decomposers and other consumers and when a fungus or other decomposer dies, its dead body becomes part of the detritus and the source of energy for yet other decomposers. It is not uncommon, for example, to see mold growing on a mushroom, one decomposer decomposing another.

In summary, despite the apparent diversity of ecosystems, they all have *structural similarity.* They can all be described in terms of photosynthetic plant producers, various categories of animal consumers, and decomposers. Some organisms do not fit neatly into a single category, but act in different roles at different times. Nevertheless, the biotic structure can still be defined in these terms.

## FEEDING RELATIONSHIPS: FOOD CHAINS, FOOD WEBS, AND TROPHIC LEVELS

In describing the biotic structure of ecosystems it becomes evident that major interrelationships among organisms involve feeding relationships. We can identify innumerable pathways where one organism is fed upon by a second organism, then a third feeds upon the second, and

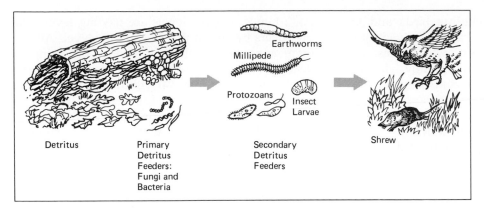

**FIGURE 1-18**
Detritus-based food web. Decomposers are not the end of the line. Fungi and bacteria, which feed on detritus, support many other organisms living in the soil and these, in turn, may be fed upon by larger consumers.

Earthworms

Millipede

Protozoans

Insect Larvae

Shrew

Detritus

Primary Detritus Feeders: Fungi and Bacteria

Secondary Detritus Feeders

so on. Each pathway is called a **food chain.** Corn eaten by a mouse that is eaten by a weasel that is eaten by a hawk is an example of a food chain. Grass to beef to humans is another. Dead plant material to mushroom to insect to frog to snake to an eagle is a third (Fig 1-19).

While it is interesting and instructive to trace these specific pathways, it is important to recognize that food chains seldom exist as isolated entities. Very few primary consumers feed upon just one kind of plant, nor are they in turn fed upon by only one kind of secondary consumer. For example, mice eat a variety of seeds and other things besides corn, and corn may be eaten by

**FIGURE 1-19**
Simple food chains. Food (nutrients and energy) is transferred from one organism to another along pathways known as *food chains.* However, food chains seldom exist as isolated entities in nature. Instead, nearly all food chains are interconnected to form a complex food web.

many different animals including birds and insects. In turn, these animals may be preyed upon by numerous species including owls, snakes, and hawks as well as weasels. Consequently, virtually all food chains are interconnected. Depicting all these interconnected feeding pathways creates a fantastic web of feeding connections (Fig. 1-20). In fact, the term **food web** is used to denote this actual, complex pattern of nutrient and energy transfer through feeding.

Despite the number of theoretical food chains and the complexity of food webs, however, all can be traced back to producers as the initial starting point. All basically lead from plant producers to herbivores to carnivores of various sorts. Or they may lead from plant producers to decomposers to various categories of consumers feeding on them.

Consequently, another important way to view feeding relationships is through the concept of **trophic levels.** Trophic literally means "feed-ing"; hence, trophic levels are feeding levels. All producers belong to the first trophic level. All primary consumers, whether feeding on living or dead producers, belong to the second trophic level, and so on. We can visualize a flow of nutrients and energy from the first to the second trophic level; from the second to the third trophic level; and so on. A diagrammatic comparison of a food chain, a food web, and trophic levels is shown in Figure 1-21.

The reason that all food chains, webs, or trophic levels are based on producers, namely photosynthetic plants, is that they produce the initial supply of organic matter. After this each heterotroph breaks down a considerable portion of what it consumes for energy; only a relatively small portion is converted into its own body. Therefore, the biomass (total combined weight) of the feeders will always be *less* than the biomass of food which they consume. This and other factors will be discussed in more detail in Chapter 2. For

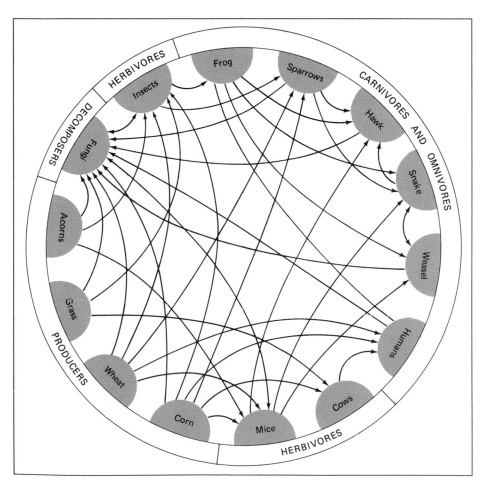

**FIGURE 1-20**
A food web. If all the actual feeding relationships in an ecosystem are represented, the picture is a fantastically complex web. Hence, the term *food web* is used to refer to the feeding relationships in nature.

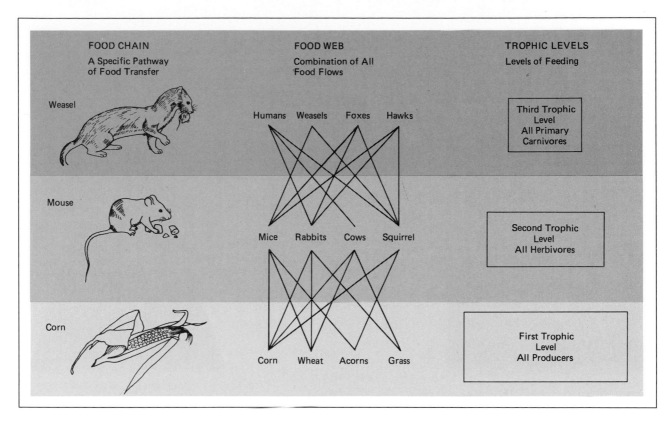

**FIGURE 1-21**
Food chain; food web; trophic levels. Three ways of representing the transfer of nutrients and energy.

now it is sufficient to recognize that as one progresses along food chains or trophic levels there is, in fact, a *diminishing biomass* (total combined weight of organisms). Depicting this graphically give rise to what is commonly called a **biomass pyramid** (Fig. 1-22).

The practical implication of this is that the amount of animal life an ecosystem can support is directly related to its primary production (growth of producers). The biomass of the animal life is only a small fraction of the biomass of production because of the diminishing effect. Can you see why animal life is much more abundant in a tropical rain forest than in a desert? Similarly can you see why there are fewer large carnivores than herbivores?

## OTHER BIOTIC RELATIONSHIPS

**Mutually supportive relationships.** In feeding relationships we generally think of one species benefiting and the other being harmed to a greater or lesser extent. However, there are many cases in which there is a mutual benefit to both species. This phenomenon is called **mutualism.** A classic example is seen in the group of plants known as lichens (Fig. 1-23). (Mosses should not be confused with lichens. Mosses have distinct stems with many minute, bright green leaves. Lichens are generally grey-green in color and are usually scaley or crusty in appearance). Lichens are actually comprised of two plants, a fungus and an alga. The fungus provides protection for the alga, enabling it to survive in dry habitats where it could not live by itself; the alga, which is a producer, provides food for the fungus, which is a heterotroph.

Another example is the relationship between flowers and insects. The insects benefit by obtaining nectar from the flowers; the plants benefit by being pollinated in the process. A third example is cleaning symbiosis observed in coral reefs. Certain fish are immune to the predatory nature of the coral and hence are able to feed on detritus in and around the coral. The fish thus benefit by having access to a food source; the coral benefits by being cleaned.

Such relationships are also called *symbiotic* relationships. However, **symbiosis** (*sym*, "to-

**FIGURE 1-22**
Biomass pyramid. A graphical representation of the biomass at successive trophic levels has the form of a pyramid.

gether," and *bio*, "live") refers to any intimate relationship between two organisms. Hence, symbiosis includes parasitism as well as mutualism.

Mutually supportive relationships, however, go far beyond the very close kinds of relationships illustrated above. For example, plant detritus provides most of the food for decomposers and soil-dwelling detritus feeders such as earthworms. Thus, these organisms benefit from plants, but the plants also benefit because the activity of these organisms is instrumental in releasing nutrients from the detritus and in maintaining soil quality, a subject that will be discussed further in Chapter 7.

In another example, many birds benefit from vegetation by finding nesting materials and places among trees. But the plant community also benefits because the birds feed on and reduce the populations of many herbivorous insects.

Even in predator-prey relationships some mutual advantage exists. The killing of individual prey, which are usually weak or diseased, may benefit the population at large by keeping it healthy and preventing it from becoming so abundant that it overgrazes the environment.

**Competitive relationships.** We have noted that different species of plants are adapted to different abiotic factors; animals, in addition, are adapted to feeding on different materials in different places at different times of day. This reduces direct competition between species for resources. Nevertheless, competition does occur. This is especially true in areas where ecosystems merge and plant species must compete with each other for nutrients, water, and/or light.

For example, consider the transition between grasslands and deserts. We mentioned that low rainfall is the primary limiting factor where grasslands give way to deserts. However, many desert species can tolerate far more than 25 cm (10 in.) of rainfall per year. What then prevents desert species from pushing into grasslands? Competition is largely responsible. Where rainfall is adequate to support grass, grass proves to be a more vigorous competitor for space than the

**FIGURE 1-23**
Lichens. The crusty-appearing ''plants'' commonly seen
growing on rocks or the bark of trees are actually comprised of
a fungus and an alga growing in a mutualistic relationship.
(USDA-Soil Conservation Services.)

highly specialized desert species. Consequently,
where there is enough water, desert species are
crowded out by grasses. Similarly, grasses can tolerate, in fact they thrive on, more than 75 cm (30
in.) of annual rainfall. However, where there is
enough moisture to support the growth of trees,
the trees prove better competitors. They shade
out grasses and the grasslands give way to forests. Likewise, we frequently find that unique
species that inhabit only very specialized situations are not found elsewhere because they are
not good competitors where other species can
grow vigorously.

## Biotic and Abiotic Interactions: A Summary

We should emphasize again that biotic factors do not act singly, although in certain situations a particular biotic factor may stand out as
being the dominant or limiting factor. For example plant populations are frequently limited by
the presence of a microbial disease, parasite,
plant-eating insect or some other species. Animal
populations are also commonly limited by disease
organisms, parasites, or larger predators as well
as by limitations of food or suitable habitat,
among other factors.

What is more, abiotic factors, in turn, affect
these biotic interactions in numerous ways. For
example, certain parasitic fungi on plants become
more virulent in humid conditions and less virulent in dry conditions. Likewise, certain voracious
insects thrive only under certain conditions of
temperature and moisture. Thus, the survival of a
species may be affected by a natural enemy that
in turn is controlled by abiotic factors. In conclusion, we find that the biomes and ecosystems
within them are far more than individual species
interacting independently with the abiotic environment. The biota (the entire community of
plants, animals, and microbes) is really an exceedingly complex, dynamic entity as all the species
interact through innumerable feeding, competitive, and mutually supportive relationships.
Thus, the ecosystem is really this entire dynamic

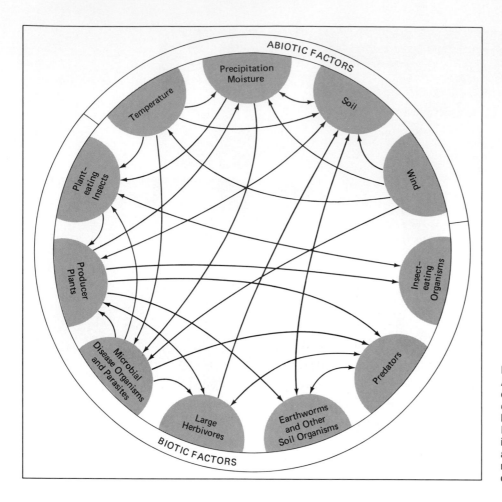

**FIGURE 1-24**
An ecosystem is an extremely complex, dynamic entity in which each species is influenced by a host of biotic and abiotic factors. In turn, each species may influence other species and also abiotic factors. Only a few of the major interrelationships are depicted here.

biota interacting with the abiotic environment (Fig. 1-24). Here, we should understand more fully what is entailed in the definition of an ecosystem—a grouping of plants, animals and microbes interacting with each other and their environment.

We also noted that ecosystems may perpetuate themselves, some more or less indefinitely. It should now be clear that as long as abiotic and biotic factors remain constant, established balances may be maintained and perpetuation or continuation occurs. However, it should also be evident that if either abiotic or biotic factors change, then balances are upset and other changes will occur. Further, once changes are set in motion, they may be exceedingly far-reaching. A domino effect occurs because all species are interconnected through various biotic relationships. In addition, changes in the biotic community may lead to alteration in abiotic factors (e.g., changes in the microclimate), which produce another round of effects. For example, removal of vegeta-

tion allows soil erosion and the change in soil affects the kinds of plants that can reestablish themselves. This, in turn, affects the rest of the biota. We shall return to a further discussion of this principle in Chapter 3 and we shall find its numerous applications throughout the text.

## Physical Barriers

A final factor limiting the invasion of one ecosystem by a species from another is the existence of physical barriers that particular species cannot cross. Oceans, deserts, and mountain ranges are examples of such barriers. Indeed it is found that when such barriers are overcome, for example by humans transporting a species from one continent to another, the introduced species frequently *does* make a successful "invasion." But a successful invasion by a foreign species is rarely a desirable event. For example, around the turn of the century a fungus that was parasitic on American chestnut trees was unwittingly introduced from Asia into the United States through the introduction of Chinese chestnuts which were resistant carriers. Over the next 50 years nearly all the American chestnuts, which were the dominant tree in eastern deciduous forests, perished. In short, an introduced species may upset the existing biotic factors and lead to gross alteration of the recipient system, including the extinction of many of its species. Additional examples of this will be explored in Chapter 3.

## IMPLICATIONS FOR HUMANS

What relevance does the study of ecosystems have to us as humans? Given our highly complex technological society organized into cities, industries, transportation systems, and various human institutions, we may often feel quite independent from nature. Fundamentally, however, we along with our domestic plants and animals comprise what is known as the **human ecosystem.** Because we maintain, so far as possible, a rigorous control over it, our human ecosystem is an artificial as opposed to a natural ecosystem. Nevertheless, it has the same attributes as natural ecosystems. Our crop plants are producers, livestock and poultry are primary consumers, and we ourselves are both primary and secondary consumers or omnivores. Furthermore, despite our attempts at control, our human ecosystem is still largely at the mercy of natural abiotic factors. Consider, for example, the effects of droughts, floods, or abnormal freezes on agricultural production. Likewise, our human ecosystem has innumerable biotic interrelationships with natural species and ecosystems, some desirable, other undesirable. Desirable interrelationships include our obtaining fish, wood, and many other valuable resources from natural ecosystems. Undesirable factors are seen in our continuing battle against parasites, rats, mice, flies, cockroaches, weeds, and many other organisms that affect both our domestic plants and animals and ourselves.

However, the human ecosystem is not limited as are natural ecosystems. We are able to (and are) invading all natural ecosystems and to a large extent supplanting them with our own agricultural ecosystem. In those that we are not totally supplanting, we are upsetting the natural balances by changing both biotic and abiotic factors in various ways. Examples, which will be discussed in more detail later in the text, are listed below.

### Altering Biotic Factors

1. Exploiting and hence drastically reducing certain species
2. Introducing species from one ecosystem into another
3. Introducing grazing animals that upset the balance between competing plant species.

### Altering Abiotic Factors

1. Diverting water for human use and thereby causing other areas such as wetlands and lakes to be depleted
2. Overfertilizing waterways through the discharge of nutrient-rich wastes
3. Changing pH through acid rain
4. Perhaps altering the climate of the entire earth through the $CO_2$ greenhouse effect

Because altering any biotic or abiotic factor will influence one or more species directly, far-reaching biological changes may occur through a domino effect.

As alterations of both biotic and abiotic factors become progressively greater under the pressure of increasing human populations and development, a serious question must be addressed: Are we managing our human ecosystem in an ecologically sound way so as to ensure the long-term survival of the human species? The answer should be unsettling to say the least. The more we learn about ecology, the more we recognize how the stability of an ecosystem is dependent on the dynamic balance provided by many interacting factors. Thus the more we upset or destroy natural ecosystems, the more we place in jeopardy the overall stability and sustainability of our own human ecosystem. The recent famine affecting Ethiopia and a number of other countries in Africa is, in many ways, the result of ecological mismanagement. Thus, it provides a vivid illustration of the tragic end-point to which such mismanagement can lead. Yet, as we shall emphasize throughout this text, we now have both the knowledge and the technology required for sound ecological management. It is only a matter of our desire and will as individuals and as a society to put what is known into practice.

Now that we have gained some understanding of what ecosystems are, we shall continue in Chapter 2 with a discussion of the principles that underlie the functioning of ecosystems.

# 2

# HOW ECOSYSTEMS WORK

## CONCEPT FRAMEWORK

| Outline | | Study Questions |
|---|---|---|

1. What are the six most significant elements that make up the chemical structure of living things?

2. Explain the chemical differences between the abiotic sphere and the structure of living things. What is the *law of conservation of matter* and how does it apply to fundamental life processes?

3. Define *matter*. Define *energy*. Contrast the two.

4. Discuss the three laws of thermodynamics and why they play an important role in determining how living organisms and ecosystems function.

5. What chemical processes make producers significant?

6. Distinguish between *producers* and *consumers*. What is *cell respiration* and how does it relate to the energy requirements of various organisms?

7. What are *decomposers* and what role do they play in the food chain?

8. Explain element recycling and its importance to the balance of the ecosystem?

9. Differentiate between the *carbon cycle* and the *phosphorus cycle*. What effects may human disruption of these cycles have on the ecosystem?

10. What makes the *nitrogen cycle* significant? What roles does it play in agriculture?

11. Why is biomass important? For what reasons may scientists want to predict biomass at various trophic levels?

12. Give several examples of how human activity has upset the ecological balance and examine the potential impact of such upsets.

If we look at all motor vehicles we see what appears to be a great difference between various trucks, buses, cars, and motorcycles. However, at the basic structural and functional level, they are essentially the same. They all have the basic structure of an engine, transmission, and wheels mounted on or in a body or chassis. Functionally, they all burn fuel in the engine, which delivers power through the transmission to the wheels.

Likewise, while ecosystems appear to be vastly different, ranging from tropical forests to deserts and tundra, we found in Chapter 1 that they all have the same basic parts or structural components: producers, consumers, and decomposers. Our objective in this chapter is to gain a better understanding of how these components function together in terms of the movement of energy and nutrients. Of particular interest is how many ecosystems, barring human interference, are able to perpetuate themselves more or less indefinitely. How do they avoid depleting resources and/or accumulating intolerable levels of pollutants, problems that seem to plague human societies?

In order to achieve an understanding of how ecosystems function, it is first necessary to have some concepts at the chemical level.

## ELEMENTS, LIFE, ORGANIZATION, AND ENERGY

A discussion of atoms of different elements and how they bond to form molecules and compounds of various gases, liquids, and solids is

**Table 2-1** | **Elements Found in Living Organisms and Their Biologically Important Locations in the Environment**

| ELEMENT (kind of atom) | SYMBOL | BIOLOGICALLY IMPORTANT MOLECULE OR ION IN WHICH THE ELEMENT OCCURS[a] | | LOCATION IN THE ENVIRONMENT | | |
|---|---|---|---|---|---|---|
| | | Name | Formula | Air | Dissolved in Water | Some Rock and Soil Minerals |
| Carbon | C | Carbon Dioxide | $CO_2$ | X | X | |
| Hydrogen | H | Water | $H_2O$ | | (Water Itself) | |
| Oxygen (required in respiration) | O | Oxygen Gas | $O_2$ | X | X | |
| Oxygen (released in photosynthesis) | $O_2$ | Water | $H_2O$ | | (Water Itself) | |
| Nitrogen | N | Nitrogen Gas | $N_2$ | X | X | |
| | | Ammonium Ion | $NH_4^+$ | | X | X |
| | | Nitrate Ion | $NO_3^-$ | | X | X |
| Sulfur | S | Sulfate Ion | $SO_4^{2-}$ | | X | X |
| Phosphorus | P | Phosphate Ion | $PO_4^{3-}$ | | X | X |
| Potassium | K | Potassium Ion | $K^+$ | | X | X |
| Calcium | Ca | Calcium Ion | $Ca^{2-}$ | | X | X |
| Magnesium | Mg | Magnesium Ion | $Mg^{2+}$ | | X | X |
| *Trace Elements*[b] | | | | | | |
| Iron | Fe | Iron Ion | $Fe^{2+}$, $Fe^{3+}$ | | X | X |
| Manganese | Mn | Manganese Ion | $Mn^{2+}$ | | X | X |
| Boron | Bo | Boron Ion | $Bo^{2+}$ | | X | X |
| Zinc | Zn | Zinc Ion | $Zn^{2+}$ | | X | X |
| Copper | Cu | Copper Ion | $Cu^{2+}$ | | X | X |
| Molybdenum | Mo | Molybdenum Ion | $Mo^{2+}$ | | X | X |
| Chlorine | Cl | Chloride Ion | $Cl^-$ | | X | X |

NOTE: These elements are found in *all* living organisms: plants, animals, and microbes. Some organisms require certain elements in addition to these. For example humans additionally require sodium and iodine.
[a]A molecule is a chemical unit of two or more atoms bonded together. An ion is a single atom or group of bonded atoms that has acquired a positive or negative charge as indicated.
[b]Elements of which only small or trace amounts are required.

given in Appendix C. You may wish to study this to gain background helpful in understanding this material.

## Organization of Elements in Living and Nonliving Systems

Many early biologists and chemists studied living things expecting to find some particular substance or vital essence that was responsible for imparting "life" to organisms. No such substance and no evidence that such a substance or element exists has ever been found. Instead they found the same **elements** or kinds of **atoms** in living things as in nonliving matter, that is, air, water, and minerals. Furthermore, of the 96 elements that occur in nature only about 20 are found in living organisms. These elements and where they occur in the environment are given in Table 2-1.

**FIGURE 2-1**
Clean dry air is a mixture of three important gases as indicated. In addition, air generally contains a certain amount of water vapor and dust. Also, numerous other compounds may be present as pollutants. It is the lack of bonding or other attractions between molecules that results in air being gaseous. When attractions between molecules or ions occur, the result is a liquid or solid.

The most significant of these are *carbon (C)*, *hydrogen (H)*, *oxygen (O)*, *nitrogen (N)*, *phosphorus (P)*, and *sulfur (S)*. You can remember them by the acronym *N. CHOPS*.

The key feature that does distinguish the chemical structure of living and nonliving things is the way these atoms are organized or bonded into molecules. A **molecule** is defined as a chemical unit of two or more atoms bonded together. In the abiotic sphere, molecular combinations are relatively simple. Clean dry air is a simple *mixture* of molecules of three important gases: oxygen ($O_2$), nitrogen ($N_2$) and carbon dioxide ($CO_2$) as shown in Figure 2-1. A few other gases that have no biological importance are also present. Water is comprised of molecules formed by the bonding of two hydrogen atoms to an oxygen atom, $H_2O$, as shown in Figure 2-2. Rock and soil minerals are made up of dense clusters of atoms of two or more elements bonded together by the attraction between positively and negatively charged atoms (Fig. 2-3).

There are interactions between air, water, and minerals: molecules from the air may dissolve in water, water molecules may enter the air as water vapor, and various minerals may dissolve

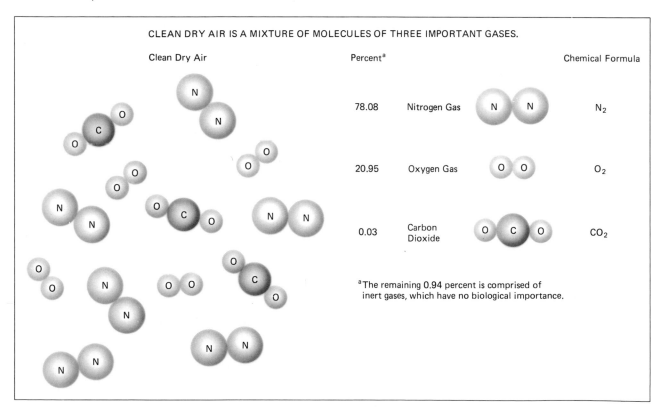

CLEAN DRY AIR IS A MIXTURE OF MOLECULES OF THREE IMPORTANT GASES.

Clean Dry Air

| Percent[a] | | Chemical Formula |
|---|---|---|
| 78.08 | Nitrogen Gas | $N_2$ |
| 20.95 | Oxygen Gas | $O_2$ |
| 0.03 | Carbon Dioxide | $CO_2$ |

[a] The remaining 0.94 percent is comprised of inert gases, which have no biological importance.

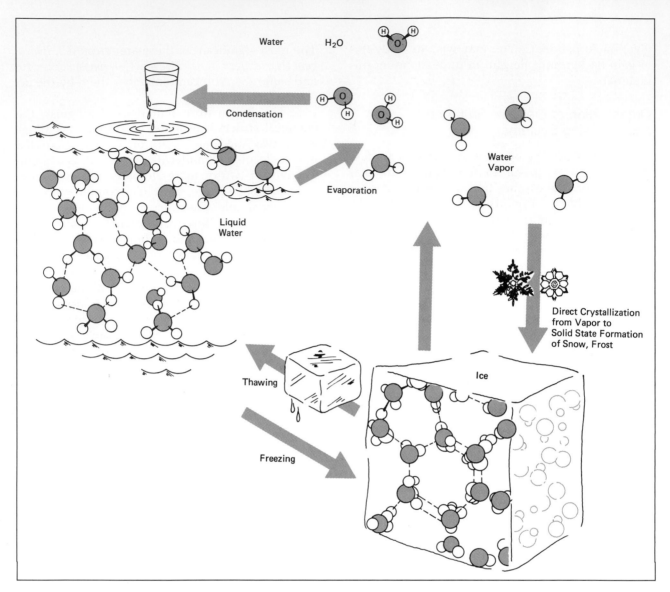

**FIGURE 2-2**
Water consists of molecules, each of which is made from two hydrogen atoms bonded to an oxygen atom. In addition, there is a weak attraction between water molecules indicated by the dotted lines. At normal temperatures, molecular kinetic energy is such that attractions break and re-form, producing the liquid state. As water is cooled, molecules lose energy and attractions become rigid, forming ice. As water is heated, molecules gain sufficient energy to break the weak attraction betweeen them and they become water vapor. As water vapor is cooled, the attractive force again holds molecules together, resulting in the re-formation of water or condensation. In all these changes in state, the water molecules themselves remain the same, $H_2O$.

into and recrystallize from water solution (Figure 2-4). Nevertheless, the molecular combinations remain relatively simple. Notably absent from the abiotic sphere are organic molecules (except where they have been introduced by humans). Chemically speaking, **organic** molecules are molecules containing carbon-carbon bonds and/or carbon-hydrogen bonds. Thus, the abiotic sphere is basically inorganic.

In contrast the chemical structure of living things is based entirely on organic molecules. Proteins, carbohydrates, fats, lipids, sugars, nucleic acids such as DNA (the genetic material), and innumerable other molecules of living organisms are constructed in large part from chains of carbon atoms bonded with each other and/or with hydrogen and certain other elements such as nitrogen, phosphorus, and sulfur. In turn, organic molecules make up the structural and functional

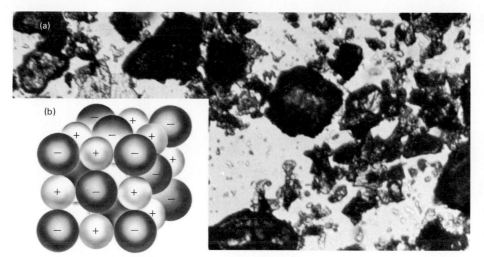

**FIGURE 2-3**
Rock and soil minerals. (a) Mineral particles of soil enlarged 100 ×. Minerals, as seen in rock or inorganic soil particles, are comprised of dense clusters of atoms of two or more elements. The atoms (or small groups of atoms) have acquired a positive or negative charge and are bonded together by the attraction between opposite charges as shown in the insert. (b) Atoms or small groups of atoms with a charge are called *ions*. Various minerals are formed by different combinations of elements, some of which form positive ( + ) ions, others of which form negative ( − ) ions. (Photo by author.)

components of living cells, which in turn make up the entire living organism (Fig. 2-5). The complexity of such molecules is fantastic—some may contain millions of individual atoms—and their potential diversity is infinite. Indeed, the diversity of living things is a function of the diversity of organic molecules.

In contrasting living and nonliving things, what we see, then, is not different elements but a profound difference in the *organization* of these elements into molecules. All living things are based on complex organic molecules while the nonliving, natural sphere is based on inorganic molecules.

Thus, on a chemical level, the fundamental life processes of growth and reproduction may be seen as a chemical process of "reorganizing" carbon, hydrogen, and other elements from their simple arrangement in inorganic molecules into the complex arrangements of organic molecules. Conversely, what occurs chemically in the process of death and decay (or burning) is a breakdown of the complex organic molecules and the rearrangement of their constituent atoms into the simple inorganic molecules. For example, as sugar is burned its carbon, hydrogen, and oxygen atoms re-form into carbon dioxide and water (Fig. 2-6). Additional oxygen is also generally used in this burning and hence this process is also referred to as **oxidation.**

The **law of conservation of matter** should be emphasized. In chemical processes, atoms are neither created nor destroyed. They are only rearranged into different molecules or compounds.

An exception to the generalization of equating "nonliving" with "inorganic" is that chemists now produce innumerable **synthetic** organic compounds. Such human-made organic compounds form the basis of all plastics and synthetic fibers such as polyesters and many solvents, drugs, and pesticides. These compounds are based on bonded carbon atoms; hence they are organic by the chemical definition, but they have nothing to do with natural life processes except that many prove to be highly toxic (e.g., pesticides), and some may benefit certain abnormal or disease conditions (e.g., medicines).

## Considerations of Energy

Chemical life processes, however, involve more than the arrangement of complex organic molecules and their subsequent breakdown. Considerations of **energy** are also crucially important. To begin this consideration we must be clear concerning the basic distinction between matter and energy.

### MATTER AND ENERGY

*Matter* is defined as anything that occupies space and has mass, that is, can be weighed when gravity is present. The particles that make up atoms—the subatomic particles known as **protons, neutrons,** and **electrons**—are the basic units of matter. Thus, matter includes all solids, liquids, and gases, and living as well as nonliving things.

In contrast, common forms of *energy* include *heat, light* and other forms of *radiation, motion,* and

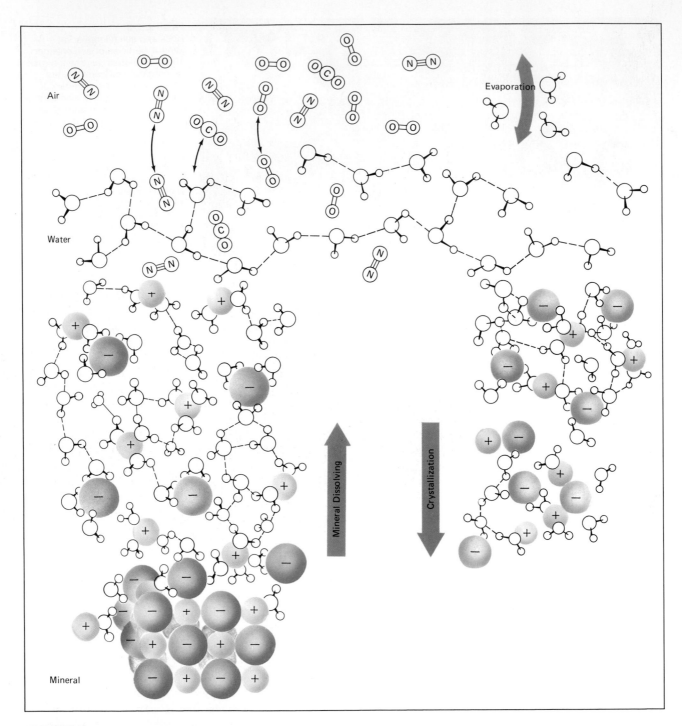

**FIGURE 2-4**
Interrelationships between air, water, and minerals. Minerals and gases dissolve
in water, forming solutions. Water evaporates into air, causing humidity. These
processes are all reversible. As water evaporates, minerals in solution
recrystallize. Water vapor in the air condenses to re-form water.

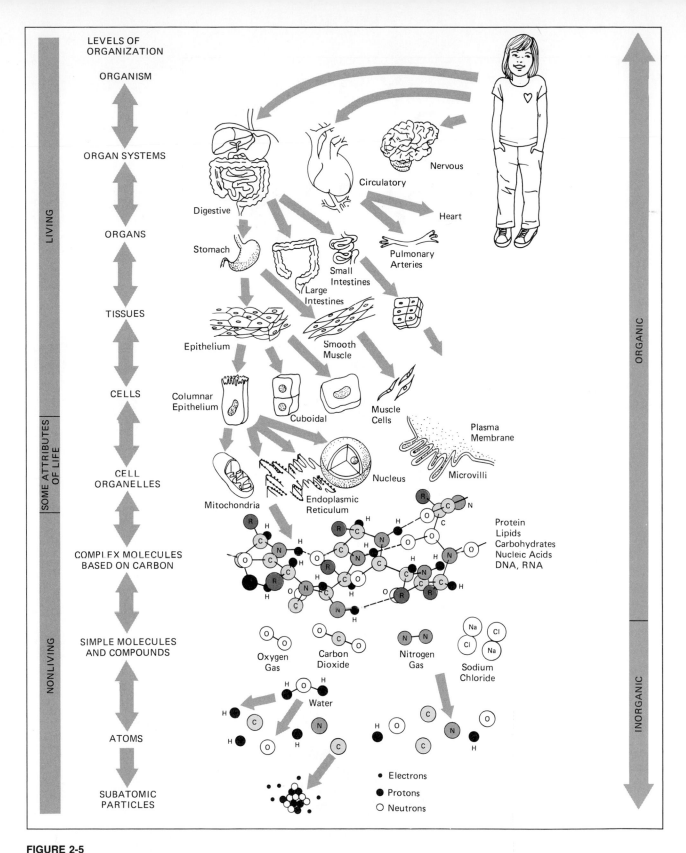

**FIGURE 2-5**
Life can be seen as a hierarchy of organization of matter. In the inorganic sphere, elements are in very simple arrangements of molecules of the air, water, and minerals. In living organisms, they are arranged in very complex organic molecules which, in turn, make up cells which comprise the whole organism.

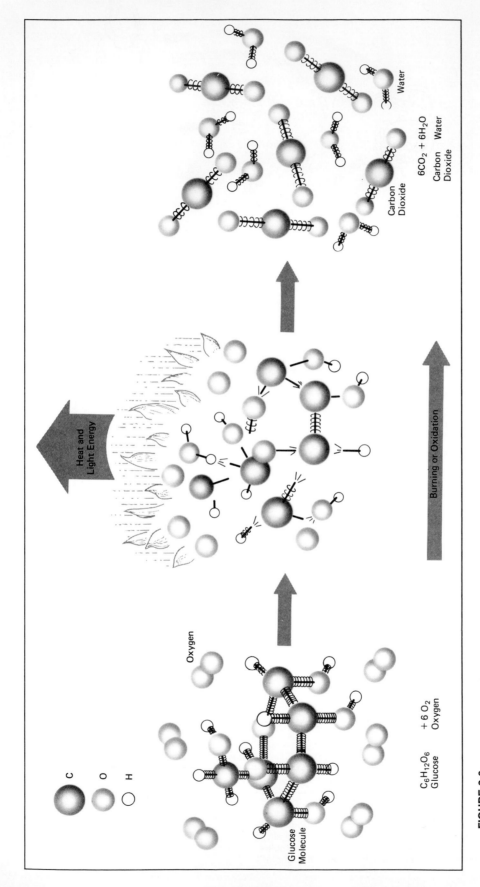

**FIGURE 2-6**

Burning or oxidation of glucose. Burning or oxidation of complex organic molecules involves their breakdown and the rearrangement of their atoms into simple molecules. Generally oxygen is used in the process, and carbon dioxide and water are the end products. Note that no atoms are created or destroyed; they are only rearranged. Also extremely important in this reaction is the release of energy, potential energy contained within the glucose molecule. All heterotrophs or consumers, including humans, derive their body energy from the oxidation of organic molecules, such as glucose, through a process called *cell respiration.*

C $C_6H_{12}O_6$
Glucose

O $+ 6 O_2$
Oxygen

H

Glucose
Molecule

Oxygen

Heat and
Light Energy

Burning or Oxidation

Water

$6CO_2 + 6H_2O$
Carbon        Water
Dioxide

Carbon
Dioxide

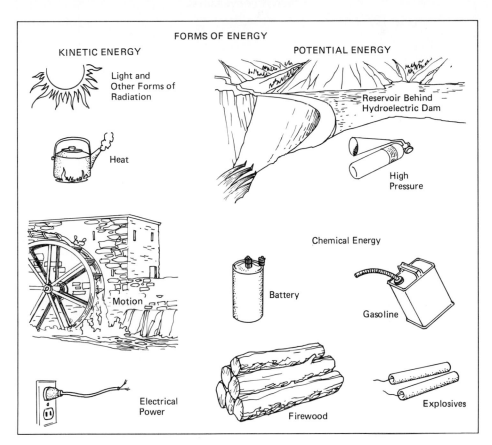

**FORMS OF ENERGY**

KINETIC ENERGY

Light and Other Forms of Radiation

Heat

Motion

Electrical Power

POTENTIAL ENERGY

Reservoir Behind Hydroelectric Dam

High Pressure

Chemical Energy

Battery

Gasoline

Firewood

Explosives

**FIGURE 2-7**
Energy. Energy is distinct from matter in that it neither has mass nor occupies space, but it has the ability to act on matter, changing its position and/or its state. Kinetic energy is energy in one of its active forms. Potential energy refers to systems or materials which may release kinetic energy.

*electrical power.* Note that these do not occupy space; that is, one cannot have a tank or balloon full of light. Nor do they have mass; one cannot weigh light or other forms of energy. But, all forms of energy have one property in common: they may cause a change in matter. The change in matter may involve a change either in *position* or *motion* (acceleration or turning) or it may involve a change in *state* (e.g., changing water to water vapor by adding heat energy). Physicists define such a change as **work;** consequently, *energy* is defined as the ability to do work.

Further, energy is commonly divided into two categories: *kinetic* and *potential* (Fig. 2-7). Kinetic means active. Thus **kinetic energy** is energy in action or motion. The kinds of energy noted above are all forms of kinetic energy. On the other hand, a substance or system with **potential energy** has the capacity or potential to release one or more forms of kinetic energy. To illustrate, gasoline has high potential energy; when it is burned, it releases heat, light, and, in an engine, its combustion drives the moving parts. The high potential energy noted in such chemicals or fuels

as gasoline, oil, coal, wood, and natural gas (methane) is also referred to as **chemical energy.**

Countless observations and experiments involving energy and energy transformations have led to the derivation of several natural laws, the **laws of thermodynamics.** These laws are extremely important to our understanding of how living organisms and ecosystems function.

The *first law of thermodynamics* states that, like matter, energy is neither created nor destroyed, but any form of energy may be converted directly or indirectly into any other form. Some familiar examples of converting energy are shown in Figure 2-8. Particularly significant is that all forms of energy can be converted to *heat*, which is measured in **calories.** Thus, any form of energy can be measured in calories. The number of calories indicates the amount of work that can potentially be performed. You are probably most familiar with food calories, a measure of the energy content of various foods. A calorie is defined as the amount of heat required to raise the temperature of 1 g (approximately 0.03 oz) of water 1 degree Celsius (1.8 degrees Fahrenheit). But, since

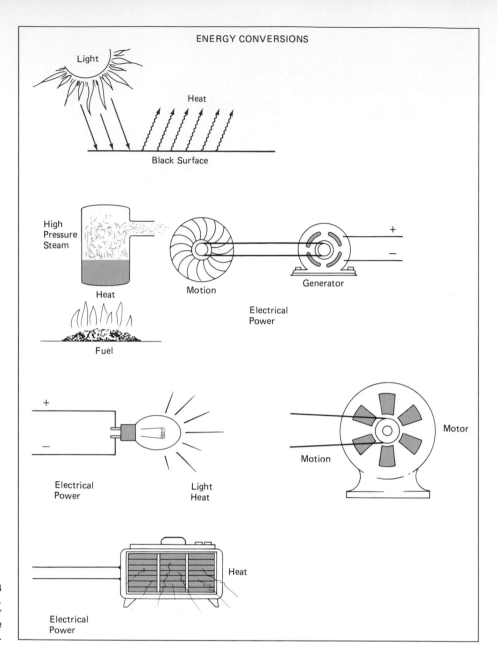

ENERGY CONVERSIONS

Light

Heat

Black Surface

High Pressure Steam

Heat

Fuel

Motion

Generator

Electrical Power

+

−

Electrical Power

Light Heat

Motion

Motor

Electrical Power

Heat

**FIGURE 2-8**
The first law of thermodynamics. Energy is neither created nor destroyed but any form can be converted into any other.

this is a very small unit, it is frequently more convenient to speak in terms of kilocalories (1 kilocalorie = 1000 calories), the amount of heat required to raise 1 ℓ (1.057 qt) of water 1 degree Celsius. Food calories (often capitalized as Calories to distinguish them from "small" calories) are actually kilocalories. In any case, the calorie content of an energy source indicates the work that can be obtained from it if it is suitably converted.

Because energy is not destroyed and because it may be converted from one form to another, it might seem that we could recycle it continuously and run our world without new energy imputs.

Unfortunately, this is impossible because here we are confronted with the *second law of themodymics,* which can be stated as follows: Energy only flows one direction, "downhill." In terms of heat, this is to say that energy always flows from a hot body to a cooler one or to cooler surroundings. For example, when we boil a pot of water we put it over a burner and the high heat from the burner flows into the cooler pot. In turn, the heat from the pot flows into the cooler surroundings. We never observe the heat from the cooler surroundings flowing into and concentrating in the pot.

Relating this to energy's ability to do work,

we find that energy can only perform work as it flows "downhill." Consequently, as energy is converted to heat and the heat reaches the temperature of the surroundings, no further work can be performed by it nor can we get it back!

Here is the reason we cannot recycle energy: in every energy conversion, at least part of the energy is converted to heat and flows out of the system towards cooler surroundings. Consequently, in every energy conversion, there is a *loss of energy*, the amount that escapes as heat. Thus the second energy law can also be stated as follows: every energy conversion involves a loss of energy (Fig. 2-9).

While energy always flows "downhill" and there is always a loss of energy, importantly, this does not preclude part of the system from gaining energy, at least temporarily. Consider the pot of water again. The cooled water can be reheated (increased in energy content) by placing it back on the burner. But if we measure the expenditure of burner calories we shall find that more calories were lost here than are gained by the water. The net loss is heat that escapes into the surroundings. Indeed, the net loss for energy conversions is frequently greater than 50 percent. Said another way, most energy conversions are less than 50 percent efficient. Many are in the range of only 1 to 10 percent efficient, the other 90 to 99 percent being lost as waste heat in the process.

In conclusion, we may see a gain in energy in one place but for every such gain there must be a greater loss of energy in some other part of the system. We shall find that this has tremendous implications concerning the way we use energy in our society as well as for ecosystems.

## ENERGY AND ORGANIC MATTER

Now, let us connect these concepts of energy to our discussion of organic matter and living systems.

Organic molecules, which make up living organisms, are really more than just the carbon, hydrogen, and certain other atoms that comprise them. They also contain *high potential energy*. On the other hand, inorganic molecules, such as carbon dioxide, water, and rock and soil minerals, are very *low* in potential energy. This is made conspicuous by the fact that organic materials such as wood, fats, and oils burn and release substantial quantities of energy, whereas carbon dioxide, water, and minerals don't burn; in fact, they may be used to extinguish fires (Fig. 2-10).

Thus, on the chemical level, the life processes of growth and reproduction involve an *increase* in potential energy as well as the net production or organic matter. Processes of food breakdown and decay (breakdown of organic molecules) involve a *release* of potential energy.

I'll use power from the generator to run the motor which will drive the generator to give me more power. . .

Heat

Heat

Motor

Generator

Why Doesn't It Work?

**FIGURE 2-9**
The second law of thermodynamics. Energy cannot be recycled because every conversion involves a loss of heat which cannot be recovered because heat only flows toward cooler surroundings. Consequently, the power output of a generator, for example, will always be less than the power input because of the loss of energy as heat.

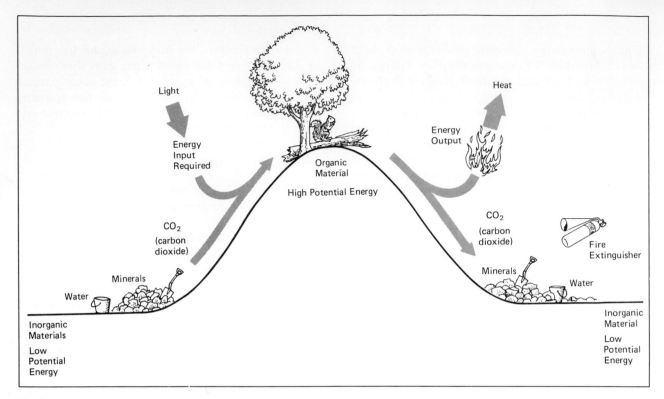

**FIGURE 2-10**
Difference in energy level between organic and inorganic states of matter. Inorganic materials are at low potential energy; organic material is at high potential energy. The change is made possible by the input of energy from light through photosynthesis. Energy is released again as organic material is again broken down to inorganic material.

Some of the released potential energy is used for necessary life functions. The remainder is lost as waste heat. You observe this heat as body heat and you are familiar with the fact that the harder you exercise, the more heat your body produces. Also you "work up an appetite," which is your body asking for more food fuel to burn.

A key question that may be asked at this point is: Where does the energy come from that must go into producing the high-potential-energy organic molecules? The answer is that it comes from light energy trapped by plants in the process of photosynthesis.

## MATTER AND ENERGY CHANGES IN PRODUCERS, CONSUMERS, AND DECOMPOSERS

With these concepts of matter and energy in mind, the functional relationships between producers, consumers, and decomposers, discussed in Chapter 1, should now be more clear.

## Producers

Producers are all those organisms that are capable of utilizing an abiotic (nonorganic) energy source for the production of organic compounds from simple inorganic chemicals in the environment. In all major ecosystems the primary producers are chlorophyll-containing plants, which use light energy to produce sugars from carbon dioxide and water, releasing oxygen as a byproduct, in the process known as *photosynthesis*. They then go on to use the sugars, plus a few additional inorganic compounds absorbed from the soil which supply nitrogen, phosphorus, sulfur and other elements to synthesize other organic compounds that make up their cells. Some sugar is also broken down to $CO_2$ and $H_2O$ to release the energy required for these conversions (Fig. 2-11).

Chlorophyll-containing plants, whether as tiny as microscopic algae or as large as trees, are generally recognizable by their "greenness," which is indicative of chlorophyll and hence the capacity to carry on the photosynthetic process. However, there are cases where the green of chlorophyll is masked by the presence of additional pigments. Japanese maples (red), purple plants, red algae, and brown algae are examples. Such

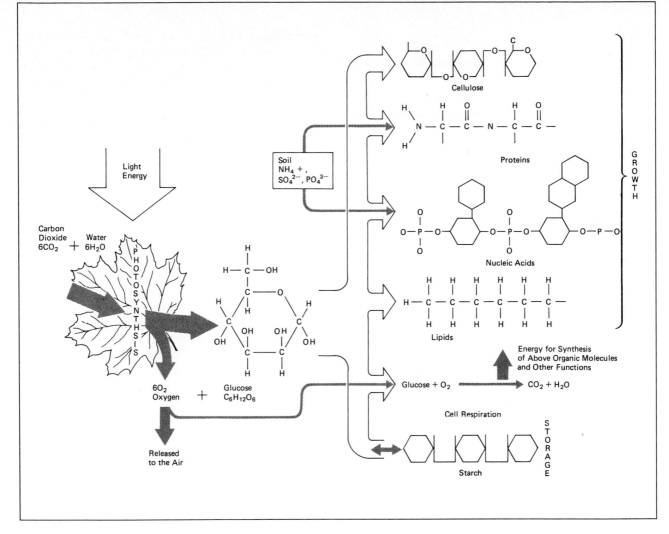

**FIGURE 2-11**
Producers. Producers are remarkable chemical factories. Using light energy, they make glucose from carbon dioxide and water. Oxidizing some of the glucose to provide additional chemical energy, they convert the remaining glucose with certain additional nutrients from the soil into other complex organic molecules used in growth.

plants are still active photosynthesizers and hence producers, as are a few photosynthetic bacteria that absorb light energy through alternative pigments. But all fungi, regardless of color, and a few higher plants as well, as noted in Chapter 1, lack photosynthetic pigments. Consequently, they are not producers.

By far the most abundant source of abiotic energy is light; hence the most predominant producers are photosynthetic organisms. However, a small number of high-potential-energy, inorganic compounds do exist, and in certain situations they are in relatively high concentrations. Hydro-

gen sulfide ($H_2S$) emitted from "sulfur" springs is an example. Certain species of bacteria are capable of using the energy from these inorganic compounds as green plants use light. This process of using chemical energy is known as **chemosynthesis.** There are a few highly specialized situations, such as in deep ocean rifts where light does not penetrate, in which chemosynthetic bacteria are the major producers of the ecosystem.

Importantly producers do not violate the second law of thermodynamics in their production of high-potential-energy organic compounds. Indeed, one finds that generally only 1 to 5 calories worth of organic chemicals are produced for every 100 calories worth of light energy falling on the plant. Thus, they are really only 1 to 5 percent efficient in converting light energy to stored chemical energy.

## Consumers

Like producers, consumers require nutrients for the synthesis of complex organic compounds which make up their cells, and they also require energy for this synthesis. Then, they require additional energy for the various active movements that typify animals. Consumers meet both of these requirements through the ingestion of organic food, some of which is converted into the body of the consumer, and some of which is broken down to release its potential energy. Also there is a certain quantity of ingested material that is not digested but passes out as fecal waste.

### ORGANIC MATTER FOR BODY GROWTH, MAINTENANCE, AND REPAIR

The chemical capability of consumers is, in general, less sophisticated than that of plants. Ingested nutrients for body growth and maintenance must include many specific organic nutrients such as amino acids (present in proteins) and vitamins, as well as less specific material. Certain elements such as calcium, iron, and sodium are also required, and may be ingested as inorganic compounds, such as salt for sodium. If any one or more of these specific nutrients are not present in the diet, various diseases of **malnutrition** will develop regardless of how much other nonspecific material may be consumed. Here you may see the problem of *junk food;* it contains few, if any, of the specific nutrients required for body growth, maintenance, and repair.

### ORGANIC MATTER FOR ENERGY

The energy requirements of all consumers are met by their breaking organic molecules down through a process called **cell respiration.** We observed earlier, in Figure 2-6, that the potential energy in **glucose** may be released by oxidizing or burning it in the presence of oxygen:

$$C_6H_{12}O_6 + 6\ O_2 \longrightarrow 6\ CO_2 + 6\ H_2O$$

$$\text{Glucose} \quad \text{Oxygen} \quad \downarrow \quad \text{Carbon} \quad \text{Water}$$
$$\text{Energy} \quad \text{Dioxide}$$

Cell respiration is the same overall reaction. The only distinction is that in cell respiration the overall reaction occurs in many small steps so that the energy is released in small "packets" appropriate for performing body functions. You may note that in its overall aspects cell respiration is the *reverse of photosynthesis;* oxygen and glucose are consumed while carbon dioxide and water are waste products.

Since oxygen consumption and carbon dioxide release are directly related to energy released through cell respiration, and these gases are exchanged through the lungs, can you see why your breathing rate increases with physical activity? It is also interesting to note that water is a waste product of energy metabolism. Can you see how certain desert mammals (e.g., the kangaroo rat) and insects survive without ever drinking water? Their bodies, being very water conserving, gain sufficient water from cell respiration. Humans and most animals require additional water because they are basically water wasteful.

Importantly, cell respiration is *not* restricted to consumers. Decomposers also require energy for growth, as do producers at times and in parts of the plant where photosynthesis does not occur. These energy requirements are, again, met through the process of cell respiration. Some decomposers, however, are able to derive sufficient energy from the partial breakdown of glucose, which may occur in the *absence* of oxygen gas. Such metabolism is referred to as **anaerobic** (without air) **respiration.** The waste products of anaerobic respiration, one form of which is fermentation, include such products as alcohol, methane gas, and acetic acid (as in vinegar).

Any natural organic compound may be broken down, releasing its potential energy. If the organic compound being broken down in cell respiration is glucose or another molecule that is made of just carbon, hydrogen, and oxygen atoms, the wastes are only carbon dioxide and water. However, if the organic molecule being broken down also contains elements such as nitrogen, phosphorus, or sulphur, these elements are left as inorganic waste compounds, or "ash." Such wastes are flushed out of the body through the urine or analogous waste in other kinds of animals. (Because these wastes are mineral in nature, they will not pass off as gas through the lungs; but they are soluble in water and consequently are removed in water solution.) These consumer functions are diagrammatically summarized in Figure 2-12.

You may note here that consumers require

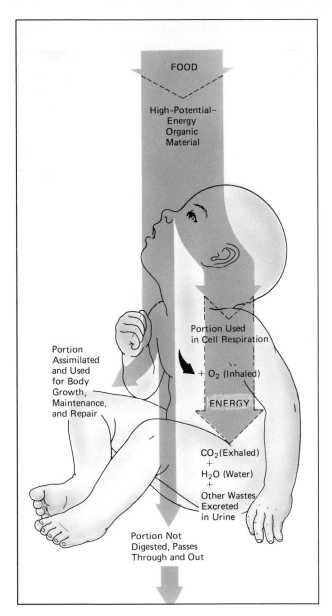

**FIGURE 2-12**
Consumers. Only a portion of the food ingested by a consumer is assimilated into body growth, maintenance, and repair. A larger amount is oxidized through cell respiration to provide energy for assimilation, movements, and other functions. Waste products of oxidation are carbon dioxide, water, and various mineral nutrients. A third portion is not digested but passes through, becoming fecal wastes.

Labels within figure:
FOOD
High-Potential-Energy Organic Material
Portion Used in Cell Respiration
Portion Assimilated and Used for Body Growth, Maintenance, and Repair
+ $O_2$ (Inhaled)
ENERGY
$CO_2$ (Exhaled) + $H_2O$ (Water) + Other Wastes Excreted in Urine
Portion Not Digested, Passes Through and Out

**FIGURE 2-13**
Animal wastes are plant fertilizer. The oxidation of food for energy leaves the inorganic nutrients needed by plants. The photograph shows a dog urine spot on a lawn. The "donut" of dark green grass is where urine has been diluted to optimal concentration; the grass in the center has been killed by overfertilization by concentrated urine. (Photo by author.)

both the organic material (food) and the oxygen produced by plants, and that the respiratory wastes, carbon dioxide and water, and the inorganic wastes are exactly the nutrients needed by producers. Indeed, urine is a good fertilizer if properly diluted (Fig. 2-13). The waste from con-

sumers being reusable by producers is a key to ecosystem function on which we shall elaborate later in this chapter.

In concluding this section, it is of interest to ask what proportion of food is required for energy versus nutritive functions of growth, maintenance, and repair. The answer varies depending on the rate of growth, the activity level, and the consumer in question. However, always more than half and frequently in the order of 70 to 90 percent of what is absorbed is required for energy. But it remains critically important to have the remaining percentage contain the specific nutrients for growth, maintenance, and repair. Herein lies a nutritional problem. Many people find it all too easy to overconsume sugars, refined starches, fats, oils, and alcohol. Such materials are commonly referred to as *empty calories* or *junk food*, implying, correctly, that they supply energy but none of the nutrients necessary for growth, maintenance, and repair of the body. As noted before, lack of nutrients leads to diseases of malnutrition despite the level of calorie consumption.

A related problem involves the need to balance calorie input and output. If calorie input is greater than output, the excess is converted to body fat, leading to excessive weight gain. Conversely, calorie output being greater than input leads to weight loss as body material is broken down to meet the deficit; if continued long enough calorie deficit leads to starvation and death. A balanced diet for humans or any other animal is one that supplies proper quantities and proportions of both calories and nutrients.

### ORGANIC MATTER THAT IS NOT DIGESTED

Finally, a portion of what is eaten is not digested but simply passes through the digestive system and comes out as fecal wastes. In consumers that eat plants, such material is largely plant cell walls and we refer to it as *fiber, bulk* or *roughage*. Some such material is necessary for the intestines to have something to push through to keep them clean and open. However, woody organic material is essentially all cell wall material known as **cellulose** and, as such, has no food value to most consumers. This brings us to decomposers.

## Decomposers

While consumers in general cannot break down cellulose (cell wall material) and gain any food value from it, certain bacteria and fungi can. These organisms, decomposers, feed on such material, giving it the appearance of rotting or decomposing. Of course decomposers may also feed on other organic material, causing the rotting of food.

Like other consumers, fungi and bacteria consume organic matter both as a source of raw material for growth and as a source of energy. About half the organic material they consume is broken down through cell respiration to release energy to assimilate the remainder into growth. Again, carbon dioxide, water, and other inorganic compounds are released and there is also a dissipation of waste heat; indeed the heat in a pile of rotting material may be quite intense.

Importantly, while most animals cannot derive food value from woody material, they can gain food value (energy and nutrients) by eating the fungi and/or bacteria that feed on the detritus. Caution must be taken, however, because some species are poisonous. Commercial mushrooms are grown mostly on horse manure. Thus, decomposers are very significant in routing detritus back

into consumer food chains as well as breaking it down. Indeed, many grazing animals such as goats and cows utilize coarse plant material (largely cellulose) through maintaining bacteria in their specialized stomachs.

## ECOSYSTEM FUNCTIONS: NUTRIENT CYCLING AND ENERGY FLOW

When one recognizes the necessary inputs and outputs of producers, consumers, and decomposers, one may be struck by how they fit together to make an integrated system. When the parts are arranged as in Figure 2-14, two fundamental principles of ecosystem functioning become evident: (1) elements are continually recycled from the environment through one or more organisms back to the environment, and (2) there is a continuous flow of energy through the system as it enters as light and exits as waste heat which cannot be reused. We shall look at these two processes in somewhat more detail.

### Cycling of Elements

The general theme of element recycling is as follows: the elements are absorbed from simple organic compounds in the environment and many are **assimilated** into complex organic compounds by producers. From here, they may be passed to almost any number of consumers and/or decomposers, depending on what is eaten by what in a food chain. At every point, however, a portion of the total is released back to the environment as a portion of organic material is broken down to release its energy. Also, combustion of organic matter will release the elements back to the environment as simple inorganic compounds. In either case, once back in the environment, the elements may be reabsorbed by producers and recycled.

Importantly, there are no obligatory "dead ends" where wastes simply accumulate. Instead, any wastes from one part are the necessary raw materials for another. Given this ability to recycle elements, an ecosystem will not deplete its reservoir of nutrients nor will it poison itself with wastes (pollution) unless, in one way or another, the recycling process is disrupted. But failing to recycle any one element could lead to collapse of the entire ecosystem. Recall the law of limiting

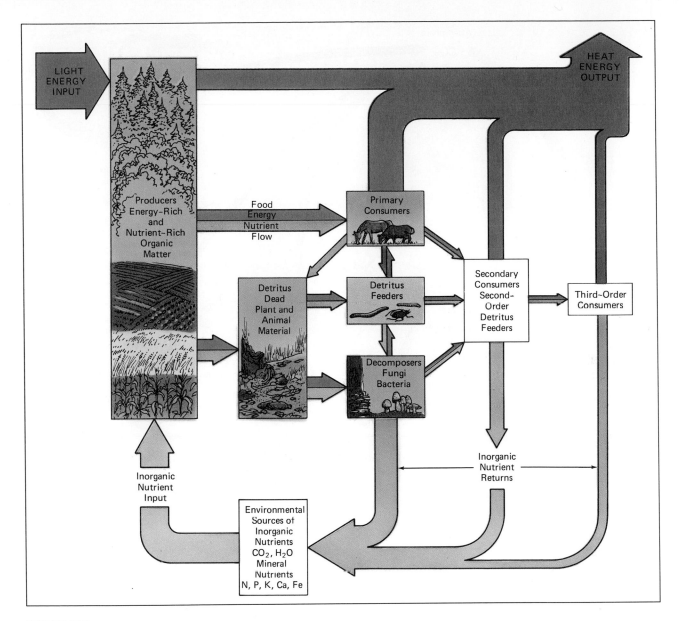

**FIGURE 2-14**
Nutrient cycling within, and energy flow through, ecosystems. Arranging organisms according to feeding relationships and depicting energy and nutrient inputs and outputs of each shows that there is a continuous recycling of nutrients within the ecosystem and a continuous flow of energy through the system.

factors. In this regard, it is important to look at cycles of particular elements because they do behave differently and present somewhat different problems in terms of recycling. Cycles of three elements are particularly important: (1) carbon, (2) phosphorus, which is representative of other mineral elements, and (3) nitrogen.

## THE CARBON CYCLE

The carbon cycle (Fig. 2-15) begins with the prime environmental "reservoir" of carbon, the carbon dioxide present in the air and/or dissolved in water. Through photosynthesis, carbon atoms from carbon dioxide are incorporated into glucose and then into other organic molecules that make up all plant tissues. They may then be passed to consumers and/or decomposers through feeding. As plants themselves or any consumer or decomposer breaks down organic molecules through cell respiration to release energy, the carbon atoms

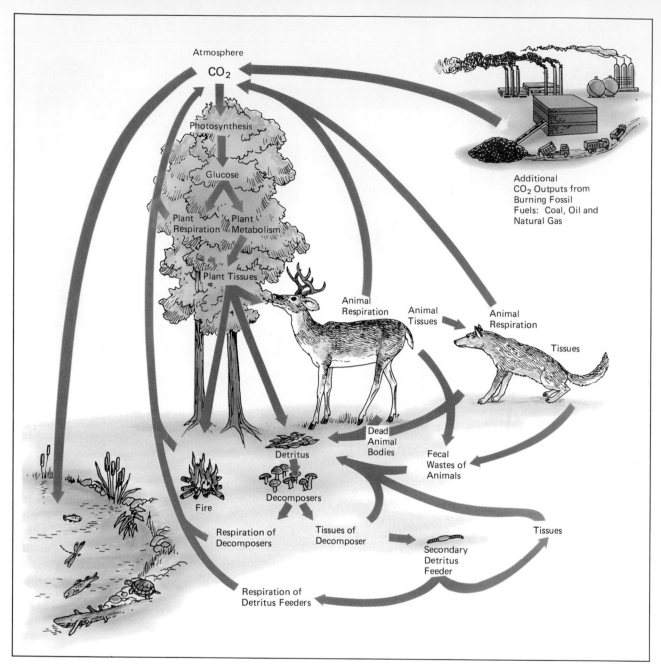

**FIGURE 2-15**
The carbon cycle. See text for explanation. The carbon cycle also operates in aquatic ecosystems as carbon dioxide dissolves in water.

are returned to the air or water solution as carbon dioxide molecules. If organic material is simply burned this also returns carbon dioxide to the air. In either case, it may be reabsorbed by plants and repeat the cycle.

Another interesting and important aspect of the carbon cycle is that in ancient geological times (hundreds of millions of years ago) much of the organic matter produced in photosynthesis was neither consumed nor decomposed; it accumulated and was buried under sediments. As a result of millions of years under heat and pressure in the earth, this detritus has been converted to crude oil, natural gas, or coal; the particular form of this **fossil fuel** depended on the plant material involved and the conditions of heat, pressure, and length of time in the earth. We are now mining or pumping these fossil fuels in huge quantities to run our industrialized society. In burning these fuels, we are, in one sense, completing the natural cycle and returning carbon dioxide to the air. But we are also increasing the concentration of carbon dioxide in the air since release is now greater than reabsorption. This has serious climatic implications (see greenhouse effect, Chapter 14).

## THE PHOSPHORUS CYCLE

Inorganic phosphorus, the basis of the phosphorus cycle (Fig. 2-16), exists in certain mineral compounds as phosphate ($PO_4^{3-}$). Phosphate dissolves in water but it does not enter the gas phase. Phosphate is absorbed from soil minerals through water solution by plant roots and is incorporated into various organic compounds by plants; phosphate in organic compounds is frequently referred to as organic phosphate. The organic phosphate is passed from plants to animals or decomposers through feeding; but, again, as organic material is oxidized for energy in cellular respiration by one organism or another, inorganic phosphate is released and exits from the animal body through urine or analogous waste. Thus, reentering the environment, it may be reabsorbed by plants and recycled.

If we contrast the carbon cycle with the phosphorus cycle, we observe a very important

**FIGURE 2-16**
The phosphorous cycle. See text for further description. The phosphorous cycle also operates in aquatic ecosystems as phosphate is soluble in water.

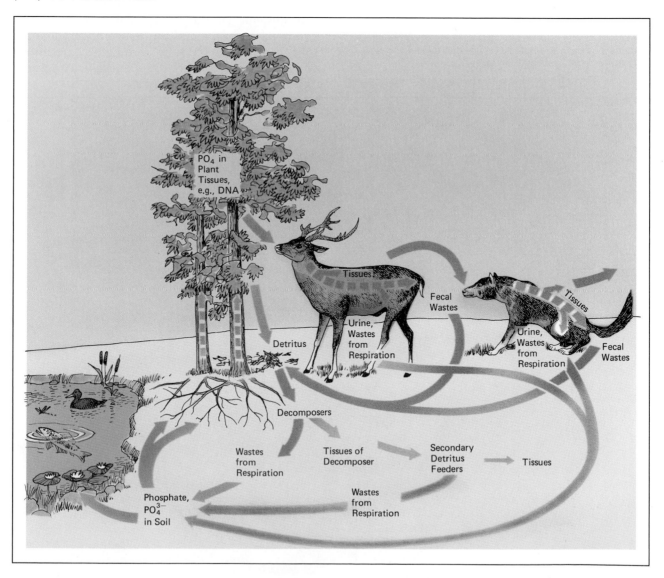

difference. The carbon cycle has a gas phase; therefore, wherever carbon dioxide is released it will mix into and move through the air where it can be reabsorbed by plants and repeat the cycle any number of times. Phosphate, on the other hand, has no gas phase in the cycle. If phosphate gets into waterways, it gradually makes its way toward the ocean where there is very limited return. Birds feeding on fish may return some to land through their droppings, but this will be a small portion of the total. Also, ocean sediments may be uplifted by geological processes, but this only occurs over millions of years.

Consequently, phosphate and other such mineral nutrients from the soil are, for the most part, only recycled in ecosystems insofar as plant and animal wastes are deposited on the soil from

which they come. In a natural ecosystem, this is basically what happens. However, we have constructed our human ecosystem such that the effluents carrying out wastes, for the most part, go into waterways. What are the long-term implications? We shall discuss this further later in this chapter and in later chapters.

### THE NITROGEN CYCLE

The nitrogen cycle (Fig. 2-17) is somewhat more complex than the carbon and phosphorous cycles. It has a gas phase so in part it is like the carbon cycle, but it also has a mineral phase, which makes it similar to the phosphate cycle.

The main reservoir of the element nitrogen is the air as about 80 percent of the air is nitrogen gas ($N_2$). Curiously, however, plants cannot utilize nitrogen gas directly from the air; the nitrogen must be in mineral form such as ammonium ($NH_4^+$) or nitrate ($NO_3^-$). Fortunately, a number of bacteria can convert nitrogen gas to the am-

**FIGURE 2-17**
The nitrogen cycle. See text for details. The nitrogen cycle also operates in aquatic ecosystems as nitrogen dissolves in water and may be fixed by blue-green algae.

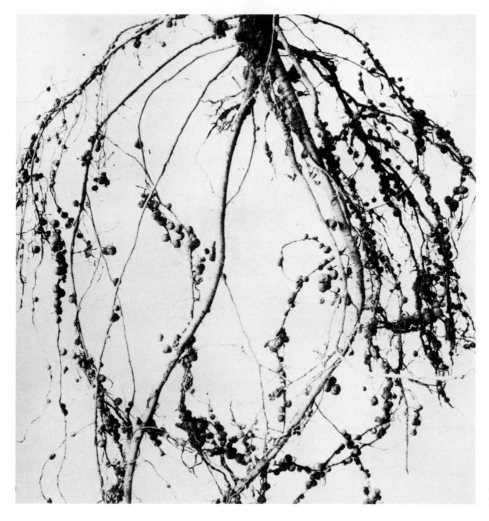

**FIGURE 2-18**
Root nodules. Nitrogen fixation is carried out by bacteria that live in the nodules. (USDA photo.)

monium form, a process called **nitrogen fixation.** Most important among these nitrogen fixers is a bacterium called *Rhizobium,* which lives within the nodules on roots of legumes, the pea-bean family of plants (Fig. 2-18). This is a good example of a mutual relationship. The legume obtains fixed nitrogen from the bacterium; the bacterium obtains food (sugars) from the legume. From the legume, fixed organic nitrogen may be passed to other organisms through feeding.

As organic compounds containing nitrogen are used in cell respiration for energy, nitrogen is returned to the environment in the ammonium form, which may be converted by bacteria to nitrate. Either form, ammonium or nitrate, may be absorbed by plants and recycled as a mineral nutrient. However, it does not remain in the mineral cycle indefinitely; another kind of bacterium in the soil gradually changes the nitrate form back to

nitrogen gas. A similar nitrogen cycle also occurs in aquatic ecosystems, but here blue-green algae are the nitrogen fixers. Some nitrogen gas is also fixed by the electrical discharges of lightning through the air and comes down with rainfall, but this is estimated to be only about 10 percent of what is fixed by the biological process.

One can see the advantage of legumes (which include peas, beans, clover, alfalfa, locust trees, redbud trees, and many others) being distributed worldwide and present in every major terrestrial ecosystem. It is also interesting to note that many of these legumes are the first plants to colonize an abandoned or burned-over area. Without them all production would be sharply impaired because of lack of available nitrogen.

Only humans have been able to bypass the need for legumes in growing nonlegume crops such as corn, wheat, and other grains. We do this by fixing nitrogen in chemical factories and applying it to crops as fertilizer. Our artifically fixed nitrogen is in the form of ammonium and/or nitrate so there is no real difference in what the plants receive. However, the high cost of artificially fixed nitrogen is causing many farmers to make more use of legumes.

## Flow of Energy and Decreasing Biomass at Higher Trophic Levels

Of all the solar energy falling on the earth, only a small fraction is trapped by photosynthetic plants in the production of plant biomass (organic matter). The rest is either reflected or absorbed by air and water and thereby serves as the driving force of the water cycle, wind, and ocean currents (Fig. 2-19). In short, solar energy is responsible for weather. Gradually all of this energy is converted to heat, which radiates from the earth into colder outer space. Also the fraction of solar energy that is trapped in plant biomass is gradually

converted to heat as various organisms in the food chain convert and utilize organic matter for energy. This heat, too, is eventually lost into outer space. Thus one can visualize a continuous flow of energy through the entire biosphere, both biotic and abiotic components.

As each heterotroph (consumer or decomposer) utilizes a fraction of the organic matter for energy, the total biomass is reduced by that fraction. Thus, there is a decrease in biomass at each higher trophic level, as noted in Chapter 1. This can be determined by direct observation as well as derived by theory.

Ecologists have actually determined the biomass of each trophic level for various ecosystems. In a grassland, for example, this is done by digging up square-meter sample plots of vegetation, separating all the dirt and other debris, then drying and weighing all the plant material. After the average biomass of producers per square meter is determined, simple multiplication gives the biomass per hectare (1 ha = 2.5 acres) or any other desired area. Similarly, trapping procedures allow one to estimate populations of herbivores and carnivores and determine their biomass per hectare. Estimates can also be made of fungi, bac-

**FIGURE 2-19**
Only a small portion of the solar energy hitting the earth is absorbed and trapped by photosynthesis, but this is then the total energy available to the rest of the ecosystem.

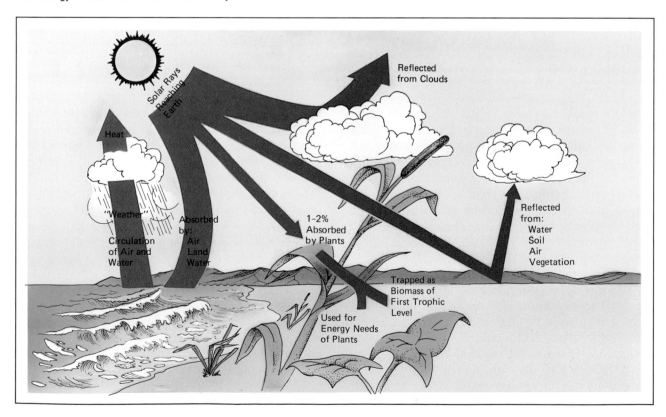

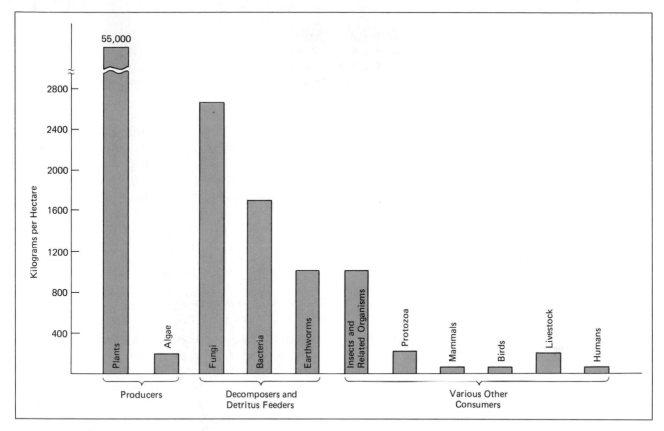

**FIGURE 2-20**
The average biomass (weight) in the United States for various groups of organisms. (Redrawn with permission, from D. Pimentel et al., "Environmental Quality and Natural Biota," *Bioscience,* 30 [November 1980], 750.)

teria, earthworms, insects, and other such organisms found in the soil. Results of such measurements and estimates are shown in Figure 2-20.

It is significant to note that producers constitute about 88 percent of the total biomass; all heterotrophs constitute only about 12 percent of the total. Further, the bulk of biomass of heterotrophs is fungi and bacteria (decomposers) and earthworms (detritus feeders). Only relatively minute portions are in larger consumers such as birds and mammals.

Two factors in addition to consumption of biomass for energy account for the distribution of biomass noted. First, any population can be looked as a **standing biomass** to which there is a yearly addition by growth and reproduction and a subtraction by death and/or consumption. If the standing biomass is to remain constant, as it does (more or less) in a stable ecosystem, it follows that yearly subtractions by death and consumption

cannot exceed yearly production by new growth and reproduction. Thus, in stable ecosystems much of what is measured as biomass of producers, for example, represents standing biomass. Only a small portion of the total, that balanced by new production, is actually consumed.

Second, of the amount of biomass that is consumed, larger animals, birds and mammals in particular, generally consume only a small portion, seeds and/or fruits for example. Further, a portion of what they do consume is not digested but simply goes through the intestinal tract of the consumer and comes out as fecal wastes.

Thus, the bulk of plant production—most leaves and wood parts—is not utilized by large consumers but becomes available to decomposers and detritus feeders such as earthworms. Consequently, the biomass of these organisms is much larger than that of large consumers.

The result of all these considerations is that, in general, the biomass of the second trophic level (decomposers and detritus feeders as well as larger herbivores) is generally no more than about 10 percent of the total biomass of producers, and the biomass of the third trophic level is generally

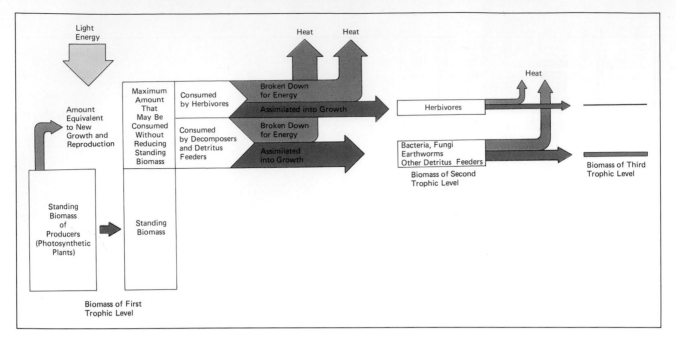

**FIGURE 2-21**
Decreasing biomass at higher trophic levels. The decrease results from the fact that much of the preceding trophic level is "standing crop" that is not available for consumption and much of what is consumed is broken down for energy.

no more than 10 percent of the second. In other words, with each transfer from one trophic level to the next, total biomass is decreased by 90 percent or more (Fig. 2-21). Said another way, to support a given biomass of herbivores requires a biomass of producers at least 10 times larger. To support a given biomass of carnivores, the biomass of producers must be 100 times larger (10 × 10). While it is possible to imagine any number of trophic levels, ecosystems generally have no more than three or four discernible levels, because after two or three steps there is insufficient biomass to support higher levels.

It is interesting that science fiction writers commonly ignore this principle and conceive huge monsters, carnivores at that, living in barren deserts. They also portray food chains that are circular and supposedly self-supporting in the absence of producers. It should be clear that the second law of thermodynamics makes such things impossible. Such systems, even if they were initiated, would quickly starve into extinction.

## IMPLICATIONS FOR HUMANS

Understanding the ecological principles of nutrient cycling and energy flow, scientists and engineers at the U.S. National Aeronautics and

Space Administration (NASA) have engaged in design studies for space stations that could maintain astronauts indefinitely without resupply of food or water. Such stations would have "greenhouses" for the propagation of food plants. The plants would use carbon dioxide, water, and mineral nutrients and would produce food material and oxygen in the course of photosynthesis. Astronauts would consume the food and oxygen and release carbon dioxide; their urinary wastes and fecal and plant wastes, after decomposition, would be used as fertilizer to replenish the nutrients needed by the plants. All water would be purified and recycled.

With a proper balance between the number of astronauts and the amount of "agriculture," such a station could maintain a space colony indefinitely without additional material input or output to the outside. The one and only necessity would be the energy source—sunlight or some artificial source of radiant energy—required for photosynthesis, and there would be a balancing energy loss of heat (Fig. 2-22).

Unfortunately, on earth we have developed and operated our human ecosystem in ignorance of these ecological principles.

In contrast to nutrient recycling, we have developed our human ecosystem on the basis of a one-directional nutrient flow. Nutrients are removed from the soil by our agricultural crops, transported to where the food is consumed, then—through wastes of both animals and humans—they are mostly discharged into natural waterways. (Sewage treatment, as it is generally practiced today, does not alter this discharge of nutrients.) Soil nutrients are replenished by mining them from various deposits and applying them as chemical fertilizers. Thus, there is an overall flow of nutrients from mines to waterways (Fig. 2-23).

There are problems at both ends of this flow. At the resource end there is a negative environmental impact of the mining process itself. Moreover, these resources may ultimately be depleted. At the discharge end, there is a problem of too much fertilization leading to gross changes in the natural biota of receiving bodies.

Similarly, our industry requires many additional elements. Here, too, we have developed a

**FIGURE 2-22**
Self-sustaining space station. Scientists and engineers are capable of building a self-sustaining space station. It would utilize the ecological principles of recycling all nutrients and water, and would operate entirely on solar energy.

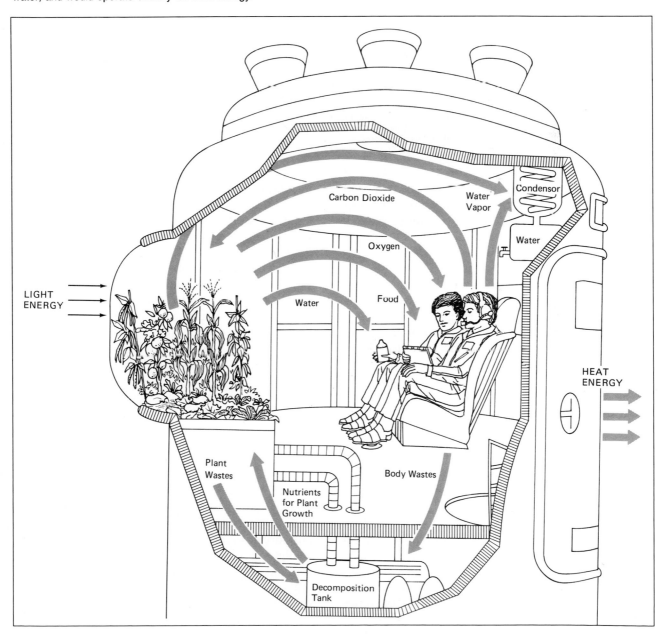

LIGHT ENERGY

Carbon Dioxide

Water Vapor

Condensor

Water

Oxygen

Water

Food

HEAT ENERGY

Plant Wastes

Body Wastes

Nutrients for Plant Growth

Decomposition Tank

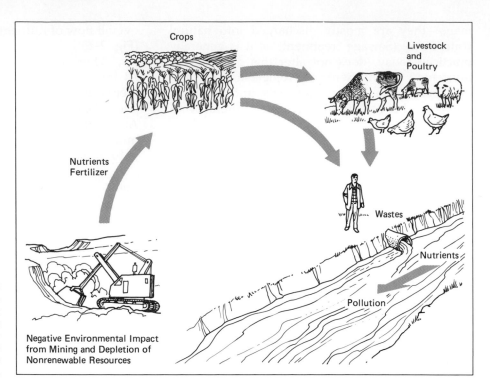

**FIGURE 2-23**
In contrast to applying the ecological principle of nutrient recycling, human society has developed a pattern of one-directional nutrient flow. There are increasing problems at both ends.

Crops

Livestock and Poultry

Nutrients Fertilizer

Wastes

Nutrients

Pollution

Negative Environmental Impact from Mining and Depletion of Nonrenewable Resources

one-directional flow rather than a recycling system. This raises again the problem of resource depletion at the input end and even more serious pollution problems at the output end since many of the elements used in industry, such as mercury and lead, are extremely toxic.

The development of synthetic organic compounds has also created problems. All natural organic compounds produced in natural ecosystems are subject to breakdown by consumers and/or decomposers. This prevents the buildup in the environment of undesirable or toxic waste products. In contrast, our technological society is now producing all manner of synthetic organic compounds that are **nonbiodegradable**; that is, organisms cannot break down these materials. Such compounds simply accumulate in the environment. Furthermore, many of these compounds are extremely toxic. Consequently they are becoming an increasing threat to both environmental and human health.

Human exploitation of fossil fuels is another source of concern. The energy source for natural ecosystems is sunlight, which is both nonpolluting and inexhaustible. (Astronomers have determined that the sun itself will "burn out" in several billion years, but this is millions of times

beyond the realm of human experience. So, for all practical purposes, we can assume solar energy is an everlasting source). In contrast, human systems currently operate to a large degree on fossil fuels (coal, oil, and natural gas). Our use of these fuels is often inefficient. In agriculture, for example, we expend an average of 10 fossil-fuel calories for every food calorie we consume (Fig. 2-24). Moreover, fossil fuels pollute when burned; most air pollution comes directly or indirectly from burning these fuels. Further, we are already facing problems of depletion of oil reserves.

The reduction of biomass that occurs at higher trophic levels has implications regarding the level at which humans feed. We are omnivores; we can do very well feeding on just plant materials (i.e., vegetarianism), but we generally prefer to eat a high proportion of meat. How does this choice affect the amount of primary agricultural production that is needed to support us and what is the environmental impact? In fact, most U.S.-grown grains and soybeans, which could support human nutritional needs directly, are actually fed to livestock and poultry to produce meat. Although it differs with the particular animal, the average efficiency of the conversion is only about 10 percent. That is, an average of 10

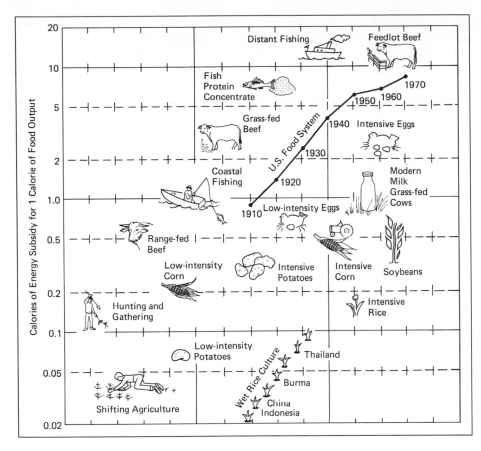

**FIGURE 2-24**
In contrast to natural ecosystems, which operate purely on solar energy, modern human society has become largely dependent on nonrenewable fossil fuels—coal, oil, and natural gas. Even food production has become more and more subsidized by fossil fuel inputs for tractors, transportation, and processing in addition to the solar energy for actual photosynthesis. (From J. S. Steinhart and C. E. Steinhart, "Energy Use in the U.S. Food System," *Science*, 184 [1974], 312.)

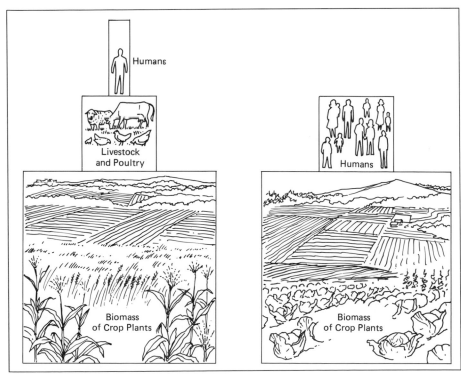

**FIGURE 2-25**
Raising livestock and poultry is a food-consuming, not food-producing, operation. Many more humans can be supported on the same agricultural base if livestock and poultry are eliminated from the food chain and humans feed on the second trophic level rather than the third.

pounds of edible grain protein is consumed by the animal to produce 1 pound of meat protein. Thus a meat diet, in contrast to a vegetarian diet, requires on the order of tenfold more primary production. In terms of actual agricultural land, about 0.2 acres (1 acre = 0.4 ha) are required to support one person on a vegetarian diet; 2 acres are required for a person on a meat diet. When factors associated with agriculture, such as soil erosion, depletion of water resources, use of pesticides, and pollution from fertilizer, are considered, the environmental impact of this food choice is tremendous (Fig. 2-25).

Finally, there are simple limits. Is it any wonder why most people in overpopulated countries eat little, if any, meat? Is meat-eating a luxury that people in developed countries can continue as the world becomes increasingly populated? Clearly, there are challenges as well as opportunities for developing a more sustainable human ecosystem on earth.

# WHAT KEEPS ECOSYSTEMS THE SAME? What Makes Them Change?

---

## CONCEPT FRAMEWORK

1. What factors influence biotic potential and environmental resistance? How do biotic potential and environmental resistance work to determine an ecosystem's population balance?

2. Why is the phenomenon of critical numbers so important?

3. Why is the predator-prey relationship exemplified by the lynx and the rabbit atypical? How does the predator-prey relationship affect the balance between herbivores and vegetation?

4. Explain why a balance between competing species depends on adaptations of the species involved.

5. Give an example of how territoriality influences population balance.

6. When is fire a positive abiotic factor?

7. Distinguish between *primary* and *secondary succession* and discuss how they contribute to the emergence of a *climax ecosystem.*

8. What factors may contribute to the succession, upset, or collapse of an ecosystem?

9. What human actions could be taken to better maintain a balance in the ecosystem?

Ecosystems are able to sustain themselves by cycling nutrients and drawing on the continuous flow of energy from the sun. Yet, each ecosystem is a dynamic structure of hundreds or even thousands of populations of producers, consumers, and decomposers interacting with each other in a complex food web and in other relationships. What prevents herbivores from devouring all the producers? What prevents carnivores from eliminating their prey? What prevents one species from pushing another completely out of the competition?

In Chapter 1 we learned the general principle that various *limiting factors* tend to keep populations from spreading beyond areas to which they are best adapted. Here, we wish to expand upon this concept to gain a clearer picture of how ecosystems maintain their integrity in spite of the conflicts among the many populations that make up the system. In so doing, we shall also become more familiar with factors that cause ecosystems to change. This has profound implications to us in terms of our desire to maintain our natural environment, as well as our own human ecosystem.

# THE KEY IS BALANCE

Studies show that there are no overriding forces, laws or rigid structures that require ecosystems to remain the same or that prevent them from changing. The only thing that maintains natural ecosystems over long periods of time is that all the dynamic relationships occurring within the system are balanced.

## Ecosystem Balance Is Population Balance

Each species found in an ecosystem is actually represented by a breeding, reproducing **population,** that is, a number of individuals in the area at one time that constitute an interbreeding, reproducing group. The term *population* may apply to plants, animals, or microbes. If an ecosystem remains stable over a long period of time, it really means that each population within it remains constant over this time.

Every population is a dynamic entity in which there is a constant input of new members through reproduction and a constant loss of members through dying or being killed. If a population remains stable over time it follows that the intro-

duction of new members is being equalled by the death of other members. The problem of ecosystem balance thus boils down to a problem of how reproductive rate is balanced with death rate for each population in the ecosystem.

## BIOTIC POTENTIAL VERSUS ENVIRONMENTAL RESISTANCE

All the factors that tend to increae the number of new individuals brought into the population are placed in one category: **biotic potential.** All the factors that cause death in one way or another are categorized under **environmental resistance.** Whether a population increases, decreases, or remains constant depends on the relative *balance* between biotic potential and environmental resistance (Fig 3-1).

The most obvious factor in a species' biotic potential is its **reproductive rate,** but that is not the only factor. Equally important is **recruitment,** which refers to the process of successfully surviving through the hazardous stages between birth, egg, or seed and actually becoming part of the breeding population. Fish and shellfish, for example, typically lay thousands or even hundreds of thousands of eggs. However, they may lack suitable genetic vigor or resistance to adverse conditions and hence fail to develop and they are extremely vulnerable to predation. The same may be said for plants, which typically set huge numbers of seeds. Thus, a very high reproductive rate may achieve nothing in terms of population growth unless recruitment is also significant. Conversely, a much lower reproductive rate can give a very sizable increase if recruitment is favorable. For example, in mammals, humans in particular, we see relative low reproductive rates, but high recruitment rates through parental care.

Still other important factors of biotic potential include the ability of animals to migrate or of seeds to disperse to similar habitats in other locations; the ability to adapt to and invade other habitats; defense mechanisms such as quills or thorns, or offensive odor or taste; the ability to avoid or survive adverse environmental conditions; and resistance to disease.

While different species have developed different means of enhancing their biotic potential, one phenomenon is invariably the case. Every species has the biotic potential to rapidly increase its population if all environmental factors are fa-

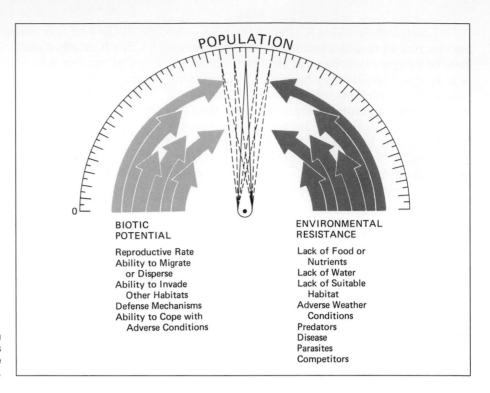

POPULATION

BIOTIC
POTENTIAL

Reproductive Rate
Ability to Migrate
   or Disperse
Ability to Invade
   Other Habitats
Defense Mechanisms
Ability to Cope with
   Adverse Conditions

ENVIRONMENTAL
RESISTANCE

Lack of Food or
   Nutrients
Lack of Water
Lack of Suitable
   Habitat
Adverse Weather
   Conditions
Predators
Disease
Parasites
Competitors

**FIGURE 3-1**
Population is a balance between factors that increase numbers and factors that decrease numbers.

vorable. In fact, the population growth may be so rapid that such a situation is commonly referred to as a **population explosion.**

The reason that populations in natural ecosystems generally do not "explode" is not because of lack of biotic potential; it is because environmental factors are seldom ideal. For instance, one or more abiotic factors such as sufficient water or suitable temperature may be limiting. Or there may be negative biotic factors such as predators, parasites, disease organisms, or lack of sufficient food. Together, these abiotic and biotic factors constitute *environmental resistance.* Importantly, the resistance typically does not generally reduce the birth rate (egg or seed production) of reproducing individuals. Rather, environmental resistance typically causes dieoffs of the young; thus, it affects recruitment. In more severe cases, it also causes dieoffs of adults, thereby reducing the breeding population.

With biotic potential pushing from one side and environmental resistance pushing from the other, so to speak, a balance is reached between the two (Fig. 3-1). This balance is *dynamic* or continuously readjusting because, as a population spreads over more area and/or becomes denser, factors of environmental resistance tend to be-

come more severe and dieoffs increase accordingly. Conversely, as the poulation becomes smaller, environmental resistance generally becomes less severe and allows the biotic potential to expand the population again. For example, as population density becomes higher, spread of parasites and/or disease becomes more severe and this may cut the population back. Also, both abiotic and biotic factors may vary from year to year. For example, a drought may cause a population to die back one year but, in following years, if conditions are normal, the population recovers.

Thus, balance is a relative phenomenon. A very stable balance fluctuates very little; an unstable balance may fluctuate widely, but so long as decreased populations restore their numbers, the system may be said to be balanced.

### CRITICAL NUMBERS

Nothing demands that populations must balance. Factors of environmental resistance may become such that populations are reduced to zero resulting in **extinction** of the species, unless there are surviving populations in other areas. Even if adverse conditions are reversed at the last moment, a population may not be able to recover if it has been reduced below a certain **critical number.** For example, the passenger pigeon was an extremely abundant bird that was exploited extravagantly for food prior to the 1900s. As populations plunged precipitously there were efforts at conservation; however, flocks continued to diminish and disappear in spite of curtailed hunting. Apparently, the species could only cope and reproduce in large flocks so that, when flocks were reduced below a critical number, extinction was inevitable.

We do not know what the critical number is for most species. However, we should be well aware of the phenomenon so that conservation efforts to protect diminished species may be started early enough. Otherwise, even massive efforts at conservation may be totally fruitless.

## Mechanisms of Population Balance

Recognizing that population balance in nature is a dynamic balance between biotic potential and environmental resistance, we turn our attention to some of the specific mechanisms of balance. We shall cite examples where a single mechanism seems to be predominant; however, you should bear in mind that an ecosystem is always a dynamic interaction between numerous biotic and abiotic factors. Therefore, a species' population balance is invariably maintained by a number of factors acting simultaneously. After examining these mechanisms we should become acutely aware of how balances may be all too easily upset by humans. Such ecological upsets often result in tragic consequences.

## *BALANCE THROUGH PREDATION, PARASITISM, OR DISEASE*

Predator-prey relationships are important in population balance. A classic example is the predation on rabbits by lynx, members of the cat family. When the rabbit population is low, each rabbit can find abundant food and plenty of places to hide and raise offspring. In other words, the environmental resistance for the rabbits is relatively low, and their population increases despite the presence of the lynx predator. However, as the rabbit population increases, there is relatively less food and fewer hiding places per rabbit. More rabbits provide easier hunting for the lynx so that, with plenty of rabbits to feed their young, the lynx population begins to increase. Decreased food and shelter and the increased predatory population mean higher environmental resistance for the rabbits, and consequently their population begins to fall. As the population falls, the amount of food and shelter available to each rabbit again increases. Also, the surviving rabbits are those that are most healthy and best able to escape from the lynx. Hunting becomes harder for the lynx; many of them starve and their population begins to fall. These factors result in a lower environmental resistance for the rabbits, and their population increases again, repeating the cycle. These events explain the fluctuating, but continuing balance found between the rabbit and lynx populations (Fig. 3-2).

While large predators such as the lynx draw attention, there are relatively few situations where they are the primary controlling factor.

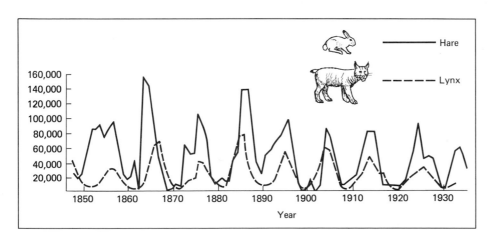

**FIGURE 3-2**
Predator-prey relationship creates balance between predator and prey populations. Data based on pelts received by Hudson's Bay Company. (From D. A. MacLulich, University of Toronto Studies, Biological Series No. 43, 1937.)

Much more abundant and ecologically important in population control is a huge diversity of parasitic organisms. These organisms range in size from tapeworms, which are many feet long (Fig. 1-15a) to smaller species of worms to microscopic disease-causing protozoans, fungi, bacteria, and viruses.

In an ecological sense, all of these parasitic organisms play essentially the same role as predators; hence, they may be thought of as specialized predators. As population density (number of individuals per unit area) of the host organism increases, parasites and their vectors such as disease-carrying insects, have little trouble finding a new host and infection rates become high. Conversely, where population density of the host is low, transfer of infection from one individual to the next is impeded and there is a great reduction in levels of infection.

While we have noted that parasites generally do not kill their host, at least not immediately, they generally weaken it and make it more vulnerable to adverse conditions and/or attack by larger predators. Indeed, it is commonly observed that the animals killed by predators are infected with parasites whereas animals killed by hunters are frequently healthy. It appears that natural predators generally do not or cannot kill healthy members of the prey population. This has important implications in terms of reproduction of the genetic traits for vigor and disease resistance, a subject we shall consider further in Chapter 4.

Finally, while parasites generally do not kill their host immediately, many kill it eventually, particularly when there are repeated occurrences of infection. Then, there are many cases where a single attack is fatal. For example, many parasitic insects lay their eggs in a host caterpillar (larval stage of another insect). When the eggs hatch, the larvae of the parasite feed on the caterpillar and destroy it. Such parasitic insects are the main factor in controlling populations of certain plant-eating insects. Indeed, considerable work is being done to utilize such parasites in the control of pest insects. Also, parasites are not limited to any particular trophic level. Parasites, including disease-causing organisms, may effectively limit producer, herbivore, or carnivore populations.

We must be wary of looking for a single predator or parasite as the controlling factor. We have noted that simple food chains are rare in nature. Any given organism is generally affected by a combination of various predators, parasites, and diseases. Consequently, the balance between predators and prey can be thought of more broadly as a balance between an organism and its **natural enemies.** But it still follows the general rule: as population increases in density, it becomes more vulnerable; lower population densities are less vulnerable. Hence, there tends to be a dynamic balance between each species and its natural enemies. Balances involving numerous natural enemies in a complex food web are generally more stable and less prone to wide fluctuation. The wide swings in population as noted in the rabbit-lynx case are generally typical of very simple ecosystems.

It is also significant to observe that an effective predator-prey or host-parasite balance implies mutually balanced adaptation on the part of both predator and prey. When the rabbits have abundant food and places to hide, they are largely able to escape the lynx, and the lynx population suffers accordingly. This is to say, rabbits are so well adapted that the lynx are unable to hunt down and kill every remaining rabbit. On the other hand, the lynx are well enough adapted to catching rabbits that they limit the population to a certain level. Similarly, a host-parasite balance implies a balance between the virulence of the parasite and the resistance of the host. The critical nature of such balances will be seen as we discuss the following examples. More will be said about how such balances develop in Chapter 4.

The predator-prey or host-parasite balance may also serve to keep an herbivore in balance with vegetation. Herbivore populations have the potential to expand to the point where they consume vegetation faster than it can regenerate, a phenomenon known as **overgrazing.** Drastic effects on the ecosystem follow. Frequently, predators and/or parasites are the principle factor that holds herbivore populations below levels that would result in overgrazing.

The importance of this balance becomes clear through observing the ecological upsets that occur when predators are removed from ecosystems or when herbivores are introduced into ecosystems where they are not controlled by natural enemies.

In 1859 rabbits were introduced from England to Australia for sport shooting. The Austra-

**FIGURE 3-3**
Results of rabbits overgrazing the Australian ecosystem. On one side of a rabbit-proof fence there is lush pasture; on the other side, it is barren. Rabbit-proof fences were built over thousands of miles but proved unsuccessful in stopping movement. (Australian Information Service photo.)

lian environment proved favorable to rabbits and their numbers increased rapidly, more rapidly than hunters could control, and they went on to devastate vast areas of rangeland by overgrazing (Fig. 3-3). Although it is often stated that the rabbit had no natural enemies in Australia, this is only partially true. Australia does have a natural predator, the dingo. Unfortunately, the dingo is not well adapted to catching rabbits—it cannot run fast enough—even when rabbits are extremely numerous. Therefore, it fails to establish an effective balance with rabbits.

Many of the most severe insect pests of agricultural crops and forests, the Japanese beetle and the gypsy moth, for example, are species from other ecosystems that do not have effective natural enemies in the ecosystem to which they were introduced.

The obvious remedy for such situations, introducing a natural enemy, is being investigated and has been used successfully in some cases, but it is not as simple as it seems on the surface. The problem is that there is no guarantee that the introduced natural enemy will, in the new ecosystem, focus its attention on the target pest. For example, foxes were introduced in Australia to control the rabbits, but the foxes also found they could catch other Australian wildlife more easily than rabbits. Hence, they went their own way and seldom bothered the rabbits. This topic of controlling pests with natural enemies is discussed further in Chapter 17.

Predator-prey or host-parasite balances have also been upset by introducing predators that are *too* effective. Domestic cats introduced into island ecosystems have often proved to be overly effective in comparison to native animals. Many unique species of island wildlife have been completely exterminated by cats.

A devastating upset caused by a disease organism is seen in the case of the American chestnut. Prior to the twentieth century, the American chestnut was the dominant tree in deciduous forests of the eastern United States. As many as one-fourth of the trees were chestnuts. Some trees were 7 feet in diameter and reached heights of 30 m (100 ft) or more. The chestnut was highly valued for its timber, its aesthetic qualities as a shade tree, and its prolific production of chestnuts, which were a major food for forest wildlife as well as a prized food for many people (Fig. 3-4). But in 1904 a chestnut blight disease was observed in New York. From then until about 1950 the disease spread in an ever-widening wave, leaving only dead trunks behind. Today only some 50 mature American chestnut trees are known and all of them are diseased. In some cases the roots of old trees continue to live and send up sprouts that may live for several years before they too are infected by the blight and killed.

The chestnut blight disease is caused by a parasitic fungus that enters the tree and causes a fast-growing canker that girdles the tree and cuts off sap flow, thus killing the tree. Spores (repro-

**FIGURE 3-4**
American chestnut. An important species of tree in eastern U.S. deciduous forests was essentially eliminated by an introduced disease, the chestnut blight. (National Archives photo No. 95-6-250527.) **Insert shows fruit of chestnut tree.** (U.S. Forest Service photo.)

ductive cells) of the fungus may be carried many miles by wind, birds, and perhaps insects.

The disease was apparently unwittingly introduced with some young Chinese chestnut trees brought from China for planting in the New York Zoological Park in about 1895. The Chinese trees are adapted to this disease, so while they carry it, they are not particularly affected. That is, in China, the disease is part of a balance between host and natural enemies. However, in the North American ecosystem, the American chestnut was not adapted to the disease; it had essentially no resistance and hence was virtually wiped out. Efforts are being made to select and breed blight-resistant chestnut trees or to find some other cure for the disease, but these efforts have not been highly successful to date.

What happened to the deciduous forest ecosystem when the American chestnut died out?

For the most part the spaces vacated by the chestnuts were filled by various species of oaks that had been present but in smaller numbers. This brings us to the next major factor of population balance, competition between species.

## BALANCE THROUGH COMPETITION BETWEEN SPECIES

Photosynthetic plants compete for light, nutrients, and water. Similarly, all herbivores compete for food, water, and shelter. Why doesn't one species win the competition and eliminate others? How is a balance between competing species maintained?

Part of the answer lies in the fact that the environment of a region is not uniform but consists of many different microenvironments resulting from differences in abiotic factors such as terrain, soil type, exposure, drainage, and so on. Biotic factors differ as well. Each combination of factors creates a slightly different environment.

In turn, species vary in their degree of **adaptation** to these microenvironments. Each species of plant, for example, is generally best adapted to one or a very narrow range of environments. If a population spreads into adjacent environments to which it is less well adapted, it faces higher environmental resistance. One or more abiotic factors may be limiting, or, more importantly, the increased environmental resistance that is imposed by abiotic factors makes the individuals more vulnerable to attack by parasites and/or various herbivores, especially plant-eating insects. The end result is that the individuals lose out to species that are better adapted to that particular environment.

The following example may be seen throughout the central portion of eastern deciduous forests: sycamore trees are found along river banks and flood plains (areas that are occasionally flooded) and oaks are found on the valley sides. The sycamore is best adapted to very wet soils and can withstand frequent flooding. Oaks, on the other hand, are best adapted to well-drained soils and cannot withstand frequent or prolonged flooding. Sycamore seeds do spread up the valley sides and, in the absence of oaks and other competing species, grow there. However, given the competition with oaks, the sycamore's poorer adaptation to drier conditions results in the sycamore usually losing out to oaks. Similarly, acorns fall on the flood plain, but oak seedlings and saplings are retarded by conditions that, for them, are too moist and they generally lose the competition to sycamores. Thus, each species is able to win on its own "home ground," and hence the competitive balance may be maintained indefinitely (Fig. 3-5).

The maintenance of such balances assumes, of course, that the underlying conditions remain unchanged. What would you expect to happen, for example, if the river were dammed and the flood plain became a drier habitat? Also, it is noteworthy that in the absence of competition, a species may inhabit an environment to which it is not particularly well suited. For example, in the absence of competition sycamores will grow in drier conditions.

The same principle holds true for animal species in a given ecosystem. For example, different species of birds may appear to be competing with each other in a great free-for-all. However, robins, which eat worms, are not competing with woodpeckers, which eat insects in dead trees, and neither is competing with sparrows, which eat seeds. Flycatchers (birds) and bats both feed on flying insects, but they are not competing because flycatchers feed during the day and bats feed at night. The giraffe is able to graze on treetops; therefore, it does not compete with other grazing animals around it. In short, animal species are adapted to feed on different foods, or in different locations, or at different times of day, as well as being adapted to the overall abiotic factors of the environment. In an ecological sense, all these factors—climate, food sources, times of activity and

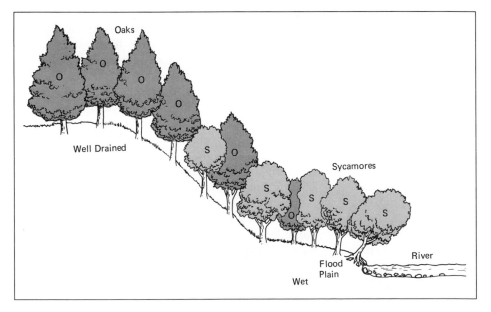

**FIGURE 3-5**
Competing species coexist in the same area by occupying different habitats. Oaks are better adapted to well-drained soils. Sycamores are better adapted to wet soils and can withstand flooding.

so on—constitute an animal's **niche.** Each animal being more or less specifically adapted to a particular niche minimizes competition, but it does not eliminate it. Niches generally overlap to a greater or lesser extent, and competition occurs in the areas of overlap.

For example, consider two bird species, one adapted mainly to feeding on large seeds; the other adapted mainly to feeding on small seeds. They may compete in feeding on intermediate-sized seeds but each wins in its own particular niche and thus maintains its place in the system. Again one generally finds that if one species is removed from the system, others will expand in population and move into its place.

As in predator-prey relationships, it is important to note that a balance between competing species depends on very precise adaptations of the species involved. Again, this is demonstrated by experiences of transferring plants and animals between ecosystems. In many cases, the introduced species does not find a niche in which it can compete with native species and it dies out. Although this seems unfortunate, it is actually good news because the receiving ecosystem is not harmed by the intrusion.

The bad news, which occurs all too frequently, is that the introduced species finds conditions that are favorable and proves to be a more vigorous competitor, a situation that results in native species being forced out. For example, in 1884 the water hyacinth, originally from South and Central America, was introduced into Florida as an ornamental flower. It soon escaped into the waterways, where it proved to be a more vigorous competitor than any of the native species; also it had no natural enemies in the Florida ecosystem. Consequently, it has proliferated to the extent of making navigation difficult or impossible (Fig. 3-6a). Millions of dollars have been spent attempting to get rid of this weed, but with little success.

Kudzu, a vigorous vine introduced into the Southeast to control erosion, is winning the competition against everything else and taking over whole forests (Fig. 3-6b). Starlings and house sparrows, introduced from Europe, have displaced native species such as bluebirds from many areas because of their more aggressive nature. Wild burros are overgrazing desert ecosystems in the West and are displacing bighorn

sheep and small mammals. Native species of fish have been displaced from many rivers and lakes by the introduction of foreign species, such as carp. Goats introduced into many islands have caused the extinction of many species by being more effective grazers. Unfortunately, the list of such examples goes on and on. Of all the introductions that have been made, very few have hit the balance of being able to survive and reproduce within the new system without upsetting it.

Finally, it is important to emphasize that in nearly every competitive situation, some part of the balance is determined by biotic factors such as predation or parasitism acting on one or both the competitors. Changing a biotic factor, then, can grossly alter a long-established balance. A prime example of this occurs in regions of semiarid grasslands where there is a tenuous balance between grass and desert species. Intensive grazing by cattle on such lands adversely affects the grass but does not affect the desert species, which the cattle do not eat. Consequently, the balance is upset and such lands are taken over by desert species (Fig. 3-7). Restoring the original condition may be exceedingly difficult or impossible because each change affects many others.

### BALANCE THROUGH AVAILABILITY OF RESOURCES AND TERRITORIALITY

We have seen that populations may be limited by any combination of abiotic and biotic factors. However, at what population density does the limitation occur? This depends on the availability of resources. The main resources required by photosynthetic plants are light, water, and mineral nutrients. Animals require food and water, suitable places for hiding, and materials and/or locations for nests, burrows, or dens. Any organism that does not obtain an adequate supply of one or more resources becomes more susceptible to attack by predators or parasites. For example, plants weakened by lack of adequate water, nutrients, or light become more susceptible to attack by insects and/or plant pathogens. The same is true for animals weakened by lack of adequate food or water. In addition, for animals, lack of suitable places for hiding and reproducing are equally important. In particular, populations of many birds are limited more by lack of suitable materials and places for nesting than by lack of food.

(a)

**FIGURE 3-6**
Introduced species don't have balanced relationships with their new partners. (a) Water hyacinth overgrowing waterways. (U.S. Department of the Army Corps of Engineers photos.) (b) Kudzu overgrowing forests. (USDA–Soil Conservation Service photo.)

(b)

Thus, what may be seen as a balance through predation or parasitism should also be viewed as a balance between the target population and the availability of resources. This has important implications for wildlife management. The population of a desired species can generally be increased more effectively by supplying additional amounts of a limiting resource than by eradicating its predator. Indeed if a resource is limiting, eradication of a predator will generally have no lasting effect on increasing the prey population; it will only promote dieoff through some other factor. Conversely, the population of an organism that is considered a pest can often be reduced most effectively by limiting some resource. For example, some pest insects can be controlled by elimination of their breeding habitat.

There are situations in which animal populations explode, deplete their food resource, then die off as a result. This cycle may be repeated when and if the food species recovers (Fig 3-8). However, many species of animals have behavioral adaptations that tend to keep their population within resource limits. This commonly ob-

**FIGURE 3-8**
Population limited by depletion of food resource. In the absence of other limiting factors, a population may grow exponentially to the point of exhausting its food resource. This is followed by a precipitous dieoff due to starvation. The cycle may be repeated when and if the food resource recovers. This type of population swerve is seen in a number of plant-eating insects that are not controlled by natural enemies.

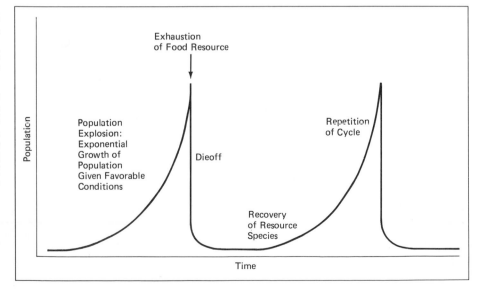

served trait is **territoriality.** For example, in many species of songbirds, the males "stake out" a territory at the time of nesting. Their song actually has the function of warning other males to keep away. The territory defended is such that it assures the male and his mate of being able to gather enough food to successfully rear a brood. One can readily see the advantages of territoriality. However, while some members are successful, other members are unable to find and/or defend a territory and are prevented from breeding. Therefore, the reproduction of the population is balanced with the environment required to support it.

Additional forms of territoriality can be observed in other animals. For instance, wolves defend a territory as a pack rather than as individuals. When the pack gets beyond a certain size it splits or individuals are forced out; if these splinter groups are not able to form a pack with a territory of their own, they will be forced into nonbreeding, if not death.

### BALANCE WITH ABIOTIC FACTORS AND FIRE

We observed in Chapter 1 that the spread of plant populations is limited by various abiotic factors to which they are not well adapted. Rainfall (amount and distribution), temperature (extremes), salinity, pH, soil type, and nutrients were noted.

One particular abiotic factor that needs elaboration, however, is fire. A raging forest fire consumes virtually every living thing and is obviously destructive (Fig. 3-9). Unfortunately, about 50 years ago forest and range managers interpreted this potential destructiveness of fire to mean that all fire is bad, and embarked on fire prevention programs that eliminated fires from many areas. Unexpectedly, the elimination of fire did not preserve all ecosystems in their existing state. Relatively dry pine forests of the western United States that were originally clear and open became cluttered with trunks and branches of trees that died in the normal aging process. These became the breeding ground for wood-boring insects that proceeded to attack live trees.

In pine forests of the southeastern United States, economically worthless scrub oaks and other broadleaf species grew up under and proceeded to displace the more valuable pines. Semi-

**FIGURE 3-9**
(a) Fires that burn from the ground to the treetops are highly destructive as shown here in the aftermath of a forest fire. (U.S. Forest Service photo.)

arid grasslands were gradually taken over by scrubby, woody species that hindered grazing.

It is now recognized that fires, since they may be started by lightning, have always been a significant, natural abiotic factor. As with all other abiotic factors, different species have varying degrees of adaptation to fire. In particular, grasses and pines have their growing buds located deep in the center of leaves or needles where they are protected from fire. Broadleaf species such as oaks have their buds in exposed locations where they are sensitive to fire damage (Fig. 3-10).

Consequently, periodic burning may be instrumental in maintaining a balance in favor of pines and/or grass by cutting back other species. In the absence of fire the broadleaf species gradually win the competition and the balance tips in their direction.

Fire may also play a valuable role in releasing nutrients from dead organic matter. In a relatively dry ecosystem where natural decomposition is slow, the absence of fire may result in significant portions of available nutrients becom-

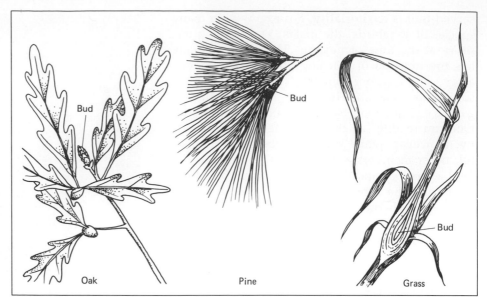

**FIGURE 3-10**
Position of growing bud gives different degrees of protection against fire. Species such as pines and grasses with protected buds are fire resistant; oaks do not have protected buds and are not fire resistant.

Oak      Pine      Grass

Bud

**FIGURE 3-11**
Fire is an important factor in maintaining grasslands. (USDA–Soil Conservation Service photo.)

ing tied up in the dead organic matter, slowing the growth of the entire system.

Fire is now being used as a tool in the management of rangelands (Fig. 3-11) and conifer forests. However, it is important to distinguish between ground fires and crown fires. Crown fires are the destructive fires that rage from the ground to the treetops, destroying everything. Ground fires simply burn along the ground, removing the dead litter, eliminating the breeding ground of wood-boring insects, and killing broadleaf seed-

lings, thus maintaining pines. Ground fires do not harm mature trees, and wildlife can generally escape them (Fig. 3-12).

Furthermore, if ground fires occur every few years, there will be relatively little accumulation of deadwood, and little danger that a ground fire will ignite a crown fire. However, in those forests in which all fire has been prevented for more than 60 years, so much deadwood has accumulated that, if it does catch fire, it will almost certainly set off a crown fire (Fig 3-13). Therefore, it re-

**FIGURE 3-12**
Ground fire. Far from being harmful, periodic ground fires as shown here are necessary to preserve the balance of pine forests. Note the absence of deadwood and competing species. (USDA Soil Conservation Service photo.)

**FIGURE 3-13**
Controlled ground fires are virtually impossible after a significant amount of deadwood has accumulated. The ground fire shown here is becoming a crown fire. (USDA–Soil Conservation Service photo.)

mains exceedingly important to prevent fires from occurring in these areas until the deadwood can be removed by other means.

It is important to emphasize that fire is important in maintaining only certain kinds of ecosystems. In deciduous forests and tropical forests that have achieved balance without fire, even ground fires can serve only to destroy the humus-rich topsoil and permit harmful erosion and loss of nutrients.

## ECOSYSTEM CHANGE

When a factor of the ecosystem is changed, some species are disfavored and their populations dwindle and may be eliminated altogether. Other species, however, are favored by the change and their populations increase. Still other species may invade the area for the first time. Overall, a change leads to one or more species being displaced by other species, a process called **succession.** Succession is a natural phenomenon although it is frequently the result of human-induced changes. Natural succession occurs when the growth of one community causes changes that make the environment more favorable to a second community and less favorable to the first. This results in a natural, gradual transition from the first to the second system. There may be additional stages before a lasting balance among species is reached.

### Primary Succession

If the area has not previously been occupied, the process of initial invasion and then progression from one ecosystem to another is referred to as **primary succession.** A classic example is the gradual invasion of a bare rock surface by what eventually becomes a forest ecosystem. Bare rock is an inhospitable environment. There are few places for seeds to lodge and germinate, and if they do, the seedlings are killed by lack of water or exposure to wind and sun on the rock surface. However, moss is uniquely adapted to this environment (Fig. 3-14). Its tiny spores, specialized cells that function reproductively, can lodge and germinate in minute cracks, and can withstand severe drying. Upon drying out, moss is not killed; it simply becomes dormant. With each bit of moisture it quickly revives and resumes its growth. As it grows, the moss forms a mat that acts as a sieve, catching and holding soil particles as they are broken from the rock or as they blow or wash by. Thus a layer of soil, held in place by the moss, gradually accumulates (Fig. 3-15a). The mat of moss and soil provides a suitable place for seeds of larger plants to lodge, and the greater amount of water held by the mat permits their germination and growth (Fig. 3-15b). The larger plants in turn serve to collect and build additional soil, and eventually there is enough soil to support shrubs and trees. In the process, the fallen leaves and other litter from the larger plants

**FIGURE 3-14**
Moss growing on bare rock.
(Photo by author.)

(a)

(b)

FIGURE 3-15
Primary succession on bare rock.
(a) Moss holds soil. (b) Seed
plants invade the moss-soil mat.
(Photos by author.)

smother and eliminate the moss and most of the smaller plants that initiated the process. Thus, there is a gradual succession from moss through small plants and finally to trees. Erosion, earthquakes, landslides, and volcanic eruptions expose new rock surfaces so that primary succession is always occurring somewhere.

Another example of primary succession is the gradual filling and invasion of lakes by forest eocsystems. As streams or rivers enter large bodies of water, their velocity slows, agitation decreases, and soil particles carried by the water settle to the bottom. When sediments build up to within a few feet of the water surface, various aquatic plants, such as waterlilies and cattails, root in the bottom and send their leaves to the surface. The roots of these plants stabilize the sediments, and the dense strand of stems and leaves acts as a filter that traps increasing amounts of sediments. Also, as the plants grow and die, their dead remains accumulate and add further to the sediments. (Decomposition in such environments is very slow; therefore, accumulation of the dead organic matter does occur.) Floodwaters and plant production continue to add sediments, finally bringing the level of the sediments higher than the average water level and providing an environment that is dry enough for grasses, then shrubs, and finally trees. Thus, the land ecosystem slowly invades a body of water and aquatic organisms are gradually elimi-

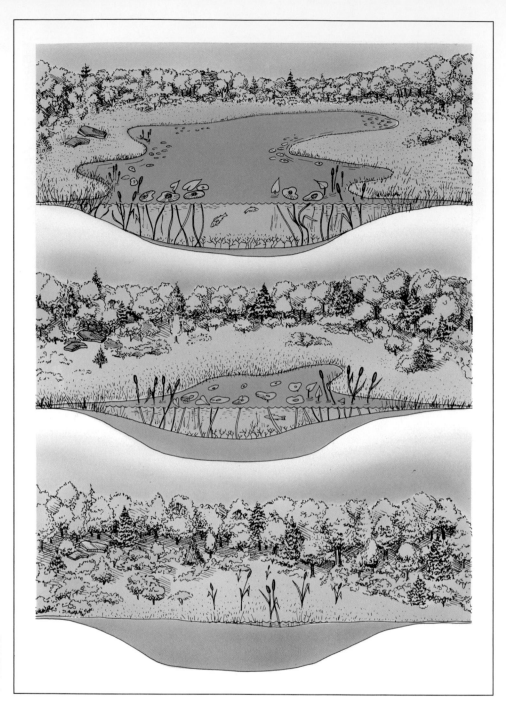

**FIGURE 3-16**
Primary succession. Ponds and lakes are gradually filled and invaded by the surrounding land ecosystem.

nated. In time a lake may become completely filled and the area can become indistinguishable from the surrounding ecosystem (Fig. 3-16).

In the course of geological history, new lakes were formed by glaciation, crustal movements, or the blockage of river courses by landslides. In recent times, humans have formed many new lakes by building dams. All these lakes are destined to be gradually filled by the process of succession.

## Secondary Succession

When an area is cleared, as for agriculture, and then abandoned, the dominant ecosystem of the area will generally return through a series of well-defined stages. Since this is the reestablishment of an ecosystem that was originally present, the process is termed **secondary succession.** A classic example is the progression from abandoned agricultural fields back to broadleaf trees that occurs in deciduous forest regions of the eastern United States (Fig. 3-17).

On an abandoned agricultural field, crabgrass is predominant among the initial invaders. Crabgrass is particularly well adapted to invading bare soil. Its seeds germinate in the spring and it grows and spreads rapidly by means of runners; moreover, it is exceptionally resistant to drought. Anyone who has had the experience of fighting crabgrass in a lawn knows that it grows vigorously over the summer months in sunbaked soil when most other plants seem to be dying from lack of water. However, in spite of its vigor on bare soil, crabgrass is easily shaded out by taller plants. Consequently, taller weeds and grasses, which take a year or more to develop, eventually take over from the crabgrass. Next, young pine trees, which are well adapted to thrive in the direct sunlight and heat of open fields, gradually develop and shade out the smaller, sun-loving weeds and grasses; eventually they form a pine forest. But this is not the end because the pine trees also shade out their own

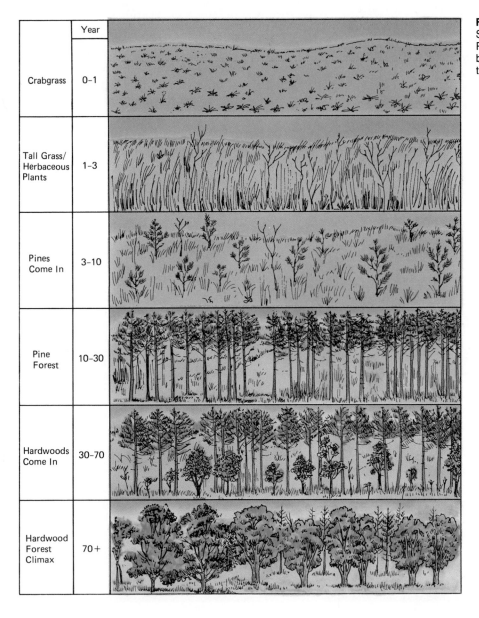

| | Year | |
|---|---|---|
| Crabgrass | 0–1 | |
| Tall Grass/ Herbaceous Plants | 1–3 | |
| Pines Come In | 3–10 | |
| Pine Forest | 10–30 | |
| Hardwoods Come In | 30–70 | |
| Hardwood Forest Climax | 70+ | |

**FIGURE 3-17**
Secondary succession. Reinvasion of an agricultural field by a forest ecosystem occurs through a series of stages.

seedlings. It is the seedlings of deciduous trees, not pines, that develop in the cool shade beneath the pine forest. Consequently, as the pines die off (their life span is 40 to 100 years), they are replaced by oaks, hickories, beeches, maples, and others that characterize eastern deciduous forests. The seedlings of the latter continue to flourish beneath the cover of the parents providing a stable balance and completing the process of succession.

Both primary and secondary succession imply that spores and seeds of the various plants and breeding populations of the various animals are present to invade the area as conditions become suitable for them. If not, succession will be blocked or altered accordingly. Also, it is assumed in secondary succession that a fertile soil base is present at the beginning. If this soil base has been destroyed by erosion or some other means, it may be necessary for the system to start over in a manner similar to primary succession (Fig. 3-18).

**FIGURE 3-18**
Copper Hill, Tenn. The forest in this area was originally killed by fumes from a copper-smelting operation. Although the fumes are no longer present, reinvasion and succession has been blocked by severe soil erosion. (U.S. Fish and Wildlife Service photo.)

## The Climax Ecosystem

Succession finally reaches a point at which all the species present continue to reproduce in proportion to each other and no further change occurs. This balanced state is called the **climax** and the system is called a **climax ecosystem.** The nature of the climax ecosystem differs according to the prevailing abiotic factors of the region. The climax in hot dry areas is a desert ecosystem; in hot wet areas it is a tropical rainforest. The major biomes described in Chapter 1 are characterized by the climax ecosystem typical of the region.

However, it is important to emphasize that even a climax system is in no way static or rigid; it is simply a system in which all the species involved have reached a balance with each other and their environment. In this dynamic balance there is continual adjustment and readjustment as conditions and populations fluctuate from year to year. Then, over the longer period (thousands of years), conditions not only fluctuate around an average, there are trends of gradual change in climate and other abiotic or biotic factors. Such changes introduce a certain degree of imbalance and consequently certain populations will increase and others will decrease accordingly in the process of establishing a new balance. Thus, even

climax ecosystems are probably always in a state of at least very gradual change or succession. For instance, some 10,000 years ago, during the last ice age, tundra existed where there are now deciduous forests in the eastern United States.

## Degree of Imbalance and Rate of Change: Succession, Upset, or Collapse

The rate of change, whether it occurs slowly over the course of many thousands of years or rapidly over the course of just a few years, is proportional to the degree of imbalance. Succession implies a slow, gradual change such that the degree of imbalance at any given time is not great. Thus, there is a gradual but orderly displacement of some species by others, but a diverse ecosystem of many interacting species is maintained throughout.

On the other hand, there may be very pronounced sudden changes that lead to a population explosion of one species at the expense of most other species in the system. This is referred to as an *ecological upset* as opposed to succession. Introduction of foreign species leading to problems such as those discussed earlier are examples of upsets. Another example is the discharge of nutrient-rich wastes into waterways, leading to an explosion of undesirable algal growth.

Finally, changes may be so drastic or of such a nature that almost nothing survives. Such an event is referred to as *collapse of the ecosystem*. After the dieoff or collapse there may be an invasion of the area by species that are able to tolerate the new condition, but this invasion of species is not the cause of the dieoff. In fact, it initiates a new round of succession. Collapse of an ecosystem may be caused by such factors as pollution with one or more toxic substances or the diversion of a waterway causing wetlands to dry up.

It is significant to note that with few exceptions such as earthquakes and volcanic eruptions, natural changes are generally gradual, leading to succession. On the other hand, human-induced changes are frequently sudden and/or drastic, leading to upsets or collapse.

## IMPLICATIONS FOR HUMANS

Examining the human population from the point of view of balancing factors provides a number of insights. For about the last 200 years the human population has been undergoing explosive growth. Since the early 1800s it has increased five-fold and continues to grow at an unprecedented rate (see Fig. II-1). What is the cause?—increased biotic potential or decreased environmental resistance? Clearly, humans have not increased their biotic potential; birth rates are actually down somewhat; however, we have reduced our factors of environmental resistance. Most significant has been the control of parasitic and disease-causing organisms that used to cause high death rates, especially among infants and children. Recruitment, that is, survival to reproductive age, used to be between 10 and 20 percent, but it is now over 90 percent (95 percent in developed countries). This results in rapid population increase despite a somewhat lower birth rate.

In terms of population balance, then, the human population explosion is not an unnatural phenomenon. It is what occurs in any species when or if controlling factors of environmental resistance are removed. But this does not make our population explosion desirable or acceptable in light of our long-term goal of survival.

As noted at the end of Chapter 1, destruction of natural ecosystems can only undercut the ecological stability of the entire biosphere which is necessary for the survival of all species including our own. Yet many of our activities are leading to further destruction and collapse of natural ecosystems. Such activities include the following:

1. Stripping forests for agriculture, highways, housing and other forms of development.
2. Allowing erosion and sediment from these activities to drain into rivers and cause the collapse of aquatic ecosystems.
3. Manipulating natural waterways to flood some areas and drain others.
4. Dumping wastes into air and water causing pollution. Acid rain is currently causing the collapse of many ecosystems, both aquatic and terrestrial.

Additionally, numerous other activities are altering natural balances in one way or another and are consequently causing ecological upsets. The most significant of these include:

1. Intentional and unintentional introduction of species from one ecosystem into another. These may be producers, herbivores, carnivores, or parasites. All are capable of causing upsets.

2. Overgrazing with domestic livestock.

3. Discharging nutrient-rich wastes into waterways.

4. Intentional elimination of predators.

5. Use of pesticides, which causes unintentional side effects, often through the elimination of natural enemies.

6. Overhunting, overfishing, or overcutting of particular plant and animal species for profit or sport.

Again, there are aesthetic and moral questions concerning these actions that science does not answer. But we can project where such actions will lead if they continue.

Currently, the human population is largely free of major controlling factors; therefore, population is expanding. But can this situation be maintained? Ultimately other factors of environmental resistance will come into play. Most importantly we can see limits of the fundamental resources of water and soil necessary to sustain our agricultural crops. Yet we are squandering these resources by allowing soil to erode and converting prime agricultural land to development. We are depleting many groundwater supplies and making other supplies unusable by pollution. We are also undercutting the long-term genetic viability of agricultural plants and animals by destroying related wild species that may be used in hybridization.

Some would like to consider the human agricultural system as a new climax (long-lasting) ecosystem taking its position on earth. However, it must be emphasized that we maintain a precarious balance here. The sustainability or perpetuation of a climax system is given only by a fundamental balance among all the biotic and abiotic factors involved. Any unbalanced ecosystem is guaranteed to be a passing phase unless balances are developed.

We can see all too clearly that the human ecosystem is grossly out of balance in many ways. But do not jump to conclusions. There is an old saying, "If you do not like where you are going, change directions." The point is we have the basic understanding required to restore balances; we need only change directions. Importantly, this does not mean abandoning modern technology and returning to a primitive condition. Continuing technological development and restoration of ecological balance are not incompatible. To the contrary, technological development will be essential in such areas as recycling materials and nutrients, harnessing solar energy, controlling pollution, and birth control. We only need to put greater emphasis on these directions.

Admittedly this task of changing directions is enormous, and bombardment with media reports of environmental tragedies makes many feel pessimistic about our ability to change course in time. Unfortunately, the drama of bad news often tends to mask good news. Actually, there is much to be optimistic about.

Considering that the environmental movement only started in the 1960s, enormous progress has been made. Numerous environmental laws have been passed. All levels of government (federal, state, and local) have formed new departments or agencies to deal with environmental issues. Hundreds of thousands of concerned people have formed and are active in various citizen organizations (see Appendix A) working toward ecological solutions. As a result of these efforts, air quality in most cities is significantly better now than in the late 1960s; pollution in many streams and rivers has been significantly reduced; recycling of wastes is advancing; and some bird populations, formerly threatened by pesticides, are now recovering.

However, many major issues remain and there must be high levels of continuing citizen understanding and support to maintain progress and prevent backsliding. Also, solutions in many areas will involve actual lifestyle changes and such changes conspicuously demand understanding and cooperation on the part of almost everyone.

# 4

# EVOLUTION

## CONCEPT FRAMEWORK

| Outline | | Study Questions |
|---|---|---|
| **I. OBSERVATIONS LEADING TO THE THEORY OF EVOLUTION** | 95 | |
| **A. The Taxonomic Tree** | 95 | 1. Define *taxonomy*. Into what categories is the taxonomic tree divided? What does the tree suggest? |
| **B. Problems in Defining a Species** | 95 | 2. What is a *total population?* What is a *species?* How does *interbreeding* relate to the taxonomic tree? |
| **C. Homologous Structures and Vestigial Organs** | 98 | 3. What are *homologous structures* and what do they imply? What are *vestigial organs* and what is their historical significance? |
| **D. The Fossil Record** | 98 | 4. What kinds of evidence are provided by fossil records? |

93

5. Explain *Lamarckianism.* Why is the theory unsupportable?

6. What three observations underlie Darwin's theory of evolution and natural selection? Why can survival of the fittest also be viewed as a process of selection?

7. How does the process of natural selection result in survival of the fittest, according to Darwin?

8. Compare and contrast artificial selection with natural selection.

9. How does the peppered moth study illustrate natural selection? Why has pesticide use failed to eliminate the problem?

10. What are *mutations?* How do mutations and sexual reproduction support Darwin's theories?

11. What is *selection pressure?* When selection pressures change, what alternatives are available to affected species?

12. Explain *speciation* and give examples.

13. How does *adaptation* relate to natural selection? What characteristics tend to lead to rapid adaptation?

14. Discuss how balanced relationships develop in ecosystems.

15. Distinguish between *evolutionary succession* and *ecological succession.* Which process is shorter? Why?

16. Discuss some of the factors of natural selection that may have led to the development of the human as we now know it. How may natural selection contribute to the development of various races?

17. How does the evolution of humankind differ from that of most other species?

18. What is considered the most significant turning point in human development?

19. What indications suggest that humans are not the culmination of the evolutionary process? What phenomenon would assure the collapse of modern civilization?

In Chapter 3 we repeatedly emphasized that the balance between populations of different species within an ecosystem depends on specific adaptations of each species to its environment and to the other species with which it usually interacts. In fact, each species can be viewed as an integrated set of adaptations which may be divided into five categories: (1) adaptations for coping with climatic and other abiotic factors; (2) adaptations for obtaining food and water in the case of animals, or nutrients, energy, and water in the case of plants; (3) adaptations for escaping from or protection against predatation, and resistance to disease-causing or parasitic organisms; (4) adaptations for finding or attracting mates in animals, or pollination and setting seed in plants; (5) adaptations for migration in animals or seed dispersal in plants. It is an instructive exercise, in fact, to select any species (plant, animal, fungus, or bacterium) and list its specific adaptations in each of these areas (Fig. 4-1).

Seeing these unique characteristics that enable each species to cope with its natural environment, one can hardly help but wonder how species came to have such marvelous adaptations. Prior to the 1800s, both scientists and lay people of the Western world generally accepted the biblical story of Genesis, which states that God created all the creatures on earth. With this belief, the adaptations of each species were simply accepted as God's wisdom in providing each species with features necessary for survival; no more needed to be said. Now, however, very few scientists hold to a literal belief in the biblical story of creation. Instead, scientists and most lay people accept the theory of **evolution,** although it is commonly misunderstood by people outside of biology. Our objective in this chapter is to examine the evidence and logic that underlie the theory of evolution and clarify our understanding of it. Seeing all life, including ourselves, as a product of biological evolution has a profound impact on how we view our interactions with ecosystems and other life on earth.

## OBSERVATIONS LEADING TO THE THEORY OF EVOLUTION

As noted in the introductory chapter, a theory is a rational explanation accounting for various observations. What are the observations pertinent to the theory of evolution?

## The Taxonomic Tree

The systematic collection, examination, and giving of scientific names to plants and animals began in the 1400s and it continues as the science of **taxonomy** or systematics. Early on, however, such study revealed that species are not entirely independent and distinct. Instead they fall into a hierarchy of groups and subgroups that are clearly related by various degrees of similarity as illustrated in Figure 4-2. Thus, classifying species according to their similiarities produces a treelike arrangement, commonly referred to as the *taxonomic tree,* which is suggestive of an evolutionary process. That is, it suggests that present species are the result of a process of change and modification of preexisting species.

## Problems in Defining a Species

The concept of species undergoing change and modification was reinforced even more when taxonomists confronted the problem of defining just what a species is. From the viewpoint of creationism, early taxonomists anticipated that each specimen of a plant or animal collected would represent a clearly definable group or species and that all members of the group should be virtually identical. However, as more and more specimens were collected and studied, more and more variation became apparent. What, then, is a species? Indeed, taxonomists still debate whether some groups of organisms should be defined as one species with a high degree of variation, or whether they should be defined as a number of separate species (Fig. 4-3).

In recent years, taxonomists have stressed the aspect of interbreeding in defining a species. A **species** is defined as the total population, all the members of which do or potentially can breed together to produce viable offspring (offspring that can, themselves, reproduce as opposed to sterile hybrids such as the mule). Distinctive variations that may be observed within a population that can interbreed may be defined as different races, subspecies, or varieties (*varieties* is a term used in reference to plants). However, even this definition does not resolve all the problems because testing for interbreeding is obviously impractical in many cases and impossible in the case of fossil specimens; hence the debate continues.

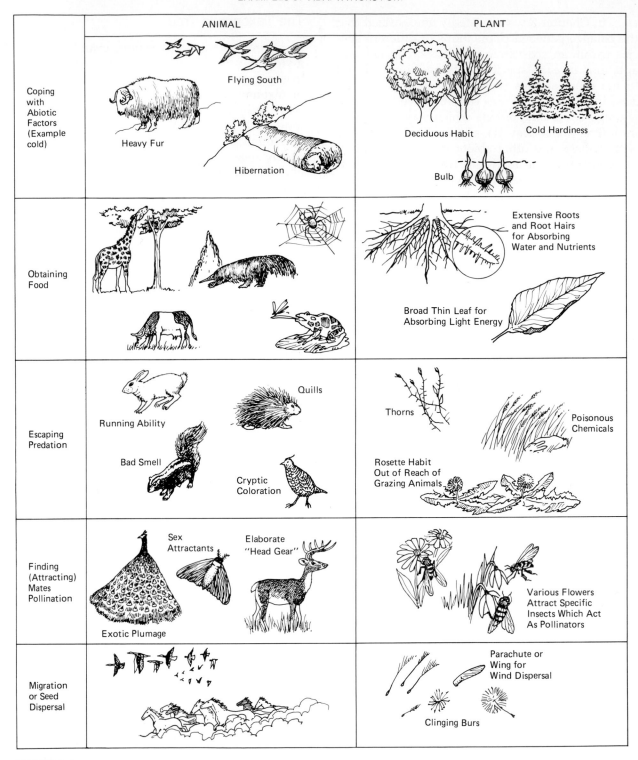

**FIGURE 4-1**

Examples of adaptations. Every species can be viewed as a complex of adaptations enabling the individuals to: (1) cope with abiotic factors; (2) obtain food and water, in the case of animals, or nutrients, energy, and water, in the case of plants; (3) escape from or gain protection against predation; (4) find and attract mates and reproduce in the case of animals or pollinate and produce seeds or spores in the case of plants; and (5) migrate or disperse seeds or spores.

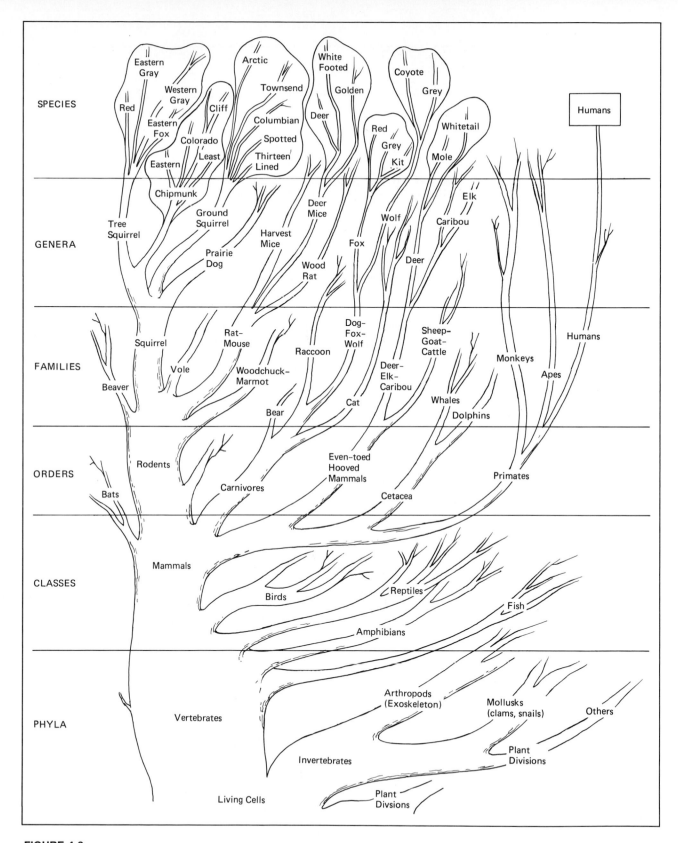

**FIGURE 4-2**
The taxonomic tree. When species are grouped according to relative similarities and differences, they fall naturally into a tree-like pattern as illustrated. This illustration depicts only a few species from selected groups of mammals, but the same tree-like arrangement can be found throughout the plant and animal kingdoms.

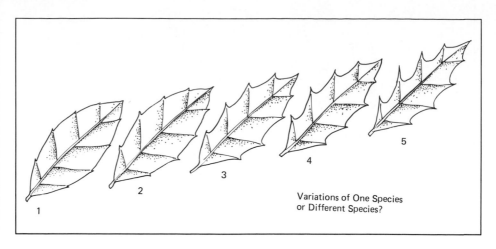

**FIGURE 4-3**
What is a species? Historically a species is defined as a population of organisms with distinctive morphology (body features). On this basis, if 2, 3, and 4 were not present in this illustration, 1 and 5 might be defined as separate, distinct species. However, given the presence of intermediate variations, the entire group may be considered one species. This kind of situation is now recognized as the process of speciation (the development of two or more species from one) in progress.

Variations of One Species or Different Species?

The only real answer to this problem must be a recognition that what is expected from the idea of individual creation, each plant and animal representing a distinct group, simply does not correspond to what is observed. On the other hand, what is observed does fit well with the idea that species are entities that have undergone and are continuing to undergo a process of change and modification.

## Homologous Structures and Vestigial Organs

At the same time taxonomists were categorizing species, anatomists were examining and describing the internal parts of organisms. They observed a striking similarity in the basic skeletal structure and internal anatomy throughout a taxonomic group in spite of apparent diversity. For example, one can observe that the front leg of a dog or a horse, the wing of a bat, and the flipper of a whale all have the same basic skeletal structure (Fig. 4-4). In the horse, for example, one can see the three central foot bones have become fused to provide a longer leg; the two outer foot bones are reduced. Such outwardly different structures, which are based on the same anatomical structure, are called **homologous structures.**

In many cases a more advanced organism may carry a remnant of a part or organ that played a significant role in a more primitive organism, but has no function in the more advanced organism. Such remnants are referred to as **vestigial organs.** The human body has several examples: the appendix, wisdom teeth, and a ru-

dimentary tail (Fig. 4-5). Again, such findings are suggestive of present species being a product of modification of preexisting species.

## The Fossil Record

The church of the eighteen and nineteenth centuries maintained a belief in a young and static world; the events of creation were calculated from lineages described in the Bible to have occurred about 5000 years ago. In contrast, the concept of evolution is based on changes occurring over many millions of years. Therefore, one of the problems faced by Darwin and other early evolutionists was to provide evidence that change had occurred in the biological and geological history of the earth, and that sufficient time had been available for these changes to take place. Fossils provided both kinds of evidence. As they were gradually accepted as representing extinct species, fossils showed that a sequence of different species had occupied the earth at different times. In addition, as geological processes that had led to entrapment of fossils in various rock layers and formations were understood, it was recognized that these time periods did encompass hundreds of millions of years (Fig. 4-6). Finally, some fossils were found that appeared to represent the transitional forms in the process of forming new species from old.

In summary, evidence from numerous different areas of study pointed to the concept that all species, extinct and living, are a product of gradual modification of preexisting species, that

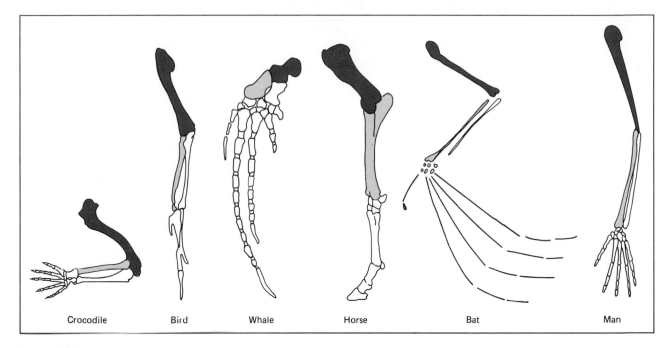

| Crocodile | Bird | Whale | Horse | Bat | Man |

**FIGURE 4-4**
Homologous structures. Observe that limbs of mammals which perform very different functions such as running (horse), flying (bat), swimming (whale), or grasping (human arm) all have the same basic skeletal structure. Only individual bones are modified in size and shape and in some cases fused together. This suggests that different species have developed by a process of modification from a common ancestral origin. (From Helena Curtis, *Biology*, 3rd ed., 1979, Worth Publishers Inc., New York.)

**FIGURE 4-5**
Vestigial tail. The human skeleton has a remnant of a taillike structure at the base of the spine. Such rudimentary structures or organs which have no apparent functional role are now interpreted as representing a vestige from our primate ancestry where tails were present and functional. (Photo by author.)

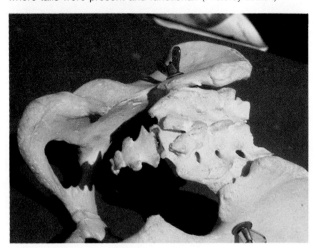

is, of evolution. Furthermore, evidence indicates that present-day species are not static but that the process is still going on.

## THEORIES OF EVOLUTION

Contrary to popular belief, the concept of evolution did not originate with Charles Darwin. The idea actually dates from ancient Greek times. What was lacking, however, was sufficient evidence to accept the theory of evolution as opposed to the idea of distinct creation. Also lacking was any understanding of a mechanism by which gradual change could occur. In the absence of such information, the concept of evolution held relatively little attraction. However, as evidence such as that cited above accumulated, theories of evolution demanded more serious consideration.

### Inheritance of Acquired Characteristics: The Wrong Theory

One theory of evolution, the inheritance of acquired characteristics, was promoted strongly by Jean Baptiste de Lamarck in the early 1800s. The idea is sometimes referred to as **Lamarckianism.** We present the theory here because, while it has been proven wrong, it is still a commonly held notion among nonbiologists.

| Species and Period | | Series and Epoch | Distinctive Features | Years Before Present | Number of Biological Families |
|---|---|---|---|---|---|
| CENOZOIC | QUATER- NARY | RECENT | Modern man. | 11 thousand | 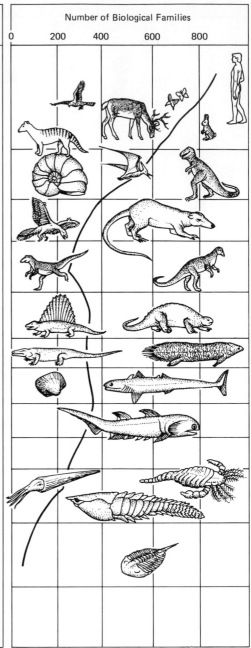 |
| | | PLEISTOCENE | Early man; northern glaciation. | 0.5 to 3 million | |
| | TERTIARY | PLIOCENE | Large carniovres. | 13 ± 1 million | |
| | | MIOCENE | First abundant grazing mammals. | 25 ± 1 million | |
| | | OLIGOCENE | Large running mammals. | 36 ± 2 million | |
| | | EOCENE | Many modern types of mammals. | 58 ± 2 million | |
| | | PALEOCENE | First placental mammals. | 63 ± million | |
| MESOZOIC | CRETACEOUS | | First flowering plants; climax of dinosaurs and ammonites. | 135 ± 5 million | |
| | JURASSIC | | First birds, first mammals; dinosaurs and ammonites abundant. | 180 ± 5 million | |
| | TRIASSIC | | First dinosaurs. | 230 ± 10 million | |
| PALEOZOIC | PERMIAN | | | 280 ± 10 million | |
| | CARBON- IFEROUS | PENNSYLVANIAN | Great coal forests, conifers. First reptiles. | 310 ± 10 million | |
| | | MISSISSIPPIAN | Sharks and amphibians abundant. Large and numerous scale trees and seed ferns. | 345 ± 10 million | |
| | DEVONIAN | | First amphibians and ammonites; fishes abundant. | 405 ± 10 million | |
| | SILURIAN | | First terrestrial plants and animals; jawless fishes first appear. | 425 ± 10 million | |
| | ORDOVICIAN | | First jawless fishes; invertebrates dominant. | 500 ± 10 million | |
| | CAMBRIAN | | First abundant record of marine life; trilobites dominant, followed by extinction of about two-thirds of trilobite families. | 600 ± 50 million | |
| | PRECAMBRIAN | | Fossils extremely rare, consisting of primitive aquatic plants. Evidence of glaciation. Oldest dated algae over 2600 million years. | | |

**FIGURE 4-6**

The fossil record. Study of fossils and geological processes reveals that the earth has been occupied by a succession of different kinds of plants and animals at different times. Mammals and birds are of relatively recent origin. Also over time there has been a gradual increase in biological diversity as indicated by the increasing number of biological families.

(Reprinted by permission from Samuel N. Luoma, *Introduction to Environmental Issues*. New York: Macmillan Publishing Company, 1984.)

One notes that, in coping with the environment, individuals may develop certain traits. Limbs or organs may be strengthened by use or atrophied by disuse; skills may be learned; changes may occur as a result of accidents. The Lamarckian hypothesis was that such characteristics acquired during the lifetime of an individual would be inherited by offspring and developed further in the next generation, thus leading to a gradual evolution of these characteristics. The classic hypothetical example is that predecessors of giraffes presumably stretched their necks in grazing on trees. The stretched necks were presumably inherited by offspring and the process repeated until finally a population of long-necked animals, giraffes, was derived. Another example, it was thought that tailless mice had evolved from mice which had lost their tails accidentally.

This was an intriguing idea; however, all the evidence points against it. Experiments clipping the tails of mice did not produce a tailless breed. The offspring of highly trained animals have no more talent than offspring of untrained individuals of the same breed. Laborers who have developed heavily callused hands don't produce children with hands that are tougher than average. The tails of sheep are cut routinely, and this practice has been carried on for thousands of generations; yet sheep are still born with tails as long as ever. In short, there was (and still is) no evidence to support the idea that traits acquired during the lifetime of an individual are inherited by its progeny. More recent evidence now clearly shows that such acquired characteristics do not change the genetic makeup of the individual and hence cannot be inherited. With no evidence to support the Lamarckian hypothesis or other hypotheses proposed prior to the 1850s, the theory of evolution remained speculative.

## Darwin's Theory: The Origin of Species by Natural Selection

Charles Darwin first contributed a tremendous body of scholarly work supporting the general concept of evolution. The sheer volume of this work began to tip the balance toward people thinking in terms of evolution. Then, and most significantly, he presented in tremendous detail a theory as to how evolution could occur, namely, the *Origin of Species by Natural Selection* published in 1854. Darwin's proposal was not only plausible; the observations on which it is based were so easily verified that they were, and still are, undeniable. This, plus the voluminous data Darwin amassed in support of evolution, precipitated general acceptance of the theory.

Darwin's proposal for the evolution of species by natural selection is based on three phenomena and logical reasoning from these observations.

### NATURAL SELECTION: A PROCESS OF OVERREPRODUCTION, VARIATION, AND SURVIVAL OF THE FITTEST

First, Darwin emphasized the observation made in Chapter 3 regarding population balance. Namely, all species have a reproductive rate such that if all the offspring (young, eggs, or seed) were to survive and reproduce there would be a population explosion in every case. Darwin noted that even the most slowly reproducing species, elephants, would flood the entire earth in a matter of a few hundred years if all the young survived and reproduced in turn. A rapidly reproducing species such as fruit flies, which can produce up to 200 offspring in about 15 days, could produce a mass of flies the size of the earth in about seven months if all survived and reproduced. The second phenomenon, closely related to the first, is the fact that population explosions rarely occur in nature and, when they do, they do not go on indefinitely.

Taken together, these two phenomena force one to admit that in nature only some of the offspring survive and reproduce. Indeed, a constant population implies an average survival to maturity of just *two* offspring per couple. These two would be sufficient to replace the parents when they die. The other tens, hundreds, or even thousands of offspring produced must die off before reproduction. Darwin referred to this as **overreproduction.** Overreproduction, however, begs the question, Which of the many offspring actually survive and reproduce? Is survival a matter of random chance or do certain offspring have some advantage over others? Darwin pointed out a third phenomenon that is critical to answering this question.

The third phenomenon is **variation:** there is variation among the individuals of all species. This is obvious in humans. We come in a wide range of different sizes, shapes, colors, and abilities and these variations may occur in any combination; they are what permit us to recognize each other. Similar variation exists among rabbits, elephants, oak trees, houseflies, and all other species. We tend to think of all houseflies, for example, as identical, but this is only because we do not examine them closely. When one looks closely—and Darwin spent many years examining species after species to verify this—one finds slight differences in size, color, and any other trait one focuses on.

Further, the differences that are visible are only part of the picture. As in humans, individual organisms exhibit different levels of tolerance to heat or cold, resistance to disease, sensitivity to various drugs, and so on. Paraphrasing Darwin: Whatever trait one wishes to examine, one finds some individuals that express it a little more and some a little less.

Recognizing that virtually every trait of an organism can be described in terms of its adaptation to obtain food or nutrients, escape from predators, find mates, and so on, one is virtually forced to admit that many variations will affect an organism's ability to cope in one way or another. For example, an animal whose color variation makes it blend into its background better is more likely to escape predation.

The conclusion, which Darwin pointed out, is that because of overreproduction, there is a struggle for survival and that individuals with advantageous variations are most likely to survive; that is, there is a **survival of the fittest.** Unfortunately, the "fittest" has often been misinterpreted to mean the strongest, biggest, or the most aggressive. It is important to recognize the meaning Darwin intended; he considered the "fittest" to be the organism best adapted to survival and reproduction in the particular environment in which the species exists. If small size improves survival and reproduction in a particular organism, it is small individuals that survive as the "fittest." If no legs enhance survival and reproduction for a species, the fittest are the ones with no legs—and we have snakes, which are remarkably well adapted for their particular niche. Also, it is extremely important to keep the idea of *reproduction*

together with *survival* because it is only through reproduction that the genes of the fittest are passed on to the next generation. The genes carried by a healthy, robust individual who does not reproduce are eliminated from the population.

Survival of the fittest can also be looked at as a process of *selection*. In effect, from a large population of offspring, nature *selects* a few of the fittest to reproduce the next generation and the rest either perish or fail to reproduce; hence, Darwin's term, **natural selection.** However, it is important to recognize that nature's selective process is by no means a form of conscious selection of a favored few. In nature, each population is subjected to predators, disease and parasites, floods, droughts, heat, cold, and whatever other biotic or abiotic factors exist in the environment. If a few individuals of each generation survive this onslaught, these are the "selected" ones. If none survives, then the population becomes extinct.

In conclusion, Darwin hypothesized that the process of nature "selecting" the "fittest" or best adapted, generation after generation, is the mechanism underlying evolution (Fig. 4-7).

### MODIFICATION OF TRAITS THROUGH SELECTION

**Modification by artificial selection.** The recognition of natural selection brings up the central question: Is the process of natural selection sufficient to explain evolution? Natural selection acts on the variations among offspring. For the most part, these differences are so slight and subtle that careful observation is required to note them at all. Can selecting on the basis of such small, seemingly trivial differences lead to the profound changes in species implied by the concept of evolution?

In addressing this question, Darwin drew heavily on the empirical evidence (evidence gained through experience) from plant and animal breeding. Plant and animal breeding, Darwin emphasized, has been carried on in an empirical way probably since the first domestication of wild plants and animals began some 10,000 years ago. As a result, we have numerous species of domestic plants and animals that are very different from their ancestors. Indeed, in some cases (e.g., corn) the domestic species has been modified so much that its wild ancestor is unknown.

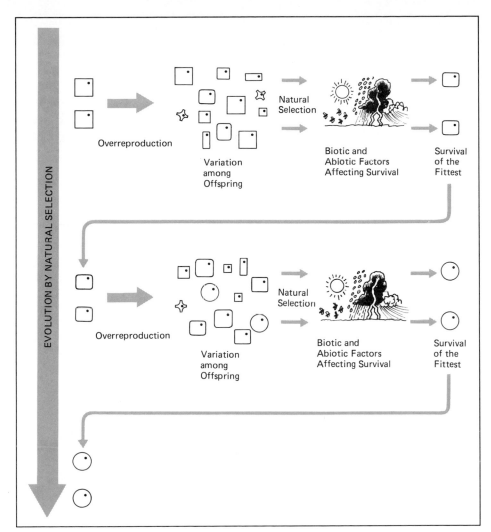

**FIGURE 4-7**
This illustration depicts the basis of Darwin's theory of evolution by natural selection. Note, offspring of "squares" have a variety of similar shapes representing over-reproduction and variation. If "roundness" survives better than other shapes, there will be a gradual evolution toward "circles."

How does the empirical process of plant and animal breeding work? Basically, as Darwin stressed, it is a matter of selection—in this case, **artificial selection.** Breeders desiring a particular trait observe that some individuals have the trait a little more, some a little less. Often, Darwin emphasized, the differences are virtually imperceptible to the inexperienced eye. Breeders select those individuals that show the trait and use them as the parents for the next generation. The offspring tend to be like the parents, but, again, the careful eye of the breeder can note that some express the particular trait more than the parents, some less. Thus, the breeder again selects those that show the trait to parent the next generation and this process is repeated over and over again.

This process is graphically illustrated in Fig-ure 4-8a. One only need note different breeds of dogs or other domestic animals to be impressed by how dramatically species may be modified in almost any way by selection of small variations over many generations (Fig. 4-8b).

There is a common misconception that new breeds or varieties are created in one step through **hybridization** (crossing between quite different parents) or through a single, large mutation (genetic change). While hybridization and mutations are extremely important factors in breeding, and occur in nature as well, they do not supplant the process of selection. For example, a breeder may cross a domestic variety with a wild relative in order to introduce a trait—enhanced disease resistance, for example—from the wild strain into the domestic variety. However, after the cross is

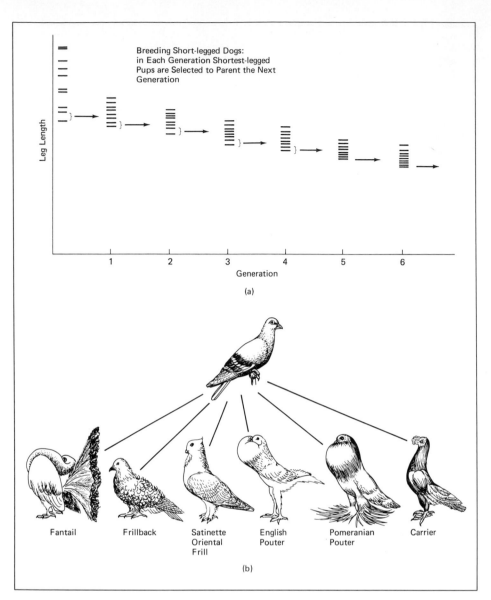

**FIGURE 4-8**
(a) Hypothetical breeding for short-legged dogs. In each generation, the shortest-legged dogs are selected to be parents of the next generation. (b) Breeds of pigeons developed by selective breeding. The wild rock pigeon of Europe, top, is thought to be the ancestor of all the domesticated breeds shown here. (Redrawn from W. W. Levi, *The Pigeon*. Sumter,.S.C.: Levi Publishing Co., 1957.)

made, the process of selection continues as the offspring with the desired resistance (and other traits as well) are chosen to parent further generations.

**Modification by natural selection.** To be sure, natural selection does not have the "critical eye" of breeders, nor does it have a specific objective in selecting for a given trait; it is based only on *survival and reproduction*. Yet, there are documented cases of natural selection leading to change in a population. A classic example is the peppered moth.

The peppered moth was a grey moth often found on lichen-covered tree trunks in nine-teenth-century England. Because the lichen-covered trees were also grey, the moth was nearly invisible to bird predators. In 1845, however, a single specimen of black peppered moth was discovered near the industrial area of Manchester. By the 1950s, the once rare black peppered moths had increased in numbers while the more common grey peppered moths had significantly declined.

A study conducted by H. B. D. Kettlewell showed that the change stemmed from industrialization, which left a layer of black soot over tree trunks in the area. As a result, the grey peppered moths were no longer camouflaged and were captured by bird predators. Kettlewell surmised that a shift in bird predation was the cause and proved it by capturing large numbers of both dark and light moths. He marked them with a spot of paint and released an equal number of each in (1) an industrialized area that contained almost exclusively dark moths and (2) an unpolluted area that contained almost exclusively light-colored moths. When he recaptured the moths, Kettlewell found that he had almost twice as many dark moths from the industrialized area and nearly twice as many light moths from the area where industry was not present.

To prove that this phenomenon was the result of natural selection, Kettlewell photographed bird predation in both places. The photographs confirmed his beliefs. The birds successfully sought out and captured the more obvious moths in each area (Fig. 4-9).

It is interesting to note that in recent years the British government has established pollution controls that have reduced the amount of soot and grime on trees and the number of grey colored moths has since increased.

Other examples of modification of species occurring in nature involve populations of pest insects developing increasing resistance to the pesticide chemicals that are used against them, and disease-causing bacteria becoming resistant to antibiotics. This occurs because the pesticide or drug kills the more sensitive members of the population while the more resistant members survive and reproduce the resistant trait (Fig. 4-10).

It is true that breeders have not produced cats from dogs, much less mammals from fish. However, in connection with this, Darwin pointed out that, regardless of how far breeders have gone in developing a particular trait, variation among offspring never disappears. Always some show the trait a little more and some a little less. Thus, there is always the opportunity to go further. In the Orient, for example, chickens have been bred that produce tail feathers up to 5 m (16 ft) long, and still there is variation!

One can reflect then that, given selection over enough time, any degree of change is possible. Has nature had enough time? Here Darwin relied heavily on geological evidence, which implied that the fossil record covers hundreds of millions of years. Modern theory based on several additional lines of evidence holds that the earth is about 4 billion (4,000 million) years old. If breeders are able to make the changes they have accomplished in a few centuries of selection, it seems quite plausible that much larger changes could take place over hundreds of millions of years. The only distinction is that breeders are selecting for particular traits desired by humans, whereas traits emphasized by natural selection involve an organism's ability to survive and reproduce in its particular environment, that is, traits

Light

Dark

Light

Dark

Peppered Moth
Population in
Rural Areas

Peppered Moth
Population in
Industrialized Areas

**FIGURE 4-9**
Modifications by natural selection. The peppered moth study illustrated here showed that predation is a prime factor of natural selection responsible for the development of cryptic (camouflage) coloration. Given the different background of trees, opposite colors of moths stand out in the two areas and are hence subject to more intense predation.

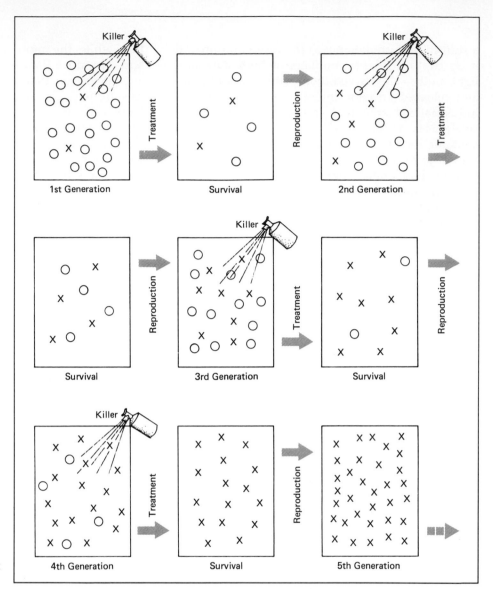

**FIGURE 4-10**
Survival of the fittest. A treatment eliminates mostly those that are sensitive (o). The few individuals that happen to be resistant (x) survive and reproduce. Eventually the entire population is made up of resistant members.

Labels within figure: Killer, 1st Generation, Survival, Treatment, Reproduction, 2nd Generation, Treatment, Survival, Reproduction, 3rd Generation, Treatment, Survival, Reproduction, 4th Generation, Treatment, Survival, Reproduction, 5th Generation

that involve coping with the various biotic and abiotic factors present. It is hardly surprising, then, that all species are highly adapted for survival and reproduction in their particular niche or habitat.

In summary, Darwin pointed out that a process of natural selection occurs in nature and that selection over many generations does indeed lead to change. Given that many millions of generations have occurred over the course of geological history, an evolutionary process seems not only possible; it is virtually inevitable.

It is interesting to compare Darwin's hypothesis with Lamarck's. Both contain the idea of selective breeding, that is, reproduction of the most adapted and elimination of the others. The critical

difference is in where the variations come from. Lamarck proposed that the variations were acquired during the lifetime of the individual, through its efforts or by accident. However, it has been proven that variations *acquired* in this way do not affect the genes of the organism. Hence, such variations prove not to be hereditary; that is, they are not passed on to offspring. Darwin, on the other hand, noted that individuals are born with variations and these variations are hereditary; selective breeding based on these innate variations does lead to gradual change of the population.

While Darwin recognized the existence of innate variations, he was at a loss to explain them. He was at even more of a loss to explain why variations did not disappear in the process of selection. Since his theory hinged on the idea that variation keeps occurring, this troubled him greatly, but he died without finding the answer. However, the answers to questions regarding the source of variations and their inheritance were not long in coming.

## SUPPORT FROM THE STUDY OF GENETICS

Starting with the work of Gregor Mendel in the 1880s, tremendous strides were made in **genetics,** the study of how traits are inherited and passed from one generation to the next. It is beyond the scope of this text to describe the specific studies and the development of our knowledge in this area. However, it is now clearly understood that inherent traits are controlled by units called genes. These units, or **genes,** are actually molecules of the chemical **DNA** (deoxyribonucleic acid) and they, along with other material, comprise the chromosomes found in the nucleus of all cells. As cells divide, it can be seen that chromosomes are replicated and copies separate so that each resulting cell receives its complement of genes (Fig. 4-11).

In general, the DNA within a cell is not altered and the replication process is exact so that genes may be passed from generation to generation without change. However, two processes are responsible for introducing and maintaining variation in a population: **mutation** and **sexual reproduction.**

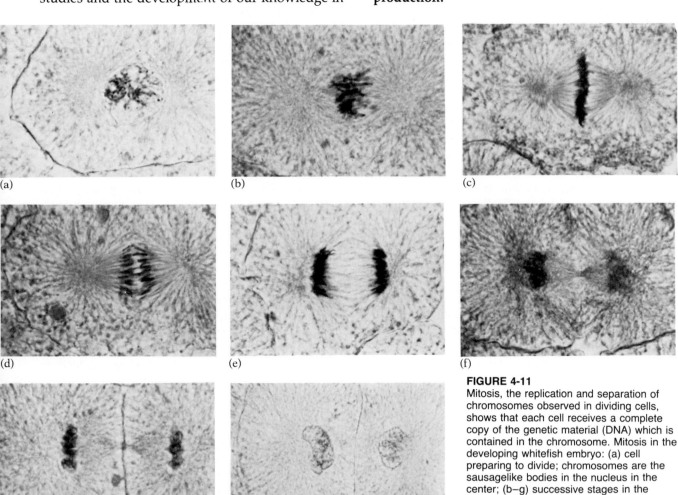

(a)　　　(b)　　　(c)

(d)　　　(e)　　　(f)

(g)　　　(h)

**FIGURE 4-11**
Mitosis, the replication and separation of chromosomes observed in dividing cells, shows that each cell receives a complete copy of the genetic material (DNA) which is contained in the chromosome. Mitosis in the developing whitefish embryo: (a) cell preparing to divide; chromosomes are the sausagelike bodies in the nucleus in the center; (b–g) successive stages in the separation of replicated chromosomes; (h) two resulting daughter cells. (Courtesy of Barrett, Abramoff, Kumaran and Millington, from *Biology,* © 1986, Prentice-Hall, Inc., Englewood Cliffs, N.J.)

Mutations are chemical alterations in the DNA that, in turn, lead to the alteration of some trait or characteristic. Chemically altered DNA may be replicated and passed on to progeny in the same way as the original DNA, so mutations are inheritable. Mutations occasionally occur spontaneously (for no apparent reason) but their frequency of occurrence is greatly increased by certain chemicals (mutagens) and/or high-energy radiation. Importantly, mutations are always random events and hence produce random, unpredictable changes that are more often harmful than beneficial. Nevertheless, they are nature's way of introducing new genes into a population.

Then, sexual reproduction promotes variation by the fact that formation of ova (eggs) and sperm, followed by fertilization (union of egg and sperm), involves a *random* segregation and recombination of the chromosomes from the two parents. Consequently, each offspring (with the exception of identical twins) has a unique combination of genes from its two parents. This is the basis for genetic variation among offspring.

In conclusion, we can visualize, in every population, a continual infusion and distribution of new genetic variations through mutation and sexual reproduction. Selection acting on these promotes change. In short, Darwin's theory of evolution by natural selection is fully consistent with and supported by the findings in genetics.

## WHERE DOES EVOLUTION LEAD?

As humans we are frequently inclined to look at evolution as having some purpose or end. Significantly, there is no evidence for any outside force guiding evolution in one way or another. The mutations and recombinations that occur through sexual reproduction are random changes. Therefore, variations in traits that result from these genetic changes are also random. In turn, whether a given variation is selected for (and thus enhanced) or selected against (and thus eliminated) depends only on whether it aids or detracts from survival and reproduction. This will be examined more thoroughly in terms of *selection pressure*.

### Selection Pressure: The Promoter of Change

**Selection pressure** refers to one or more environmental factors that result in individuals with a certain trait being more likely to survive and re-

produce than individuals without that trait. One can use this concept to visualize the development of almost any trait seen in wild plants or animals. For example, many offspring are lost to natural enemies; individuals that chance to have some protective trait are more likely to survive and produce the next generation. Using our new term, there is a strong *selection pressure* for traits that enable organisms to escape or avoid predation and resist attack by parasitic or disease-causing organisms.

The variation does not have to involve any particular characteristic; the bottom-line question in natural selection is: Does it work? If it works, it will tend to be selected for and thus enhanced. If it doesn't work, those individuals will perish and the respective genes will be eliminated from the population. Thus, among different animals, we see numerous traits that confer a significant degree of protection against natural enemies. For example, skunks have a noxious odor, porcupines have sharp quills, rabbits can run very swiftly, armadillos have incredibly tough skin, certain insects have offensive tastes or odors, many animals have coloration that enables them to blend into their background, and both plants and animals have some degree of resistance to diseases.

As organisms become better adapted, however, selection pressure and, consequently, the rate of change may be expected to diminish. For example, when natural selection has led to the point where all skunks emit an odor so noxious that would-be predators can't tell which one smells worse, then there is no longer selection pressure for increasing the offensiveness of the odor. When ancestors of giraffes started grazing on trees, there would be strong selection pressure for the ability to graze higher and higher until all the population can graze on most of the trees available. After that, there is no longer selection pressure for increasing neck length. Indeed, some counterselection pressure comes into play because the very tall stature of giraffes presents severe problems in equalizing blood pressure from the head to the body when the animal changes positions, for example, when it bends over to drink water. The same may be said of any number of other traits. As long as an organism is well adapted to its habitat or niche, there is little selection pressure in any specific direction and, hence, little if any change. Indeed, in a well-adapted species, selection pressure probably acts to preserve

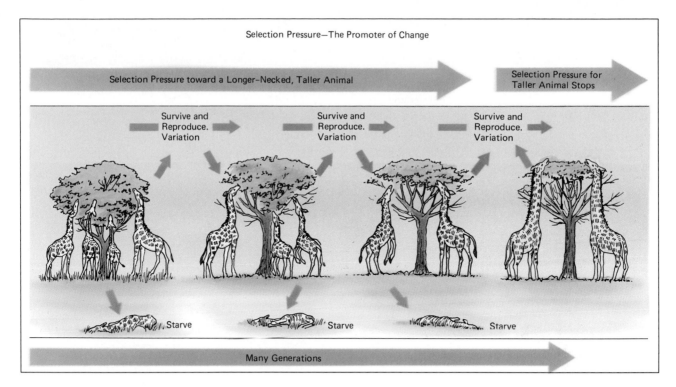

**FIGURE 4-12**
Selection pressure. As long as a situation persists in which a taller, longer-necked animal gets more food and hence enhances its survival advantage, there is selection pressure and hence evolution toward the taller animal. When a longer neck is of no additional benefit, selection pressure and further modification stop.

the status quo because, as in the case of the giraffe's neck length, once a trait has reached an optimum, any further increase or decrease has a negative effect on survival (Fig. 4-12).

However, environmental factors do not stay constant indefinitely. For example, it is well known that the earth has experienced cool periods, the Ice Ages, and various regions have fluctuated between wetter and drier climates. Also, invasion by new species may shift any number of biotic factors. When any one or more factors change, selection pressures change accordingly. In response, a species must do one of three things: (1) it may migrate to similar conditions in another area as, for example, species have generally moved southward as climate cools and northward as climate warms; (2) through natural selection, its population may evolve and adapt to the new conditions; or (3) if it fails to do either of the above, it will become extinct (Fig. 4-13).

You may note that human activities are now creating a tremendous selection pressure for nearly all species. Some, such as rats and numerous insects, have adapted to new human environments all too well. However, countless others are being pushed into extinction.

## Speciation

In the process of adapting to new conditions, a species may change so significantly that it becomes a different species. Also, separate populations of the same species, if they are exposed to different biotic and/or abiotic factors, may evolve differently, giving rise to two or more species. This process of a single species evolving into two or more different species is called **speciation.**

A classic example of a species adapting to changing conditions is seen in the evolution of the horse. According to the evidence that has been pieced together from the fossil record, the modern horse is descended from a collie-sized animal, *Eohippus,* which inhabited forests at the beginning of the Tertiary Period some 65 million years ago. *Eohippus* was well adapted to browsing on leaves in the forest habitat. However, through the Tertiary Period, the climate gradually became drier and forests gradually gave way to grasslands. This changing habitat created numerous selection pressures. With decreasing amounts of

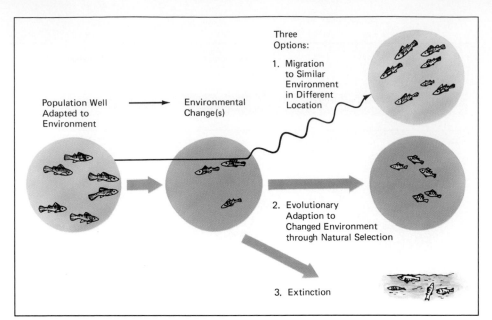

**FIGURE 4-13**
Species may be well adapted to their environment, but what occurs if one or more abiotic or biotic factors change?

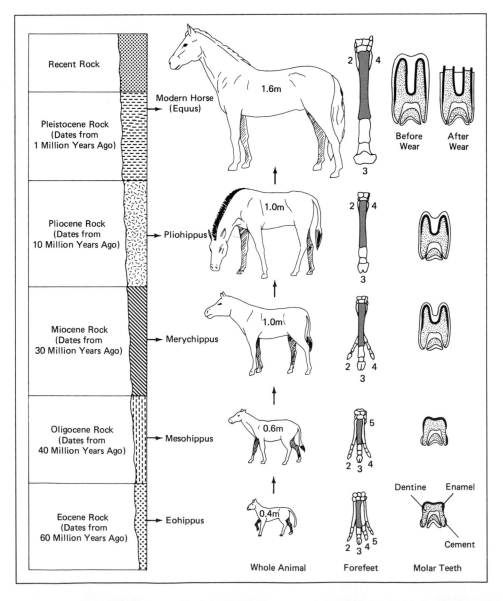

**FIGURE 4-14**
Evolution of the horse. Gradual change brought about by the enhanced survival and reproduction of those individuals that chanced to have traits that made them better adapted to an altered environment. (Redrawn from G. deBeer, *Atlas of Evolution*, p. 49. New York: Thomas Nelson, Inc., 1964.)

browse available, individuals with teeth that were better for grinding grass, which is more coarse, were favored. Also, without the extensive shelter of forests, the larger, faster animals that were able to outrun their predators were favored. Finally, variations that allowed the animal to easily traverse the increasing distances between grazing areas and sources of water aided survival. Thus, one can visualize how the selection pressure of the changing environment led to the transition we see in the fossil record from the forest-adapted *Eohippus* to the prairie-adapted horse (Fig. 4-14).

The evolution of single species into two or more species can be easily visualized by observing related species of the arctic and temperate regions. Take foxes, for example; in the arctic, selection pressures favor individuals that have variations such as heavier fur and shorter tail, legs, ears, and nose, all of which help conserve body heat. In southern regions, selection pressures regarding adaptation to temperature are the reverse; the above variations would actually be harmful because in warmer climates the fox needs to dissipate excessive body heat. If fox populations in the two regions interbreed, the resulting genetic mixing will preserve them as one species. However, if they are separated in such a way that interbreeding does not occur, natural selection

may eventually produce differences great enough that the populations of the two areas become different species. Observe the arctic fox and gray fox shown in Figure 4-15.

A classic example of speciation, observed by Darwin, involved what he described as a "most remarkable group of finches" that inhabit the Galapagos Islands. There are 14 species, each with a unique form and behavior adapting it to a food supply different from those of the others. Yet the similarities among the species reveal that they are all related members of the finch family. It is hypothesized that the Galapagos Islands, being of relatively recent geological origin and remote from other landmasses, had no land birds until a flock of one species of finch chanced to arrive and invaded each of the islands.

In general, finches are seed eaters but, on the Galapagos Islands, there was an abundance of other food materials available and no competition from other bird species. We can presume that the original population fed on seeds to which they were best adapted. As the population grew, however, there was increasing competition among individuals striving for the same food supply. This situation would have created strong selection pressure toward feeding on other available foodstuffs. In turn, natural selection would select from

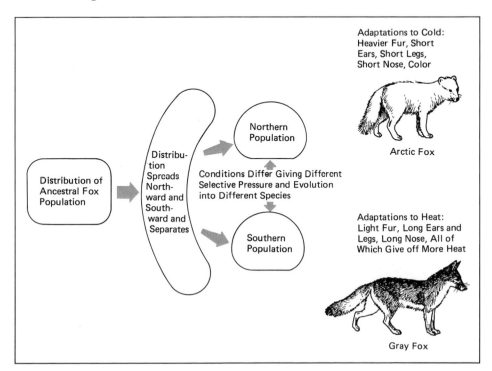

**FIGURE 4-15**
A population spread over a broad area faces different selective pressures in different regions. Eventually this may result in development of different species. Drawing shows a hypothetical distribution of the ancestral fox population that speciated into the arctic fox and the gray fox.

among random variations those that enabled more efficient feeding on those different kinds of materials. Over the course of time, this natural selection resulted in the development of distinct species, each adapted to feed in a particular way (Fig. 4-16).

Numerous examples of speciation in progress can be seen. Many plant species that inhabit mountains are comprised of different varieties, some of which are better adapted to low altitudes, others to high altitudes (Fig. 4-17). Many species of animals differ in distinctive ways between the northern and southern extremes of their range. Interbreeding throughout the entire population will keep the group as one species; however, if interbreeding is somehow blocked between populations occupying different areas, we can imagine that, given many generations, these different varieties may become distinct species.

## Adaptation or Extinction

Up to this point, we have stressed species adapting through natural selection to new or changing conditions. However, there is no rule that species will adapt; changing conditions or new factors may also lead to extinction. The fossil record is replete with extinct species; dinosaurs are among the best known. In addition, the history of the past few centuries documents the extinction of a hundred or so more, and close to 1000 additional species are now recognized to be close to extinction. These are referred to as **endangered species** and the list of such species grows yearly.

What factors determine whether a species will be able to adapt to changing conditions or whether it will be come extinct? Before addressing this question directly, we should reemphasize that changing conditions or new factors do not create or cause the variations on which natural selection acts. The variations and the genetic differences which cause them are already present in the population *before* the new condition(s) arise. Thus, adaptation through natural selection does not involve the change of living individuals; it involves individuals with certain traits being better able to cope with the new conditions and subsequently to reproduce, while individuals without those traits generally die out. Hence, the change that occurs in any single generation is likely to be small, and hundreds to thousands of generations

**FIGURE 4-16**
Four of the 13 different species of Darwin's finches. The birds are all small, dusky brown or blackish, with stubby tails. Differences between them are mainly in their bills, which adapt them to feed on different food sources. *Camarhynchus pauper* (a) is an insect eater. *Geospiza scandens* (b) feeds on cactus blooms and fruit. *Geospiza fuliginosa* (c) and *Geospiza fortis* (d) are both seed eaters. The large bill of *Geospiza fortis* enables it to crack larger seeds. On seeing these variations in one small, intimately related group of birds, Darwin wrote, "One might really fancy that . . . one species had been taken and modified for different ends." It is now accepted that this is exactly what did occur through separation of the ancestral population into different areas and subsequent natural selection to different niches.

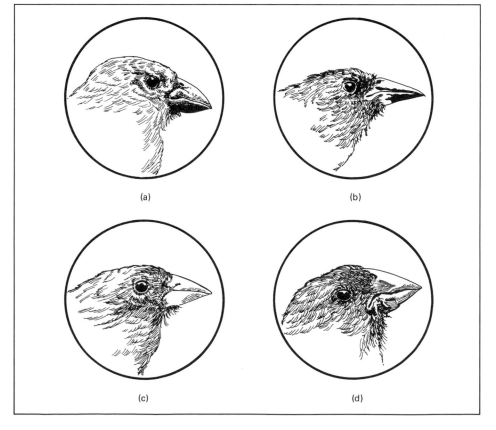

(a)

(b)

(c)

(d)

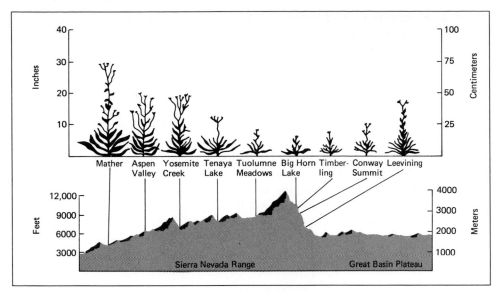

**FIGURE 4-17**
Speciation in progress. The plants shown in the illustration are still one species as demonstrated by the fact that they are all capable of interbreeding. But environmental factors clearly favor different variations in different areas. Given time (1000+ generations), the high-altitude and low-altitude plants may diverge enough to become separate species. (Redrawn from J. Clausen, D. Keck, and W. Hiesey, Carnegie Institute of Washington Publication No. 581, Washington, D.C., 1948.)

will probably be required to achieve a significant degree of modification.

It follows that, if any biotic or abiotic factor is changed so suddenly or drastically that none of the individuals of the species is able to cope, the species will become extinct, as has effectively occurred in the case of the American chestnut. Short of this, however, there are several factors that make some species much more likely than others to survive and adapt to changing conditions. The characteristics that enhance survival and adaptation are as follows:

1. A large initial population with a high degree of variation. This makes it more probable that there will be a substantial number of individuals that may have variations that enable them to tolerate and cope with the new biotic and/or abiotic factors.

2. A very high reproductive rate (hundreds, thousands, or even more offspring per mating). The more offspring, the more rapidly the entire population may be replaced by offspring from the few surviving "fit" individuals.

3. A short generation time (such as is seen in many insects, which may produce several generations per year). New variations and the selective process occur with each generation. Hence, the more rapid the turnover of generations the more rapidly adaptive modifications may develop.

4. A small size. A small animal is more likely to find suitable habitat even though the environment at large has been drastically altered. For example, a garbage container can maintain a breeding popu-

lation of flies whereas large animals, such as panda bears, need a very large, relatively undisturbed habitat to maintain a breeding population.

5. A wide geographic distribution. Changes or new factors seldom affect a very wide area uniformly. A species with a wide geographic distribution may be wiped out in one area but its survival in other areas increases the chance that adaptive variations will occur and develop.

6. Generalization. An organism that is generalized to survive in a number of different habitats is much more likely to survive than one that is so specialized that it requires very specific conditions. For example, an animal that is generalized to feed on many different things is more likely to survive than one that requires a very specific food.

In considering this list, we should be struck by the fact that large, exotic reptiles, birds, and mammals, which we would most like to keep on earth for their aesthetic beauty and interest, have those characteristics that make them *least* able to survive and adapt to changing conditions. Consequently, as we humans alter the earth in various ways, these species are increasingly threatened with extinction. Especially at the rate at which tropical habitats are being destroyed or altered, some scientists estimate that several thousand exotic species of plants and animals may be extinct by the turn of the century.

On the other hand, we would like to exterminate certain insects and plants that are serious agricultural pests or disease-causing organisms.

Unfortunately, these organisms have all the characteristics that enable species to adapt most quickly. It should hardly be surprising then that the use of poisonous chemicals to control insects and weeds has, in many cases, threatened exotic wildlife, but populations of the pests themselves have bounced back in forms that are adapted, that is, resistant, to the pesticides.

## Evolution of Ecosystems

In discussing adaptation through natural selection, it is necessary to focus on one species at a time. However, one should always bear in mind that selection pressures are working simultaneously on all the species of an ecosystem. For example, in a predator-prey relationship, natural selection is selecting variations in the prey that give the prey greater protection against the predator. However, at the same time, natural selection is selecting variations in the predator that enhance its ability to capture its prey. The same is true in all biotic relationships. Therefore, all the species in an ecosystem may gradually evolve without upsetting the overall balance among the biotic components in the ecosystem. If a balance is upset, it follows the rule given in Chapter 3; because unbalanced relationships cannot be sustained, sooner or later they are eliminated from the ecosystem leaving it with *only* balanced relationships. Thus, it is not surprising that we find climax ecosystems that show remarkable balance among all the biotic factors.

A feature that profoundly affects the evolution of ecosystems, however, is what species happen to be present as evolution of the ecosystem commences. For example, when the Galapagos Islands formed and their ecosystems developed, mammals were not present but tortoises were. Consequently, natural selection, acting on variations in tortoises, resulted in tortoises evolving into the niche of being the prime herbivore (Fig. 4-18a). In Australia, again mammals were not present but marsupials (mammallike animals that carry their young in a pouch) were. Thus, marsupials, namely kangaroos, evolved as the prime grazing animal (Fig. 4-18b).

Another example that illustrates this principle of ecological similarity but specific difference involves the niche of obtaining food (insects) from under the bark of trees. On the continents, this niche is occupied by woodpeckers; however, on remote islands that woodpeckers never reached, other species have become adapted to this niche. In the Hawaiian Islands, a bird of the honey-creeper family has evolved a woodpeckerlike beak; on the Galapagos Islands, a bird of the finch family uses a cactus thorn to probe under the bark; in Madagascar, the aye-aye, a member of the monkey-ape group, has evolved a grotesquely elongated third finger that enables it to extract insects (Fig. 4-19). Thus, through natural selection, whatever species are present in an area become gradually adapted to the various niches present in the ecosystem. In each case, a balanced relationship develops between these animals and the rest of the ecosystem.

Recognizing how these balanced relationships have evolved in various ecosystems over millions of years should reinforce our understanding of the dangers involved in transferring organisms from one ecosystem to another as described in Chapter 3. It is most unlikely that a species' adaptations that developed in relation to one ecosystem will balance in new ecosystems. Ultimately, the evolutionary process will lead to new balances, but massive change and the extinctions of countless species may occur in the process. It should be noted that extinctions can occur far more rapidly than speciation.

## Evolutionary Succession

The process of speciation and extinction together constitute the process of **evolutionary succession.** For example, consider a hypothetical species with a broad range of tolerance which invades a wide geographic region. Through different pressures of natural selection in different parts of the region, it gradually undergoes speciation. As some of these new species become more and more highly adapted and specialized to different niches, they become less tolerant of change and may be forced into extinction by the invasion of a new species or by other environmental changes. Thus, the species that inhabit the earth at any one time are gradually replaced or succeeded by other species. The fossil record is re-

**FIGURE 4-18**
Evolution of ecosystems. Natural selection can only act on and modify species that are present. Therefore, species that fill particular niches differ depending on what species were present as evolution of the ecosystem began. (a) On the Galapagos Islands where mammals were not present but tortoises were, tortoises evolved into the niche of being the prime grazing animal. (Photo by author.) (b) Similarly, in Australia, with the absence of mammals, the marsupials present (kangaroos) evolved as the prime grazing animal. (Australian Information Center photo.)

**FIGURE 4-19**
Different occupants of the woodpecker niche: (a) Woodpecker finch, Galapagos; (b) Honeycreeper, Hawaiian Islands; (c) Striped opossum, New Guinea; (d) Aye-aye, a primate from Madagascar. Both mammals have evolved long fingers that enable them to extract insects from under the bark of trees.
(Adapted from Robert M. May, "The Evolution of Ecological Systems," *Scientific American*, September 1978. Copyright © 1978 by Scientific American, Inc. All rights reserved.)

plete with examples of new species emerging, spreading, and diverging into many more species, many or all of which may fade into extinction (Fig. 4-20).

Evolutionary succession should not be confused with ecological succession, described in Chapter 3, although the two processes may go on together. Ecological succession involves a species from another area invading a new habitat to which it is already adapted. Since ecological succession need not involve species changing or adapting, it may occur at a relatively rapid rate, over tens or hundreds of years. Evolutionary succession, which involves the development of

new species through natural selection, occurs at a much slower rate, requiring many thousands or even millions of years.

## HUMAN EVOLUTION AND IMPLICATIONS

### Human Evolution

One can hardly look at the theory of evolution without contemplating how we, as humans, fit into and become part of the picture. We fit into the taxonomic tree as a family of the order primates. Our skeletal structure is conspicuously homologous with that of the ape (Fig. 4-21). Our upright posture is brought about by relatively minor modifications of the skull, feet, legs, pelvis, and lower back. Indeed, the modifications to upright posture are much less than perfect, giving our species a legacy of problems including lower back pains, hemorrhoids, hernias, sinus problems, and fallen arches. These problems all derive from an upright posture being imposed on an anatomy

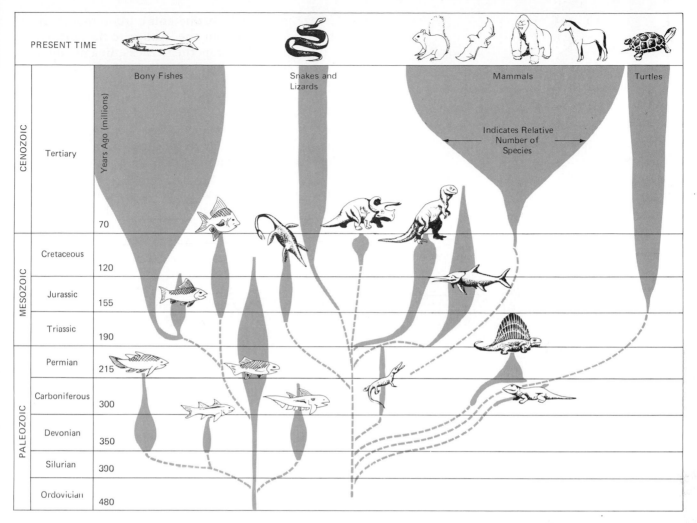

**FIGURE 4-20**
Survival of no species is assured. The fossil record is replete with examples of new species emerging, spreading, and diverging into many more species, many or all of which may fade into extinction.

that is basically designed for a more horizontal position.

These traits, and the fact that the oldest humanlike fossil showing erect posture dates from about 3 million years ago, lead anthropologists to believe that the evolutionary transition from true apes to human form occurred over the previous 5 to 15 million years. The major question is, what factors of natural selection, that is, what *selection pressures*, could have brought about this gradual change? Of course, no one was there to observe and record the process; however, many hy-

potheses have been put forth. It is worth considering what now seems to be the most plausible hypothesis consistent with the existing evidence.

## SELECTION PRESSURES THAT MAY HAVE LED TO HUMAN EVOLUTION

Fifteen million years ago, ape populations similar to present-day chimpanzees already existed. Therefore, the story starts from here. Two behavioral traits that distinguish apes such as the chimpanzee from primitive human cultures of hunter-gatherers are *food-sharing* and *division of labor*. Apes live in social groups but they are basically individual foragers. After weaning, the young must gather food and feed themselves. Young may gain food from their parents' efforts,

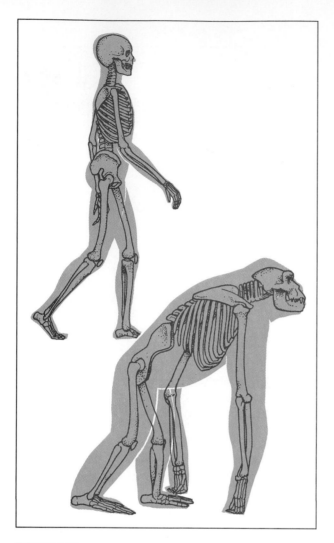

**FIGURE 4-21**
Comparison of human and ape skeletons. The human upright posture is brought about by relatively minor modifications in skull, feet, legs, pelvis, and lower back. Many of the modifications are less than perfect, resulting in frequent back, foot, and other problems. (From *Contemporary Biology*, 2nd ed. Mary E. Clark. Copyright © 1973/1979 by W.B. Saunders Co., Philadelphia. Reprinted by permission of CBS College Publishing.)

forage individually as the entire troop moves. As the troop moves from one location to the next, the young cling to their mothers' backs. In a not infrequent accident, however, a baby loses its grip, falls and is injured, and consequently perishes. The alternative to individual foraging is that some members of the group hunt and/or gather food at some distance away and bring it back to a "home base" where they share it with others who stay and care for the young.

The survival advantage of these two traits is pronounced. Monkeys and apes successively rear to adulthood only about one in three babies. Primitive human hunter-gatherers, even without the benefit of modern medicine, successfully raised one out of every two children born. Since the former case would give a "return" of 100 for every 300 births and the latter gives 150 for every 300 births, food-sharing behavior enabled a 50 percent increase in survival!

Therefore, if the behavioral traits of food-sharing and division of labor developed in a population of apes, it would create a strong selection pressure for other characteristics that would support and enhance this behavior.

The most conspicuous trait that would enhance gathering and sharing is bipedal motion (walking on two legs), which frees the arms for carrying. Chimpanzees are exceedingly comical as they attempt to gather bananas, for example, and carry them away. They tuck them under the chin, between the thighs, under the arms, and hold them in their mouths, hands, and feet; then they drop them all over the place as they try to run off on all fours. Given apes with a food-sharing behavior, one can readily see that a parent that happens to have an anatomical variation that makes it a little more able to move on two legs and carry food in the hands is more likely to successfully rear its offspring and thus enable this genetic, anatomical trait to be passed on to the next generation. Allowing that such natural selection occurred over many thousands of generations, one can picture the gradual transition from the ape structure, which uses all four limbs for locomotion, to the erect posture using just legs for locomotion.

This hypothesis is consistent with the fossil evidence, which shows that erect, bipedal motion developed in hunter-gatherer groups of early hu-

but this is more "tolerated scrounging" than true sharing. A young chimpanzee that is too weak from injury or illness to gather its own food will die before its mother's eyes. Gathering and saving food for the young is not in the parents' behavioral repertoire. On the other hand, the human parental trait of collecting food and sharing it with each other and their young is conspicuous even in the most primitive human cultures.

The second behavioral trait, division of labor, is closely associated with food-sharing. Apes

manlike primates some 2 million years before there was any significant increase in brain size.

What selection pressure could lead to increasing intelligence after the ability of bipedal motion and carrying? It was formerly thought that humans were the only animals that used tools. However, it has now been observed that a number of animals use what are in effect tools. For example, the woodpecker finch, as we have noted, uses a cactus thorn to probe for insects. Chimpanzees, in particular, use a wide variety of "tools" such as grass blades or sticks to probe for ants and termites in their nests, leaf sponges to collect water or to wipe dirt from the body, stones for cracking nuts, leafy twigs for whisking away flies, and an assortment of items used in display of aggression. The strange thing to us as humans is they always leave their "tools" when they are finished and start afresh the next time; they never seem to develop these crude tools into more effective implements. The reason, anthropologists speculate, is their inability to carry implements with them. Thus, they never really get to compare two tools and reflect on which is better.

Our food-sharing primate ancestors probably had the same capacity as present-day chimpanzees to use "tools." As they gathered food to carry it back to home, one tool that came into use was probably a slab of bark which could be used as a tray to aid in carrying nuts, berries, or seeds. In the process of carrying a bark tray, reusing it, and comparing it with others, there was more time to reflect on what made a better carrying device. Of course, given the wide variation in intelligence seen in all mammals, some would be able to carry this reflection to a productive conclusion and some would not. However, those that could were able to carry more home and have more to share; this would increase the chances of survival for their young, and thus would perpetuate those genes for intelligence. Here, then, is natural selection for intelligence. Along with intelligence, greater dexterity must also have been selected because the ability to think of a basket does not increase survival unless there is also the dexterity to weave the basket.

Another trait that is much more highly developed in humans is the capacity of complex verbal communication. Survival of individual apes such as chimpanzees depends largely on the group as a whole for protection. Yet, as they are basically individual foragers, there is little need for complex verbal communication. However, as food sharing becomes more developed and members rove farther from home base to hunt and/or gather, one can see that there is an increasing need and survival advantage in communicating more complex information such as where the good gathering places are, what dangers exist in the area, and so on. Even more complex verbal communication is necessary, along with intelligence and dexterity, in the process of developing better tools. The bright idea frequently cannot be realized unless it can be communicated sufficiently to gain cooperation and help from other members of the group.

In conclusion, beginning with the behavioral trait of food sharing, there would be a strong selection pressure for all the major human traits: walking and running erect, intelligence, manual dexterity, and verbal communication. Given several million years between the start of the process and the age of the oldest humanlike fossils (3 million years old), it is not hard to conceive that natural selection could have brought about the differences observed between modern humans and primate ancestors (Fig. 4-22).

Further, some racial differences may also be explained on the basis of natural selection. Skin color is one possibility. Sunlight on the skin is needed for the body to synthesize vitamin D; however, long, full-body exposure to sunlight over a period of time can lead to skin cancers. Dark skin, which blocks the damaging ultraviolet portion of sunlight, is therefore advantageous in tropical climates where such exposure is common. However, in northern latitudes, light skin appears to have been an advantage because it enables synthesis of vitamin D even in modest amounts of sunlight; protection against intense sunlight is generally not needed in these regions. Consistent with this are the facts that dark-skinned people originated in equatorial regions, while light-skinned people originated in the northern regions which receive less intense sunlight.

As humans spread from their center of origin and settled in different parts of the earth, different forces of natural selection may have resulted in the divergence into different races. However, the process did not go far enough to

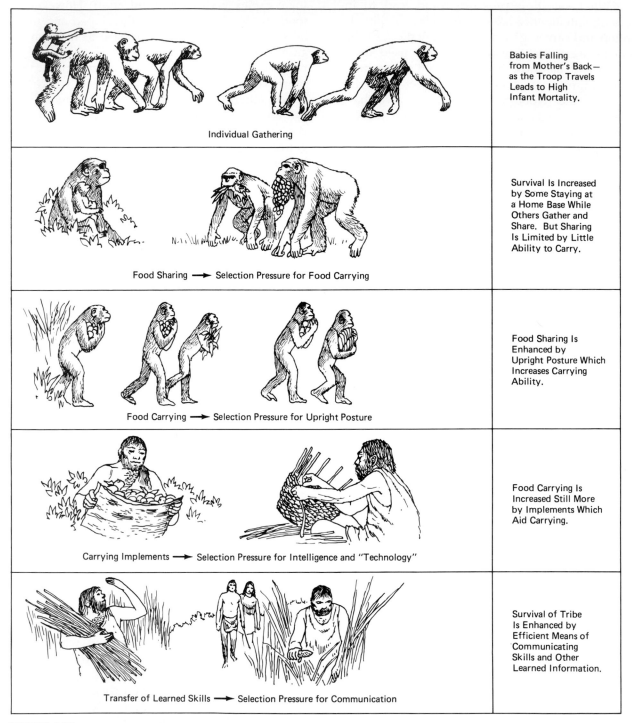

| | Babies Falling from Mother's Back— as the Troop Travels Leads to High Infant Mortality. |

Individual Gathering

Food Sharing ➝ Selection Pressure for Food Carrying

Survival Is Increased by Some Staying at a Home Base While Others Gather and Share. But Sharing Is Limited by Little Ability to Carry.

Food Carrying ➝ Selection Pressure for Upright Posture

Food Sharing Is Enhanced by Upright Posture Which Increases Carrying Ability.

Carrying Implements ➝ Selection Pressure for Intelligence and "Technology"

Food Carrying Is Increased Still More by Implements Which Aid Carrying.

Transfer of Learned Skills ➝ Selection Pressure for Communication

Survival of Tribe Is Enhanced by Efficient Means of Communicating Skills and Other Learned Information.

**FIGURE 4-22**
Hypothetical model for human evolution from an apelike
primate ancestral population. Beginning with the trait of food
sharing, which greatly increases survival of young, there is a
strong selective pressure for other traits that support and
enhance food-sharing capacity.

Survival of the Fittest

An Inefficient Predator Allows Survival of the Fittest and Provides a Factor of Population Balance

Passenger Pigeons

Overkill

Extinction

Humans Are Such Efficient Predators That the Fittest May Not Survive and Extinction May Occur

**FIGURE 4-23**
In most cases, evolutionary adaptation has led to such specialization that a predator can exploit only a few species, and even in these cases, it is not capable of eliminating all its prey. Hence there is a continuing survival of the fittest. In the case of humans, however, evolutionary adaptation led to the generalized ability to exploit almost any species and with such efficiency that extinction may easily result.

create different species of humans, for all races can and do interbreed. Presently, the races are gradually mixing together again.

### GENERALIZATION VERSUS SPECIALIZATION

The evolution of humankind does differ in one unique way from that of most other species. In other species, natural selection has generally led to **specialization** as well as adaptation to a particular niche. For example, the giraffe is marvelously adapted to grazing on treetops but, as such, it is also *specialized* and hence restricted to grazing on trees and shrubs. Only with great difficulty can it bend down to graze on the ground. Similarly, the anteater is remarkably well adapted to eating ants but, being so adapted, it is incapable of catching and/or eating other prey, and such is the case for countless other species. Thus, in general a single species is unable to exploit more than a small portion of an ecosystem's resources.

For humankind the reverse is true. The traits selected have led to very *generalized* capability. Humans with highly developed intelligence and manipulative capability can do virtually anything. Far from evolving into a specialized niche in balance with natural enemies, competitive species, and other environmental factors, humankind

evolved in such a way that it is capable of moving into and exploiting virtually every environment on earth and even in space (Fig. 4-23). No natural predator or competitor offers significant resistance, and other natural enemies such as diseases are controlled, at least to the extent that they no longer provide effective population control.

Humans undoubtedly caused ecosystem upsets even in early times by being overly effective predators. For example, there is considerable evidence that the extinction of the giant mammals such as the saber-toothed tiger and the hairy mammoth some 10,000 years ago was largely brought about by humans. However, as long as humans were hunter-gatherers, they were basically a part of natural ecosystems and subject to the limiting biotic and abiotic factors of those ecosystems.

### DEVELOPMENT OF THE HUMAN ECOSYSTEM

Archaeologists have determined that humans existed exclusively as hunter-gatherer societies until about 10,000 years ago. About this time, there are indications from the kinds of tools and plant remains found at sites of habitation that humans started propagating specific plants for food as opposed to relying exclusively on hunting and gathering in the wild. This was the advent of agriculture and it is now recognized as a most significant turning point in human development.

Agriculture provides a much more abundant and reliable food source than a natural ecosystem. To illustrate, the notion of going into the woods and "living off the land" as do wild animals is sometimes appealing. However, any attempt to do this quickly reveals that during most of the year the number of calories of food palatable to humans in a natural ecosystem is relatively small and hard to come by even if one knows what to look for. In fact, it is estimated that a natural, terrestrial ecosystem will support no more than two or three human hunter-gatherers per square mile (3 km$^2$). On the other hand, intensive agriculture in a good climate can support as many as 2000 humans per square mile. It is hardly surprising that agricultural societies flourished, spread, and displaced hunter-gatherer societies in all but the most remote regions.

With the advent of agriculture, we as humans effectively began the development of our own distinct ecosystem, which has continued through the present and now involves not just humans and distinct varieties of domestic plants and animals, but all kinds of technological, industrial, and social activities as well.

## Overview and Implications

Viewing humans in an evolutionary context shows that our species is a very recent phenomenon. Just how recent can be better appreciated by the following: Suppose the entire time period of organic evolution (some 4 billion years) is condensed into a single year where each day represents about 11 million years (Fig. 4-24). In this time frame, most of the year, up to mid-September, is taken up by the evolution of primitive bacterialike organisms. However, about September 1, 1.4 billion years ago, the first complex cells typical of present-day plants and animals were formed and then the pace quickened. All the major invertebrate phyla of marine organisms developed through September and October and the first vertebrates (forerunners to fish) developed during the first part of November (some 450 million years ago). The Ages of fish and amphibians, each last-

ing roughly 100 million years, follow in two-week intervals through November and, by the first of December, amphibians are giving way to the giant reptiles (dinosaurs) that dominate the earth during the first half of December. During the third week of December, dinosaurs fade into extinction and mammals and birds become the dominant animals.

Finally, the first humans appear about 4:00 P.M. on December 31 (3 million years ago) but agriculture is only established in the last two minutes and the explosion of technology and knowledge of the last 200 years is represented by the last two seconds.

The entry of humans into the evolutionary scene can definitely be considered a new evolutionary age as the human ecosystem now dominates large sections of the earth and continues to displace or alter natural ecosystems at a rapid pace. However, is anything revealed in evolutionary theory that indicates that humans are the culmination or the end point of the evolutionary process? Unfortunately, the answer is definitely no! To the contrary, we see that all the processes of evolution are ongoing.

The one factor that determines longevity of a species on earth is its reaching and maintaining a stable balance with other species in an ecosystem that has efficient recycling of nutrients and sustainable energy flows. Recognizing this, we see that modern humans could be a mere flash on the evolutionary scene—a few seconds compared to two or three weeks for dinosaurs—because we have developed a system based not on balance but on extensive and continuing exploitation of soil, energy, nutrient, water, and numerous biological resources. Moreover, in using pesticides we maintain a very tenuous balance between our agriculture crops and pests. Our technological ability to exploit resources has certainly been responsible for our phenomenally rapid rise on the evolutionary scene, but it is also rapidly bringing us to the brink of ecological collapse as critical resources become limited, or the already tenuous balances break down.

Many have the notion that sacrifice of natural environments and ecological systems is the price we must pay for human progress. It is desperately important that we change this thinking

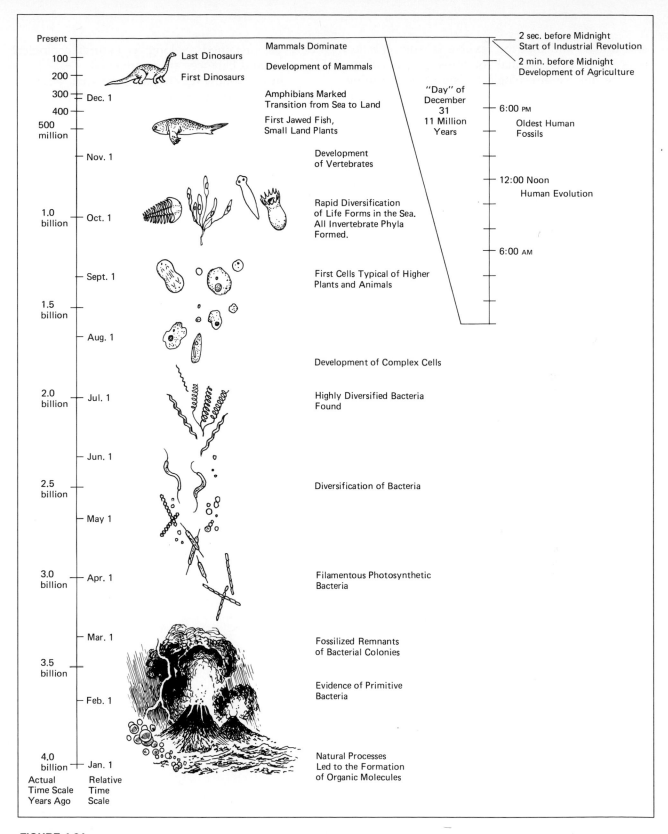

**FIGURE 4-24**

Contrasting the geological time scale with a single year gives one a better appreciation for the relative amount of time taken for various evolutionary stages. Note that two-thirds of the time is taken in the development of cells; then the pace quickens. The origin and development of humans occurs only in the last 8 hours of the last day, and progress since the Industrial Revolution occupies only the last 2 seconds of the "year."

and understand that the reverse is true. *Balanced ecosystems* are essential for life-support. Allowing the demise of ecological systems is one thing (aside from nuclear war) that can assure the collapse of modern civilization. Continuing technological, economic, and social development can only be sustained as long as natural ecological systems and balances are maintained.

In the following chapters, we shall examine particular problem areas in more detail and see what is being done and what remains to be done to achieve lasting, sustainable balances.

# POPULATION

We have noted that we, as humans, evolved as a distinct species some 3 million years ago. For most of our 3 million–year history, our population grew slowly and suffered frequent setbacks due to epidemics of disease and intermittent famines. It was roughly 1830 before world population reached the 1 billion mark. In the period from the 1700s through the 1800s, however, a remarkable turnabout occurred. Human population changed from a condition of slow growth punctuated by setbacks to one of explosive growth. In 1930, just 100 years after reaching the first billion, the pop-

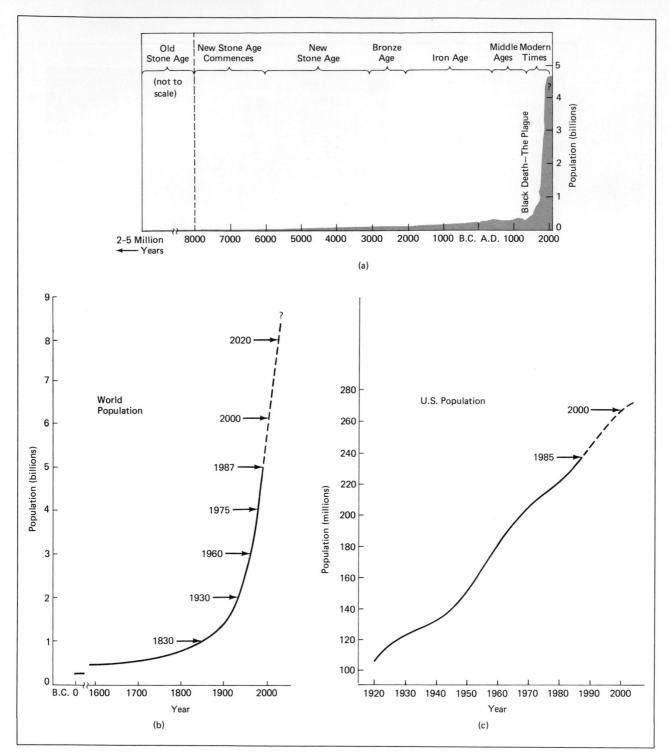

**FIGURE II-1**
The population explosion. (a) For most of human history the human population
grew very slowly, but in modern times it has suddenly "exploded" as shown.
Data prior to 1800 are estimates. (Reprinted by permission from E. J. Kormondy,
*Concepts of Ecology,* 3rd ed., Englewood Cliffs, N.J.: Prentice-Hall, Inc., 1984.) **(b)** Recent
world population data (solid line) and estimates (broken line). (Population Reference
Bureau, Inc., Washington, D.C.) **(c)** Recent U.S. population data and estimates. (Ibid.)

ulation increased to over 2 billion. Barely 30 years later, in 1960, it reached 3 billion, and in only 15 more years (1975) it had climbed to 4 billion. Then, 11 years later, 1986, it crossed the 5 billion mark and the 6 billion mark is expected to be crossed in 1997. This growth pattern could go on well into the twenty-first century (Fig. II–1).

As pointed out in Part 1, such a population explosion is a basic ecological imbalance that cannot be sustained. How many people can Spaceship Earth support? How close are we to the limits? Studying the situation does not give a clearcut answer but it reveals two basic facts.

First, in many respects the human population is supporting itself through exploitation and/or degradation of essential soil, water, and biological resources. If such degradation continues these resources will not support even the present world population, much less additional billions.

Second, some three-fourths of the world's population live in various degrees of poverty, many with barely adequate or inadequate food and/or shelter (Fig. II–2). The famine in Africa which attracted worldwide attention in 1985 is a case in point. Yet, most of the 80 to 90 million new arrivals each year are being born into impoverished conditions in less-developed countries where they can look forward to little more than mere survival, if that. As the population explo-

sion continues, then, it seems destined to increase the ranks of the poor, while the affluent in developed countries become a smaller and smaller minority. If we are to be human we must be concerned with *quality of life,* not just numbers.

In a public address, Philip Handler, past president of the United States National Academy of Sciences, expressed it this way: "I cannot believe that the principal objective of humanity is to establish experimentally how many human beings the planet can just barely sustain. But I can imagine a remarkable world in which a limited population can live in abundance, free to explore the full extent of man's imagination and spirit."

If Handler's ideal is to be our goal, we must stop thinking in terms of theoretical maximums and move ahead as vigorously as possible on three fronts: (1) stemming the human population explosion, (2) improving the quality of life for the existing population, and (3) implementing better ecological management, particularly in areas of soil, water, and wildlife conservation and in land management.

These fronts are interrelated in complex ways. Improving quality of life and implementing sound environmental management becomes increasingly difficult and, ultimately, all such efforts may be overwhelmed by the pressures of increasing population. Elevating quality of life may be

**FIGURE II-2**
Scene from a village in Benin, West Africa. Cou Provence rural Development project. (Photo by Joseph Hadar/World Bank.)

necessary to reduce population growth and improve ecological management. But as affluence increases, a population places greater demands on resources, making sound ecological management even more imperative. This interrelatedness means that we must work on all three fronts simultaneously.

In Chapter 5 we shall examine the causes of the population explosion. We shall also explore why rapid population growth tends to be linked to the social and environmental conditions of poverty in less-developed countries. In Chapter 6 we shall study methods of dealing with these complex problems. Subsequent chapters will focus on the population and resource conservation issues that must be addressed if we are to attain a sustainable society.

# 5

# POPULATION AND POVERTY

## CONCEPT FRAMEWORK

3. State the formula for finding the crude birth rate necessary for population stability.

4. In developing countries why are children considered an economic asset. Discuss other reasons why people in developing countries may choose to have large families. What are the potential consequences of such actions?

5. How does poor education in regard to the use of contraceptives and of infant aids such as milk formula affect the birth rate in developing countries?

6. What is the *demographic transition?* Discuss its three phases and factors that influence them.

7. How have advances in modern medicine added to population problems in developed countries?

8. How does the pyramid illustrate population momentum? Explain why reducing fertility rates is not enough to curtail population growth.

9. How does increasing population lock moderate- and low-income countries into a perpetuating cycle of poverty and high birth rates?

10. What factors have influenced people in developing nations to move to the cities? What are the results?

11. Is it in the best interests of developed countries to deal with the population problems of developing nations or to separate themselves from such problems? Why?

In viewing the world population situation it is first necessary to recognize that we live in a world where tremendous economic disparity exists between nations.

The world is commonly divided into three main economic categories (Fig. 5-1):

1. *Highly developed*, industrialized, or high-income countries: mainly the United States, Canada, Japan, Australia, and the countries of Western Europe and Scandinavia.

2. *Moderately developed* or middle-income countries: mainly those in Latin America, North and West Africa, and East Asia.

3. *Less developed* or low-income countries: mainly the countries of East and Central Africa, India, and the People's Republic of China. (Although China is still placed in this category, recent developments suggest it may soon move into category 2.)

Socialist countries, with the exception of China, are generally considered in a separate category as are a few countries like Saudi Arabia where most people are poor, but the national income is high due to exporting oil.

The highly developed nations are also commonly referred to as develop*ed* countries while middle- and low-income countries are referred to as develop*ing* nations.

The disparity in distribution of wealth among these nations and their people is mind-boggling. Highly developed nations hold just 25 percent of the world population, but they control 79 percent of the world's wealth. Thus, developing countries have 75 percent of the world's population, but only 21 percent of the world's wealth. This disparity is illustrated in the following analogy: Imagine the world's economic wealth as a pie sitting on a table. Twenty people, each representing about 5 percent of the world's population sit before the pie, which is cut into 20 equal slices. Four people, representing the population of the highly developed nations, take 16 slices or about four-fifths of the pie leaving only 4 slices (one-fifth) for the other 15 people. But even these 15 did not divide the remainder equally. Five people, representing populations of moderately developed countries, take 3 of the remaining 4 slices. Only 1 slice, about 5 percent of the total, is left for the remaining 10 people who represent the less-developed countries.

Of course, the distribution of wealth is also disproportionate within each country. Between 10 and 15 percent of the people in highly developed countries are recognized as poor and about 10 percent of those in moderate and less-developed countries are wealthy. Nevertheless, the economic concerns of the average person in a highly developed country centers on owning a comfortable home, a car, several television sets, and other amenities. Food is thought of only in terms of what is wanted to eat, not whether there *is* anything to eat (Fig. 5-2). On the other hand, large portions of the population in moderately and less-developed countries have inadequate food and/or shelter and virtually no amenities. Their primary economic concern is simply day-to-day survival. At least a billion people in developing countries fit into the latter category and in less developed nations, particularly Africa, malnutrition and starvation-related deaths are common, especially among infants and children.

Yet it is within the less-developed and moderately developed countries that the population explosion is most intense; they are growing at rates that, if sustained, will double their present populations, in 25 to 35 years. In contrast, barring immigration, populations in highly developed nations are approaching stability (no growth). Contrast the stylized world map in Figure 5-3a, where countries are scaled according to their share of economic wealth, with the map in Figure 5-3b, where they are scaled according to number of births and birth rates.

Our objective in this chapter is to study the causes behind the population explosion and why it continues in developing countries while it has tapered off in high-income nations. These factors must be understood before meaningful strategies for stabilizing populations can be considered.

## PRINCIPLES OF POPULATION INCREASE OR DECREASE

### The Population Equation

Considering the world as a whole, any change in population is a function of just two factors: number of births and number of deaths. Indeed, literally hundreds of other factors may affect births and deaths—disease, war, family and social traditions, economics, religious traditions,

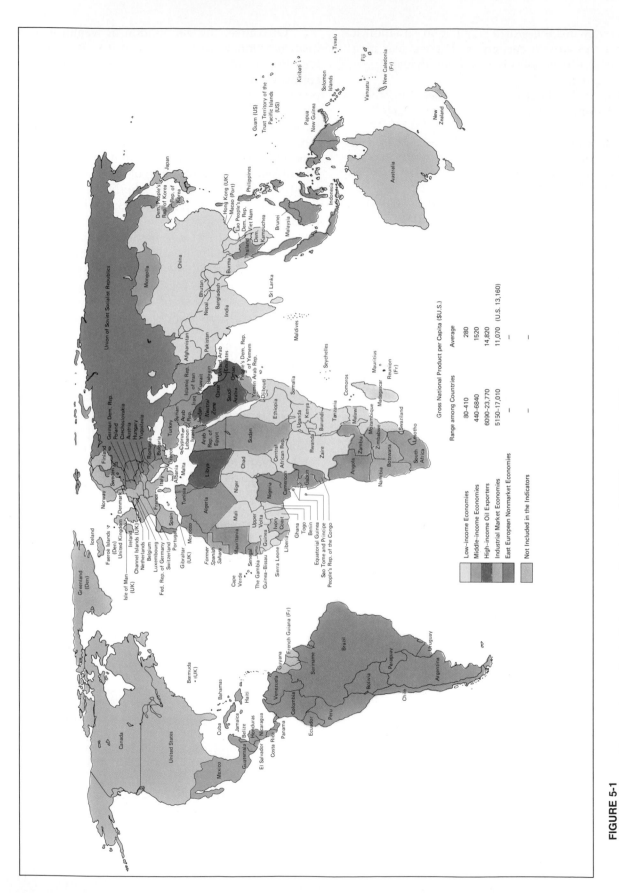

**FIGURE 5-1**

Nations of the world grouped according to their gross national product (GNP) per capita. GNP per capita is a general indicator of standard of living. (From the World Development Report 1984. Copyright © 1984 by the International Bank for Reconstruction and Development/The World Bank. Reprinted by permission of Oxford University Press, Inc., New York.)

Gross National Product per Capita ($U.S.)

| | Range among Countries | Average |
|---|---|---|
| Low-income Economies | 80–410 | 280 |
| Middle-income Economies | 440–6840 | 1520 |
| High-income Oil Exporters | 6090–23,770 | 14,820 |
| Industrial Market Economies | 5150–17,010 | 11,070 (U.S. 13,160) |
| East European Nonmarket Economies | — | — |
| Not Included in the Indicators | — | — |

**FIGURE 5-2**
(a) In highly developed countries food is thought of in terms of what is wanted to eat. (Photo by Bob Gambarelli.) (b) In low-income nations the question often is, Is there anything at all to eat? Nor is there adequate shelter, clothing, or other amenities. (FAO photo by W. Williams.)

(b)

and morals. Still, population changes are figured by subtracting the number of deaths in a given time period from the number of births during that same time period:

$$
\begin{array}{ccc}
\text{Number of} & \text{Number of} & \text{Change} \\
\textit{Births} & -\ \textit{Deaths} & =\ \text{in Population} \\
\text{per Year} & \text{per Year} & \text{per Year}
\end{array}
$$

Immigration and emigration also affect population in individual nations. Consequently, the change resulting from the number of births minus the number of deaths is called the **natural increase** (or decrease). For the time being, we shall consider only *natural* increases or decreases.

Of course number of births and deaths will be a function of total population size. To compare

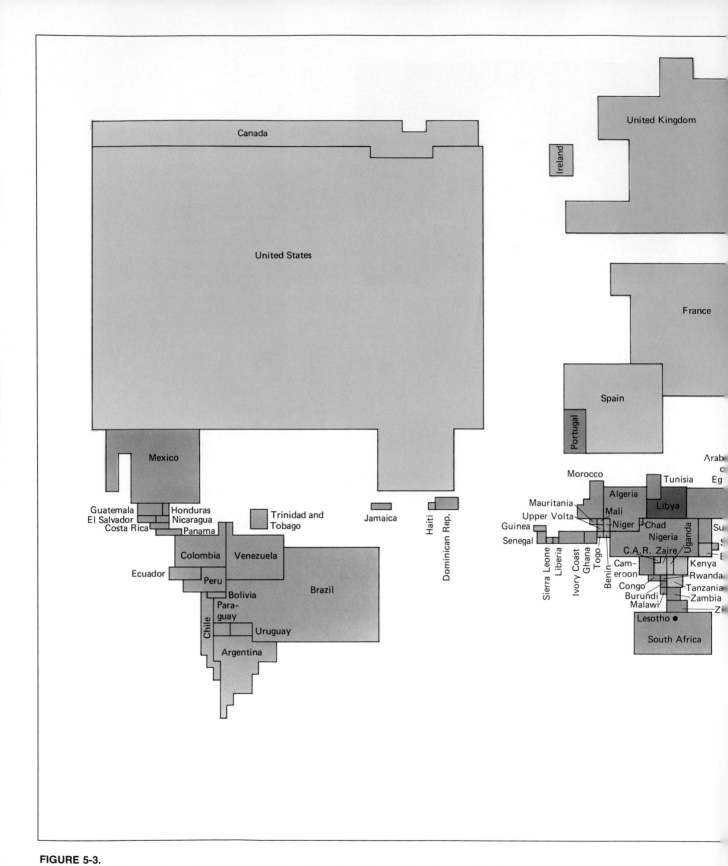

**FIGURE 5-3.**
(a) Distribution of economic wealth among selected countries. Countries are drawn in proportion to
their share of total GNP. Those for which data on GNP are not available, as well as those with
fewer than 1 million inhabitants, are excluded.

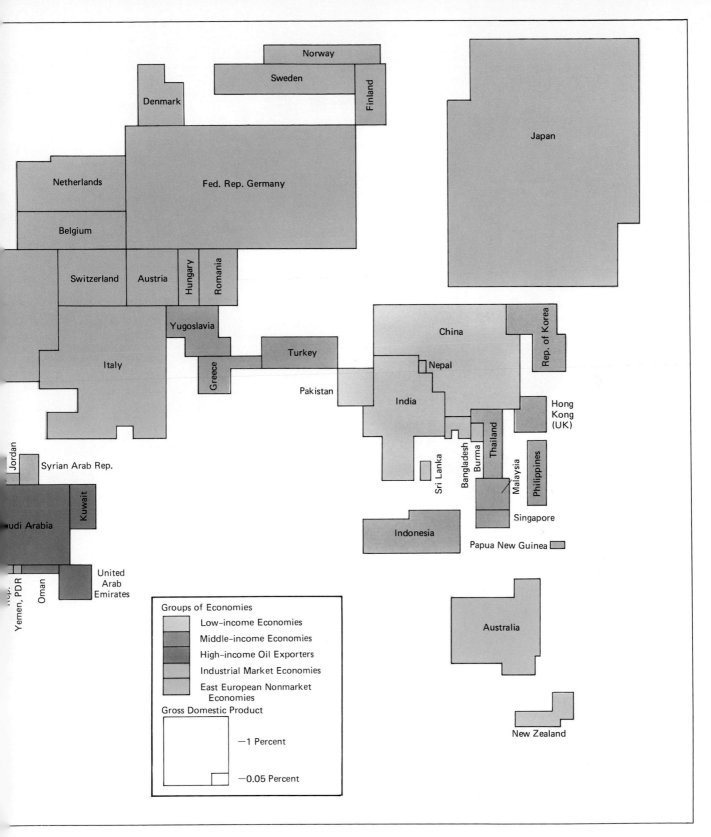

Groups of Economies

Low-income Economies
Middle-income Economies
High-income Oil Exporters
Industrial Market Economies
East European Nonmarket Economies

Gross Domestic Product

—1 Percent

—0.05 Percent

(a)

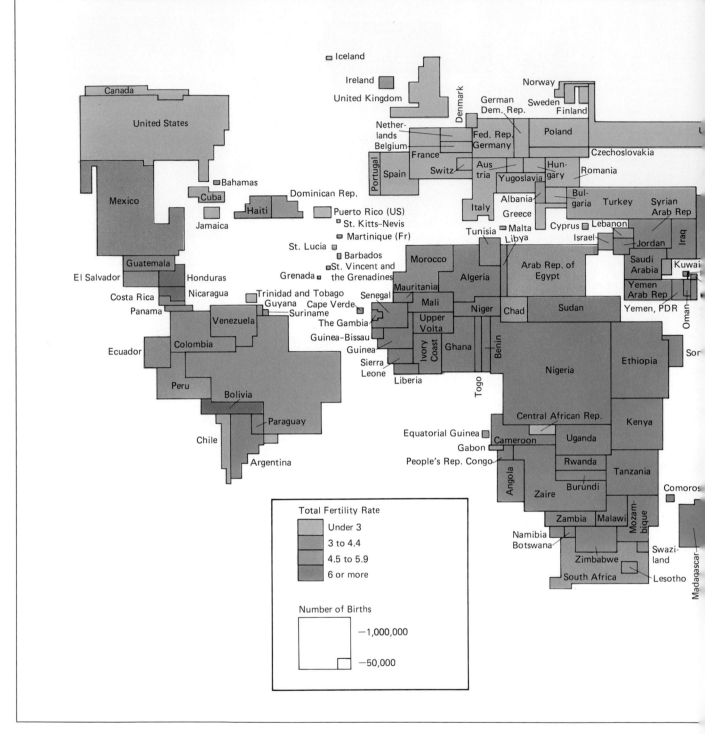

**FIGURE 5-3** (cont.)

(b) Distribution of births among selected nations. Countries on this map are drawn in proportion to the number of births in 1982. Thus India, with more than 24 million births, appears as the largest country, and China, with about 21 million births, is slightly smaller. The colors represent each country's total fertility rate, which corresponds to the number of births a woman would have if, during her childbearing years, she were to bear children at each age in accord with current age-specific fertility rates. (Both maps from the World Development Report 1984. Copyright © 1984 by the International Bank for Reconstruction and Development/The World Bank. Reprinted by permission of Oxford University Press, Inc., New York.)

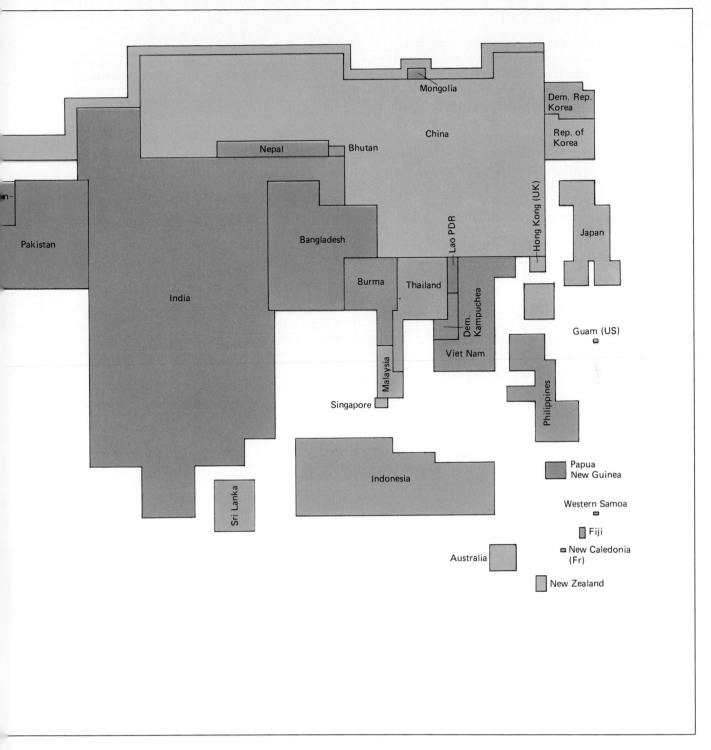

relative rates of population growth in countries of different size it is customary to divide the population into groups of 1000 and present the data as births and deaths per 1000 people per year. This is known as the **crude birth rate** and **crude death rate,** respectively. (The term *crude* is used because there is no consideration given to population structures, that is, what proportion of the population is old or young, male or female, and so on).

The *rate* of population increase (or decrease) is found by subtracting the crude death rate (CDR) from the crude birth rate (CBR):

Rate of population change = CDR − CBR

and then changing the result to a percent.

For example, the crude birth rate for the United States in 1984 was 16 per thousand and the crude death rate was 9 per thousand. Thus the natural increase for 1984 was

16 − 9 = 7 per thousand, or 0.7 percent

Again, this is the *natural* rate of increase. It does not consider immigration or emigration.

It is also common to express population growth rates in terms of **doubling time,** that is, how long it will take the population to *double* at its current rate of natural increase. This is found by dividing the percentage rate of increase into 70. Thus, for the United States, increasing at the current rate of 0.7 percent per year, it would take 100 years to double (70 ÷ 0.7 = 100).

By comparison, many low-income nations have natural increases that will cause their populations to double in just 23 to 24 years. For example, the average crude birth rate for countries of East Africa is 47 and the crude death rate is 17. This results in a natural increase of 3 percent and a doubling time of 23 years. (47 − 17 = 30 per 1000, or 3 percent; 70 ÷ 3 = 23). The natural increase in Kenya, an East African nation, is one of the most severe. The country's crude birth rate is 53 and its crude death rate is 13, resulting in a natural increase of 40 per thousand, or 4 percent per year. Unless changed, this will double the population in just 17 years. Crude birth and death rates for some other areas of the world are presented in Table 5-1. You can practice calculating the doubling times by dividing the annual percent rate of increase into 70.

**Table 5-1**

| Population Data for Selected Regions or Countries in 1984 | | |
|---|---|---|
| *REGIONS* | *BIRTH RATE* | *DEATH RATE* |
| World | 28 | 11 |
| Highly Developed Nations | 16 | 9 |
| Developing Nations (excluding China) | 37 | 13 |
| Northern Africa | 41 | 12 |
| Western Africa | 49 | 18 |
| Eastern Africa | 47 | 17 |
| Middle Africa | 45 | 18 |
| Southern Africa | 36 | 10 |
| Asia (excluding China) | 34 | 12 |
| North America | 15 | 8 |
| Latin America | 31 | 8 |
| Northern Europe | 13 | 11 |
| Western Europe | 12 | 11 |
| Eastern Europe | 16 | 11 |
| Southern Europe | 14 | 9 |
| *Countries* | | |
| Canada | 15 | 7 |
| China | 21 | 8 |
| Cuba | 16 | 6 |
| Egypt | 38 | 11 |
| Ethiopia | 47 | 23 |
| France | 15 | 10 |
| India | 34 | 14 |
| Israel | 24 | 7 |
| Japan | 13 | 6 |
| Kenya | 53 | 13 |
| Mexico | 32 | 6 |
| Nicaragua | 47 | 11 |
| Nigeria | 51 | 22 |
| Philippines | 32 | 7 |
| Switzerland | 12 | 9 |
| United Kingdom | 13 | 12 |
| United States | 16 | 9 |
| USSR | 20 | 10 |
| West Germany | 10 | 12 |

Source: Population Reference Bureau, Inc., *1984 World Population Data Sheet,* Washington, D.C.

## Reasons for the Population Explosion

From the population equation, it is evident that if the death rate and birth rate were equal, the net change would be zero and the population would be stable. An increase may result from either an increase in the birth rate or a decrease in the death rate. It may surprise you to know that the population explosion which continues today is basically due to a decrease in the death rate. We shall examine this phenomenon in more detail.

Prior to 1800, human birth rates were high (CBR in the range of 40 to 50). But death rates were similarly high, primarily because of widespread disease and periodic famine. Consequently, population grew slowly, and during serious epidemics deaths were so numerous that population setbacks occurred. Then a profound decrease in death rate commenced, setting off the population explosion whose features are explained below.

### CHANGE IN INFANT AND CHILDHOOD MORTALITY

Prior to 1800, death rates among infants and children were particularly severe. It was not uncommon for parents to have seven to ten children, of whom only one to three made it to adulthood. Severe epidemics, such as the diphtheria epidemic that plagued western Canada at the turn of this century, occasionally left families with no children at all. These high mortality rates among infants and young children contributed most to the fact that the overall death rate was nearly equal to the birth rate.

It is interesting to point out that similarly high reproductive rates and high death rates are typical of nearly all populations of plants and animals. Recall from Chapter 3 that a high reproductive capacity is a basic biological phenomenon,

but the population is generally held in check by a similarly high dieoff of the young caused by one or more factors of environmental resistance.

For humans, however, these high death rates began to be reduced in the 1800s as a result of improvements in disease prevention, control, and cure. Particularly significant innovations were vaccinations against certain diseases and improvements in sanitation to keep sewage-borne pathogens from contaminating water and/or food supplies. The most profound result of disease prevention was a dramatic reduction in infant and childhood mortality. In developed countries today, it is unusual for parents to lose a child to disease, whereas 100 years ago it was uncommon not to.

This decrease in infant and childhood mortality can also be shown as an increase in **survivorship**. Survivorship is a graphic expression of how an average group of 1000 newborns may be expected to decline as it ages. It is constructed from census information and death statistics, which give the number of people in each age group and the age of each person who dies. Curve A in Figure 5-4 is typical of the situation before the advent of modern medicine. Note that it shows a rapid drop during the years of infancy and childhood; only about half the population reaches 20 years of age. Then, the decline is gradual to the limit of longevity. Curve B in Figure 5-4 expresses the modern situation. Note that it

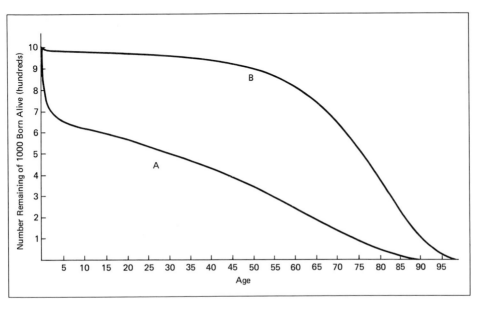

**FIGURE 5-4**
Survivorship. Starting with a population of 1000, survivorship curves indicate the number that will reach each age. (a) Survivorship of American nonwhite males (1902), typical of human population prior to modern health care. (b) Survivorship of U.S. population (1976). (Data from U.S. Department of Health, Education and Welfare, Public Health Service.)

shows that over 95 percent of newborns survive to age 40 or beyond; a rapid decline does not begin until after age 50.

### CHANGE FROM PREREPRODUCTIVE TO POSTREPRODUCTIVE DEATH

Why is the decrease in infant and childhood mortality or the increase in survivorship so significant? The answer lies in whether a person dies *before* having children (prereproductive death) or *after* having children (postreproductive death).

To illustrate, consider two cases in which parents traditionally have four children. In the first case, suppose that two of the four children in each generation die before they reach reproductive maturity (Fig. 5-5a). The original parents then have just two children who reach maturity. These children produce four grandchildren, but only

two of these reach maturity, so there can only be four great-grandchildren, and so on. As the older generation dies, a stable (nongrowing) situation results.

On the other hand, if all the children live to sexual maturity and reproduce, the four children grow up to produce eight grandchildren and these produce 16 great-grandchildren, and so on (Fig. 5-5b). Even as the older generation dies, this situation causes the population to double each generation. Thus, population growth is strikingly affected by whether people die before or after they reproduce.

### THE POPULATION EXPLOSION

Here is the essence of the population explosion. Before the 1800s, birth rates were high but infant and childhood mortality were also high, so that, on the average, only two offspring survived to reproduce, and the population remained essentially constant. Then the infant and childhood mortality rate dropped while birth rates remained

**FIGURE 5-5**
Effect of prereproductive and postreproductive death on population growth. (a) Prereproductive death holds growth in check. (b) Postreproductive death allows population to grow.

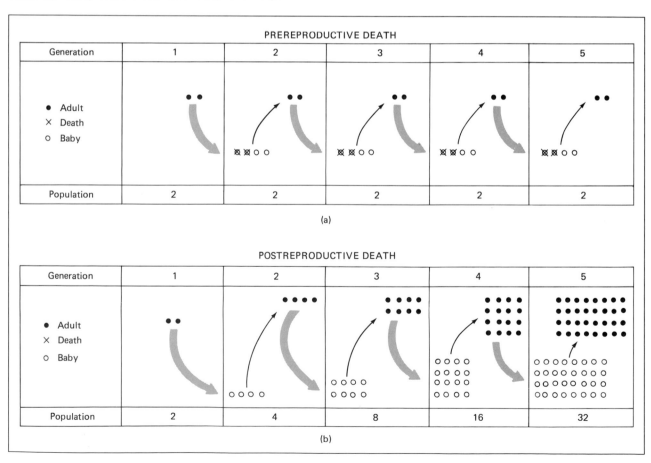

high, putting population on the rapidly multiplying course that continues today in developing countries (though it has been moderated in developed countries by reduced birth rates).

### THE MINOR IMPACT OF LONGEVITY

A common but mistaken belief is that increased longevity is a major factor in the population problem. This notion is false. Increasing longevity has only a slight impact on population growth and even that impact is temporary. Figure 5-6 illustrates the point. To simplify the picture, it is assumed in the figure that there are four children per family and that the generation time is 20 years. Life expectancies of 60, 80, and 40 years are compared. In each case, the population starts with a total of 14 people: two grandparents, four parents, and eight children. With a life expectancy of 60 years, the oldest generation is removed from the graph with the addition of each new generation, keeping three generations in the picture (Fig. 5-6a). Note that the population doubles each generation: from 14 to 28 to 56, and so on. In Figure 5-6b, the life expectancy is increased to 80 years; thus, four generations are kept in the picture. Old people remaining in the population for another generation initially increase the total population by about 7 percent, but after this the population still only doubles each generation: 30 to 60 to 120, and so on. Finally, in Figure 5-6c, the life expectancy is decreased to 40 years, thus removing a generation of older people from the population. This initially results in a population about 15 percent less than when the life expectancy is 60 years (24 compared to 28); but after this reduction, the population still proceeds to double each generation.

Consequently, whether diseases of maturity such as cancer are cured so that people live another 20 to 30 years or worsen, causing more people to die in their 40s and 50s will have little long-term impact on population growth. It will, however, have a social and economic impact.

### THE MINOR IMPACT OF ACCIDENTS AND NATURAL DISASTERS

It is sometimes suggested that war, accidents, and natural disasters replace the other mortality factors that humans have overcome. Again, such factors are not directed at prereproductive death and, despite the social and economic impact of such losses, they have a relatively small and temporary effect on the overall population increase. For example, in the Vietnam War, about 45,000 American lives were lost. Given the natural increase for the U.S. population of about 100,000 per month, such a loss is made up in about two weeks, or one month if men alone were considered. The total losses of U.S. soldiers in all wars combined, 650,000, is made up within six months by our current natural increase. Similarly, our yearly loss in car accidents, about 50,000, is replaced within two weeks. Several years ago a tidal wave swept a coastal rice-growing region of India and about half a million persons were killed, a tremendous loss in terms of a natural disaster. However, India's natural rate of increase replaced this loss in about 30 days. Some wars have had a significant population impact on certain countries and a wide-scale nuclear war directed at population centers would significantly reduce population. However, we cannot say that wars, accidents, or natural disasters of the recent past have been significant in this respect. Even the current loss of 3 to 6 million people per year to famine and malnutrition is small when compared to the world's natural increase of about 80 million per year. The loss again represents about two weeks of natural increases.

## Decreasing Population Growth

Population growth could obviously be curtailed by allowing death rates to increase again. However, this could not happen without a corresponding increase in human misery and suffering. If our overall objective is to improve the human condition, then any thought of increasing death rates must be totally dismissed.

Therefore, the only course is to reduce birth rates to a level that will stabilize population. The crude birth rate necessary for population stability is found by dividing 1000 by the average life expectancy. For developed countries, average life expectancy is 73 years; therefore, a stable population will be achieved with a crude birth rate of 13.7 (1000 ÷ 73). Crude birth rates in developed countries, particularly those of Western Europe, are close to or below this level; the average CBR for developed countries is 16. Hence, populations of developed countries are stabilizing, at least in terms of natural increase. In developing countries, however, the picture is quite different. Av-

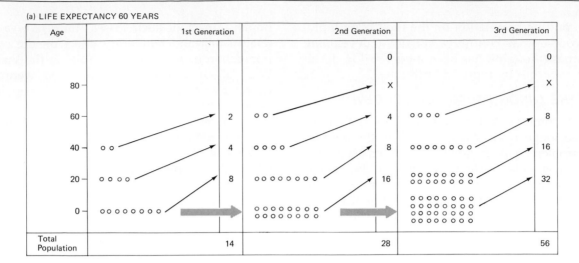

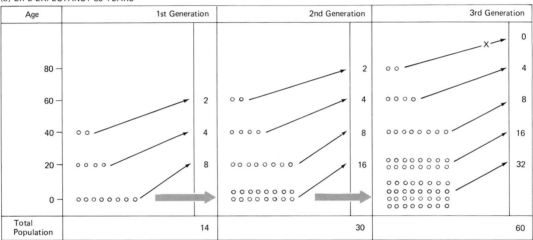

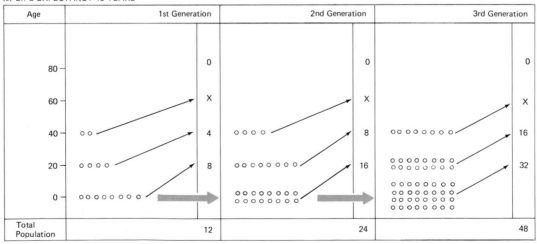

**FIGURE 5-6.**
Effect of longevity on population growth. Four-child families and a 20-year generation time are assumed in each case. (a) Assumes a 60-year life span; note that population doubles each generation. (b) Assumes an 80-year life span; note that population is only slightly larger and just doubles every generation. (c) Assumes a 40-year life span; population is 15 percent smaller but still doubles every generation.

**Table 5-2** | **Total Fertility Rates (TFR) for Selected Countries**

| HIGH-INCOME COUNTRIES | GNP PER CAPITA[a] | TFR | LOW- AND MIDDLE-INCOME COUNTRIES | GNP PER CAPITA[a] | TFR |
|---|---|---|---|---|---|
| Switzerland | 17,010 | 1.6 | Chad | 80 | 5.9 |
| Norway | 14,280 | 1.7 | Bangladesh | 140 | 7.0 |
| Sweden | 14,040 | 1.6 | Ethiopia | 140 | 6.7 |
| United States | 13,160 | 1.8 | Pakistan | 170 | 6.5 |
| Denmark | 12,470 | 1.4 | Uganda | 230 | 6.1 |
| West Germany | 12,460 | 1.4 | India | 260 | 4.8 |
| France | 11,680 | 1.9 | Haiti | 300 | 5.5 |
| Canada | 11,320 | 1.8 | Kenya | 390 | 8.0 |
| Australia | 11,140 | 1.9 | Indonesia | 580 | 4.4 |
| Belgium | 10,760 | 1.6 | Egypt | 690 | 5.3 |
| Japan | 10,080 | 1.9 | Philippines | 820 | 4.8 |
| United Kingdom | 9660 | 1.8 | Nicaragua | 920 | 6.4 |
| Italy | 6840 | 1.6 | Ecuador | 1350 | 5.5 |
| Ireland | 5150 | 3.0 | Mexico | 2270 | 4.7 |
| Average | 9190 | 2.0 | China | 310 | 2.6 |
| | | | Average (including China) | 750 | 4.4 |
| | | | (excluding China) | 940 | 5.1 |

[a]Gross national product per capita in U.S. dollars.
Source: Population Reference Bureau, Inc., *1984 World Population Data Sheet*, Washington, D.C.

erage life expectancy is about 55 years, requiring a crude birth rate of 18.2 to stabilize the population. But the average crude birth rate for developing nations is 37.

The comparison may be seen even more clearly in terms of **total fertility,** the average total number of children being born per woman. The total fertility in developed countries is 2.0; in less-developed countries, excluding China, it is 5.1 (Table 5-2). The key question is: What factors are responsible for the low **fertility rate** in highly developed countries and the continued high fertility rate in moderately and less developed countries?

Two factors stand out. The first concerns the number of children a couple choose or desire to have. The second factor depends on the availability of contraceptive techniques to achieve the desired end. In turn, both of these factors are related to economic development, as we shall see in the following section.

## RELATIONSHIP BETWEEN FERTILITY AND ECONOMIC DEVELOPMENT

### Factors Influencing Desire Regarding Number of Children

Along with the conspicuous emotional factors of love and fulfillment that influence a person's decision to have children, there are a num-ber of economic factors that affect how many children may be desired. The following are particularly significant.

### OLD-AGE SECURITY

Highly developed countries offer an elaborate array of government-sponsored social security, employer pension programs, and health care plans for the elderly. With these, each individual is expected and usually able to provide for his or her own retirement and old-age care. Most people do not expect to become dependent on their children when they reach old age; consequently, they are not motivated to have children for this reason.

The opposite situation exists in most middle- and low-income nations where there are no government social security plans and most people, particularly the poor, are not employed by organizations that offer pensions and health care programs. Along with these economic considerations, social tradition in many developing countries calls for children to take care of their elderly parents. Hence, the more children born to a couple, the greater their old-age assurance, particularly in countries where high death rates among children are common. In these areas, couples desire more children so that they are assured some survivors for their old-age security.

## CHILDREN: AN ECONOMIC ASSET OR LIABILITY?

In highly developed countries where most of the population lives in cities, it is taken for granted that children must be housed, clothed, fed, educated, and entertained, all of which are demanding on both income and time. Hence, children are an economic liability; they make little, if any, contribution to the family income.

In developing countries, however, much of the population is still rural, and agricultural work is done by hand. A child as young as 4 years old can offer significant help with lighter chores and 12-year-olds often do the same work as adults. Thus, in poor, rural settings, children are desired because they are economic assets.

## EDUCATIONAL AND CAREER OPPORTUNITIES

The availability of educational and career opportunities in highly developed countries, particularly for women, frequently supplants the desire for children. At the very least, these opportunities can lead people to postpone marriage and children until later in life. Consequently, the most fertile years, which for women are between 19 and 24, pass without children, and, in the end, fewer children are conceived (Fig. 5-7). In some cases, pursuit of career goals essentially eliminates the desire for any children. In highly developed countries, this phenomenon has become more pronounced in recent years because of the many new jobs now open to women.

In contrast, in moderately and less-developed nations, unemployment is extremely high (50 percent or more) and higher educational opportunities, especially for women, are rare. The lack of alternatives increases the desire for children because they offer a means of self-satisfaction and fulfillment.

## TRADITION AND/OR RELIGIOUS BELIEFS

Culture and religion influence a couple's decision to have children, and in many cases both favor large families. Here again, the poor are more likely to adhere to such traditions and beliefs while the affluent are more likely to break away. For example, the Roman Catholic religion

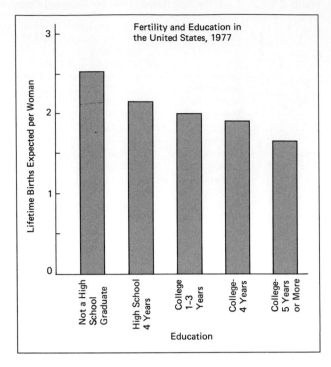

**FIGURE 5-7**
Total fertility rate tends to decrease as years of education increase. (U.S. Department of Commerce, Bureau of the Census.)

speaks against the use of artificial contraception and encourages large families. But surveys show that Catholics in the United States and other developed countries generally pay little attention to this edict; they plan their families around their own economic circumstances and desires. In moderately or less-developed countries, however, religion may act as a considerable barrier to reducing population growth.

## RURAL VERSUS URBAN LIVING

A factor that tends to incorporate all of the preceding factors is rural versus urban living. Rural families tend to be outside the network of social security or pension systems, even where these exist. Children can provide real assistance with farm chores. Eductional opportunities are more limited in rural settings. Finally, rural areas are often cut off from new ideas and social values that spread rapidly in cities. In industrialized countries only a small percent of the population remains rural. In developing countries, however, nearly two-thirds of the population is still rural, though a large-scale urban migration is occurring.

## SUMMARY

It is erroneous to assume that people in developed countries have children for the same reasons as those in developing countries. The underlying emotional desire may be similar, but economic factors and goals such as education and careers vary between the two groups. In developing nations there are strong economic reasons favoring many children and few conflicting goals. But in highly developed countries the opposite situation exists. Economic factors and conflicting goals promote the desire for fewer children. This is graphically illustrated in Figure 5-8.

Of course couples in developing countries may find that having a large number of children does not achieve the intended objectives. In fact, more children may lead to additional hardship and suffering. Unfortunately, this situation often is not realized until after the children are born.

**FIGURE 5-8**
Reasons for having children: a comparison between developed and developing nations. In less-developed nations, in addition to emotional reasons, there are strong economic incentives favoring having children, and few conflicting goals. As a country develops, economic incentives for having children decrease and conflicting goals increase. (From the *World Development Report 1984*. Copyright © 1984 by the International Bank for Reconstruction and Development/The World Bank. Reprinted by permission of Oxford University Press, Inc., New York.)

## Availability of Contraceptive Techniques

Even if couples desire to limit the number of children they have, **contraceptive** or **birth control** techniques must be available to make it possible. Here again, a disparity between the affluent and poor exists, particularly when one compares highly developed nations with less-developed nations. In high-income countries a great variety of contraceptives has been freely available for many years and is widely used. In low- and moderate-income countries, on the other hand, owing to a host of economic, political, religious, and social reasons, contraceptives are seldom regularly available to the masses of poor people; and even when they are available, the people are frequently not given the information necessary to use them correctly. This also tends to be true for the poor of industrialized nations.

During the past ten years abortion has been increasingly used to terminate unwanted pregnancies among both the affluent and the poor. However, the poor are generally dependent on public funds and services to obtain abortions and recent backlashes against abortion have sharply limited these funds and services. Regardless of how one feels about abortion, it is clear that such

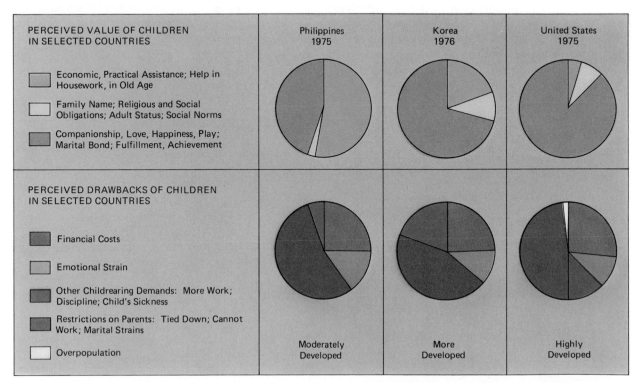

actions increase the disparity in births between the well-to-do and the poor.

In addition to the limited availability of abortions and artificial contraception among the poor, an important natural contraception method may be lost. A woman usually does not ovulate during lactation (milk production). Therefore, by nursing a baby as long as possible, a woman may space her children two to three years apart. Conversely, not nursing allows ovulation to start again and another baby may be conceived shortly after the birth of the preceding child. This situation has been aggravated in recent years in developing countries by promotion of baby formula and bottle feeding by certain multinational food corporations. Since other contraceptive methods are not available (or used) this practice has significantly increased the birth rate in some areas. Adding to the problem, closely spaced children in poor families are less likely to receive adequate care; a woman caring for an infant can hardly care for a one-year-old. Consequently both children and an overworked, exhausted mother are more likely to suffer from ill health.

Ill health is further aggravated by use of bottles and formulas because the people are not always taught how to use the materials or do not have the means to adequately sterilize water and bottles. Hence, the formula may become contaminated or parents may attempt to stretch the supply by overdiluting it. The result is increased malnutrition, misery, and infant mortality. Ironically, poor health and mortality of infants and children lead parents to desire still more children in order to achieve their initial objectives. Consequently, the whole cycle of poverty and overpopulation is further perpetuated.

## The Demographic Transition

### DEFINITION

In light of what has been discussed, we can observe that death rates and birth rates change in predictable ways as a society or nation develops from a low-income to a high-income situation. Demographers, people who study human population trends, term these changes the **demographic transition.** As graphically illustrated in Figure 5-9, the demographic transition is divided into three phases.

*Phase I* is the condition before industrial development in which the birth rate is high but infant and childhood mortality is correspondingly high so that the population grows slowly, if at all.

*Phase II* is the period in which society learns to control the diseases that are primarily responsible for high infant and childhood mortality. In this phase, childhood mortality rates drop dramatically, but the birth rate remains high, giving rise to rapid population growth and the population explosion.

*Phase III* is the period in which social and/or economic changes decrease the desire to produce many children and birth rates come down. At the end of this phase, population stability is again achieved because low infant and childhood mortality is balanced by low birth rates.

**FIGURE 5-9**
Demographic transition. The three phases of the demographic transition are shown in this idealized graph. Phase I is characterized by a stable population with a high birth rate and high death rate. Phase II is characterized by a dropping death rate but continued high birth rate, leading to an increasing rate of population growth. Phase III is characterized by a falling birth rate, which restores population stability. Highly developed countries are nearing the end of Phase III. Developing countries are still in the early stages of Phase III.

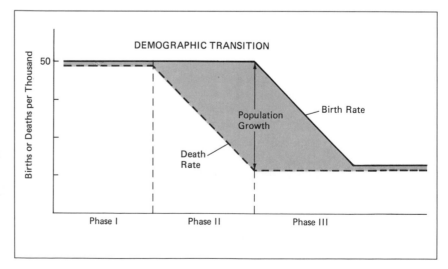

## COMPARISON OF HIGHLY DEVELOPED COUNTRIES AND LESS DEVELOPED COUNTRIES

Simply speaking, we can say that highly developed countries have passed through all three phases and are now at or almost at the new stable end point. Moderately and less-developed nations, however, are still in the latter stages of Phase II or the early stages of Phase III where death rates have dropped, but birth rates remain high. Consequently, they are in the most rapid period of population growth.

Realistically speaking, however, the phases of demographic transition are not as clear-cut, nor do they follow as automatically as this picture indicates. In present high-income countries, such as the United States, Phases II and III occurred more or less simultaneously over the past 100 years. Disease control developed gradually with industrialization so that jobs and other opportunities became available in cities as people left the rural way of life. Thus, economic development and urbanization led to a lowering of birth rate that was not far behind the decline in death rate. Consequently, present highly developed countries never experienced a severe population explosion (Fig. 5-10a).

Countries like the United States grew extremely rapidly, but this was due to immigration

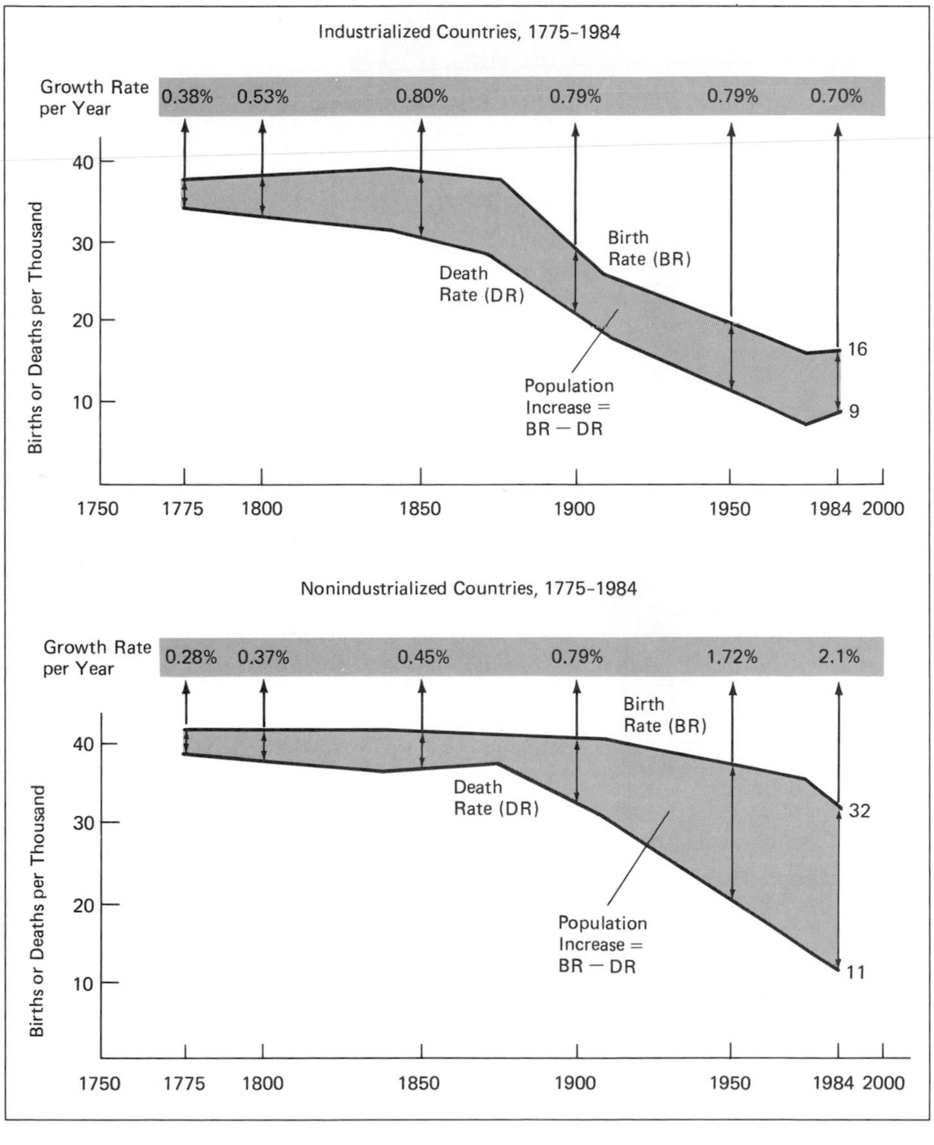

**FIGURE 5-10**
The increase in world population has been due to a drop in death rate, not to an increase in birth rate. In fact, birth rates have also declined somewhat. The increasing rate of population growth is seen in the increasing gap between birth rate and death rate in nonindustrialized countries. (Redrawn with permission of Population Reference Bureau, Inc., Washington, D.C.)

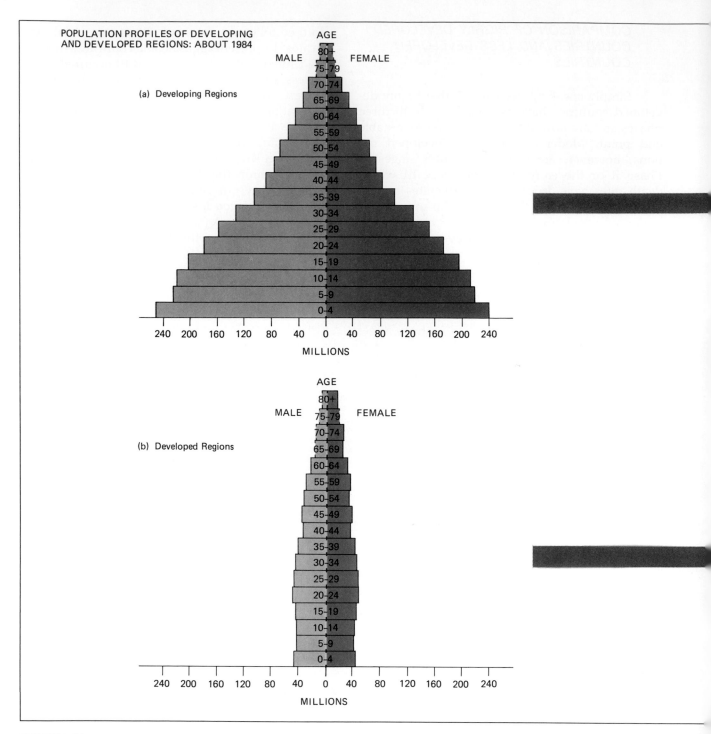

**FIGURE 5-11**
Population profiles for developing and developed regions, 1984 and estimates for 2024. (a) All developing countries combined. (b) All developed countries combined. Note that (a) and (b) are drawn on the same scale; each bar indicates actual numbers of people. (From Leon Bonvier, "Planet Earth, 1984–2034: A Demographic Vision," *Population Bulletin,* 39, No. 1 [Population Reference Bureau, Inc.] Washington, D.C. 1984.)

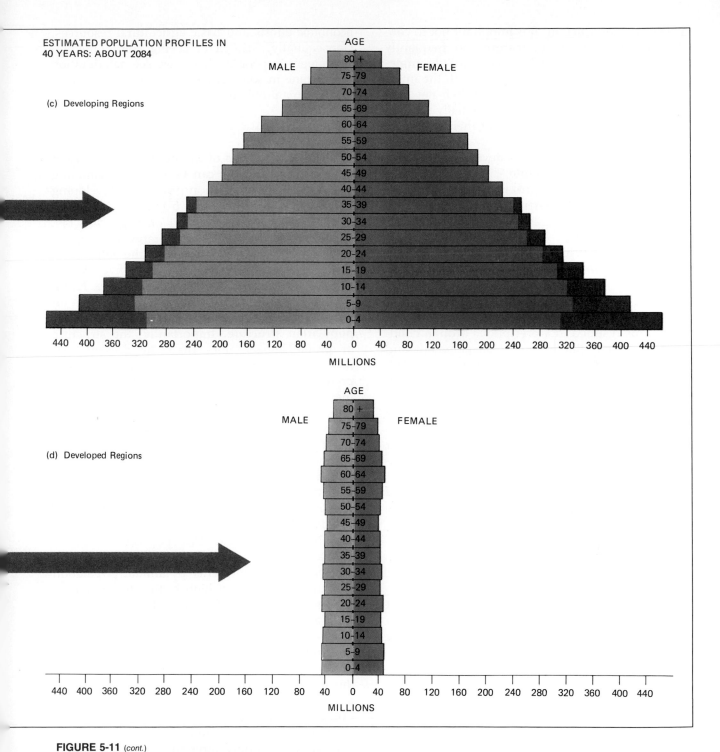

ESTIMATED POPULATION PROFILES IN
40 YEARS: ABOUT 2084

(c) Developing Regions

AGE
MALE    FEMALE

80 +
75-79
70-74
65-69
60-64
55-59
50-54
45-49
40-44
35-39
30-34
25-29
20-24
15-19
10-14
5-9
0-4

440 400 360 320 280 240 200 160 120 80 40 0 40 80 120 160 200 240 280 320 360 400 440

MILLIONS

(d) Developed Regions

AGE
MALE    FEMALE

80 +
75-79
70-74
65-69
60-64
55-59
50-54
45-49
40-44
35-39
30-34
25-29
20-24
15-19
10-14
5-9
0-4

440 400 360 320 280 240 200 160 120 80 40 0 40 80 120 160 200 240 280 320 360 400 440

MILLIONS

**FIGURE 5-11** (cont.)
(c) The population of less developed regions in 40 years (2024). Total length of bars is assuming
current birth rates continue. Lighter portion only is assuming fertility rates decline to 2.0 by 2004, a
highly optimistic assumption. (d) Population of developed regions in 2024 assuming fertility
continues at current rate 2.0. Note that even with favorable assumptions the population of less
developed regions will increase enormously while that of developed regions will increase little,
placing a much greater burden on the developed world.

as well as reproduction. It is important not to confuse the introduction of mature and frequently highly skilled workers who can contribute immediately to the industrial labor force with the introduction of infants who, because of their need for care, education, and job training, actually subtract from the industrial labor force for their first 15 to 20 years of life.

In moderately and less-developed nations, however, the story is quite different. In these countries, Phase I, high birth rates and high childhood mortality rates, generally prevailed well into the 1900s and in cases as late as the post–World War II period. Then, in a short period of time in the late 1940s and early 1950s, industrialized countries introduced modern disease control methods, most importantly improved sanitation, antibiotics, vaccinations, and the use of DDT and other pesticides to control diseases spread by insects. This resulted in a precipitous drop in childhood mortality. But economic development and social change did not keep pace. Therefore, birth rates remained high. The 1984 *World Development Report*, recently published by the World Bank, aptly describes the situation: "In Kenya (Africa) which has the world's highest birth rate, the average woman used to have eight babies and four living children. Now, with improved health measures, she has eight children."

The result, again, is the sharply widened gap between birth and death rates and consequently the population explosion (Fig. 5-10b).

## DIMENSIONS OF THE PROBLEM

Based on the demographic transition, one might predict that development of low-income nations would cause their birth rates to decline and the population problem to be resolved in this way. Consequently, emphasis may be placed on promoting development rather than on reducing fertility per se. This approach has been used over the past several decades and it has improved the quality of life for many people. It is also at least partly responsible for some decline in birth rates, though they are still high as seen in Figure 5-10b.

However, in recent years, the population explosion in developing regions has gained a momentum such that it is threatening to overwhelm economic efforts and basic agricultural resources. Consequently, instead of seeing continuing improvement in peoples' lives, we see deterioration and famine in some areas of the world. We shall explore this problem more fully.

### Momentum of Increasing Population

The momentum of the population explosion in developing regions can be seen by examining and contrasting population profiles of developing and developed regions (Fig. 5-11).

As shown in Figure 5-11 on pages 148–149, a **population profile** is a bar graph, constructed from census data, which shows the number of persons in each age group, usually at five-year intervals, from birth through the end of life. Numbers of males are shown on one side, females on the other. The profile is also referred to as the **population structure.** Importantly, the structure or profile shows more than the present age-makeup of the population; it also can be used to predict future population trends. Simply imagine that every five years the bars move up one position. As elderly people die, the uppermost bar is removed from the top and other upper bars are slightly reduced. Then a new bar representing the number of births for this period is added at the bottom.

When we look at the population profile for developing regions (all developing countries combined), we see that it forms a pyramid (Fig. 5-11a). Nearly 40 percent of the population is below 15 years of age; a very small portion (about 7 percent) is 60 or older. This pyramidal population structure is a result of high birth rates combined with relatively low infant and childhood mortality. Each new age group entering the reproductive years is larger than the preceding one, and, in turn, has produced a new group of babies two to three times larger than itself. At the same time, the population in the older age groups, where most deaths occur, remains proportionately small. Consequently, the number of deaths remains small in proportion to the number of births.

Thus, if fertility rates in less developed regions stay at their current level (average 4.4 including China; 5.1 excluding China) we can visualize this population pyramid expanding enormously in the years ahead as each new bar added at the bottom will be about 15 percent larger than the one above. Indeed, in the next 60 years it would expand about fourfold but would keep the same general shape.

Even if fertility rates in developing regions dropped to 2.1 immediately (an extremely unlikely event) population would continue to grow for some time before stabilizing. We can see the reason for this if we consider the movement of the current population of 0- to 14 year olds into their reproductive period. Even with the "ideal" fertility rate of 2.1, they will still produce their own number of offspring and this number will greatly exceed the dieoffs among the relatively small number of people in older age groups. This situation will persist for 60 to 70 years, until the present young people reach old age and begin dying off. At that time, the profile will have roughly the shape of a straight column, the population of old people roughly equaling the population of younger people. With this population structure, the deaths among the old will balance the reproduction of the young. But at this point, the population will be two to three times larger than it is now (Fig. 5-11c).

Thus, even using the most optimistic projection, an immediate decline in fertility rate to 2.1, population growth in developing countries has such a tremendous momentum that it will con-

tinue for at least 60 years. If nothing is done to curtail fertility, it will continue indefinitely until environmental limits are reached.

In contrast, the population profile for developed regions (all industrialized nations combined) presents a very different picture (Fig. 5-11b). It already has close to a straight-column shape. The number of people in the 20- to 39 year-old age group is slightly larger as a result of the temporary high birth rate following World War II. Now, however, these people have a fertility rate of 2.0 or less, so that they are not quite even replacing their own numbers (Fig. 5-12). Nevertheless, the population is still growing because there are still relatively small numbers of people in the old-age groups. Hence, births still outnumber deaths despite low fertility. We can see, however, that population growth in developed regions will stabilize and even begin to decrease as the 20- to 39-year-olds move into old age in another 30 to 40 years (Fig. 5-11d). Of course, this assumes that current low fertility rates are maintained.

Finally, it is necessary to emphasize the relative population sizes seen in comparing the population profiles. The profiles emphasize the fact, stated earlier, that developed nations account for only 25 percent of the world's present population; 75 percent is in developing regions. Further, as the population of low-income regions expands

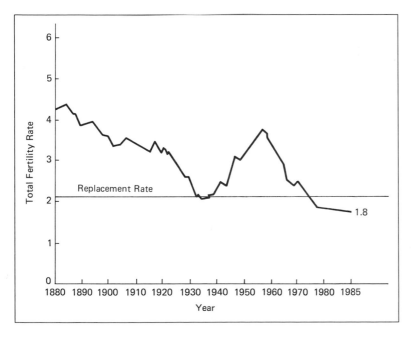

**FIGURE 5-12**
Fertility rate in the United States. Note the sharp increase in fertility following World War II, then resumption of the downward trend. Where does it go from here? (From Population Reference Bureau, Inc., Washington, D.C.)

and that of developed regions stabilizes, this discrepancy will become more and more pronounced.

## Poverty and Population Growth as Mutually Reinforcing

To advance the average standard of living, economic development must be divided among the number of people. This means that, for a country whose population is doubling in 30 years, all facilities and services (housing, schools, jobs, industry) must also double during the same period of time in order to just maintain the status quo. Nevertheless, with the aid of industrialized nations, many developing nations have achieved economic growth rates above their population growth rates. However, since populations are growing more rapidly than ever, the need for further development aid has increased.

But critical problems have emerged. For the aid to be effective, there must be an experienced work force in the recipient nation so that it is used efficiently. As noted above, about 40 percent of the population in developing countries is under 15. Another substantial percentage is occupied in bearing or caring for the children. Finally, there is the percentage that is too old, sick, or for other reasons cannot hold jobs. The result is that only a small portion of the population, in many cases about 25 percent, is economically active and able to perform jobs that contribute to the economy. And of this 25 percent, the largest portion is in the 15 to 25 age group and handicapped by lack of experience and training. By contrast, in developed countries, close to 50 percent or more of the population is economically active.

A second problem in developing nations stems from the fact that most past development was funded with borrowed money. Many developing countries are now struggling under mounting debt burdens and most of their national income must go toward interest payments. The situation makes industrialized countries increasingly reluctant to loan additional money to less-developed nations.

Finally, in many developing nations resources are being stretched to the limit. Agricultural production, for example, all too often has been increased at the expense of erosion, desertification, salinization, and depletion of water resources. In short, it appears doubtful that development efforts can stay abreast, let alone move ahead, of population. And if rising population outpaces development, backsliding will occur, resulting in more poverty, less education, and all of the other circumstances that tend to increase fertility rates. Hence, despite the demographic transition theory, it appears that increasing population may lock moderate and low-income countries into a perpetuating cycle of poverty and high birth rates.

## Migration and Urbanization of the Poor

Although the population explosion in developing countries has been going on for several decades, a relatively new phenomenon has recently complicated the situation—a large-scale migration from the countryside to cities that are ill prepared, due to inadequate housing, water and sewer systems, and so on, to handle the influx. The result is producing slums and squalor of unbelievable proportions in cities of unprecedented size (Fig. 5-13). We shall look at this in a little more detail.

In preindustrialized societies, 90 to 95 percent of the population is rural because, due to a lack of mechanization and other modern agricultural methods, farmers produce little more than enough to feed their own families. As industrialization and mechanization of farms occurs, the need for hand labor diminishes and there is increasing migration from rural areas to cities. In the United States and other developed countries, this phenomenon took place gradually over about 150 years during which jobs in industry became available at the same time as labor needs on farms decreased. And, because this migration occurred slowly over a long period, most cities were able to keep pace in providing needed housing and services.

Again, a very different situation is occurring in developing countries. Traditionally, farms were passed to and subdivided among sons. As long as population growth was slow and new land was available, this process continued. However, due to the population explosion, family plots have become so small that further subdivision is impractical; essentially all of the land that can be farmed easily is now under cultivation. Bringing new land into cultivation involves high development costs for such practices as drainage and irrigation. In addition, rapid mechanization promoted by industrialized nations has drastically reduced labor

**FIGURE 5-13**
Slum area of Bombay, India. Coupled with growing population, modernization of agriculture in nonindustrialized countries is causing a massive migration from rural areas to cities. But cities can provide neither enough jobs nor housing. The result is slums, the extent and squalor of which are difficult for Americans to imagine. Homeless thousands sleep in the streets. (Photo by J. P. Laffont—SYGMA.)

needs. Finally, in some areas, farmers are being forced off the land because soil deterioration, combined with economic factors, has undercut profitability to the point of no return.

These circumstances, along with the continuing population explosion, are causing an unprecedented migration of people from rural areas to cities where they are seeking a better life. But cities in low- and moderate-income countries are ill prepared to provide for the massive influx. In 1984 the United Nations listed 22 cities in developing nations with populations greater than 5 million. Mexico City, Mexico, and São Paulo, Brazil, are the giants with about 16 million each (in comparison, the population of the New York metropolitan area is about 15 million). These mush-

rooming cities in developing countries are already unable to provide adequate services. Many people live in sprawling slums and shanty towns with inadequate water supplies, poor sewage removal, crowded or even no housing, inadequate schools, and few if any other services. Meanwhile, jobs remain scarce and unemployment rates of 50 percent are not uncommon.

Despite these deplorable conditions, the United Nations projects that by the year 2025 some 16 cities in Latin America, Africa, and Asia will have populations that range between 20 and 30 million. Another 74 cities are predicted to have over 5 million people.

## CONCLUSION

It is clear that the world population problem is intertwined with the world poverty problem. The socioeconomic conditions of poverty tend to keep birth rates high, and high birth rates tend to perpetuate poverty. Since the population problem is centered in less-developed countries, some believe that overpopulation is "their problem" and that developed nations should not burden themselves with it. However, such an attitude is similar to affluent people in the bow of a ship laughing at poor people in the stern as they say, "Your end of the boat is sinking."

The growing disparities between over-consumption of the rich and deprivation of the poor constitute an increasing threat to peace. It is important to remember that all parts of the world are economically and politically connected. Poor nations are not going to "sink" without profoundly affecting the economic, social, and political concerns of the rest of the world. The poor's demand for a better life while the rich seek to maintain their affluence is likely to produce military conflicts such as those occurring in Central America in the 1980s. At the very least, the situation will stimulate massive migrations as people seek a better life elsewhere. Already, illegal immigration is a vexing problem for the United States because many "illegal aliens" accept very low wages, which undercut the rest of the labor force. Also many of these people require social services, thus placing additional burdens on systems that are already hard-pressed to meet the needs of citizens. If collapse should occur in developing countries, such immigration could become massive and uncontrollable.

Even if the desire existed, the United States and other developed nations cannot accommodate unlimited immigration and still maintain a semblance of the standard of living to which their citizens are accustomed. It is important to remember that developed nations represent only 25 percent of the world population and this percentage is shrinking as moderate- and low-income nations swell. If just the natural increase in world population migrated to the United States, the population here would double in less than three years.

The urgency of the population problem cannot be overstated. But it is dangerous to attack the problem simplistically. For example, in India, Indira Gandhi's government was thrown from power in the 1970s largely because of the overly zealous promotion of sterilization. Mrs. Gandhi was returned to power a few years later, but the sterilization program was abandoned. Hence, attempts to coerce people to use birth control methods may result in a social backlash that defeats birth control and causes the upheavals we wish to avoid.

In Chapter 6 we shall discuss the approaches for managing the population problems.

# 6

# ADDRESSING THE POPULATION PROBLEM

## CONCEPT FRAMEWORK

Outline                                          Study Questions

4. How may large-scale centralized projects aggravate poverty? What is the *debt crisis?*

5. Explain *appropriate technology.*

6. What are some of the key aspects of appropriate technology?

7. Based on the examples cited in the text, how does appropriate technology boost not only local people, but the economy in general? What are some of its drawbacks?

8. What role does average income play in determining fertility? Aside from economics, what key factors have been found to influence fertility rates?

9. What three points do health care programs emphasize in an effort to slow population growth? If adhered to, how do they succeed? How does education help slow population growth?

10. What incentives and deterrents have been used to successfully curtail population growth?

11. What are some of the key elements in family planning services?

12. A number of factors influence fertility. Think of a hypothetical situation where some, but not all, key elements of population control are available, and discuss how these deficiencies may work against factors that are already present.

13. Why are family planning services so important in developing countries? How are they influenced by politics?

14. In what ways are the affluent as large a threat to global stability as the poor?

In Chapter 5 we saw that population growth is integrally connected with socioeconomic conditions and that the two issues are mutually reinforcing. Therefore, any effort to reduce the fertility rate must include an attempt to improve the socioeconomic conditions of the poor. The objective of this chapter is to look at past efforts to control population growth and stress those approaches that show the most promise.

## CAN THE WORLD PRODUCE ENOUGH FOOD?

The first prerequisite for providing a better life for the very poor is to assure that they obtain an adequate diet. Discussion of education, jobs, improved health care, and other opportunities are meaningless unless this prerequisite is met. We must ask: Can the world produce enough food?

Malnutrition, if not starvation, is widespread among the extremely poor. **Malnutrition** refers to consumption of inadequate amounts of one or more nutritional factors, for example, protein, vitamins, or minerals. Depending on the particular deficiency, malnutrition can lead to health problems ranging from lack of physical energy to retarded and abnormal growth and development and increased susceptibility to various diseases. **Starvation** refers to inadequate calorie consumption such that the body wastes away. Both malnutrition and starvation can cause death or lead to it because the body cannot fight secondary infections such as pneumonia. In today's world, about a billion people suffer from some degree of malnutrition and/or starvation.

The very existence of malnutrition and starvation suggests that the world may not be able to produce enough food to support its population. But starvation and malnutrition are not new; to some degree, they have existed throughout history. Nearly 200 years ago, in 1798, when the world population was still less than 1 billion, British economist Thomas Malthus pointed out that population has the capacity to increase geometrically (i.e., increasing more and more rapidly, as was shown in Figure II-1) while agriculture, which is dependent on a finite amount of arable land, is limited. Malthus predicted that, in the absence of checks, the world was headed toward catastrophic famines as human populations outpaced agricultural capacity. Numerous others have echoed this cry.

Since Malthus's time, the world population has increased some six- to sevenfold. But the percentage of people affected by malnutrition or starvation (10 to 20 percent) is probably no higher now than it was 200 years ago. Remarkably improved farming methods, including fertilizing, irrigation, and new varieties of plants and animals, have exponentially increased agricultural production so that the amount of food per person has remained roughly constant or improved slightly. The improvement in developed countries has been dramatic, but production has also increased in developing countries (Fig. 6-1). Nevertheless the Malthusian echo is still heard as many argue that population is about to overtake food production and/or that food production cannot be sustained by current methods. Others maintain, however, that past trends in food production can

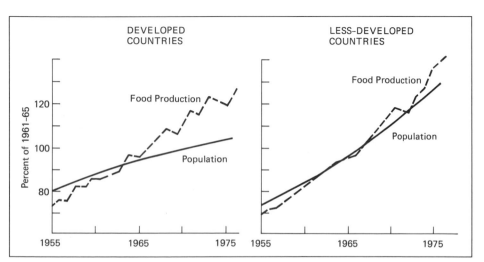

**FIGURE 6-1**
Food production and population in developed and less-developed countries. (Redrawn with permission from S. Wartman and R. W. Commings, Jr., *To Feed This World*. Baltimore: Johns Hopkins University Press, 1978. Based on data from USDA, Economic Research Service.)

**Table 6-1**  Positive and Negative Factors Affecting Food Production

| KEY FACTORS AFFECTING FOOD PRODUCTION | POSITIVE ASPECTS | NEGATIVE ASPECTS |
|---|---|---|
| Available Land | New land may be brought into production. | Virtually all good land is in production. High development costs and/or sacrifice of other natural values are involved in developing additional land. Good land is being taken out of production by urban development, highways, etc. |
| Soil Quality | Soils may be improved and made more productive. | On balance, soils are being degraded and made less productive by erosion, overgrazing, and accumulation of salts. |
| Water | Irrigation can greatly increase production in dry regions. | Sufficient water may be unavailable or too costly to obtain. Water supplies are being depleted in many irrigated areas. Irrigation may destroy soil by adding salts (Fig. 6-2). |
| Fertilizer | Crop production per acre has been increased dramatically by adding fertilizer. Fertilizer is underutilized in developing nations. | Fertilizer is often too expensive for poor farmers to afford or unavailable when it is needed in developing nations. |
| Plant and Animal Breeding | More productive varieties of plants and animals may be developed (Fig. 6-3). | Destruction of natural ecosystems and consequent loss of their biota is depleting genetic resources required for breeding. |
| Pest and Disease Control | One-third to one-half of potential production is lost to pests such as insects, rats, and disease organisms, which attack both before and after harvest. | Present control of pests relies largely on the use of toxic chemicals that may endanger both human and environmental health. Pests become resistant, which demands greater use of chemicals. |
| Food from the Sea | Harvest from the sea may be increased. | Traditional stocks are already depleted by overfishing (Fig. 6-4). |
|  | Fish farming may be developed. | Because it is very costly, fish farming does not provide food the poor can afford. |
| Pollution Control | Crop production in some areas is retarded because of pollution and/or acid rain. Better pollution control would have corresponding benefit. | Certain widespread pollution problems (e.g., ozone and acid rain) are worsening. |
| Climate | Drought-stricken areas may get relief. | Drought-stricken areas may spread. |

**FIGURE 6-2**
A farmer attempting to irrigate the desert finds that there is not enough water, and soil erosion is severe. This land should have been left in range. (USDA-Soil Conservation Service photo.)

**FIGURE 6-3**
Hybrid corn compared with previous variety. Yields per hectare have been greatly increased by the development of hybrids. (USDA photo.)

be maintained, at least for the forseeable future. They argue that the current problem of world hunger is an economic problem of insufficient buying capacity on the part of the poor, not a problem of insufficient production capacity. We shall examine these arguments in more detail.

## The Hungry: A Problem of Food-Producing Capacity

Agricultural production may be affected positively or negatively by numerous factors. Table 6-1 lists these factors and we shall examine them in detail in later chapters. For now, it is important to see that the picture is not entirely optimistic. Many of the efforts that increased production in the past are the result of exploitation without conservation. Consequently, potential gains are now being undercut by losses. For example, increases in the amount of land in agricultural production is being offset by loss of land to urbanization, degradation of soil by erosion, and depletion of water resources. Furthermore, we are sacrificing some of the natural ecosystems that produce the very genetic resources that are necessary for con-

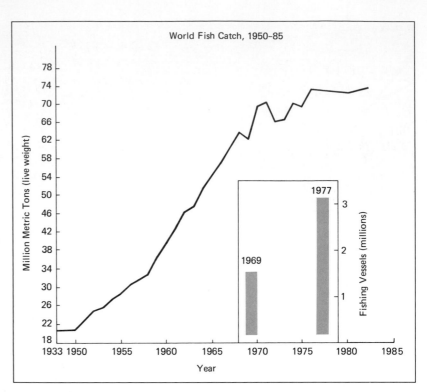

**FIGURE 6-4**
The sea is not an unlimited source of food waiting to be harvested. World catches have not increased significantly since 1970, despite increasing fishing effort, indicating that the seas are already being overfished, at least in terms of familiar species. (Data from National Marine Fisheries Service, National Oceanic and Atmospheric Administration, U.S. Department of Commerce.)

tinued development and maintenance of more productive varieties of plants and animals. The outbreak of famine in Africa in 1984 was not only the result of increasing population; it was also due to a failure to adequately maintain soil and water resources which undercut production as well. Better ecological management is needed in developing as well as developed nations in order to sustain agricultural production.

## The Hungry: A Problem of Food-Buying Capacity

Despite potential limits of food production looming sometime in the future, there is ample evidence that current world hunger problems are basically problems of **economics.**

### SURPLUSES VERSUS SHORTAGES

The conspicuous fact is that surpluses exist in some regions while hunger exists in others. The United States, for example, has chronic problems with agricultural surpluses and periodically the government has used various incentives to keep land out of production. Although the United States has been world leader in supplying food aid to countries in need, food supplies, like all

commodities, tend to flow according to economic buying power. Food staples are frequently diverted from the poor in favor of food luxuries for the affluent.

### EATING HIGH ON THE FOOD CHAIN

Grains, particularly wheat and corn, and soybeans are excellent foods for direct human consumption. Yet developed nations regularly feed these crops to animals to produce meat products. About nine-tenths of the grain's food value is lost in this process; only about one-tenth is converted into edible portions of the animal (see Chapter 2).

It is estimated that if people in developed countries eliminated meat from their diet and ate grains directly, as do most of the people in developing countries, there would be enough grain to feed another 3 to 5 billion people. However, because the poor cannot afford to pay 30¢ a pound for grain, it goes toward producing $3-a-pound meat for the affluent.

### LUXURY FOOD AND CASH CROPS

During the era of colonialism, the territories, which are now developing countries, were dominated by such European powers as England, France, Spain, and Portugal. A major goal of colonialism was to exploit the agricultural base of the colony to produce luxury food and/or cash crops for the mother country. Thus, a large portion of the best agricultural land in the colonies was put into the production of crops such as coffee, tea, citrus, bananas, cacao, cotton, tobacco, and sugarcane. While these territories have become independent countries, the legacy remains; a large portion of the agriculture in Africa, South and Central America, and Asia remains devoted to cash crops for export that benefit a relatively few affluent individuals while production of food crops is inadequate or barely adequate to feed the population at large.

## COSTS OF INCREASED PRODUCTION AFFORDED ONLY BY THE AFFLUENT

Increasing production involves costly investments, whether they be for fertilizer, pest control, clearing forests, draining marshland, irrigating arid lands, or mariculture (farming the sea). Although such investments can create innumerable opportunities for increasing food production, they will not be made unless the products can be sold at prices that provide a suitable economic return for the producer. The poor simply cannot afford the prices that underwrite such investments, but the affluent can. Consequently, we see tropical rainforests in developing countries being converted to raise cattle for beef export to developed nations. And, we see hitherto untapped marine resources such as krill (a shrimplike species rich in high-quality protein), which could alleviate world hunger, being exploited instead for cattle and pet food.

These examples show that, at least for the present, world hunger is a problem of food-buying capacity rather than food production capacity. If the poor had more money to spend on food, the food would be forthcoming.

Why do we not overcome the economic problem by simply *giving* the poor the food they need? This brings us to the next point.

## Food Aid: The Utterly Dismal Theorem

In times of social strife or natural disasters (wars, droughts, earthquakes) people desperately need help until they can resettle and begin anew.

The United States has led the world in providing such relief and the benefits are without question.

However, routinely supplying food aid in an attempt to alleviate chronic hunger in developing countries has been self-defeating. "The food generosity of industrial countries, whether in their own self-interests (disposing of food surpluses) or under the mantle of alleged distributive justice, has probably done more to sap the vitality of agricultural development in the developing world than any other single factor," write authorities S. Wortman and R. Cummings in their book, *To Feed This World.* This situation has occurred because the poor (and everyone else for that matter) will not pay more than they have to for food. Therefore, provision of foreign food for free or at any price less than that of domestically produced food, undercuts the local market. In effect, local farmers must compete economically with free or low-cost imported food. When they cannot earn a profit they stop producing and eventually enter the ranks of the poor. The cycle continues in that the people who sell goods to the farmer also suffer when the farmer loses buying power. In the long run, the entire local economy deteriorates. Hence, the availability of free food, while well-intended, often aggravates the very conditions of poverty and hunger that it is meant to alleviate. Meanwhile, population pressures continue to build and the magnitude of the entire problem increases.

Economist Kenneth Boulding observed and described this phenomen in 1957 and called it *the utterly dismal theorem.*

Interestingly, the governments in some developing nations have led themselves into the trap of the utterly dismal theorem with their internal policies. The problem is particularly evident in newly independent countries of Africa where leaders have mandated that food prices remain low in an effort to maintain the favor and support of large urban populations. The policy has made producing food unprofitable and forced farmers out of business. As a result these countries are increasingly short of food. This phenomenon also contributes to the migration of rural farm families to urban centers as described in Chapter 5.

In conclusion, we must refocus our attention on the issue raised by Philip Handler in the introduction to Part II: Are we working to experimentally establish how many humans the planet can barely sustain, or are we working to establish a remarkable world in which a limited population can live in abundance, free to explore the full extent of the human imagination and spirit?

Assuming the latter, humanitarian efforts to provide aid are still in order. But they must go beyond simplistic acts that offer bare sustenance. They must focus on mitigating the socioeconomic conditions that trap people in the cycle of high birth rates and poverty. We must help people become self-sufficient and strive for ecologically balanced management of resources.

## DEVELOPMENT TO MITIGATE POVERTY

Economic development aid has long been recognized as a means both to improve quality of life and to reduce fertility rates. However, economic development aid may be targeted in two ways: (1) large-scale centralized projects or (2) local diversified projects.

### Large-Scale Centralized Projects

Examples of large-scale centralized projects include construction of hydroelectric dams, nuclear power plants, modern steel or textile mills, and new, high-speed highways. Such projects are advantageous because they are relatively easy to administer; lenders or donors need only interact with a few people in the recipient nation. In addition, progress on these projects is easy to measure and monitor, and they result in very conspicuous end products that both grantors and recipients can point to with pride.

These projects may also substantially increase the gross national product (GNP) of the recipient nations so that, on the books, it looks as if the country grew wealthier. Unfortunately, the appearance may be deceiving. Too often there is little change in the general condition of poverty. For example, a new power plant may enable the affluent in cities to buy and use additional electrical appliances. But the poor in slums and rural areas remain without electricity because they cannot afford the installation of power lines or the cost of the electricity. The new highway is an obvious advantage to those with cars but rural farmers will continue to struggle over mud-rutted roads with oxcarts. The new road also increases

hazards for the majority of the population who use it, but on foot.

In many cases, these centralized projects actually aggravate poverty. A modern textile mill, for instance, may provide some well-paying factory jobs, but it may also put hundreds of individual weavers out of business. This problem has been especially acute in connection with agriculture. A common thought in industrialized nations is that poor countries would prosper if they modernized agriculture. Thus, industrialized nations have promoted the use of modern agricultural machinery and techniques in low-income countries. However, the multitude of peasant farmers in these countries have small plots of land and cannot make use of large tractors and other machinery. Therefore, these products go to already wealthy farmers with large land holdings. This enables these farmers to cut costs by dismissing laborers. It also enables them to acquire more land, pushing small farmers into unemployment. At the same time, such mechanization generally has not increased food production in developing countries because the highest production per hectare is achieved on small plots where the land is carefully tended and harvested by hand.

Finally, mechanization undercuts both diversification and varied timing of food supply. When numerous independent farmers tend small plots, diversity of crops and variation in times of planting and harvest tend to occur, especially in the tropics where few seasonal restrictions exist. Individual farmers then sell and trade fresh produce in large, open markets (Fig. 6-5). With this direct farmer-to-consumer exchange, there is little need for middlemen and storage facilities. As farming becomes centralized, however, it becomes impractical to raise a diverse variety of crops and harvesting takes place at one time. This creates the need for expensive storage facilities and refrigeration. The situation also creates transportation problems because people must travel further to get the food. All of these costs are ultimately reflected in higher food prices and a further strain on the poor.

Large centralized projects also add financial problems. They involve large, up-front investments, and payback is extended over 30 to 40 years or more. Thus, in the short run they provide little if any real economic benefit. Many of

**FIGURE 6-5**
Open-air market in Equador. Numerous small, independent farmers bring their produce and crafts to these huge market places for sale and trade. The system encourages individuality and creates tremendous diversity. (Photo by author.)

the developing nations that borrowed money for such projects are now finding that, with high interest rates and reduced markets for their products, it is increasingly difficult, if not impossible, to meet their loan obligations. Indebtedness of developing nations reached $700 billion in 1984. Defaults on loans of this magnitude could have economic repercussions around the world. Many economists refer to the situation as the *debt crisis.*

Despite these problems, some industrialized countries still promote large-scale projects in developing countries as a means of benefiting themselves. Consider a bank president who owns a large amount of stock in a company that manufactures earth-moving equipment. If he can convince a Third World country to build a large hydroelectric dam, he can lend it money to buy the equipment, then collect both interest on the loan and increased stock dividends.

## Decentralized Projects—Appropriate Technology

By the early 1960s the problems inherent in large centralized projects were apparent and it was recognized that to effectively alleviate poverty, projects must be on a scale such that the poor can immediately benefit. A number of terms were coined for this new approach, the most common being **appropriate technology**.

Appropriate technology cannot be defined in terms of specific programs because what is appropriate in one situation may be totally inappropriate in another. A listing of attributes and examples best explains this concept.

## ATTRIBUTES OF APPROPRIATE TECHNOLOGY

1. Most importantly, appropriate technology is directly relevant to the economic and social needs of the people in need of help, but it should not upset the existing social structure.

2. Appropriate technology should not require a high degree of training or skill. It should involve the kinds of techniques that one person can teach his or her friends and neighbors.

3. It should utilize local resources that are plentiful and inexpensive.

4. It should not require expensive, centralized workplaces; existing homes, farms, and small shops should suffice.

5. It should allow many individuals to use their own talents and minds, express their creativity, and develop self-reliance.

## EXAMPLES OF APPROPRIATE TECHNOLOGY

**Handloom versus textile mill.** We noted above that building a modern textile mill was inappropriate because it put many local people out of work and provided only a few jobs. In contrast, introduction of an improved handloom proved very appropriate. Women could still produce cloth in their own homes while caring for children. The social structure was not upset and these women could produce more and increase their earning power.

**Handmade brick dwellings versus a large apartment complex.** Inappropriate technology was almost employed in a situation where the government planned to clear a poor slum area and build a huge complex of apartments. This project would have drastically altered the existing social structure. The appropriate technology involved teaching people to make bricks from the local mud. The technique spread quickly and soon nearly every family in the area was making bricks and building itself a simple, but adequate dwelling.

**Leg power versus gasoline power.** The ideas of appropriate technology are perhaps most applicable to agriculture. For instance, in a dry area where crops would benefit from additional water, gasoline-powered irrigation pumps were introduced. This method proved inappropriate because few farmers could afford them and even fewer could afford the gasoline to run them. In addition, spare parts were seldom available. The appropriate alternative was a simple pump constructed from local materials that could be leg-powered by harnessing it to a bicycle. Nearly every farmer already had a bicycle and enough family members to pump. Further, an abundance of repair shops and spare bicycle parts existed.

**Better wheels versus better roads.** In another situation food was being produced, but it was not getting to market because the roads were so bad that the wheels on the farmers' oxcarts broke. One solution would be to bring in modern earth-moving machinery and begin a massive road-building program. However, this would drastically impact the environment and the way of life of the people. The appropriate solution proved to be the introduction of a relatively simple device with which farmers could bend iron straps and to make metal rims for their wheels to keep them from breaking.

**Sophisticated techniques.** Frequently appropriate technology is simple, as the foregoing examples illustrate. However, at times it can also be highly sophisticated. For example, agricultural production in many regions has been greatly increased by the introduction of new plant varieties (Fig. 6-6). Some of these varieties require new tillage techniques and additional fertilizer and water. When farmers recognize the production advantages, however, they are usually willing to adopt the new methods and invest in fertilizer. Farmers may also be amenable to adopting cultural methods of natural pest control (Chapter 17). In China individual farmers are building anaerobic digesters to dispose of agricultural wastes and produce methane gas for cooking. The Chinese are also world leaders in utilizing agricultural wastes for high-density fish culture. Other developing nations are becoming leaders in mariculture (Fig. 6-7).

Such techniques not only improve the productivity and income of numerous farmers, but they boost the entire local economy. When the farmers have more money, they buy more products made by townspeople; thus others also increase their incomes, and they in turn can now buy more from the farmer. The cycle of economic exchange perpetuates general improvement.

(a)

**FIGURE 6-6**
Comparison of an old variety of wheat (a) (photo by Phi Farnas/ Photo Researchers) with a new short-stemmed variety (b) (photo by J. Latta/Photo Researchers). The first of the more productive short-stemmed varieties was developed under the direction of Norman Borlaug. This advance was heralded as the beginning of the "Green Revolution."

(b)

**FIGURE 6-7**
Mariculture in the Philippines. Natural wetlands have been made into fishponds. (FAO photo by P. Boonserm.)

In conclusion, the key to appropriate technology is not simplicity or sophistication; it is aid that enables greater productivity and economic returns within the framework of the existing social and economic system. Administering appropriate technology, however, is often difficult. Empathetic individuals must live and interact with the community for some time to determine how the people see their problems and what solutions are practical. In addition, community leaders must be identified. To efficiently spread new ideas or techniques, it is necessary to introduce these concepts to the right people and to respect established lines of communication. Finally, a mechanism of providing small loans to numerous individuals should be available.

The concept of appropriate technology, however, is not entirely without drawbacks. Most seriously, it may be viewed by people of developing nations as an effort to prevent them from entering the modern industrialized world.

In fact, some mix of centralized projects and appropriate technology is probably in order. One example that is commonly cited is the need for power plants and electrification. These are essen-tial for development because without them people are handicapped simply by the lack of light when it comes to education, reading, studying, and learning new techniques and skills. Whether projects are small or large, however, it is now recognized that conservation of the environment and protection of basic soil and water resources must be a primary consideration. In fact, an entire field of study called **technological assessment** has grown around this issue.

## DEVELOPMENT AND FERTILITY REDUCTION

Development to improve the quality of life of the poor is a desirable goal in itself. But here we return to the question of whether this, in turn, will lead to a reduction in fertility as predicted by the demographic transition theory. Considerable data on changes in fertility in relation to changes in average income have been collected over the past 15 years. Figure 6-8 shows that in nearly every developing country where average income has increased, fertility has dropped. However, it is also significant to note that a vast difference between average income and fertility exists in different countries. For example, average income is about the same in Kenya and China; yet Kenya's aver-

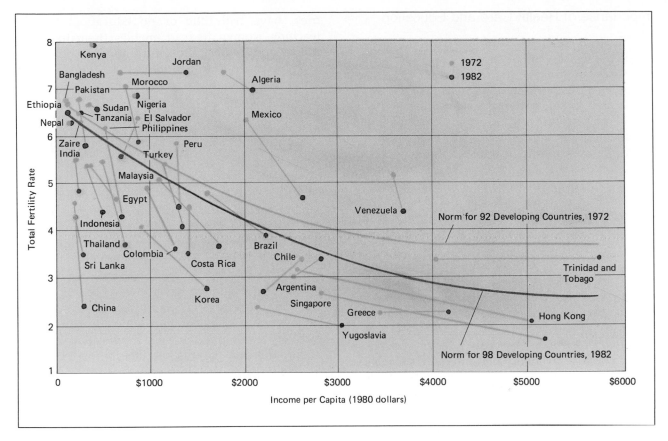

**FIGURE 6-8**
Fertility in relation to income in developing countries, 1972 and 1982. There is a correlation between rising income and decreasing fertility rates, but it is a weak correlation. Additional significant factors involved are improvements in education, health care, and promotion of family planning services. (From the *World Development Report 1984*. Copyright © 1984 by the International Bank for Reconstruction and Development/The World Bank. Reprinted by permission of Oxford University Press, Inc.)

age fertility is 8.0 while China's is down to about 2.6. Similarly, a substantial drop in fertility is seen in some countries with little increase in income—China, Thailand, and the Philippines, for example. In other countries a significant increase in income has occurred with little effect on fertility.

Some of the apparent discrepancies are caused by unequal income distribution. For instance, the fact that Venezuela has a high fertility rate despite a high average income reflects the fact that high profits from oil sales are not well distributed; much of the population is still very poor. But enough knowledge and understanding has been gained to know that much more than average income is involved.

Economists define poverty in terms of income. However, poverty is a psychological condition as well as an economic condition and the two aspects, although often correlated, may vary independently.

The psychological condition of extreme poverty involves a feeling of hopelessness and helplessness. Under these circumstances, children seem to offer the benefits, rewards, and security discussed in Chapter 5, and hence we find high birth rates. Conversely, when people see opportunities to improve their lives along avenues that entail having fewer children, they begin to act accordingly, though actual incomes in the two situations may be the same.

In fact, declines in birth rate have been found to be more closely related to adult literacy, particularly of the mother, and life expectancy than with gross national product per capita. Thus, improved education and health care along with economic opportunity are key ingredients in the development aimed at reducing fertility. Contraceptive techniques come into play next.

## Importance of Health Care and Education

In industrialized countries we tend to take excellent health care of mothers and infants for granted; we think of improved health care primarily in terms of extending the lives of the aged. But in developing nations, the situation is different. Poor health and mortality are common among mothers and children and frequently involve poor nutrition. In some developing countries, for example, women are typically married by the time they are 16 or 17 and they have several children by their early 20s. Because the teenage body is not fully developed, pregnancy at this age places stress on both the mother and child. When children are closely spaced, the stress becomes greater. As a result, mothers are often in poor health and a sickly mother cannot adequately care for children, regardless of income. Hence, the children are likely to also be in poor health.

Since mothers bascially desire the best health possible for their children and themselves, health care programs that promote better maternal health and fewer but healthier children are well received and effective. These programs emphasize several major points:

1. Delay childbearing until the early 20s.
2. Practice good nutrition during pregnancy and lactation.
3. Nurse the baby 12 to 15 months or more.
4. Practice good sanitation, hygiene, and disease-control techniques.

The first three points have a specific effect in slowing population growth. Delaying the time of childbearing lengthens the time between generations. Better nutrition enables an extended nursing period, and lactation acts as a natural contraceptive by preventing ovulation in most cases, thus leading to greater spacing between children. Finally, parents with healthier, faster growing, and more vigorous children are frequently content with fewer children because, if for no other reason, a few healthy children are more likely to provide security in old age than many unhealthy children.

Better educational opportunities are also connected with reducing fertility rates. Many of the poor, especially women in developing countries, have had little or no education. Simply teaching women to read enables them to better utilize and understand nutritional, health care, and child care information. As illustrated in Figure 6-9, studies show that in every case examined, a reduction in the number of children corresponds to the increased education of the mother. Note that we are not talking about college education. The figure shows years of *elementary* education.

Further, when schools are available and parents see them as a way for their children to achieve better lives, they are willing to invest more of their resources in sending their children to school. However, it becomes clear that if there are a lot of children in one family, the parents will not be able to send them all to school. Again, this promotes decisions to have fewer children.

Of course, it is important that education lead somewhere, so economic opportunities (jobs) must be developing, as well. If education ends in unemployment, parents and children will become disillusioned with its virtue.

## Additional Incentives and Deterrents

When opportunities for improving standard of living are available, governments may provide additional incentives and/or deterrents for reducing the number of children per couple. *Incentives* are defined as payments given to individuals, couples, or groups to delay or limit childbearing. *Deterrents* are the withholding of certain benefits from those whose family size exceeds the desired norm. Some examples include:

- Limiting tax deductions or child allowances after a prescribed number of children have been born.
- Limiting maternity benefits after a given number of births.
- Giving priority in school admission to first and second children. School admission is limited after a certain number of children.
- Allocating housing independently of family size, so that large familes do not receive more space than small familes. (Recall that housing in many developing countries is in short supply and, as new housing is built, it is allocated.)

China, the largest developing country, with over 1 billion people, currently has the most strin-

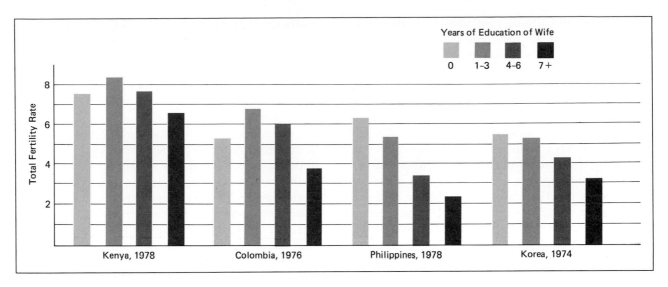

**FIGURE 6-9**
Total fertility rate by years of education of the woman in selected developing countries. Even a few years of elementary education has a significant effect on reducing total fertility. The apparent increase between 0 and 1–3 years seen in some cases probably indicates that the group of women with no education tends to include more women that are unable to bear children for other physical and/or mental reasons. (From the *World Development Report 1984*. Copyright © 1984 by the International Bank for Reconstruction and Development/The World Bank. Reprinted by permission of Oxford University Press, Inc.)

gent and comprehensive birth control policies of any country. The policy is based on recognition by the country's leaders that unless population growth is controlled, the country will be unable to live within its limits. Under the plan, a one-child program has been instituted along with an elaborate array of incentives and deterrents. The prime ones are as follows:

- Paid leave to women who have fertility-related operations, namely sterilization or abortion procedures.
- A monthly subsidy to one-child famiies.
- Job priority for single children.
- Additional food rations for single children.
- Housing preferences for single-child families.
- Preferential medical care to parents whose only child is a girl. (There is a strong preference for sons in China and parents generally wish to have children until at least one son is born.)

Penalties for an excessive number of children in China include these:

- Repayment of bonuses for the first child if a second is born.

- Payment of a "tax" for a second child.
- Payment of higher prices for food for a second child.
- Denial of maternity leave and paid medical expenses after the first child.

Along with improving economic opportunities, these incentives and deterrents have helped China achieve a precipitous drop in its total fertility rate from about 4.5 in the mid-1970s to just about 2.6 in 1982. Because of this unique achievement, and its tremendous size, China is frequently excluded from the demographic categories of developing countries.

## Family Planning Services

All of the factors discussed above may bring about the desire for fewer children, but it is doubtful that any of them decrease sexual activity. To make the desire for fewer children a reality, individuals must have easy access to family planning services. The most effective family planning services have broad-based health care objectives. They provide a wide range of information, guidance, and services regarding pre- and postnatal care of both infants and mothers. Then, consistent with the health care objective, they provide information, devices, and/or procedures that may be used to avoid unwanted pregnancies.

Organizations such as Planned Parenthood, Parenthood Federation of America, and International Planned Parenthood Federation, all of which are financed by public and private donations, have been making such efforts for many years. They have been instrumental in calling attention to the problem and paving the way for government programs, many of which were established in the 1960s and 1970s. Today, some 85 countries in the developing world, which represent 95 percent of the population, have family planning programs.

Since the introduction of these services, a sharp drop in birth rates has been seen in many countries. This emphasizes that the socioeconomic factors that lead people to desire fewer children are often already present. The only lacking ingredient is family planning services. Thus, the introduction of family planning services has had more effect than has development on the decline in birth rates seen in Figure 6-8.

## Summary

In summary, a woman's fertility, the number of children she bears during her lifetime, is primarily determined by her age at the time of marriage, by whether she breast-feeds, by the use of contraceptives by herself and her partner, and perhaps by abortion. These aspects of fertility are influenced by family decisions regarding the timing of marriage, the number of children, and whether both parents must or may choose to work outside the home. These family decisions are made within the framework of the socioeconomic environment, which includes educational opportunities, availability of family planning and health services, women's status, and financial and labor markets. Finally, governments often manipulate this socioeconomic environment with policies and laws, by funding for education, health, and family planning services, and by providing various incentives and deterrents. These factors and their relationships are summarized in Figure 6-10.

Thus, the general concept of the demographic transition remains valid; birth rates drop with increasing development. However, one must recognize that "development" here is not just a simple factor of average income per capita; it is a complex matrix of many interrelated socioeconomic factors. If any one factor is missing, all other efforts to reduce fertility rates may be ineffective. By the same token, determining the weakest part of the structure and applying efforts in this area may markedly reduce fertility rates with minimal capital expenditure.

**FIGURE 6-10**
Summary of factors influencing fertility. (From the *World Development Report 1984*. Copyright © 1984 by the International Bank for Reconstruction and Development/The World Bank. Reprinted by permission of Oxford University Press, Inc.)

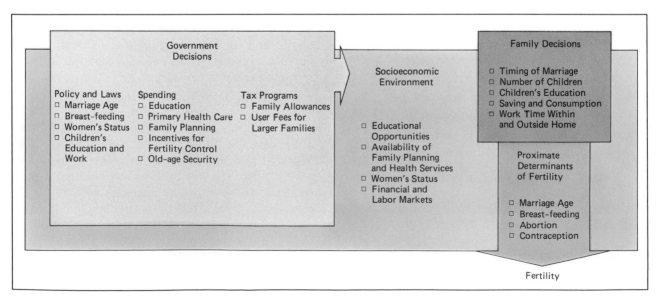

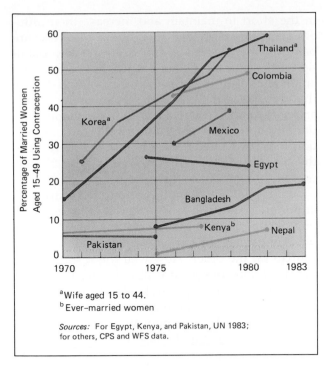

**FIGURE 6-11**
Trends in contraceptive prevalence, 1970–83, in selected
countries. (From the *World Development Report 1984*. Copyright ©
1984 by the International Bank for Reconstruction and Development/
The World Bank. Reprinted by permission of Oxford University Press,
Inc.)

## POLICY DECISIONS

After surveying the population growth and development situations of developing nations, the World Bank, a leader in development efforts, found a critical, unmet need in family planning services: In developed countries, contraceptive use among married women of childbearing age is nearly 100 percent. In developing nations, however, while contraceptive use is generally increasing, it is still below 50 percent in most and even below 10 percent in some (Fig. 6-11). "Nearly all programs still fail to reach most rural people; even in towns and cities, the quality of services is often poor and discontinuation rates are high. . . . Twenty-seven countries have yet to introduce family planning programs," according to the World Bank, *World Development Report, 1984*.

The survey also found that in virtually all countries many of the women who say they do not want more children do not use contraceptives or they use relatively ineffective methods. In some cases, nearly half the women of childbearing age fell into this category. Therefore, lack of adequate family planning services seems to stand out as an important contributor to population growth and one that could be corrected with minimal expense.

Despite this unfilled need for family planning services, progress in meeting this need has slowed in recent years as a result of a shift in United States policy under the Reagan Administration. Assistance for family planning, adjusted for inflation and population increases, turned sharply down after 1979 and in 1981 was no higher than in 1973.

Developing countries themselves recognize the problem. A World Population Conference in Mexico City in 1984 called for a fourfold increase in assistance for family planning services. The United States, however, maintained that population growth itself is not a problem, and advocated large-scale development as the solution. As noted earlier, this position was essentially discredited in the 1960s.

Another factor that aggravates the situation is a policy instituted in 1984 in the United States that denies family planning where such aid is associated with abortion. Abortions make up a very small portion (less than 10 percent) of family planning activities and they are generally restricted to cases where full-term pregnancy would jeopardize the health or life of the mother and the baby. Denial of abortions in these cases seriously undercuts the credibility of health care services. It also leads women to attempt self-induced abortions and abets the activities of unskilled practitioners who take on exceedingly dangerous activities in unclean conditions. This often results in permanent injury or death and again undercuts the efforts of health care services to improve the health and well-being of the people.

## CONCLUSION

In conclusion, evidence indicates that if intelligent policies are pursued and proper environmental management is employed, population growth may level off within the bounds of what the earth can sustain. However, this will not happen automatically. Much greater effort must be made to recognize, understand, and accept the problems as they are and to construct policies accordingly.

But we must reemphasize the dangers of exploding population. The situation represents much more than just the present reality and potential increase of human poverty and suffering in developing countries. As populations in poor countries continue to expand and those in developed nations level off, the affluent will become a smaller and smaller minority; yet they will continue to control the bulk of the world's economic wealth (see Fig. 5-3). The political, social, and economic stability of the world is threatened by this growing disparity. Hence, humanitarianism is not the only reason why developed nations must use their economic wealth and skills to help people in developing nations obtain economic self-sufficiency and reduced birth rates.

It is also a mistake to view the population explosion in the developing world as the whole of the problem. Whether or not there is an overpopulation problem will always depend on the relative balance between humans on one side and their necessary resources on the other. This is a key consideration for high-income, developed nations such as the United States, as well as for low-income developing nations. While developed nations have achieved a fertility rate that will stabilize their populations, they have not achieved a balance between population and resources. They are continuing to exploit soil, water, mineral, biological, and energy resources at an alarming rate

in the effort to maintain and increase their affluent standards of living. In addition, resources are being further degraded and destroyed by pollution. This extreme lack of balance between population and resources leads many experts to declare that developed nations such as the United States are even more overpopulated than developing nations. Indeed, the point is made by the observation that the United States, with just 5 percent of the world's population, consumes about 25 percent of the world's energy resources.

But here we reemphasize another extremely important point: at any level of affluence, the population-resource balance is not given by a simple comparison of people on one side and resources on the other. How well the resources are managed and conserved is also a crucial factor. A small population can carelessly destroy its resources as well as a large one and it is also possible for a relatively large population to live well by carefully managing and conserving its resources.

Consequently, every effort must be made to achieve sound ecological management of. resources and pollution control as well as population stabilization. With this in mind, we turn our attention to the issues of resource management and pollution control.

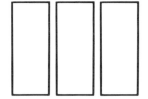

# SOIL AND WATER

Fertile soil and adequate water are the basis of almost all food, livestock, feed, forage, timber, and natural fiber production. Therefore, as world population grows we can anticipate an increasing need for fertile soil and water resources. Unfortunately, poor management has and continues to undercut these resources. Millions of hectares (1 ha = 2.5 acres) of formerly productive land have been turned into barren, desertlike wastes, and productivity has declined accordingly. For example, largely because of soil losses per capita, grain production in Algeria dropped 61 percent between the early 1950s and 1970s, greatly increasing that country's dependence on imported grain. Statistics are little better for many other nations. The United States is one of the few countries that still produces quantities of food significantly beyond the needs of its own population. But ominously, soil degradation is occurring here as well.

Likewise, water resources are overdrawn in many areas and exhaustion of water supplies for irrigation is likely to impose even more severe constraints on agriculture than will soil deterioration.

Our objective in Part 3 is to examine fundamental aspects of soil and water so that we can both appreciate the nature of the problems and understand the kinds of management procedures that need to be adopted to correct past problems and avert future problems.

# SOIL AND THE SOIL ECOSYSTEM

## CONCEPT FRAMEWORK

| Outline | | Study Questions |
|---|---|---|
| **I. CRITICAL ASPECTS OF THE SOIL ENVIRONMENT** | **178** | |
| **A. Mineral Nutrients and Nutrient-holding Capacity** | **178** | 1. How are rocks broken down so that the nutrient ions are released for plants? What is *leaching?* What two factors are important regarding nutrients? Why? Define *eutrophication.* |
| **B. Water and Water-holding Capacity** | **179** | 2. How do plants hold water? What is *transpiration?* How does it affect plant survival? Describe the three aspects of soil that determine the size of the capillary water reservoir. |
| **C. Oxygen and Aeration** | **180** | 3. How do roots derive energy? What is soil *aeration* and why is it significant? Give examples. |

4. What is pH?

5. What is *water balance*? What is *osmosis*? What determines water balance?

6. What is *soil texture*? How are mineral particles classified? How does soil texture affect water-holding capacity, infiltration, and nutrient-holding capacity? How does it affect agriculture?

7. Describe *humus*. How does it affect soil structure? Where does humus formation occur? What are the advantages of humus?

8. Discuss how soil organisms may influence a plant's nutrient absorption.

9. Describe the mutual relationship between plants and soil. In what other ways do plants protect the soil environment? What is *mineralization*? Summarize the importance of establishing a balance between plants and soil and discuss the consequences of poor soil management.

10. What does soil *erosion* involve? Describe *splash* erosion and its effects. What are *surface, rivulet, gully,* and *wind* erosion? What are their effects? What protects soil from erosion?

11. What are the benefits and disadvantages of plowing?

12. Describe *contour farming, strip cropping, shelter belts,* and *terracing.* Why have they not been adopted as a regular part of agricultural practices?

13. Why was no-till agriculture not used until the mid-1960s? What are its advantages and disadvantages?

14. How could a shift in crops from annual to perennial species protect against soil erosion?

15. Discuss the differences between and the effects of *inorganic* and *organic* fertilizers. How are they best used?

16. How can the agricultural principles discussed thus far be employed by home gardeners?

17. How do erosion and overgrazing lead to desertification? Why is the desertification plaguing world agriculture not a natural process?

18. What is *clearcutting?* What effects does it have on the soil? How would selective cutting improve the situation? Why is it not widely practiced?

19. What is the most common limiting factor in agricultural production? What are the negative effects of irrigation? Can they be alleviated?

20. Give an example of one way in which pesticides may affect the soil.

21. Cite examples of how soil can be regenerated and discuss the role of degradation and regeneration in agriculture.

The surface area of a plant's roots has been calculated to be even greater than the surface area of its branches and leaves. Thus, what is below the surface of the ground constitutes more than half the plant's complex environment.

On examining a rich topsoil one finds a complex combination of mineral particles plus organic matter from dead vegetation and animal wastes in various stages of decomposition. In addition, one finds a host of living organisms, ranging from the decomposers (fungi and bacteria) to larger detritus feeders (earthworms, snails, and insects), which form a complex food web based on the detritus. Thus, a soil that supports the productive growth of plants is much more than "dirt." It is a complex ecosystem in its own right (Fig. 7-1). The combination of bacteria, fungi, earthworms, and other organisms that feed on the detritus in or on the soil constitute a biomass that is very much larger than that of all "above ground" feeders, including humans (see Fig. 2-20).

As with other ecosystems, altering any one or more abiotic or biotic factors may upset or even cause the collapse of the entire ecosystem. This, in turn, may profoundly affect the kinds and quantities of plants the soil will support.

Our first objective in this chapter is to become familiar with the major aspects of the soil environment and to understand how they relate to the sustenance of plants. Secondly, we shall look at how these aspects vary in different kinds of soil. This information will provide the background necessary to understand how various human practices degrade the soil environment and how such practices must be modified to assure continued productivity.

## CRITICAL ASPECTS OF THE SOIL ENVIRONMENT

In order to sustain plants, the soil environment must supply the plants' needs for mineral nutrients, water, and oxygen. Additionally, the pH (relative acidity) and salinity (salt concentration) of the soil environment are particularly important.

### Mineral Nutrients and Nutrient-Holding Capacity

The needs of plants for mineral nutrients such as nitrate ($NO_3^-$), phosphate ($PO_4^{3-}$), potassium ($K^+$), and calcium ($Ca^{2+}$) were described in Chapter 2. Except for nitrogen compounds, which may be derived from the atmosphere through the nitrogen cycle, all mineral nutrients are originally present as part of the chemical composition of rocks along with nonnutrient chemicals such as silica, aluminum, and oxygen (see Fig. 2-3). However, plants cannot absorb the mineral nutrients as long as these nutrients are bound tightly in the mineral structure. They can only be absorbed after the rock is broken down and the nutrient ions are released into water solution or a loosely bound state (see Fig. 2-4).

In nature, rock, referred to as **parent material,** is gradually broken down into finer soil particles by a combination of factors collectively called **weathering.** These factors include physical forces such as freezing and thawing, heating and cooling, and the abrasive action of sand particles carried by wind or water; biological forces such as pressure exerted by roots growing into small cracks; and processes involving various chemical reactions.

As nutrient ions are released they may be absorbed by roots or they may be washed away by water percolating through the soil, a process called **leaching.** Consequently, two factors are important regarding nutrients: (1) the initial supply of nutrients, and (2) the ability of the soil to

**FIGURE 7-1**
A productive soil is more than just "dirt." Maintaining soil involves a dynamic interaction between mineral particles, detritus, and detritus feeders and decomposers. Altering any of the three main factors may have drastic effects on the soil.

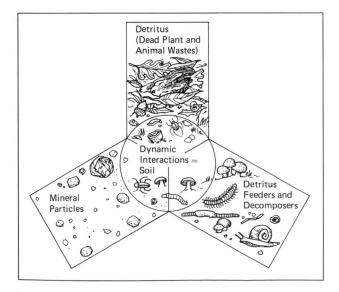

Detritus (Dead Plant and Animal Wastes)

Dynamic Interactions = Soil

Mineral Particles

Detritus Feeders and Decomposers

loosely bind and hold the nutrient ions such that they may be absorbed by roots and prevented from leaching. This latter aspect is referred to as the soil's **nutrient-holding capacity** or **ion exchange capacity.**

The importance of nutrient-holding or ion exchange capacity deserves special emphasis. First, one must recognize that weathering is a slow process. Even if the parent rock material is rich in nutrients, its rate of weathering will not release nutrients fast enough to support optimum growth of plants. If plants were dependent on rate of weathering for nutrients, their growth would be severely limited. In most ecosystems, the major source of nutrients that support current vegetation is not the weathering of parent material but rather the decay of organic matter. In other words, the growth of the current generation is dependent on recycling nutrients from past growth as discussed in Chapter 2. This is especially true on geologically old soils, which exist in moist tropical regions. Here, essentially all the weathering that can take place has already happened. The remaining inorganic portion of the soil consists of highly stable, insoluble minerals that are virtually devoid of nutrient ions.

In agricultural systems there is an additional removal of nutrients through the harvesting and removal of crops. Nutrients contained in the harvest are routinely replaced and natural supplies are supplemented by the addition of **fertilizer. Inorganic (chemical) fertilizers** are appropriate mixtures of necessary mineral nutrients. **Organic fertilizers** are plant and animal wastes such as manure that release nutrients upon their decay.

Even if nutrients are added via fertilizer, the nutrient-holding capacity of the soil is important. First, it is an obvious economic waste to add fertilizer only to have it leach away. Second, and more importantly, nutrients that leach from the soil must go somewhere; they may leach into groundwater, making it hazardous to drink, or they may end up in lakes and other bodies of water, causing a serious pollution problem of overfertilization known as **eutrophication.** These aspects of nutrients are summarized in Figure 7-2.

## Water and Water-Holding Capacity

The plight of plants wilting and dying for lack of water is well known; however, the reason for the water demand is of interest. Plants use a small amount of water as a nutrient in photosynthesis, but a 100-fold larger amount is required because the stomas or pores in the leaves, which enable the absorption of carbon dioxide ($CO_2$) and the release of oxygen ($O_2$) in photosynthesis, also allow the escape of water vapor, which evaporates from the moist cells inside the leaf (Fig. 7-3). Such loss of water vapor is called **transpiration.** It has been estimated that 0.4 hectare (1 acre) of corn may transpire as much as 2 million liters (0.5 million gallons) of water in a single growing season; this is equivalent to a layer of water 43 cm (17 in.) deep over the field. The water lost from the top of the plant must be continually replaced by absorption through the roots from a reservoir held in the soil. Without adequate water, dehydration, wilting, and death occur.

A major factor in the distinction between desert and forest tree species is the amount of water they lose in transpiration. Trees lose a great deal; therefore, support of forests requires large amounts of water. Conversely, desert species have adaptations such that they lose relatively little water and can thus survive in much drier situations.

Between rains, most plants depend on a reservoir of **capillary water,** which is water held between soil particles against the pull of gravity, as water is held in a sponge. The roots of most plants are within a meter (about 3 ft) of the soil surface; water that trickles down through the soil more than 1 m or so becomes unavailable to most plants.

In addition to the amount and frequency of precipitation (or irrigation), three aspects of the soil are critical in determining the size of the capillary water reservoir or **field capacity.** First is the nature of the surface: it must enable **infiltration;** that is, it must be such that water can soak in readily. Water that runs off the surface cannot benefit plants. Second, and generally most important, is the **water-holding capacity** of the soil. There are spongy soils that can hold sufficient capillary water to sustain plants over a considerable dry period. Conversely, there are soils that are sand that hold very little water—most of it percolates beyond the reach of roots. Without frequent rainfalls or irrigation, plants on such soils may easily exhaust the small reservoir of capillary water and suffer from drought. Frequently, soils with low water-holding capacity can only support drought-tolerant species.

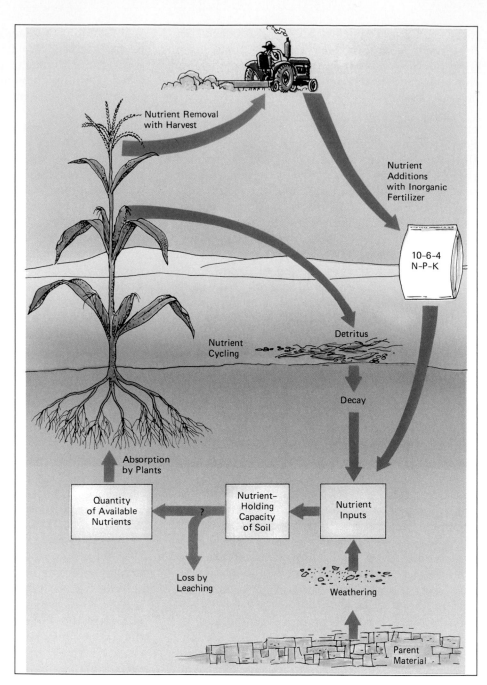

**FIGURE 7-2**
Plant-soil-nutrient relationships. Soil nutrients may come from decay of detritus, additions of fertilizer, and very slowly from weathering of parent material. Whether these nutrients remain in the soil until they are absorbed by plants or are lost from the system depends on the nutrient-holding capacity of the soil.

The third point to consider is that the reservoir of capillary water may be significantly diminished by evaporation from the surface. Therefore, a surface layer that reduces evaporation while still permitting good infiltration is highly significant, particularly in situations where capillary water may be limited by other factors. These aspects of soil water are summarized in Figure 7-4.

## Oxygen and Aeration

Roots require energy for growth and chemical functions including the absorption of nutrients from the soil. Roots derive this energy from the oxidation of glucose in the process of **cell respiration.** Hence, they require oxygen and produce carbon dioxide as a waste. The top of the plant

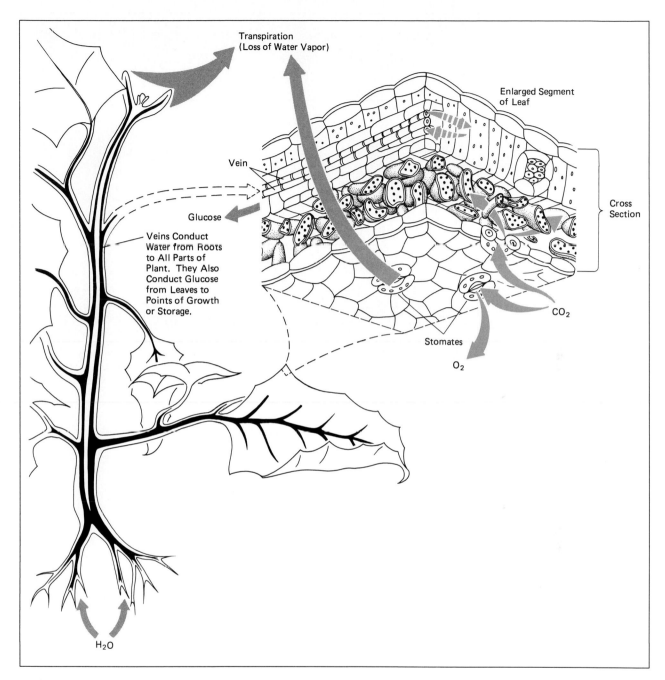

**FIGURE 7-3**
Transpiration water loss. Pores in the leaf surface called *stomas*, which allow the exchange of $CO_2$ and $O_2$ for photosynthesis, also allow the escape of water vapor, which evaporates from the moist cells inside the leaf. This loss of water is called *transpiration.* If the water lost in transpiration is not replaced, the plant will wilt and die.

Labels in figure:
- Transpiration (Loss of Water Vapor)
- Enlarged Segment of Leaf
- Cross Section
- Vein
- Glucose
- Veins Conduct Water from Roots to All Parts of Plant. They Also Conduct Glucose from Leaves to Points of Growth or Storage.
- $CO_2$
- Stomates
- $O_2$
- $H_2O$

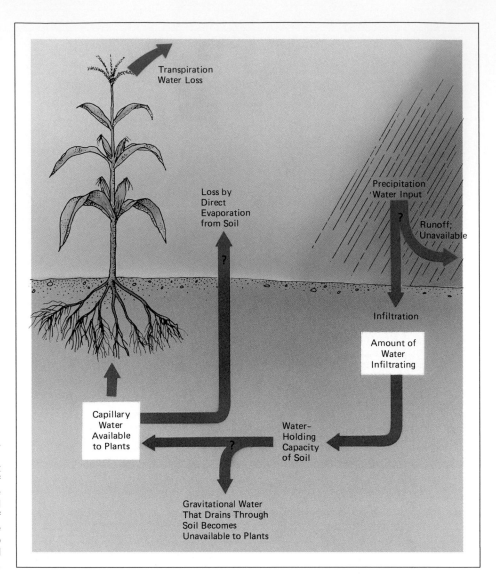

**FIGURE 7-4**
Plant-soil-water relationships. Water lost in transpiration must be replaced from a reservoir of capillary water held in soil. The size of this reservoir, beyond amount and frequency of precipitation, depends on the soil's ability to allow water to infiltrate, hold water, and minimize direct evaporation.

Within the figure:
Transpiration Water Loss
Loss by Direct Evaporation from Soil
Precipitation Water Input
Runoff; Unavailable
Infiltration
Amount of Water Infiltrating
Capillary Water Available to Plants
Water-Holding Capacity of Soil
Gravitational Water That Drains Through Soil Becomes Unavailable to Plants

may be carrying on photosynthesis and producing oxygen; however, most plants have no means of transporting gases from one part of the plant to another. Therefore, respiration in the roots is dependent on their ability to absorb oxygen from and dispose of carbon dioxide into the soil environment. Consequently, the ability of soil to allow the diffusion (passive movement) of oxygen from the atmosphere into the soil and the reverse movement of carbon dioxide is another important aspect of the soil environment (Fig. 7-5). This property is referred to as soil **aeration.**

A loose porous soil will have good aeration; a dense compacted soil will have poor aeration and hence may significantly retard the growth of plants (Fig. 7-6). If aeration is blocked, the resulting suffocation of roots may cause death of the plant. Such a situation can result not only from **compaction,** but also from overwatering or flooding. **Drowning** of plants occurs when soils become so waterlogged that gases cannot readily diffuse through them. Some plants, such as the bald cypress, mangrove, and Spartina (marsh grass) sedges and reeds have anatomical adaptations that enable oxygen to diffuse through the stem from the top down into the roots. Hence, these plants can and do thrive in very wet situations.

## Relative Acidity (pH)

Early observers noticed that when soil is touched to the tongue it may taste sour, it may have no taste, or it may taste sweet. If soils are

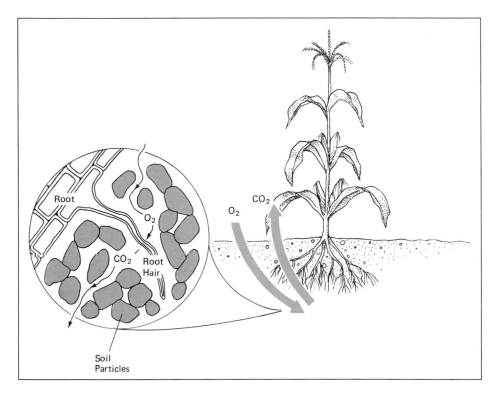

**FIGURE 7-5**
Soil aeration. Plant roots are dependent upon an exchange of oxygen and carbon dioxide in respiration. In turn, these gases must be able to diffuse through the soil between soil particles.

**FIGURE 7-6**
Poor soil aeration reduces plant growth. The cut through the soil shows that the upper layer has been severely compacted. For the soybean plant on the left, the compacted layer was broken and loosened by tillage; for the plant on the right it was not. The small size of the plant on the right shows that compacted soil severely retards growth, an effect largely due to poor soil aeration. (USDA–National Tillage Machinery Laboratory photo.)

too sour or too sweet they fail to support plant growth. The terms *sour* and *sweet soils* are still occasionally used; however this taste assay has been quantified in terms of **pH,** the symbol used to designate the relative degree of *acidity* (sourness) or *alkalinity* (sweetness). The pH scale extends from strongly acidic at pH = 0 through neutral at pH = 7 (the pH of pure water) to strongly alkaline or basic at pH = 14 (see page 364).

Most plants require a soil or water solution close to neutral and most natural environments provide this. Unfortunately, under the onslaught of acid precipitation, many environments are becoming more acid, a situation that is profoundly damaging both plant and animal life. This problem is of such magnitude that Chapter 14 is devoted to it. Here, it is only necessary to note that pH is one of the important soil water parameters.

## Salt and Osmotic Pressure

To function properly, all cells of living organisms must maintain a certain amount of water, a feature referred to as **water balance.** Interestingly, however, cells have no way of directly pumping water in or out. Instead, water balance is determined primarily by the relative salt concentration inside versus outside the cell membrane. Basically, water molecules are attracted to salt ions. The cell membrane inhibits the passage of salt, but water moves rapidly through the membrane toward the higher salt concentration, a phenomenon called **osmosis** (Fig. 7-7).

Thus cells control their water balance by adjusting their internal salt concentration, and water follows passively by osmosis. However the capacity of cells to do this is limited. Consequently, if the salt concentration outside the cells is too high, the cell cannot absorb water; indeed water may diffuse out of the cells by osmosis, resulting in dehydration and death of the plant. This is why freshwater species cannot tolerate seawater, which has a salt concentration of 3.5 percent. Marine species have various adaptations so they can maintain water balance in seawater; however, the principle remains the same.

Most land plants require fresh water. As soil becomes salty, its ability to support plants decreases. We shall see that this is a problem because irrigation gradually increases the salt content of soil.

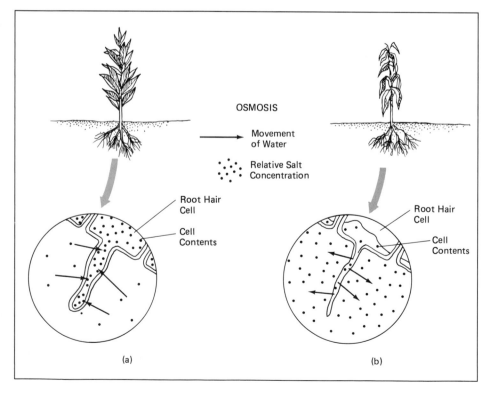

**FIGURE 7-7**
Soil salt and movement of water by osmosis. (a) Cells absorb potassium salts but exclude sodium salts. When soil water is fresh, roots absorb potassium such that salt concentration inside cells is greater than outside. The result is movement of water into cells, keeping the plant turgid (nonwilted). Cell walls keep cells from bursting. (b) If there is a high concentration of salt outside, especially sodium salts, which cells exclude, water is drawn out of the cells, causing wilting and death of cells.

OSMOSIS

Movement of Water

Relative Salt Concentration

Root Hair Cell

Cell Contents

Root Hair Cell

Cell Contents

(a)

(b)

# ABIOTIC AND BIOTIC FACTORS AFFECTING THE SOIL ENVIRONMENT

The important question now is: How are the critical aspects of the soil environment affected by various abiotic and biotic factors? Two factors in particular influence nutrients, water, and aeration: (1) *soil texture* or size of the mineral soil particles, and (2) the amount and nature of the *organic matter* present, both living organisms and dead material.

## Soil Texture—Size of Mineral Particles

The mineral particles of soil that result from rock weathering are classified into three main categories based on size: sand, silt, and clay. Relatively coarse particles are **sand;** somewhat finer particles are **silt;** extremely fine particles are **clay** (Table 7-1). The proportions of sand, silt, and clay present in a soil define its **texture.** For example, if most of the mineral particles are in the silt size range, the soil is said to have a silty texture.

Table 7-1

| USDA Classification of Soil Particles | |
|---|---|
| NAME OF PARTICLE | DIAMETER (mm) |
| Very Coarse Sand | 2.00–1.00 |
| Coarse Sand | 1.00–0.50 |
| Medium Sand | 0.50–0.25 |
| Fine Sand | 0.25–0.10 |
| Very Fine Sand | 0.10–0.05 |
| Silt | 0.05–0.002 |
| Clay | Below 0.002 |

Some soils are virtually pure sand, silt, or clay, but commonly there is a mixture. One can easily determine the proportions of differently sized particles by shaking a small amount of soil with water in a glass cylinder and then allowing it to settle. Sand particles, which settle first, are thus separated from silt and clay, which settles last.

The soil texture triangle shown in Figure 7-8 allows one to plot any combination of sand, silt, and clay and determine its texture classification.

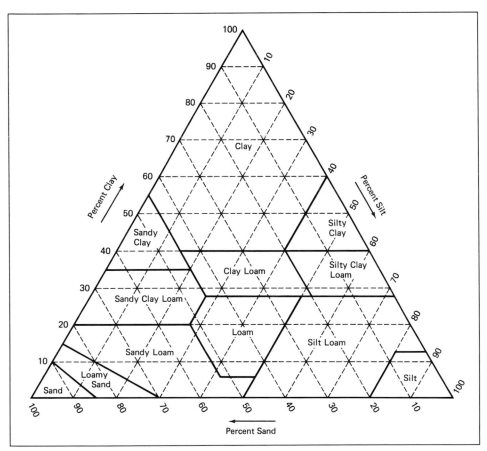

**FIGURE 7-8**
Soil texture classes. Soils are classified according to texture on the basis of the percentage of sand, silt, and clay that they contain. (USDA.)

A classification of considerable importance is **loam** in which the mineral portion of the soil is roughly 40 percent sand, 40 percent silt, and 20 percent clay.

Soil texture has a profound effect on infiltration, water-holding capacity, nutrient-holding capacity, and aeration. Larger particle size also means larger spaces between particles. Consequently, infiltration and aeration increase with *increasing* particle size. Sandy soils are excellent in terms of infiltration and aeration; clayey soils are poor in these respects because of very small pore size; silty soils are intermediate.

However, water-holding and nutrient-holding capacity have the opposite relationship to particle size. Both of these factors are very low in sandy soils and improve with *decreasing* particle size.

Water- and nutrient-holding capacity improve with decreasing particle size for two reasons. First, both water molecules and nutrient ions tend to stick to surfaces. Second, relative surface area increases as particle size decreases. This can be visualized by imagining a rock being broken in half again and again. Every time it is broken, two new surfaces are exposed, one on each side of the break, but the overall weight or volume of the rock is not changed (Fig. 7-9). Thus, in a given volume of soil, the smaller the particle size, the greater the overall surface area and the greater the nutrient- and water-holding capacities (Fig. 7-10).

Soil texture also affects **workability,** the ease with which a soil can be cultivated. This has an important bearing on agriculture. Clayey soils are very difficult to work. With modest changes in

**FIGURE 7-9**
Relative surface area increases as particle size decreases. Every time the cube is cut, it exposes more surface area (color) without changing the volume of material. Therefore, the surface/volume ratio increases as particle size decreases.

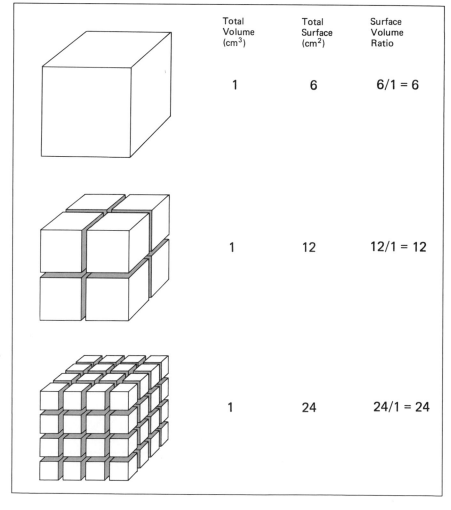

| | Total Volume (cm³) | Total Surface (cm²) | Surface Volume Ratio |
|---|---|---|---|
| | 1 | 6 | 6/1 = 6 |
| | 1 | 12 | 12/1 = 12 |
| | 1 | 24 | 24/1 = 24 |

| Table 7-2 | Relationship Between Soil Texture and Physical and Chemical Properties | | | | |
|---|---|---|---|---|---|
| SOIL TYPE | WATER INFILTRATION | WATER-HOLDING CAPACITY | ION EXCHANGE CAPACITY | AERATION | WORKABILITY |
| Sand | Good | Poor | Poor | Good | Good |
| Silt | Medium | Medium | Medium | Medium | Medium |
| Clay | Poor | Good | Good | Poor | Poor |
| Loam | Medium | Medium | Medium | Medium | Medium |

moisture content, they alternate rapidly from being sticky and muddy to being hard and bricklike. Sandy soils are very easy to work; they do not become muddy when wet, nor do they become hard and bricklike when dry.

These relationships between soil texture and various properties are summarized in Table 7-2. Which is the most productive kind of soil? Considering the principle of limiting factors, the poor attributes of sandy or clayey soils may severely curtail the productivity of a natural ecosystem and may preclude agriculture altogether. The most productive mineral soils (those with no organic matter) are silt or loam in which there is a compromise or mixing between the favorable and unfavorable properties of sand and clay. Note here, however, that good qualities are diminished just as poor qualities are improved.

Is it possible to maximize all factors? Yes, this occurs with the addition of organic matter, as will be described in the next section.

## Detritus, Detritus Feeders, Humus, and Soil Structure

A truly remarkable improvement in all the aspects we have been discussing comes with the addition of a small percentage of organic material called **humus**. The process of adding humus to soil begins with the accumulation of dead leaves, dead roots, and other detritus on and in the soil. This detritus supports a complex food web consisting of several trophic levels. Organisms include numerous species of bacteria, fungi, protozoans, mites, insects, millipedes, spiders, centipedes, earthworms, snails, slugs, moles, and

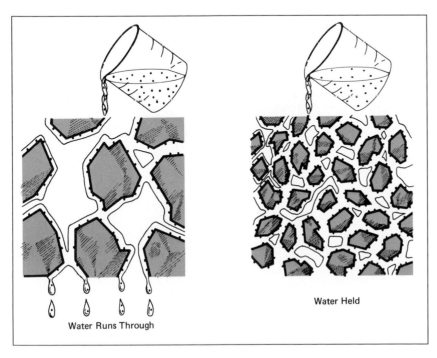

Water Runs Through

Water Held

**FIGURE 7-10**
Soil water- and nutrient-holding capacity increase as particle size decreases. Both water and nutrient ions (represented by dots) tend to cling to surfaces, and smaller particles have relatively more surface area.

other burrowing mammals (Fig. 7-11). As these organisms feed on the detritus and each other in a complex food web, most of the biomass is oxidized through their cell respiration to carbon dioxide, water, and the mineral nutrients as described in Chapter 2. However, each organism leaves "fecal pellets," the portion of its total intake that is resistant to digestion. This residue of organic material is called *humus*. You have probably observed pure humus as the residue of dark-colored, spongy material inside an old log after most of the interior has rotted away (Fig. 7-12). As feeding of soil organisms, particularly fungi and bacteria, continues, the humus too is gradually decomposed at a rate of 20 to 50 percent per year. It is only somewhat more resistant to breakdown than the bulk of the detritus.

Soil organisms feeding on detritus not only decompose it to humus; but their activity also intimately mixes it with the mineral particles and develops what is called **soil structure**. For exam-

**FIGURE 7-12**
Humus is organic material that is more resistant to decomposition and therefore remains after the major portion of decomposition has occurred, as in this tree stump. Gradually humus also decomposes to inorganic materials. (Photo by author.)

**FIGURE 7-11**
Soil organisms. The major groups of soil organisms are depicted in the illustration. It is the action of these organisms that reduces detritus to humus, intimately mixes the humus with soil, and in the process develops soil structure. (From Robert Leo Smith, *Ecology and Field Biology*, 2nd ed., [Figure 17-5, p. 540]. Copyright 1966, 1974 by Robert Leo Smith. Reprinted by permission of Harper & Row, Publishers, Inc.)

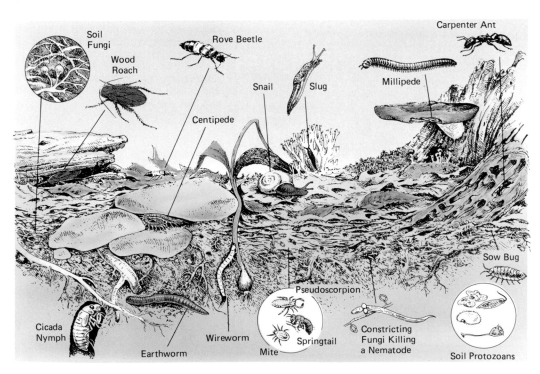

ple, as earthworms feed on detritus they ingest many inorganic soil particles as well. As the mineral particles go through the gut, they become thoroughly mixed and "glued" together with the nondigestible organic compounds. As much as 40 tons of soil per hectare (2.5 acres) may pass through earthworms each year. Also, slimy secretions from earthworms, which aid in lubricating their burrows, add to the humus and tend to glue soil particles together. The result is that the mineral particles of soil become bound together with humus into larger clumps and aggregates. At the same time, the burrowing activity of various or-

ganisms keeps the clumps loose. This loose clumpy characteristic is the soil structure (Fig. 7-13).

This process of humus formation and development of soil structure occurs mainly in the upper 2 to 3 dm (8 to 12 in.) of soil. Thus a layer of dark-colored (from humus addition) soil with clumpy structure develops on top of the lighter-colored, humus-poor, compacted soil. This layer of humus-rich soil is called **topsoil**; the soil below is **subsoil** (Fig. 7-14).

The attributes of topsoil in contrast to subsoil are legendary. Humus has phenomenal nutrient- and water-holding capacity, as much as 100-fold greater than clay on the basis of weight. Thus, humus added to a soil of any texture will greatly increase its water- and nutrient-holding

**FIGURE 7-13**
Humus and the development of soil structure. (a) On the left is a humus-poor sample of loam. Note that it is a relatively uniform, dense "clod." On the right is a sample of the same loam but rich in humus. Note that it has a very loose structure composed of numerous aggregates of various sizes. (Photo by author.) (b) A diagrammatic illustration of the difference.

(a)

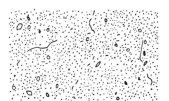

Lack of Structure
Gives Compacted Soil
with Poor Aeration
and Poor Infiltration

Addition of
Organic Matter
and Humus–forming
Process Involving
Numerous Organisms

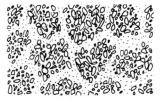

Structured Soil with
Excellent Aeration and
Excellent Infiltration

(b)

**FIGURE 7-14**
Soil profile. A cut through soil generally reveals a layer of loose, dark topsoil overlying light-colored, compacted subsoil. The topsoil layer results from the addition of organic matter and the activity of soil organisms. (USDA photo.)

in clayey soils. Added to loam or silt soils, humus markedly enhances all aspects (Table 7-3).

The gardener or farmer may see the greatest difference in workability. When wet, the humus-rich sample breaks cleanly and easily into small clumps, whereas the humus-free sample is a muddy mass. When dry, the humus-rich sample still breaks easily into smaller clumps, whereas the humus-lacking sample is a bricklike clod. As water percolates through humus-rich soil it remains relatively clean whereas water percolating through humus-poor soil is muddied by clay particles. This means that the humus-rich soil will maintain a loose clumpy structure when watered (whether by natural or artificial means) while humus-poor soil settles into a compacted muddy mass with poor infiltration and aeration.

## Direct Interactions between Roots and Soil Biota

Some very important direct interactions between many plants and certain soil organisms exist. One is a mutually beneficial symbiotic relationship between the roots of some plants and certain soil fungi. Masses of fungal filaments called **mycorrhizae** surround the roots. Some of the filaments penetrate and draw some nourishment from the plant, but they also ramify through the soil and help the plant absorb nutrients (Fig. 7-15). Another important relationship is the role of certain soil bacteria in the nitrogen cycle discussed in Chapter 2.

Not all soil organisms are beneficial, however. There are certain small worms known as nematodes that feed on living roots and are highly destructive pests to a number of agricultural crops. In a flourishing soil ecosystem, however, nematode populations may be controlled by other soil organisms. Interestingly, there is another fungus, the *hyphae* which actually form little snares that catch nematode worms and proceed to feed on them (Fig. 7-16).

capacity. This is particularly important in sandy and silty soils. Moreover, the development of soil structure promoted by humus greatly enhances infiltration of water and aeration. This is essential

| Table 7-3 | Comparison of Humus-Poor and Humus-Rich Loam | | | | |
|---|---|---|---|---|---|
| LOAM TYPE | WATER INFILTRATION | WATER-HOLDING CAPACITY | ION EXCHANGE CAPACITY | AERATION | WORKABILITY |
| Humus-poor | Medium | Medium | Medium | Medium | Medium |
| Humus-rich | Excellent | Excellent | Excellent | Excellent | Excellent |

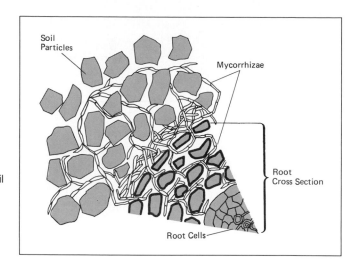

**FIGURE 7-15**
Mycorrhizae. In many plants, soil fungi, or mycorrhizae, aid in the absorption of nutrients. Drawing shows cross section of rootlet with mycorrhizae forming a connection between the root cells and soil particles.

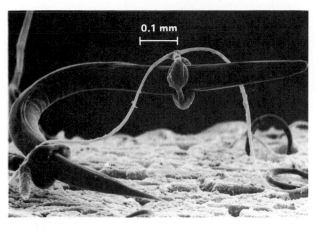

**FIGURE 7-16**
Soil nematode (roundworm) captured by the constricting rings of the predatory fungus *Arthrobotrys anchonia*. (Courtesy of Nancy Allin and O. L. Barron, University of Guelph.)

## MUTUAL RELATIONSHIP BETWEEN PLANTS AND SOIL

You should now be able to visualize the mutual relationship that exists between plants and soil. Plants are dependent upon soil for nutrients, water, and aeration of roots. However, the soil's ability to provide nutrients, water, and aeration of roots is largely dependent on humus and soil structure, which in turn are dependent on the activity of soil organisms feeding on detritus provided by plants (Fig. 7-17).

Plants play two other roles in protecting the soil environment. Most prominently, a vegetative cover protects the soil from erosion, a topic that we shall address in detail in the next section. Sec-

ondly, the cover of dead leaves or other litter protects the soil from excessive evaporative water loss and it is also the source of organic material for humus formation. It also provides additional protection from erosion.

In short, plants play a tremendous role in developing and maintaining a soil environment that is most conducive to their own support. However, it must be emphasized that, as in all ecological relationships, this is a dynamic balance, not a fixed situation. Humus is resistant to digestion and oxidation; it does oxidize at a rate of 20 to 50 percent per year. Therefore, the humus content will diminish unless there is sufficient input of new detritus to preserve the balance. With the loss of humus, the soil structure and those attributes dependent on soil structure deteriorate and one is left with only the mineral portion of the soil. This process is commonly spoken of as **mineralization** of soil (Fig. 7-18).

An irony of soil degradation is that it tends to start a vicious cycle that becomes ever more difficult to halt or reverse. Degraded soil supports less abundant plant production. In turn, this means less detritus and humus production and less protection from erosion. This leads to further soil deterioration and even less plant production. Indeed, productive land can become totally barren in the end (Fig. 7-19).

Recognizing the possibility of this vicious cycle underscores the imperative that we manage our endeavors in ways that preserve the soil environment. This will be the topic of the following sections.

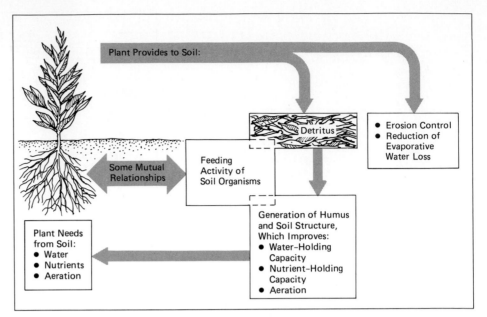

**FIGURE 7-17**
A mutual relationship exists between plants and the soil environment. Attributes of plants are enhanced by inputs of detritus derived mainly from dead plant wastes.

Plant Provides to Soil:

Detritus

- Erosion Control
- Reduction of Evaporative Water Loss

Some Mutual Relationships

Feeding Activity of Soil Organisms

Plant Needs from Soil:
- Water
- Nutrients
- Aeration

Generation of Humus and Soil Structure, Which Improves:
- Water-Holding Capacity
- Nutrient-Holding Capacity
- Aeration

**FIGURE 7-18**
Humus and corresponding soil structure is the result of a dynamic balance between organic additions leading to humus formation and loss of humus through decomposition and oxidation. Soil characteristics important in supporting plants will vary accordingly.

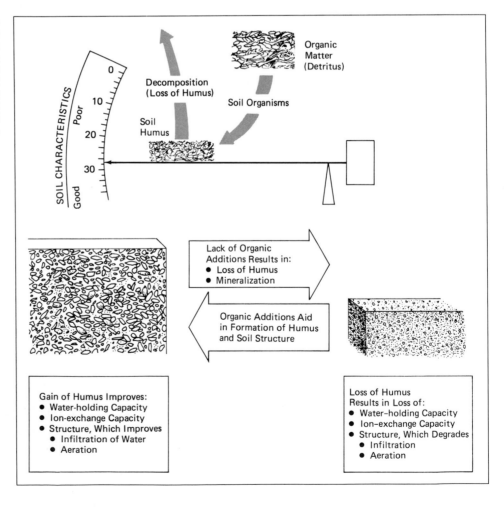

SOIL CHARACTERISTICS
Poor
0
10
20
30
Good

Decomposition (Loss of Humus)

Organic Matter (Detritus)

Soil Organisms

Soil Humus

Lack of Organic Additions Results in:
- Loss of Humus
- Mineralization

Organic Additions Aid in Formation of Humus and Soil Structure

Gain of Humus Improves:
- Water-holding Capacity
- Ion-exchange Capacity
- Structure, Which Improves
  - Infiltration of Water
  - Aeration

Loss of Humus Results in Loss of:
- Water-holding Capacity
- Ion-exchange Capacity
- Structure, Which Degrades
  - Infiltration
  - Aeration

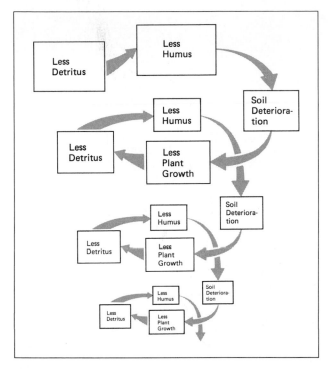

**FIGURE 7-19**
Soil degradation may become a vicious cycle. Degraded soil supports fewer plants, permitting further soil degradation.

## HUMAN IMPACTS ON THE SOIL ECOSYSTEM

Maintaining the soil ecosystem is crucially important in maintaining plant productivity. Yet, in many cases, traditional practices connected with agriculture and forestry along with recent phenomena like acid precipitation negatively impact the soil ecosystem. Not only are soil and plant productivity affected by these occurrences; but increased erosion and leaching from soil adversely affects water quality and aquatic ecosystems, as well. If we wish to sustain the human ecosystem, we must recognize these negative impacts and adopt practices that will control them.

Our objective in the following sections is to understand how the soil ecosystem is degraded by various traditional practices and look at some of the alternatives for corrective action.

### Bare Soil and Erosion

By far the most destructive force on soil is **erosion.** Since erosion results from many practices

and because it is a complex process, we begin with a look at soil erosion itself.

*Soil erosion* is the physical process of wind or water picking up and removing soil particles. Thus, we speak of *wind erosion* and *water erosion* respectively. Normally, a vegetative cover protects soil from erosion. However, whenever soil is laid bare, as it often is in agriculture, forestry, mining, and construction activities, it becomes subject to erosion. Erosion is more complex than just removal of "dirt"; it involves breakdown of the soil's aggregate structure and a selective removal of clay and humus that seriously decreases the ability of the remaining soil to support plants.

Water erosion starts with **splash erosion,** the effect of raindrops splashing against bare soil. When a raindrop hits bare soil, the effect is a miniature explosion that blasts the soil particles apart (Fig. 7-20). Recall that the importance of topsoil is its structure of clumps glued together with humus for aeration and infiltration. Splash erosion does more than destroy the structure of the surface layer. As aggregates disintegrate, the soil particles wash into and clog pores and spaces between other aggregates, thereby decreasing both aeration and infiltration. On level land, decreased infiltration leads to water standing in puddles; hence, the effect is referred to as **puddling.** Plants on puddled soil frequently suffer and die from lack of sufficient aeration, an effect commonly called *drowning.* Even more serious, however, water that cannot infiltrate readily begins to run off the surface, causing further stages of erosion.

As water flows over the surface, it picks up and carries along soil particles. A uniform loss of soil from the surface is called **sheet erosion.** Frequently, small stones protect the soil under them from the combined effects of splash and sheet erosion. Such stones are left sitting on pedestals of soil where they give mute testimony to the amount of soil lost (Fig. 7-21). As further runoff occurs, the water converges into rivulets and streams, which have greater volume and velocity and hence greater energy. The increased energy provides the water with greater capacity to pick up and carry soil particles, so that the water gouges out channels of various sizes. Many small channels are called **rivulet erosion;** large channels are called **gully erosion** (Fig. 7-22).

As sheet, rivulet, and gully erosion occur, the texture of the remaining soil becomes progres-

**FIGURE 7-20**
Splash erosion. Raindrops hitting bare soil have a destructive effect. They break up soil structure and seal soil pores. In turn, aeration and infiltration are decreased and surface runoff is increased. (USDA–Soil Conservation Service photos.)

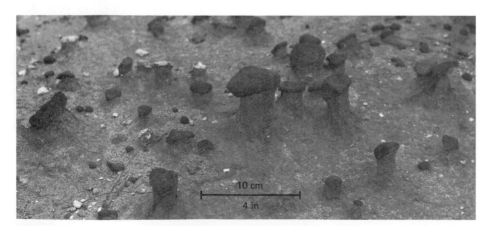

**FIGURE 7-21**
Sheet erosion. Stones protect the soil under them against splash and sheet erosion. As soil is eroded away, stones are left on pedestals, the height of which shows how much soil has been removed. (USDA–Soil Conservation Service photo.)

**FIGURE 7-22**
Rivulet and gully erosion. Water flowing over a surface tends to collect in rivulets of increasing volume, velocity, and energy. Soil is accordingly eroded in a pattern of converging rivulets and gullies of increasing size. (USDA–Soil Conservation Service photo.)

**FIGURE 7-23**
Erosion removes small soil particles more easily than large ones; therefore, soil subjected to erosion becomes progressively more sandy and stony. (a) Coarse, stony soil remaining after wind erosion; overgrazing removed vegetation allowing erosion (Russ Kinnel/Photo Researchers)

(a)

sively more coarse, that is, more sandy and stony. This occurs because soil particles are moved according to weight. Previously we noted that if loam, which consists of sand, silt, and clay, is shaken up in a glass cylinder with water, sand settles first and clay last. Humus particles may actually float. Conversely, as aggregates are shattered by splash erosion and sheet erosion occurs, humus, clay, and silt particles are carried away most readily, leaving the coarser sand and stones behind. Thus, rocky, stony soils are frequently a reflection of past erosion (Fig. 7-23).

People commonly believe that clay will not erode easily because they visualize clay as hard, bricklike clods. It is worth the effort of pouring water over a clod of dry clay and observing that the water running off is very muddy in appearance. This attests to how easily individual clay particles are separated from the clod and brought into water suspension.

Thus, the selective removal of humus, clay, and silt caused by erosion can readily degrade a rich loam topsoil into a layer of coarse sand and stones overlying compacted subsoil. Since nutrient ions are generally bound to humus and clay particles, a great loss of nutrients also occurs. From the discussion in previous sections you should be well aware of what this means in terms

**FIGURE 7-23** (cont.)
(b) Coarse soil remaining after water erosion. (USDA photo)

of changes in the soil environment and how this affects the soil's ability to support plant growth.

Rivulets and gullies appearing in a plowed field after a sudden rainstorm (see Fig. 7-22) attest to the seriousness of water erosion. However, the gradual, but pervasive effect of wind erosion may be more serious in drier climates. Wind erosion has the same effect of preferentially removing humus and clay from topsoil and leaving behind the coarser sand and stones. Did you ever wonder why deserts are full of sand? It is because all the finer-sized particles have been blown away. In low rainfall regions the most serious consequence of loss of fine particles and humus by erosion is the loss of water-holding capacity. Frequently this loss can make the difference between a soil's ability to support grass and its ability to support only a few desert species or even nothing at all (Fig. 7-24). This phenomenon is so conspicuous and widespread that it has been given the name **desertification** and it is a problem of worldwide concern. We shall discuss desertification in more detail later in the chapter.

In considering erosion, one should also be aware that soil degradation is only one aspect of the total picture. The soil particles that are removed from one place must be deposited somewhere else. Sediment (soil particles) washing down streams and rivers and filling channels and reservoirs also creates serious problems, which will be addressed in Chapters 9 and 11.

The key to protection against erosion is to keep the soil covered. Vegetation provides a tremendous protection against all forms of erosion. When there is a vegetative or litter cover (cover of dead leaves and other detritus) the energy of falling raindrops is dissipated against such material. The water then trickles gently from the leaves and infiltrates into the soil pores without disturbing the soil structure. As infiltration is preserved, runoff is reduced. Further, the velocity of runoff that does occur is slowed by its movement through the vegetative or litter mat so that it does not have sufficient energy to pick up soil particles. Grass is particularly good for erosion control because when volume and velocity of runoff increase, well-rooted grass simply lies down, making a smooth mat over which the water can flow without disturbing the soil underneath. Similarly, vegetation slows the velocity of wind near the ground so that soil particles are not blown away (Fig. 7-25).

In summary, any practice that exposes bare soil to wind and rain may result in serious erosion. The cures lie in shifting to practices that keep the soil covered or at least minimize its ex-

**FIGURE 7-24**
Erosion causes loss of soil productivity. The rocky, stony soil to the left of the fence has such poor water-holding capacity that it will no longer support the growth of grass. The condition results from initial destruction of the protective grass cover by overgrazing, but then erosion has removed fine material leaving only coarse material. Note the drop of 10–15 cm (4–6 in.) at the fence line. Since the eroded soil now only supports desert species, the process is frequently referred to as *desertification.* (USDA photo.)

**FIGURE 7-25**
Erosion. Bare soil is extremely vulnerable to erosion. The splash of falling raindrops breaks up soil aggregates into individual particles. The finer particles of humus, clay, and silt, are readily carried away by runoff or wind, leaving only a layer of coarse sand, stones, and rocks, which are resistant to erosion on the surface. A vegetative cover protects soil from all forms of erosion.

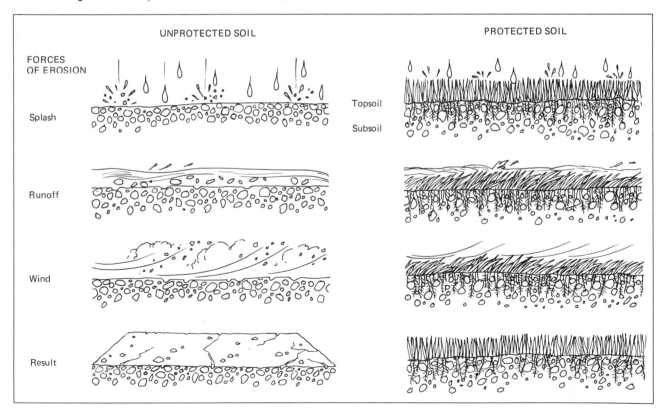

posure. We shall see this principle applied repeatedly as we discuss various practices connected with food and forest production in the next section.

## Impacts of Certain Agricultural Practices

### PLOWING AND CULTIVATION

Traditionally the first step in growing crops has been, and to a large extent still is, to plow the soil. The basic purpose of plowing is weed control. Plowing picks up and turns upside down the first 20 to 40 cm (8 to 16 in.) or so of soil. This buries and smothers weeds, which would otherwise compete with the crop plants for water and nutrients. However, in addition to controlling weeds, plowing and subsequent cultivation may lead to serious deterioration of the soil.

Most significantly, plowing exposes bare soil to erosion and it may remain exposed for a considerable part of the growing season because some time passes before the planted crop forms a complete cover; later, after harvest, the soil again may be left largely exposed. Also, at least in the short run, these activities cause excessive aeration and warming of the soil under direct rays from the sun, which enhance the activity of decompos-

ers and cause an increase in the rate of oxidation of humus. Unless the increased rate of oxidation is balanced by equal or greater inputs of organic matter, mineralization and collapse of the aggregate soil structure occurs. Thus, between erosion and increased mineralization, a serious deterioration of soil quality occurs.

It is ironic that plowing and cultivation are frequently deemed necessary to "loosen" the soil to improve aeration and infiltration when all too commonly the effect is the reverse. Ensuing splash erosion and mineralization destroy the soil's aggregate structure so that it collapses and becomes more compacted than before cultivation. The weight of tractors adds to the serious compaction of some soils.

In terms of water, the compaction impedes the infiltration and promotes runoff and erosion. This, in turn, decreases the soil's water-holding capacity. Also, bare soil under the sun becomes very warm, increasing evaporative water loss and further reducing the amount of water available to plants. As compacted soils with high clay content dry, they also shrink, causing the formation of wide, deep cracks (Fig. 7-26). Such cracking also aggravates evaporative water loss.

Deterioration of agricultural land because of erosion is not a new problem. Some historians feel that a significant factor in the fall of some ancient civilizations was soil erosion, which undercut the agricultural base. In the United States some 20 percent of our original agricultural land

**FIGURE 7-26**
Bare soil that has been puddled (beaten down by rain) often forms deep cracks as it dries. The cracks permit excessive evaporative water loss. A mulch cover prevents this. (USDA–Soil Conservation Service photo.)

has been rendered nonproductive by erosion. China has lost about one-third of its land to erosion and the story is similar or even worse for many other countries. In the United States, the Soil Conservation Service (SCS) was formed in 1932 in an attempt to prevent soil erosion. Yet in spite of many billions of dollars spent by SCS in this effort, the problem persists.

It is estimated that the processes of weathering and soil formation will produce a layer about 0.4 mm (1/64 in.) thick or about 12.5 tons per hectare (5 tons per acre) per year. This is a very rough estimate because the actual rate varies greatly depending on parent material and climatic factors. But, based on this estimate, soils could sustain 12 tons of erosion per hectare and still remain in balance. Unfortunately, recent surveys indicate that very significant proportions of U.S. croplands are experiencing at least two to three times this amount of erosion (Fig. 7-27 on pp 200-201). Currently the Midwest loses about two bushels of topsoil for every bushel of corn harvested, obviously an unsustainable practice.

## METHODS OF CONTROLLING EROSION

**Traditional soil conservation techniques.** When the Soil Conservation Service was formed in 1932 there was no practical substitute for plowing to control weeds. However, several techniques were promoted that greatly reduced the amount of erosion and these techniques are still used with advantage today. They include contour farming, strip cropping, shelter belts or windrows, and terracing.

**Contour farming** is the technique of plowing and cultivating at right angles to the slope so that runoff water is caught between furrows rather than running down their length. This practice gives the water more time to soak in and hence checks runoff and erosion (Fig. 7-28). **Strip cropping** is the technique of cultivating alternate strips, generally leaving strips of grass or hay between strips of a cultivated crop like corn. Erosion from the cultivated strip is caught and held in the grass strip, thus preventing overall erosion. Strip cropping is especially effective in reducing wind erosion. Wind velocity near the ground is lessened by the grass strip; hence it does not tend to pick up so many soil particles from the cultivated strip (Fig. 7-29). Strip cropping and contour farming are often used in conjunction with each other

(Fig. 7-28b). Shelter belts, rows of trees planted around fields, have also been used to lessen wind erosion. **Terracing,** or grading the slope into a series of steps, is also effective (Fig. 7-30). Water is caught on the steps and does not run down the slope. Very steep slopes were elaborately terraced by ancient civilizations, but such terracing is no longer compatible with modern machinery (Fig. 7-30b).

Currently, erosion occurs not because these techniques do not work, but because they have been abandoned, especially in the Midwest. In the economic climate of modern farming they are seen as not cost-effective. To meet land and machine costs and make a slight profit, farmers have been effectively forced to maximize production on all available land. This has meant sacrificing windrows and protective strips of less profitable hay and cultivating slopes that should be left in grass.

Compounding the irony of this situation is the fact that over 90 percent of the most severe erosion stems from the production of corn and soybeans which are fed to livestock to support a high-meat human diet. Most nutritionists regard such a diet as neither necessary nor desirable. Government economists are constantly engaged in manipulating the economic system in one way or another to achieve desired ends. Adjusting the system so as to provide more incentive for soil conservation needs their attention.

**No-till agriculture.** Although traditional techniques of soil conservation greatly reduce erosion they do not prevent it nor do they prevent the other undesired side effects of plowing, such as increased evaporative water loss and oxidation of humus.

One way to avoid all these adverse effects is to stop plowing in the first place; that is, to leave the soil undisturbed with a vegetative cover and simply plant the desired crop. The virtue of this idea was recognized in the 1930s; the concept is called **no-till agriculture.** Until the 1960s, however, no-till agriculture was a practical impossibility because if weeds are not controlled they compete with the crop plants. In competition with weeds, crop plants are severely reduced in size and yield, if not totally eliminated. Thus, plowing and cultivation remained necessary until the advent of specific chemical herbicides (weed killers) in the mid-1960s.

Herbicides have enabled no-till agriculture to

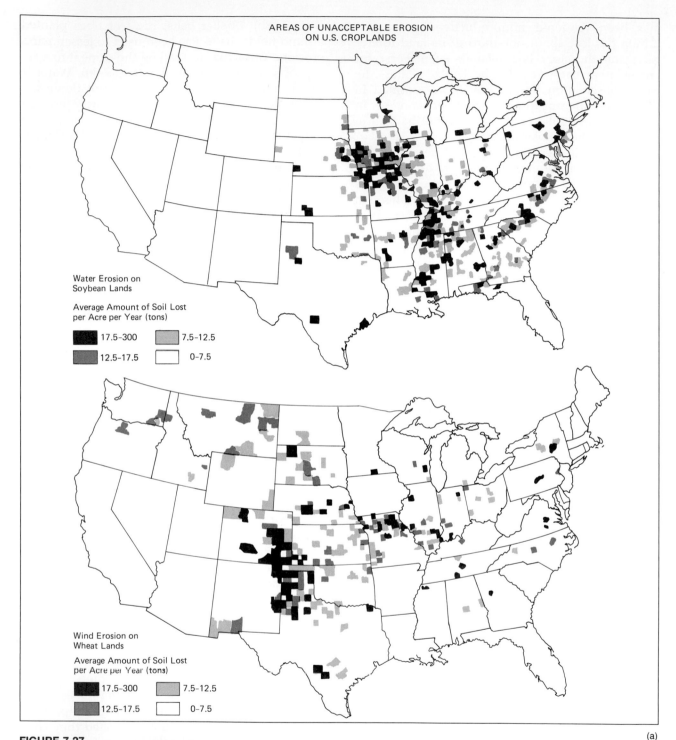

AREAS OF UNACCEPTABLE EROSION
ON U.S. CROPLANDS

Water Erosion on
Soybean Lands

Average Amount of Soil Lost
per Acre per Year (tons)

17.5–300    7.5–12.5

12.5–17.5    0–7.5

Wind Erosion on
Wheat Lands

Average Amount of Soil Lost
per Acre per Year (tons)

17.5–300    7.5–12.5

12.5–17.5    0–7.5

(a)

**FIGURE 7-27**
Soil erosion in the United States. Losses of more than 5 tons per acre per year
are unsustainable. (a) As these maps show, a very significant portion of U.S.
cropland is suffering from unacceptable rates of erosion from wind and/or water.
The data is based on a county-by-county survey. The shading indicates the
average determined for each county. (From Sandra C. Batie and Robert G. Healy,
"The Future of American Agriculture," *Scientific American*, February 1983. Copyright ©
February 1983 by Scientific American, Inc. All rights reserved.)

(b)

FIGURE 7-27 (cont.)
(b) the poor cover given by corn
allows very significant erosion.
(J.P. Jackson, © 1984, Photo
Researchers, Inc.)

(a)

 (b)

FIGURE 7-28
Contour farming. (a) Cultivation
up and down a slope promotes
water running down furrows and
may lead to severe erosion as
shown. (b) The problem is
reduced by plowing and
cultivating along the contours at
a right angle to the slope as
shown. In this case strip
cropping is also being utilized.
(USDA photo.)

**FIGURE 7-29**
(a) Soil deposited at fenceline attests to wind erosion occurring on the open cropland to the right.
(Dean Krakel/Photo Researchers.)
(b) Wind erosion is most severe when wind is unimpeded. Stripcropping (leaving alternate strips in grass) causes wind eddies near the surface and greatly reduces erosion.

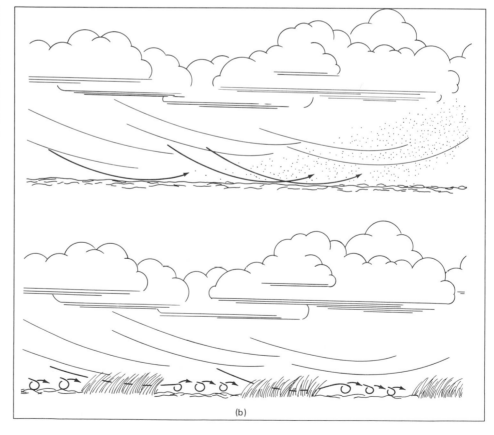

(b)

**FIGURE 7-30**
Terracing. (a) Slopes are graded into "stair steps" to reduce runoff. Only the steps are cultivated; the rises between are kept in grass. (Georg Gerster/Photo Researchers.) (b) Terraced rice fields in the mountains of the Philippines. These terraces were constructed by ancient civilizations and are only compatible with hand agriculture. (Philippine Ministry of Tourism.)

(b)

(a)

**FIGURE 7-31**

No-till planting. (a) The no-till planting machine is pulled behind a small tractor and performs several operations at once. (USDA–Soil Conservation Service photo.) (b) Schematic diagram of the apparatus. The direction of movement of the machine in this drawing is to the left. The first wheel is a fluted coulter that opens a narrow band in the untilled soil, usually cutting also through the residue of the preceding crop, which is left on the ground as a mulch. The coulter is followed by a disc that applies fertilizer. Here the fertilizer is dry, but in some machines it is applied as a liquid. (Herbicide can also be applied at about the time of planting to control weeds.) Next comes a disc, offset from the fertilizer disc by about 2 inches, that opens the furrow in which the seed is to be planted. It is followed by the planting unit, which receives the seed from the container above it. The seed is forced into the slots in the planting disk by air from a blower; the air also holds each seed in place until the slot approaches the ground and the seed drops into the furrow. The last wheel presses the soil down over the seed. Several units of this kind are normally ganged together in one machine. (From Glover B. Triplett, Jr., and David M. Van Doren, *Agriculture Without Tillage.* Copyright © Jan. 1977 by Scientific American, Inc. All rights reserved.)

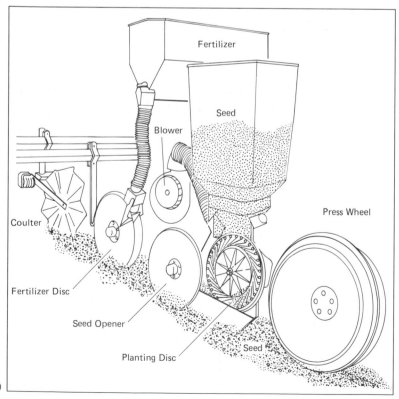

(b)

become a common practice. Herbicides kill unde-sired plants without disturbing the soil or harm-ing the desired plants. Moreover, the no-till planting apparatus pulled behind a tractor accom-plishes several operations at once. A steel disc cuts a furrow through the mulch, drops in seed and fertilizer, and then closes the furrow (Fig. 7-31). The dead plants remain as a mulch (detritus cover), protecting the soil from erosion and reduc-ing evaporative water loss. Crop plants may grow as well or better than in a plowed and cultivated seedbed, presumably because of the better water conservation under the mulch (Fig. 7-32).

As the mulch decays, it provides a source of humus that supports the soil structure. Immedi-ately after harvest, or even slightly before, a new cover such as rye grass may be planted. Rye grass seeds scattered on top of the soil germinate rap-idly in the fall and provide a new vegetative cover that protects the soil over winter. The grass may be mowed for hay in the spring, the remainder killed with herbicides, and the new crop planted. Wastes from the harvested crop, such as leaves and stems of corn, also may be left in the field to add to the protective cover and to supply organic matter.

**FIGURE 7-32**
Crop plants planted by no-till methods grow as well as or better than on fields cultivated in the traditional way. (USDA–Soil Conservation Service photo.)

In short, no-till agriculture does not disturb the topsoil structure and thus enables a vegetative and/or detritus cover to remain at all times. As a result, erosion may be reduced to no more than what occurs in a natural ecosystem, and organic inputs balance the decomposition of humus. Thus, with no-till agriculture topsoil may be maintained indefinitely.

No-till agriculture has several advantages to the farmer in addition to minimizing soil erosion. First, the farmer accomplishes in one operation what formerly required at least three—plowing, discing, and planting. Thus, there is a tremen-dous savings in time (or labor expense) and en-ergy (tractor fuel). A second advantage is that the single pass with a tractor minimizes soil compac-tion. Third, a farmer practicing no-till is able to get his crop in at the optimal time in the spring; formerly he was often delayed by having to wait for the soil to become dry enough to plow. The early planting and the ability to plant a second crop immediately after harvesting the first may al-low double cropping or the growing of two suc-cessive crops on the same field in the same year. Formerly, there was only time for one. These fac-tors generally give no-till methods an economic advantage in addition to controlling erosion.

However, no-till agriculture is not without disadvantages. Most prominent is its reliance on chemicals to control weeds and pests. More is in-

volved than just the herbicides used to kill the initial vegetative cover. A number of serious insect pests and plant pathogens may overwinter in the old crop residues left on the surface. Also, certain pests, including slugs and mice, thrive in undisturbed soil. The traditional practice of plowing under crop residues effectively controlled such pest organisms. With no-till methods, further chemical treatment is required to control them. In all, no-till requires two to four times more pesticides than conventional farming techniques and the demand for chemicals tends to become even greater because the use of chemicals tends to result in the natural selection of strains with increasing resistance as described in Chapter 4. At the very least, the cost of chemicals and spraying equipment is eliminating the economic advantage of no-till. In addition, we must seriously question the ecological and human health side effects of using pesticides.

It is neither logical or intelligent to conclude that use of all chemicals is bad. Scientists are now well aware of the potential hazards such as those caused by DDT. Thus, chemicals are being developed that are effective but safer. However, it will probably never be possible to state with absolute certainty that a given herbicide will not have undesirable side effects in certain situations. Therefore, regular monitoring procedures to test safety, to make sure herbicides are kept to prescribed uses, and to check for undesired side effects should continue. In the meantime, farmers must aim for a sensible balance between the practices of no-till and traditional agriculture with soil conservation practices.

**Other crops.**  Another approach to decreasing soil erosion involves a reconsideration of the kinds of crops that are grown. In the course of history, humans have focused on developing and culturing a very small number of plants as the major food base of society. One of these is corn. Unfortunately, growing corn leads to more soil erosion than does almost any other crop because it covers the soil for only a short period of the year.

Scientists recognize that our dependence on corn is more a happenstance of history than a basic necessity. It may be possible to develop other kinds of plants that will produce equally abundantly. Under consideration are perennial species that regrow from their roots year after year. Shifting our agricultural base from annual to perennial species would provide enormous protection from soil erosion, to say nothing of additional savings of the labor and energy now spent in replanting seed each year.

### INORGANIC VERSUS ORGANIC FERTILIZER

A bushel of corn contains about 0.5 kg (1 lb) of nitrogen and lesser amounts of all the other mineral nutrients as well. Therefore, harvesting corn or any other crop amounts to the removal of these nutrients from the soil. If they are not replaced, the soil's nutrient supply will gradually be exhausted and plant growth will be limited accordingly. Such nutrient-depleted soils are frequently described as "worn out." In Colonial America, soils were frequently farmed until worn out, then abandoned. Much of the great westward migration occurred when farmers moved on to find new fertile land.

However, it was also recognized that soil fertility could be maintained by adding animal manure, and that growing and plowing in a legume crop such as clover or alfalfa every second or third year restored fertility. Recall that bacteria in the root nodules of legumes fix nitrogen from the air (Chapter 2) and nitrogen is frequently a limiting nutrient. This practice is called **crop rotation.** Importantly, both manure and crop rotation provide organic matter as well as nutrients, thus regenerating humus and maintaining topsoil structure.

Through the first half of the 1900s, agricultural practices were changed by the introduction of inorganic fertilizer. Inorganic fertilizer is produced by mining minerals that are rich in particular nutrients. For example, phosphate is mined as phosphate rock, large deposits of which are found in Florida. Nitrogen is an exception in that nitrogen compounds are produced from nitrogen gas in the air. With the ability to provide optimal levels of nutrients with inorganic fertilizer, farmers could make more money growing a crop such as corn every year; hence crop rotation was largely abandoned.

The result is predictable: in many cases there was no longer an adequate balance between organic inputs forming new humus and the oxidation of existing humus. Consequently, some agricultural soils have mineralized and thus lost nutrient- and water-holding capacity and other favorable characteristics even without erosion. Unfortunately, as mineralization leads to a collapse of soil structure, infiltration is decreased and this may intensify erosion which, of course, compounds the problem.

Noting the effect of mineralization stemming from the use of inorganic fertilizer has led some people to proclaim that inorganic or "chemical" fertilizers "poison" the soil. It is important to recognize that the fault is not in the inorganic fertilizer itself; it is in the failure to add sufficient organic matter to sustain the production of humus and to protect the soil from erosion. Indeed, exclusive use of organic material as fertilizer can have its problems, too. Many organic wastes simply do not contain enough of certain nutrients to support the needs of modern high-yielding crops. It is necessary to recognize the distinct roles of both organic matter and inorganic nutrients and to provide both.

In response to the problems of mineralization, some farmers are returning to growing legume cover crops particularly as the use of no-till methods allows double cropping. Also, treated sewage sludges are increasingly recognized as potential organic fertilizer. However, their use may be precluded in some cases because of contamination with toxic heavy metals. This will be discussed further in Chapter 10.

### HOME GARDENING

The agricultural principles we have discussed apply equally well to home gardening. Amateur gardeners often start with a conviction that they must "work" the soil. Rototilling, the home substitute for plowing, is followed by an arduous regime of periodic hoeing and raking to keep the soil "clean and loose." Can you see the effects of this? First the churning action of rototilling is highly destructive to soil structure. After one good rain, the soil settles into a highly compact mass. In addition, hoeing hastens evaporative water loss, oxidation of humus, and maximally exposes soil to the effects of erosion. I have seen ardent, amateur gardeners totally mineralize a good topsoil in the course of a single year with these procedures.

On the other hand, the concept of no-till agriculture can be applied with advantage. The use of chemical herbicides is not practical on a small

scale, nor should they be handled by amateurs. However, the generous use of an organic mulch will hold down weeds, maintain soil structure, and provide for regeneration of humus and soil structure. Keeping a mulch on the soil is also much less work than all the digging and cultivation.

A mulch of fresh grass clippings is most desirable in that it is also nutrient rich. However, mulches of dead leaves, woodchips, sawdust, or even shredded newspaper may be used. Such mulches are claimed by some to be undesirable because they "rob" the soil of nutrients. The "robbing" effect occurs because the latter mulches are virtually pure cellulose; they contain almost none of the necessary mineral nutrients. Fungi and bacteria that feed upon this material absorb other necessary nutrients from the soil, thus robbing it. Nevertheless, such materials still provide all the advantages of a litter cover and an addition of organic matter; they only need to be supplemented with additions of inorganic fertilizer.

### OVERGRAZING AND DESERTIFICATION

As noted in Chapter 1, there is a transition in ecosystems from forests to grasslands to deserts which is mainly determined by decreasing amounts of rainfall. The semiarid (low-rainfall) grasslands bordering deserts are too dry to support crops, but they have traditionally been used for grazing livestock. Unfortunately such lands are frequently subjected to **overgrazing.** That is, the grass is grazed unto death, leaving areas of barren soil exposed. If grazing were halted at this point, one might think that the system would quickly regenerate. Unfortunately, wind erosion in such areas is often severe and selectively removes humus, clay, and silt components and the nutrients bound to these particles; only coarser material with reduced water-holding capacity is left. Given the relatively low rainfall of such areas, this makes the difference between the land being able to support grass and its being able to support only drought-resistant desert species. Thus, as a result of erosion following overgrazing, semiarid grasslands are for all practical purposes converted into desert, a process called **desertification** (Fig 7-33).

The U.S. Agency for International Development has estimated that 650,000 km² (250,000 mi²) of once-productive grazing land along the south-

(a)

**FIGURE 7-33**
Desertification. Desertification is a two-stage process. (a) First, overgrazing removes the protective cover of grass. (USDA–Soil Conservation Service photos.) (b) Then erosion reduces the water holding and nutrient content of the soil by selectively removing humus and clay. Thus the potential for recovery and/or reclamation is drastically reduced, as the eroded soil can only support a few desert species as seen on this hillside originally overgrazed by sheep. (Yva Momatiuk/Photo Researchers.)

(b)

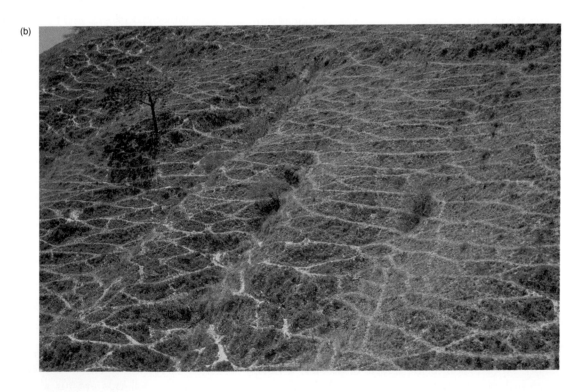

ern edge of the Sahara have become desert in the last 50 years. What is more disturbing is that overgrazing and consequent desertification is continuing in many regions of the world, including the United States (Fig. 7-34).

Why should humans fall into this tragedy? Actually, more is involved than just introducing excess numbers of animals and letting them graze the land into barrenness. Ranchers may think their herds are within the carrying capacity of the

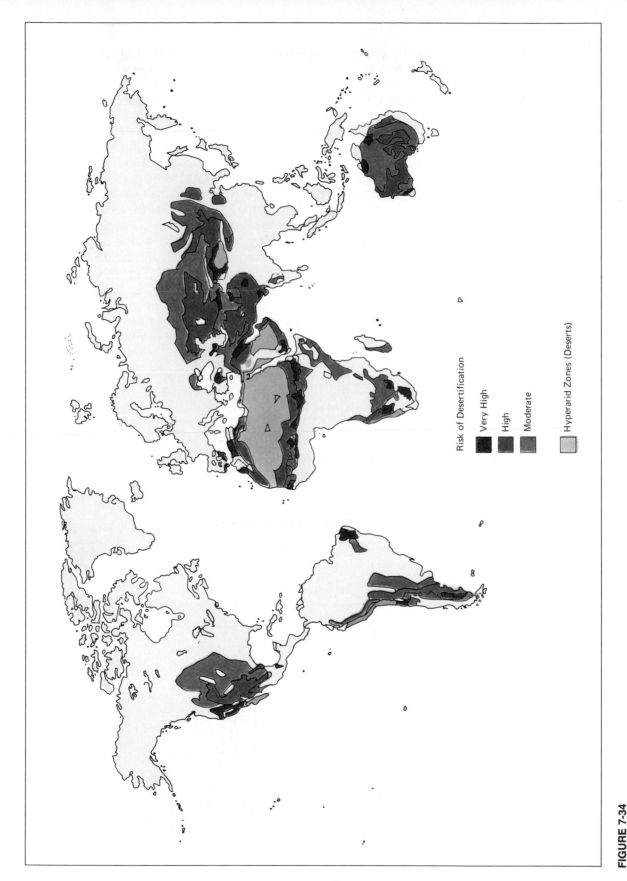

**FIGURE 7-34**
Deserts and areas subject to desertification. Throughout the world, overgrazing and/or deforestation is causing vast low-rainfall areas to degenerate into deserts. Reprinted with permission from "Desertification: Its Causes and Consequences," United Nations Conference on Desertification, Nairobi, 1977. Copyright 1977, Pergamon Books Ltd.

Risk of Desertification

Very High

High

Moderate

Hyperarid Zones (Deserts)

land, that is, within the number of animals that the land can support without a deterioration of the ecosystem. However, with intensive grazing there is less detritus to generate soil humus. In addition, the physical weight of the cattle compacts the soil. Thus, there may be a subtle decline of soil quality over many years that is not readily recognized. Then, because of gradual deterioration of soil, and perhaps an abnormally dry year, grass production drops below the level of what is being eaten. At this point continued grazing cuts directly into the regenerative capacity of the grass and creates a rapidly accelerating cycle—the less the grass grows, the more the animals feed on and destory the remaining grass. The land is grazed into desertlike barrenness with surprising and often catastrophic suddenness (Fig. 7-35).

Given the suddenness with which the final stage of desertification can occur, it is hardly en-

couraging that the U.S. Bureau of Land Management reported recently that of land under its care, 20 million hectares (50 million acres), an area equivalent to the size of Utah, was in "poor" or "bad" condition mostly because of overgrazing. Another 32 million hectares (80 million acres) was reported to be in only fair condition. Stricter monitoring and control of grazing is definitely needed.

In less-developed countries more than regulations will be required to halt overgrazing. Here overgrazing may be forced by the pressures of overpopulation (see Chapter 5). People struggling for an existence by grazing a few cattle on arid lands may have no alternative until desertification and starvation take their toll (Fig. 7-36).

Another factor contributing to desertification in many less developed countries is deforestation for firewood. Throughout developing nations, the poor often gather firewood or other burnable ma-

**FIGURE 7-35**
Overgrazing begins with a slow deterioration of rangeland, gradually reducing the production of grass. The end may come with extreme suddenness as production of grass fails to keep up with what is being eaten and cattle graze the land into barrenness.

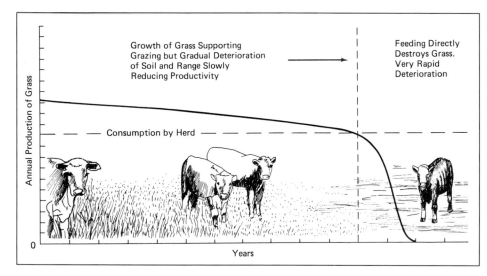

**FIGURE 7-36**
Overpopulation and desertification. People struggling for survival by grazing a few cattle on arid lands may have no alternative until desertification and starvation take their toll.
(FAO photo.)

terial such as animal dung for cooking. In many cases, areas for several miles around villages have been picked clean of all combustible material, and erosion and desertification are ensuing.

(a)

**FIGURE 7-37**
Erosion is extremely severe on forested slopes that have been clearcut. (a) Deforestation led to erosion seen here. (Photo by Victor Englebert/Photo Researchers.) (b) A stream filled with sediment eroded from clearcut slopes. (Photos by David Van de Mark; courtesy of Save-the-Redwoods League.)

## FORESTRY: CLEARCUTTING VERSUS SELECTIVE CUTTING

Exploitation of forests for wood products, mainly lumber and paper, runs a close second to agriculture in amount of land used, and increasingly forests are being treated and managed like other agricultural crops. Such management usually involves **clearcutting;** that is, all the trees are cut so that the land is clear. Then new trees may be planted in pure stands, fertilized, and sprayed to control pests. When the trees reach maturity the stand is again clearcut and the process is repeated. The practice of clearcutting has serious implications for the soil.

Forest soil generally has little vegetative cover other than trees because the limited amount of light beneath the trees prevents the growth of most other plants. There may be a natural litter cover, but this is generally mixed into the soil or pushed about irregularly by machinery in the process of cutting and removing logs. Therefore, when trees are removed, bare soil is exposed to erosion, which is frequently intensified by the fact that timbered slopes are often much steeper than those under cultivation (Fig. 7-37). There is no shortage of grim reminders as to where erosion from clearcut slopes can lead. The now barren rocky hills of Lebanon, for example, once supported huge stands of great cedar trees, the legendary "Cedars of Lebanon." Again, the problem goes beyond losing the soil. Many streams, rivers, and reservoirs are choked with the sediments (Fig. 7-37b).

(b)

Even without severe erosion, clearcutting may lead to a dramatic loss of soil nutrients. Investigators at Hubbard Brook Forest in New Hampshire, found that after clearcutting an experimental section, the loss of nitrates in runoff increases as much as 45-fold. The interpretation of this data is that when the trees were not present to actively reabsorb nutrients as they are released from decomposing detritus, the nutrients were susceptible to leaching.

The alternative to clearcutting is **selective cutting,** wherein only selected mature trees are periodically removed and the rest remain, thereby protecting the soil. Selective cutting is a more tedious, time-consuming, and expensive procedure. Thus, timber interests are usually vigorously opposed to and have successfully defeated legislation aimed at restricting clearcutting. This debate will probably continue for some years; however, if the interest is in long-term preservation of the soil as opposed to short-term economic gain, the choice is clear.

## IRRIGATION AND SALINIZATION

Given the availability of nutrients through fertilizer, water is the most common limiting factor in agricultural production. Thus, tremendous gains have been achieved through **irrigation,** that is, artificially applying additional water. Irrigated land in the United States now totals about 60 million acres. Traditionally water has been diverted from rivers through irrigation canals (Fig. 7-38) and flooded through furrows in fields (Fig. 7-39). In recent years, **center pivot irrigation** has become much more popular. Here water is pumped from a well in the center through a gigantic sprinkler system that slowly pivots itself around the well (see Fig. 8-18).

In either case irrigation may lead to an intolerable increase in soil salinity (saltiness), a problem called **salinization.** This occurs because irrigation water invariably contains at least trace amounts of salts dissolved from the earth. When irrigation water escapes by evaporation or by transpiration, only water molecules leave; salts in solution remain behind. This can be easily demonstrated by pouring salty water on some soil in a bowl. As the soil dries you will observe the formation of salt crystals on the surface of the soil. As salt accumulates, the soil becomes less and less able to support plants because of the osmotic pull of water exerted by the salt. It is estimated that 25 to 30 percent of the irrigated croplands in the United States already have excessive salinity. The problem is especially acute in the lower Colorado River Basin area and in the San Joaquin Valley of California where some 160,000 ha (400,000 acres) have been rendered nonproductive, representing an economic loss of more than $30 million per year (Fig. 7-40).

**FIGURE 7-38**
Irrigation channel. Such channels are constructed to carry water from where it is plentiful to where it is needed for irrigation. (Gary M. Handsher/Photo Researchers.)

**FIGURE 7-39**
Flood irrigation. A traditional method of irrigation is to periodically flood fields with water. (USDA photo.)

**FIGURE 7-40**
Many acres of irrigated land are now worthless because of the accumulation of salts left behind as the water evaporates, a phenomenon known as *salinization*. (Photo courtesy of Texas Agr. Exp. Sta., El Paso.)

Salinization can be avoided, or even reversed, if sufficient water is applied to leach the salts down through the soil. However, this is only possible where there is suitable natural drainage or where artificial drainage is installed to allow the salty water to escape (Fig. 7-41). Installing artificial drainage is very expensive, but without drainage the field can become a salty quagmire.

In another approach to the problem of salinization, plant breeders are attempting to develop varieties of crop plants that are more salt tolerant. The potential of this approach is demonstrated by the fact that countless marine and seashore species are adapted to such conditions. If similar adaptations can be bred into commercial species, then perhaps salinized areas can again become somewhat productive.

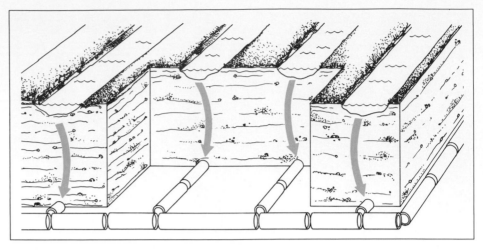

**FIGURE 7-41**
Drainage of irrigated land. To prevent salinization, excess water with salts must be drained away. This may necessitate a system of tiles as shown here and additional environmental problems may be caused by the discharge of such high-mineral-content drainage into natural waterways.

An even more pronounced constraint on irrigation in the future, however, is likely to be lack of sufficient water. Under the demands for irrigation, water resources in many areas are being depleted. This will be discussed in more detail in Chapter 8.

### PESTICIDES

Application of various insecticides, herbicides (weed killers), and other pesticides is a routine practice in the production of many crops. They are also widely used in forests. Frequently, however, less than 1 percent of the amount applied reaches the target pest and much of the remainder enters the soil as toxic pollutants. Recall that the soil environment houses a highly complex ecosystem with countless organisms that are instrumental in the formation and maintenance of topsoil. Many soil organisms have mutual relationships with plants; recall, for example, mycorrhizae and nitrogen-fixing bacteria. Various studies show that pesticides have marked effects on the soil biota. Unfortunately, soil ecosystems have not been sufficiently studied for us to understand what the long-term implications of such upsets may be. The absence of such knowledge, however, hardly provides confidence. Alternatives to the use of chemical pesticides are discussed in Chapter 17.

### Regenerating Soil and Artificial Soil

Soil formation occurs in weathered and eroded soils, but it is a very slow process and continuing erosion may erase any progress that is made. It is not impossible, however, to regenerate even very badly degraded soils by digging ample quantities of organic matter, liming to adjust the pH, and applying mulches to stabilize the surface against erosion. Given this help, natural processes will work in concert to develop a good topsoil. Many people have made it a labor of love to buy an old farm with badly eroded soil and gradually bring it back to a productive state. Israel has achieved spectacular success in bringing the desert, which in that region is really a product of desertification, back to agriculture.

In another approach, plants can be grown without soil, a technique known as **hydroponics.** Plants are rooted in a gravel or sand bed and irrigated more or less continually with a circulating nutrient solution (Fig. 7-42). Plants cultured hydroponically do exceedingly well and tests indicate that they are just as nutritious as those grown on good soil.

Why worry about soil degradation then? The problem becomes one of economics. In order for people to have enough to eat, food must be more than available; it must also be affordable. The la-

214 Part III: Soil and Water

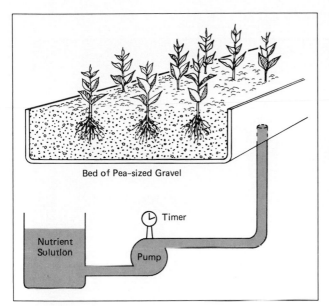

Bed of Pea-sized Gravel

Timer

Nutrient Solution

Pump

**FIGURE 7-42**
Hydroponics, growing plants without soil. One method of hydroponic culture is to grow plants in a bed of gravel. A solution containing all the necessary mineral nutrients is pumped up into the gravel every eight hours and allowed to drain back to the storage tank. The gravel periodically wetted with nutrient solution provides an environment of water nutrients and air that is ideal for roots. Costs, however, preclude hydroponics being used for general food production.

bor and material inputs entailed in soil restoration or hydroponics are such that food produced must command a very high price to justify the costs. For example, to support growing wheat hydroponically would require bread to sell for about $10 a loaf. (This is why I mentioned above that restoring eroded soil is a labor of love; the economic payoff for the labor involved will be minimal.)

The only way that we can have food and other agricultural or forestry products at a price most of the population can afford is to preserve the integrity of the natural system, which provides the service of soil maintenance free of charge.

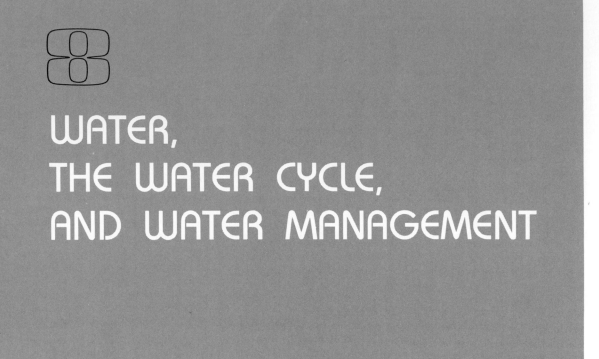

# WATER,
# THE WATER CYCLE,
# AND WATER MANAGEMENT

4. Discuss the five general methods of water purification. What is *disinfection?* What is the most common disinfection technique? Differentiate between disinfection and purification.

5. Describe the basic hydrological cycle.

6. From where do water molecules evaporate? Why are evaporation and/or transpiration the purification stages of the water cycle? What is *relative humidity?*

7. Compare air movements in desert and equatorial regions.

8. What determines the *infiltration runoff ration?* Why is it important? What is the role of *capillary water?* What happens to infiltrated water after the soil is at field capacity? What are *aquifers?* Why is groundwater generally considered pure? How may leaching affect groundwater?

9. Discuss the differences between *consumptive* and *nonconsumptive* uses of water.

10. What factors of current use affect water supplies? Why can't we depend on groundwater supplies when surface water supplies decline?

11. What effect will water depletion have on the Great Plains states?

12. What is *land subsidence?* At what rates does it generally occur?

13. Why and where does *saltwater intrusion* occur? What are its effects?

14. How does water depletion affect natural waterways in general and estuaries in particular?

15. How much did water consumption in the United States increase between 1950 and 1975? Discuss some of the solutions that have been proposed to increase supplies.

16. What water conservation methods are being used?

17. What effects have developments had on streams? What are the effects of unnatural flooding?

18. What are some of the techniques employed in *stormwater management?* Why must the total watershed be considered in stormwater management?

Water is one of the most vital of all resources both for humans and for natural ecosystems. But two problems increasingly plague water supplies. First is the problem of *quantity*. Supplies are being depleted and shortages are becoming critical in some areas because of the increasing demands of expanding populations, industry, and agriculture. The second problem is one of *quality*. Is water sufficiently pure? Too many cases exist where pollution is degrading water quality so much that natural ecosystems may be destroyed and human health endangered.

Our objective in this chapter is to gain a better understanding of water, the water cycle, and human impacts on the water cycle.

## WATER: SOLUTIONS, POLLUTION, AND PURIFICATION

Before we can fully appreciate problems of water quantity and quality, we must understand certain physical and chemical aspects of water itself.

### Physical States of Water

Water may exist in any of three distinct physical states with which you are familiar: solid (ice), liquid (water itself), or gaseous (water vapor). These three states are the result of different degrees of interaction between water molecules. Two forces are involved: one is a weak attraction between water molecules called **hydrogen bonding.** Hydrogen bonding tends to hold water molecules together. The second is **kinetic energy,** the energy of vibrational movement which is inherent to all atoms and molecules. This kinetic energy tends to separate molecules from one another. Consequently, the degree to which molecules are held together, and hence the physical state of water, depends on the relative balance between hydrogen bonding and the kinetic energy of the molecules (Fig. 8-1). Hydrogen bonding is a constant attraction; kinetic energy, however, varies with temperature: the warmer the temperature, the greater the kinetic energy.

Below freezing, 0° C (32° F), the kinetic energy of water molecules is small compared to the attraction of hydrogen bonding. Thus, the hydrogen bonding holds the molecules in a rigid position with respect to one another. This gives water

its solid state, ice. As temperature increases above freezing, the increasing kinetic energy of the molecules literally shakes the structure apart and the result is melting. However, as a hydrogen bond breaks at one point it re-forms at another. Thus, molecules "slide" about one another but basically remain held together. This results in the liquid state of water.

Finally, at the boiling point, 100° C (212° F), molecules gain sufficient kinetic energy to break the hydrogen bonds altogether and they go off into the air as free (unattached) water molecules, a process we observe as **evaporation.** We speak of water molecules in the air as water vapor, and the amount of water vapor in the air is commonly measured as **humidity.**

All these processes are reversible. **Condensation** is basically a reversal of evaporation. As temperature and corresponding kinetic energy decrease, water molecules in the vapor state may be held together by hydrogen bonding when they come in contact. If temperature is low enough, water vapor may go directly from the vapor state to the solid or crystalline state as in the formation of frost in nature or in a freezer (see Fig. 2-2).

These changes in the physical state of water are the basis of the water cycle of the earth. Water is constantly entering the atmosphere by evaporation and then returning to earth through condensation and precipitation. Before discussing the water cycle in detail, however, we should first understand how water carries materials in solution and how it is purified.

### Solutions, Suspension, and Pollution

When we add substances such as sugar or salt to water they seemingly "disappear"; however, we can tell by taste or perhaps color that they are still present. Thus, we speak of such substances not as disappearing but as dissolving or going into **solution.** What really happens is that water molecules are attached around individual molecules or ions of other substances such that they are "floated" away among water molecules. Substances differ as to how readily they dissolve in water. Some are very soluble—that is, relatively large quantities may dissolve and be carried in solution. Other substances are relatively insoluble—water will carry very little in solution.

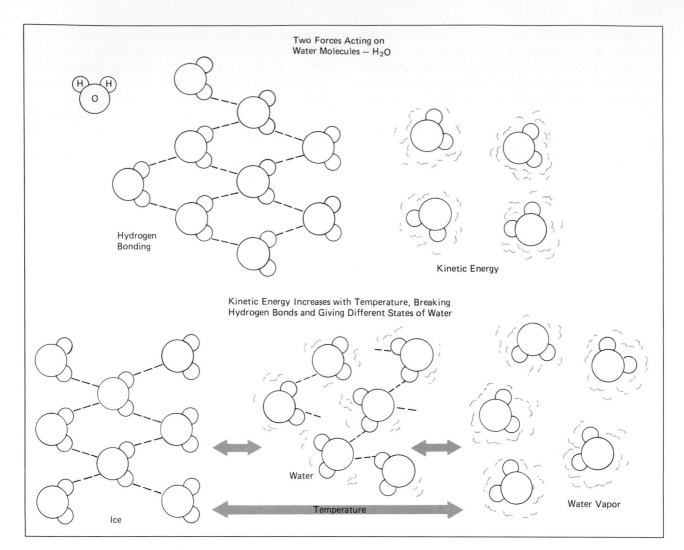

Two Forces Acting on
Water Molecules — $H_2O$

Hydrogen
Bonding

Kinetic Energy

Kinetic Energy Increases with Temperature, Breaking
Hydrogen Bonds and Giving Different States of Water

Ice

Water

Temperature

Water Vapor

**FIGURE 8-1**
Physical states of water. Water molecules are affected by two
forces: hydrogen bonding, which tends to hold them together,
and kinetic energy, which tends to break them apart. As kinetic
energy, which increases with temperature, overcomes
hydrogen bonding, different states of water result.

When much larger particles such as clay particles, bacterial and algal cells, and particles of detritus are carried in water, we speak of them as
being in water **suspension** as opposed to solution.
The term solution implies that the material is dispersed as individual ions or molecules of the substance. (Note that a bacterial cell, for example
while microscopic in size still contains in the order of trillions of atoms.)

Terms used to describe water such as good
quality, poor quality, fresh, brackish, and salty refer to the relative presence or absence of materials

in solution or suspension. These and other terms
used to describe water are defined in Table 8-1.

Many substances in solution may be beneficial rather than harmful. For example, aquatic life
depends on adequate amounts of carbon dioxide,
oxygen, and mineral nutrients in solution. It is
also desirable that drinking water contain some
minerals. Indeed much of the water that is drunk
by people in the United States is in the form of
sodas which are laden with sugar, carbon dioxide
for the fizzy effect, and various flavorings, all in
solution.

In contrast, *polluted water* is water that carries
one or more substances in solution or suspension
that make it undesirable or unfit for a given use.
Excessive heat can also be harmful and is known

**Table 8-1**

| Terms Commonly Used to Describe Water | |
|---|---|
| *TERM* | *DEFINITION* |
| Water Quantity | The amount of water that is available to meet desired demands. |
| Water Quality | The degree to which water is pure enough to meet desired uses. |
| Fresh Water | Water with a salt concentration of less than 0.01 percent. As a result of purification by evaporation, all forms of precipitation are fresh water, as are lakes, rivers, groundwater, and other bodies of water which have a through-flow of water from precipitation. |
| Salt Water | Water, typical of oceans and seas, that contains at least 3 percent salt (30 parts salt per 1000 parts water). |
| Brackish Water | A mixture of fresh and salt water, typically found where rivers enter the ocean. |
| Hard Water | Water that contains minerals, especially calcium and/or magnesium, that cause soap to precipitate, producing a scum, curd, or scale in boilers. |
| Soft Water | Water that is relatively free of those minerals that cause soap to precipitate causing scale buildup in boilers. |
| Polluted Water | Water that contains any one or a combination of impurities that make it unsuitable for a desired use. |
| Purified Water | Water that has had pollutants removed or rendered harmless. |

as thermal pollution. Importantly, **pollution** is relative to the desired use of the water or the type of aquatic ecosystem it supports. For example, normal seawater contains about 30 parts salt per 1000 parts water. This concentration of salt is intolerable for drinking and for freshwater ecosystems. In such situations, the salt would be considered a pollutant.

## Purification

Because materials in solution or suspension are simply carried among water molecules, the water molecules themselves are not affected in any way. Thus, water purification may be achieved by removing or separating the undesirable materials from the water. When this is done, even the most polluted water may become pure water of the highest quality. There are four general methods for removing materials from solution or suspension: (1) evaporation and recondensation or distillation, (2) filtration, (3) chemical adsorption, and (4) biological oxidation.

### EVAPORATION AND RECONDENSATION

When water evaporates only water molecules leave the surface; salts and other materials in solution remain behind (except in the case of dissolved gases or volatile liquids such as alcohol). When the water vapor is recondensed it is pure water. Chemically pure water for use in laboratories is generally obtained by this method (Fig. 8-2). This process is called **distillation.** The same process takes place on a gigantic scale between the earth and the atmosphere. As noted previously, water evaporates from the surface and recondenses in the atmosphere; this is the basic source of all fresh water on earth.

### FILTRATION

The principle of **filtration** is the passage of water through a screen; particles larger than the holes in the screen are thus removed. Filtration has limits in terms of water purification; it involves the relationship between the size of particles to be removed and the size of the holes in the filter (Fig. 8-3). Common filters or material such as sand or soil which may act as natural filters only remove relatively large particles such as algal and bacterial cells in suspension. Their pore size is much too large to remove materials in solution, which are dispersed as individual ions or molecules. With modern technology, filters with a pore size sufficiently small to strain out dissolved materials have been produced. But such filters also impede the passage of water and, hence, require very high pressures. This makes the method very expensive. Finally, filters of all dimensions become clogged with the material being removed and must be changed or cleaned, adding to their cost.

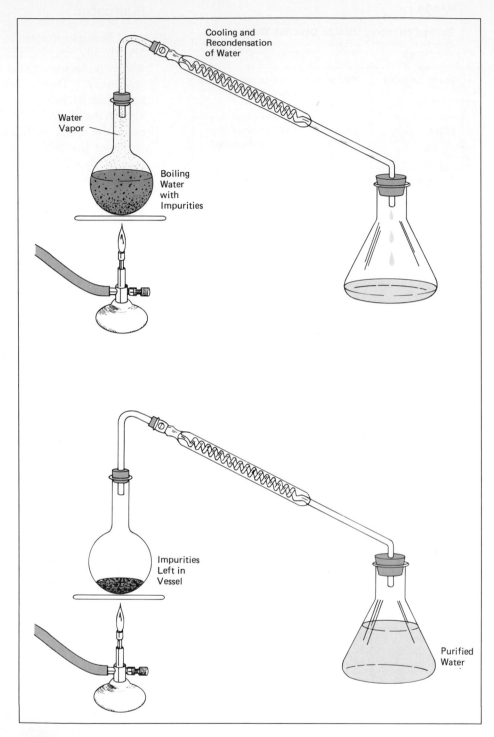

**FIGURE 8-2**
Purification of water by distillation. When water evaporates,
only water molecules leave the surface; impurities remain
behind. Hence, water may be purified by distillation, a process
of boiling the water and recondensing the vapor.

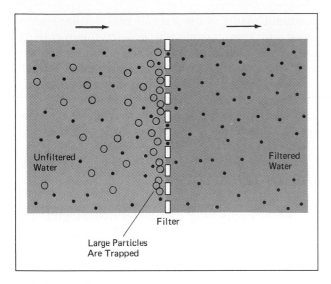

**FIGURE 8-3**
Filtration. Filtration will only remove particles that are larger than the pore size of the filter. Materials in solution are not removed by usual filters.

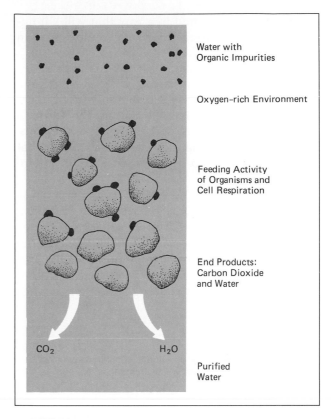

**FIGURE 8-4**
Biological oxidation. Many organic impurities may be removed from water by utilizing the feeding activity and cell respiration of bacteria and other organisms, a process known as *biological oxidation.* This method will not remove nonbiodegradable organics or inorganics, and such systems may be inhibited by toxics that affect organisms.

## CHEMICAL ADSORPTION

Certain materials such as activated carbon actually bind and hold various other molecules. We refer to this as **adsorption,** the noncarbon molecules being adsorbed to the carbon. If water (or air) carrying certain foreign molecules is passed through activated carbon, the foreign molecules may be adsorbed, held, and thus removed. Devices containing activated carbon or other such materials are commonly referred to as *chemical filters.* In nature, clay and particularly humus have considerable adsorptive capacity. In Chapter 7 we spoke of this as the nutrient-holding capacity of these materials. Importantly, chemical adsorption differs from the filtration described above in that it does not depend entirely on pore size; it depends on a chemical binding. Thus materials in solution may be removed by suitable chemical filters without greatly impeding the passage of water. Again, chemical filters are expensive and must be changed frequently.

## BIOLOGICAL OXIDATION

In **biological oxidation** natural organic material in water may be fed upon and thus broken down through a series of trophic levels in the aquatic ecosystem (Fig. 8-4). This is the "self-purification" process of natural waterways. This method is set up artificially for sewage treatment, which will be described in Chapter 10.

Since this method involves the feeding of organisms, it can only break down and remove natural organic materials. It will not affect inorganic compounds or synthetic organic compounds which are nonbiodegradable. It also requires the maintenance of an environment such that organisms can thrive. Any factor that adversely affects the organisms will reduce or negate the process.

## DISINFECTION VERSUS PURIFICATION

The potential presence of disease-causing microorganisms is a pollution hazard of particular significance. In order to eliminate this hazard, water is commonly disinfected. **Disinfection** means that the water is treated in a way that will kill microorganisms; nothing is actually removed as in purification.

Disinfection is used because it is both an inexpensive and a positive way of eliminating the disease threat. The most commonly used technique of disinfection is **chlorination,** the addition of chlorine gas ($Cl_2$) or some other chlorine compound to water. Quantities are adjusted so that they are harmless to humans yet still kill microorganisms.

Unfortunately, disinfection, particularly chlorination is commonly referred to as "purification." The distinction between the two terms should be clarified. Far from removing anything, chlorination adds a toxic compound. This has further environmental implications as heavily chlorinated water from swimming pools or sewage

treatment plants is discharged into natural waterways and may negatively impact the receiving body. Also, chlorine atoms may bond with organic compounds, producing **chlorinated hydrocarbons,** which tend to be long-lived compounds with many adverse biological effects.

With these principles in mind, let us now turn our attention to the water cycle of the earth.

## THE WATER CYCLE

The **water cycle,** also called the **hydrological cycle,** is represented in Figure 8-5. Basically, the cycle consists of an alternation of evaporation and condensation. Water molecules enter the air by evaporation and transpiration. In the atmosphere, the water molecules condense and return to earth as precipitation. The water drains over or through the ground, and eventually reevaporates, thus repeating the cycle. Let us examine this cycle in more detail.

**FIGURE 8-5**
The water cycle. The water on earth is continuously recycled through evaporation or transpiration, condensation, and precipitation.

## Water into the Atmosphere

Since oceans cover about 70 percent of the earth's surface, it is not surprising that the largest amount of evaporation occurs from the ocean surfaces. However, water molecules also evaporate from lakes, rivers, moist soil, and other wet surfaces.

Over vegetated land, large amounts of water enter the atmosphere by transpiration from plants, the process by which water evaporates from moist cell surfaces in the leaf and diffuses out through the stomas. The combination of both evaporation and transpiration is called **evapotranspiration.**

Evaporation and/or transpiration are the main purification stages of the water cycle because only water molecules leave the surface; ions and molecules that may be in solution remain behind, as in distillation. We have already noted in Chapter 7 how salt may be deposited on soil as irrigation water evaporates. Natural salt deposits such as the salt flats surrounding Great Salt Lake in Utah and other lakes that have no outlet are also the result of this process. Indeed, the ocean itself remains salty for the same reason.

The amount of water vapor that air can hold is dependent on temperature: the warmer the air, the more water vapor it can hold; the cooler the air, the less water vapor (Fig. 8-6). **Relative humidity** is a measure of how much water vapor is in the air compared to the maximum amount the air can hold at that particular temperature. For example, a relative humidity of 50 percent means that the air has half as much vapor as it can hold at that temperature. When the relative humidity is low, the air is said to be dry and evaporation is rapid. When the relative humidity is high, the air is said to be moist and evaporation is slow.

Since the amount of moisture that air can hold changes with temperature, relative humidity also changes with temperature. Given a constant amount of water vapor in air, the relative humidity will drop as the air is warmed. This is why cold winter air, warmed to room temperature, is very dry. Conversely, relative humidity increases as air is cooled. Suppose that high-humidity air is cooled so that the relative humidity reaches 100 percent, then still further cooling occurs. The air now cannot hold all the water vapor it contains, so the excess water vapor condenses into tiny droplets, creating fog, mist, or clouds. If the temperature increases again, the droplets may evaporate, resulting in the fog "clearing." However, if further cooling occurs, condensation leads to precipitation (Fig. 8-7).

**FIGURE 8-6**
Capacity of air to hold water vapor. The amount of water vapor air can hold varies with temperature. Graph is for standard pressure—1000 millibars.

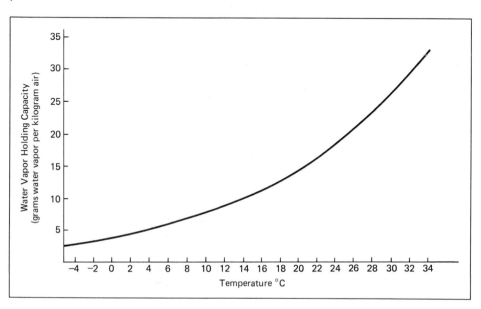

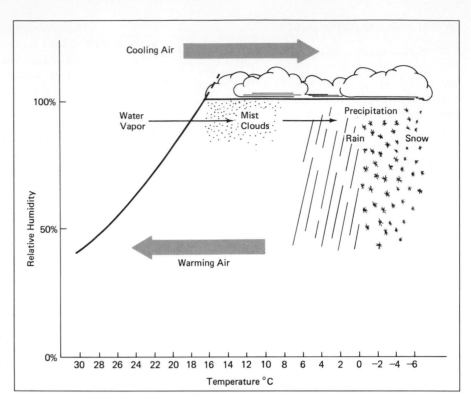

**FIGURE 8-7**
Given a certain amount of water in the air, relative humidity increases as temperature drops. When air cools below the point of saturation (100%), condensation and precipitation result. When air warms, there is a drop in relative humidity.

## Water out of the Atmosphere: Precipitation

The term **precipitation** is used to refer to any form of moisture, such as rain, snow, sleet, and hail, that falls from the sky. Frequently temperatures in clouds are such that ice crystals form and grow until they are heavy enough to fall. If they melt on the way down, the result is rain; if not, the result is snow. In addition, when moist air comes in contact with a cold surface, water vapor may condense directly on the cold surface. Thus, we observe the formation of dew or the "sweating" of a cold glass. If the surface is below freezing, we observe the formation of frost. Importantly, condensation and precipitation are the source of all fresh water on earth since impurities were left behind in evaporation.

We noted in Chapter 1 that the amount of precipitation is a primary factor in determining the type of ecosystem that an area can support. The distribution of precipitation over the earth, which ranges from near zero in some areas to more than 3 m (120 in.) per year in other areas, is basically dependent upon patterns of heating and cooling of the earth's atmosphere. The high rain-fall observed in equatorial regions comes about as follows: Solar heating produces currents of rising air over the equatorial regions—air expands with increasing temperature, hence it becomes less dense and rises. The rising air cools because of the further expansion allowed by the lower pressure at higher altitudes and the loss of heat into outer space. Finally, the cooling air cannot contain its water vapor, and rainfall results.

On the other hand, subtropical regions (25° to 35° north and south of the equator) are typified by deserts. The Sahara of Africa is the prime example. Such regions have low rainfall because the air has risen and dropped its moisture over equatorial regions before it descends over subtropical regions (Fig. 8-8). As this air descends, it warms and its relative humidity drops. Far from producing rain, these descending air currents produce an extreme drying effect on the land below. Such air (or water) currents, which result from unequal heating, are called **convection currents.** Figure 8-9 shows other convection currents to the north and south.

Convection currents are influenced by the rotation of the earth so that in the completion of the convection loop there is a general west-to-east movement of air in both northern and southern temperate regions, and an opposite movement in tropical regions. These prevailing winds blowing across mountain ranges produce even more striking contrasts between high- and low-rainfall regions. As moisture-laden air encounters a mountain range, it is deflected upward. The rising air cools and moisture is precipitated on the windward slopes, (those facing toward the wind). As the wind crosses the range and descends on the other side, it becomes warmer and increases its capacity to pick up moisture. Hence, deserts occur on the leeward sides (those facing away from the wind) of mountain ranges. The dry area to the leeward side of mountains is referred to as a **rain shadow** (Fig. 8-10). The most severe deserts in the world are caused by the rain shadow effect. For example, the westerly winds, full of moisture from the Pacific Ocean, strike the Sierra Nevada

**FIGURE 8-8**
Equatorial tropical rainforests and subequatorial deserts. Solar radiation causes maximum heating in equatorial regions and produces rising currents of moist air. As the moist air cools, there is heavy precipitation over the equatorial regions supporting tropical forests. The air then descends over subequatorial regions. As it descends, it becomes warmer and drier, resulting in subequatorial deserts.

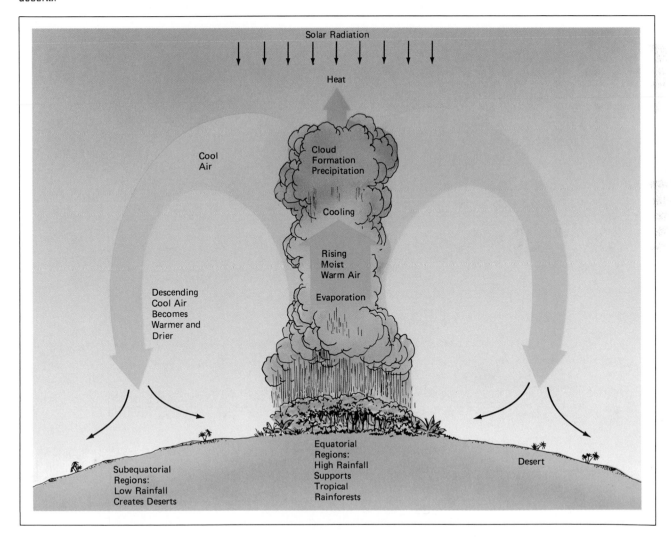

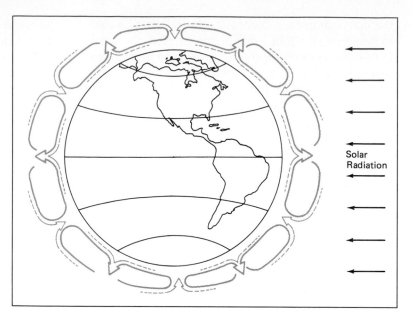

**FIGURE 8-9**
Idealized circulation of the earth's atmosphere. Solar heating of the atmosphere causes air to rise over equatorial regions, setting in motion other successive convection currents. (Adapted from L. Battan, *The Weather,* p. 38. Englewood Cliffs, N.J.: Prentice-Hall, Inc., 1974. Reprinted by permission of Prentice-Hall, Inc.)

**FIGURE 8-10**
Rain shadow. Moisture-laden air cools as it rises over a mountain range, resulting in high precipitation on the windward slopes. Desert conditions arise on the leeward side as the descending air warms and tends to evaporate water from the soil. (Redrawn from *Trees: The Yearbook of Agriculture, 1949.* Washington, D.C.: USDA.)

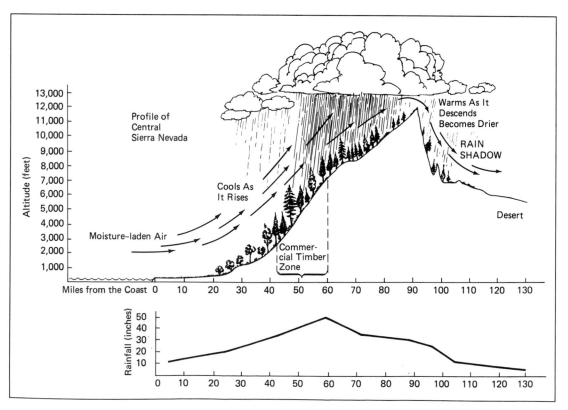

mountain range in California. As the winds rise over the mountains, large amounts of water precipitate out, supporting the lush forests on the western slopes. Immediately east of the Sierra Nevada range is the desert, Death Valley, produced by the rain shadow. Further north in Oregon and Washington, a similar effect is produced by the Cascade ranges.

## Water over and through the Soil

As precipitation hits the ground there is a fork in the pathway of the water cycle. Water from rainfall or snowmelt may infiltrate or soak into the ground, or it may run off the surface. Thus we speak of **infiltration** and **runoff.**

Runoff flows directly over the surface into streams, which in turn make their way to rivers, lakes, and eventually the ocean. The land area that drains into a particular waterway is called the **watershed** of that waterway. The water of all lakes and rivers and other open bodies is referred to as **surface water** as opposed to **groundwater,** which is found below ground.

The relationship between the amount of infiltration and the amount of runoff is like the two ends of a seesaw: when one is high the other must be low. This is commonly expressed as the **infiltration-runoff ratio.** The infiltration-runoff ratio is influenced by the character of the surface, the nature of the soil, the slope, the rapidity with which rain falls (or snow melts) and the total amount of precipitation as shown in Figure 8-11.

The infiltration-runoff ratio is of great significance because if runoff is excessive it causes several problems: (1) it is responsible for erosion; (2) it is responsible for a large portion of pollution as it picks up and carries all sorts of dirt, grime, and chemicals from the surface into streams and rivers; and (3) it is responsible for flooding as large volumes of water flow directly into streams and rivers. On the other hand, infiltration is of much value: it nourishes plants and recharges the reservoir of groundwater.

As water infiltrates into soil there is another fork in the pathway. Some water is held between soil particles by capillary action as water is held in sponge. This water, referred to as **capillary water,** is available to plants between rains as discussed in Chapter 7. Capillary water completes the water cycle by being absorbed and given off by plants

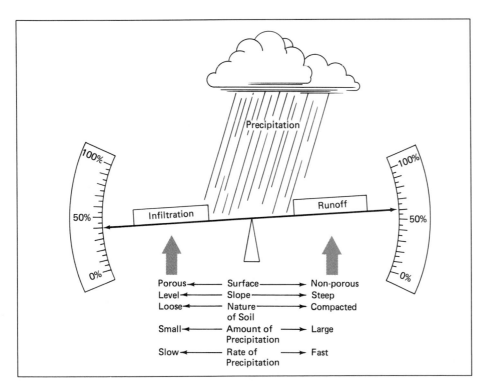

**FIGURE 8-11**
Infiltration-runoff ratio. Precipitation must either infiltrate or run off the surface. Several factors, as listed, influence which direction the water goes.

in transpiration, or some of it may evaporate directly from the soil surface.

When the soil absorbs all the capillary water it can hold, it is said to be at **field capacity.** Additional infiltrated water percolates down through the larger cracks, pores, and spaces between soil particles under the influence of gravity and is hence called **gravitational water.** Gravitational water moves down beyond the reach of plants until it comes to a layer of rock or dense clay, which it cannot penetrate. Water then accumulates above such an impervious layer, completely filling all the cracks, pores, and spaces. This accumulated water is called groundwater; its upper surface is the **water table** (Fig. 8-12). Wells are dug or drilled below the water table; groundwater seeping into the well provides a source of water.

Underground rock layers frequently slope. Consequently, groundwater proceeds to percolate more or less horizontally through various layers. The layers of porous material through which water moves are called **aquifers.** The actual location of aquifers is complex. Layers of porous rock are often found between layers of impervious material and the entire formation may be folded or fractured in various ways. Thus groundwater may be found at various depths between layers of impervious rock. Also, it should be noted that the **recharge area,** the area where water actually enters an aquifer, may be many miles from where it is withdrawn. If the recharge area is at a higher elevation it may result in high pressure, which forces the movement of water through the aquifer. An aquifer with water under pressure is called an **artesian aquifer** (Fig. 8-13).

In nature, groundwater moves through aquifers until it finds some opening to the surface. We observe such natural exits of groundwater as springs or seeps. A **seep** is where water seeps out over a relatively wide area. In turn, springs and seeps feed lakes, streams, and rivers, which make their way to the ocean or occasionally end in evaporation lakes such as the Great Salt Lake.

As water percolates through soil and porous rock, microorganisms and other suspended impurities are filtered out. Thus free of disease-causing organisms, groundwater is generally of such quality that it is used without treatment for drinking purposes. New wells are tested by the county health department to be sure they are free of bacterial contamination and old wells may be retested if there is any question about their purity.

**FIGURE 8-12**
Pathways of infiltrated water. Water that infiltrates the soil and is held against further movement is called *capillary water.* When soil reaches its capacity of capillary water (field capacity), additional infiltrated water percolates downward under the pull of gravity and is called *gravitational water.* Water that accumulates above an impervious layer, saturating all the pore spaces in the soil, is called *groundwater.* The upper surface of the groundwater is called the *water table.* Groundwater is withdrawn by drilling wells to below the water table.

However, the picture is very different with respect to dissolved materials, which, you may recall, are not removed by simple filtering. Indeed, as water percolates through the ground it may dissolve and carry along in solution any number of compounds. This process is called **leaching.**

Minerals leached from the earth impart various characteristics to the water. Some combinations of minerals in water give a particularly good taste and such water is highly valued for drinking. Sulfur springs exist where groundwater has

**FIGURE 8-13**
Artesian aquifer. Aquifers may lie between two impermeable layers and the recharge area may be at a higher elevation, so that water in the aquifer is under pressure from the higher head in the recharge area. Such an aquifer with water under pressure is an artesian aquifer.

leached sulfur-containing minerals, and the poison springs of the Southwest occur because the groundwater has leached arsenic or other poisonous minerals. Limestone (calcium carbonate) is particularly subject to leaching, and such leaching has produced the huge limestone caves that exist in many parts of the United States. Stalactites and stalagmites are formed by the leaching and redeposition of limestone within the cavern (Fig. 8-14).

**FIGURE 8-14**
Limestone caverns such as this are the result of groundwater leaching away limestone. Formations develop as limestone from seeping groundwater recrystallizes. (National Park Service photo by Fred E. Maug, Jr.)

## Summary of the Water Cycle

In summary, the water cycle always consists of evaporation, condensation, and precipitation, but in completing the cycle there are three principal "loops": (1) the *surface runoff loop,* in which water runs off the surface and becomes part of the surface water system; (2) the *evaporation-transpiration loop,* in which water enters the soil and is held as capillary water and then returns to the atmosphere by way of evaporation from soil or through absorption by plants and transpiration; and (3) the *groundwater loop,* in which water enters and moves through the earth, finally exiting through springs, seeps, or wells, thus rejoining the surface water system.

It should be evident that all the water we use must come from this cycle at one point or another and all the waste water we discharge must go into it. Can you begin to visualize how human activities affect the water cycle and its ability to sustain modern civilization? Specific problems and solutions will be addressed in following sections and chapters. However, points of major concern are listed in Figure 8-15 so that you can begin to appreciate their number and diversity.

## WATER MANAGEMENT

Humans have traditionally taken whatever amounts of water they needed (or could get) from ground and surface waters and carried on other endeavors with little thought toward impacts on the water cycle. Indeed, as long as human populations were relatively small, impacts were relatively minor. However, as population grows and

**FIGURE 8-15**
Human impacts on the water cycle. (1) Groundwater depletion. (2) Air pollution adds impurities to precipitation. (3) Surface runoff carries pollutants into waterways. (4) Wastewater discharges add pollutants. (5) Leaching from landfills and leaking storage tanks pollutes groundwater. (6) Hard surfacing increases runoff and causes downstream flooding. (7) Groundwater depletion causes drying up of springs, land subsidence, and saltwater intrusion.

demands for water increase, we are finding increasingly that traditional supplies are inadequate to meet desires, and overuse of supplies is having a number of adverse impacts. Clearly, a sustainable future will depend on our exercising better water conservation and management practices. We shall investigate this further in the following sections.

## Overall Patterns of Water Use

### CONSUMPTIVE VERSUS NONCONSUMPTIVE USES

A person need only consume about 2.3 ℓ (1 ℓ = 1.05 qt) of water per day to survive. However, in the United States, agriculture consumes another 2700 ℓ (700 gal) per person per day, primarily for irrigation; electrical power production uses an additional 2300 ℓ (600 gal) and industry and homes use still another 2000 ℓ (520 gal) for flushing and washing. In total, about 7000 ℓ (1820 gal) of water per person are used each day in the United States. This does not account for the water needed to maintain the natural state of lakes, streams, and rivers.

When addressing these uses of water and the adequacy of supplies it is necessary to recognize that some water uses are *consumptive* and some are *nonconsumptive*. Irrigation in agriculture and watering home gardens and lawns is the major consumptive use of water. It is called *consumptive* because most of this water goes back to the atmosphere through direct evaporation or transpiration and is not available for further use. On the other hand, most of the water used in the production of electrical power and in homes and industry is used to wash and/or flush away wastes. (In the production of electricity water is used to carry away waste heat). These water uses are *nonconsumptive* because the water itself remains available, albeit polluted, for further use. This presents problems as well as opportunities. Pollution may preclude further use and/or it may damage ecosystems where it is discharged. However, if the pollutants are removed, and they can be, the water remains available for reuse. Thus, in principle, at least, water can be reused any number of times before it returns to the atmosphere through evaporation. In long river systems, such as the Mississippi, it has been traditional practice to withdraw, use, and put water back many times before it

finally reaches the ocean. But because purification of the water discharged into the Mississippi is significantly less than 100 percent, the pollution load is substantial at the end.

### MAJOR TRENDS IN WATER USE

Fresh water is withdrawn from rivers or lakes or from groundwater through wells. The problem in drawing water from rivers is that, while average flow may be more than adequate, flow varies tremendously through the year and from one year to the next. These variations in river flow seldom correspond to fluctuations in human needs, which are frequently greatest (e.g., for irrigation), when river flow is lowest. Therefore, it has been common practice to construct dams and create reservoirs which accumulate water at times of high flow and are drawn down at times of low flow.

A dam and reservoir may have several functions beyond providing water. Water allowed to flow through the dam may be used to generate electricity. The reservoir may support fishing and recreation. When there is exceptional runoff, the storage capacity of the reservoir may prevent downstream flooding. In some cases, one of these latter functions is the primary objective.

It is estimated that as long as no more than 30 percent of the average flow is withdrawn, the downstream ecology of a river will not be greatly affected and the water supply will be sustainable even during periods of drought. However, demands for water continue to increase, especially for irrigation purposes (Fig. 8-16). The problem is aggravated by the fact that in some cases normally nonconsumptive uses have become consumptive. For example, water from one river is piped so far for municipal water use that it is impractical to return it to the river or lake from which it was withdrawn. The problem of water **diversion** is particularly severe in the West where water from the Colorado River is piped to Los Angeles and then discharged into the ocean. Consumptive demands on the Colorado River now exceed 100 percent of its average flow. Little and sometimes no fresh water reaches the mouth of the river; hence, the natural ecology of the upper Gulf of California has been destroyed and human desires for water from the river often go unmet.

The Colorado River is not the only problem area. Withdrawal of water from Texas to Califor-

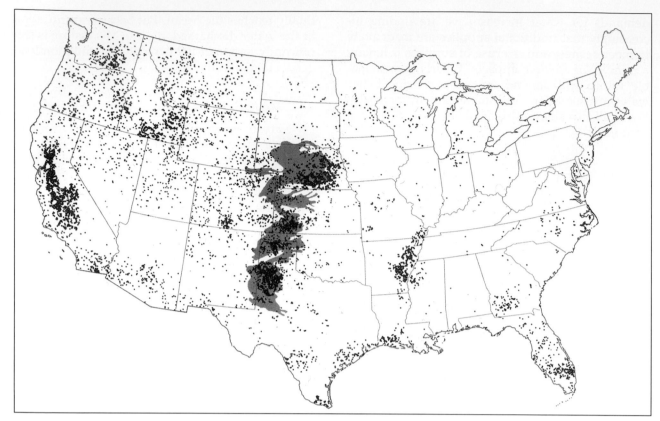

**FIGURE 8-16**

Irrigated land in the United States totals about 60 million acres. Each dot on the map represents 8000 acres where irrigation facilities existed in 1977, the latest year for which information of this kind is available. The colored area shows the lands underlain by the Ogallala aquifer, where much of the postwar growth in irrigated acreage has been concentrated. Crops are raised here mainly to feed cattle. (From Sandra S. Batie and Robert Healy, "The Future of American Agriculture," Copyright © February 1983, Scientific American, Inc. All rights reserved.)

nia and northward into the Mountain States is moving beyond the 30 percent level and increasing the potential for critical water shortages in the future (Fig. 8-17). As the 30 percent level is passed, people witness the severe drawdown, if not total exhaustion, of reservoirs in periods of drought. This has severe impacts on downstream ecosystems to say nothing of the human inconveniences. Yet there is little room for expanding surface water supplies because dams and reservoirs have already been developed at most sites where it is practical to do so.

As a result of pressing the limits of surface water supplies, attention has turned to the pumping of groundwater. Improvements in drilling technology has aided this alternative. Particularly

significant is the advent of center pivot irrigation systems in agriculture. With this system, water is withdrawn from a central well and applied to the field through a gigantic sprinkler system that moves itself in a circle about the well (Fig. 8-18). The use of such systems has increased tremendously in the last 20 years. Indeed it is largely responsible for an increase in agricultural production in recent years. However, it consumes huge amounts of groundwater. A single system may use as much as 40,000 $\ell$ (10,000 gal) of water per minute. In addition, vast numbers of new homes in suburban areas and, in some cases, entire municipalities depend on groundwater.

The prodigious use of groundwater amounts to treating it as an infinite and everlasting re-

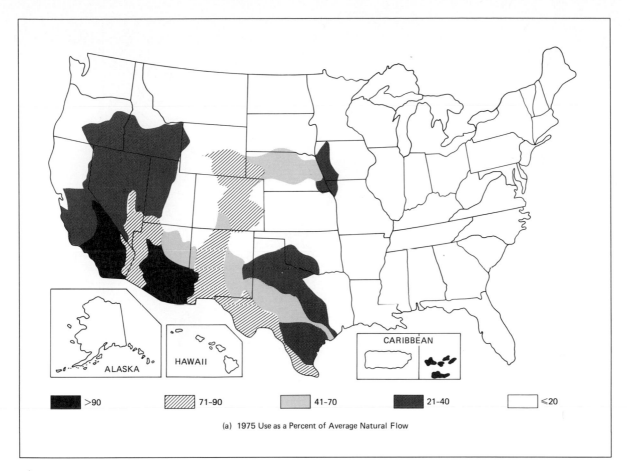

| >90 | 71-90 | 41-70 | 21-40 | ≤20 |

(a) 1975 Use as a Percent of Average Natural Flow

**FIGURE 8-17**
Surface water use for an average year. No more than 30
percent of the average level of surface water flow can be
counted on to be available in 95 out of 100 years. (a) Large
areas of the country are already depleting water at or above
this level.

source. Yet, groundwater is indeed a reservoir
with limits. It is a very large reservoir; it is estimated that in the United States there is tenfold
more fresh water stored as groundwater than is
stored in surface lakes and reservoirs. However,
the replacement of groundwater is a very slow
process, less than 1 percent per year on the average. This means that one cannot withdraw more
than 1 percent per year without gradually depleting this resource. The hard fact is that over much
of the United States, especially in the Plains States
and in the Southwest, groundwater is being withdrawn considerably above this rate. Of course this
means eventual exhaustion of this resource, but
there are also other effects, as will be described in
the following section.

## Problems of Excess Withdrawal of Water

### DEPLETING GROUNDWATER

Since irrigation is the greatest consumer of
water, depletion of groundwater will have its
greatest impact on agriculture. It is predicted that
3.5 million acres in the Great Plains will be converted to dryland farming (ranching and production of forage crops) by the year 2000 because of
depletion of the great Ogallala aquifer (see Fig. 8-
16). As a result, yields of food crops, such as corn
and cotton, which are now supported by irrigation, will drop by about 60 percent in these regions. Some areas have already gone out of production because of groundwater depletion. This
experience is by no means unique to the United
States.

Before actual exhaustion of groundwater, irrigation will become increasingly costly as deeper
wells must be drilled to compensate for the falling
water table and as more energy is required to

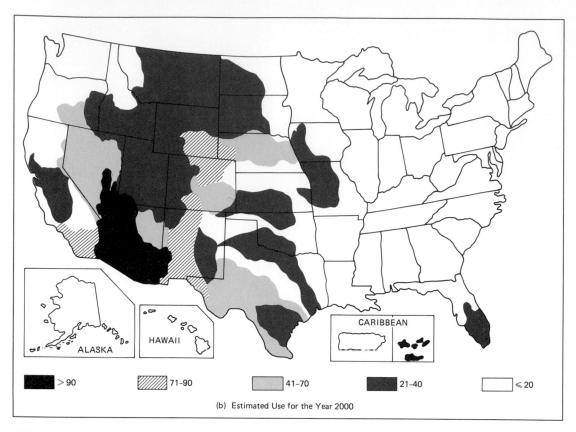

| | | | | |
|---|---|---|---|---|
| ■ >90 | ▨ 71–90 | ▨ 41–70 | ■ 21–40 | □ ≤ 20 |

(b) Estimated Use for the Year 2000

**FIGURE 8-17** *(cont.)*
(b) By the year 2000, much larger areas will be above the 30
percent level. Thus severe water shortages are inevitable.
(U.S. Water Resources Council.)

**FIGURE 8-18**
Center pivot irrigation. Water is
pumped from a center well. (a) A
self-powered boom rotates
around the well, spraying water
as it goes.

FIGURE 8-18 (cont.)
(b) The resulting circular irrigated fields are as much as 1 mile in diameter. (Earl Roberge/Photo, Researchers.)

pump the water from the greater depths. The rising costs may actually cause farmers to abandon irrigation before all the water is exhausted or at least to reserve irrigation for high-value crops. It is interesting to note that present depletion of groundwater in the Plains States is the result of irrigation used on crops that go to cattle feed. Recall again that about 90 percent of the food value is wasted in such conversions (Chapter 2).

### LAND SUBSIDENCE

Over the ages, groundwater has leached spaces and cavities in the earth, and the water it-self has played a role in supporting the overlying rock and soil. Consequently, as groundwater is withdrawn there may be a gradual settling of the land, a phenomenon known as **land subsidence.**

The rate of sinking may be on the order of 15 to 30 cm (0.5 to 1 ft) per year. In some areas of the San Joaquin Valley, California, land has settled as much as 9 m (30 ft) because of groundwater removal. Under cities, land subsidence leads to cracking of building foundation, roadways, and water and sewer lines (Fig. 8-19). In coastal areas it leads to flooding unless levees are built for protection. For example, a 10,000 km$^2$ (4000 mi$^2$) area

FIGURE 8-19
Land subsidence. Removal of groundwater may allow ground to settle, resulting in severe property damage, such as the cracking foundations seen here. (USDA–Soil Conservation Service photo.)

in the Houston–Galveston Bay region of Texas is gradually sinking because of removal of groundwater, and coastal properties are being abandoned as they are gradually inundated by the sea. Land subsidence is also a serious problem in New Orleans, in sections of Arizona, and in many other places throughout the world.

Another kind of land subsidence, the occurrence of a sinkhole, may be sudden and dramatic (Fig. 8-20). A sinkhole results when an underground cavern, drained of its supporting groundwater, suddenly collapses. Sinkholes may be 100 m or more across and as much as 50 m deep.

Formation of sinkholes is particularly severe in the southeastern United States where ancient beds of underlying limestone are highly leached. An estimated 4000 sinkholes have occurred in Alabama alone, some of which have "consumed" buildings, homes, livestock, and sections of highways.

### SALTWATER INTRUSION

Another problem resulting from groundwater removal is **saltwater intrusion,** also called **saltwater encroachment.** In coastal regions, springs and seeps of outflowing groundwater may lie under the ocean. As long as the water table on land is higher than the ocean level and the pressure in the aquifer is maintained, there is a net flow of fresh water into the ocean. Thus wells near the ocean yield fresh water (Fig. 8-21a). However, a lower water table and/or rapid rates of groundwater removal may reduce the pressure in the aquifer, permitting salt water to push back into the aquifer and hence into wells (Fig. 8-21b). Saltwater intrusion is problematic at many locations along the Atlantic Coast from New York to Florida, the Gulf Coast, and the coast of California.

### EFFECTS ON NATURAL ECOSYSTEMS

We have noted that up to 30 percent of the average flow of a river may be removed without causing undue damage to natural ecosystems. However, this figure has been exceeded in many cases. Throughout the southwestern United States, 70 percent depletion of instream flow is common. The flow of the lower Colorado River has been depleted by nearly 100 percent. The depletion largely results from direct removal; however, removing groundwater also contributes to the problem because the falling water table results in many springs and seeps going dry. Consequently, a river may be deprived of many of its natural inputs. This is particularly significant in dry periods when runoff is nil and a river's only source is springs.

In addition, the large surface areas of reservoirs and irrigation canals may significantly increase the amount of water lost through evaporation. In a related effect, dams on the Colorado

**FIGURE 8-20**
Sinkhole. Removal of groundwater may drain an underground cavern until the roof, no longer supported by water pressure, collapses, resulting in a sinkhole such as this one in Alabama. (Department of the Interior, U.S. Geological Survey photo.)

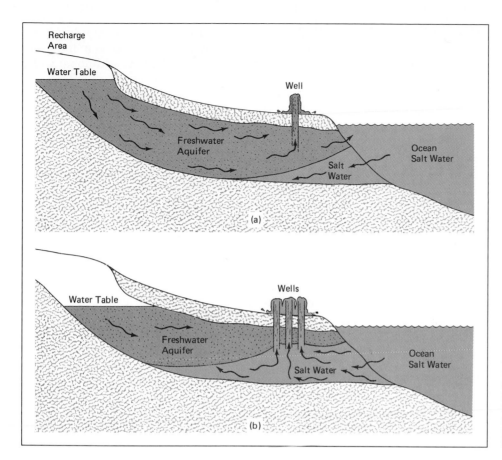

**FIGURE 8-21**
Saltwater encroachment.
(a) Where aquifers open into the ocean, fresh water is maintained in the aquifer by the head of fresh water inland. (b) Excessive removal of water may reduce the pressure so that salt water moves into the aquifer.

River have eliminated flooding and this has allowed an increase in riverside vegetation, which in turn withdraws water through transpiration.

As a river approaches dryness, the effect on aquatic life is severe; however, effects on ecosystems go beyond the river itself. Marshlands along many rivers, no longer nourished by occasional overflows, have dried up, resulting in tremendous dieoffs of ducks and other wildlife that depend on these habitats. A particular case is the Everglades National Park at the tip of Florida. This area is effectively one gigantic marsh nourished by freshwater overflow from Lake Okeechobee in central Florida. As this fresh water has been diverted, there has been an estimated 90 percent decline in populations of wading birds, many of them unique species found nowhere else in the world. Belatedly some efforts are now being made to return more water to the Everglades, but in other cases the outlook is increasingly dim.

This problem also impacts estuaries. Estuaries are bays along the ocean where there is an inflow of fresh water from a river gradually mixing with the salt water of the sea. This creates an environment of brackish water—water with a salt concentration between fresh water and seawater. Estuaries are among the most productive aquatic ecosystems; they are rich breeding grounds for many species of fish and shellfish that are adapted to the particular intermediate salt concentrations. As river flow is reduced there is less fresh water flushing of the estuary and its salt concentration increases. This may profoundly affect the ecology of the estuary. For example, the entire ecology in the upper end of the Gulf of California has been altered by the reduced flow of fresh water from the Colorado River.

A similar situation is occurring in Mono Lake, a 100 mi$^2$ lake in east central California. Mono Lake has no outlet other than evaporation but a substantial inflow of fresh water from snowmelt off the Sierra Nevada Mountains immediately to the west has kept its water at a level of modest salinity. Thus, it is a rich habitat supporting numerous species of wildlife; especially impressive are its huge flocks of aquatic birds. But now, much of the freshwater inflow is being diverted to support the profligate water consumption of Los Angeles. As a result, Mono Lake is losing more water in evaporation than it is receiving; therefore, it is shrinking rapidly and becoming more salty (Fig. 8-22). If this situation is not altered, it may become a small, dead lake (too salty to support life) surrounded only by many square miles of barren salt flats.

In addition to fresh water, a river also carries nutrients that are essential to the support of aquatic plant life. The reduced input of nutrients may be another factor in ecological collapse. For example, commercial fisheries that existed at the mouth of the Nile River in Egypt were virtually eliminated when the Aswan High Dam cut off the normal flow of nutrients and fresh water into the Mediterranean Sea.

## Getting More Water versus Using Less

### LIMITS OF INCREASING SUPPLIES

Based on past trends, which show that per capita water consumption in the United States increased by about 50 percent between 1950 and

1975, economists project another significant increase in water consumption between now and the end of the century. Where is additional water to come from?

We noted that most practical sites for dams and reservoirs have already been developed. However, this does not prevent some people from proposing and promoting ever more grandiose schemes for damming and diverting surface waters from water-rich to water-poor areas. For example, a proposal has been made to dam the Yukon River in Alaska, and rechannel its water all the way to the southwestern United States, Mexico, and parts of the Midwest, distances of up to 5000 km (3000 mi).

But this is where we must emphasize the word *practical.* The cost of construction would be astronomical to say nothing of the ecological costs. The proposed reservoir would flood some 10 percent of the state and would inundate a large share of the land that supports most of Alaska's wildlife. In other cases, the construction of new dams and reservoirs would entail relocating substantial communities or would inundate more farmland than would be opened by additional water for irrigation. In short, the economic, social, and ecological costs of further water diversion projects are becoming increasingly unacceptable. Similarly, pumping additional groundwater in

**FIGURE 8-22**
Mono Lake. (Americ Higashi/Stockpile.)

water-rich areas and transporting it to water-short areas is economically prohibitive.

Purification of seawater is feasible and some water-short coastal cities have built desalinization plants. However the energy consumption and consequent costs are high, making this source of fresh water impractical beyond supplying limited municipal needs.

Development of salt resistant plants (see Chapter 7) may permit certain crops to be irrigated with seawater. This would alleviate the demand for fresh water but only in low-lying areas near coasts such as southern California and Florida. Beyond such areas pumping costs would become prohibitive.

Given these limitations of water supplies on the one hand and projections of rising demands on the other, it does not take great insight to foresee that shortfalls are increasingly probable. When? How great? The great unknown in this equation is climate. The climatic record shows that great fluctuations in rainfall have occurred in the past and should be anticipated for the future. During some drought periods yearly flow of surface water may drop 70 percent or even more below average. Hence, the maxim follows that no more than 30 percent of the average flow should be counted on to be available in 95 out of 100 years. Therefore, as long as human demands are

below this level, supplies generally will be adequate. However, as demands increase above this amount, and to an increasing extent they are, shortages can only become more common and severe (Fig. 8-23). Also, more and more of the country will be affected.

## WATER CONSERVATION

Such dire predictions of water shortages are based on extrapolation of current trends. They are not meant to predict the future so much as to point out the need to change directions. In addressing water shortfalls it is necessary to examine the question: Do our lives really depend on present and projected levels of water use and consumption? We should address each of the three major areas: irrigation, municipal water supplies, and industrial uses.

**Irrigation.** **Drip irrigation** systems may be substituted for present methods. Drip irrigation is a system of pipes and tubes that literally feeds water drop by drop to each individual plant (Fig. 8-24). This is impractical for crops such as wheat, but it has been feasible for larger, individual plants. The capital costs of installing the system are obviously high but these may be largely offset by reduced pumping costs and a reduction of 80 to 90 percent in the amount of water used. With less water applied, the rate of salinization (salt buildup) is also reduced.

Further, the extent of irrigation itself should be reexamined. Increasing irrigation through the 1970s led to very significant increases in agricultural production. This sounds good, but much of

**FIGURE 8-23**
This hypothetical diagram shows natural variations in surface water flow and increasing human demands on water. Do water shortages occur because of droughts (periods of low flow) or because of excessive demands on the system?

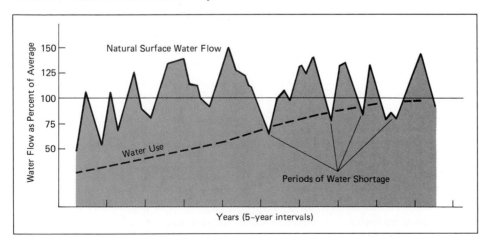

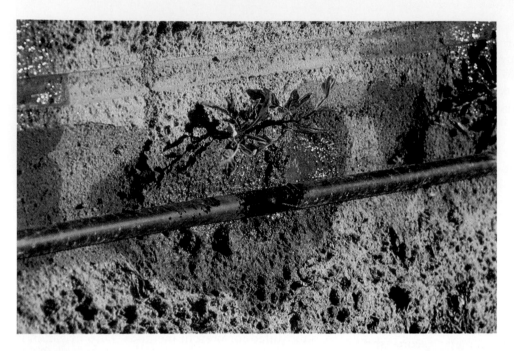

**FIGURE 8-24**
Drip irrigation. Irrigation is the most consumptive water use. Drip irrigation offers a conservative method of applying water, dripping it on each plant through a system of plastic pipes. (Lowell Georgia/Photo Researchers.)

the increase in the United States simply piled up as surpluses. In turn, the surpluses depressed grain prices, causing economic hardship for many farmers and, in 1983, leading the United States government to reinstitute a program of effectively paying farmers not to produce. There is more than a little irony in overdrawing water supplies to increase production on the one hand, while paying farmers not to produce on the other.

**Municipal water supplies.** Regarding municipal use, there are dozens of "how to" manuals on ways to use less water, and numerous devices are on the market which will conserve water in showers, toilets, and other devices. Also, many cities and industries are considering means of purifying and recycling waste water (see Chapter 10). In some areas, particularly in water-short periods, watering lawns and gardens is a major use of municipal water. Beautiful gardens can be achieved with much less or no watering, by landscaping with more drought-resistant species, and by increased use of mulches to reduce evapora-

tion of water from soil. Also, **gray water,** water from sinks and tubs, may be used for lawn and garden water if drains are separated from those of toilets and a holding reservoir is installed (Fig. 8-25).

**Industrial uses.** Another significant factor in the projected increase of demand for water is in the development of coal and oil shale energy resources. This is particularly problematic because these energy resources are primarily located in the already water-short western states (Colorado, Utah, Wyoming, and Montana). Similarly, nuclear power plants require large amounts of water for cooling. Developing other energy resources (Chapter 23) would alleviate these conflicts.

Managing and conserving water resources demand focusing attention on more than direct uses. Water resources also may be affected greatly by how we use land and dispose of wastes. Therefore, we address these topics in the following sections.

## Problems of Changing Land Use

Many human activities change the nature of the land surface such that the infiltration runoff ratio is shifted toward less infiltration and greater runoff. Most conspicuously, urban and suburban

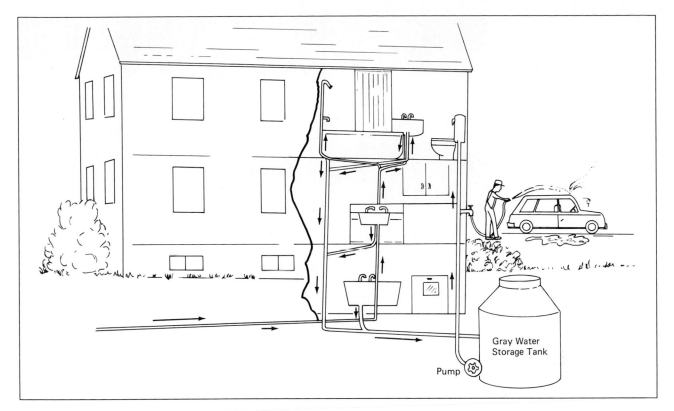

**FIGURE 8-25**
Gray water use. It is extravagant to use drinking-quality water for such uses as toilet flushing and lawn watering. Significant water conservation could be achieved by trapping waste water from tubs and sinks (gray water) and recycling it for such purposes.

development greatly increases runoff by creating innumerable hard, impervious surfaces such as roadways, parking lots, and rooftops. Even the soil of suburban lawns is generally compacted so that runoff is significantly increased. Agricultural practices that cause the soil to become puddled and/or compacted, as well as clearcutting of forests and overgrazing by livestock also lead to increased runoff. Whatever its cause, shifting the infiltration runoff ratio in this way has numerous and far-reaching effects.

### CHANGE FROM STEADY STREAM FLOW TO ALTERNATION BETWEEN FLOODING AND DRYNESS

In the water cycle, the groundwater system regulates stream flow. Infiltration recharges the reservoir of groundwater, which combined with water trickling out of springs keeps streams flowing at more or less uniform rates in spite of fluctuations in rainfall. With a steady flow, streams support a rich variety of aquatic species and amphibians such as frogs and salamanders.

When development increases runoff and decreases infiltration, the situation is changed dramatically. Rainfall flows from hard surfaces in a massive sheet funnelled into storm drains. Storm drains are simply pipes that lead directly to the nearest convenient stream channel (Fig. 8-26). Thus, a large portion of rainfall enters stream channels almost immediately, causing a huge surge if not a flood (Fig. 8-27a). Just as importantly, springs and seeps go dry as the groundwater reservoir receives less water from infiltration and the water table drops. Deprived of the continual seepage from springs, streams may be totally dry between rains (Fig. 8-27b).

In short, hard surfacing, which is so much a part of development, causes nearby streams to change from a steady flow that is ecologically rich to a flood-drought situation that is ecologically

**FIGURE 8-26**
Storm drains generally are direct
pipes to the nearest convenient
stream bed. Storm drains from
residential area in background
empty here into a stream bed.
(Photo by author.)

(a)

(b)

**FIGURE 8-27**
Effect of development on stream flow. Before
development, this stream maintained a
generally modest flow of water throughout the
years. Now, after development of the
surrounding area with suburban homes, the
flow fluctuates sharply between (a) high
surges of runoff and (b) dryness. Photo (a)
was taken during an average summer
thunderstorm; photo (b), a few hours later.
Note the bank erosion as well as the dryness.
This situation is a typical result of
development. (Photos by author.)

destitute (Fig. 8-28). Such streams are little more than open storm sewers; indeed, they have frequently been incorporated into the storm drain system by laying sewer pipe in the stream bed and covering it over (Fig. 8-29). This is a particularly tragic loss when contrasted with our need and desire to preserve and protect more of the natural ecology within the increasingly urban-suburban landscape.

If the stream channels are left open, however, the surges from each rainfall lead to economic and ecological loss. An increase in the frequency and severity of flooding is the most conspicuous result. It is true that floods have always been a part of nature. However, with increased runoff, a very modest storm may lead to a flood-producing volume of water, whereas under natural conditions only a rare, very severe storm would produce such a volume. Countless communities, many of them expensive new, suburban developments, have experienced flooding with increasing frequency and severity as expand-ing development has paved more and more of the upstream watershed; thus, flood damages have generally increased despite flood control measures (Fig. 8-30). Paradoxically, while flooding is increasing, wells may be going dry because decreased infiltration causes the water table to drop.

Another major problem is **stream bank erosion.** There is a common misconception that surges of water deepen a stream channel; actually, the opposite is the case. Stream banks are stabilized by vegetation against normal flows. However, high flows cut under and around tree roots, causing the trees to topple (Fig. 8-31). Fallen trees and other debris divert the water and cause even further erosion of the banks. Rocks, stones, and coarse sand, which are too heavy to be carried away, collect in the bottom of the channel and cause the water to spread out and erode the banks even more (Fig. 8-32). Consequently, the channel becomes broader and shallower. Additional sources of sediment will accentuate this effect. Gradually, a narrow, tree-lined stream may

**FIGURE 8-28**
Changes in stream flow occur with development. Curves are for similar storms on Brays Bayou in Houston, Texas, before, during, and after development. Note the increasing height of the surge occurring with the storm and also the decreasing volume of flow that occurs later in the cycle. (From D. Van Sickle, in *Effects of Watershed Changes on Stream Flow*, ed. W. Moore and C. Morgan. Austin, Texas: University of Texas Press, 1969.)

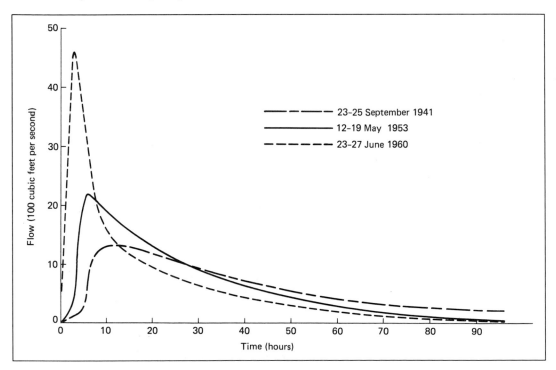

23–25 September 1941
12–19 May 1953
23–27 June 1960

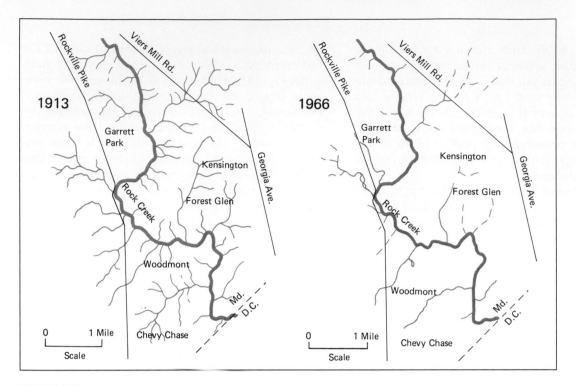

**FIGURE 8-29**
With urbanization, streams become storm drains. This 26 square mile section of the Rock Creek watershed in Maryland, now a heavily populated surburb of Washington, was rural in 1913, with many small tributaries fed by springs and seeps. Ensuing development, carried out in ignorance of natural processes, covered most of the old acquifer recharge areas with pavements and rooftops, so that more precipitation ran rapidly off the land instead of soaking in and flowing out gradually into streams. Flooding during storms and loss of flow at other times caused most of the tributaries to be covered over as storm sewers; of 64 miles of natural-flowing stream channels that existed in 1913 in this section, only 27 miles could be found above ground in 1966. (National Park Service.)

**FIGURE 8-30**
Property losses caused by flooding. Increased runoff and clogging of stream channels with sediments results in increased flooding. The weather, and consequently flooding, is highly variable from year to year; however, since 1960 the trend toward flooding has been distinctly upward. (Data from U.S. Water Resources Council.)

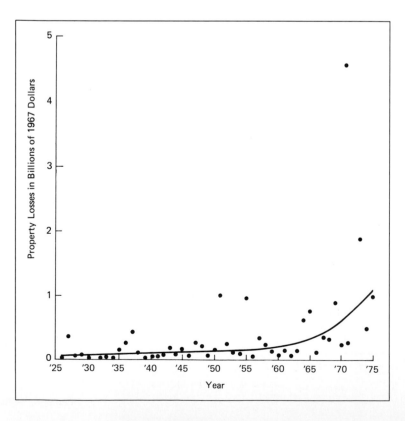

**FIGURE 8-31**
Stream bank erosion. The surges of increased runoff greatly accelerate erosion of stream banks. Note the many trees that have fallen because of undercutting. Fallen trees block the channel and cause still further erosion and undercutting. (Photo by author.)

be converted into a broad area of fallen trees and drifts of sand and gravel sediment. The ecological devastation is self-evident.

Another aspect of the urban runoff problem is that storm drain outlets are frequently located high on the side of the valley rather than in the stream bed. Consequently, gullies are eroded as water flows from the storm drain exit to the stream, both defacing the valley side and adding to the burden of sediment (Fig. 8-33).

Along with these problems, it should be re-emphasized that the surge of increased runoff is not only quantitatively undesirable; it is also of poor quality. The list of pollutants that may be picked up from the surface includes soil particles; nutrients from lawn and garden fertilizers; pesticides and herbicides applied to lawns and gardens; fecal wastes and associated bacteria from pet droppings; oil and grease leaked from cars onto streets and parking lots or poured down storm drains by do-it-yourself auto mechanics; salt and other chemicals used in de-icing streets; chemicals and other materials stockpiled in the open; soot, grime, and various chemicals that are expelled from smokestacks or exhaust pipes and then settle to the ground; human litter such as paper, plastic, bottles, and cans; natural litter such as leaves and twigs; and heat absorbed from hot pavements and rooftops. From rural areas the most important categories in addition to sediments are fecal wastes from barnyards and feed-

(a)

(b)

(c)

**FIGURE 8-32**
Change in stream channel.
(a) Accelerated erosion of banks.
(b) The stream channel becomes broader and shallower. (c) The main channel of this stream has been filled in, and the water is washing and eroding the woodland floor over a broad area. (Photo by author.)

lots, fertilizers, pesticides, and various agricultural wastes.

Two techniques, relocation of people and channelization, address the problems of flooding but do nothing for the root cause of increased runoff. Yet they are frequently used, so they should be understood.

Flood damages are at least partly covered by public disaster relief funds. As frequency of flooding has increased, political jurisdictions have found it expedient to use such funds to actually buy and demolish flood-prone properties and relocate residents rather than pay for recurring flood damages. Generally, this is not the best solution because, as stream banks erode and channels fill with sediment, the areas flooded gradually become broader. The situation is aggravated as further development continues to increase the

amount of runoff. In fact, average yearly flood damages are increasing, as was shown in Figure 8-30, and may be expected to increase further until runoff is controlled.

Second, since flooding is aggravated when stream and river channels are clogged with sediment, dredging the channels is often advocated. But unless sources of sediment are controlled, dredged channels fill rapidly. In addition to dredging, however, the channel may be straightened or made into smooth, sweeping curves and lined with rock or cement, a procedure known as **channelization** (Fig. 8-34). The smooth-sided channel carries water more swiftly and efficiently so that the excess water does not overflow on the banks. Also the cement or stone sides prevent further erosion.

Still, channelization is highly controversial.

**FIGURE 8-33**
Erosion from storm drain outlets. Frequently storm drains open onto valley sides, and water is allowed to find its own way down the slope. The gully seen here is the result of erosion by water coming from a storm drain outlet in the left background. (Photo by author.)

It may alleviate the problem of flooding and bank erosion, but only in the immediate area of the channel. More efficient movement of water and sediments may simply transfer the problems downstream. Ecologically, a channelized stream has little if any semblance to a natural stream. Beyond the alternation in water flow between surges and dryness, all the crevices and pools that might house aquatic life are eliminated. Further, the channel may be a hazard to wildlife in that animals that fall in may not be able to get out. Aesthetically and functionally, as you can see in Figure 8-34, a channelized stream is effectively an open storm drain.

### STORMWATER MANAGEMENT

How to cope with the related problems of excessive runoff, stream bank erosion and flooding has become a subject of study and a profession in its own right. It is called **stormwater management.** Many government jurisdictions now

**FIGURE 8-34**
Channelized stream. To decrease flooding and erosion, stream channels may be dredged, straightened, and lined with concrete, a process that may simply transport the problem downstream; it also destroys all semblance of the natural stream ecosystem. (USDA–Soil Conservation Service photo.)

have departments of stormwater management addressing these problems. The issues are many and complex, but we shall discuss some of the general principles.

Traditionally, the only attention given to stormwater when designing developments was to have it drain away as quickly as possible. For example, in a heavy rainfall a 1 hectare (2.5 acre) parking lot may generate runoff at the rate of 6000 $\ell$ (1500 gal) per minute. The storm drain was designed to handle double this flow, so as to meet even the most demanding situations. Of course, the more efficient the drainage, the greater the runoff surge and related downstream problems.

Stormwater management promotes the reverse philosophy—hold stormwater at or near where it falls and let it drain away slowly or infiltrate. The techniques of stormwater management include building parking lots with porous surfaces, and designing rooftops, parking lots, and other large flat surfaces so that they "pond" the water and let it trickle away slowly. Also, runoff may be funnelled into holding reservoirs built under or beside such areas. From these reservoirs, the water can drain away slowly or infiltrate. Such reservoirs are called **stormwater retention** (or detention) **reservoirs.**

Stormwater retention reservoirs may addi-

**FIGURE 8-35**
Stormwater reservoirs. Rather than letting the excessive runoff from developed areas cause environmental damage, it may be funnelled into such reservoirs from where it may drain away slowly or infiltrate. The design of the standpipe allows this reservoir to hold and slowly drain away excess water while retaining some to create a useful pond.

tionally provide pockets of wildlife habitat, thus enhancing the otherwise urban-suburban setting (Fig. 8-35), or they may provide a practical source of water for nondrinking purposes such as car washing or lawn watering, thus lessening the demand on the municipal system (Fig. 8-36).

Large stormwater retention reservoirs can double as recreational facilities (Fig. 8-37). During a storm the recreation area around the reservoir can store floodwater. Recreational equipment can be made so that it will not be damaged by the high water and there is little inconvenience to people since most would leave the area at the time of a storm whether it was flooded or not.

Another technique under serious consideration is the funnelling of excess stormwater runoff into huge "reverse wells" which may replenish groundwater supplies. This has been done in some areas and seems to have merit. But it seems destined to pollute groundwater because surface water is often highly polluted.

On a broad scale stormwater management demands **total watershed planning.** This means that the total watershed is considered in making plans that both permit development and preserve the ecological integrity of the waterways. Different types of control may be developed for different segments of the watershed. For example,

some areas of the watershed may be left as permanent open space so that the soil remains capable of absorbing water. Another area might be developed and the increased runoff from it controlled by a large retention reservoir that also provides recreational uses. In a commercial-industrial area, on-site retention structures might be required in development.

The legal impetus for coping with these problems of runoff has been established by Section 208 of the Clean Water Act of 1972. Under this law, most regions are currently attempting to promote various control procedures. Since actions on these projects are at the local level, it is an area where individual citizens can become highly involved and play a significant role.

## Conclusion

In conclusion, the problems associated with assuring adequate water supplies and preserving the ecological integrity of natural waterways demand a much greater awareness of the water cycle, its limits, and the results of human impacts on it than has been afforded in the past. To come to terms with these problems, there must be a much greater emphasis on both conservation and better management of available water. Of course, here we have concentrated on just the quantitative aspects. Qualitative aspects are equally important and in following chapters we shall address various aspects of pollution.

**FIGURE 8-36**
Stormwater may be a cheap source of water for many uses; here, runoff from a car dealership is being stored and used to wash cars.

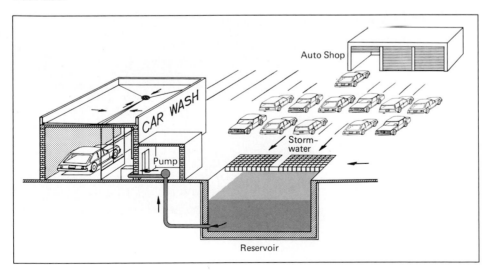

(a)

(b)

(c)

**FIGURE 8-37**
Lake Needwood flood control project. (a) This lake and
surrounding area is a favorite recreational area in Montgomery
County, Maryland, a suburban area of Washington, D.C. In
addition, the area is a flood control reservoir. (Air Photographics,
Inc., Silver Spring, Maryland.) (b) In the event of a storm, the dam
can hold up to an additional 40 feet of water and allow it to
drain down gradually, preventing flooding and erosion
downstream. (c) Recreational equipment is designed to
withstand flooding. (Photos by David Hunley.)

# IV
# POLLUTION

Pollution has become a household word that calls up images of despoiled water, air, or land. Yet pollution is actually much more complex. It eludes simple definition because it may involve hundreds of factors that stem from numerous sources. The Environmental Pollution Panel of the President's Science Advisory Committee in its 1965 report, *Restoring the Quality of Our Environment,* defined environmental pollution as "the unfavorable alteration of our surroundings, wholly or largely as a by-product of man's action. . . ."

The alteration may occur as a direct or indirect result of changes in energy levels, radiation levels, chemical factors, physical factors, or populations or organisms. Some changes, such as contamination of air or drinking water may directly affect human health and well-being. Changes in agriculture or production of other biological products, corrosion of materials, and altered opportunities for recreation and appreciation of nature may affect humans less directly, but nevertheless affect them.

The key point is simply that an "unfavorable alteration of our surroundings" occurs. Some of the more significant sources or categories of water pollution include sediments eroding from soil that adversely affect aquatic ecosystems; disease-causing organisms from sewage wastes that contaminate drinking or recreational water supplies; organic wastes from sewage or other sources that cause depletion of dissolved oxygen and suffocation of aquatic life; oversupply of nutrients from sewage and fertilizer runoff that causes overgrowth of undesirable algae and upsets the existing ecosystem; toxic chemicals from wastes and pesticides that contaminate water supplies; acid precipitation that causes acidification of lakes and rivers and consequent upset of existing aquatic ecosystems; waste heat from electric power plants that upsets existing aquatic ecosystems; and radioactive wastes and materials from the production of nuclear power and weapons that may potentially contaminate air, water, and/or soil.

Some of the major sources of pollution affecting air include direct and indirect products from the combustion of fuels, and fumes from toxic chemicals that pollute the air we breathe. All air pollutants eventually settle on the earth where they may present water and soil pollution problems as well.

Finally, certain human activities may affect the climate on a global scale and hence affect the entire biosphere. For example, an increased concentration of carbon dioxide in the air as a result of combustion of fuels and deforestation is causing a global warming trend known as the *greenhouse effect*. Certain chemicals can cause a depletion of the ozone layer, which protects the earth from damaging ultraviolet radiation. Dust and smoke particles emitted into the air by a nuclear war would cause a drastic cooling effect on the earth referred to as a *nuclear winter*.

As these examples illustrate, the maxim "don't pollute," which is taught to elementary school children, is a gross oversimplification of the pollution problem. As ecologist Edward Kormondy points out: Pollutants are

> normal by-products of people as purely biological organisms and as creative social beings. They are the organic and inorganic wastes of metabolic and digestive processes and of creativity in protecting and augmenting the production of crops, of warming homes, clothing the body and harnessing the atom. . . . Solutions do not and cannot lie solely in removing the cause because as long as humanity exists, it will have by-products. Rather, answers lie in intelligent management of that production and through regulating the unfavorable alteration of our surroundings.*

Indeed, every organism in a natural ecosystem produces potentially polluting waste products. What makes natural ecosystems sustainable is that the wastes from one kind of organism become the food and/or raw materials of another. In balanced ecosystems, wastes do not accumulate to produce "unfavorable alterations"; they are broken down and recycled.

Through much of their history humans have relied on the same natural processes to dispose of their wastes. But the situation has become extremely unbalanced. Exploding human population coupled with increasing use of materials and energy has led to enormous volumes of wastes and other materials being discharged into the environment. Even when materials are **biodegradable,** that is, of a kind that can be assimilated and recycled by organisms, sheer volumes overwhelm the capacity of natural systems to cope. Aggravating the problem is the production of increasing amounts and kinds of wastes and materials that are *not* readily broken down and assimilated by natural processes. Such materials and chemicals are referred to as **nonbiodegradable.**

Regaining the balance, however, does not mean abandoning modern technology. The solution lies in enlisting technology to regulate and recycle byproducts so that they become new resources.

---

*Kormondy, Edward J., *Concepts of Ecology*, 3rd ed. Prentice-Hall, 1984, p. 247.

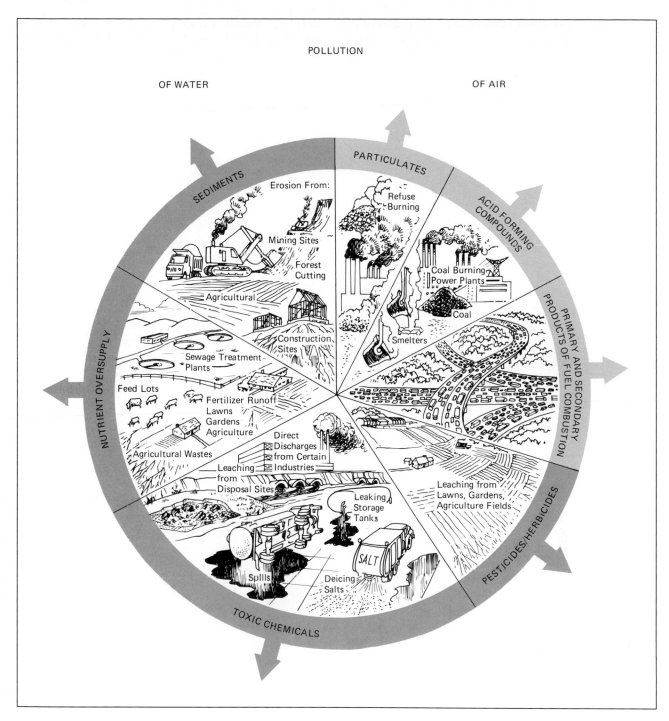

**FIGURE IV-1**

Pollution may be defined as any chemical or material out of place. Thus, chemicals and materials that are useful in one place cause pollution as they are discarded in or migrate to places where they are not wanted and where they may cause damage to environmental and human health. The figure shows the major categories of pollution and illustrates their most important sources.

Clearly, pollution involves so many different factors from so many different sources that there is no single or simple remedy (Fig. IV-1). In each situation the pollutant(s) causing the problem must be identified, sources determined, and then appropriate control strategies developed and implemented. This is a complex and difficult task. It should hardly be surprising that despite efforts and significant progress in some areas, huge problems remain. In Part IV we shall address pollution issues that are of greatest and most widespread concern and show what is being done and what remains to be done to solve the problems. Additional problems such as those involving pesticides, thermal pollution, and radioactive wastes will be considered in later chapters.

# 9

# WATER POLLUTION DUE TO SEDIMENTS

---

## CONCEPT FRAMEWORK

| Outline | | Study Questions |
|---|---|---|
| **I. PROBLEMS CAUSED BY SEDIMENTS** | **259** | |
| **A. Damage to Aquatic Ecosystems** | **259** | 1. Explain how sediment movement affects the ecosystem. How is aquatic life affected by sand and silt when they are not suspended? Why are estuaries particularly vulnerable to sediment problems? |
| **B. Filling of Channels and Reservoirs** | **260** | 2. What are the advantages and disadvantages of dredging? How does the Baltimore Harbor illustrate the situation? |
| **II. SOURCES AND CONTROL OF SEDIMENTS** | **263** | 3. What is the main source of sediments? |
| **A. Agriculture and Related Activities** | **263** | 4. Why do many farmers ignore soil conservation procedures? What are the long-term effects? How could regulations help the situation? |

5. Why are sediments resulting from development likely to erode? How do sediment traps work? What measures could be employed to better control erosion resulting from development?

6. Why are the opportunities for erosion greater at mining sites than at construction sites?

7. How can individuals help prevent sediment erosion in urban areas?

Sediment pollution is not a glamorous new issue; it does not receive the headlines given to toxic wastes, ozone, or acid rain. Yet, sediment often tops the list of water quality problems in streams, rivers, and estuaries. What is sediment?

**Sediment** is the name given to particles which are carried by flowing water and eventually deposited as the flow of water slows. Most sediment consists of sand, silt, and clay particles originating from soil erosion, but water transport invariably separates the particles and deposits them in different locations. Thus, sediments in a given location bear little if any similarity to the soil that was eroded.

Some erosion occurs even in undisturbed natural ecosystems. For example, undisturbed forest land loses about 0.25 tons per hectare per year (0.1 ton per acre per year). Streams and rivers are able to handle this load without difficulty. But human activities such as plowing, logging, mining, and the bulldozing that occurs with land development frequently cause erosion to increase as much as 100- or even 1000-fold. The resulting increase in sediment load frequently has a devastating effect on the waterway and the problems magnify as sediments are deposited at its end. Then human activities are so widespread that few waterways anywhere in the world escape the impact of sediments. Our objective in this chapter is to understand what the impacts of sediments are and to learn how they can be controlled.

## PROBLEMS CAUSED BY SEDIMENTS

### Damage to Aquatic Ecosystems

When erosion is slight, streams and rivers draining in the area run clear. They maintain ecosystems in which producers are mainly algae and other aquatic plants attached to rocks or rooted in the bottom as submerged aquatic vegetation. These producers support a complex food web including bacteria, protozoans, worms, insect larvae, crayfish, and fish. These organisms, in turn, maintain themselves in flowing water by attaching to rocks or, as in the case of fish, seeking shelter behind or under rocks. Even fish that maintain their position by active swimming occasionally need such shelter.

Sediment movement has compound effects upon the ecosystem. Clay and organic particles in suspension not only make the water look muddy; they reduce light penetration and the rate of photosynthesis. As the sediment settles, it coats everything and continues to block photosynthesis. It also kills the animal organisms by clogging their gills and feeding structures. Eggs of fish and other aquatic organisms are particularly vulnerable to smothering because of sediment.

Equally destructive is the **bedload** of sand and silt not readily carried in suspension, but gradually washed along the bottom. As particles roll and tumble along, they scour organisms from the rocks. They also bury and smother much of the bottom life and fill in the hiding and resting places for fish and crayfish. Aquatic plants may be prevented from reestablishing themselves because the bottom is a continually shifting bed of sand. Little of the natural ecosystem can survive in a stream subjected to high sediment loads (Fig. 9-1).

When high sediment loads are coupled with increased runoff, bank erosion occurs. As a result, the river channel becomes wider, but shallower. During times of low water, the river may be little more than a wide expanse of sediment deposits with a minor rivulet meandering through (Fig. 9-2).

The unfavorable effects may extend beyond the collapse of natural stream ecosystems. Bacteria and detritus entering from sewage outfalls or runoff tend to persist rather than be consumed by the aquatic ecosystem. Therefore, bacterial pollution may become a greater health hazard in sediment-loaded waterways.

As rivers and streams carry sediments into lakes, bays, and estuaries, the same ecological damage occurs. Fine clays and organic material are particularly troublesome because wave action and very slight currents maintain them in suspension and distribute them throughout the entire body. You have probably seen a lake or bay turn "muddy" after a storm. After about a week most of the sediment settles and the water clears, but in the meantime light penetration and photosynthesis have been reduced to a tiny fraction of normal. This may cause a significant dieoff of producers which affects the rest of the ecosystem.

Heavier sand and silt particles settle relatively close to the mouth of the waterway, but the ecological impact can be disproportionately large.

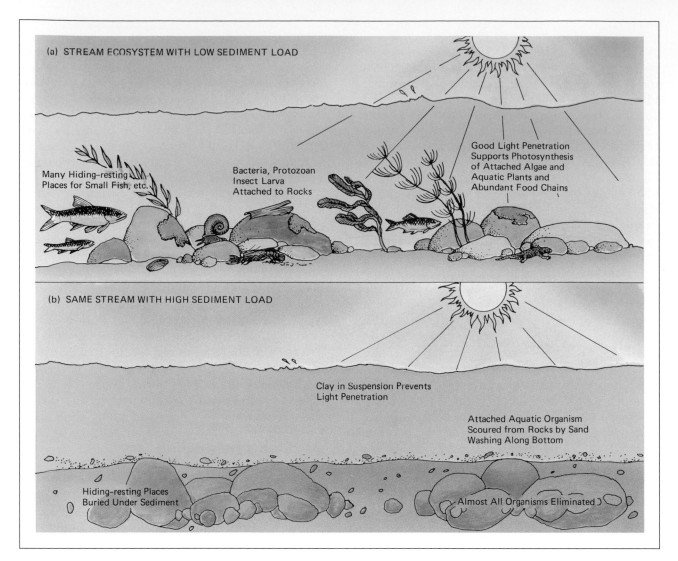

**FIGURE 9-1**
Negative impacts of sediments on the aquatic ecosystems of streams and rivers. (a) The ecosystem of a stream that is not subjected to a large sediment load. (b) The changes that occur when there are large sediment inputs.

Streams and rivers normally carry new supplies of nutrients and detritus; thus, the area where a stream or river enters a larger body is often the most productive area for aquatic plants and animals. This is particularly true of estuaries, bays where rivers enter the sea. Many fish and shellfish are restricted to these regions and additional species utilize them for breeding and nursery grounds. Consequently, excessive sediment loads which bury aquatic life at the river's mouth may dramatically affect organisms throughout the receiving body, especially if the problem occurs during times of spawning.

## Filling of Channels and Reservoirs

Sediment also causes serious economic problems. In Chapter 8 we discussed how sediments fill and clog stream channels and aggravate problems of stream bank erosion and flooding. Additionally, water supply reservoirs are filled, shipping channels are made impassable, and irrigation canals are clogged.

In the United States, many millions of dollars are spent yearly in dredging sediments to alleviate these problems. The task is unending because dredged areas soon fill in with new sediments, and present dredging efforts do not address all the problem areas. Beyond the ecological damage, many millions of cubic meters of water storage capacity in reservoirs are lost each year because of sedimentation (Figs. 9-3 and 9-4). This loss intensifies the problem of predicted water shortages.

**FIGURE 9-2**
A river channel choked with sediment from erosion upstream. Large sediment loads have combined with flooding and stream bank erosion to make this river channel a broad expanse of unstable sediment deposits that shift with the increased runoff of each additional storm. How much of the natural ecosystem is left? Platte River at Lexington, Nebraska. (Charles R. Beliuky/ Photo Researchers.)

**FIGURE 9-3**
Lake Como, Minnesota was entirely filled with sediments in the ten-year period from 1926 (a) to 1936 (b). The sediments came from the erosion resulting from timber clearing and plowing up and down slopes. (USDA–Soil Conservation Service photos.)

(a)

(b)

261

(a)

(b)

**FIGURE 9-4**
(a and b) Before and after photos of a reservoir rendered worthless because of filling by sediments. (USDA−Soil Conservation Service photos.) (c) The filling process. The bottom portion of this photo shows a growing deposit of sediment which is being brought into the lake by a small stream entering at the lower left. (Photo by author.)

(c)

Since there are few or, in many cases, no new locations that are suitable for reservoirs, one can anticipate a need to redouble dredging efforts in the future. However, costs will be more than just financial. The process of dredging inevitably stirs up and redistributes the fine sediment, thus initiating another round of sediment pollution. Then there is the problem of disposing of the dredged material. One might think that it could be trucked back to the land from which it came. However, it is not the same as the soil that originally eroded from the land. Recall that sand, silt, and clay settle out at different rates and hence in different locations. The magnitude of the problem is illustrated by Baltimore Harbor in Maryland. Baltimore Harbor needs to be dredged, but the project has been held up for 15 years because of controversy over what to do with the dredged material. In this case, the material is a "mucky ooze" of fine clay containing about 90 percent water and a generous mixture of sewage and industrial chemical wastes that accumulated in the harbor over years.

In all, the Soil Conservation Service estimates that the impacts of sediments cost the United States over $6 billion each year but much of the damage goes uncorrected. Clearly, a solution to all these problems lies in practices that maintain soil on the land where it belongs.

## SOURCES AND CONTROL OF SEDIMENTS

The source of sediments is erosion. The major points of erosion are plowed croplands and over-grazed rangelands, clearcut forests, construction sites, and surface mining sites. Erosion also occurs from stream and river banks because of increased stormwater runoff resulting from development, as discussed in Chapter 8. Which of these is most important will depend upon the activities in the given region. In each case practices can be adopted or amended to minimize erosion and/or hold the sediments on the site.

### Agriculture and Related Activities

In Chapter 7 we discussed erosion from plowed fields, clearcut forest land, and over-grazed rangeland as well as control procedures.

Though soil conservation practices exist, landowners do not always adopt and maintain these practices. It has generally been considered that if a farmer, for example, lets his soil erode, it is his problem. However, both the damage caused by sediments and the long-term decline in soil productivity become problems for all of society. Hence, many believe that farmers should take it upon themselves to be more conscientious about soil conservation practices. It is important to note, however, that such practices can be costly, and farmers who adopt them may not be able to compete economically with those who produce crops less expensively, albeit at the long-term ecological expense of reduced soil fertility and polluted aquatic ecosystems.

Consequently, pressure is mounting for regulations that will require farmers and other large landowners to adopt suitable soil conservation measures. Under such regulations, all farmers would be on the same economic footing, and increased costs could be uniformly passed on to the consumer.

### Construction Sites

Approximately 0.6 million hectares (1.6 million acres) of land are affected each year by development of housing tracts, highways, and other construction. The subsoil exposed by construction activity has little capacity for infiltration, and slopes created by such activities are frequently steep and unstable. Therefore, there is a high percentage of runoff and resulting erosion is often severe. Water-gullied embankments along highways under construction are probably familiar to everyone (Fig. 9-5). Losses of 2500 tons of soil per hectare (1000 tons per acre) are not uncommon and may be as high as 25,000 tons per hectare (10,000 tons per acre). In other words, the soil lost from a construction site during one year may exceed what would be lost over 20,000 to 40,000 years under natural conditions. Since the entire runoff from a construction site almost always goes into a single stream, the impact from sediment may be tremendous.

Many techniques may be used to reduce the loss of soil from construction sites. Most commonly a sediment trap is constructed at the lower end of the site. The sediment trap is essentially a pond into which runoff from the site is chan-

**FIGURE 9-5**
The gullies in the embankments of this highway under construction attest to severe erosion. Such construction activities may be the most significant source of sediments entering waterways. (Photo by author.)

**FIGURE 9-6**
A sediment trap. Sediments eroding from construction sites may be trapped by making a "pond" and channelling the runoff into the pond. Sediments settle in the pond while the water overflows through the standpipe. Sediments may be periodically dredged from the pond and redistributed on the site. (USDA—Soil Conservation Service photo.)

nelled. As water enters, its velocity is reduced and sediment settles in the pond. Sediment-free water flows out over a rock dam or through a standpipe (Fig. 9-6). On small sites where the slope is gentle, the lower perimenter of the site may be diked with bales of straw (Fig. 9-7), which filter the runoff and remove sediment. These techniques do not serve to prevent erosion itself, but they trap and hold at least the heavier sand and silt sediments—the clays tend to wash on through. The trapped sediment may then be redistributed over the site and stabilized with sod.

A practice that aggravates the sediment problem is that the entire site is often left open and erodes during the entire period of construction. Alternatively much of the site can be resta-

**FIGURE 9-7**
Trapping sediments with hay bales. A row of hay bales placed across the path of runoff may serve to filter sediments from the water. This technique may be used where the amount of runoff and/or erosion is relatively small. (Photo by Tommy Noonan.)

**FIGURE 9-8**
Stabilizing earth with vegetation. In this photograph note that little if any erosion occurred where the bank was protected by grass, but severe erosion occurred where it was left bare. Leaving the sowing of grass until the very end of the construction process allows the maximum amount of erosion. Alternatively, erosion can be largely controlled by completing the final grading and sowing grass as soon as possible. (USDA–Soil Conservation Service photo.)

bilized with grass immediately after grading without interfering with later construction. This technique is particularly appropriate in highway construction (Fig. 9-8).

Unfortunately, having techniques available and getting developers to use them conscientiously are two different things. Unlike a farmer, a developer does not have a vested interest in saving the soil. Eroded soil makes as good a base upon which to lay concrete as does good soil. Grass sod can be laid over almost any base and, with watering, will survive. To many developers, soil conservation practices are just another nuisance, an added expense they want to reduce as much as possible. Erosion control on construction sites must therefore be promoted through legislative action and legal enforcement. The current trend is to add sediment control to other building codes or construction requirements so that developers submit sediment control plans along with other plans for site development. These plans are reviewed by the government and modified if necessary before a building permit is issued or a contract is let. Then, the site is inspected periodically to be sure sediment traps or other devices are installed and maintained properly.

Inspection and enforcement are frequently the weak link in this procedure. It is not uncommon to find sediment traps that are nonfunctional for lack of proper installation or maintenance. This is an issue where local citizen action can be effective.

Save Our Streams and Trout Unlimited are two citizen organizations whose members take it upon themselves to act as "inspectors" overseeing construction projects in their areas. When they find uncontrolled erosion threatening waterways they report it to the authorities. If the authorities fail to take action they report it to the local media. Media publicity frequently brings action when all else fails.

## Mining Sites

In the United States some 1.2 million hectares (3 million acres) have been disturbed by surface mining; furthermore, mining activities increase each year, especially as we turn increasingly to coal to meet our energy needs. Erosion from mining sites is similar to that from construction sites, but sites are generally larger and remain open for longer periods of time (Fig. 9-9). A federal law requiring reclamation of strip-mined areas was passed in 1977, but it has been unevenly enforced and attempts to weaken it persist. In the meantime, some 0.8 million hectares (2 million acres) of old mined soils still remain more or less exposed to ongoing erosion. The sediment control procedures that may be used on mining sites are similar to those described for construction sites but must be employed on a larger scale.

## Miscellaneous Sources

As we go about our lives we may notice patches around our schools, homes, and shopping centers where the earth has been denuded or seeded after construction did not succeed. Such areas often do not revegetate by themselves but remain open and erode indefinitely (Fig. 9-10). These patches may not be very large individually, but taken together they are the source of a significant portion of the sediment problem.

Government bureaucracies tend to be too large and cumbersome to deal with such small areas effectively. The best, and in many cases only, solution is for a few people to get together and do some raking, mulching, and reseeding. The Soil Conservation Service, which maintains

**FIGURE 9-9**
Erosion from mining sites. Strip mining leaves massive amounts of earth exposed to the forces of erosion. As a result, sedimentation of waterways draining such areas is often severe. Further pollution of waterways results from the leaching of various chemicals from the disturbed rock formations. (Francis Current/Photo Researchers.)

**FIGURE 9-10**
Following construction there are frequently numerous patches, such as this one on a highway embankment, that fail to be adequately stabilized. The continuing erosion from such patches adds significantly to the sediment burden of waterways. (Photo by author.)

an office in every county of the United States, does not have the work force to do the work itself, but is generally enthusiastic about providing support with advice and technical assistance.

In conclusion, we should emphasize again that sediment problems stem from the removal of natural vegetation. While impacts may be reduced by various measures such as sediment traps and brought under control by revegetation, it would obviously be much simpler if the land were not disturbed in the first place. Perhaps at first this strikes one as being ridiculous, but once we begin to ask if the new highway, shopping center, agricultural land, or whatever is really needed, more thorough investigation frequently reveals that the answer is no. What is promoted as a "need" is often simply an attempt to support an underlying but outmoded policy of profligate use of resources, energy, and land to enhance the profits of developers. Thus, much of the solution ultimately lies in changing these policies, as will be discussed in Chapter 24.

# 10

# WATER POLLUTION DUE TO SEWAGE

## CONCEPT FRAMEWORK

Outline                                                         Study Questions

1. What are *parasitic organisms* and how do they enter the environment? Why are humans more vulnerable to the spread of disease than animals? What measures have been taken in developing countries to interrupt disease cycles caused by sewage-borne pathogens? What is the situation in less-developed countries?

2. What is *dissolved oxygen* and what causes the depletion of dissolved oxygen? What are the effects of this depletion? What is *biological oxygen demand*?

3. What is *eutrophication*?

4. What are the two phases of sewage handling and treatment?

5. Prior to the adoption of modern sewage treatment methods, how were wastes handled? What changes were made as a result of Louis Pasteur's discovery of bacteria? Why did the Thames River in London become a public health threat? What did the Clean Water Act do to improve sewage treatment? What are some of the more pressing sewage problems that need to be addressed?

6. What are the three stages of sewage treatment? What is *tertiary treatment?*

7. Describe *pretreatment.*

8. How is *primary treatment* conducted? What does it accomplish? What effects does pretreatment have on the ecosystem when it is not followed up with primary treatment?

9. Why is secondary treatment referred to as *biological treatment?* Describe *trickling filters* and *activated sludge systems* and the advantages of each. According to the Environmental Protection Agency, how much sewage in the United States is discharged with only primary treatment or no treatment at all?

10. What steps may be taken to halt eutrophication? Why is chlorination controversial? What are the alternatives?

11. How could water conservation help the sewage treatment process?

12. What is *raw sludge?*

13. Describe *anaerobic digestion* and the sludge that remains when the process is complete.

14. How is sludge composted?

15. What has prevented the widespread use of treated sludge?

16. Name the heavy metals that are particularly dangerous when discharged into public systems. What problems do they cause?

17. How do septic tanks and drain fields operate? What are the advantages and disadvantages? Describe the *Clivus Multrum method.*

18. What is *E. coli* and how is it used as a measure of contamination?

While sediments are the number one pollution problem in terms of volume, raw sewage—that is, untreated excrements from humans and other animals—is the greatest threat to public health. Further, discharge of sewage wastes into waterways, even after treatment to reduce disease hazards, may cause severe ecological upsets. Nevertheless, sewage wastes are nutrient-rich materials, which, when handled properly, may be used as a resource to achieve the ecological ideal of nutrient recycling.

The objective of this chapter is to understand sewage and sewage treatment and to examine the techniques that may solve problems associated with these wastes.

## POLLUTION PROBLEMS RESULTING FROM SEWAGE WASTES

### Disease Hazard

Raw sewage presents a significant public health hazard becuase humans and other animals infected with pathogens—disease-causing bacteria, viruses, and other parasitic organisms—generally discharge large numbers of these organisms or their eggs in their fecal excrements. When discharged into the environment, these parasitic organisms may infect other individuals and cause the spread of disease (Fig. 10-1).

The spread of disease may occur directly if sewage wastes contaminate drinking water, food supplies, or water used for swimming or wading; or disease may be spread indirectly through various food chains. For example, animals may become infected with parasites from human excrement; humans may then become reinfected if they eat the animals without sufficiently cooking them.

Humans are not the only animals vulnerable to this problem. It also occurs in the natural ecosystem. Recall from Chapter 3 that parasitic organisms are among the most important biotic factors holding natural populations in check. In general, pathogenic organisms survive only a few days outside of a host, and the number of organisms ingested or otherwise contracted by the host is an important factor in determining subsequent infection. When populations of host animals are sparse, levels of contamination remain low and

more time may elapse between the elimination by one host and contact by the next. Thus, relatively little transfer of the pathogenic organisms occurs. As populations become denser, however, the reverse is true. Humans tend to live and work in high-density urban situations making themselves vulnerable to the spread of disease.

In the mid-1800s, before the connection between disease and sewage-borne pathogens was discovered and suitable public health measures were instituted, disastrous epidemics were common in cities. Occasionally such epidemics killed as much as a third of the population of a city. Typhoid fever, for example, killed hundreds of people and was common in the United States until the turn of this century. Today, public health measures that interrupt the disease cycle have been adopted throughout most countries. These measures involve (1) disinfection of public water supplies with chlorine or other agents, (2) personal hygiene, cleanliness, and sanitation, especially in relation to the preparation, handling and serving of food, cooking and eating utensils, and bathroom functions, and (3) proper collection and treatment of sewage wastes.

Despite these control measures, however, problem areas remain. Some communities still discharge raw sewage into natural waterways. Many private septic systems are overflowing and may contaminate water with sewage wastes. In many cities, sewage collection and treatment facilities are old and inadequate to handle loads of increasing populations. Hence, overflows and leaks of raw or poorly treated sewage occur at least intermittently.

In less-developed countries, much more severe problems exist. Most cities do not have sewage systems capable of meeting demands. In these countries, inadequate collection, disposal, and/or treatment of sewage wastes is a primary factor in drinking water contamination. Hence, there is an extremely high rate of sickness and mortality, especially among infants and children. And the problem is getting worse because of rapid growth and urbanization of populations. With increasing amounts of raw sewage going uncollected or being discharged directly into natural waterways without treatment, public health officials are concerned that widespread, uncontrollable disease epidemics as well as ecological problems may result.

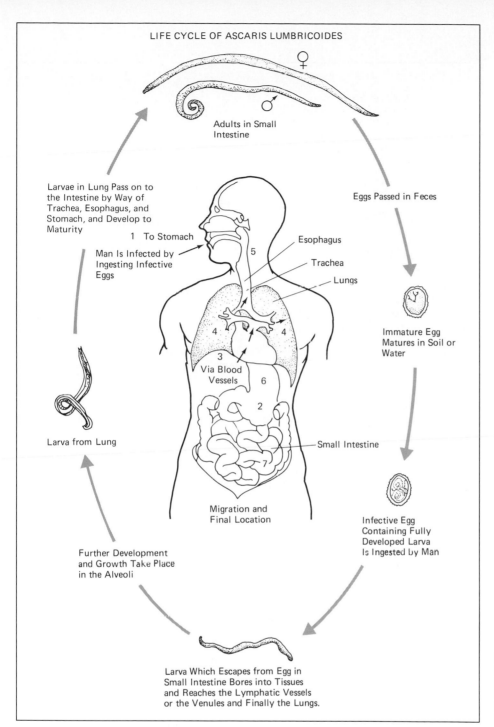

LIFE CYCLE OF ASCARIS LUMBRICOIDES

Adults in Small Intestine

Larvae in Lung Pass on to the Intestine by Way of Trachea, Esophagus, and Stomach, and Develop to Maturity

1 To Stomach

Man Is Infected by Ingesting Infective Eggs

5

Esophagus

Trachea

Lungs

4    4

3
Via Blood Vessels

6

2

Larva from Lung

Small Intestine

7

Migration and Final Location

Eggs Passed in Feces

Immature Egg Matures in Soil or Water

Infective Egg Containing Fully Developed Larva Is Ingested by Man

Further Development and Growth Take Place in the Alveoli

Larva Which Escapes from Egg in Small Intestine Bores into Tissues and Reaches the Lymphatic Vessels or the Venules and Finally the Lungs.

**FIGURE 10-1**
Life cycle of a parasitic roundworm *Ascaris lumbricoides.* (Courtesy of Gary E. Kaiser, Catonsville Community College.)

## Depletion of Dissolved Oxygen in Waterways

The discharge of raw sewage into waterways not only leads to disease, it may also cause the collapse of aquatic ecosystems because the raw sewage depletes dissolved oxygen in the waterways. As a result, fish and shellfish suffocate.

The problem occurs because the undigested organic matter present in fecal material or other wastes is readily digested by decomposers and detritus feeders. These organisms also consume **dissolved oxygen** through respiration. Because water holds only a minute amount of dissolved oxygen compared to the atmosphere, and because

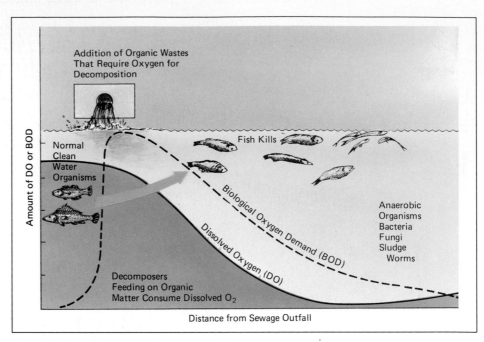

**FIGURE 10-2**
Dissolved oxygen (DO) and biological oxygen demand (BOD). Fish and other aquatic organisms depend on dissolved oxygen for their respiration. Additions of organic matter create an additional biological oxygen demand because bacteria (decomposers) that feed on the organic matter consume oxygen as well. Thus, additions of organic wastes to waterways may result in fish kills through depletion of dissolved oxygen. After the organic wastes are fully decomposed, recovery of dissolved oxygen will occur as atmospheric oxygen slowly dissolves in the water.

entry of more oxygen from the atmosphere or by photosynthesis is a slow process, loading an aquatic system with too much organic matter from sewage or other sources causes an excessive proliferation of aquatic decomposers. These organisms consume oxygen faster than it enters the system, thus, depleting the oxygen supply (Fig. 10-2).

The depletion of dissolved oxygen does not eliminate the decomposers themselves because some are capable of surviving by means of anaerobic respiration. They consume oxygen only when it is available. Thus, dissolved oxygen may remain at or near zero as long as there is an overload of organic matter supporting the excessive population of decomposers. Such an anaerobic body of water is not only incapable of supporting fish and shellfish, it also smells bad because many of the waste products of anaerobic metabolism are compounds with an unpleasant odor. Indeed, this process is what gives sewage its characteristic smell.

In addition, depletion of dissolved oxygen may aggravate microbial pollution. Pathogenic organisms tend to die off quickly or are consumed in an oxygen-rich environment. They survive much longer in an anaerobic or oxygen-poor environment. Consequently, such organisms are more likely to be present in water with low dissolved oxygen.

Clearly, dissolved oxygen, commonly abbreviated **DO,** is an extremely important and frequently measured parameter of water quality. The potential for organic materials to absorb dissolved oxygen through the decomposition process is an equally important consideration. Wastes and organic materials present in natural waterways are commonly measured in terms of their **biological oxygen demand,** the amount of oxygen that will be consumed by their biological decomposition. The biological oxygen demand is often abbreviated and referred to as **BOD.**

## Eutrophication

Another environmental problem resulting from the discharge of sewage into waterways is **eutrophication.** Eutrophication literally means true- or overfeeding. The eutrophication process occurs when the addition of nutrient-rich material from sewage or other sources stimulates the growth of algae, especially undesirable species. The eventual death of the algae creates an oversupply of detritus, which leads to depletion of the dissolved oxygen and resultant fish kills.

Eutrophication is part of the normal aging process or succession that occurs in natural ponds and lakes (see Chapter 3). However, the tremendous acceleration of the process that results from human causes is obviously an unfavorable altera-tion that we need to avoid. We shall examine this further in Chapter 11.

## SEWAGE HANDLING AND TREATMENT

Handling and treatment of raw sewage involves two distinct phases: collection and removal of sewage from the immediate human proximity, that is, from homes and workplaces, and suitable treatment and/or disposal of wastes to minimize the threat of diseases and ecological problems. These two phases are independent. Adequate treatment and ultimate disposal does not result from clean bathrooms and the flushing of wastes away from houses and workplaces. This becomes clear with an understanding of the development of sewage systems.

### Background

Prior to the mid-1800s, the general means of sewage disposal was the outdoor privy or collec-tion of the "night soil" from pots kept in the home and dumped (Fig. 10-3). In some areas, es-pecially in developing countries, these methods are still practiced. Historically, this situation re-sulted in contamination of drinking water sup-plies and disease, especially in cities where pri-vies and wells were located near each other.

The discovery by Louis Pasteur and others in the mid-1800s that sewage-borne bacteria were responsible for many infectious diseases led to in-tensive efforts to rid cities of wastes as expedi-ently as possible. Many cities already had drain-age systems to remove stormwater, but these systems were not used to remove human wastes. In fact, until the turn of this century, many cities had ordinances prohibiting the dumping of wastes into the storm drains. However, thinking quickly changed. The flush toilet was developed and sewers were tapped into storm drains. Storm drains, thus, became sanitary sewers carrying raw wastes into streams, rivers, or other receiving bodies (i.e., any body of water that receives wastes).

As long as the receiving body is large com-pared to the input of sewage, and as long as the wastes are adequately mixed and diluted, the nat-ural ecosystem can handle additional inputs with-out upset. However, as waste loadings increase, a critical point is reached. Dissolved oxygen is de-pleted and the existing ecosystem collapses into one with high levels of disease organisms, obnox-ious odors, and other undesirable qualities.

**FIGURE 10-3**
Outhouses were the widely used means of sewage disposal through the turn of the century and remain in use in many rural locations. This photograph also shows that dropping wastes into streams was accepted as a means of flushing them away. (USDA–Soil Conservation Service photo.)

By the late 1870s, many bodies of water including the Thames River in London were so noxious and presented such a public health menace, it was clearly evident that more had to be done. Construction of sewage treatment facilities began at major sewage outfalls. However, progress has been uneven. In the United States, the first municipal sewage treatment facilities were built near the turn of this century. However as late as the early 1970s a considerable number of cities were still discharging raw sewage into receiving bodies. Also, many small outfalls of raw sewage remained where sewers had not been tied into the central systems. Increasing population and urbanization added to the problem by overburdening existing systems. Signs reading "No Swimming! Polluted Water!" proliferated along formerly clean rivers and beaches.

The Clean Water Act of 1972 offered a tremendous impetus toward cleaning up. The act provided federal money for new interceptor sewers to collect wastes and for the construction, expansion, and upgrading of sewage treatment plants. As a result, many waterways are much cleaner now than they were in 1972. Still, many problems remain.

Continuing expansion of urban areas and increasing water use (and hence sewage output) in modern homes increase loads on sewer lines and treatment facilities, many of which are more than 100 years old and nearing a state of collapse. Building codes now require the installation of separate sanitary and storm drain systems, but because of past practices of tapping sanitary sewers into storm drains, large amounts of stormwater still enter sanitary sewer systems, especially in older cities. As a result, leaks, breaks, backups, and overflows of raw sewage still occur and the need to expand and replace sewer lines and treatment facilities continues.

Meanwhile, other problems persist. What to do with wastes that are removed from sewage effluents in the treatment process and how to handle the nutrients that cause eutrophication are two of the problems that will be explored when we discuss the process of sewage treatment itself. Another problem that cannot be ignored is the situation in developing countries, many of which have no central sewer system, let alone treatment plants. In many cities in these countries, sewage from homes empties into roadside gutters, moves into storm drains, and is discharged into rivers which are becoming increasingly polluted.

## Traditional Processes of Sewage Treatment

Raw sewage is about 1000 parts water for every part of polluting material because the typical flush toilet uses 3 to 5 gallons of water to flush away small amounts of waste. In addition, gray water—waste water from sinks and tubs that is not contaminated with sewage—empties into the same collection system. Still, the polluting materials that do contaminate waste water are important.

The major pollutants considered here are *microorganisms* (especially pathogens), *detritus* dispersed as fine particles, and *dissolved nutrients* such as nitrogen compounds and phosphate which are contained in urine (Table 10-1). The task of sewage treatment is to remove or destroy these polluting materials before releasing effluent waters into the environment.

Sewage treatment as it is commonly practiced in the United States consists of three stages: **pretreament, primary treatment,** and **secondary treatment.** These stages remove most of the organic matter including microorganisms, but they do not remove dissolved nutrients. Additional treatment, referred to as **advanced treatment** or **tertiary treatment,** is now being practiced to remove dissolved nitrate and phosphates.

### PRETREATMENT

Pretreatment, as the name suggests, involves the steps required before removal of actual pollutants can begin. The sewage stream may carry large pieces of debris such as rags and plastic bags or coarse sand or grit. If these materials remain, they will clog or hinder later stages. They are removed by letting the water flow through a **bar screen,** a row of bars mounted about 2.5 cm (1 in.) apart (Fig. 10-4). Debris is mechanically removed from the screen and taken to be incinerated. Coarse sand and grit, which are carried along sewer lines, are removed by passing the water through **grit chambers** where its velocity is slowed just enough to permit such material to settle (Fig. 10-5). The settled material is mechanically removed from these tanks and generally put in landfills.

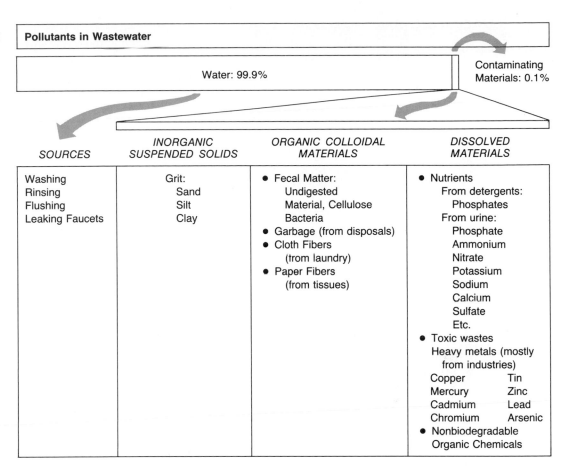

| Table 10-1 | **Pollutants in Wastewater** | | | |
|---|---|---|---|---|
| | Water: 99.9% | | | Contaminating Materials: 0.1% |
| *SOURCES* | *INORGANIC SUSPENDED SOLIDS* | *ORGANIC COLLOIDAL MATERIALS* | *DISSOLVED MATERIALS* | |
| Washing<br>Rinsing<br>Flushing<br>Leaking Faucets | Grit:<br>  Sand<br>  Silt<br>  Clay | • Fecal Matter:<br>  Undigested<br>  Material, Cellulose<br>  Bacteria<br>• Garbage (from disposals)<br>• Cloth Fibers<br>  (trom laundry)<br>• Paper Fibers<br>  (from tissues) | • Nutrients<br>  From detergents:<br>    Phosphates<br>  From urine:<br>    Phosphate<br>    Ammonium<br>    Nitrate<br>    Potassium<br>    Sodium<br>    Calcium<br>    Sulfate<br>    Etc.<br>• Toxic wastes<br>  Heavy metals (mostly<br>    from industries)<br>  Copper      Tin<br>  Mercury    Zinc<br>  Cadmium   Lead<br>  Chromium  Arsenic<br>• Nonbiodegradable<br>  Organic Chemicals | |

**FIGURE 10-4**
Bar screen. Channelling the waste water through a bar screen is the first step of pretreatment. The screen removes large pieces of debris. Such debris is removed from the screen by a mechanical rake. (Washington Suburban Sanitary Commission photo.)

**FIGURE 10-5**
Grit chamber, the second step of pretreatment. The velocity of flow through these chambers is slowed to 1 to 2 feet per second, which allows sand and other coarse grit to settle to the bottom while the water flows over the top edge (foreground). The grit is removed by mechanical plows scraping the bottom.
(Photo by Tommy Noonan, Washington D.C.)

## PRIMARY TREATMENT

After pretreatment, the water moves on to primary treatment where it flows very slowly through large tanks called **primary clarifiers** (Fig. 10-6). While flowing slowly through these tanks, the water is nearly motionless for several hours. This permits the heavier particles or organic matter, about 30 to 50 percent of the total, to settle to the bottom where they can be removed. At the same time, fatty or oily material floats to the top and is skimmed off. All the material removed is known as **raw sludge,** and its treatment and disposal will be considered shortly.

It is intersting to note that the term *treatment* suggests a complex chemical or technological pro-

cess. At this point treatment is nothing more than letting material sink to the bottom of a tank. Nevertheless, it accomplishes significant removal at minimal cost.

In some cases, pretreatment is the only treatment provided. When this occurs, water from the clarifier is chlorinated to kill pathogens and discharged into natural waterways. Conspicuously, such waste water still carries a large amount of organic matter and hence its high biological oxygen demand may deplete the dissolved oxygen in the receiving waterway. Also, the presence of chlorine may cause other problems, which will be considered later.

However, under pressure of the Clean Water Act, sewage treatment in most communities of the United States has been upgraded to include at least secondary treatment which is designed to remove most of the organic material remaining after the primary process. Where secondary treatment is included, the water flows directly from the primary to the secondary process.

## SECONDARY TREATMENT

Secondary treatment is also called **biological treatment** because it employs organisms that literally consume the organic matter and break it down through their cell respiration to carbon dioxide and water. Either of two types of systems may be used: *trickling filters* or *activated sludge systems*.

In **trickling filter** systems, the water is sprayed onto and allowed to trickle through a bed of fist-sized rocks (Fig. 10-7). The environment is similar to a natural stream. The well-aerated water, rich in organic matter, supports several trophic levels of organisms attached to the rocks. Bacteria are the first organisms to attack and digest organic particles but the food chain continues through protozoans, rotifers, and various small worms so that biomass is reduced (Fig. 10-8). Organisms and clumps of detritus that wash from the trickling filters are removed when the waste water moves from trickling filters into secondary clarifiers, tanks similar to the primary clarifiers. Material that settles here is handled as raw sludge. Through primary treatment and the trickling filter system, 85 to 90 percent of the total organic matter is removed from the waste water.

The alternative secondary treatment method, the **activated sludge system,** passes the

(a)

Barrier
Blocks Overflow
of Scum

Clarified
Water

30-50 Percent of Organic
Material Settles

Sludge ← | ← Waste Water

(b)

**FIGURE 10-6**
Primary clarifiers used for primary treatment. (a) The water
enters these tanks at the center and exits over the wires at the
edge. The slow velocity (1–2 inches per minute) of flow
through the tanks permits 30 to 50 percent of the colloidal
organic matter to settle while oil and grease rise. Settled
organic material, raw sludge, is pumped from the bottom while
oil and grease are simultaneously skimmed from the surface.
(Photo by Tommy Noonan, Washington, D.C.)
(b) Cross section of clarifier.
(Courtesy of Walker Process Division of C.B.I.)

water from primary treatment into a tank that
could hold several tractor trailer trucks parked
end to end (Fig. 10-9). As water moves slowly
through the tank, it is vigorously aerated. Again,
an environment ideal for the growth of organisms
is created. A mixture of organisms, referred to as

**activated sludge,** is added to the water entering
the tank. As these organisms feed, the biomass of
organic matter is reduced. When the waste water
leaves the aeration tank, it still contains the rich
mixture of the feeding organisms. Therefore, from
the aeration tank, the water goes into a secondary
settling tank. Since the feeding organisms are
usually clumped on bits of detritus, settling is relatively efficient. The settled organisms are continuously pumped from the bottom of the settling
tank back to the entrance of the aeration tank, assuring a continuing high population of active organisms (Fig. 10-10).

Through the settling process of primary
treatment and the activated sludge process, 90 to

(a)

(b)

**FIGURE 10-7**
Trickling filters, the old method of secondary treatment. (a) The water from the primary clarifiers is sprinkled onto and trickles through a bed of 6 to 8 feet of rocks. (b) Various bacteria and other detritus feeders adhering to the rocks consume and digest the organic material in the water as it trickles by. The water is again collected at the bottom of the filters. (Photos by author.)

**FIGURE 10-8**
(a) Some of the organisms that are active in secondary treatment. (b) These organisms form a biomass pyramid of detritus feeders. Through this pyramid the biomass of organic matter entering the system is reduced by up to 90 percent.

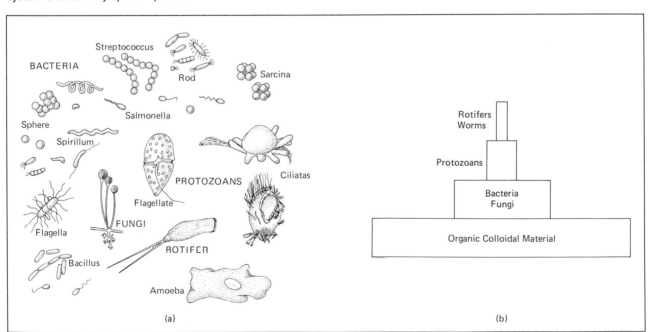

(a)

(b)

**FIGURE 10-9**
Aeration tank used in activated sludge treatment. As waste water from the primary clarifiers moves through the tank it is vigorously aerated by air forced up from the tubes along the bottom. (Photo by author.)

**FIGURE 10-10**
Activated sludge treatment. (a) In the oxygen-rich environment of the aeration tank, microorganisms consume the organic matter. Organisms (activated sludge) settle out in the secondary clarifier and are returned to the aeration tank while the clarified water flows on. (b) Aeration tank in operation. The discharge from the pipe in the foreground is the return of activated sludge. (Photo by author.)

(b)

95 percent of the organic matter that was present in the raw sewage is removed.

In construction of new wastewater treatment plants or in upgrading old ones, the activated sludge system is generally installed because it offers several advantages. First, it removes a somewhat higher proportion of the organic material, 90 to 95 percent versus 85 to 90 percent for the trickling filter. Second, it is a more compact system, which saves limited space. Third, it is less subject to clogging. It has one notable disadvantage, however. Activated sludge systems require large amounts of energy for the aeration pumps. With rising energy costs, activated sludge systems are becoming increasingly expensive to operate. Trickling filter systems, on the other hand, are generally gravity feed systems, and their operation requires little if any additional energy. It is important to note that neither secondary treatment system removes dissolved nutrients, the major cause of eutrophication.

After secondary treatment the water is generally disinfected with chlorine to kill any pathogenic organisms that may persist and is then discharged into a natural waterway (Fig. 10-11). A summary of the sewage treatment process is shown in Figure 10-12 on pages 282–283.

In 1972 the U.S. Environmental Protection Agency **(EPA)** estimated that nearly a third of the sewage in the United States was discharged with only primary treatment or no treatment at all. It is hardly surprising that many lakes and rivers were notoriously polluted. As mentioned earlier, the Clean Water Act of 1972 attempted to improve the situation by establishing a framework for cooperation between federal, state, and local agencies. Under the act, state and local governments prepare plans and the federal government provides most of the financing for new or improved sewage treatment facilities. Over the years, small, inefficient plants have been phased out in favor of larger, more efficient plants. Moratoriums were and still are declared on new developments if adequate sewage collection and treatment are unavailable. Now, most of the sewage in the United States does receive at least secondary treatment. This has markedly improved water quality and many formerly polluted beaches have been reopened.

## Remaining Problems and Advanced Treatment

While the combination of primary and secondary treatment effectively removes organic matter, and chlorination greatly reduces the disease hazards, problems remain. First, the nutrients remaining in the water may cause eutrophication; second, the use of chlorine as a disinfecting agent may be hazardous.

Where eutrophication is a problem, the Clean Water Act mandates that major sewage treatment plants be upgraded to advanced treatment. Advanced treatment follows secondary treatment and removes one or more of the nutrients from the water. Numerous methods are available for this task. For example, the water could be 100 percent purified by distillation, chemical filters, or microfiltration. The problem is cost. The total flow is about 500 $\ell$ (about 150 gal) per person per day, 150 million gallons per day for a population of 1 million. Purifying this amount of water by these methods would be very expensive. However, the costs might be justified in water-short areas as the quality of the treated water would be so high it could be recycled back into the municipal water supply.

Many people pale at the idea of recycling waste water in this way, but we should remember that all water is recycled by nature. These methods provide better-quality water than is obtained by drawing water from a river into which upstream cities have discharged sewage. This occurs along the Mississippi and other major rivers of the world.

**FIGURE 10-11**
Tanks of chlorine gas used in disinfecting waste water. Is this purification, or the addition of one more toxic substance? (Photo by author.)

However, less expensive alternatives to solve the eutrophication problem are available. In most natural water, growth of undesirable species of algae is held back by lack of phosphate ($PO_4$). Therefore, removal of phosphate alone may suffice. One method involves passing the water through a lime filter in which phosphate combines with calcium and becomes insoluble. Another process is similar to the activated sludge system. Microorganisms that absorb phosphate are introduced and then removed. Or one can choose not to discharge the nutrient-rich waste water into a natural waterway and use it instead for irrigation or culturing desired aquatic plants. Finally, as much as two-thirds of the phosphate present in waste water comes not from human excrement, but from detergents. The substitution of low- or no-phosphate detergents would mitigate the problem of eutrophication. All of these alternatives will be discussed further in Chapter 11.

The use of chlorine gas ($Cl_2$) for disinfection also presents environmental problems that need consideration. First, chlorine is used because it is both effective and relatively inexpensive. But chlorine gas is extremely toxic; transporting tank car loads from points of manufacture to water and sewage treatment plants presents a serious accident hazard. Several accidents involving tankers of chlorine have required the evacuation of large numbers of people. It has been more luck than good management that has kept many people from being killed during these accidents. Second, chlorine is exceedingly toxic to some fish. Levels that are too low to measure have been found to inhibit the hatching of trout eggs and development of the embryos. Furthermore, chlorine may upset the entire ecosystem by modifying food chains.

Finally, it has been found that to some extent chlorine reacts spontaneously with organic compounds to form chlorinated hydrocarbons, organic molecules with chlorine atoms attached. Many of these compounds are toxic and nonbiodegradable, and some have been identified as compounds that cause cancer, abnormal development, and reproductive problems.

Hence, disinfecting agents other than chlorine may be more satisfactory. One is ozone ($O_3$). Ozone is extremely effective in killing microorganisms, and in the process it breaks down to oxygen gas, which improves water quality. However, ozone is not only toxic, it is explosive. Thus, ozone must be manufactured at the point of use, a step that demands considerable capital investment and energy. With improved technology, however, costs might be comparable to the use of chlorine, and safety hazards might be reduced. Another suggestion is to pass the water under an intense source of ultraviolet or other radiation that would kill microorganisms but would not otherwise affect the water. Also, after chlorine disinfection, other chemicals, such as sulfur dioxide, may be added that react with chlorine to form inactive, harmless compounds.

## Water Conservation and Sewage Treatment

In all phases of wastewater treatment, cost is proportional to the volume of water being treated. Further, both primary and secondary treatment processes are time dependent—time is needed for settling and for biological decomposition. The more water forced through a treatment plant, the faster it must move. Hence, there is less time for settling and decomposition and the quality of treatment diminishes accordingly. Hence, water conservation, while generally promoted on the basis of saving water resources, would also significantly reduce the costs and increase the efficiency of sewage treatment processes.

The fear of reduced water flow causing sewer line clogging is not realistic. Cutting water use in half would, in effect, double the pollutant concentration. However, since raw sewage is only about 0.1 percent solids, such doubling would still make raw sewage only 0.2 percent solids (99.8 percent water). This would not significantly effect its flow through the system.

## SEWAGE SLUDGE: WASTE OR RESOURCE

Recall that 30 to 50 percent of the organic matter present in raw sewage is removed in primary treatment. This material is called *raw sludge*. Pumped from the bottom of the primary settling tanks, raw sludge is a black, foul-smelling, syrupy liquid of about 98 percent water and 2 percent organic matter that includes many pathogenic organisms. With more widespread and efficient sewage treatment, the volume of raw sludge has grown and its disposal has presented problems. In the past, it was incinerated, landfilled, or simply barged to sea and dumped. All of these prac-

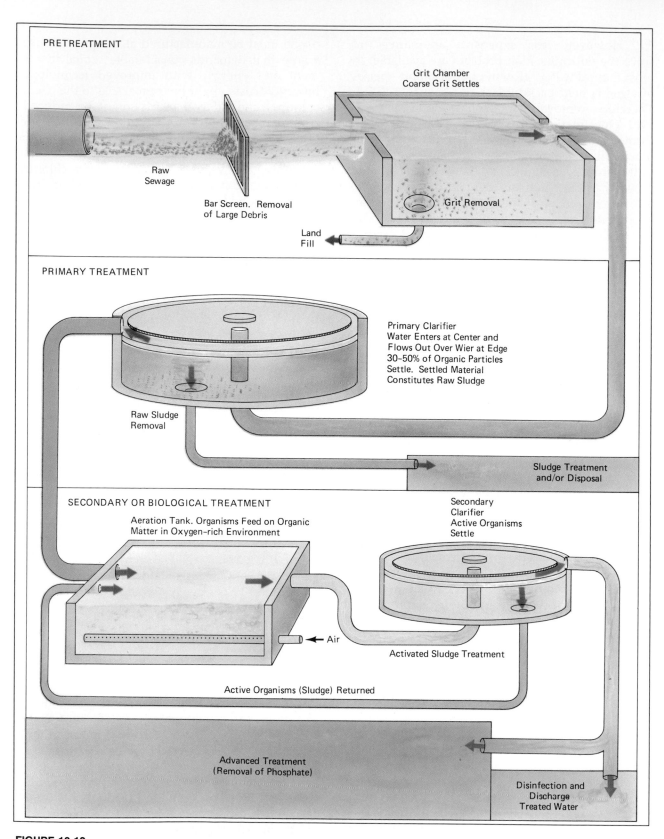

**FIGURE 10-12**
Diagram (a) is a summary of wastewater treatment through secondary treatment. Further purification processes, referred to as advanced treatment, may be added on after secondary treatment.

**FIGURE 10-12** (cont.)
Aerial view of a large sewage treatment plant. Indicated on the photograph: (b) grit settling tanks; (c) primary clarifiers; (d) trickling filters; (e) sludge digesters. (Baltimore Environmental Center, Inc., photo.)

tices create other pollution problems that can be avoided. Raw sludge can be treated so that it is not a waste at all, but a valuable resource. The two basic sludge treatment methods are **anaerobic digestion** and **composting.**

## Anaerobic Digestion

In anaerobic digestion the raw sludge is put into large airtight tanks called **sludge digesters** (Fig. 10-13). Bacteria feed on the organic matter

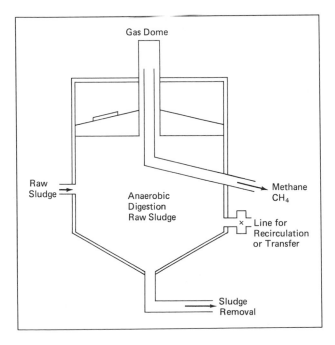

**FIGURE 10-13**
Treatment of raw sludge by anaerobic digestion. The raw sludge withdrawn from primary clarifiers (primary treatment) may be treated by placing it in airtight tanks for a period of 6 to 8 weeks. In the absence of oxygen, bacteria feed on the sludge (anaerobic digestion), producing biogas (about 60 percent methane) as a waste product and leaving a reduced amount of stable humuslike sludge (treated sludge) as a byproduct. (EPA.)

but, in the absence of air, they carry on anaerobic respiration. An important waste product of anaerobic respiration is **biogas,** which is a mixture of gasses about two-thirds of which is **methane.** Methane is the same as the natural gas used widely in heating and cooking. Thus, the biogas from digesters, can be burned. In fact, some 50 percent of it is generally used to heat the digesters themselves since their efficient operation requires that they be maintained at about 38° C (100° F). However, the gas from the digester is generally considered a low-quality fuel. In the past it was not deemed cost effective to purify it for other uses. Therefore, common practice has been simply to burn the excess gas in the air as a means of disposing of it (Fig. 10-14).

With the rise in natural gas prices, however, the economic balance has shifted. Raw sludge is now viewed as an economical source of methane and its use is being promoted. However, the amounts involved are relatively small when compared with our total energy consumption. The quantity of methane that may be obtained from a

**FIGURE 10-14**
Flaring of biogas. The biogas produced by anaerobic digestion is flammable. Flares are burning, but the clear, smokeless flame is not visible. Biogas is commonly used to heat the digesters to keep them at an optimum temperature, but the excess is frequently simply burned off (flared). However, more attention is now being given to its potential use as a "free" fuel.
(Photo by author.)

**FIGURE 10-15**
Use of treated sludge. The treated sludge remaining after anaerobic digestion is a humus- and nutrient-rich liquid that is an excellent soil conditioner. The vehicle shown is specially designed for applying sludge to soils. (Courtesy Big A Equipment Co.)

**FIGURE 10-16**
Filtering of treated sludge. (a) A vacuum inside the rotating drums sucks water from the sludge as it is applied to the belts. (Photo by Tommy Noonan, Washington, D.C.) (b) A semisolid sludge cake of humuslike material results. This material also has great value as a soil conditioner but unfortunately most is either dumped in landfills or burned. (Photo by author.)

sewage treatment plant is about 0.03 m³ (1 ft³) per day per person served. By comparison, through a gas heater, an average individual uses 0.7 to 0,8 m³ (25 to 30 ft³) per day just for personal hot water needs. It is interesting to note, however, that in China many small farmers maintain individual digesters into which they put agricultural wastes. They supply their own fuel needs for cooking in this way.

The sludge remains in the tanks from four to six weeks during which time digestion is more or less completed. What remains of the raw sludge is now called **treated sludge.** It is still a liquid, black with organic matter, but this organic matter is now relatively stable and odorless, and pathogenic organisms have been greatly reduced if not eliminated so they no longer present any great hazard. The end product is essentially the same as the humus resulting from decay in natural ecosystems. Like natural humus, it can improve soil quality. However, a note of caution must be added. Treated sewage sludges are frequently contaminated with toxic heavy metals such as copper from various sources. These can cause problems as will be discussed shortly.

Treated sludge may be applied directly to lawns and agricultural fields in the highly liquid state in which it comes from the digesters, thus providing benefit from both the humus and the nutrient-rich water (Fig. 10-15). Alternatively, the treated sludge may be filtered (Fig. 10-16a), leaving a semisolid humus **sludge cake** (Fig. 10-16b). Such a cake is easier to handle, but most of the nutrients are lost with the water, which is generally put back into the wastewater stream.

When sewage sludge is treated in this way, the resulting "humus" is usually available to the public free of charge. Numerous communities, especially in the Southwest, apply their sludge production to agricultural fields. Chicago is presently using much of its sludge to reclaim soils ravaged by strip mining (Fig. 10-17). Milwaukee, which has a particularly rich sludge resulting from the brewing industry, pasteurizes, bags, and sells it under the trade name "Milorganite."

## Composting

Raw sludge may also be treated by composting. In this process, raw sludge is filtered, mixed with wood chips or other material to improve aeration, and piled in windrows (Fig. 10-18). As

**FIGURE 10-17**
Soil conditioning value of treated sewage sludge. Plants on the right were grown on strip-mined soil treated with 10 percent sludge. Plants on the left were grown in untreated soil. (USDA photo.)

**FIGURE 10-18**
Treatment of raw sludge by composting. (a) Raw sludge is mixed with wood chips, which absorb the excess water, and the mixture is placed in piles through which air is drawn. (b) Diagram of the piles shown in (a). In the aerated piles, bacteria reduce the sludge to a nutrient-rich, humuslike material that can be used as a soil conditioner. Heat produced by the aerobic respiration of bacteria is sufficient to kill pathogenic organisms. (USDA photo by Robert C. Bjork.)

(a)

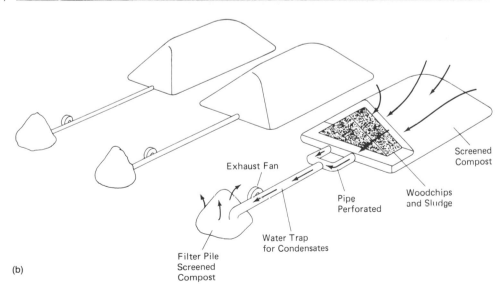

Exhaust Fan

Screened Compost

Woodchips and Sludge

Pipe Perforated

Water Trap for Condensates

Filter Pile Screened Compost

(b)

**FIGURE 10-18** (*cont.*)
(c) Woodchips are screened out
and reused. (© Peter Arnold, Photo
by Ray Pfortner, NYC.)

shown, additional air may be drawn through the piles to further increase aeration. As long as the system is kept well aerated, obnoxious odors that come from waste products of anaerobic respiration are negligible. Aerobic respiration of organisms breaks down organic material, pathogenic organisms lose out in the competition, and a nutrient-rich humuslike material is produced which again may be used for soil improvement.

### Using Treated Sludge

Unfortunately, many people still have the impression that treated sludge differs little, if at all, from the raw "stuff" that went down the drain. Consequently, not only do they refuse to use it themselves, they go so far as to bring court action against individuals or municipalities who do use it. Also, some people in government, especially those in the sewage treatment branch, believe they should not involve themselves in the "fertilizer business." The use of sludge for fertilizer is also handicapped by economics. Compared to inorganic fertilizers, the nutrient value of sludge is low on a per-ton basis. Consequently, the cost of energy required to haul it long distances defeats its cost-effectiveness.

It is interesting that while we in America generally reject putting treated sludge on lawns or agricultural fields, we have tolerated untreated sewage going into lakes and rivers that serve as water supplies. This attitude not only aggravates pollution; it prevents the use of a valuable organic fertilizer that could improve degraded soils.

### PROBLEMS OF TOXIC WASTES CONTAMINATING SEWAGE

In the past, and still continuing to some degree, it was common practice for industries to flush toxic chemical wastes into the municipal sewer system. The chemical wastes of particular concern are **heavy metals** such as lead and mercury and nonbiodegradable organic compounds such as chlorinated hydrocarbons.

Sewage treatment is in no way designed to handle such wastes, which do not settle and are not broken down by biological oxidation. Indeed, they poison the organisms used in secondary treatment and thus impede the process. They cause other problems in the environment when they are discharged with the water. Furthermore, they tend to bind with organic material and to this extent they end up contaminating treated sludge. Hence, the treated sludge from many cities has been unusable as a soil conditioner because of unacceptable levels of heavy metals.

Under mandate of the Clean Water Act, industries are now being required to pretreat their waste water, that is, to remove such toxic pollutants, *before* discharging into municipal sewer systems, or to find alternative means of disposal.

However, if the wastes are not put down the drain, what is to be done with them? This question will be addressed in Chapter 12.

## INDIVIDUAL SEPTIC SYSTEMS

In areas that do not have a centralized sewer system, a common practice is to install a septic tank and drain field for each individual home or commercial building (Fig. 10-19). The effluent from the home flows into a large tank where the heavier particles settle to the bottom. The water, still carrying much of the fine organic particles and nearly all the dissolved material, overflows into a system of drain tiles buried in the ground, and gradually percolates into the soil. Bacteria gradually digest the organic material that settles in the tank, reducing it to a stable humus that must be pumped out every two to three years. Likewise, soil bacteria decompose the organic material that comes through the drain tiles.

Septic systems seem more ecologically sound than centralized sewer systems in that they do return nutrients to the soil. In fact, some peo-

**FIGURE 10-19**
Individual sewage treatment.
Septic tank and drain field.
(USDA–Soil Conservation Service.)

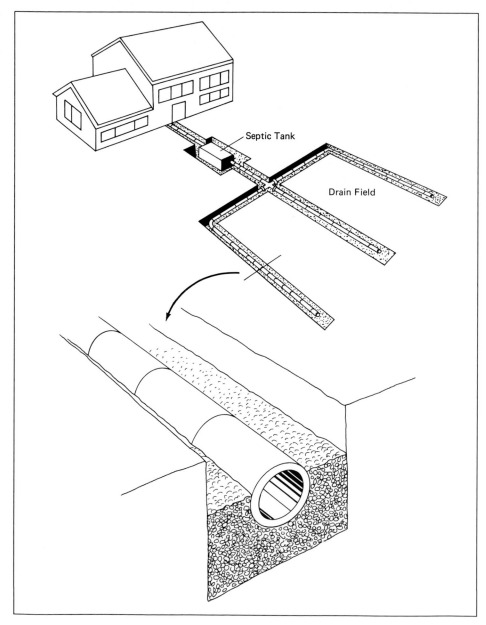

Septic Tank

Drain Field

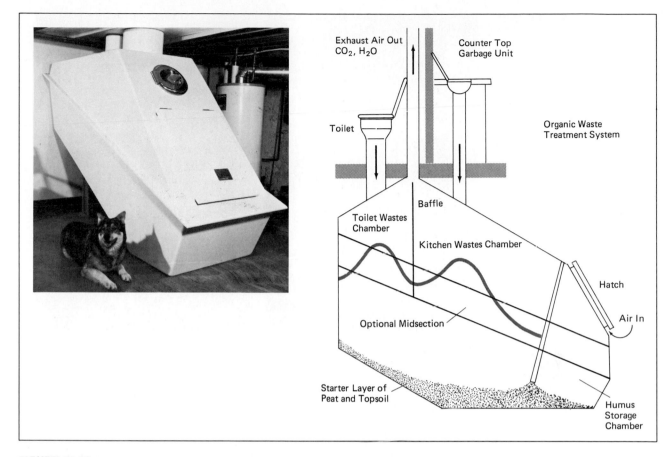

**FIGURE 10-20**
Clivus Multrum, a dry waste treatment system. Sewage and garbage wastes deposited in the Clivus Multrum decompose aerobically. A dry humus- and nutrient-rich compost is removed from the Clivus Multrum as a byproduct of treatment. (Courtesy of Clivus Multrum, W. Pa., Inc.)

ple establish successful vegetable gardens over septic drain fields, thus using and recycling the nutrients from wastes rather than using commercial fertilizer from nonrenewable resources.

If circumstances are ideal, a septic system from which the humus material is periodically removed from the tank may function indefinitely. However, organic material frequently enters the soil faster than it decomposes, gradually clogging the soil pores and preventing percolation of the water. With nowhere else to go, effluent with the objectionable organic material forces its way to the surface where it is unsightly, causes objectionable odors, contaminates surface water, and is a general health hazard. If the lot is not large enough to relocate the drain field, little can be done about this problem except to try to get centralized sewage as soon as possible. In many

areas of the country, expensive suburban developments using individual septic systems have become quite obnoxious as the septic systems have failed and lot sizes have not permitted relocation. How long individual septic systems function depends to a large extent on how intensively and how conscientiously they are maintained. Disposal of paper diapers, sanitary products, and kitchen garbage, and failure to pump it regularly will clog a septic system in a relatively short time. The average life expectancy of a system is 7 to 14 years.

An interesting alternative for the handling of personal wastes in individual homes is a **composting toilet.** An example is the Clivus Multrum developed in Sweden (Fig. 10-20). The multrum receives only personal excrements and food wastes—no water other than urine. These wastes pass through a series of chambers as they decompose and after two to four years they arrive at the final chamber as a stable, nutrient-rich humus that is suitable for application on lawns and gardens. Bath, dishwashing, and other gray water must still be disposed of, but since it is not con-

**FIGURE 10-21**
The fecal coliform test. The number of fecal coliform bacteria in a sample of water is determined by: (a) inserting a filter disc in a filtering apparatus, and (b) drawing the water sample through the filter. Bacteria in the sample are entrapped on the filter disc. (c) The filter disc is then placed on a medium for growing bacteria and incubated at 38° C for 24 hours. In this period each *E. coli* bacterium entrapped on the disc multiplies to form a colony that is visible to the naked eye. (d) The number of bacteria that were present in the water sample is thus given by the number of colonies (seen in the photograph as spots) on the disc. Fecal coliform bacteria are identified by colonies that have a distinctive metallic-green sheen. (Photos by Bob Hudson.)

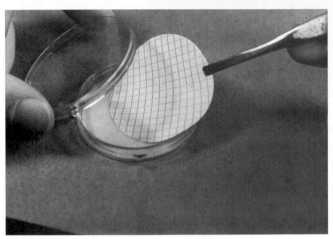
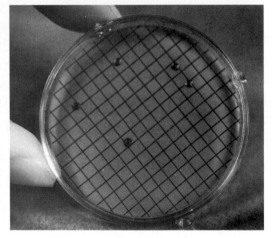

taminated by human excrements, this offers relatively little problem. In fact, keeping gray water separate from sewage and using it for irrigation of lawns and gardens has been advocated as a means of water conservation.

## MONITORING FOR SEWAGE CONTAMINATION

In spite of general improvement in sewage collection and treatment, there are and always will be unrecognized leaks, breaks, or overflows in sewage systems. Therefore, the need to monitor for

sewage contamination continues. It is worth understanding how this is done.

It would be exceedingly difficult, time consuming, and costly to test for each specific pathogen that might be present. Therefore an indirect method called the **fecal coliform test** has been developed. This test is based on the fact that huge populations of a bacterium called *E. coli (Escherichia coli)* normally inhabit the lower intestinal tract of humans and other animals, and large numbers are invariably excreted with fecal material. *E. coli* is not found in the environment except when it enters from animal excrement. It is not a pathogen; in fact our bowels would not function properly without it. However, since *E. coli* is invariably part of fecal wastes, its presence reveals a persistent source of raw sewage and, consequently, the potential presence of other sewage-borne pathogens. Conversely, the absence of *E. coli* organisms indicates that the water is probably free of sewage-borne, microbial pathogens.

The fecal coliform test, which is described in Figure 10-21, is a means of detecting and counting the actual number of coliform bacteria in a sample of water, usually 100 m$\ell$ (about 1/2 cup). Thus, the test shows how much sewage pollution is present and the relative degree of hazard. For example, it is considered that for water to be safe for drinking the *E. coli* count should be zero, that is, absolutely no sewage contamination. However, water with as many as 100 to 200 *E. coli* per 100 m$\ell$ is still considered safe for swimming. Beyond this level a river may be posted as polluted and contact recreation should be avoided. By comparison, raw sewage water itself (99.9 percent water/0.1 percent wastes), has counts in the millions.

When water is too polluted for a desired use, two possible approaches exist: disinfection and/or reduction of sewage sources. Drinking water supplies and swimming pools are generally disinfected with chlorine. However, there is obviously no way of disinfecting natural bodies of water without also killing everything else in the system. Therefore, there is no substitute for continuing and improving the quality of sewage treatment.

Unfortunately, however, we still tend to disregard the importance of sewage treatment. As long as something disappears down the drain, the general public believes it has been taken care of. Consequently, the question of whether adequate sewage treatment facilities are needed frequently becomes a matter of political debate about how to spend the taxpayers' money. It is important to recognize the connection between our sanitary bathroom and polluted beaches and waterways. Sewage treatment is not a political issue. It is a vital public health and environmental concern.

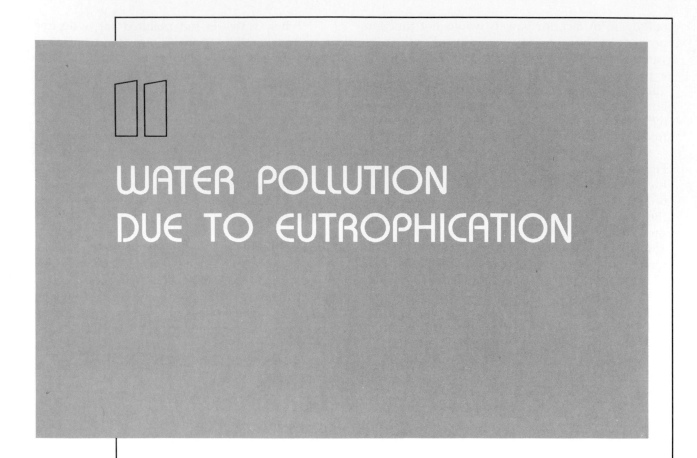

# WATER POLLUTION DUE TO EUTROPHICATION

---

## CONCEPT FRAMEWORK

| Outline | | Study Questions |
|---|---|---|

1. What are *benthic* and *planktonic* plants? Describe the appearance of water when phytoplankton increases. What is the *euphotic zone*? Where is the lower limit of the euphotic zone?

2. In nutrient-poor water, food chains depend on what? Are dissolved oxygen levels high or low in these systems? Why?

3. How do nutrient-rich waters differ from nutrient-poor waters in terms of dissolved oxygen levels, growth of benthic plants, and the organisms such as environment supports? Why are eutrophic lakes often called "dead" lakes?

4. Discuss the linear manner in which the eutrophication process takes place.

5. When is eutrophication a natural process? What causes it to become an unnatural process?

6. What nutrient is the most common limiting factor in eutrophication?

7. Why were herbicides used in an attempt to counter eutrophication? Why did they fail?

8. How does aeration help the eutrophication problem? Where is it practiced?

9. What is the most efficient and cost-effective way to control nutrient overload?

10. Define *point* and *nonpoint sources.*

11. What is one of the most common point sources?

12. Why were phosphates added to detergents?

13. What is the most significant nonpoint source of nutrients? What are other nonpoint sources? How could they be controlled?

14. What problems plague the diversion of nutrient-rich waters?

Imagine a lake with clear water that is a favorite area for swimming, boating, and fishing. Gradually, it is plagued by bursts of **algae** or **algal blooms** that turn the water green. Over the years the blooms become more intense and last longer. The fish start dying; whole schools of fish turn up dead, floating among the scum of algae on the surface. Eventually, the fish are gone. The lake is not good for recreation because the water is "thick" with algae from spring until fall when they are killed by the cold. Biologists test the water. It is rich in fertilizer nutrients such as phosphate and nitrogen compounds; no toxic chemi-

cals are found. But the water is low in dissolved oxygen. Deeper areas, where fish tend to hide, are even devoid of dissolved oxygen at times. This is what killed the fish; they suffocated!

Unfortunately the above story describes what has happened or is happening to many lakes, reservoirs, bays, and other bodies of water throughout the United States and the rest of the world. Technically, as we learned in Chapter 10, the process is called **eutrophication** (pronounced yoo-traf-i-ká-shun).

In some notable cases eutrophication has been controlled or even reversed. However, in

**FIGURE 11-1**
Two basic categories of aquatic plant life. (a) Benthic or bottom-rooted plants, also referred to as submerged aquatic vegetation. (b) Phytoplankton, various species of plants that exist as single cells or small groups or filaments of cells that float freely in the water. Drawings of phytoplankton are enlarged 500 to 1000-fold. Benthic plants withdraw nutrients from the sediments and hence do well in nutrient-poor water, while phytoplankton depends on nutrients dissolved in the water.
From: *Fundamentals of Ecology*, 3rd ed., by Eugene P. Odum. Copyright © 1971 by W.B. Saunders Co. Copyright 1953 and 1959, W.B. Saunders Co. Reprinted by permission of CBS College Publishing.

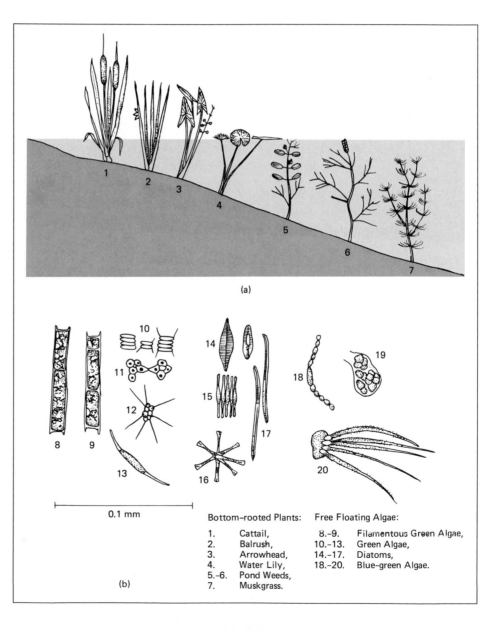

0.1 mm

Bottom-rooted Plants:

1. Cattail,
2. Balrush,
3. Arrowhead,
4. Water Lily,
5.–6. Pond Weeds,
7. Muskgrass.

Free Floating Algae:

8.–9. Filamentous Green Algae,
10.–13. Green Algae,
14.–17. Diatoms,
18.–20. Blue-green Algae.

large part, it is still a growing problem and it will demand increasing attention in the years ahead to develop and implement effective countermeasures. Our objective in this chapter is to gain a more thorough understanding of eutrophication, the sources of nutrients responsible for it, and potential means of controlling these sources.

## EUTROPHICATION IN DETAIL

### Two Kinds of Aquatic Plants

To understand eutrophication more thoroughly, we need to consider two distinct life forms of aquatic plants. The first category includes plants that are rooted in the bottom and that have a stem-and-leaf structure. These are called **benthic plants** (from *benthos*, "deep") or rooted plants (Fig. 11-1a). The second category includes microscopic plants consisting of single cells or small groups or "threads" of cells that maintain themselves suspended in the water (Fig. 11-1b). These are referred to as **planktonic plants** (from *plankton*, "floating") or **phytoplankton** (*phyto*, "plant") and includes various groups of algae. In a lake with relatively clear water, phytoplankton populations are low but rarely absent, as can be seen by dragging a fine net through the water and observing the catch under a microscope. As phy-

toplankton populations increase, the water appears increasingly turbid (cloudy) and greenish. In extreme situations water may become literally pea-soup green and a scum or mats of phytoplankton float on the surface (Fig. 11-2). The algae also frequently impart an unpleasant fishy odor.

The location of mineral nutrients determines which of the two groups of plants dominates. Benthic plants, being rooted in the bottom, may obtain their mineral nutrients from bottom sediments. However, they depend on the penetration of light through water for photosynthesis. The layer of water between the surface and the area where light is barely adequate to support plant growth is known as the **euphotic zone.** Water absorbs light passing through it. The lower limit of the euphotic zone is generally the area where light is reduced to 1 percent of full sunlight. In very clear water, this area may be located 30 m (about 100 ft) below the surface. However, in places where turbulence upsets sediments and/or plankton, the euphotic zone may be reduced to less than 1 m.

Planktonic algae, on the other hand, are able to maintain themselves in the euphotic zone near the surface, but mineral nutrients must be dissolved in the water. Can you see how the amount of dissolved nutrients will affect the relative growth of these two groups of plants?

**FIGURE 11-2**
In nutrient-rich water, phytoplankton may grow so thickly that it forms a dense scum over the surface. However, even without forming a scum, phytoplankton may grow so thickly that benthic plants are severely shaded. (Photo by author.)

## Two Kinds of Ecosystems: The Concentration of Dissolved Nutrients

### THE OLIGOTROPHIC (NUTRIENT-POOR) LAKE

The water of a lake that is clear and aesthetically pleasing for swimming and boating is characterized by *nutrient-poor* water. It is the lack of dissolved nutrients that limits the growth of phytoplankton and leaves the water clear.

Lakes with nutrient-poor water nevertheless support productive ecosystems that include numerous species of snails, clams, aquatic insects, crayfish, and the game fish we find most desirable. The food chains supporting these organisms are based on benthic plants, which grow abundantly wherever the bottom is located less than a few meters below the surface. In addition, the benthic plants in shallow water provide important habitats and shelter for reproduction of fish and shellfish.

Finally, and most importantly, dissolved oxygen in these systems generally remains high. This is particularly significant in deeper areas frequented by game fish. The maintenance of a high dissolved oxygen level in deeper regions results from the fact that the few decomposers there demand little oxygen. In the absence of such demand, the entrance of oxygen from the atmosphere combined with a normal mixing process provides a sufficient level of dissolved oxygen. Why is the oxygen demand of decomposers low? Recall the deeper areas are at the lower end of, or below, the euphotic zone. Hence, photosynthesis and a corresponding production of detritus is low. With little detritus, populations of decomposers and detritus feeders are small. Thus, little oxygen is consumed.

In summary, an oligotrophic lake is one with nutrient-poor, but oxygen-rich, clear water. It supports a diverse ecosystem based on rooted plants (Fig. 11-3a).

### THE EUTROPHIC (NUTRIENT-RICH LAKE)

A lake with water that is rich in dissolved nutrients is profoundly different from an oligotrophic lake. Not only is there a tremendous amount of phytoplankton supported by the availability of dissolved nutrients; there are far fewer benthic plants because they are shaded out by the turbidity caused by the phytoplankton and sediment. The difference in producers profoundly affects the rest of the ecosystem. Algae neither support the same food chains nor provide the same habitats as benthic plants.

A depletion of dissolved oxygen, especially in deeper areas, is even more significant here. We noted that in a nutrient-poor lake there is little production of biomass and detritus in the deeper areas. However, in the nutrient-rich lake, phytoplankton grows abundantly at the surface. Phytoplankton has a high turnover rate; that is, at high densities rapid growth is balanced by equally high dieoff. Dead phytoplankton sinks to the bottom, resulting in a rapid accumulation of detritus. Hence, populations of decomposers multiply quickly. Their demand for oxygen for respiration increases proportionally. Dissolved oxygen is thus consumed faster than it is replenished. As a result, it is depleted or held at very low levels.

Depletion of dissolved oxygen is aggravated by the loss of benthic plants, which serve to restore dissolved oxygen through photosynthesis. Phytoplankton also produce oxygen in photosynthesis; however, because they grow at or near the surface, much of the oxygen they produce escapes to the atmosphere. On a calm sunny day, one can often observe bubbles of oxygen entrapped by filamentous algae being released at the surface.

As dissolved oxygen is depleted, of course, fish and other organisms that depend upon oxygen for respiration die. Game fish are especially vulnerable because they require large amounts of oxygen and tend to inhabit the deeper water that is the first to be depleted. Some fish, like carp, survive or even thrive because they can tolerate lower concentrations of dissolved oxygen, inhabit shallow water, and have a behavioral adaptation of gulping air from the surface.

In summary, eutrophication is the process of nutrient enhancement and alteration of the ecosystem (Fig. 11-3b). A eutrophic lake supports a tremendous growth of planktonic algae, but is low in dissolved oxygen and therefore lacks or has small populations of aquatic life other than the phytoplankton and the decomposers which maintain the low level of dissolved oxygen (Fig. 11-3c and d).

Eutrophic lakes have been termed "dead." Biologically, the term dead is a definite misnomer because the total production of biomass is many times greater in a eutrophic lake than in an oligotrophic lake. However, the lake is dead in terms

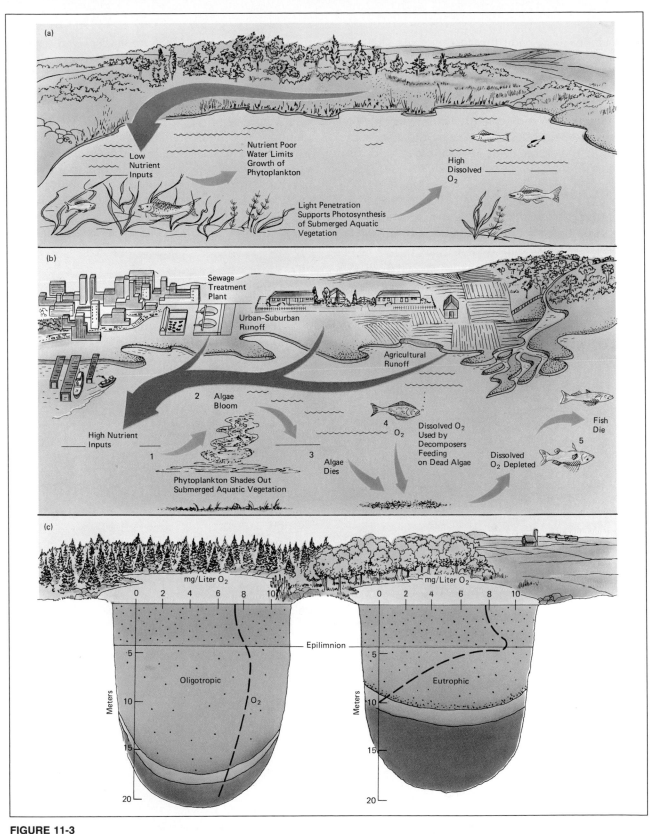

**FIGURE 11-3**

The process of eutrophication. (a) Natural drainage is poor in nutrients, maintaining lakes in nutrient-poor condition. Fish are supported by primary productivity of wetlands and marshes along the shore and of bottom algae. (b) Dissolved nutrients from various human sources (1) permit an overgrowth of planktonic (floating) algae (2). As algae die (3), they are attacked by decomposers (4) which utilize the dissolved oxygen, causing fish to suffocate (5). (Based on material from *The World Book Encyclopedia*. © 1979 World Book-Childcraft International, Inc.) (c) Dissolved oxygen levels typical of oligotrophic and eutrophic conditions are shown. (Redrawn by permission: Robert Leo Smith, *The Ecology of Man: An Ecosystem Approach*, 2nd ed., p 281, Harper and Row, Pub., Inc., 1976.)

of its ability to provide the aesthetic pleasures of swimming, boating, and sport fishing. Also, if the lake is a source of drinking water, its value may be greatly impaired because algal cells rapidly clog water purification filters; thus the filters need to be changed more frequently. Moreover, the offensive taste and odor imparted by the algae may remain in the water even after filtration and disinfection.

### EUTROPHIC RIVERS

In rivers, the process of eutrophication may spread in a linear manner, depending on the rate of river flow. Nutrients discharged into the river are carried downstream and development of an algal bloom occurs some distance from the nutrient source. Decomposing algae continue to move downstream and decrease dissolved oxygen further from the nutrient source. Further downstream, the river may recover if there are not more additions of nutrients and/or organic wastes (Fig. 11-4). However, even relatively small areas of eutrophication can profoundly affect fish populations. A school of fish may swim into a eutrophic area and suddenly die from the lack of oxygen, or a eutrophic segment drifting in a river or bay may entrap schools of fish. This may be the cause of many "mysterious" fish kills. The decaying bodies of fish create an additional oxygen demand that compounds the cycle of eutrophication.

## An Oligotrophic Lake Gradually Turns Eutrophic

It is important to recognize that eutrophication is part of the normal aging process of lakes. As discussed in Chapter 3, lakes tend to become

**FIGURE 11-4**
Eutrophication of a river. When nutrient-rich wastes are discharged into a river, the stages of eutrophication may be spread out in a linear fashion as shown.

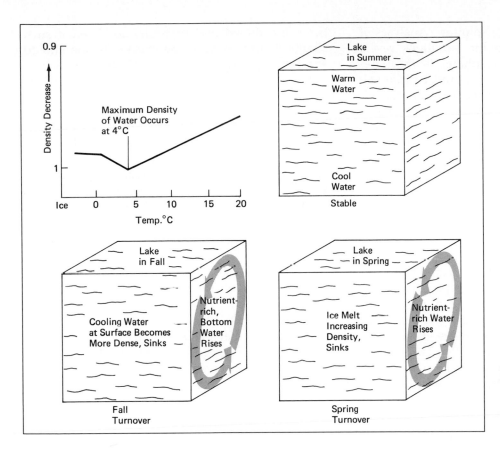

**FIGURE 11-5**
Temperature changes in the spring and fall result in a turnover of the water in a lake as shown. The nutrient-rich water brought to the top may stimulate a natural algal bloom at these times.

increasingly nutrient-rich, and hence eutrophic, as they fill with sediments. Also, eutrophication is not an all-or-nothing process. Even in lakes that are generally oligotrophic, algal blooms may occur in the spring when cold water from melting ice sinks to the bottom, and bottom water rises to the top, bringing along nutrients from the sediments. A similar turnover occurs in the fall as surface water cools, sinks, and forces nutrient-rich, bottom water to the surface (Fig. 11-5). In lakes that are relatively short of nutrients, however, the algal bloom is reduced in terms of duration and intensity. All the available nutrients are soon absorbed by the growing phytoplankton and are carried to the bottom where the cells die.

Furthermore, an entire body of water, unless it is a small pond, does not become eutrophic at once because complete and uniform mixing does not occur. In oligotrophic lakes evidence of eutrophication—the algal scum and low levels of dissolved oxygen—is usually first observed in secluded coves around stream outlets where nutrients enter.

Human-induced eutrophication causes a tremendous speedup of the natural process. This speedup occurs because of greatly increased rates of nutrient loading. In most undisturbed terrestrial ecosystems, efficient recycling of nutrients occurs and leaching of nutrients from the soil into aquatic ecosystems is slight. Further, nutrients that enter the system are largely bound to clay particles that settle and take the nutrients to the bottom. Very small concentrations of nutrients are found in the free, dissolved state. Various human endeavors, however, especially those associated with agriculture and sewage treatment, tend to flush large amounts of nutrients into waterways.

## COMBATTING EUTROPHICATION

There are two general approaches to combatting the problem of eutrophication. One is to attack the symptoms—use chemicals to inhibit the algal blooms and/or use artificial means to restore dissolved oxygen. This approach has been less than successful, and we should be aware of the shortcomings so that we are not tempted to repeat mistakes.

The second approach is to control the input of nutrients that are at the root of the problem.

This should be the ultimate solution. Analyses of nutrient levels in oligotrophic situations reveal that a single nutrient, phosphate, is most frequently the limiting factor. Other nutrients are already present in abundance. Hence, it is the addition of phosphate that stimulates the excessive growth of phytoplankton. Thus, controlling eutrophication may center on controlling sources of phosphate.

## Attacking the Symptoms

### CHEMICAL TREATMENT

**Herbicides** have been used very successfully in agriculture to eradicate unwanted plants. Therefore, it was an obvious extrapolation to believe that unwanted algae also could be eliminated by chemical treatment (Fig. 11-6) and thousands of tons of chemicals were spread on ponds and lakes in the 1960s and 1970s. The general finding was that chemicals do not work. The planktonic algae, especially blue-green species which are the most obnoxious, prove to be among the most resistant of all organisms. If enough chemical is added to kill them, it also kills everything else in the body of water. When herbicide concentrations dissipate, the algae are among the first species to reappear because the root cause, high nutrient levels, has not been affected.

To date, no chemical has been found that will kill algae and not harm other aquatic plants and animals. Thus, this method has generally been abandoned and rightfully so. Even if such chemicals did work in the short run one should remain skeptical of long-term effects.

### AERATION

Installing a mechanical aeration system in a eutrophic lake or pond will keep the dissolved oxygen high and at least prevent the fishkills due to suffocation (Fig. 11-7). In addition, aeration may also reduce the level of insoluble phosphate. When the dissolved oxygen level is low, phosphate remains more soluble. Since phosphate is often the limiting nutrient in growth of algae, decreasing phosphate solubility by means of aeration may significantly reduce algal blooms.

However, aeration does not solve the underlying problem of nutrient loading. Though it does not produce any of the undesirable ecological side effects or health hazards of herbicides, aeration systems are expensive to install and energy costs for operation are high. They are practical only for small lakes and reservoirs where a very high value is placed on controlling eutrophication.

### HARVESTING ALGAE

Residents at times have taken to removing algae by hand from community lakes that they value. By the way, if you undertake such an effort, a raft made by lashing a few planks across two canoes, as shown in Figure 11-8, is remarkably effective. This does remove unsightly mats of algae and other litter that collects in various places and greatly improves the appearance.

**FIGURE 11-6**
Spraying water with herbicides to inhibit undesired plants. What will be the side effects? (FAO photo by S. Baron.)

**FIGURE 11-7**
Lakes and reservoirs may be aerated to avoid the
consequences of eutrophication. Aeration also aids in
stabilizing phosphates in sediments. (USDA–Soil Conservation
Service photo.)

**FIGURE 11-8**
To fight the use of chemical
herbicides to control algae, these
residents in Columbia, Maryland,
resorted to harvesting algae by
hand. The city has now
purchased a mechanical
harvester. (Photo by Bruce Fink.)

However, it does little to improve the underlying problem. Beyond what has collected in mats, plankton, because it is primarily single cells in suspension, does not lend itself to harvesting. Even attempting to filter it does not work because it clogs filters so quickly.

## Identifying and Controlling Nutrient Inputs

It should be clear that real control of eutrophication requires decreasing nutrient inputs. However, accomplishing this task is difficult. Sources of nutrients must be identified and evaluated and then suitable cost-effective methods of control must be implemented. Since many nutrient sources result from human practices, the most cost-effective method of control lies in changing these practices.

### POINT AND NONPOINT SOURCES OF NUTRIENTS

Nutrients come from both point and nonpoint sources (Fig. 11-9). **Point sources** are major, specific discharges that can be pinpointed, such as the discharge from a sewage treatment plant or a particular industrial discharge. **Nonpoint sources** include the general runoff and/or leaching from urban, suburban, and agricultural areas.

### CONTROLLING POINT SOURCES

Sewage treatment plants are most frequently identified as the major point sources. Hence, as discussed in Chapter 9, there is a major thrust in many areas to upgrade these plants to include advanced treatment that will at least remove phosphate. Also, industries with nutrient-rich discharges of wastes are now required to install similar systems of treatment.

### BANNING THE USE OF PHOSPHATE DETERGENTS

Another way to limit the amount of phosphate discharged from sewage treatment plants is to limit what goes down the drains in our homes. It has been estimated that only about 30 percent of the phosphate in waste water comes from human excrements. Most comes from detergents.

Traditional soaps combine with calcium in water to form an insoluble precipitate or "curd"—the ring around the bathtub is an example. When modern washing machines, which use the spin-dry principle, were introduced, the curd remained on the clothing. Therefore, soaps were replaced by detergents. Detergent molecules consist of a long chain of hydrocarbons at one end. These combine with grease and oil. At the other end of the detergent molecule is a highly water-soluble group (Fig. 11-10). Detergent molecules lift fatty or oil substances and are not precipitated by minerals such as calcium in hard water.

The first detergents used a sulphonate group for the water-soluble end. However, such detergents were bactericidal—they stopped the functioning of sewage treatment plants and led to banks of foam on the rivers where they were discharged. To provide for decomposition, sulphonate was replaced by phosphate, which is nontoxic and allows the hydrocarbon portion to be degraded. Hence, such detergents were called **biodegradable,** but importantly, after degradation the phosphate group remains. Phosphate is a basic group of atoms that cannot be broken down or decomposed. Hence, such detergents aggravate the problem of eutrophication.

Now, most manufacturers have made substitutions and offer low- or no-phosphate products along with traditional high-phosphate brands. Unfortunately, phosphate content is not always readily identifiable from the label. Consumers must check the fine print to determine phosphate content. Importantly, the term *biodegradable* has no relevance to phosphate content.

Attempts to legislatively ban the use of detergents containing phosphate have been made in some states or counties, but such legislation is vigorously opposed by the detergent industry and has generally been defeated. However, such bans are now in effect in the Great Lakes region, on Long Island, and in Maryland. The ban in the Great Lakes area was implemented in the mid-1970s when Lake Erie was becoming eutrophic as a result of phosphate inputs from the many cities around its shoreline. Since the ban, the lake has improved considerably. The ban in Maryland was enacted in 1985 after eutrophication was shown to be a major factor in the decline of aquatic life in Chesapeake Bay. Unfortunately other states bordering the bay have not yet followed suit.

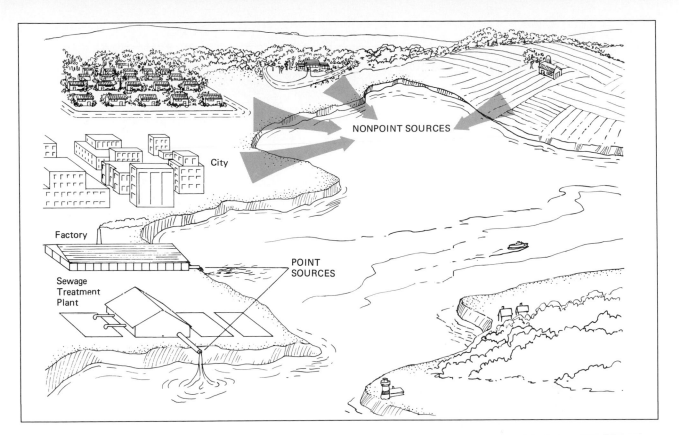

**FIGURE 11-9**
Point and nonpoint sources of pollution. Pollution may come from both point sources, such as identifiable drains and outlets, and nonpoint sources, such as the generalized runoff from cities, suburbs, and farm lands.

**FIGURE 11-10**
Detergent molecule. Its action results from the hydrocarbon end dissolving in fat while the ionic end dissolves in water. Fats are thus brought into solution. The hydrocarbon portion may be biodegradable, but phosphates, at the ionic end, will remain in solution.

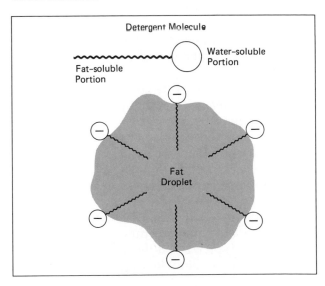

## CONTROLLING NONPOINT SOURCES (LAND MANAGEMENT)

Unfortunately, problems of eutrophication have not declined in proportion to the reduction of major point sources. This has made us more aware of the significance of nonpoint sources. Runoff from agricultural areas is often very high in nutrients from leaching of fertilizer and/or animal wastes. Likewise, urban-suburban runoff carries significant amounts of nutrients from fertilizer used on lawns and gardens and from pet droppings.

The most significant agricultural sources of runoff are controllable. Most important is erosion. Phosphate, more than other nutrients, binds to particles of clay and humus. Thus, control of erosion, as discussed in Chapter 7, is as important in the prevention of eutrophication as it is for the maintenance of soil productivity. Other agricultural sources are also significant and controllable. Washings from dairy barns or cattle feeding areas frequently drain directly into streams and then into rivers and lakes. Farmers are now encouraged to construct holding ponds to collect such washings. The water from these ponds can be used for irrigation providing some benefit in re-

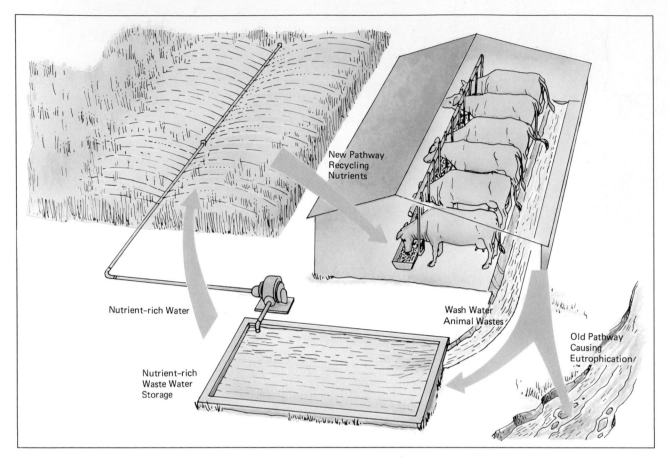

New Pathway
Recycling
Nutrients

Nutrient-rich Water

Wash Water
Animal Wastes

Old Pathway
Causing
Eutrophication

Nutrient-rich
Waste Water
Storage

**FIGURE 11-11**
A collection pond for dairy barn washings. If washings from
animal facilities such as dairy barns are flushed into natural
waterways, they may contribute significantly to eutrophication.
This may be avoided by collecting the flushings in ponds from
which both the water and nutrients may be recycled. Many
farmers use such flushings for irrigation.

turn (Fig. 11-11). Likewise, improved soil management, which increases the soil's ability to hold
nutrients, benefits the farmer and is a deterrent to
eutrophication.

Where fields are directly adjacent to streams,
rivers, or lakes, farmers are encouraged to plant
buffer strips of trees between their fields and the
waterways in order to catch and reabsorb the nutrient-rich leachate (Fig. 11-12). However, this advice is not so readily accepted by farmers because
it means taking some land out of cultivation and
putting it into what, for the farmer, is a less productive use.

Note that farmers are only encouraged in
these respects. How farmers manage their land is
still considered their own business. Since runoff

has been recognized as a major cause of pollution,
sentiment is growing to make "best management
practices" mandatory.

Runoff per hectare from urban-suburban
areas frequently contains even higher levels of
nutrients and fecal bacteria than that from agricultural areas. Fertilizer used on lawns and gardens and pet droppings are the major sources, although failing septic systems may be a problem
in some areas.

The problem of failing septic systems is
solved by municipal sewer hookups. However,
other aspects of the urban-suburban runoff problem present a tremendous challenge. How does
one approach millions of home and pet owners,
for example, and educate them in terms of better
fertilizer management on lawns or in cleaning up
after pets?

Furthermore, the amount of nutrients coming from urban-suburban runoff is increasing as
development spreads. At the root of the problem
are our policies of land development (see Chapter

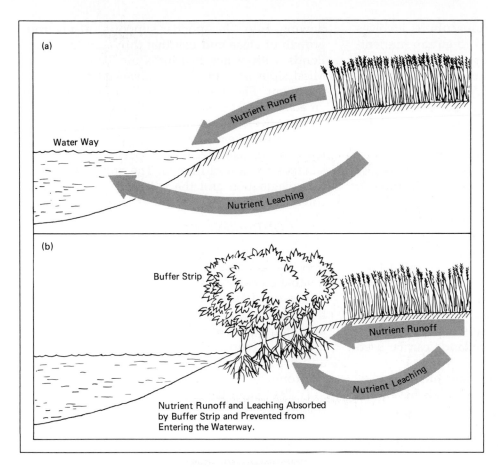

(a)

Nutrient Runoff

Water Way

Nutrient Leaching

(b)

Buffer Strip

Nutrient Runoff

Nutrient Leaching

Nutrient Runoff and Leaching Absorbed
by Buffer Strip and Prevented from
Entering the Waterway.

**FIGURE 11-12**
Mitigation of eutrophication by
buffer strips of trees. Maintaining
a buffer strip of trees along
waterways helps to absorb and
hold nutrients that would
otherwise leach from agricultural
fields into the waterway.

21). Development along waterways and particularly around reservoirs should be restricted and buffers of trees maintained to protect water quality.

### DIVERTING NUTRIENT-RICH WATER TO USEFUL PURPOSES

Another approach that may be used in connection with both point and nonpoint sources is to find a use for the nutrient-rich water. After all, nutrient-rich water is not bad in itself; it is only bad when it causes the growth of the wrong kinds of plants in the wrong places. Why not use it to grow useful plants in desired places? With this thought in mind, many communities have undertaken or are contemplating projects in which waste water, after secondary treatment, is used for irrigation of parks or agricultural lands.

Straightforward as this idea seems, it does have some limitations. First, there must be adequate land nearby and preferably downhill to receive the water. Otherwise prohibitive costs are involved in piping and pumping the water. Second, storage capacity must be built into the system to hold the waste water when irrigation is not possible. For example, when the ground is frozen or when it is saturated with rainfall it will not accept additional water. Such conditions may prevail for several months of the year. Another problem is the potential presence of heavy metals as described in the section on sewage sludges in Chapter 10.

Similarly, water from urban-suburban runoff may be trapped in reservoirs and then used for irrigation. (This also has benefits in terms of controlling flooding and stream bank erosion; see Chapter 8). Here again, however, heavy metals may be a problem. Urban runoff, in particular, tends to be contaminated with lead from leaded gasoline which escapes from exhaust pipes and then settles on the ground. Zinc, which is used in tires, presents a similar problem.

Aquaculture is a related but alternative approach to using nutrient-rich waste water for useful biological production. Aquaculture involves constructing an artificial pond to entrap nutrient-rich water from point and/or nonpoint sources. These ponds may be used for the intensive culture of one or more species. Any number of useful aquatic plants can be grown. For example, water hyacinths and water lilies can be used for livestock feed. Cattails and other such "reeds" may be used in weaving "straw" mats, baskets, and so on. Or such plants may be anaerobically decomposed to produce methane, fermented to produce alcohol, or burned directly as a boiler fuel. If the water is suitably disinfected, a second trophic level could be shellfish, waterfowl, or any combination (Fig. 11-13). The producers absorb most of the nutrients so that water exiting the ponds is relatively pure. In effect, one is using solar energy through photosynthesis to remove nutrients, thus purifying the water while obtaining biological products. Such ponds require considerable amounts of land, but wetlands can be used in many coastal cities.

In the Philippine Islands and other parts of the world, sections of coastal wetlands have been "fenced off" to create ponds. Nutrient-rich water flowing into the ponds supports a vigorous growth of algae and fish that thrive on the algae. Ponds with water hyacinths are being used in Mississippi as a means of advanced wastewater treatment (Fig. 11-14). A warm sunny climate is an advantage, but some of the climatic limitations may be overcome by judicious choice of species. Also, waste heat from power plants might be used to maintain desired temperatures. The possibilities are almost infinite. The only limitation is the imagination and the will to see the problem solved.

### CONTINUING PROGRESS

In spite of all the problems with eutrophication, the good news is that it is reversible. Some notable cases have shown that when nutrient inputs are reduced, existing nutrients are gradually flushed from the system and/or become stabilized in bottom sediments. Algal blooms cease, organic matter decomposes, dissolved oxygen returns, and the lake again provides an attractive, healthy environment.

The legal framework for addressing water pollution problems was set up under the Clean

**FIGURE 11-13**
Biological advanced treatment. Algae or various other water plants may be used to absorb the nutrients from waste water after secondary treatment. The result is plant and/or animal products as well as relatively clean water.

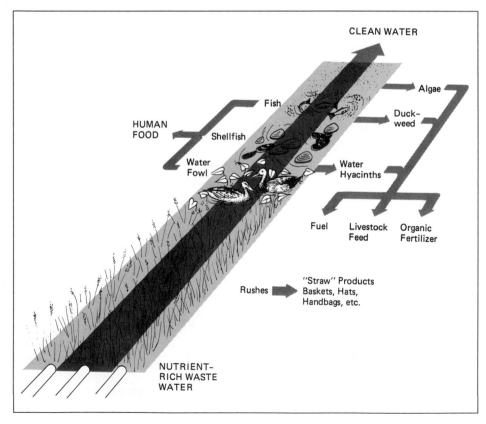

**FIGURE 11-14**
Natural wetlands may be converted into aquaculture using waste water as the primary source of nutrients. Photograph shows a water hyacinth farm in Lucedale, Mississippi. (Photo by N. D. Vietmeyer, National Space Technology Laboratory.)

Water Act of 1972. However, the mere passing of a law cannot guarantee that its objectives will be met. If the public becomes apathetic, we can expect the government to behave likewise. Planning and construction schedules will fall behind, compliance will not be enforced, and laws themselves may be changed or relaxed. If we desire healthy rivers and lakes, we must become involved and find out what is being done and what is not being done to maintain or restore good water quality. We need to support and encourage public officials who are doing what they can and prod those who are not.

# 12

# GROUNDWATER POLLUTION AND TOXIC WASTES

## CONCEPT FRAMEWORK

| Outline | Study Questions |
|---|---|

As discussed in Chapter 8, groundwater is a tremendously important resource. The volume of known groundwater supplies in the United States is about 50 times larger than surface flow. Moreover, in most cases, the quality (purity) of groundwater is such that it meets safe drinking water standards; purification or treatment is not usually required. Thus, the trend to use groundwater has increased. Between 1950 and 1980, groundwater withdrawals nearly tripled. Pumped groundwater currently provides about half of the population with water and fills about 40 percent of the nation's irrigation needs. In addition, about 750,000 new wells are drilled each year.

As long as withdrawal rates do not exceed recharge rates, there is no theoretical reason why this practice cannot continue indefinitely. Unfortunately, however, in many cases formerly high-quality groundwater has become polluted with toxic chemicals. Serious illnesses have occurred as a result, and wells have been closed, forcing people to bring in water from other sources or to resort to costly purification processes (Fig. 12-1). The threat of such pollution is becoming increasingly widespread. Indeed, groundwater pollution is recognized as one of the most serious public health issues of the 1980s. And it is a problem complicated by the fact that once groundwater has been contaminated, it is often permanently lost. Due to its underground location and its way of percolating through the earth, purification and treatment of polluted groundwater is economically and/or technologically impractical.

Hence, it is extremely necessary to take steps to protect groundwater from pollution. The objective of this chapter is to examine the various sources of groundwater pollution and to provide some understanding of what is being done and what remains to be done to prevent its further contamination.

## MAJOR SOURCES OF GROUNDWATER CONTAMINATION

The most important fact concerning groundwater is that it is recharged by water that infiltrates at the earth's surface and percolates down through the earth to the water table. While percolating, particles including bacteria and other microbes are generally filtered out. This is why most groundwater is suitable for drinking. However, the "soil filter" is not fine enough to remove individual ions or molecules. Thus, the water tends to carry materials that dissolve (go into solution) as it percolates through the earth. The general principle to remember is that any chemical or material that is used, disposed of, or stored on or in the earth has the potential to leach (dissolve and be carried by percolating water) into and contaminate ground-

**FIGURE 12-1**
Residents in McKeesport, Pa, getting army-purified water because the municipal supply was contaminated. As groundwater is contaminated with toxic materials, more and more people must seek other sources which may be inconvenient, costly, and limited in quantity. (Grapes Michaud/Photo Researchers.)

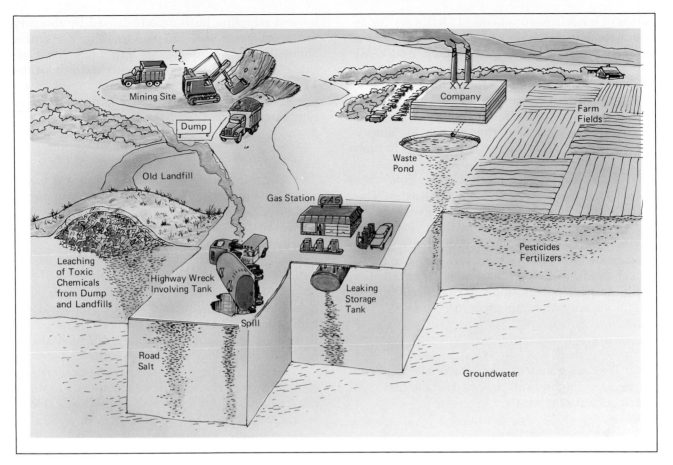

**FIGURE 12-2**
Any chemical used, stored, spilled, or disposed of on or near the surface may leach into the groundwater. This illustration depicts some of the most significant sources of groundwater contamination.

Labels within figure: Mining Site; Dump; Old Landfill; Gas Station; GAS; Leaching of Toxic Chemicals from Dump and Landfills; Highway Wreck Involving Tank; Spill; Road Salt; Leaking Storage Tank; XYZ Company; Waste Pond; Farm Fields; Pesticides Fertilizers; Groundwater

water. With this principle in mind, it is not difficult to recognize that there are innumerable sources of chemicals threatening groundwater (Fig. 12-2). The sources of toxic chemicals currently recognized as the most threatening include the following:

1. Inadequate landfills or other unprotected facilities where chemical wastes have been dumped or disposed of and from which they may leach or escape into the ground.

2. Leaking underground storage tanks or pipelines. The leakage of gasoline from service station storage tanks is a particular problem.

3. Pesticides and fertilizers used in agriculture or on lawns and gardens.

4. De-icing salt used on roads.

5. Waste oils used on dirt roads to keep dust down.

6. Overapplication of sewage sludges or waste water.

7. Transportation spills.

While all of the above are significant sources of groundwater pollution, illegal dumping and/or inadequate disposal of chemical wastes is considered the most serious problem. Hence, we shall devote most of our discussion in this chapter to the disposal of toxic chemical wastes.

## THE TOXIC CHEMICAL WASTE PROBLEM

### Major Sources of Chemical Wastes

Chemical wastes come from three main sources. The first source includes chemical processes, all of which produce certain wastes. For example, ores are refined into metals and raw materials such as crude oil are made into numerous synthetic organic chemicals used in plastics, synthetic fibers, coatings, pesticides, drugs, and other products. Such processes generate thousands of miscellaneous chemical "leftovers" or by-products that are either of no use or are economically impractical to separate, purify, and reuse. The second source includes the numerous chemicals used as solvents, cleaning fluids, or other agents in various processes; these "process materials," are also discarded after use. The third source includes residues that remain in "empty" containers and drums along with unused portions of chemicals that are discarded.

Many of the chemical wastes from these sources are classified as hazardous wastes. They include materials that are (1) explosive or highly reactive; (2) flammable; (3) corrosive, such as acids and bases; and (4) toxic, poisonous, or that may cause long-term physiological problems such as cancer, birth defects, and other abnormalities. With respect to groundwater pollution, toxic chemicals are our major concern.

## Toxic Wastes and Their Threat

Two categories of toxic waste chemicals are particularly significant: **heavy metals** and **synthetic organic chemicals.**

### HEAVY METALS

Heavy metals, as the name implies, are metallic elements that in pure form are literally heavy (dense). Lead, mercury, arsenic, cadmium, tin, chromium, zinc, and copper are the main examples. Heavy metals are primarily contained in the wastes of metal-processing industries and industries producing pigments; heavy metal compounds are frequently intensely colored. However, many other industrial processes use heavy metals. As ions or in certain compounds, these metals are soluble in water and may be ingested and absorbed into the body.

The refining and use of heavy metals goes far back in human history. Consequently, the unfortunate experiences of those who came in contact with heavy metals has provided a great deal of information about their toxicity. For example, the expression "mad as a hatter" is derived from the fact that people who made hats in the early days were, indeed, frequently insane. The insanity, it was later found, was caused by the mercury used in the production process.

Relatively small amounts of heavy metals can have insidious health effects. The mental retardation caused by lead poisoning and the insanity and crippling birth defects caused by mercury are particularly well known. Investigations now show that heavy metals are extremely toxic because atoms tend to combine with and inhibit the functioning of particular enzymes. Thus, very small amounts can have severe physiological or neurological consequences.

### SYNTHETIC ORGANIC CHEMICALS IN GENERAL

Organic chemicals, by definition, are based on various arrangements of bonded carbon atoms with hydrogen and perhaps other elements attached to the carbon structure. The number of possible arrangements is infinite. However, relatively few basic arrangements are found in the natural organic compounds that make up proteins, nucleic acids, lipids, and other molecules found in living organisms. On the other hand, organic chemists have learned to make hundreds of thousands of unique arrangements, which are referred to as *synthetic organic chemicals.*

Unlike heavy metals, synthetic organic chemicals are relatively new; chemists produced the first ones in the latter half of the 1800s. However, production has grown phenomenally as these compounds have become the basis of all plastics, synthetic fibers, paintlike coatings, resins, innumerable drugs, pesticides, detergents, solvents, and countless other products. Moreover, production, which is already in the hundreds of millions of tons per year, continues to grow rapidly, and thousands of new compounds are produced each year.

Many synthetic organic compounds are similar enough to natural organic compounds that they interact with particular enzymes or other chemicals in the body and cause altered functions. This results in some synthetic organic compounds being acutely poisonous while others may serve as beneficial drugs. However, altered functions have also been shown to include mutagenic (mutation-causing), **carcinogenic** (cancer-causing), and teratogenic (birth defect-causing) effects. In addition, they may cause serious liver and kidney dysfunction, sterility, and numerous lesser physiological and neurological problems.

A significant problem in assessing the health risks of synthetic organic compounds is that information concerning their physiological effects is far from complete. Compounds are so numerous and their potential effects are so diverse that few (if any) have been tested adequately. Many compounds found in chemical wastes have not been tested at all.

## HALOGENATED HYDROCARBONS

Even in the absence of adequate testing, experience has shown that one subgroup of synthetic organic compounds is particularly dangerous. These are the **halogenated hydrocarbons,** compounds of hydrogen and carbon in which one or more of the hydrogen atoms have been replaced by an atom of chlorine, bromine, or iodine. These latter three elements are classed as halogens; hence the name halogenated hydrocarbons.

Among all the halogenated hydrocarbons, those containing chlorine and referred to as **chlorinated hydrocarbons** are by far the most common. Such compounds are widely used in plastics (e.g., polyvinyl chloride), pesticides (e.g., DDT, kepone, and mirex), solvents (e.g., carbon tetrachloride), wood preservatives (e.g., pentachlorophenol), electrical insultation (e.g., PCBs or polychlorinated biphenyls), flame retardants (e.g., TRIS), and many other products. Consequently, such compounds are common in the wastes of industries that produce or use these products. PCBs and **dioxin** are examples of chlorinated hydrocarbons that are notorious for their pollution hazard. Others are given in Table 12-1.

## THE PROBLEM OF BIOACCUMULATION

Both heavy metals and halogenated hydrocarbons are particularly insidious because they tend to *bioaccumulate*. **Bioaccumulation** refers to the accumulation of higher and higher concentrations of such toxic chemicals in the body. In effect, the body acts like a chemical filter; consequently, what may seem like trivial, harmless quantities ingested with food or water over a long period of time may gradually build up in the body until, finally, toxic levels are reached (Fig. 12-3). Bioaccumulation occurs for two reasons. First, these chemicals are very stable; they do not break down readily. Heavy metals, being elements, cannot be broken down or destroyed by any chemical process. Halogenated hydrocarbons may break down slowly under certain conditions such as very high temperature. However, they generally do not break down readily under normal environmental conditions or in biological systems—that is, they are nonbiodegradable. As a result, these chemicals may linger more or less indefinitely in the environment.

| Table 12-1 | Examples of Toxic, Synthetic Organic Compounds Frequently Found in Chemical Wastes |

| Chemical | Known Health Effects | CAUSES MUTATIONS | CARCINOGENIC | BIRTH DEFECTS | CAUSES STILL BIRTHS | CAUSES NERVOUS DISORDERS | LIVER DISEASE | KIDNEY DISEASE | LUNG DISEASE |
|---|---|---|---|---|---|---|---|---|---|
| Benzene | | X | X | X | X | | | | |
| dichlorobenzenes | | X | | | X | X | X | | |
| hexachlorobenzene | | X | X | X | X | X | | | |
| Chloroform | | | X | X | X | | X | | |
| Carbon tetrachloride | | | X | | X | X | X | X | |
| Ethylene | | | | | | | | | |
| chloroethylene (vinyl chloride) | | X | X | | | X | X | | X |
| dichloroethylene | | X | X | | X | X | X | X | |
| tetrachloroethylene | | | X | | | X | X | X | |
| trichloroethylene | | X | X | | | X | X | | |
| Heptachlor | | X | X | | X | X | X | | |
| Polychlorinated biphenyls (PCB's) | | X | X | X | X | X | X | | |
| Tetrachlorodibenzo dioxin | | X | X | X | X | X | X | | |
| Toluene | | X | | | | X | X | | |
| chlorotoluenes | | X | X | | | | | | |
| Xylene | | | | X | X | X | | | |

Source: Adapted from "Hazardous Waste in America" by S. Epstein, L. Brown, and C. Pope. Copyright © 1982 by Samuel S. Epstein, M.D., Lester O. Brown and Carl Pope. Reprinted with permission of Sierra Club Books.

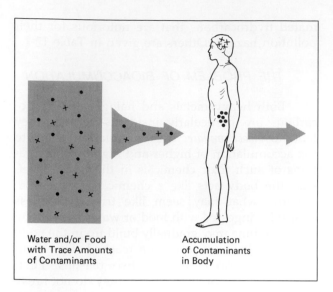

**FIGURE 12-3**
Bioaccumulation. Some of the chemicals that may be ingested with food or water are neither broken down nor excreted readily. Instead, they accumulate in the body, concentrating in the liver or other specific organs until they reach damaging levels. This phenomenon is called *bioaccumulation.* Heavy metals and halogenated hydrocarbons are two categories of chemicals that are particularly likely to bioaccumulate.

Second, these chemicals are readily absorbed into the body, but they are excreted very slowly if at all. Heavy metals bind tenaciously with proteins, and halogenated hydrocarbons are very fat-soluble and sparingly soluble in water. These characteristics prevent these chemicals from being flushed out of the body by the usual excretory (urinary) process.

Bioaccumulation may be compounded in the food chain, as shown in Figure 12-4. Organisms at the bottom of the food chain absorb the chemical from the environment and concentrate and accumulate it in their tissues. In feeding on these organisms, animals at the second trophic level receive a higher dose and accumulate still higher concentrations in their tissues, and so on up the food chain. It is hardly surprising that organisms at the top of the food chain receive lethal concentrations, which are as much as 100,000 times higher than environmental concentrations. This concentrating effect that occurs through a food chain is referred to as **biomagnification.**

One of the most distressing aspects of bioaccumulation and biomagnification is that the diseases which they cause—cancer, liver and kidney dysfunction, and birth defects—may not appear until many years after the initial exposure. By that time it is extremely difficult, if not impossible, to determine the real cause of the disease and, of course, it is too late for the individual concerned to do much if anything about it.

The danger of bioaccumulation and biomagnification of chlorinated hydrocarbons became clear in the 1960s when it was discovered that diebacks in populations of many species of predatory birds, such as the bald eagle, were due to bioaccumulation of DDT, a chlorinated hydrocarbon pesticide. Numerous sport and commercial fishing areas have been shut down because fish accumulated what are considered to be dangerous levels of PCBs and other chlorinated hydrocarbon compounds.

The danger of heavy metals was brought to public attention in the early 1970s by **Minamata disease.** The disease is named for a small fishing village in Japan where the episode occurred. In the mid-1950s, cats in Minamata began to show spastic movements, followed by partial paralysis and, later, coma and death. At first this was thought to be a peculiar disease of cats, and little attention was paid to it. However, concern escalated quickly when the same symptoms began to occur in people; additional symptoms such as mental retardation, insanity, and birth defects were also observed. Scientists and health experts eventually diagnosed Minamata disease, as it came to be known, as acute mercury poisoning.

But what was the source of the mercury? A chemical company near Minamata was discharging waste containing mercury into a river that flowed into the bay where the Minamata villagers fished. Investigation revealed that mercury deposited in the sediments was first absorbed by bacteria feeding in the sediments and then was passed and further concentrated by one organism after another in the food chain, from bacteria through fish to cats and humans. Cats had suffered first and most severely because they fed almost exclusively on the remains of fish. By the time the situation was brought under control, some 50 people had died and 150 had suffered serious bone and nerve damage. Even now, the tragedy lives on in the crippled bodies and retarded minds of Minamata descendants.

### SYNERGISTIC EFFECTS

Complicating the situation is the problem of synergistic effects. Toxic chemicals seldom occur singly. Often, as many as a dozen or more occur

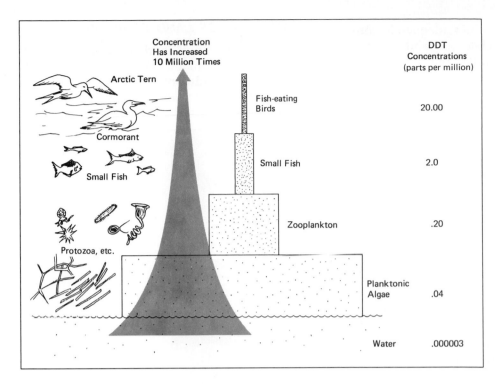

Concentration
Has Increased
10 Million Times

| | DDT Concentrations (parts per million) |
|---|---|
| Fish-eating Birds | 20.00 |
| Small Fish | 2.0 |
| Zooplankton | .20 |
| Planktonic Algae | .04 |
| Water | .000003 |

Arctic Tern

Cormorant

Small Fish

Protozoa, etc.

**FIGURE 12-4**
Biomagnification. When the phenomenon of bioaccumulation is put into the context of a food chain, each successive consumer receives a more contaminated food supply and, in turn, accumulates the contaminant to yet a higher level. For example, scientists have observed that the concentration of the pesticide DDT was magnified some 10 million-fold as it passed through the food chain shown.

simultaneously, and frequently two or more chemicals act together to produce an effect that is greater than the summation of the effects caused by the two acting separately. This is known as a **synergistic effect.** We shall discuss it further in connection with air pollution in Chapter 13.

## Background of the Toxic Waste Problem

### SHIFT FROM WATER AND AIR TO LAND DISPOSAL

Historically, chemical wastes, including all categories of hazardous wastes, have been disposed of as expediently as possible. Prior to the environmental laws of the 1960s and 1970s, it was common practice to exhaust fine particles and gases into the air and flush liquids and wash water with all kinds of wastes into natural waterways. The pollution that resulted is hard to believe by today's standards. Many streams and rivers in industrial cities were essentially open chemical sewers. They were hazardous and devoid of life. For example, in the 1960s the Cuyahoga River, which flows through Cleveland, Ohio, carried so much flammable material that it actually caught fire and destroyed seven bridges before it burned itself out. Other pollutants caused waterways to turn all colors of the rainbow. And,

of course, many discarded pollutants were not visible, but nevertheless affected biota. Wastes were also disposed of on land, as will be discussed below.

The Clean Air Act of 1970 and the Clean Water Act of 1972 were enacted in response to the public outcry against these highly visible problems. As a result of this legislation, industry spent billions of dollars on pollution control equipment to remove chemical wastes from its air and water effluents. Direct discharges of wastes into air and water were greatly reduced, and water quality in many areas improved dramatically, and fish returned to many formerly polluted waterways. Nevertheless, it is essentially impossible to control all minor sources and no removal system is 100 percent efficient. Therefore, small inputs continue and residues of toxic chemicals from past discharges remain in the sediments of many lakes and rivers where they continue to biomagnify through the food chain, tainting fish if not inhibiting their reproduction.

It is important to remember, however, that removing wastes from discharges into air and water does not make them disappear; they must be put somewhere else. In the early 1970s, wastes were largely redirected toward land disposal. Unfortunately, the potential shortcomings and dangers of this redirection were not adequately considered. Consequently, while air and surface water quality improved during the 1970s, land disposal and the potential for groundwater pollution increased enormously.

## METHODS OF LAND DISPOSAL

Three methods of land disposal of hazardous wastes predominate: (1) deep well injection, (2) surface impoundments, and (3) landfills. As will be discussed, each of these methods may be

DISPOSAL OF HAZARDOUS WASTES BY DEEPWELL INJECTION

THEORY

A Well Is Drilled into a Dry Porous Layer and Wastes Are Pumped in. Contamination of Groundwater Is Prevented by Casing and Seal around the Portion of the Well That Penetrates Groundwater

PRACTICE

1. Spills or Leaks of Wastes at Surface
2. Corrosion of Casing Allows Waste Escape
3. Inadequate Seal Permits Wastes to Back-flow
4. Fractures, Existing or Caused by Earthquakes or the Introduction of Fluids, Allow Wastes to Escape into Groundwater

**FIGURE 12-5**
Injection well, a technique that is used for disposal of large amounts of liquid wastes. The supposition is that toxic wastes may be drained into dry porous strata within the earth, where they may reside harmlessly "forever." Precautionary measures to make this method safe are listed on the left. Often the installation of these measures is inadequate and, even when it is adequate, potential for failure remains (right). (Adapted with permission from *Environmental Action* magazine, 1525 New Hampshire Ave. NW., Washington, D.C. 20036.)

safe if ideally constructed and managed; however, in practice the ideal is seldom realized.

**Deep well injection.** Some 57 percent of hazardous wastes are presently disposed of by deep well injection. This method involves drilling a well into porous material below the groundwater (Fig. 12-5). In theory, hazardous waste liquids pumped into the well soak into the porous material and remain isolated from groundwater by impermeable layers. However, it is impossible to guarantee that fractures in the impermeable layer will not eventually permit wastes to escape and contaminate groundwater. Indeed, stresses produced by the introduction of wastes may even cause such fractures. Also, there are a number of

**FIGURE 12-6**
Surface impoundments, an inexpensive technique that is used for disposal of large amounts of lightly contaminated liquid wastes. (a) The supposition is that only water leaves the impoundment, by evaporation, while wastes will remain and accumulate in the impoundment indefinitely. Precautionary measures required to make this method safe are given on the left. Frequently these measures are not installed or are inadequate, but even when they are adequate, potential for failure remains (right).

other ways in which wastes can escape into groundwater as shown in Figure 12-5.

**Surface impoundments.** Another 38 percent of hazardous wastes are deposited in surface impoundments. This is the least expensive way to dispose of large amounts of water carrying relatively small amounts of hazardous materials (wash water for example). The waste is discharged into a sealed pit or pond and evaporates while wastes settle and accumulate (Fig. 12-6). If the bottom is well sealed and evaporation at least equals input, such impoundments may receive and hold wastes indefinitely. However, seals may break down over time and/or exceptional storms may cause overflows.

**Landfills.** When hazardous wastes are in a concentrated form, they are commonly put into drums and buried in **landfills.** If a landfill is properly lined, covered, and supplied with a means to remove drainage as shown in Figure 12-7, it is presumed safe; it is called a **secured landfill.**

(a)

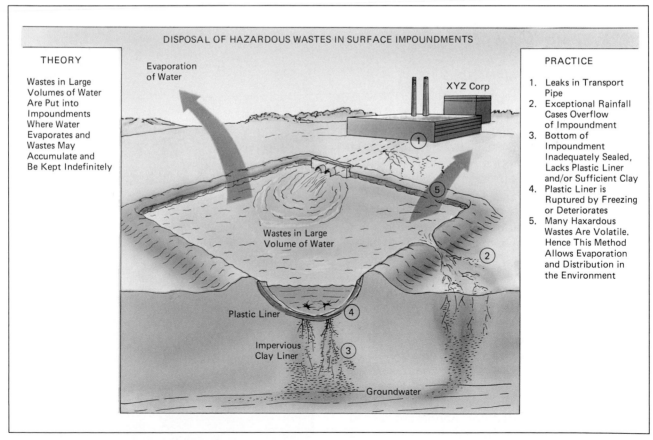

DISPOSAL OF HAZARDOUS WASTES IN SURFACE IMPOUNDMENTS

THEORY

Wastes in Large Volumes of Water Are Put into Impoundments Where Water Evaporates and Wastes May Accumulate and Be Kept Indefinitely

Evaporation of Water

XYZ Corp

Wastes in Large Volume of Water

Plastic Liner

Impervious Clay Liner

Groundwater

PRACTICE

1. Leaks in Transport Pipe
2. Exceptional Rainfall Cases Overflow of Impoundment
3. Bottom of Impoundment Inadequately Sealed, Lacks Plastic Liner and/or Sufficient Clay
4. Plastic Liner is Ruptured by Freezing or Deteriorates
5. Many Haxardous Wastes Are Volatile. Hence This Method Allows Evaporation and Distribution in the Environment

**FIGURE 12-6** *(cont.)*
(b) Surface impoundment pond (left foreground) is the main feature of this waste disposal facility. (Photo by Hal Yeager, Alabama Journal.)

**FIGURE 12-7**
Secured toxic waste landfill, a technique for disposal of modest quantities of concentrated hazardous materials. The supposition is that hazardous materials may remain isolated in the ground indefinitely. (a) Chemical dumping in missile silos near Grandview, Idaho. (© David Frazier, Photo Researchers, Inc.)

However, as noted in the figure, the various barriers are subject to damage or deterioration. Many experts feel it is only a question of time before the contents will leach from even the most secure landfills.

### PROBLEMS IN MANAGING LAND DISPOSAL

Two problems are inherent in land disposal. The first is ensuring that wastes get to disposal facilities. The second is ensuring that disposal facilities are properly constructed, managed, and sealed. Problems in both areas became strikingly apparent during the 1970s.

**Midnight dumping.** When legislation prohibited discharging hazardous wastes into air and water, many companies needed help with alternative means of disposal. Some highly reputable individuals entered the business of handling and disposing of wastes properly. However, there were also unscrupulous "operators" who formed "fly-by-night" companies and said: For a price, let us dispose of your wastes—no questions asked. As it turned out, they simply pocketed the money and left stacks of drums of hazardous wastes in old warehouses, piled or crudely buried them on vacant lots or farms, sneaked them into sanitary landfills (those intended for nonhazardous domestic wastes), or even drained them along road-

**FIGURE 12-7** (*cont.*)
(b) Secured hazardous waste landfill under construction in Alabama. (Nancy Shute.)

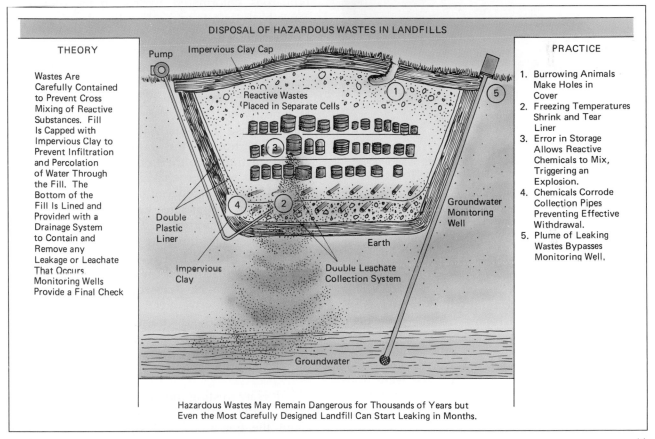

## DISPOSAL OF HAZARDOUS WASTES IN LANDFILLS

### THEORY

Wastes Are Carefully Contained to Prevent Cross Mixing of Reactive Substances. Fill Is Capped with Impervious Clay to Prevent Infiltration and Percolation of Water Through the Fill. The Bottom of the Fill Is Lined and Provided with a Drainage System to Contain and Remove any Leakage or Leachate That Occurs. Monitoring Wells Provide a Final Check

### PRACTICE

1. Burrowing Animals Make Holes in Cover
2. Freezing Temperatures Shrink and Tear Liner
3. Error in Storage Allows Reactive Chemicals to Mix, Triggering an Explosion.
4. Chemicals Corrode Collection Pipes Preventing Effective Withdrawal.
5. Plume of Leaking Wastes Bypasses Monitoring Well.

Pump

Impervious Clay Cap

Reactive Wastes Placed in Separate Cells

Groundwater Monitoring Well

Double Plastic Liner

Impervious Clay

Double Leachate Collection System

Earth

Groundwater

Hazardous Wastes May Remain Dangerous for Thousands of Years but Even the Most Carefully Designed Landfill Can Start Leaking in Months.

**FIGURE 12-7** (*cont.*)
(c) Precautionary measures to make this method safe are listed on the left. Frequently the execution of these measures is inadequate, but even when they are installed properly, many potentials for failure remain. (Adapted from an illustration by Rick Farrell, copyright ©. All rights reserved.)

**FIGURE 12-8**
Midnight dumping. In the past, large quantities of hazardous wastes were simply dumped and abandoned with no precautionary measures whatsoever. Frequently dumping was done under the cover of darkness to escape notice, hence the term *midnight dumping*. Thousands of such "midnight dumps" remain scattered about the country. Illegal disposal of parathion barrels by a group of aerial sprayers in Sumner Co., Kans. © 1984 Larry Miller, Photo Researchers, Inc.

sides in the dark of night (Fig. 12-8). The operators then disappeared without a trace. This practice has been called **midnight dumping.** Even where disposal was accepted, results have been tragic (Fig. 12-9).

**Nonsecure disposal.** The potential threat posed by accepted disposal sites that were nonsecure was dramatically brought to the public attention by Love Canal, an abandoned canal bed near Niagara Falls, New York. In the 1930s and 1940s it served as a convenient, acceptable burial site for thousands of drums of waste chemicals—over 20,000 tons in all. After the canal was filled and covered, the area seemed ripe for development. The land was transferred to the city of Niagara and to developers. Homes and a school

were built on the edge of what had been the old canal and the area of covered chemicals became a playground. Over the years, parents observed that children who attended the school and came in contact with the black gooey stuff oozing out of the ground had health problems ranging from chemical burns and skin rashes to severe physiological and nervous disorders. Even more alarming, residents began to note that an unusually high number of miscarriages and birth defects were occurring. In particular, it was noted in one neighborhood that out of 16 newborns, 9 had abnormalities. The average rate of occurrence for such abnormalities in the population at large is about 1 percent. In this particular community, the rate was 56 percent. The situation climaxed in 1978 when health authorities finally came in and identified the black ooze as a potent mixture of numerous chlorinated hydrocarbons known to cause birth defects and other disorders in experimental animals.

It is not hard to imagine the desperate feelings that were aroused. Some $3 billion in health claims were filed against the City of Niagara, several hundred times the city's operating budget, and nearly 600 families demanded relocation at state expense. Eventually the state did purchase about 100 homes, but most people only had added frustration to their other problems.

Along with the episode at Love Canal, numerous lesser incidents of groundwater contamination from various sites have occurred (Fig. 12-10). The EPA reported in 1984 that contaminated groundwater has shown up in some 2800 wells in 20 states in the last five years. Many of these incidents were discovered only after people experienced unexplainable illnesses over prolonged periods. While they were not as widely publicized, they were nevertheless tragic and financially devastating for the people involved.

### SCOPE OF THE PROBLEM

It is interesting to note that, prior to the mid 1970s, little public information was available about the quantities and kinds of toxic wastes generated, where they were being disposed of or whether sites posed a threat to human health. The recognition of midnight dumping, the episode at Love Canal, and various instances of groundwater contamination provided the impetus to initiate investigations to answer these ques-

(a)

(b)

**FIGURE 12-9**
Nonsecure disposal of toxic wastes. In addition to midnight dumping, huge quantities of toxic wastes were just deposited on or in sites with few if any of the features required for security. Thousands of such nonsecured sites containing toxic wastes remain a serious threat to groundwater.
(a) Hazardous waste stored in barrels near Chatfield Reservoir, Denver, Colorado, area. (Kent & Donna/ Photo Researchers.) (b) Site clean up, Chemical Management, Elizabeth, N.J. (Courtesy of J. B. Moore, $CH_2M$ HILL.)

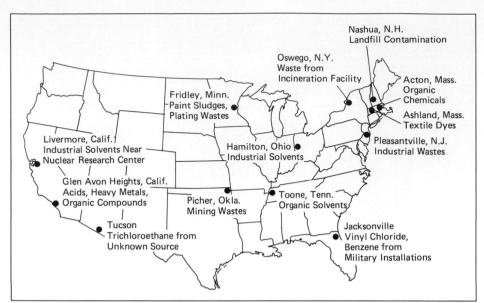

**FIGURE 12-10**
Examples of endangered water. The map depicts just a few of the many cases of water contamination that have been identified. Much more serious is the potential for future contamination. (Reprinted from *U.S. News and World Report*, Jan 16, 1984. Copyright, 1984, U.S. News and World Report.)

Map labels:
- Nashua, N.H. Landfill Contamination
- Oswego, N.Y. Waste from Incineration Facility
- Acton, Mass. Organic Chemicals
- Ashland, Mass. Textile Dyes
- Fridley, Minn. Paint Sludges, Plating Wastes
- Pleasantville, N.J. Industrial Wastes
- Livermore, Calif. Industrial Solvents Near Nuclear Research Center
- Hamilton, Ohio Industrial Solvents
- Glen Avon Heights, Calif. Acids, Heavy Metals, Organic Compounds
- Picher, Okla. Mining Wastes
- Toone, Tenn. Organic Solvents
- Tucson Trichloroethane from Unknown Source
- Jacksonville Vinyl Chloride, Benzene from Military Installations

tions. The results, which only now are becoming clear, show that the threat of groundwater contamination by toxic wastes is extremely pervasive and widespread.

The Environmental Protection Agency now estimates that there are close to 150 million tons of hazardous wastes generated each year in the United States (close to 2/3 ton per person). In the late 1970s they estimated that as much as 90 percent was being disposed of improperly. Surveys indicate that there are between 32,000 and 50,000 landfill sites that contain hazardous wastes and as many as 15,000 are estimated to need remedial action to prevent groundwater contamination. Some sites have been found where wastes are actually immersed in groundwater. Injection wells have been discovered where wastes are deposited above or even into aquifers. Numerous surface impoundments are not sealed at all and are allowing wastes to leach into groundwater. In one survey of 20 landfills licensed to receive toxic wastes, the EPA found that none had groundwater monitoring systems adequate to detect leakage should it occur. Another survey in the 1970s showed that the drinking water supplies in many cities were tainted with synthetic organic chemicals from various sources, although levels were generally considered not dangerous.

Despite the widespread nature of the problem, however, the actual threat to human health remains controversial for several reasons. First,

groundwater moves in complex patterns; therefore, it is difficult to predict what wells may be contaminated from a given site. Second, movement of groundwater is generally very slow, in the order of a few centimeters per day, and toxic chemicals tend to bind to soil particles so that they move even more slowly. However, these factors differ tremendously with different rock and soil types so that, again, it is difficult to predict how rapidly wastes will spread from a given disposal site even if it is leaking. Finally, the actual health effects of the chemicals involved is extremely controversial. Many are known to have severe effects and have been shown to be carcinogenic if doses are high. However, the effects of low doses over long periods of time are largely unknown, and many of the chemicals involved have not been tested.

The slow movement of chemicals and the absence of observable health effects at low doses is not comforting, however, because we understand that these chemicals may persist in the environment indefinitely. Even if their movement is slow, they may eventually reach water supplies. Then, given the capacity of toxics to bioaccumulate and cause disease many years after initial exposure, low doses have little relevance. The urgency of the situation is further illustrated by the fact that once groundwater is polluted with toxic wastes, it is difficult or impossible to reverse the situation. In some cases, the contaminated water

**FIGURE 12-11**
When water supplies become seriously contaminated, there may be no alternative but to abandon the property, even though it may be financially ruinous to do so. (Photo by Thomas Busler.)

can be pumped out, but the costs of this process can be prohibitive. Nor does such action address the questions of what should be done with the contaminated water, and if adequate supplies remain. If nature is left to follow its own course, it may be 100 years or more after the source of pollution is removed before pollutants are flushed through and out. And while research and debate over this problem go on, toxic chemicals continue to be found in or migrating toward increasing numbers of water supplies. Some people, in the face of unknown hazards and frustrations, are driven to abandon their homes (Fig. 12-11).

## CLEANUP AND MANAGEMENT OF TOXIC WASTES

From the preceding discussion, it should be clear that there are three major aspects to the toxic waste problem. First and overall is the need to assure safe drinking water and irrigation water supplies. Second is the need to clean up thousands of existing sites from which toxic materials are already or may soon be leaching into the groundwater. Third is the need to provide for proper management and disposal of hazardous wastes being generated now and those that will be generated in the future. At least the beginning stages

of all three have been mandated by Congress, and numerous states and municipalities are also addressing the issue.

### Assuring Safe Drinking Water Supplies

The Safe Drinking Water Act of 1974 was passed to protect the public from the risk of toxic chemicals contaminating drinking water supplies. It mandates the monitoring of water supplies; if specified toxic chemicals are found, supplies may be closed until adequate purification procedures are adopted or until pollutants are flushed out. However, in practice several shortcomings remain. Municipal supplies are monitored, but there is no systematic monitoring of groundwater or private wells. Therefore, groundwater contamination is seldom recognized until people with private wells experience ''unexplained'' illness, and report ''funny'' tasting or smelling water. Even after analyzing the water and verifying the presence of synthetic organic chemicals, action may not be taken because standards as to what constitutes ''unsafe'' versus ''safe'' pollutant levels have not been set for many of the chemicals involved. Finally, closing wells—and as of 1982 over 500 wells, some in every state, had been closed—hardly solves the problem. When wells are closed, people must bring in water from another source; in some cases the hardship has meant abandoning property. Clearly, there is a need to upgrade the requirements under the Safe Drinking Water Act as well as to identify and eliminate sources of pollution.

## Cleaning Up Existing Toxic Waste Sites

A major federal program aimed at identifying and cleaning up existing waste sites was initiated by the Comprehensive Environmental Response, Compensation, and Liability Act of 1980, popularly known as the **Superfund.** This legislation, through a tax on the chemical industry, provides a fund to clean up those sites where parties responsible for the dumping cannot be found. Where responsible parties can be found, the law forces them to clean up the site.

However, the enormity of the task precludes any quick solution, and there are many difficult choices or tradeoffs to make. First, the Superfund, even at $1.6 billion over the first five years, was sufficient to clean up only about 20 percent of the more than 500 sites that have been identified as needing prompt attention. The EPA has established a priority list, but it is evident that any such list is open to question and political influence. Second, the question arises of whether Superfund money should be used to clean up a site immediately or whether the slow process of litigation should be engaged in to get a responsible (but reluctant) party to do it. Third, cleaning up a site presupposes that there are facilities for properly disposing of the wastes and contaminated soil; without proper facilities, the problem is just transferred from one place to another. Hence, another dilemma emerges. Should one wait (and risk further contamination) until facilities offering a permanent solution are available, or should one act immediately with interim solutions, such as the installation of drainage pumps to remove leachate from the old site? Again, the latter course of action leaves less money for the former, and presupposes an acceptable way of disposing of the leachate. Given these questions, it is hardly surprising that there has been much controversy concerning the EPA's administration of the Superfund program, and such controversy is likely to continue. As of 1985 only about a dozen of the more than 500 sites needing attention had been cleaned up. Still, the Superfund does represent movement in the right direction.

## Management of Wastes Currently Produced

Indiscriminate disposal of toxic wastes is addressed through both the Clean Drinking Water Act of 1974 and its amendments of 1977 and the Resource Conservation and Recovery Act of 1976 (RCRA). The Clean Drinking Water Act provides for proper location and construction of injection wells and for setting standards for and monitoring drinking water supplies. However, few standards have been set because of the great number of chemicals involved and their unknown health effects.

The RCRA requires records to be kept on the handling and transfer of all hazardous wastes from their point of origin to their ultimate disposal. Furthermore, all landfills or other facilities receiving hazardous wastes must be authorized by the issuance of a permit. A permit implies that requirements for proper construction and operation of the site are met. While this law requires a great deal of paperwork, its intent is to prevent the practice of midnight dumping and to assure that wastes are finally disposed of in a secured site or in other ways that have minimum risk of causing pollution problems.

Controversy also surrounds the EPA's administration of the RCRA. **Interim permits** have been granted and toxic wastes continue to go into thousands of facilities that are far from ideal. This procedure is necessary initially because it is not practical to shut down industry until proper disposal facilities are developed; in the meantime, the wastes must go somewhere. Requirements under interim permits can demand rapid upgrading, but this process has proceeded very slowly. As of 1983, a survey conducted by Congress showed that only 24 of an estimated 8000 facilities had met the full requirements. Moreover, investigators also found that nearly 80 percent of the facilities were violating interim permits by not monitoring for possible leaks of toxic chemicals into groundwater. It is clear that a policy of strong enforcement is required in addition to the law.

Even with strict enforcement of the RCRA, however, significant quantities of hazardous and toxic wastes may escape regulation because the law addresses only operations that generate more than 100 kilograms (220 lb) per month of hazardous wastes. Operations that produce less than this amount are exempt from the law, as are homeowners and farmers. Surveys also show that the quantity of pesticides and other toxic and hazardous materials thrown away by homeowners, farmers, and small businesses is considerable.

## Future Management of Hazardous Wastes

It should now be clear that many shortcomings concerning regulation, management, and disposal of hazardous materials still exist. Fortunately, these shortcomings are recognized and are gradually being addressed in some of the ways described below.

### REDUCING PRODUCTION OF HAZARDOUS WASTES

Strict regulation and enforcement of hazardous waste disposal has motivated many industries to actually reclaim and recycle materials that were formerly discarded. Also, the recording of all hazardous wastes enables the creation of waste inventories that often show that one company's waste is another company's raw material. Further, many companies now research new products and/or production methods in an effort to adopt procedures that generate fewer byproducts. In some cases, a changeover has been mandated by legislation that bans further production and/or use of certain chemicals. The banning of PCB production is one such case, but large amounts of PCBs remain in old transformers.

However, it is not always possible to make substitutions in products or processes, nor is it reasonable to expect society to abandon the benefits of technologies and products that have toxic materials as waste or byproducts. Therefore the toxic waste disposal problem will remain.

### NEW TECHNOLOGIES FOR TOXIC WASTE DISPOSAL

Listed below are a number of methods that may curb disposal problems:

1. Nonbiodegradable organic compounds can be broken down at high temperatures. For example, breakdown of chlorinated hydrocarbons could yield carbon dioxide, water, and harmless chlorine compounds. Some incinerators have already been constructed for this purpose. Some are being constructed aboard vessels so that burning can be done at sea.

2. Heavy metals may be chemically precipitated (made insoluble) from solutions. The resulting sludges are rich "ores" that can be reprocessed.

3. In the future, techniques of genetic engineering may enable the development of bacteria that have the capability of decomposing organic compounds that are presently nonbiodegradable.

4. Future chemical technology may also develop catalysts (artifical enzymes) that will break down halogenated hydrocarbons.

With the implementation of these new technologies, simple disposal in facilities such as landfills will become unnecessary. In the meantime, however, the various techniques of land disposal are still unfortunately perceived as the least-cost method; alternative techniques listed above are perceived as more expensive. Other factors being equal, economic forces dictate that the least-cost method will be pursued. Importantly, we must come to recognize that the "other factors" are not equal. If we consider the costs of guarding against the perpetual uncertainties and risks entailed in storing hazardous wastes in the ground, much less the health costs if such facilities break down, land disposal is a most expensive long-term option.

Our future assignment should be to phase out land disposal (as well as air and water disposal) of hazardous wastes and to hasten the development and implementation of the alternative technologies that reuse such wastes or render them harmless. Vigorous protest by citizens against the location of hazardous waste landfills near their communities is making it all but impossible to find new disposal sites and is effectively promoting this change. However, direct public pressure on the political process to amend basic laws and regulations in this regard is also in order. In reauthorizing the RCRA in 1984, Congress did include provisions to ban land disposal of many of the most offending chemicals over the next five years.

### CLOSING LOOPHOLES IN CURRENTLY UNREGULATED TOXIC WASTE DISPOSAL PRACTICES

In 1985 the RCRA was amended to include many previously unregulated small-quantity generators, especially those that produce between 100 and 1000 kg (220 to 2200 lb) of hazardous wastes per month. It is estimted that this action will nearly double the quantity of hazardous wastes under regulation.

Another action with potential for reducing toxic waste pollution is to prevent homeowners from disposing toxic substances with their trash. It is impractical to try to place all individual homeowners and farmers under regulation because monitoring and enforcing such regulations would be infeasible and intolerable. However, studies show that if special trash pickups are provided, most homeowners cooperate. Currently, homeowners have no practical alternative other than the trash can for disposal of empty paint or pesticide cans. Citizen pressure may prompt local authorities to offer special pickups.

## OTHER PROBLEMS

In closing, we remain aware that disposal of hazardous and toxic wastes is not the only source of groundwater contamination (see the list on p. 311). Other sources also require attention and appropriate action. In particular, there is a need to check, repair, and replace underground fuel storage tanks in most service stations. About 1.5 to 2 million of these tanks exist and experts estimate that 75,000 to 100,000 of them leak.

The problem of fertilizer contamination may be addressed by better soil management as discussed in Chapter 7. We shall look at the problem of pesticide contamination in Part 5.

# 13

# AIR POLLUTION AND ITS CONTROL

## CONCEPT FRAMEWORK

Volcanoes and natural fires have emitted smoke and other substances into the atmosphere for thousands of years. However, with the discovery of fire about 100,000 years ago, humans began adding to these natural pollutants. Our early forebears probably did not consider the potential harmful effects of pollution, nor did their primitive technology provide any practical alternative. Thus, venting fire wastes into the atmosphere became an accepted practice. It was simply assumed that the atmosphere would assimilate these pollutants in one way or another.

This assumption was not totally false. Given adequate space, pollutants do disperse and dilute in the atmosphere. Further, pollutants will settle or come down to earth via precipitation, and studies show that many poisonous gases are converted to harmless products by soil microorganisms (Fig. 13-1). For example, carbon monoxide and sulfur dioxide, both highly poisonous gases generated by combustion, are converted to carbon dioxide and sulfate, which are plant nutrients.

The amount and type of pollutants that the atmosphere can handle, however, are limited.

With the advent of the Industrial Revolution, the load of pollutants discharged into the atmosphere increased exponentially. Yet, the assumption that the atmosphere could take care of the problem prevailed. In fact, belching smokestacks became a symbol of progress. It was not until the decades following World War II that people recognized that this assumption had been carried too far. It became increasingly commonplace to see cities enshrouded in a brownish haze referred to as smog, or more properly, **photochemical smog** (Fig. 13-2). On some occasions, the "bad air" spread far beyond cities. In one notorious episode in August 1969 a conspicuous buildup of smog covered almost the entire eastern United States from the Great Lakes to the Gulf Coast. And an increase in atmospheric turbidity (haze) was observed globally (Fig. 13-3).

The poor visibility and hazy sky resulting from the widespread pollution not only seriously degraded the aesthetic quality of life, but produced other adverse effects as well. Many people found the smog irritating to eyes and throat and aggravating to preexisting respiratory conditions

**FIGURE 13-1**
Dilution and assimilation of pollutants. As a pollutant mixes with a larger air mass, its concentration diminishes by dilution. With sufficient dilution, concentration is reduced to a threshold level, a level at or below which it is assumed to cause no ill effects. Further, soil microorganisms or other natural processes may absorb and assimilate pollutants, removing them from the system entirely. Thus, there is an assumption that "nature will take care of pollutants." Unfortunately modern civilization produces pollutants in such quantities or of such kinds that these assumptions do not hold.

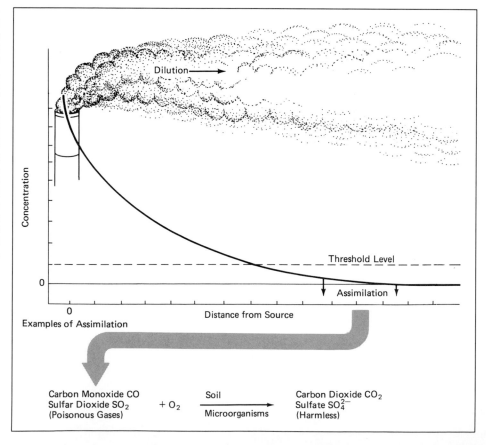

330

(a)

(b)

**FIGURE 13-2**
A typical episode of photochemical smog. (a) Early in the morning, the air is clear (Tom McHugh/Photo Researchers.) (b) Midmorning of the same day, the air very hazy with smog. (Georg Gerster/Photo Researchers.) The haze is the result of pollutants from the exhaust of rush-hour traffic reacting in the atmosphere. The reactions are promoted by sunlight. Photos are close to the same view over Los Angeles.

**FIGURE 13-3**
Increasing atmospheric turbidity. Measurements of atmospheric turbidity (haziness), taken at Mauna Loa, Hawaii, showed an increasing trend apart from natural events. Since Hawaii is far removed from human sources of pollution, this was taken to mean that human-produced pollution was affecting the entire atmosphere. (Redrawn by permission from Wilfred Back, *Atmospheric Pollution,* p 37, McGraw-Hill Book Company, 1972.)

Mauna Loa
Observatory

Linke Turbidity

Mt. Agung Eruption

Year

331

| Table 13-1 | **Major Air Pollution Episodes** | | |
|---|---|---|---|
| *DATE* | | *PLACE* | *EXCESS DEATHS* |
| Feb. | 1880 | London, England | 1,000 |
| Dec. | 1930 | Meuse Valley, Belgium | 63 |
| Oct. | 1948 | Donora, Penn., U.S. | 20 |
| Nov. | 1950 | Poca Rica, Mexico | 22 |
| Dec. | 1952 | London, England | 4,000 |
| Nov. | 1953 | New York, U.S. | 250 |
| Jan. | 1956 | London, England | 1,000 |
| Dec. | 1957 | London, England | 700-800 |
| Dec. | 1962 | London, England | 700 |
| Jan./Feb. | 1963 | New York, U.S. | 200-400 |
| Nov. | 1966 | New York, U.S. | 168 |

Source: Wilfred Bach, *Atmospheric Pollution*, 1972, McGraw-Hill Book Co., reprinted by permission.

such as asthma and emphysema. In some cities pollution reached such high levels that mortality significantly increased (Table 13-1).

In addition to fearing more disasters, the public became concerned about the potential, long-term health effects of exposure to ambient air. The possibility of serious long-term effects was emphasized in 1963 when the U.S. Surgeon General issued the first report linking long-term smoking and lung cancer. It is difficult not to associate smoking and air pollution and wonder if exposure to air pollution may lead to similar effects.

In addition, farmers near cities such as Los Angeles and New York reported damage or even total destruction of crops because of air pollution (Fig. 13-4). A conspicuous acceleration in the rate of metal corrosion and deterioration of rubber, fabrics, and other materials was noted. In short, we were overloading the natural system, especially in cities. Our rapidly expanding industry and, particularly, our mushrooming use of cars for everyday transportation had resulted in the discharge of pollutants at a faster rate than natural processes could handle.

By the 1960s it was evident that discharges into the air had to be controlled. Various air pollution laws were enacted, most notable of which were the Clean Air Act of 1970 and its amendments of 1977. These laws remain the foundation

**FIGURE 13-4**
Sweet corn growing near Los Angeles about 1968. Injury was caused by the existing levels of air pollution, particularly photochemical smog. Such examples became increasingly common through the 1960s. (Photo courtesy of Ray Thompson, Statewide Air Pollution Research Center, Univ. of California, Riverside, Calif.)

of air pollution control efforts. They designate the most widespread pollutants, call for setting standards as to how much of them may be present in the air, and establish goals as to when these standards should be met. As a result of the Clean Air Act, air quality in most cities is markedly better now than it was in the early 1970s. However, not all problems have been solved and some have become worse.

The objective of this chapter is to provide an understanding of the main aspects of air pollution, show what has been done, and most importantly, what remains to be done to control air pollution.

## THE FOUR-PHASE ATTACK ON AIR POLLUTION

Ideally, we may wish to eliminate all air pollution. However, it is technologically impractical to prevent emissions from all sources and it is economically wasteful to spend large sums of money controlling an emission that is harmless or insignificant. Further, industry is increasingly reluctant to adopt and government is reluctant to impose control measures when there is not adequate economic justification. Therefore, the attack on air pollution involves four major phases: (1) identifying the pollutants present in the air, (2) determining their sources, (3) demonstrating which pollutants are responsible for particular adverse health and/or environmental effects, and (4) developing and implementing suitable controls on the major sources of the most harmful pollutants. These four phases are not separate; all are ongoing processes. As new information regarding adverse health and/or environmental effects of pollutants are discovered, requirements and strategies for control must be modified and adjusted. However, for clarity, we shall consider these phases in the order given.

## MAJOR OUTDOOR AIR POLLUTANTS AND THEIR SOURCES

### Major Pollutants

It is important to recognize that air pollution is not a single entity; it is an alphabet soup of solid and liquid particles in suspension along with numerous chemicals present as gases or vapor.

These are all mixed with the normal constitutents of air (Fig. 2-1). The amount of each pollutant present varies tremendously depending on proximity to the source and various conditions of wind and weather. Eight pollutants or pollutant categories are of major concern. These are generally present in greater or lesser amounts in ambient air. The categories are as follows:

1.  Particulates. Particulates are solid or liquid particles that are so tiny they remain suspended in the air for long periods of time. We see these particles in mass as smoke or haze; other pollutants present as individual molecules or ions are not visible except in the case of nitrogen dioxide, which is a brownish gas. Chemically, particulates may contain any or all of the other pollutants.

2.  Hydrocarbons and other volatile organic compounds. Includes materials such as gasoline, paint solvents, and organic cleaning solutions which evaporate and enter the air in a vapor state.

3.  Carbon monoxide ($CO_2$).

4.  Nitrogen oxides ($NO_x$).

5.  Sulfur oxides, mainly sulfur dioxide ($SO_2$).

6.  Lead and other heavy metals.

7.  Ozone and other photochemical oxidants.

8.  Acids, mainly of sulfur and nitrogen.

### Sources of Pollutants

In large measure, the above pollutants are direct or indirect products of combustion. Therefore, emission may occur wherever fuel or other materials are burned.

#### DIRECT PRODUCTS OF COMBUSTION

The primary fuels, including coal, liquid fuels refined from crude oil, and natural gas, and most trash such as wastepaper, are organic. When such material is burned, the major waste products are carbon dioxide and water vapor (see Chapter 2). However, in burning, oxidation is seldom complete. Particles consisting mainly of carbon are emitted into the air; these are the **particulates** seen as smoke. In addition, there are various unburned or partially oxidized fragments of fuel molecules. These contribute to the **hydrocarbon** or volatile organic category of pollutants. Partially oxidized carbon atoms result in **carbon monoxide** as opposed to carbon dioxide.

Furthermore, combustion almost always uses air as the oxidizing medium rather than pure oxygen. Recall that air is only about 20 percent oxygen. It is 78 percent nitrogen. At the high temperatures of combustion, some of the nitrogen gas ($N_2$) is oxidized to form the gas, nitric oxide (NO). In the air, nitric oxide immediately reacts with additional oxygen to form nitrogen dioxide ($NO_2$), and/or nitrogen tetroxide ($N_2O_4$). These compounds are collectively referred to as **nitrogen oxides** and given the designation $NO_x$. Nitrogen dioxide absorbs light and is largely responsible for the brownish color of photochemical smog.

Finally, fuels or trash generally contain various impurities or perhaps additives and these are also emitted into the air with the exhaust. Coal, in particular, contains from 0.2 to 5.5 percent sulfur as an impurity. In combustion, the sulfur is oxidized, giving rise to the gas **sulfur dioxide.** Coal also may contain heavy metal impurities and, of course, trash contains an endless array of impurities. Lead is added to gasoline as an inexpensive way to prevent engine knock. As such fuels or materials are burned, atoms of these pollutants are emitted and may remain airborne and travel great distances before settling (Fig. 13-5).

**FIGURE 13-5**
Lead levels in Greenland glaciers. Age of ice samples is related to their depth in the glacier. Lead content of samples, while variable, is clearly correlated with amount of lead emitted into the air by industry and automobiles. (Reprinted with permission from M. Murozumi, T.J. Chow, and C. Patterson, "Chemical Concentrations of Pollutant Lead Aerosols, Terrestrial Dusts, and Sea Salts in Greenland and Antarctic Snow Strata," *Geochimica and Cosmochimica Acta*, 33. Copyright 1969, Pergamon Press, Ltd.)

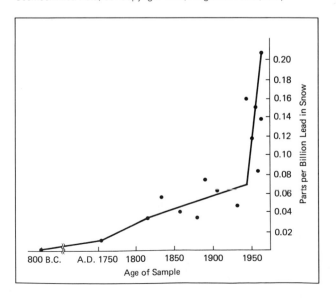

Sulfur dioxide, nitrogen oxides, and carbon monoxide are known to be highly poisonous gases. Heavy metals and hydrocarbons are also well-known toxic chemicals as described in Chapter 12, and are known to aggravate the respiratory system. They have also been linked to cancer. These pollutants are bad enough in themselves; however, even more damaging pollutants form in the atmosphere, as will be discussed below.

## INDIRECT PRODUCTS OF COMBUSTION

Analysis of smog-laden air reveals significant concentrations of ozone, highly reactive synthetic organic compounds, and acids. These pollutants, especially ozone and acids, have been shown to be the most irritating to humans and the most damaging to crops and forests. These substances are known as indirect products of combustion because they are formed by various reactions of the primary products after combustion has occurred.

Intensive research revealed that **ozone** and numerous reactive organic compounds are formed as a result of chemical reactions between nitrogen oxides and volatile hydrocarbons, with sunlight providing the energy necessary to cause the reactions to occur. Since sunlight provides the energy, these products are collectively known as **photochemical oxidants.** The major reactions involved are shown in Figure 13-6. Light energy is absorbed by and causes nitrogen dioxide to split to form nitric oxide and free oxygen atoms. The free oxygen atoms combine with oxygen gas ($O_2$) to form ozone ($O_3$). Interestingly, these reactions are spontaneously reversible; that is, if other factors were not involved, ozone and nitric oxide would react to reform the nitrogen dioxide and oxygen gas and there would be no appreciable accumulation of ozone (Fig. 13-6a).

But, when hydrocarbons are present, the nitric oxide reacts with them and has two diabolical effects. First, the reaction between nitric oxide and hydrocarbons leads to highly reactive and damaging synthetic organic compounds known as peroxyacetyl nitrates, or PANs. Second, with the nitric oxide tied up in this way, the ozone accumulates (Fig. 13-6b). We may note here that a strategy for controlling ozone has been the control of hydrocarbon emissions.

Acids of sulfur and nitrogen are also indirect products of combustion. In the atmosphere both

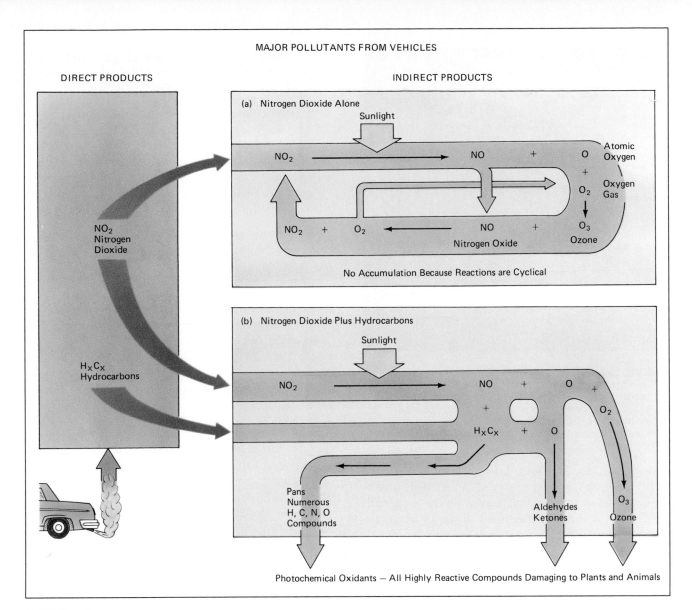

MAJOR POLLUTANTS FROM VEHICLES

DIRECT PRODUCTS

INDIRECT PRODUCTS

(a) Nitrogen Dioxide Alone

No Accumulation Because Reactions are Cyclical

(b) Nitrogen Dioxide Plus Hydrocarbons

Photochemical Oxidants — All Highly Reactive Compounds Damaging to Plants and Animals

**FIGURE 13-6**
Formation of photochemical oxidants, the most injurious ingredients of photochemical smog. (a) Nitrogen oxides, by themselves, would not reach damaging levels because reactions involving them are cyclic. (b) When hydrocarbons are also present, however, reactions occur which lead to the accumulation of numerous damaging compounds, most significantly ozone.

sulfur dioxide and nitrogen oxides react with water vapor to form sulfuric and nitric acids, respectively. These acids are the primary factor in acid rain, which is now recognized as one of our most serious environmental problems (see Chapter 14).

Finally, fine particles also take on additional potency as they sift through the atmosphere. Activated charcoal makes an excellent chemical filter because all sorts of compounds are adsorbed by its surface. The same occurs with the fine particles

in the atmosphere; they become heavily laden with other pollutants. Indeed, much of the chemistry leading to the formation of more noxious compounds may take place on the surface of these particles (Fig. 13-7).

*SPECIFIC SOURCES OF ATMOSPHERIC POLLUTANTS*

The prime sources of the above kinds or categories of pollutants are summarized in Figure 13-8a and are discussed below.

**Ozone and photochemical oxidants (hydrocarbons and nitrogen oxides).** To find the source of ozone and photochemical oxidants, we must look to the sources of hydrocarbons and ni-

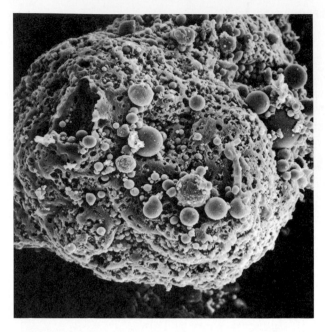

**FIGURE 13-7**
Fly ash from a coal-fired power plant (magnified 2000 times).
A particle of ash is an adsorptive surface for many other
pollutants, as shown by this photograph taken under a
scanning electron microscope. (Photo by Roger J. Cheng,
Atmospheric Sciences Research Center, The State University of New
York at Albany.)

trogen oxides. The gasoline engines of cars and
other vehicles are by far the most widespread
and, given numbers, the largest emitters of both
(Fig. 13-8b). Indeed, the onset of the photochemical smog problem is directly related to the development of the suburban lifestyle with all its commuting by car. Cars are also responsible for
hydrocarbon vapors given off from oil in engines,
from gasoline spilled in the process of fueling,
and from evaporation of gasoline from the tank.
Diesel engines have lower hydrocarbon emissions, but greater emissions of nitrogen oxides.
Recently it has been recognized that evaporation
of organic solvents used in paints, cleaning fluids,
and other processes may also contribute a significant share of hydrocarbons to the atmosphere.

**Sulfur dioxide.** Much of the electricity
used in the United States is generated by the
burning of coal to produce steam to drive turbogenerators. Consequently, such coal-burning
power plants are the major source of sulfur dioxide. In addition, metal ores generally contain considerable amounts of sulfur; therefore, metal
smelters are also potent emitters and may be the
major source of sulfur dioxide in areas where they

exist. Because gasoline, natural gas, and paper
wastes usually contain very little sulfur, combustion of these fuels is not a major source.

**Particulates.** Before the pollution controls
of the 1970s, trash and refuse incineration and
various industrial smokestacks were the major
emitters of particulates, but these sources have
largely been curtailed. Now diesel vehicles are a
significant and growing source as is the private
use of wood and coal stoves for home heating.

**Carbon monoxide.** Motor vehicles are the
major source of carbon monoxide.

**Lead and other heavy metals.** The most
significant source of lead is vehicles that burn
leaded gasoline. Coal-burning power plants are
prime emitters of other heavy metals, as are certain industries and incinerators.

## Concentration of Air Pollutants by Topography or Weather Conditions

One must always bear in mind that it is not
the absolute presence or absence of pollutants,
but the dose (the concentration times the length
of exposure) that determines the end result.

Two factors determine the final concentration of pollutants in the air. One is obviously the
output of pollutants. The second, equally important, is the amount of space into which the pollutants can disperse and dilute. What really instigated air pollution disasters and continues to
cause high-pollution conditions today is not sudden increases in the output of pollutants; it is
weather conditions that create a "closed-room effect," which prevents pollutants from dispersing
into the atmosphere at large.

In addition to still air, the weather condition
that is most significant in this regard is a **temperature inversion.** Normally, air temperature decreases with increasing elevation. In this situation, the warm air near the ground rises (because
warm air is lighter than cold air), carrying pollutants upward and dispersing them at higher altitudes (Fig. 13-9a). In a temperature inversion, a
layer of cold air at the ground is covered by warm
air above. This situation develops because the influx of cooler air moves in under the warm air.
With a temperature inversion, the upward movement of air carrying pollutants is blocked and pollutants accumulate in the cool air near the ground

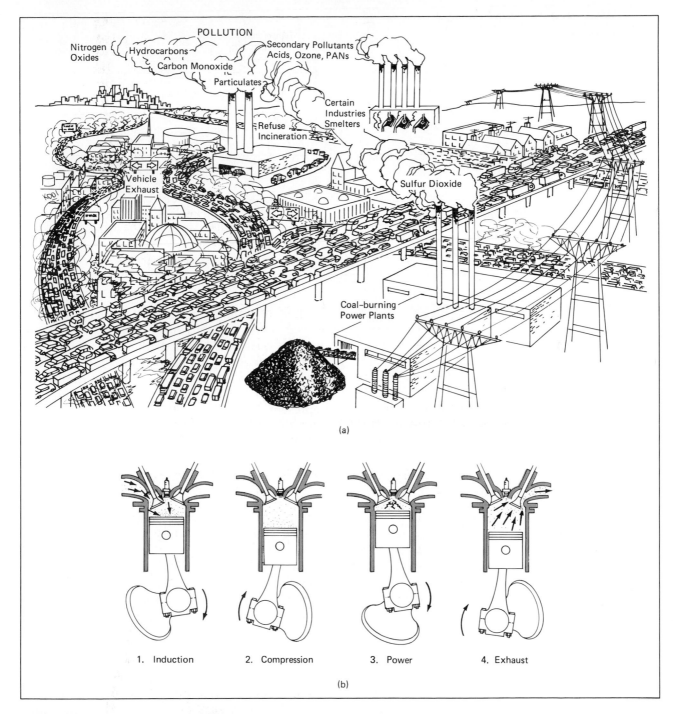

**FIGURE 13-8**
(a) The prime sources of the major air pollutants are depicted in this illustration.
(b) Vehicles are the prime source of hydrocarbons and nitrogen oxides because
in each cylinder of a gasoline engine a fuel-air mixture is (1) taken in, (2)
compressed, (3) ignited, and (4) exhausted about 50 times each second during
normal operation. The high-pressure combustion causes the production of
nitrogen oxides; moreover, combustion is often incomplete, causing emissions
of hydrocarbons and carbon monoxide.

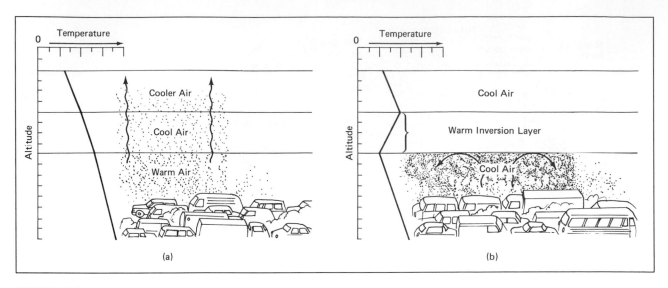

**FIGURE 13-9**
A temperature inversion may cause episodes of high
concentrations of air pollutants. (a) Normally air temperatures
are highest at ground level and decrease at higher elevations.
Since the warmer air rises, pollutants are carried upward and
diluted in the air above. (b) A temperature inversion is a
situation in which a layer of warmer air overlies cooler air at
ground level. This blocks the normal updrafts and causes
pollutants to accumulate like cigarette smoke in a closed room.

(Fig. 13-9b). Local topography may intensify the
effects of a temperature inversion, as in Mexico
City and Los Angeles where surrounding hills or
mountains prevent pollutants from moving horizontally.

These factors suggest that pollution problems can be solved by increasing dilution. "Dilution is the solution to pollution," is a phrase frequently heard. Indeed, this has been a major
control strategy. But we shall see later that it has
serious shortcomings; even much wider dispersal
is not sufficient to avoid certain adverse effects.

## Natural Sources of Pollution

As noted in the introduction of this chapter,
nature itself may produce substantial amounts of
air pollution. Volcanic eruptions and natural fires
may produce all the primary products of combustion. Dust storms and winds that blow clay particles from the soil into the air are very significant
sources of particulates. Vegetation gives off certain volatile organic compounds. Such facts are
sometimes cited as a way to justify pollution from
human sources. However, it is estimated that the
human contribution to air pollution is 10 to 100
times greater than nature's contribution. The total

loading of pollutants from all sources must be
considered in comparison to the capacity of systems to assimilate them. If the human contribution is overloading these systems (and it is), humans must assume responsibility for controlling
these pollutants regardless of the proportion they
represent.

Of course, one category of natural pollutants
is unique—pollens and mold spores. While these
are not generally considered pollutants, those
who have allergies know only too well how much
suffering they cause.

## ADVERSE EFFECTS OF AIR POLLUTION ON HUMANS, PLANTS, AND MATERIALS

We tend to be most concerned about the effects
of air pollution on human health. However, effects on crops, forests, natural ecosystems, and
inanimate materials are equally or potentially
more serious. Therefore, each of these areas must
be addressed when assessing the full impact of air
pollution.

Ideally, of the many pollutants to which we
are exposed, we would like to be able to identify
the one that causes a particular adverse effect. We
could then direct our attention and financial resources efficiently towards controlling that particular pollutant. Unfortunately, real-life situations
are seldom this simple. We and the rest of the
environment are not exposed to individual pollutants. Invariably, we are exposed to the whole
mixture, which varies in makeup and concentration from day to day, even from hour to hour,

and from place to place. Consequently, the effects we feel or observe are rarely (if ever) the effects of a single pollutant; they are the combined effect of the whole mixture of pollutants acting over the total life span. Increasingly, scientists are finding that the end result is not a simple addition or averaging of the effects of the single components. A phenomenon knowns as **synergism** or **synergistic interaction** is frequently involved.

## Synergistic Effects

A synergistic effect occurs when two or more factors interact together to cause an effect that is more severe than one would anticipate from their separate effects. You have probably heard of this in connection with the danger of taking certain drugs in combination with alcohol. For instance, small doses of tranquilizers have a relatively mild effect, as do modest amounts of alcohol. However, taking these drugs and drinking alcohol at the same time may be fatal. Another way of looking at synergism is that in a biological system, the effect of 2 + 2 often adds up to considerably more than 4.

Synergistic effects go far beyond the interaction of various air pollutants. There are countless other factors that may be interacting as well. For any organism, including humans, such additional factors may be nutritional state, presence of other toxic substances in food or water, presence of disease organisms, intrinsic genetic sensitivity or resistance, and other factors causing stress. For plants, additional factors may include insect pests, drought, and toxic substances in soil.

Given the complexity of this situation, it is extremely difficult to determine the role of any particular pollutant in causing the observed result. Consequently, as the following description shows, there is still a conspicuous lack of exact information concerning effects of various pollutants. Nevertheless, research is going forward and more information is becoming available.

## Effects on Human Health

During periods when pollution reaches high levels, many people complain of headaches; irritation of eyes, nose, and throat; nausea; and general ill feeling. More people call in sick to work and more asthmatics have attacks. Ozone seems to be the predominant factor in irritation of mu-cous membranes. Acid particles, particularly those of sulfuric acid, correlate most closely with the increase in asthma attacks, and carbon monoxide levels may be responsible for weakened judgment, drowsiness, and headaches. High levels of particulates over prolonged periods correlate with respiratory disease and lung cancer. However, all the factors may contribute to various degrees.

In some cases, as shown in Table 13-1, air pollution has reached levels such that an increase in the death rate was observed. Although this increase is attributed directly to air pollution, it should be noted that these deaths occur among people already suffering from severe respiratory and/or heart disease. While the gases present in air pollution are known to be lethal in high concentrations, such concentrations are not reached even in air pollution disasters. Therefore, deaths attributed to air pollution are not the direct result of simple poisoning. However, there is no question that air pollution puts an additional stress on the body and, if one is in an already weakened condition, this additional stress may be fatal.

The big question for most people is: Does ambient air pollution cause the really serious respiratory diseases such as lung cancer and emphysema? The heavy metal and organic constituents of air pollution include many chemicals that are known to be carcinogenic in high doses. Therefore, there is a strong feeling that the presence of trace amounts of these chemicals in ambient air may be responsible for a significant portion of the cancer observed in humans. However, the actual evidence to support this contention is far from clear. Initial studies did show a higher rate of lung disease among those living in cities with high levels of air pollution. However, as studies have progressed, the only pollution factor that clearly and indisputably correlates with serious lung disease is cigarette smoking (Fig. 13-10). When studies were repeated separating smokers and nonsmokers, it was found that nonsmokers living in polluted city air had little if any more lung disease than those living in clean air. However, smokers living in polluted air had a much higher incidence of lung disease than smokers living in clean air (Fig. 13-11).

What we see then, is a very strong synergistic interaction between smoking and air pollution. Ambient air pollution by itself does not seem to increase the risk significantly. However, when

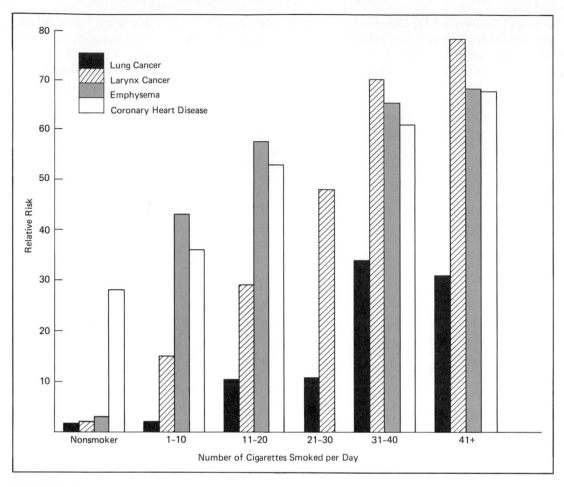

**FIGURE 13-10**
Many diseases and disease conditions are correlated with smoking, including cancer of the lung and larynx, emphysema, and coronary heart disease. (Note that data are not given for emphysema and heart disease at the 21–30 cigarettes-per-day level.) (Data from the U.S. Department of Health, Education and Welfare.)

combined with the effect of smoking, which is a significant risk in itself, the risk greatly increases. Certain diseases typically associated with occupational air pollution show the same synergistic relationship with smoking. For example, black lung disease, the dread of coal miners, is seen almost exclusively among miners who are also smokers. The same is true in the case of asbestos. Lung disease among those exposed to asbestos predominates in smokers. In a 1983 article in *Science* Dr. Bruce Ames, Chairman of the Department of Biochemistry, University of California at Berkeley, stated, "despite numerous suggestions to the contrary, there is no convincing evidence of any generalized increase in the United States (or United Kingdom) of cancer rates other than what could plausibly be ascribed to the delayed effects of previous increases in tobacco usage."

The synergistic effect between smoking and air pollution may occur because one of these effects of smoking is its tendency to deaden the action of the cilia (tiny active hairs) on the cells lining the lung passages. The cilia normally serve to remove foreign particles from the lungs. Also, small particles tend to attract other pollutants from the air as noted previously (Fig. 13-7). Thus, smoking effectively increases the lung's exposure to a host of additional air pollutants.

However, the fact remains that trace amounts of the heavy metals and organic compounds are present in ambient air pollution. Other experts feel that dismissing these as insignificant is not justified in view of the fact that a number of these substances in higher concentrations are known to increase cancer risks. The lungs comprise some 60 m² (about 650 ft²) of tis-

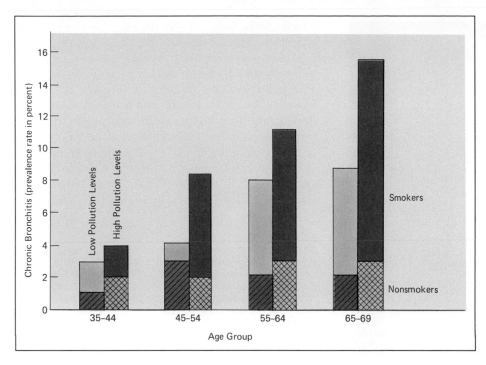

Chronic Bronchitis (prevalence rate in percent)

Low Pollution Levels

High Pollution Levels

Smokers

Nonsmokers

Age Group

35–44 45–54 55–64 65–69

**FIGURE 13-11**
Incidence of chronic bronchitis seen among people living in "low pollution" rural areas and those living in "high pollution" city areas. Among nonsmokers no significant difference is seen between the two groups. But, the increased incidence caused by smoking is increased even more by air pollution, a clear synergistic effect. (From: P.M. Lambert and D.D. Reid, "Smoking, Air Pollution, and Health," *Lancet*, April 25, 1970.)

sue exposed to air and all its pollutants. No other organ of the body is so continually exposed to known and unknown chemicals. The final outcome of this long-term exposure is simply not known at present. In a sense, we are in the midst of a vast experiment in which we are the unwitting subjects.

Also, while it remains questionable whether air pollution apart from smoking increases the risks of lung cancer, there is little question that air pollution causes or aggravates a host of lesser ailments. The correlation between air pollution and increased rates of headaches, nausea, and chronic bronchitis is seen in nonsmokers as well as smokers.

Lead is also a pollutant that deserves special attention. Lead poisoning is a well-known cause of mental retardation, a fact that has been recognized for several decades. It was formerly thought that ingestion of peeling paint chips that contain lead as a white pigment was the main source of lead contamination. In recent years, however, blood tests show that significantly elevated lead levels are much more widespread than previously expected and they plague adults as well as children. In turn, a number of ailments, including learning disabilities in children and, more recently, high blood pressure in adults, have been correlated with high levels of lead in the body.

The source of this widespread contamination is attributed to the use of leaded gasoline. The lead comes out with the exhaust and may be inhaled directly or may settle on food, water, or any number of items that are put in the mouth. This knowledge led the Environmental Protection Agency to mandate a 90 percent reduction of lead in gasoline by the end of 1985 and a total phaseout by 1988.

## Effects on Agriculture and Forests

We cannot ignore the effects of air pollution on plants, which are the basis of agriculture, forests, and all natural ecosystems. Plants, in fact, are generally more sensitive to air pollutants than humans. It was common in the past (but less so now with emission controls) to see wide areas of totally barren land or severely damaged vegetation downwind from smelters. Obviously, workers at least tolerated the emissions that devastated the vegetation (Fig. 13-12). Many species of plants cannot be grown in cities now becaue of poor air quality.

The pollutants responsible for the damage to vegetation are experimentally determined by growing plants in lighted chambers where they can be subjected to any desired concentration of single or mixed pollutants and the results com-

pared with field observations. Also, open-top chambers are set up in the field which enable plants in one chamber to receive filtered air while plants in an adjacent chamber receive ambient air, and pollutants are monitored (Fig. 13-13). Problems still occur in sorting out synergistic interactions from other factors such as drought, damage from insects, plant pathogens, and acid rain. Also tree species that grow 20 to 40 years or more present obvious difficulty. Nevertheless, through such experiments it is possible to gain insight into which pollutants cause damage and to extrapolate

**FIGURE 13-12**
Death of vegetation caused by air pollution. That air pollution can cause severe damage to vegetation has been long known. It is common to find vegetation killed back for considerable distances around various industrial operations. This forest was killed by various pollutants emanating from an open pit coal mine in British Columbia, Canada. (Paolo Koch/Photo Researchers.)

**FIGURE 13-13**
Testing the effects of pollutants in ambient (existing) air on plant growth. Open-top chambers as shown are placed around sample plots of vegetation, and filtered or unfiltered air is blown through. Comparing the growth of vegetation within the two chambers reveals the effect of ambient pollution levels under natural conditions. Many such experiments show that pollution-free air leads to a 10 to 40 percent increase in growth, which is to say that pollution levels in ambient air may be causing this degree of growth reduction. Ozone seems to be most significant in causing this effect.

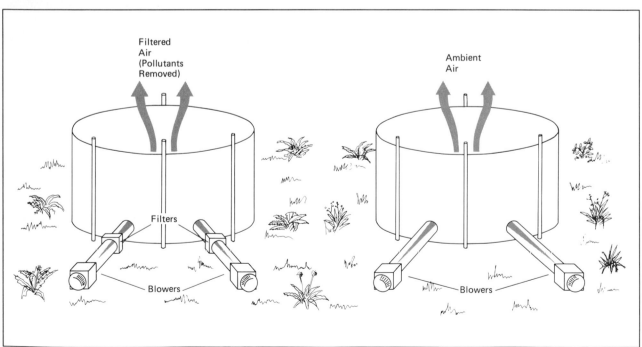

(a)

(b)

**FIGURE 13-14**
Effect of air pollution on plant growth. Various air pollutants such as ozone (a) and sulfur dioxide (b) cause conspicuous damage to plants. (c) Pollution also causes a general reduction in growth without other conspicuous damage. Potato plant at right was grown in prevailing air at Beltsville, Maryland; plant at left was grown at the same location, but air was filtered to remove pollutants. (USDA photos.)

(c)

dioxide (Fig. 13-14). Sulfur dioxide is the factor most often responsible for killing vegetation around smelters. There have been innumerable instances of severe crop damage caused by polluted city air that is carried into agricultural areas. Indeed, pollution-induced damage has become so routine that it has forced the complete abandonment of citrus growing in certain areas of California and vegetable growing in certain areas of New Jersey—areas that were among the most productive regions in the country.

Even more insidious, it has been revealed by open-chamber experiments that plants in clean (filtered) air grow considerably larger than plants in unfiltered ambient air. This shows that the pollution in ambient air is responsible for a general reduction of growth without conspicuous signs of damage or abnormality. Ozone is by far the most significant factor in this effect. A recent assessment of crop production in the United States estimates that without ozone pollution, crop yields would increase as follows: corn, 3 percent; wheat, 8 percent; soybeans, 17 percent; and peanuts, 30 percent. This represents about $3 billion worth of agricultural productivity (1978 prices) or about 10 percent of total farm income.

It is important to emphasize that these fig-

from this information to a broader picture of the effects of air pollution on agriculture, forests, and ecosystems in general. The picture that is obtained should be a concern of top priority.

Experiments show that plants are extremely sensitive, much more so than humans, to ozone and other photochemical oxidants and to sulfur ures represent an average across the United States while most agriculture is located in what are considered clean air regions; thus the effects in badly polluted areas are far more severe. Further, the impacts of air pollution on agriculture are being felt despite the control measures adopted during the 1970s. The conclusion that can be drawn is that, while the pollutants that invade the countryside are greatly diluted and the air appears clean by human health standards, the ozone concentration is still high enough to have a significant negative impact in plant production.

The negative impact of air pollution on wild plants and forest trees may be even greater than that noted for agricultural crops. Open-chamber experiments in the Blue Ridge Mountains in northwestern Virginia have shown that the growth of various wild plants was reduced between 32 and 57 percent even though ozone concentrations remained below standards set for human health.

Many tree species are also highly sensitive to ozone, although in the following examples acid precipitation may be a major factor in reduced growth. Significant damage to economically valuable ponderosa and Jeffrey pines is occurring along the entire western slope foothills of the Sierra Nevada Mountains in California. In the San Bernardino Mountains of California, an area that receives air pollution from Los Angeles, production has been reduced by 75 percent. A study plot of white pine near a roadside in western Virginia showed a 40 percent decrease in wood production.

Forests under pollution stress are more susceptible to damage by insects and other pathogens. For example, death of ponderosa and Jeffrey pines in California is generally attributed to western pine beetles, which invade pollution-weakened trees. Even normally innocuous insects may cause mortality when combined with pollution stress.

Another concern is that if widespread air pollution worsens, reductions in production could occur with disastrous suddenness. The effect may be sudden, because reduction in growth (or any other manifestation of pollution damage) is not a linear function of the pollution level. That is, one unit of pollutant does not produce one increment of damage; two units, two increments; and so on. Instead, plants will tolerate a certain level of pol-

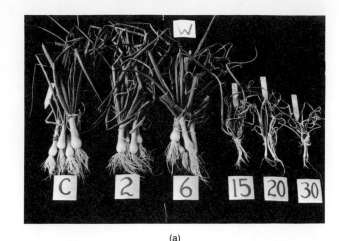

(a)

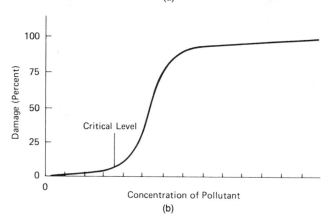

(b)

**FIGURE 13-15**
Critical level of pollution. (a) Control plants at left were grown in clean air; others were exposed to indicated concentration of sulfur dioxide (parts per 100 million). Note that the reduction in growth is not linear but that there is a sharp drop between 6 and 15 parts per 100 million. (USDA photo.) The concept of the critical level is shown graphically in (b).

lution stress with very little, if any, noticeable effect. But just a small increase in concentration or duration of exposure may push the plant beyond its capacity to cope with the pollution. This point is known as the **critical level** (Fig. 13-15). Even without drastic dieoffs, the decrease in primary productivity must ultimately affect the rest of the ecosystem, including soils because they depend on primary production as a source of humus forming detritus. Also, as sensitive species die out, they are replaced by more resistant species in the process of ecological succession. Where this will lead is uncertain, but numerous foresters and ecologists have little doubt that large-scale biological changes are already underway in the landscape as a result of current levels of air pollution.

## Effects on Materials and Aesthetics

Walls, windows, and all other surfaces turn grey and dingy as particulates settle on them. Paint and fabrics deteriorate more rapidly, and the sidewalls of tires and other rubber products become hard and checkered with cracks because of oxidation by ozone. Corrosion of metals is dramatically increased by sulfur dioxide and acids derived from sulfur and nitrogen oxides (Fig. 13-16), as is weathering and deterioration of stonework. These and other effects of air pollutants on materials increase the costs for cleaning and/or replacement of products by hundreds of millions of dollars per year.

In addition, a clear blue sky and good visibility—in contrast to the haze of "smog"—have aesthetic value and a psychological impact that few would deny. Hence, the impacts of air pollution on materials and aesthetics should not be overlooked.

## CONTROL STRATEGIES

Everyone will agree that air pollution ideally should be stopped. However, given the multitude of sources, it cannot be stopped overnight short of halting virtually all industry, electrical power production, and transportation. The final phase of our attack on pollution, the actual control of problem pollutants, itself has three distinct steps: (1) setting standards or goals one hopes to achieve, (2) developing strategies and/or technologies to meet the standards, and (3) implementing and enforcing these strategies.

Again these are all ongoing processes. New information on health or environmental effects may warrant new standards. Some strategies or technologies for control may prove to be unworkable, ineffective, or in conflict with other goals. This situation demands a return to the drawing board and delays implementation.

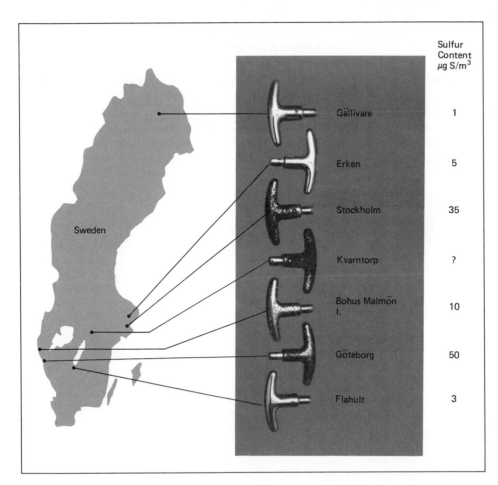

**FIGURE 13-16**
Material corrosion and air pollution. The nickel-plated handles shown were exposed to the prevailing air conditions for 4.5 years. The association between corrosion and level of air pollution is conspicuous.
(Swedish Corrosion Institute, Stockholm, Sweden.)

| Location | Sulfur Content $\mu g\ S/m^3$ |
|---|---|
| Gällivare | 1 |
| Erken | 5 |
| Stockholm | 35 |
| Kvarntorp | ? |
| Bohus Malmön I. | 10 |
| Göteborg | 50 |
| Flahult | 3 |

## Setting Standards

The Clean Air Act of 1970 mandated setting standards for the following five pollutants, which, at the time, were recognized as the most widespread and objectionable: (1) total suspended particulates, (2) sulfur dioxide, (3) ozone, (4) carbon monoxide, and (5) nitrogen oxides. These are now known as **criteria pollutants.** The **primary standard** for each is based on the highest level that can be tolerated by humans without noticing any ill effects plus a 10 to 50 percent margin of safety (Table 13-2).

In turn air quality, which is now commonly given along with weather reports, is based on these standards. The pollutant having the highest concentration relative to its primary standard determines the index (Fig. 13-17). For example, if the highest of any of the criteria pollutants is 25 percent of its primary standard, the Pollution Standards Index (PSI) is 25 and air quality is in the "good" range; if the highest pollutant is 150 percent of its standard, the Index is 150 and air quality is in the "unhealthful" range, as shown in Figure 13-17.

In the early 1970s, many cities were experiencing more than 100 days a year of pollution in the "unhealthful" range or above; Los Angeles was having in the order of 300 such days per year. The original goal was to achieve the "good" air quality on all days by 1975. It hardly needs to be said that this was not accomplished; however, marked progress has occurred in recent years

(Fig. 13-18). Given adequate support, perseverance, and experience, there is every reason to believe that further progress can be made.

## Techniques for Reducing Pollution: Successes and Failures

### CONTROL OF OZONE AND CARBON MONOXIDE

**Emission controls on cars.** Since cars are the major source of the hydrocarbons and nitrogen oxides that lead to the formation of ozone, other photochemical oxidants, and carbon monoxide, the Clean Air Act of 1970 mandated a 90 percent reduction of these emissions from 1970 levels to be achieved by 1975. This goal proved overly optimistic, and it still has not been achieved entirely. The Clean Air Act has been amended several times to give the car industry more time. However, efforts have been made and cars are now equipped with a variety of pollution control devices (Fig. 13-19c). Computerized control of the fuel mixture and ignition timing also help reduce engine emissions.

The most significant control device on cars, however, is the **catalytic converter** (Fig. 13-19b). As the exhaust passes through this device, a chemical catalyst, which consists of platinum-coated beads in this case, completes the oxidation of carbon monoxide and hydrocarbons to carbon dioxide and water. Cars with catalytic converters, which include all 1975 and later U.S. cars and

| Table 13-2 | National Ambient Air Quality Standards for Criteria Pollutants | | |
|---|---|---|---|
| | POLLUTANT | AVERAGING TIME[a] | PRIMARY STANDARD |
| | Particulates ($\mu$g/m$^3$)[b] | 1 year | 75 |
| | | 24 hours | 260 |
| | Sulfur oxides (ppm) | 1 year | 0.03 |
| | | 24 hours | 0.14 |
| | | 3 hours | — |
| | Carbon Monoxide (ppm) | 8 hours | 9 |
| | | 1 hour | 35 |
| | Nitrogen Dioxide (ppm) | 1 year | 0.05 |
| | Ozone (ppm)[c] | 1 hour | 0.12 |

[a]Averaging time is the time period over which concentrations are measured and averaged.
[b]A (microgram) is one-millionth of a gram.
[c]Revised January 26, 1979.
Source: After Council on Environmental Quality, *Seventh Annual Report,* 1976, p. 215.

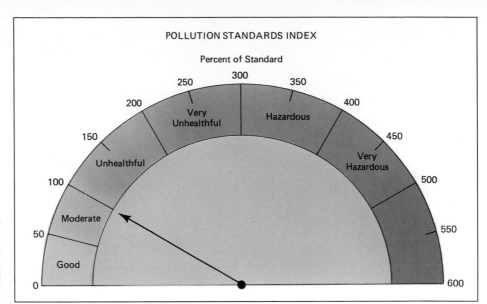

FIGURE 13-17
Pollution Standards Index (PSI) of air quality. Air quality is designated as shown according to the percentage of the highest of any of the criteria pollutants relative to its primary standard.

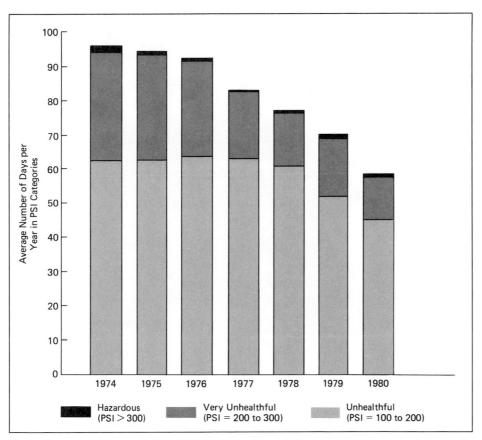

FIGURE 13-18
Air quality in 23 metropolitan areas as measured by the Pollution Standards Index (PSI). In general, air quality in metropolitan areas improved markedly in the late 1970s, as shown by the decrease in the number of days that the PSI went into the unhealthful range or worse. This shows pollution abatement strategies have had a significant positive effect, although room for improvements still remains. Source: Council on Environmental Quality. Reprinted with permission from The Conservation Foundation, *State of the Environment: An Assessment at Mid-Decade*, 1984.

many foreign cars, should not use leaded gasoline because the lead atoms quickly coat the platinum catalyst and render it ineffective. Of course, leaded gasoline should not be burned in any case because it is a serious pollutant itself and its use is being phased out for this reason.

Nitrogen oxides are a major factor in the formation of ozone and other photochemical oxidants, and an important factor in acid precipitation, as well. However, control of nitrogen oxide from cars has been much more problematic. The first control measure consisted of engine adjust-

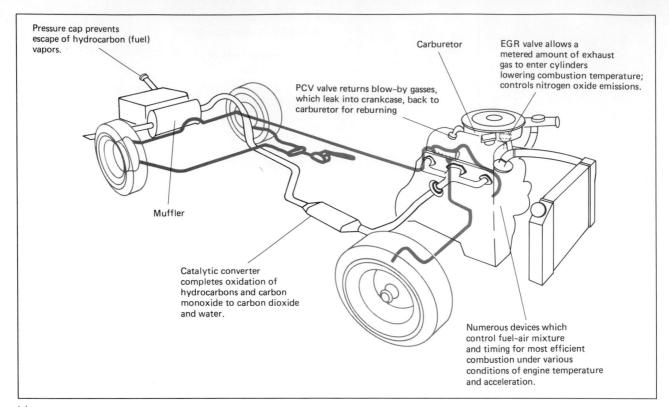

Pressure cap prevents escape of hydrocarbon (fuel) vapors.

PCV valve returns blow-by gasses, which leak into crankcase, back to carburetor for reburning

Carburetor

EGR valve allows a metered amount of exhaust gas to enter cylinders lowering combustion temperature; controls nitrogen oxide emissions.

Muffler

Catalytic converter completes oxidation of hydrocarbons and carbon monoxide to carbon dioxide and water.

Numerous devices which control fuel-air mixture and timing for most efficient combustion under various conditions of engine temperature and acceleration.

(a)

**FIGURE 13-19**
Pollution control features on cars. (a) New cars are now equipped with numerous pollution control devices. (b) The most significant device is the catalytic converter in the exhaust system. As exhausts pass through the converter, the catalyst, made of platinum-coated beads, causes hydrocarbons and carbon monoxide to react with more oxygen to form harmless carbon dioxide and water vapor. (General Motors photo.)

(b)

ments, which reduced combustion temperatures and pressures and, in turn, resulted in less oxidation of nitrogen. Unfortunately, such adjustments also reduce engine efficiency and, consequently, increased fuel consumption. Thus we see pollution control in conflict with another extremely important goal—energy conservation. This conflict has become even more intense with the introduction of diesel cars, which, while more fuel efficient, produce more nitrogen oxides and more particulates. As a result of the lack of nitrogen oxide control, this air pollution problem has increased since 1975.

An increase in nitrogen oxides might not be harmful if hydrocarbons were well controlled because both are required in the formation of ozone.

Indeed the smog problem has diminished greatly with catalytic converter control of hydrocarbons (see Fig. 13-18). Because of this, authorities have not taken nitrogen oxides seriously and have granted automotive companies delays in meeting goals. However, we now recognize that ozone damage to vegetation is increasing despite hydrocarbon control. This has led us to recognize that there are other very significant contributors of hydrocarbons such as solvents and even vegetation itself. These will be very difficult or impossible to control. Consequently, much more attention must be focused on reduction of nitrogen oxide emissions.

Technologists are working to develop a catalytic converter that will turn nitrogen oxides back to nitrogen gas. This effort, which has near-term practicality, could use more support and pressure from a concerned public. Ideas such as new kinds of engines that may be less polluting or the use of hydrogen as the "ultimate clean fuel" are still too far in the future and too speculative to be of value in addressing current problems.

In the meantime, the importance of keeping engines well tuned and pollution control devices functioning properly is critical. It is discouraging to note that a recent survey showed that in one of five cars pollution control devices do not work because of improper maintenance or tampering. Two of the most common problems were catalytic converters that did not function because of the use of leaded gasoline (a single tankful is enough to damage the converter) or catalytic converters that were disconnected in the mistaken belief that this improves fuel economy.

In areas that fail to meet pollution standards, the federal government mandates car inspection programs under the threat of loss of federal highway funds (Fig. 13-20). This program should help ameliorate the problem, but it has met considerable resistance from the public and has become a political issue concerning states' rights in some regions.

**Public transportation.**    It is also obvious that if people drove less there would be less pollution from cars and less fuel use as well. Planners in many cities have devised schemes to increase the use of public transportation, since this would move people more efficiently than private cars. In general the results of such efforts have been extremely disappointing. Except for brief reversals during the gasoline shortages of the 1970s, the general trend in ridership of public transportation continues downward. Conversely, the trend in number of cars and miles driven continues upward, tending to offset progress in reducing emissions per car. The trend of living farther and farther from places of work and using regional shopping malls contributes to the increase in driving (see Chapter 24).

**FIGURE 13-20**
Auto emissions inspection station. A sample of gas is withdrawn from the tailpipe and automatically analyzed for carbon dioxide, carbon monoxide, and hydrocarbons. Each lane of the station can test up to 40 cars per hour. Repairs are required of cars not meeting standards. Where air pollution levels remain problematical, such inspections are mandated under the Clean Air Act. (Photo by author with permission of Systems Control, Inc.)

Indeed the reduction in air pollution observed in cities may in part represent a shift from city driving to rural driving as new shopping centers and industrial parks are built in exurban areas. Unfortunately, air pollution has not been carefully monitored outside cities, but judging from crop and forest damage, concentration of photochemical oxidants in rural areas is not decreasing as might be expected with the improvement in city air. Sadly, as we withdraw from public transportation, these systems become defunct so that we gradually lose this option entirely.

## CONTROL OF PARTICULATES

The Clean Air Act placed strict limits on "visible emissions," that is particulates, and great success has been achieved in this regard (Fig. 13-21). Two factors contribute to this success. First, most particulates came from easy-to-identify point sources, such as waste incinerators or industrial smokestacks. Second, particulates can be removed relatively easily by filters and/or electrostatic precipitators, as shown in Figure 13-22, which were already available in the 1950s. However, these wastes, which frequently include heavy metals and other toxic substances, must still go somewhere. Ironically, removal of wastes from exhaust gases has added to the burden of toxic wastes going into landfills from which they may be leached into groundwater (see Chapter 12).

In addition, open burning of trash was banned in cities and some states. Smoking trash incinerators were shut down and in large part municipalities turned to landfilling of municipal wastes. Landfilling municipal wastes is not without its own problems, however, particularly those involving groundwater pollution. Therefore, the trend is again moving toward burning municipal wastes, but in new facilities that both control particulates and utilize the heat produced (see Chapter 20).

Importantly, precipitators and filters do not remove gaseous pollutants nor is it practical to install such devices on each one of 130 million cars in the United States or on home wood-buring stoves. Therefore, if the trend to use diesels and woodstoves continues, particulates could become a growing problem again. For this reason, California recently banned the sale of new diesel cars.

## CONTROL OF SULFUR DIOXIDE

Measurements of sulfur dioxide in the ambient air of cities indicate that there has also been great success in controlling this pollutant (Fig. 13-21). Actually these results are highly misleading because control strategies have led to other problems that are as bad if not worse.

The major source of sulfur dioxide in the 1960s was, as it is now, coal-burning electric power plants and smelters. In the late 1960s and early 1970s a move was made to shift from coal to

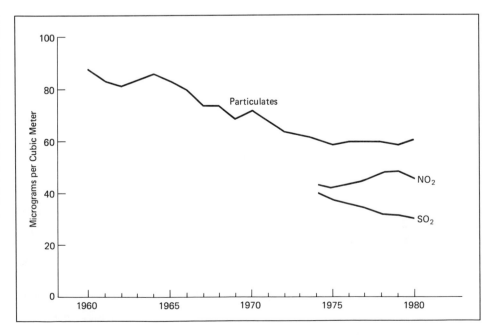

**FIGURE 13-21**
National ambient concentrations of total suspended particulates, nitrogen dioxide, and sulfur dioxide. The data show significant reduction in levels of particulates and sulfur dioxide, which attests to the effectiveness of control strategies for these pollutants. But nitrogen oxide levels have increased somewhat, attesting to the lack of attention to this pollutant. Source: U.S. Environmental Protection Agency. Reprinted with permission from The Conservation Foundation, *State of the Environment*, 1982.

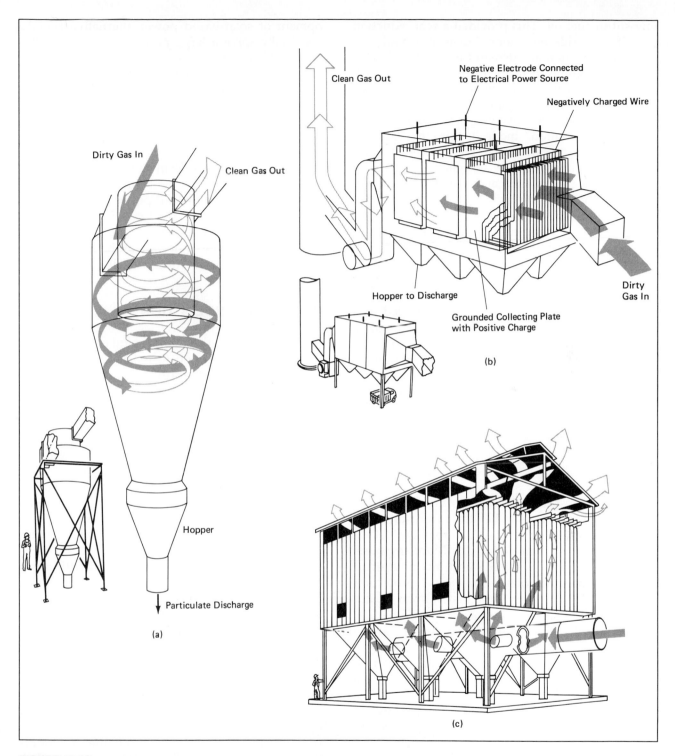

**FIGURE 13-22**
Devices to remove particles from exhaust gases. (a) Cyclone precipitator.
Particles are removed by centrifugal force as exhausts are swirled.
(b) Electrostatic precipitator. Particles are electrically ( − ) charged, then attracted
to plates of the opposite charge ( + ). (c) Bag house. Exhaust gases are forced
through giant vacuum cleaner bags. Note that none of these devices removes
very fine particles or polluting gases such as sulfur dioxide. (American Lung Association.)

low-sulfur fuel oil. This provided a real reduction in sulfur dioxide emissions, but it also markedly increased our nation's demand for crude oil. This became a significant factor in the "energy crisis" of the mid-1970s when demands, at least in the short run, outstripped supplies (see Chapter 21).

Other strategies followed the "dilution is the solution to pollution" mode of thinking. Utilities and industries located new plants well outside cities. This improved city air quality, but it degraded rural air quality and has increased the negative impact of air pollutants on agricultural crops and forests.

The third short-sighted and unsuccessful strategy was to build very tall smokestacks that would eject pollutants above the inversion layer. The idea was that, if pollutants were dispersed into the upper atmosphere, they would dilute and effectively disappear. The tall stacks enabled utilities to go back to burning coal with more sulfur and thus reduce the demand on crude oil. Far from disappearing, however, the sulfur dioxide from these tall stacks has now been shown to be the major source of acid precipitation (see Chapter 14). However, this problem, too, is solvable; we need only to recognize the failing of old assumptions.

## Toward the Future

The widespread damage to crops and forests, as well as to human health, as a result of air pollution shows that we can ill afford any let-up in efforts to control air pollutants. More emphasis should be placed on settling and attaining standards that provide total *environmental* protection and a margin of safety, not just human health protection. This demands an *absolute* reduction of emissions. No longer can we accept the axiom that dilution is the solution to pollution, particularly in connection with the hydrocarbons and nitrogen oxides that lead to ozone production and sulfur oxides. Hence, we should not permit further delays in meeting nitrogen oxide emission standards for cars and we must maintain and enforce hydrocarbon emission standards on cars and other sources.

For the long term we must place more emphasis on changing our basic source of energy. We note again that the source of most air pollution is the burning of fossil fuels. An energy policy directed toward more conservation and development of solar-based power alternatives would eventually solve a large portion of our air pollution problems as well as giving us an everlasting source of energy (see Chapter 23).

To accomplish these goals, citizens must make officials aware of their concern and their desire for action. Pollution control is not only a health and environmental issue; it is an economic and business issue. Politics have and will continue to play a key role in the adoption of measures that address the problem on both a local and national level. Without public involvement in the political process, budget cuts in antipollution programs, less stringent enforcement of already weak standards, and a decline in the research and development of new controls will certainly prevail.

## INDOOR AIR POLLUTION

After recognizing the many pollutants in outside air one may be inclined to remain indoors to escape the hazards. Unfortunately, severe and direct threats to human health are also posed by exposures to hazardous or toxic materials in the home and workplace. In fact, the ambient (average) air inside homes and workplaces often contains much higher levels of hazardous pollutants than outdoor air. In a recent report, the National Academy of Sciences called indoor air pollution an issue that "has been largely overlooked" and a matter "of immediate and great concern."*

Traditionally, hazardous indoor air pollution has been associated with blue-collar workers employed in places heavily contaminated with dusts and/or fumes from various industrial processes. It is now recognized that homes and offices may also contain significant air pollution hazards. Researchers suspect that an increasing number of pollutants inside office buildings and homes may be responsible for lethargy, lack of productivity, chronic sinusitis, headaches, and other illnesses that are usually attributed to other causes.

The overall indoor air pollution problem is threefold. First, increasing numbers and kinds of products and equipment used in homes and/or offices give off potentially hazardous fumes. Second, in the interest of rigorous temperature con-

---

*Indoor Pollutants* by the Committee on Indoor Pollutants, Board of Toxicology and Environmental Health Hazards, Assembly of Life Sciences, National Research Council, 1981.

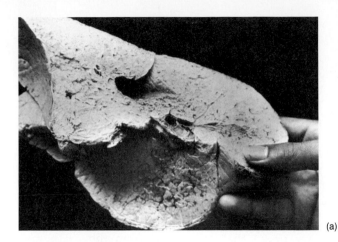

(a)

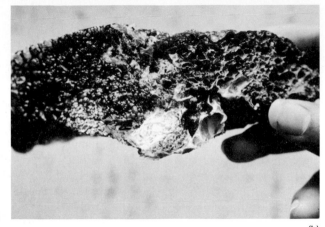

(b)

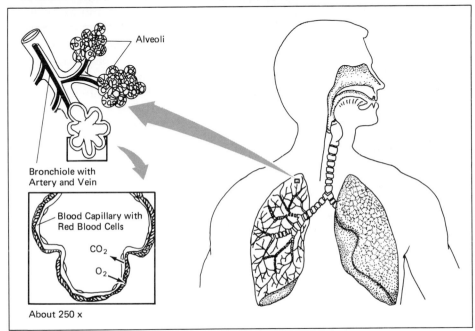

Alveoli

Bronchiole with
Artery and Vein

Blood Capillary with
Red Blood Cells

$CO_2$

$O_2$

About 250 x

(c)

**FIGURE 13-23**
In the lungs, air passages branch
and rebranch and finally end in
millions of tiny sacs called
alveoli. Alveoli are surrounded by
blood capillaries. As blood
passes through these capillaries,
oxygen diffuses from the alveoli
into the blood, and carbon
dioxide diffuses in the reverse
direction. Lung disease.
(a) Normal lung tissue. (b) Lung
tissue from person who suffered
from emphysema, a chronic lung
disease in which some of the
structure of the lungs has broken
down. Cigarette smoking is
associated with the development
of emphysema as well as other
lung diseases. (EPA–Documerica,
photos by Leroy Woodson; courtesy
of the U.S. Environmental Protection
Agency.) (c) Diagram of lung
structure.

trol and energy efficiency, buildings and homes
have become increasingly well- insulated and
sealed; hence, pollutants that would otherwise escape through leaks and cracks and dissipate in the
outdoor environment, are now trapped inside
where they accumulate to potentially dangerous
levels. Third, exposure to indoor air pollution is
longer in duration than exposure to outdoor pollution. The average person spends 70 to 80 percent of his or her time indoors. And the people
who spend the most time inside are the people
who are most vulnerable to the harmful effects of
pollution: small children, pregnant women, the
elderly, and the chronically ill. The sources of indoor air pollution are numerous. We shall look at
some of those that have been recognized as most
significant.

## Sources

### SMOKING

Tobacco smoke is rife with all the primary
pollutants of combustion in addition to the notorious tars and nicotine, and the concentrations inhaled are hundreds of times greater than those in
even badly polluted air. A striking comparison is
that it would be necessary to breathe smoggy Los
Angeles air for one to two weeks to equal the intake of volatile organic matter received by smoking a single cigarette containing 20 mg of tar. The
many adverse health effects of smoking, such as
lung cancer, emphysema, and coronary heart disease are well known (Fig. 13-23). However, it is
becoming increasingly clear that smoking indoors
also adversely effects nonsmokers. Surveys show

that nonsmokers report an increase in eye irritation, sore throats, coughing, sneezing, and generalized stress when continually exposed to cigarette smoke. In addition, studies indicate that children who are exposed to their parents' cigarette smoke are more vulnerable to respiratory illnesses. Likewise, nonsmoking spouses are more likely to suffer from cardiovascular disease if their husband or wife is a smoker. The potency of secondhand smoke may be attributed to the fine smoke particles that scavenge other pollutants out of the air and increase the potential for synergistic effects when they lodge in the lungs. The secondhand smoker may inhale as many of these as the original smoker.

In short, smoking stands out as the single most significant factor in human disease and premature deaths and the effects of smoking are not limited to the smoker.

## FORMALDEHYDE AND OTHER SYNTHETIC ORGANIC COMPOUNDS

New construction as well as home and office furnishings utilize large amounts of press or particle board, synthetic fibers, and plastics. Formaldehyde and/or other organic compounds are used as binders in these products and, over time, these compounds tend to be released into the air. The release is most pronounced during the first few months after installation, but the fumes may continue, often unnoticed, for several years.

A particularly serious episode involving formaldehyde occurred in the mid-1970s when the energy crisis was on everyone's mind. Urea formaldehyde foam, a very effective insulating material, was introduced and used by a half million consumers. When installed properly, the foam is pumped into exterior walls where it hardens and provides excellent protection from the heat and cold. Unfortunately, however, many users began complaining of headaches, dizziness, rashes, and eye irritation after the insulation was installed. The cause of these illnesses was eventually shown to be formaldehyde toxicity, usually resulting from improper installation.

The Consumer Product Safety Commission banned the use of urea formaldehyde foam insulation in 1982, but the ban did not offer recourse to homeowners who had already installed it. (Re-

moving the product usually involves tearing down the walls into which it was pumped). And the ban did not motivate other government agencies to establish health standards for formaldehyde, which is contained in many other products. Hence, homes may be plagued with formaldehyde poisoning, but homeowners and health officials have no guidelines on which to measure or trace the problem.

### ASBESTOS

Asbestos occurs naturally in fiberlike crystals (Fig. 13-24) that are mined from the earth and used as a heat and fire retardant and as an insulating material in numerous products. Until recently it was frequently installed as wrapping on steam heating pipes and as a ceiling covering in many schools and other public buildings. In the 1960s, it was determined that the inhalation of **asbestos fibers** is associated with a unique form of lung cancer that develops as many as 20 to 30 years after exposure. Clearly, the microscopic asbestos fibers that flake from asbestos-containing materials present a significant health risk.

The Environmental Protection Agency began regulating asbestos in the mid-1970s and initiated intensive campaigns to remove it from schools and other public places. However, this program has moved forward slowly and many schools and other buildings still contain asbestos products. The substance also continues to be found in such products as pot holders and ironing board covers as well as some paints and roofing materials.

### RADON

Radon is a radioactive element found in the atmosphere, soil, rocks, concrete, brick, and groundwater. It is part of the natural background radiation to which all individuals are exposed. However, in tightly constructed buildings, concentrations of radon often exceed outdoor levels. Again, increased ventilation is usually sufficient to eliminate problems.

### INDOOR COMBUSTION

The use of wood, coal and kerosene stoves has increased dramatically during the last decade. These items, along with unvented gas stoves, generate particulates, nitrogen oxide, and carbon monoxide. Carbon monoxide, which is colorless

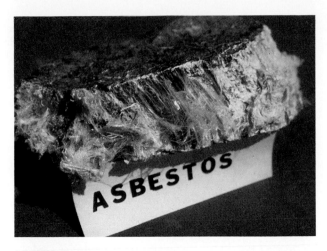

**FIGURE 13-24**
Asbestos. Asbestos is a natural mineral, the crystals of which are in the form of fine fibers. Since asbestos will not burn, it was used extensively in materials to insulate heating pipes and ducts and deaden sound throughout schools and other public buildings. Then it was discovered that asbestos fibers, when inhaled, may cause lung cancer. (Photo by author.)

and odorless, can be lethal. Nitrogen oxide and particulates irritate the eyes and upper lining of the respiratory system. Environmental and health officials consider these products to be among the most pervasive indoor pollution threats, noting that their danger has been accentuated by decreased ventilation. The experts point out that these items are especially hazardous when used inside some of the newer mobile homes because a tightly constructed mobile home is like a sealed tin can.

### LEAD

Problems of leaded gasoline have already been discussed. However, lead was also widely used in paints and piping. Most of these products no longer contain lead, but in older homes, high levels of lead have been found in water and peeling paint. Young children are apt to play with and eat peeling paint, and when ingested, lead can cause serious kidney and neurological problems. Studies have linked lead poisoning with learning disabilities and other brain disorders.

### FIBERGLASS

Fiberglass is suspected of causing cancer and has been connected with respiratory problems and skin irritation. It is found in ceramics, scouring soaps, talcum powder, insulation, furnace filters, and roofing materials.

### PESTICIDES

The ingredients that make pesticides effective against bugs are also dangerous to humans. For example, some pesticide strips and flea collars contain extremely toxic DDVP, a nerve poison and suspected carcinogen. Other pesticides contain other extremely toxic chemicals. Even when used outdoors, these products are dangerous because they can filter through open windows and contaminate food and air.

### HOBBIES

Heavy metals are key ingredients in paints and pottery glazes. Abnormally high levels of mercury, found in some water-based paints, can cause the same problems as lead. Cadmium, another toxic metal, is frequently found in orange and yellow pigments and has been linked with kidney malfunctions. Paint strippers, photographic developing materials, and other hobby supplies also generate dust particles and noxious vapors that can cause headaches, scar lung tissue, impair motor response, and even lead to death. These products are especially dangerous when used in unventilated areas, as they are in many homes where the activities associated with them are conducted in amateur settings. Further, these toxic materials can easily be spread to other family members if clothing is not washed after exposure.

### PLASTICS

Many plastics contain vinyl chloride and polyvinyl chloride, known carcinogens. Use of both chemicals has been restricted, but small amounts can still be found in upholstery, wall coverings, garden hoses, toys, records, food packaging, credit cards, and floor tiles.

### AIR FRESHENERS AND DISINFECTANTS

Many air fresheners only mask odor; they do not eliminate it. Often, these products contain chemicals far more hazardous than the odors they are designed to cover up. Many contain cresol, which can affect the central nervous system, liver, and kidneys. Paracresol, occasionally found in fragrances, soaps, detergents, creams, and lotions in the United States, is prohibited in these same products in some parts of Europe.

## OVEN AND DRAIN CLEANERS

Oven and drain cleaners are potent cleansers that usually contain lye and other caustic substances; hence, they can be very dangerous in heavy doses. A common problem with these and other household cleaners is that consumers do not heed warnings and use them in unventilated spaces. Other problems occur when consumers mix products. For example, when combined, ammonia and bleach give off dangerous, toxic fumes.

## Seeking Protection

Gaining protection against the hazards of indoor air polution is an extremely complex problem with many hurdles to overcome.

### IDENTIFICATION OF HAZARDS

The difficulty in identifying hazardous compounds and their degree of toxicity revolves around the fact that most problems associated with them result from long-term, low-level exposure. Moreover, most people who are exposed to dangerous chemicals are also exposed to a wide variety of other chemicals simultaneously. Hence, any number of synergistic effects may occur, and most of these effects do not show up until many years after exposure. Finally, the sensitivity of individuals to substances varies.

In one of its first major attempts to address the toxicity problem, Congress enacted the Toxic Substances Control Act in 1976. Known as TSCA, this law directed the Environmental Protection Agency to inventory and investigate all chemicals manufactured in the United States. The inventory, which lists thousands of substances, was completed, but as of 1986, the agency has not begun testing. Holding up test procedures is controversy over the extent and type of testing that should be conducted. Given the problems cited above, you can see the difficulty. At the very best, testing will be extremely expensive and time consuming. Even then, the information gained may be of dubious value because most studies are based on animals, but not all substances that are carcinogenic or dangerous to animals are risky for humans and not all chemicals that are hazardous to humans affect animals.

In the absence of critical animal studies we gradually identify hazards by means of harsh reality. We, ourselves, act as guinea pigs. Epidemiological studies, studies examining the incidence of particular diseases or illnesses on various populations, lead to identification of various hazards. The dangers of smoking and of asbestos were identified in this way.

### REGULATION

Even after hazards are identified, adopting regulations is a difficult practice. The Occupational Safety and Health Administration (OSHA) established pollution standards for industrial workplaces in 1968. Because no other laws exist to protect workers, these regulations are occasionally used to assess problems in workplaces outside of industry. However, these standards cover only about 20 chemicals common to industrial workplaces. They do not take into account the types of pollutants found in office buildings or in the home. Also, they do not account for the sensitivities of susceptible individuals like the elderly or pregnant women. Thus, they are of little help when it comes to problems outside of the areas for which they were specifically designed. And some labor and occupational health officials claim that they are not even adequate to protect industrial workers because federal funding cuts have kept them from being updated and adjusted in accordance with new information.

The Consumer Product Safety Commission is empowered to remove from the market specific products that are shown to be particularly hazardous. We noted the removal of urea formaldehye foam and restriction of asbestos use, for example. However, such curtailment cannot be exercised without "clear evidence" and "just cause." Given the inherent ambiguities of terms such as "clear evidence" and "just cause," large industries with vested interests often use protracted legal debates and successfully delay proposed bans or regulations on their products.

Such proposed bans or regulations must also be scrutinized at public hearings. Here, the matter of individual rights comes to the forefront. For example, even if the tobacco industry were willing to accept a ban, it is doubtful that members of the smoking public would allow it to be imposed because they consider it their right to smoke. Most people do not want the government telling them what products they should or should not buy, despite the fact that some may be hazardous. Certainly, most people would consider government monitoring of the air inside their homes an invasion of privacy.

The compromise often reached in these situations is to require that manufacturers provide consumers with more information about the hazards of their products, then let the buyer exercise her or his own choice. Warnings on cigarette packages are an example.

Another aspect of the individual rights question in connection with hazardous materials is legislation called "right to know." Right-to-know laws are aimed primarily at workers who handle hazardous materials. Many of these people are unaware of the hazards of the chemicals they come in contact with. Unless one of the chemicals they handle is a designated toxic substance regulated by TSCA, employers are not required to reveal the generic name of the substances to employees. However, employers are required to provide workers with information about the designated chemicals in the form of material safety data sheets—complete dossiers about the composition, health effects, and other peculiarities of the chemicals—but these documents are not always updated or supplied without a specific request.

This system was established primarily to protect company trade secrets. Workers in many states are now challenging the law and asking that they be supplied with information, including generic names, of every chemical they come in contact with. Firefighters, who contend that they often do not know what substances are inside a building when they are called to an emergency, are staunchly behind the workers. Many citizen action groups have also joined in, campaigning not only for better information for workers, but for more information for the public as well.

Under heat, the federal government issued a regulation in 1984 requiring employers to provide workers with more information about the substances they are exposed to. But the law is vague and does not cover all workers who handle toxics. Manufacturing firms are adamantly against even this legislation, believing that it will lead to trade secret leaks and ruin their businesses.

Some states and municipalities, including New Jersey and Philadelphia, have responded to the problem by passing their own right-to-know laws.

## IMPROVING VENTILATION

Since indoor pollution largely results from pollutants that are trapped inside relatively small spaces, the most straightforward and cost-effective way to deal with the problem may simply be to improve ventilation.

During the last few decades, however, the trend has been to markedly reduce such ventilation for two reasons. First, the desire to maintain a constant temperature throughout the year has been achieved with "climate control" systems that heat or cool and recirculate inside air. These systems do not remove pollution and they substantially reduce the amount of air that flows through buildings. In fact, windows in many new buildings cannot be opened. Second, concern for energy conservation has resulted in improved insulation and weatherstripping which seals leaks and cracks.

Because of these measures, experts estimate that air exchanges in new buildings is about 70 percent less than in older buildings. This is particularly dangerous when leaks or spills occur because these toxics remain inside the building where they are quickly circulated. Numerous incidents have occurred where large numbers of employees got sick as a result of this problem and in some cases entire buildings had to be shut down.

Still, temperature control and energy efficiency remain desirable goals. In view of the indoor air pollution problem, however, it is clear that a balance between achieving these goals and providing ventilation needs to be reached. Some countries such as Japan, Sweden, and West Germany have adopted ventilation regulations, but legally enforceable standards do not yet exist in the United States. The adoption of such standards could markedly improve the situation here, according to most environmentalists.

## EXPOSURE AS A RESULT OF ACCIDENTS

Despite the best intentions, accidents always have and always will occur. Unfortunately, as modern society uses increasing amounts and kinds of toxic materials, the stage is set for very simple mistakes or accidents to become wide-scale disasters. A few actual examples illustrate the far-reaching effects accidents can have.

## Examples

In December 1984 a leak at a pesticide manufacturing plant in Bhopal, India, resulted in the release of what some reports called a "small quantity" of methyl isocyanate, an extremely poisonous gas. Over 2000 people died and another 150,000 were injured, many seriously, as the gas spread through the community (Fig. 13-25a).

This incident is only the latest and most serious of a long list of accidents or "mistakes" involving toxic chemicals.

In 1973, a few sacks of a fire retardant chemical got mixed up with an animal feed additive by a distributor in Michigan. If the chemical had been of low toxicity, the amounts that were fed to the animals would have had little if any effect. However, the fire retardant chemical was PBB, a

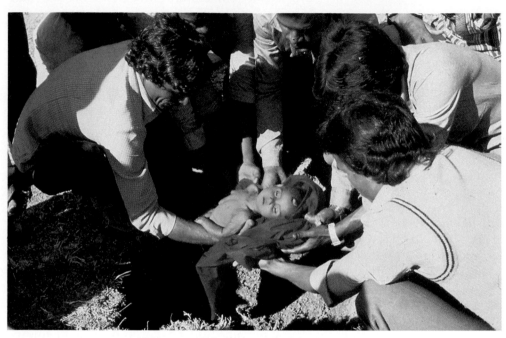

(a)

**FIGURE 13-25**
Many of the most disastrous "pollution" episodes are the result of accidents and/or carelessness. (a) In 1984 a leak of methyl cyanate from a pesticide manufacturing plant in Bhopal, India, killed some 2000 persons and injured and blinded some 20,000 more. (Copyright Baldev-Sygma.) (b) The town of Times Beach, Missouri, had to be evacuated and destroyed when it was found that roads had been treated with oil contaminated with dioxin and a flood had spread it throughout the entire town. (Environmental Protection Agency.)

(b)

**FIGURE 13-25** *(cont.)*
(c) The wreck of the supertanker *Amoco Cadiz* in March, 1978, polluted 200 miles of French coastline. Increasing demands for goods and fuels make us increasingly vulnerable to widespread pollution from such accidents. (United Press International photo.)

highly stable, bioaccumulating halogenated hydrocarbon closely related to PCB, but some five times more toxic. As a result, numerous people, mostly farm families, became sick, suffering varying degrees of nervous disorders; some 500 farms had to be quarantined; 30,000 cattle, 1.5 million chickens, thousands of sheep and hogs, and tons of cheese, milk, and eggs had to be destroyed because of the contamination. The damage was estimated at about $100 million, not including compensation for individual human suffering. Moreover, the chemical remained in the Michigan ecosystem. Several years after the initial incident, reaccumulation from "unknown" sources still caused sporadic occurrences of PBB poisoning.

In another incident, in Séveso, Italy, in 1976, a safety valve in a chemical plant malfunctioned and about a kilogram (2.216) of dioxin, a chlorinated hydrocarbon and one of the most toxic substances known, was released into the air. The entire town of 100,000 residents had to be evacuated; hundreds of people suffered severe skin ailments; animals died by the thousands; and consumption of all local food was banned. A year later an area around the factory was still uninhabitable and there is much concern that birth defects may occur in the next generation.

In still another incident, the entire town of Times Beach, Missouri, had to be evacuated and relocated and all existing structures destroyed after roads were unwittingly sprayed with waste oil contaminated with dioxin (Fig. 13-26b).

Transport of toxics is an activity that is particularly prone to accidents. Over the course of a year, about 4 billion tons of chemicals cross over roads, railroad tracks, and waterways in the United States. Numerous accidents involving tankers filled with hazardous chemicals have resulted in the evacuation of residents. Many experts speculate that the fact that such accidents have not led to wide-scale injury and death is more a matter of good luck than good management.

Even relatively nontoxic materials take on disaster potential if the volume is large. Oil is a case in point. Crude oil is a mixture of natural organic compounds that are, in modest quantities, broken down by organisms and assimilated. However, the huge amounts that may be spilled in an accident involving a supertanker can result in enormous ecological disasters. In March 1978 the supertanker *Amoco Cadiz* went aground off the French coast, spilling 220,000 tons of crude oil (Fig. 13-25c). The accident affected some 200 miles of one of Europe's most picturesque coastlines; wiped out 20,000 birds, including a whole colony of rare puffins; made 9000 tons of oysters inedible and ruined their culturing ground; obliterated marine worms, which are essential in the food chain for commercial fish; and cancelled out tourism in the region. And these were just the immediate effects. The long-term effects are still being seen.

## Protection against Accidents

The possibility of accidents cannot be eliminated, but the probability of their occurrence can be greatly reduced. Some of the measures that could reduce the likelihood of accidents include the following:

1. Development of quality standards for equipment used in the handling of hazardous materials
2. Timely inspection, maintenance, and replacement of such equipment
3. Installation of suitable monitoring and alarm systems to provide warnings of malfunctions
4. Installation of safety backup systems or devices to provide protection if the primary component fails
5. Clear labelling of all hazardous materials
6. Better training of both management and technical personnel regarding safety precautions

Some, though not enough, action has been taken in all of the areas listed above. For example, the U.S. Department of Transportation has issued a lengthy list of regulations regarding the transportation of hazardous materials. The program also calls for federal inspection of vehicles that carry hazardous materials and requires that the vehicles be properly labelled. Proper labelling means labels that can be seen on the side and rear of vehicles, and that use codes that reveal the nature of the substances carried and the type of precautions that must be taken if accidents occur and a cleanup is necessary. Additionally, many police and fire departments maintain special "emergency response units," teams specially trained and equipped to cope with spills of hazardous materials should they occur.

Yet critics point out that, because of lack of staff and money, inspections and enforcement of safety standards are extremely lax and many substandard vehicles are in use. There is little question that the margin of safety could be increased significantly. The incident at Bhopal recently spurred interest in the need for more rigorous safety controls in the production and handling of hazardous materials. Whether or not this interest materializes into concrete action depends on the amount of pressure citizens bring to bear.

# 14

# ACID PRECIPITATION AND FACTORS AFFECTING CLIMATE

---

## CONCEPT FRAMEWORK

4. Describe a *buffer*. What is one of the most common buffers?

5. What is the normal pH of rain? What is the pH of acid precipitation?

6. What causes acid fog?

7. How may increased acidity affect an aquatic ecosystem? How may acidic conditions affect an ecosystem even if the pH of the water remains in the normal range? How does acid precipitation affect the entire food chain? Why are some lakes healthy despite acid precipitation?

8. Discuss some of the positive and negative effects of acid precipitation on forests? How may these problems be catastrophic in the long term?

9. How is acid precipitation affecting buildings and monuments.

10. How may acid precipitation affect human health?

11. What is the major source of $SO_2$ emissions in the United States? How was the "dilution is the solution" approach implemented? Why did this approach fail?

12. Discuss the five options for control of $SO_2$.

13. Discuss the problems and steps being taken to address the acid rain problem.

14. Why is the carbon dioxide effect called the *greenhouse effect*? What benefits does it provide?

15. Why wasn't carbon dioxide a threat before the Industrial Revolution? Why is it a threat now?

16. What temperature changes do climatologists predict as a result of the greenhouse effect? Where will the changes be most and least significant? How may these changes affect the earth as a whole? How do particulates help shield us from the carbon dioxide effect?

17. What policy decisions may help avert the carbon dioxide effect?

18. How does atmospheric ozone differ from the ozone associated with air pollution? How is the oxygen-ozone balance upset? What is the most significant ozone catalyst?

19. What is the major source of free chlorine atoms? Name other ozone catalysts.

20. How may ozone depletion affect human health?

21. Discuss the problems and alternatives for controlling ozone depletion.

22. Why would a nuclear winter follow a nuclear conflict? How would the climatic changes affect the environment both within and beyond the target area?

As noted in Chapter 13 nitrogen oxides ($NO_x$) and sulfur dioxide ($SO_2$) are largely by-products of fossil fuel combustion. Recall that nitrogen oxides are the result of burning fuels in air, which is 78 percent nitrogen gas. At high temperatures some of the nitrogen is oxidized along with the fuel. Sulfur dioxide forms because fossil fuels, especially coal, contain some sulfur as an impurity; in combustion the sulfur is oxidized.

In the atmosphere nitrogen oxides and sulfur dioxide may react with water vapor to become, respectively, nitric acid and sulfuric acid. Finally, these acids are cleansed from the atmosphere in the condensation of water vapor and brought to earth by precipitation (Fig. 14-1). While this is commonly termed **acid rain,** these acids may be contained in any form of precipitation: rain, snow, hail, or fog. Thus, **acid precipitation** is a more accurate term. Indeed, there is some fallout of acid molecules as such without additional water. A term that covers this dry acid fallout as well as the acid precipitation is **acid deposition.** We shall use the term *acid precipitation* except where we wish to make special reference to the dry fallout.

Acid precipitation has received so much publicity in recent years that you are probably already aware that it is responsible for the loss of fish from many hundreds of lakes in the northeastern United States, Canada, northern Europe, and other industrialized regions of the world. Furthermore, thousands of additional lakes, rivers, and forests are being damaged or threatened. Indeed, many environmentalists consider acid precipitation to be the most severe pollution problem we face. Yet there is much controversy concerning acid precipitation, some even claiming that it is beneficial. The objective of this chapter is to provide a more thorough understanding of acid precipitation and its effects, to explore some of the controversy concerning it, and to consider options available for reducing it. Additionally we shall consider certain factors that may upset the climatic balance of the earth.

## MEASUREMENT OF ACIDITY AND ITS BIOLOGICAL IMPORTANCE

To understand the impacts of acid precipitation on the environment, it is first necessary to study the nature of acids and their measurement.

### Acids, Bases, and Neutralization

The chemical reactions of acids and bases are so interrelated that the two must be discussed together. The acidic properties of any solution are due to the presence of hydrogen ions ($H^+$), which are highly reactive. Therefore, an **acid** is any chemical that will release hydrogen ions on dissolving in water. Chemical formulas of a few common acids are shown in Table 14-1. Note that all ionize (the chemicals separate) to give hydrogen ions as well as the negative ion of the particular acid. The stronger an acid, the more readily it releases hydrogen ions; and the higher the concentration of hydrogen ions in a solution, the greater its acidity.

In a similar way, the caustic properties of all bases or alkaline solutions are due to presence of $OH^-$ ions called hydroxyl; hence, a **base** is any chemical that will release $OH^-$ ions (Table 14-1).

| Table 14-1 | Common Acids and Bases | | | | | |
|---|---|---|---|---|---|---|
| | *ACID* | *FORMULA* | *YIELDS* | *$H^+$ ION(s)* | *PLUS* | *NEGATIVE ION* |
| | Hydrochloric Acid | $HCl$ | $\rightarrow$ | $H^+$ | $+$ | $Cl^-$ chloride |
| | Sulfuric Acid | $H_2SO_4$ | $\rightarrow$ | $2H^+$ | $+$ | $SO_4^{2-}$ sulfate |
| | Nitric Acid | $HNO_3$ | $\rightarrow$ | $H^+$ | $+$ | $NO_3^-$ nitrate |
| | Phosphoric Acid | $H_3PO_4$ | $\rightarrow$ | $3H^+$ | $+$ | $PO_4^{3-}$ phosphate |
| | Acetic Acid | $CH_3COOH$ | $\rightarrow$ | $H^+$ | $+$ | $CH_3COO^-$ acetate |
| | *BASE* | *FORMULA* | *YIELDS* | *$OH^-$ ION(s)* | *PLUS* | *POSITIVE ION* |
| | Sodium Hydroxide | $NaOH$ | $\rightarrow$ | $OH^-$ | $+$ | $Na^+$ sodium ion |
| | Potassium Hydroxide | $KOH$ | $\rightarrow$ | $OH^-$ | $+$ | $K^+$ potassium ion |
| | Calcium Hydroxide | $Ca(OH)_2$ | $\rightarrow$ | $2OH^-$ | $+$ | $CA^{2+}$ calcium ion |
| | Ammonium Hydroxide | $NH_4OH$ | $\rightarrow$ | $OH^-$ | $+$ | $NH_4^+$ ammonium ion |

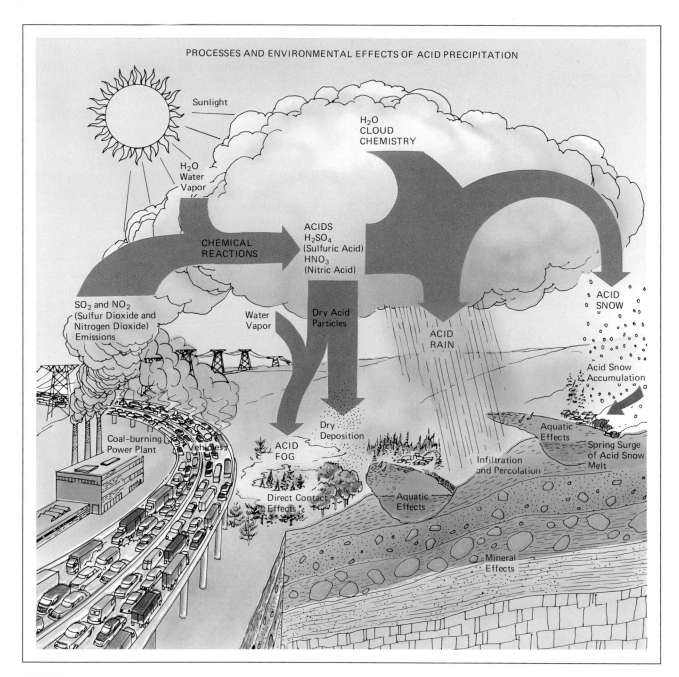

**FIGURE 14-1**
Acid precipitation. Emissions of sulfur and nitrogen oxides react with water vapor in the atmosphere to form their respective acids which come back down as dry acid deposition or mixed with water, causing the precipitation to be abnormally acidic.

In addition to possessing highly reactive properties, acids and bases neutralize each other. Recognizing that an acid is essentially $H^+$ ions and a base is $OH^-$ ions should make the process of neutralization clear. The $H^+$ and $OH^-$ ions simply come together to form HOH (i.e., $H_2O$) or water. The other ions associated with the acid and base remain as a salt, but this does not affect acidity or alkalinity (Fig. 14-2).

Thus, pure water is the balance or neutral point between an acidic solution on the one hand and a basic solution on the other. If one starts with an acidic solution and adds a base, acidity will gradually decrease until the neutral point associated with pure water is reached. Only then

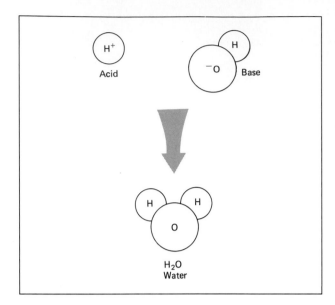

**FIGURE 14-2**
Neutralization of acids and bases. Neutralization occurs because the acid unit (H$^+$) reacts with the base unit (OH$^-$) to give water (H$_2$O).

will there be an increase in alkalinity. Thus there is a continuum from acidity through neutrality and then toward increasing basicity or alkalinity (Fig. 14-3).

In pure water there are actually trace but equivalent concentrations of H$^+$ and OH$^-$ ions. Substantial quantities of both cannot be present simultaneously because whenever the two ions come together they combine to form water molecules.

## pH: The Measurement of Acidity and Alkalinity

The actual concentration of hydrogen ions along the continuum from acid to base is expressed as pH. The pH scale goes from 0 (highly acidic) through 7 (neutral) to 14 (highly basic) (Fig. 14-4). The numbers on the pH scale actually stand for the negative powers of 10 of the hydrogen ion concentration in grams per liter. For example pII = 1 means that the concentrate of hydrogen ions in the solution is $10^{-1}$ or 0.1 g/ℓ, pH = 2 means that the hydrogen in concentration is $10^{-2}$ or 0.01 g/ℓ, and so on. At pH = 7 the hydrogen ion concentration is $^{-7}$ but here the OH$^-$ concentration is equivalent. Thus, this is the neutral point and expresses the tiny but equivalent

amounts of H$^+$ and OH$^-$ present in pure water. The pH numbers above 7 continue to express the negative exponents of hydrogen ion concentration, but more importantly they also represent an increase in OH$^-$ ion concentration. For example pH = 13 means that the hydrogen ion concentration is $10^{-13}$ (a decimal followed by 12 zeros and a one) grams per liter. However, the OH$^-$ concentration at this point is equivalent to $10^{-1}$.

It is significant to note that since numbers on the pH scale represent powers of 10, there is a tenfold difference between each unit. For example, pH 5 is ten times more acid (has ten times more H$^+$ ions) than pH 6; pH 4 is ten times more acid than pH 5, and so on. There is a 1000-fold difference between pH 6 and pH 3 (tenfold for each unit: $10 \times 10 \times 10 = 1000$). The same is true for OH$^-$ ion concentration as pH units progress above 7.

**FIGURE 14-3**
Adding a base to an acid. OH$^-$ of the base combines with H$^+$ of the acid until the neutral point, pH 7, is reached. Further addition of base increases the OH$^-$ concentration, making the solution basic.

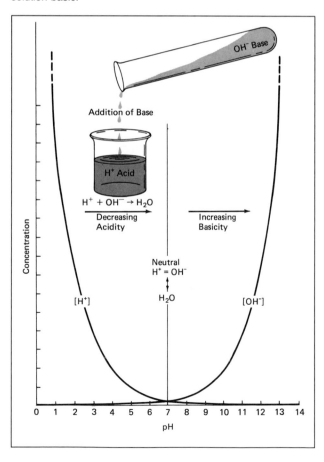

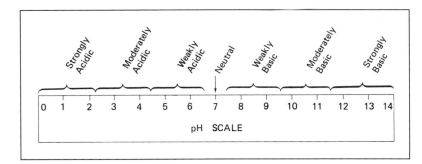

**FIGURE 14-4**
The pH scale.

While the pH scale sounds complex, pH is actually quite easy to measure. One can buy rolls of pH indicator paper from any laboratory supply house. When a strip of this paper is dipped in a solution it turns a certain color depending on the pH of the solution. One determines the pH by comparing the dipped paper with a color chart provided with the paper. Such pH indicators work because they are comprised of pigments that readily absorb or release hydrogen ions and change color as a result. The degree to which they absorb or release hydrogen ions depends on the $H^+$ ion concentration in the solution. There are also various electronic instruments for measuring pH, but the indicator papers are by far less expensive, easier to use, and, with practice, give results that are sufficiently accurate for most purposes.

## Effects of pH Change on Organisms

If pH becomes too high or too low, organisms suffer innumerable physiological malfunctions, any of which may lead to death. All these malfunctions may be understood in terms of the effect of pH on proteins. Proteins play a multitude of roles in all organisms including humans. They serve as the enzymes that direct and control virtually all the chemical reactions of metabolism. They are the carriers that move materials in or out of cells. Some proteins serve as hormones. Others control development. Still others serve as structural components of cells, and thus they are responsible for the basic characteristics of skin, bone, hair, muscle, and other tissues. Changes in pH cause changes in the structure of protein molecules and hence may affect virtually every aspect of an organism. Thus pH of water is an extremely critical environmental parameter, and no organism is immune to being affected by its change.

An important distinction exists, however,

between the pH of internal body fluids of an organism and the pH of the external environment. A significant aspect of metabolism for all organisms is the control of internal pH. The blood in our own bodies, for example, is maintained at pH 7.2 in spite of our eating and drinking things that are more acidic or more basic. Also our skin protects us from the external environment so that we may swim in water that is somewhat more acidic or basic without harm. Of course, strong acids or bases are so potent that they destroy the proteins of the skin and cause severe burning.

Aquatic and soil organisms, and especially their eggs, sperm, and developing young, have little protection and therefore are particularly dependent on an appropriate environmental pH.

Various environments on earth do have different pHs. For example, peat bogs are known for their acidity (pH about 4) and some organisms have adapted to these conditions; certain plants, for example, are known as "acid-loving." By and large, however, both fresh water and soil (actually soil-water) environments are within a range from pH 5.5 to 7.5. Correspondingly most organisms have an optimal pH within this range and can tolerate changes within this range. However, if pH drops below 5.5 they become severely stressed; and for the most part, they cannot tolerate pH below 5. Of course, as with all environmental parameters, some organisms are much more sensitive than others, some being severely stressed as pH goes below 6.

The same may be said about pHs above 7. Some natural conditions do produce a high pH. But as the pH rises above 9 or 10, most organisms cannot survive.

In addition, pH may affect organisms indirectly by acting on certain minerals. Increasing acidity causes some minerals to dissolve, thereby releasing toxic elements into solution. Conversely, increasing alkalinity can also cause certain nutrients to become insoluble. As a result, organisms may suffer from a lack of these nutrients. This will be discussed further later in the chapter.

## Maintaining pH with Buffers and Buffering Capacity

While organisms are sensitive to the pH of their environment, both artificial systems such as aquaria and natural ecosystems may be protected from pH change by *buffers.* A **buffer** is a substance that has a high capacity to absorb (or release) hydrogen ions at a given pH. For example, if acid is added to a system containing a buffer, the additional hydrogen ions will be absorbed by the buffer and the pH will remain relatively constant.

Many natural bodies of water and soils are effectively buffered by the presence of limestone, which chemically is calcium carbonate ($CaCO_3$). As shown in Figure 14-5, hydrogen ions from the addition of acid react with the carbonate ($CO_3^{2-}$) to form water and carbon dioxide, which may be released to the atmosphere. This is the "fizzing" that is observed when a drop of acid is put on limestone. Farmers have long used the addition of lime (pulverized limestones) to neutralize acid soils. Other carbonate minerals may be used as well. Shells of clams and oysters, for example, consist of calcium carbonate and are frequently used by organic gardeners.

Any buffer, however, is limited in capacity. Lime, for instance, is simply used up by the neutralizing reaction. Thus, we speak of the **buffering capacity** of a system. When the buffering capacity is exhausted, additional hydrogen ions will remain in solution and there will be a corresponding drop in the pH.

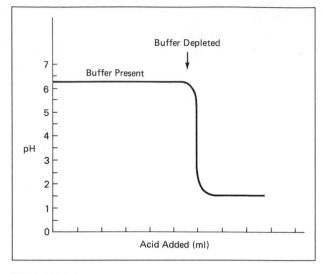

**FIGURE 14-6**
Exhaustion of buffering capacity. As the buffering capacity of a system is exhausted, the change in pH is *not* gradual. With the addition of very little more acid there is a sudden precipitous drop to the pH of the acid being added. Consequently, if we wait until buffers are depleted, disastrous consequences may occur before there is time for remedial action.

The rapid shift in pH that occurs as buffering capacity is exhausted deserves emphasis. It is graphically shown in Figure 14-6. Note that when a buffer is present, the addition of many units of acid has no appreciable effect on the pH. However, as the buffering capacity is exhausted, there is a precipitous drop in pH with the addition of just a few more units of acid. Can you see the implications of this regarding natural ecosystems under the impact of acid precipitation? Lakes with little or no buffering capacity have already turned acid, and all their aquatic organisms have died. The life in others is threatened as their modest

**FIGURE 14-5**
Buffering. Acids may also be neutralized by certain nonbasic compounds called *buffers.* A buffer such as limestone (calcium carbonate) reacts with the hydrogen ions as shown. Hence the pH remains close to neutral despite the additional acid. Note, however, that the buffer is consumed by the acid. Limestone is the most widespread natural buffer.

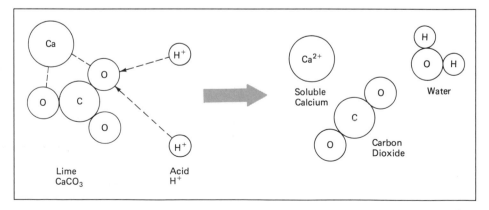

buffering capacity nears depletion. Still others with a great deal of buffering capacity may support a healthy ecosystem for a long period of time in spite of acid precipitation.

## ACID PRECIPITATION AND ITS EFFECTS

### Definition and Background

In the absence of any pollution, rainfall is normally slightly acidic, pH 5.6. this is because carbon dioxide in the air readily dissolves in and combines with water to produce a weak acid, carbonic acid (Fig. 14-7). What we refer to as *acid precipitation* has a pH that is lower, pH 5.5 or less. Values are commonly in the range of 4.5 and values as low as 1.5 have been recorded. Since, as we noted earlier, each pH unit represents a factor of 10, this means that acid precipitation is commonly 10 times, and may be as much as 10,000 times, more acid than normal precipitation. In addition, analysis of such precipitation reveals the presence of nitric acid and/or sulfuric acid, which shows that the increased acidity is derived from sulfur and nitrogen oxide pollutants.

Acid precipitation is not a new phenomenon. Acid rain and its adverse effects in causing rusting of metals and fading of dyed materials was first reported in 1872. Until relatively recently, however, acid precipitation was thought to be limited to areas where smelters or other industries that produce large amounts of sulfur dioxide are located. Beyond this, it was felt that occurrences of acid precipitation were unusual if they were observed at all, which is to say scientists were not systematically looking at the pH of precipitation.

About 35 years ago, however, anglers started noting precipitous declines in fish populations in many lakes in Sweden, Ontario, Canada, and the Adirondack Mountains of upper New York State. Many hypotheses were suggested and tested to determine the reason. Scientists in Sweden were the first to identify the cause as increased acidity, and to link this with abnormally low pH of precipitation.

This discovery revealed that acid-forming pollutants can be carried very long distances and can impact ecosystems hundreds of miles from their source. Sweden, for example, is several hundred miles downwind from industrial centers in England and Western Europe. When this fact was brought to light, scientists all over the world began monitoring the pH of precipitation and investigating its effects on ecosystems. The results obtained are cause for serious concern, to say the least.

### Intensity and Distribution

Results of monitoring show that acidity of precipitation has increased enormously both in terms of area affected and in the lowness of pH. The present distribution and intensity of acid precipitation for the United States and Canada is shown on the map in Figure 14-8. Importantly, this map does not represent a single unusual weather event; it shows the averages compiled over several years of monitoring. As you can see from the map, all of the eastern United States and Canada receive rain and/or snowfall with an average pH of 5 or less. For the northeastern United States and southeastern Canada the average is 4.5

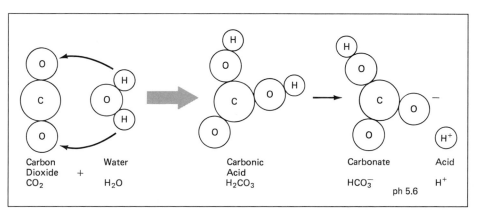

**FIGURE 14-7**
The pH of normal rainfall is 5.6. Normal precipitation is slightly acidic because carbon dioxide dissolved in water produces a weak acid as shown.

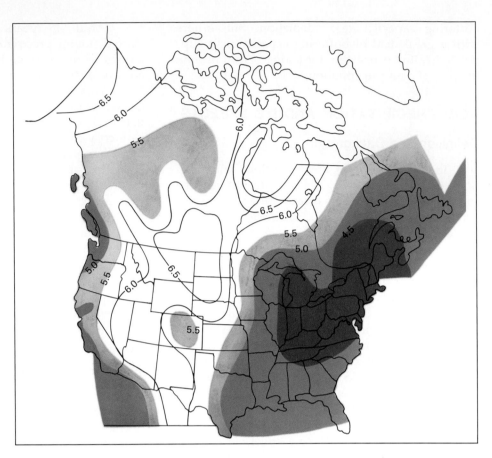

**FIGURE 14-8**
Regions receiving acid precipitation. Monitoring the pH of precipitation now reveals that acid precipitation is occurring over most of the United States and Canada. It is especially severe in the Northeast and along the West Coast. (Ontario Ministry of the Environment.)

and many areas within this region regularly receive precipitation of pH 4.0. The West Coast is also affected. Again, these are averages; individual events may be considerably lower.

At first acid precipitation was primarily associated with rain and snowfall, simply because these are most easily collected and monitored. Recently, however, fog in California was measured, and it was found to be much more acid than either rain or snow. Researchers at the California Institute of Technology have found that fog in the Los Angeles area typically has a pH value between 2.5 and 3. In forests in the mountains east of Los Angeles, scientists found fog water with a pH of 2.8 dripping from pine needles.

Why should fog be so acid? Scientists speculate that dry acid particles found in smog act as centers on which the condensation of water vapor occurs. Further, the tiny water particles may absorb more sulfur and nitrogen oxides and aid their conversion to acids. Thus the acids are highly concentrated in fog particles. In the for-

mation of snow or rain, the addition of much more water dilutes the acids.

The impact of dry acid deposition is also receiving more attention. It is now calculated that the amount of acid falling as dry particles is about equal to that falling with precipitation. Furthermore, particles may settle and accumulate on vegetation. When these become wetted with a small amount of water, as occurs with the condensation of dew, the resulting solution may be very strongly acidic.

## Effect of Acid Precipitation on Ecosystems

With the recognition of the widespread occurrence of acid precipitation, research on its effects on biological systems intensified. It has been found to affect ecosystems in many diverse ways. What is more, all negative impacts are becoming more widespread and more pronounced as acid precipitation continues and the buffering capacity of various systems is exhausted.

## IMPACT ON AQUATIC ECOSYSTEMS

The most straightforward impact of acid precipitation on aquatic ecosystems is a decrease in pH. Different species within a lake have differing sensitivites to pH (Fig. 14-9). Dropping pH from 7 to 6 may alter community composition as some species find it more difficult to compete; a few species may be eliminated, but there is still a complex, healthy ecosystem. As pH drops to 5.0 a large number of species is eliminated and others are placed under increasing stress. For example, reproduction and development of young may be impaired. Many aquatic plants are eliminated as well. As pH drops toward 4.0 virtually all organisms die off except certain species of algae that cover the bottom and grow along the edges. In some lakes these, too, disappear.

**FIGURE 14-9**
Effects of pH on the survival of various organisms. Some organisms are more tolerant than others to low pH, but very few can survive below pH 4.5, and upsets in ecosystems will occur even as the most sensitive organisms die off. (Office of Technology Assessment.)

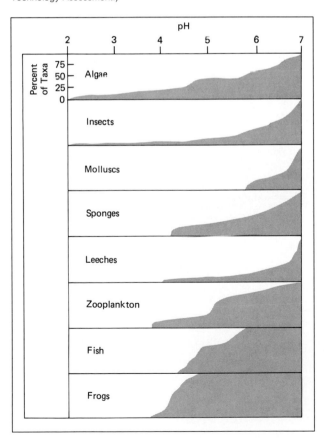

The appearance of such lakes is deceiving. From the surface, they are clear and blue, the outward signs of a healthy oligotrophic condition. However, a view under the surface is eerie. In spite of ample light shimmering through the clear water, there is not a sign of life; the lake is totally barren.

In some cases, there has been a dieoff of fish even though the pH of the lake has remained within normal range. In many cases this has been traced to *snowmelt*. Snow containing the acids accumulates over winter. The acid is then released quickly over a short period in the spring when the snow melts. It is analogous to a smoker quitting the habit for four months but continuing to stack all the packs he might have smoked on a shelf. Then he loses his resolve and smokes them all in a ten-day period. Even if natural buffers are present, the acid from snowmelt comes into streams and lakes faster than it can be neutralized, and the pH drops at least temporarily. Making a bad situation worse, this pH drop coincides with the time many fish, frogs, and other organisms are spawning and their eggs and developing young are most vulnerable.

A second cause of dieoffs of fish and other organisms in water with a normal pH has been traced to aluminum. Aluminum is toxic to fish, causing numerous abnormalities in the development of embryos. Where does the aluminum come from? Aluminum is an extremely common element; substantial quantities are present in many rock and soil minerals. Normally, these minerals are very insoluble and thus harmless, but under the attack of acid they break down and release the aluminum into solution. The process is frequently referred to as **mobilization;** in this case, mobilization of aluminum. Other toxic elements, such as lead and mercury may be mobilized as well. Synergistic interactions between toxic elements and low pH may also occur.

The loss of wildlife does not stop with the dieoff of fish and other aquatic organisms. Through the complexity of the food web, nearly all wildlife depends at least to some extent on the productivity of lakes. For example, naturalists report that in areas of the Adirondacks where the lakes no longer support fish, there is a total absence of loons and other waterfowl. Populations of many birds that feed on insects breeding in the water are also down as are populations of raccoons and many other mammals. If the situation continues it is not hard to imagine what will happen to geese and other migrating birds that use lakes as places to stop and feed (Fig. 14-10).

While research progresses, unfortunately so does acid precipitation and the dieoff of lakes. In Norway and Sweden the fish have died in some 6500 lakes and seven Atlantic salmon rivers. In Ontario, Canada, some 1200 lakes are now dead, and in the Adirondacks, a favorite recreational region for New Yorkers, more than 200 lakes are without fish.

Why are these particular areas so severely affected while we have noted that acid precipitation is occurring over a much broader region? Even within these regions one lake may be severely acidic while an adjacent lake remains normal. The answer is buffering capacity. If the watershed (the land area draining into a lake) contains adequate limestone, most of the water entering the lake will be well buffered and the pH

will be maintained. However, in the absence of buffering minerals, the lakes may quickly turn acidic. In short, it is found that lakes that still support healthy ecosystems despite acid precipitation have adequate buffering capacity. Those lacking in this capacity have become acidic. What does this portend for the future if acid precipitation is allowed to continue? The unfortunate fact is that large portions of the United States and Canada that receive acid precipitation have rock and soil types that are low in buffering capacity (Fig. 14-11) and what capacity remains is rapidly being depleted. A recent study of the United States east of the Mississippi River conducted by the Congressional Office of Technology Assessment showed that of a total of 17,059 lakes, 2993 (about 16 percent) are already showing signs of damage and another 6430 (about 38 percent) are endangered as indicated by very little remaining buffering capacity. Of a total 117,423 miles (187,876 km) of streams and rivers, 24,688 miles (21 percent) already show damage from acid precipitation and another 21 percent are endangered. Although the problem is worse in the Upper Midwest and Northeast where up to 80 percent of lakes and streams are threatened, nearly every state from

**FIGURE 14-10**
Acidification of lakes and rivers has a profound effect on terrestrial wildlife because numerous birds and land animals are supported by food chains that have their origin in aquatic systems.

**FIGURE 14-11**
Acid-sensitive regions. Large areas of North America, particularly in Canada, are especially sensitive to acid precipitation because these regions have granitic rock formations that have little buffering capacity. (Ontario Ministry of the Environment.)

## IMPACT ON FORESTS AND AGRICULTURE

The impact of acid precipitation on forests and agriculture may also be severe. Between 1963 and 1973, the growth rate of spruce trees in the Green Mountains of Vermont declined by 50 percent. More recently, scientists have observed similar slowdowns in the growth of forests over wide areas of the eastern United States and in California. An unprecedented number of trees is falling prey to insect and disease attack, and large-scale dieoff of forest species has begun to occur in New England, California, and other regions of high acid precipitation (Fig. 14-12). Many scientists are convinced that acid precipitation is a prime factor behind all of these effects because they have found that it may affect vegetation in the following ways.

**Direct contact effects.** Through simulated acid rain studies in greenhouses it has been shown that the acid damages the cuticle, the wax-like protective layer of leaves, making plants more vulnerable to attack by insects, fungi, or other plant pathogens. This effect may be particularly severe when crops and forests are exposed to intensely acid fogs or the accumulation of dry acid deposition.

**Effects on nutrients.** Analysis of water draining from various natural areas under differ-

Maine to Florida reports some damage. In Canada, Ontario alone stands to lose an additional 48,000 lakes within the next 20 years if acid precipitation continues at the 1983 rate.

**FIGURE 14-12**
Dieback of forests. A retardation of growth and a dieback of a number of species is being observed in many areas impacted by acid rain. Acid precipitation in addition to other factors of pollution is apparently stressing forests to the critical point. Shown here: Dieback of forest on Mount Mitchell, western North Carolina (1985) (Courtesy of Dr. Dwight Billings, Duke Univ.)

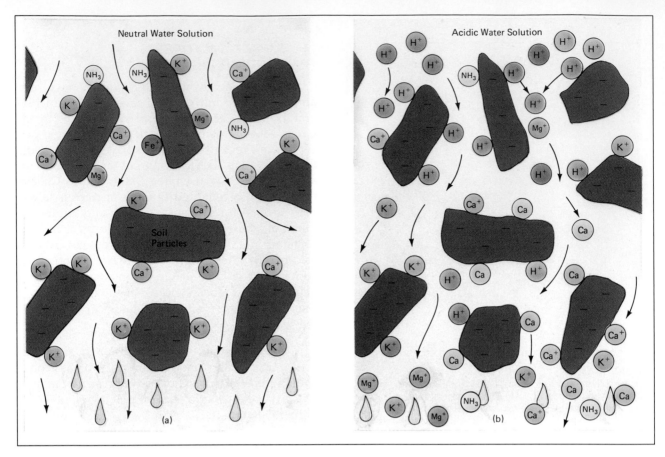

**FIGURE 14-13**
Acid precipitation leaches soil nutrients. Clay and humus particles tend to be negatively charged and hence bind and hold positively charged nutrient ions such as potassium ($K^+$), ammonium ($NH_3^+$), and calcium ($Ca^{2+}$). (a) The attractive force is strong enough to hold ions despite water percolating through the soil. (b) Acidic solutions cause leaching because hydrogen ions ($H^+$) displace nutrient ions.

ent conditions has shown that acid precipitation greatly increases the leaching of nutrients. Hydrogen ions effectively displace nutrient ions from their places in the soil and humus (Fig. 14-13). In addition, low pH also retards the activity of decomposers and nitrogen-fixing organisms, causing even further nutrient shortages. Finally acidic precipitation washing over vegetation has been shown to leach nutrients and other metabolites from leaves. Considering all these events together we can see that acid precipitation may cause nutrient deficiencies in a number of ways. Such deficiencies may be responsible for the reduced growth that is observed in the field and under simulated acid precipitation conditions in greenhouses.

**Mobilization of aluminum and other toxic elements.**  Lastly, many plants are highly sensitive to aluminum. The mobilization of aluminum and certain other elements may thus have a serious toxic effect on terrestrial plants as well as on aquatic organisms.

The major cause for concern is the fact that all of these acid-related effects may intensify with precipitous suddenness as buffering capacities are exhausted (Fig. 14-6). An example from Europe is noteworthy. In a region of high acid precipitation centered on the border between East Germany and Czechoslovakia, the woods are dying at an alarming rate. A 1980 study showed that 60 percent of the fir trees were healthy; two years later 98 percent were dead or dying. A similarly rapid dieback of forests throughout West Germany is occurring. According to the German government surveys, forests affected by damage as a result of pollution increased from 8 percent to 50 percent in just the two years between 1982 and 1984. German researchers warned that symptoms observed in United States forests in 1985 resembled those seen in 1980 in West Germany.

As changes occur in forests, you can surmise the effect on other wildlife populations. If a sudden collapse of the forest ecosystems occurs, the ramifying effects of soil erosion, sedimentation of waterways, flooding, and deterioration of water supplies will be catastrophic. At the very least, we can expect a succession in which the dying trees are replaced by acid-loving species. But the variety of such plants is very limited, and most are mosses, ferns, and other scrubby plants that are economically worthless even for grazing.

## IMPACTS ON MATERIALS

As noted previously, the impact of acid precipitation on materials was the first effect observed, and this impact has not diminished. Limestone and marble, which is a form of limestone, are favored materials for the outside of buildings and monuments. The limestone acid reaction shown in Figure 14-5 is causing these materials to weather and erode at a tremendously accelerated pace. Monuments and/or buildings that stood for hundreds of years with little change are now literally dissolving and crumbling before our eyes (Fig. 14-14). Concrete also contains lime, and acid precipitation may be hastening the deterioration of roadways and bridges; however, road salt seems more significant in this regard. Acid

precipitation and deposition also have a negative impact on paint and they accelerate the corrosion of metals. Here again, however, the effects of acid precipitation are difficult to separate from those of other air pollutants.

## IMPACTS ON HUMANS

Many state officials are concerned that the ability of acid precipitation to mobilize aluminum and other toxic elements may result in contamination of both surface and groundwater supplies. It is noteworthy that aluminum has recently been implicated as a cause of Alzheimer's disease, a form of premature senility. Increased acidity of water mobilizes lead used in some old plumbing systems and from the solder used in copper systems; this is a problem in some areas.

Some scientists believe that inhalation of the highly acidic fog particles and dry acid particles is the major cause of the breathing and respiratory problems that are experienced by some people during smog events. In addition, there is evidence that inhalation of such particles renders lung tissues more susceptible to the carcinogenic effects of other pollutants. However, if acid precipitation is allowed to continue at the present rate, by far the greatest impact on humans will be in the deterioration and loss of lakes and forests and their associated economic, ecological, and aesthetic values.

 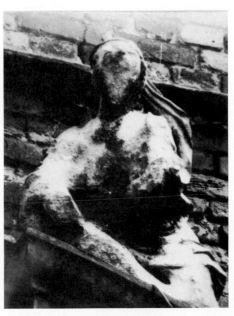

**FIGURE 14-14**
Corrosive effects of acid rain are highly destructive to stonework. The statue shown is in the Ruhr, an industrial district of West Germany. (a) Appearance in 1908; (b) appearance in 1969. Erected in 1702, it has suffered more erosion in the last 60 years than it did in the previous 200. (From E. M. Winkler, *Stone: Properties, Durability in Man's Environment*. Photos supplied by Schmidt-Thomsen, Landesdenkmalamt, Westfalen-Lippe, Muenster, Germany.)

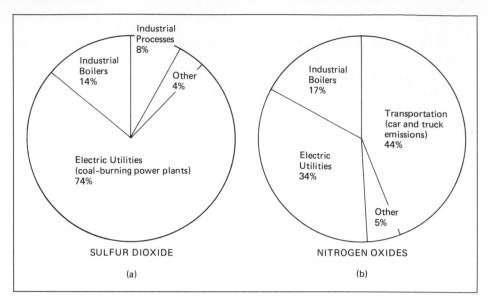

FIGURE 14-15
Sources of acid-forming pollutants in the 31 eastern states. (a) Sulfur dioxide. (b) Nitrogen oxides. (Office of Technology Assessment; data for 1980.)

**FIGURE 14-16**
Standard smokestacks of this coal-burning power plant have been replaced by new 1000-foot stacks to aid in the dispersion of pollutants into the atmosphere. They may alleviate local problems, only to create more widespread ones. (Farrell Grehan/ Photo Researchers.)

# IDENTIFICATION OF SOURCES AND CONTROL OF ACID PRECIPITATION

## Identification of Sources

Total emissions of sulfur dioxide and nitrogen oxides can be estimated from the amounts and kind of fuels that are consumed for various purposes. Thus, emissions from just the states east of the Mississippi River are placed at some 22.4 million tons of sulfur dioxide and another 14 million tons of nitrogen oxides per year. A further breakdown as to sources is shown in Figure 14-15. You can see from this figure that 74 percent of the $SO_2$, about 16.5 million tons, comes from the burning of fossil fuels by electric utilities. Since this forms sulfuric acid ($H_2SO_4$), which carries two $H^+$ ions per molecule, this represents about 55 percent of the total acid produced from both sulfur and nitrogen oxides. Further, through tracking cloud movements by satellite, the source can be largely traced to the tall stacks of some 50 huge, coal-burning power plants (Fig. 14-16). The same plants are also significant emitters of nitrogen oxides. The locations of these plants are shown in Figure 14-17. In the process of generating steam to drive generators, such plants consume about 10,000 tons of coal per day. If one assumes the coal has an average sulfur content of 3 percent, a plant will emit some 600 tons of sulfur dioxide per day or 220,000 tons per year.

Early on it was recognized that sulfur dioxide was a serious air pollutant and that coal-burning power plants were major contributors. As the

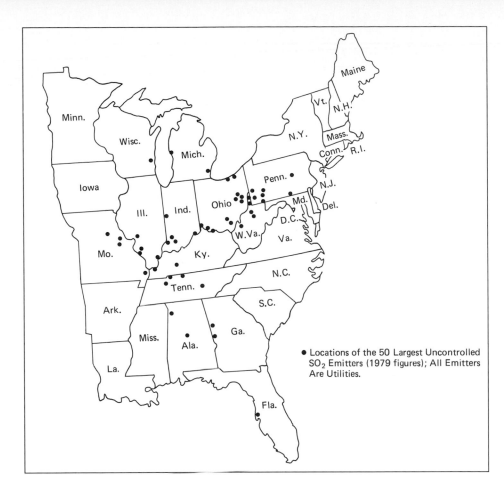

● Locations of the 50 Largest Uncontrolled SO₂ Emitters (1979 figures); All Emitters Are Utilities.

**FIGURE 14-17**
Locations of the 50 largest sulfur dioxide emitters, all of which are utility coal-burning power plants. They account for 74 percent of the sulfur dioxide and a large portion of the nitrogen oxide emissions as well. ("Acid Rain, a Major Threat to the Ecosystem," Conservation Foundation Letter, December 1982.)

Clean Air Act of 1970 was being written, environmentalists called for stringent controls on such plants. However, the electric utility industry lobbied hard for and won permission to take the "dilution is the solution to pollution" approach. Their claim was that huge, tall smokestacks would eject pollutants high in the air where they would be carried far away and would eventually fall "harmlessly" to earth. Thus, huge stacks over 300 m (1000 ft) high were built to replace smaller stacks (Fig. 14-16). The installation of the tall stacks *did* improve local air quality around the plant. However, winds carry and distribute the pollutants over the eastern United States, especially the Northeast, and into eastern Canada where they come down as acids (Fig. 14-18).

At the same time tall stacks were built, precipitators were installed to control particulates. But these seem to aggravate the problem still more. The particulates contain alkaline materials that formerly helped neutralize the acids, although they also presented their own pollution problems.

## Options for Control

Scientists calculate that a 50 percent reduction in present acid-causing emissions would effectively prevent further acidification of the environment. This would not correct the already bad situation, but natural buffering processes are estimated to be capable of preventing further deterioration. Since we know that about 50 percent of the acid-producing emissions come from the tall stacks of coal-burning power plants, control strategies center on these sources. Five main strategies have been proposed: (1) alternative power plants, (2) fuel switching, (3) coal washing, (4) scrubbers, and (5) fluidized bed combustion. A brief examination of each strategy leads to the conclusion that scrubbers are the only practical solution for the near term, and acid precipitation must be controlled in the near term as well as the long term.

### ALTERNATIVE POWER PLANTS

The only current alternative that could produce electricity on a scale equivalent to fossil fuel

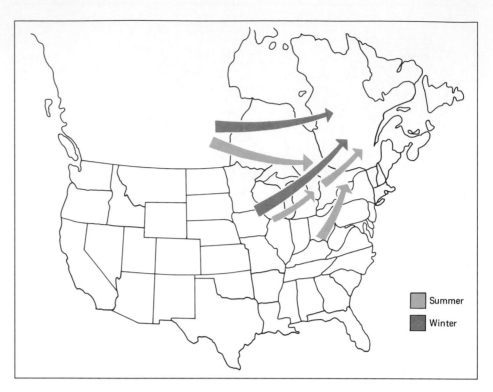

**FIGURE 14-18**
Major summer and winter storm paths. Acid-forming pollutants emitted from the tall stacks of power plants located in the Ohio Valley tend to fall out over New England and eastern Canada.
(Ontario Ministry of the Environment.)

Summer
Winter

power plants is nuclear power plants. However, it is economically prohibitive to scrap the huge coal-fired plants, which are relatively new, and embark on building 50 nuclear plants that cost several billion dollars each, require 10 to 12 years to construct, and, while they don't emit acid-forming pollutants, do present other problems (see Chapter 22). The tremendous cost of nuclear power plants and the public reaction against them have, in fact, promoted the recent trend to cancel plans for nuclear plants and build more coal-fired plants.

### FUEL SWITCHING

The present plants would not emit sulfur dioxide if the fuel were not contaminated with sulfur. Is it feasible to switch to low- or no-sulfur fuels? The United States does have abundant reserves of low-sulfur coal, but most are located in the western states (Montana, Wyoming, Colorado, and Utah). The logistics and economic dislocations of shifting the mining and transportation of some 200 million tons of coal per year from eastern operations to the West is horrendous to contemplate. It has also been pointed out that western low-sulfur coal has a lower heating value. Therefore, while there is less sulfur per ton, one

would have to burn more tons; thus the net reduction in $SO_2$ emissions might be small. If low-sulfur fuel oil or natural gas were to be substituted, these fuels would have to come from increased imports. This would be highly counterproductive in terms of U.S. efforts to become less dependent on foreign fuel supplies. Indeed, a significant factor in decreasing reliance on foreign supplies in recent years has been the utilities' switch from oil and gas to coal.

### COAL WASHING

Simply washing coal with water before it is burned will remove a portion of the sulfur from high-sulfur coal. However, most high-sulfur coal is already being washed to some degree, so further opportunities in this regard are limited. Several chemical techniques for removal of sulfur have been proposed, but none of these has yet been developed on a commercial scale (10,000 tons per day).

### SCRUBBERS

*Scrubber* is the term for a filter that consists of a liquid spray. In this case, exhausts from coal combustion are drawn through a spray containing lime (Fig. 14-19). The sulfur dioxide reacts with

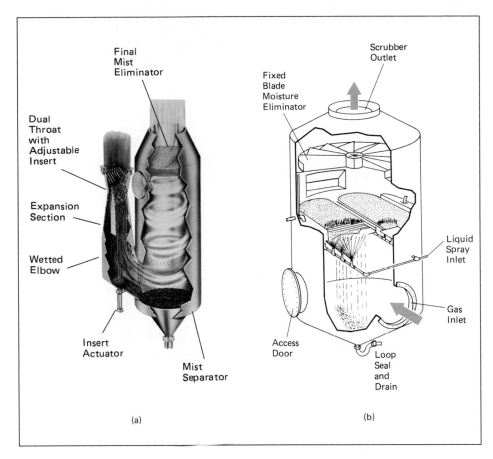

Final
Mist
Eliminator

Dual
Throat
with
Adjustable
Insert

Expansion
Section

Wetted
Elbow

Insert
Actuator

Mist
Separator

(a)

Scrubber
Outlet

Fixed
Blade
Moisture
Eliminator

Liquid
Spray
Inlet

Gas
Inlet

Access
Door

Loop
Seal
and
Drain

(b)

**FIGURE 14-19**
Scrubbers. Sulfur dioxide may be removed from flu gases by passing the furnace exhaust through a spray of lime and water. The sulfur dioxide reacts with the lime and a calcium sulfate precipitate is removed. Two designs for such devices, called *scrubbers*, are shown here. [(a) Courtesy of FMC Corporation-Air Quality Control. (b) A cutaway view of a SLY IMPINJET℠ Gas Scrubber (The W. W. Sly Manufacturing Company, Cleveland, Ohio)]

the lime and is precipitated as calcium sulfate ($CaSO_4$). Scrubbers are expensive to install and operate, and disposal of the toothpastelike calcium sulfate sludge may present logistical problems (although it is not toxic). However, plants in various parts of the world, including some in the United States, have demonstrated that scrubbing is practical and also highly effective.

### FLUIDIZED BED COMBUSTION

Fluidized bed combustion is an alternative method of burning coal (or other fuels) in which sulfur dioxide may be removed in the actual combustion process (Fig. 14-20).

Of course it is not necessary to choose a single approach; any combination of these options might be used.

## Muddling Toward Action

The seriousness of the acid precipitation problem and the involvement of coal-burning power plants were clearly evident by the end of the 1970s. Indeed, in 1981 the National Academy of Science after studying the problem, concluded that the circumstantial evidence linking power plant emissions to the production of acid rain was "overwhelming." Other studies completed since have added to the already "overwhelming evidence." This analysis, combined with the fact that scrubbers are a proven technology, would lead one to expect swift action. Quite the opposite has been the case.

The electric power industry has raised every argument and excuse imaginable to confuse the issue, to disclaim any responsibility, and to avoid or delay installing scrubbers. Many of the arguments they present are totally spurious. For example, they make the categorical statement, "All rain is acid," attempting to make the public believe that rainfall of pH 4.0 or less is no more acid than normal rain of pH 5.6. They also attempt to show that acid precipitation has nothing to do with the situation by citing that adjacent lakes receiving the same rainfall have different pH values. Of course, this ignores the well-known facts about differing buffering capacities. They disclaim

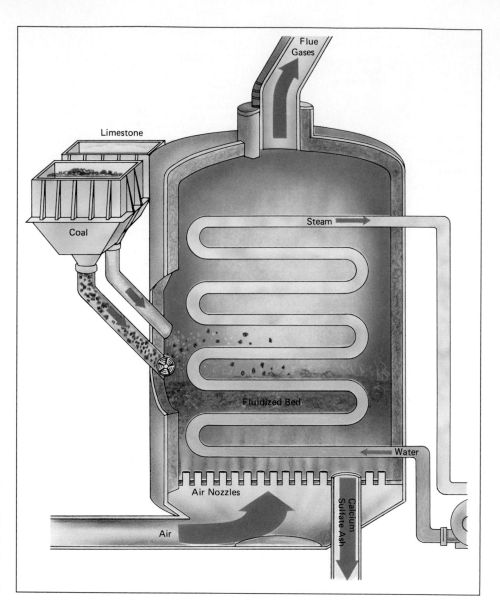

**FIGURE 14-20**
Fluidized bed combustion. Sulfur dioxide may be removed in the combustion process itself by burning the coal in a mixture of sand and lime, all of which is kept fluid by forcing air up from the bottom. (Redrawn from an illustration by Rick Farrell, copyright ©. All rights reserved.)

all evidence showing that acid precipitation has played a role in forest damage.

One of their most common arguments is that scientists are unable to show any case in which emissions from plant X have caused lake Y to become acid. Unless or until there is such definitive evidence, they claim that they cannot and should not be held responsible. This argument is analogous to saying that we should not have pollution controls on cars because we cannot prove which car produced the smog that gave John Doe bronchitis. Clearly, this is absurd. Lakes and ecosystems are suffering from a general loading of the atmosphere with acid-producing pollutants, and it makes sense to reduce the major sources of this

loading. It is both ridiculous and totally unnecessary to demand that lines be drawn from specific power plants to specific lakes.

Finally, the electric power industry claims the costs of installing scrubbers would be unaffordable. It is interesting to note here that the industry promotes nuclear power as an answer to the acid rain problem. The nuclear power option, however, is at least 25 to 50 times more expensive than scrubbers. All 50 of the coal-fired power plants in question could be fitted with scrubbers for the cost of one or two nuclear power plants. In fact, several nonutility organizations have analyzed prospective costs of installing and operating scrubbers and have estimated an increase in the range of 2 to 8 percent. Actual experience from the Tennessee Valley Authority (TVA), a quasi-government utility company, shows that installing and operating a scrubber increases costs by only 3 to 5 percent.

Blatantly absent in any case is discussion concerning costs of environmental destruction that will be incurred if emissions are not controlled. In the United States, acid damage could be disastrous to the $49 billion forest and forest products industry which employs 1.2 million people. This is not to mention the impacts on recreation and water supplies. The National Academy of Sciences warns that allowing emissions of sulfur and nitrogen oxides to continue unchecked in view of the known hazards is taking a grave risk with human health and biosphere protection. The Academy goes on to strongly recommend prompt tightening of the controls on acid forming emissions.

Unfortunately, much to the frustration of those who understand the seriousness of the problem, particularly the Canadians who receive a large portion of acid precipitation from us, the U.S. government has not yet imposed regulations to curtail present emissions from power plants. During its first term in office, the Reagan Administration persistently echoed and supported the argument of the electric power industry and exercised delaying tactics by calling for more studies. In early 1986 the administration did agree with Canadian officials to support a joint 5 billion-dollar program to determine the most economical way to clean up coal burning. Yet this still delays any substantive reductions to 1990 or beyond. Such inaction is allowed to persist largely because the public has remained strangely silent on the issue. We often find that government responds to the loudest voice, and if that voice is not the public's, then it is most likely to be that of an industry that is never apathetic regarding its self-interest.

## FACTORS AFFECTING CLIMATE

Humans have tried to influence the weather for millennia. Now we are at the threshold of major human-caused changes in climate. Unfortunately, these changes are neither planned nor directed. They will be caused by unwitting changes in the earth's atmosphere.

The changes can be attributed to two major sources: (1) the increased concentration of carbon dioxide in the air, and (2) the addition of materials to the atmosphere that cause a depletion of stratospheric (very high altitude) ozone. There is still much uncertainty regarding the specific impacts of these processes. Yet the predicted impacts are severe and once they occur they will be irreversible. Therefore, it is not too soon to make policy decisions to deal with the pollutants involved. In addition, a nuclear conflict, beyond its immediate destructive effect, could drastically alter the climate for a period of several years. We shall examine these issues in more detail.

### The Greenhouse or Carbon Dioxide Effect

#### DESCRIPTION OF THE PROBLEM

The **carbon dioxide effect** is also commonly called the **greenhouse effect** because it is analogous to the solar heating that occurs in a greenhouse, or in your car when it is left parked in the sun. Sunlight (energy) enters through the glass and is absorbed by the surfaces it strikes. As the surfaces become warm, they reradiate the energy as infrared or heat radiation. The nature of glass is such that, while it is highly transparent to light, it tends to block infrared radiation. Therefore, the energy that enters as light is trapped and causes the temperature to rise (Fig. 14-21a).

On a global scale, carbon dioxide in the atmosphere plays a role analogous to the glass in a car or a greenhouse (Fig. 14-21b). The atmosphere, apart from particulates, is largely transparent to incoming light energy; however, outgoing infrared radiation is absorbed by carbon dioxide, leading to a warming of the atmosphere. It follows that the greater the amount of carbon dioxide in the atmosphere, the more infrared energy absorbed and the greater the warming of the atmosphere. The present atmospheric concentration of carbon dioxide seems slight, 0.034 percent or 340 parts per million (ppm). But this absorbs sufficient infrared radiation to give the moderate temperatures we enjoy over much of the earth. Space probes to the planet Venus show what the extreme can be like. Venus's atmosphere is 98 percent carbon dioxide and its resulting temperature is 477° C (nearly 900° F), hot enought to melt lead. If Venus's atmosphere were the same as ours, its temperature would be comparable to ours; it would be only a few degrees warmer owing to Venus being closer to the sun.

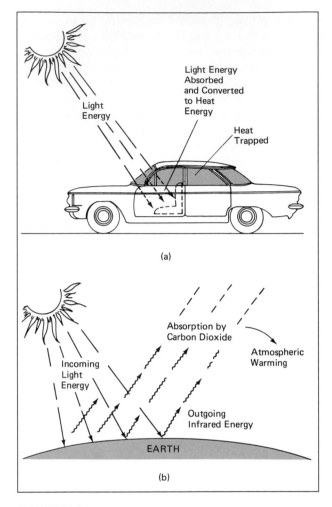

**FIGURE 14-21**
The greenhouse effect. Analogous to the heating of your car when it is in the sun, the earth's atmosphere is heated by outgoing infrared radiation being absorbed by carbon dioxide molecules. The greater the concentration of carbon dioxide, the greater will be the absorption and heating of the atmosphere.

There is no danger of the earth's atmosphere ever approaching that of Venus. Yet we are increasing the carbon dioxide concentration of our atmosphere, and if this trend continues there is little question that it will lead to significant warming over the next 40 to 70 years. This will trigger many other climatic changes.

### SOURCES OF INCREASED CARBON DIOXIDE

In our study of the carbon cycle in Chapter 2 we observed that carbon dioxide is absorbed from the atmosphere and is converted into or-ganic matter by photosynthesis. In turn, carbon dioxide is returned to the atmosphere by the oxidation of organic matter by organisms and by burning. Thus, in balanced ecosystems carbon dioxide inputs and outputs are equal.

However, human activities have created a distinct imbalance in the carbon cycle. Most importantly, since the beginning of the Industrial Revolution (1760), we have burned fossil fuels (coal, oil, and natural gas) at an accelerating rate. The major waste product of such combustion is carbon dioxide. Another factor, which some scientists feel is almost equally important in releasing carbon dioxide into the atmosphere, is widescale deforestation and soil mineralization, particularly that now occurring in tropical regions around the world. As trees are cut and burned and soil humus oxidizes, all of the carbon they contain is converted to carbon dioxide.

Not all of the carbon dioxide produced remains in the atmosphere. Carbon dioxide is soluble in water; therefore the oceans act as sinks aborbing a portion of the excess carbon dioxide. Some of this carbon dioxide is then converted to calcium carbonate by shell-forming organisms, thus removing it from the system (Fig. 14-22). However, a substantial portion of the excess carbon dioxide is accumulating in the atmosphere as shown by precise monitoring since the 1950s (Fig. 14-23). The current mean level is about 340 ppm. This is a 30 percent increase over the estimated 260 ppm at the start of the Industrial Revolution.

Given present trends there is little prospect for curtailing the carbon dioxide buildup. The use of crude oil has levelled off because of shortages and higher prices, but coal, which is much more abundant, is being used in increasing amounts. Even more carbon dioxide is produced from coal per unit of energy because coal is a less efficient fuel. Likewise deforestation and oxidation of soil organic matter show little sign of abating. Unless there is a dramatic change in policy regarding energy and deforestation in the near future, one can project that between the years 2025 and 2075 the carbon dioxide concentration in the atmosphere will reach the level of 600 ppm, double its level before the Industrial Revolution. The year 2025 is well within the expected lifetime of most students reading this text.

Carbon dioxide is not the only gas that absorbs infrared radiation. Methane, hydrocarbons,

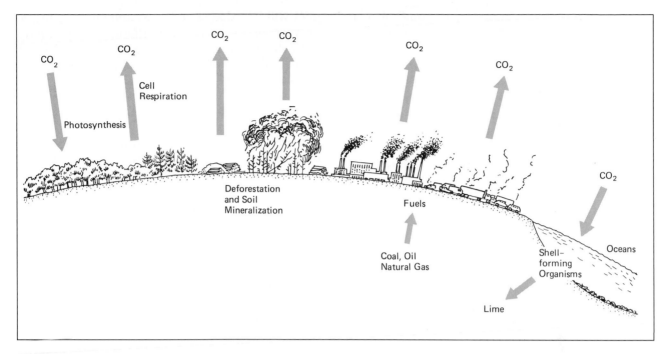

**FIGURE 14-22**
The carbon dioxide level in the atmosphere is the result of a balance between processes that add $CO_2$ and processes that remove it. Burning of fossil fuels, destruction of forests, and mineralization of soils make current additions much greater than removals.

nitrogen oxides, and freons are also strong absorbers. Therefore these pollutants will also contribute to the greenhouse effect.

### DEGREE OF WARMING AND ITS PROBABLE EFFECTS

Lack of good information in the past has led to tremendous speculation about the greenhouse effect. If a doubling of the carbon dioxide level to 600 ppm is used as a reference point, estimates have ranged from "no effect" to a total melting of the polar ice caps and a rise in sea level of about 25 m (80 ft). This would cause massive flooding of all coastal and low-lying regions. Unprecedented heat waves causing massive dieoffs or at least reproductive failure of many species including humans have also been postulated. However, recent advances in computer modeling of climatic factors has reduced the uncertainty a great deal.

Climatologists are now in general agreement that the overall warming will be between 1.5° C and 4.5° C (3° F to 8° F). Warming is likely to be more pronounced in polar regions, as much as

10° C (18° F), and less pronounced in equatorial regions, 1–2° C (2–4° F). This degree of warming is not likely to produce the catastrophes noted above; however, the effects may still be severe. Because the difference in temperature between the poles and the equator is a major driving force for atmospheric circulation, the greater heating at the poles will reduce this force. This will change atmospheric circulation patterns, which will, in turn affect the distribution of rainfall. Some regions of the world are likely to see an increase in rainfall; other areas, a decrease, as shown in Figure 14-24.

North Africa, which is largely desert at present, is likely to profit by increasing rainfall. However the United States and Canada are likely to be losers. The central portion of North America is a major "breadbasket" of the world, producing huge amounts of wheat and corn. Rainfall, already minimal for these crops, is likely to become much less. Quoting Dr. Walter Orr Roberts, former director of the National Center for Atmospheric Research,

> The Dust Bowl of the middle 1930s in the United States was the greatest climatic disaster in the history of our nation. . . . [However] the Dust Bowl of the 30s may seem like children's play in comparison to the Dust Bowl of the 2040s. Because of the effects of the warming . . . , natural rainfall may decline by as much as 40%, and the sum-

mers will be hotter, increasing the evaporation of soil moisture. The soils will desiccate, and the winds will lift them to the skies.*

Nor can irrigation be expected to provide much relief. Recall that the water table is already being drawn down to support agriculture through much of the region; by 2040 and probably considerably before then, most of the groundwater for the re-

---

*Roberts, Walter Orr, "It is Time to Prepare for Global Climate Changes," *Conservation Foundation Letter*, April 1983.

**FIGURE 14-23**
Atmospheric carbon dioxide concentration. Carbon dioxide concentration in the atmosphere fluctuates between winter and summer due to seasonal variation in photosynthesis. But the *average* concentration is gradually increasing owing to human activities, namely, burning fossil fuels and oxidation of soil organic matter. This trend may lead to an increase in global temperatures, which will result in other widespread effects. (M. C. MacCracken and H. Moss. "The First Detection of Carbon Dioxide Effects," *Bulletin of the American Meteorological Society* 63:1165, October 1982.)

gion will be exhausted or prohibitively expensive because of the need for drilling and pumping from deep levels. Coastal regions of the United States will probably retain favorable rainfall but, ironically, this is where large amounts of agricultural land have been and continue to be sacrificed for suburban development (see Chapter 24).

Why are we not already observing a warming effect due to the 30 percent increase in carbon dioxide that has already occurred? In fact, we may be. Weather conditions fluctuate so much from year to year that it is difficult to observe gradual trends. However, glaciers in many parts of the world are receding, as is sea ice. This indicates a gradual warming trend. Still, it is difficult to know whether the warming is due to carbon dioxide buildup or is the result of other climatic cycles that are not well understood.

Another factor that influences the situation is the number of particulates in the atmosphere,

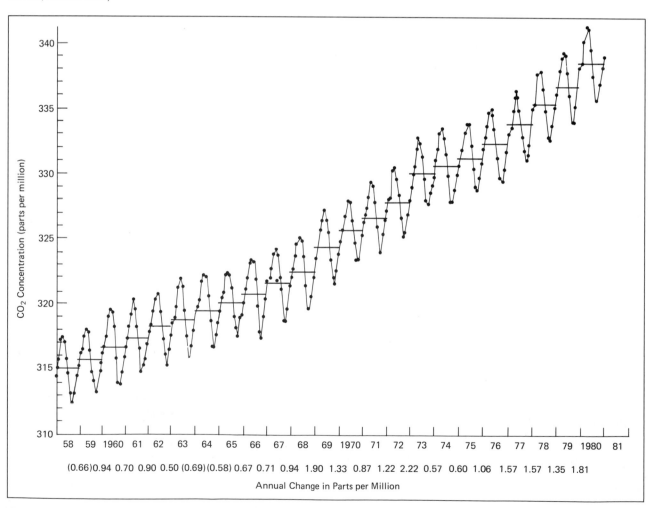

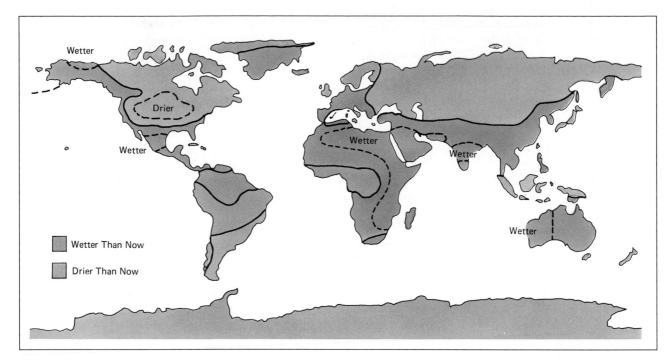

**FIGURE 14-24**
Regions of the world that are likely to become wetter or drier than now under the influence of the carbon dioxide effect. Predictions may not be entirely accurate, but areas demarked by a dotted line have a higher probability of being correct.
(William W. Kellogg and Robert Schware, 1981. *Climate Change and Society: Consequences of Increasing Atmospheric Carbon Dioxide,* Westview Press; Boulder, Colorado.)

which may somewhat offset the carbon dioxide effect. Particulates in the upper atmosphere tend to reflect sunlight. This reduces the energy input and the earth's atmosphere cools as a result. Particulates both from human activities and from recent volcanic eruptions—particularly the eruption of El Chichon in Mexico in 1982, which ejected a huge amount of material into the upper atmosphere—may be countering the carbon dioxide effect now. As particulates from these volcanic eruptions subside over the next few years, however, climatologists feel that the carbon dioxide warming trend will become noticeable.

One factor may tend to counter the carbon dioxide buildup. Experiments have shown that increasing carbon dioxide concentration increases net photosynthesis in most plants and it also leads plants to use water more efficiently. The latter occurs because the plants can still gain sufficient carbon dioxide even though their stomas are largely closed, thus reducing transpiration. However, potential increases in growth may not be realized unless we reverse trends in deforestation, soil degradation, air pollution, and acid rain.

## COMING TO GRIPS WITH THE CARBON DIOXIDE EFFECT

It will be difficult to come to grips with the carbon dioxide effect because it will creep up slowly over the next 40 to 70 years and it will remain hidden in normal year-to-year variations in weather. Abnormally wet periods and abnormally dry periods have always been part of nature and this situation will continue. Who is to say that a given drought is caused by or made more severe by the carbon dioxide until we find ourselves with the Dust Bowl conditions mentioned above? At that point, there will be no chance to reverse the situation. Even if carbon dioxide emissions were sharply reduced at that point it would probably be 1000 years or more before natural processes would "clean" the atmosphere of the excess carbon dioxide.

There is some thought that agriculture in North America might be moved north, taking advantage of the longer growing season. This may occur to some extent but this option is limited because soils in a large portion of northern North America are very shallow and poor.

Even with recognition of the carbon dioxide effect there is no straightforward technological way to reduce carbon dioxide emissions from burning fossil fuels. The only effective strategy involves basic changes in energy use and agriculture. Since such changes involve large sectors of the economy, very long lead times are required. Basic policy decisions should be made now. What policy decisions are in order?

First and foremost, continued and increased emphasis must be placed on energy conservation and development of alternatives to fossil fuels (see Chapter 22 and 23). Reducing the use of fossil fuels would reduce other forms of air pollution and acid rain, as well as carbon dioxide output. Emphasis should also be placed on agricultural practices that maintain or increase soil organic matter and on reforestation. This would shift carbon dioxide from the atmosphere back into biomass. Finally, development of more drought-resistant and/or salt-tolerant crops would be highly beneficial in making agriculture more resilient to current situations of salinization and desertification as well as to potential droughts brought on by the carbon dioxide effect. It is significant to note that policies needed to mitigate the carbon dioxide effect are the same as those needed to cope with other environmental problems.

## Depletion of the Ozone Shield

### DESCRIPTION OF THE PROBLEM

In Chapter 13 we learned that pollutants that cause ozone to form in the lower atmosphere are a threat because ozone is highly poisonous to both plants and animals. Paradoxically the opposite situation exists in the stratosphere (the layer of atmosphere 10 to 15 km or 6 to 10 mi above the earth). Here, the presence of ozone is essential to life on earth because it screens out damaging ultraviolet radiation. Some pollutants threaten to deplete this ozone. The two problems, excess ozone in the lower atmosphere and depletion of ozone in the stratosphere, exist simultaneously and are essentially independent because there is negligible mixing between the stratosphere and the lower atmosphere.

In order to understand the problem of ozone depletion in the stratosphere, the hazard of ultraviolet (UV) radiation should be stressed first. Ultraviolet is a part of the natural radiation from the sun; the wavelengths are just slightly shorter and have higher energy content than those of visible

**FIGURE 14-25**
Ultraviolet, visible light, infrared, and many other forms of radiation are, in fact, just different wavelengths of the electromagnetic spectrum. (By permission from Joe R. Eagleman, *Meteorology: The Atmosphere in Action*, D. Van Nostrand copyright 1980 by Litton Educational Publishing, Inc. Reprinted by permission of Wadsworth Publishing Co.)

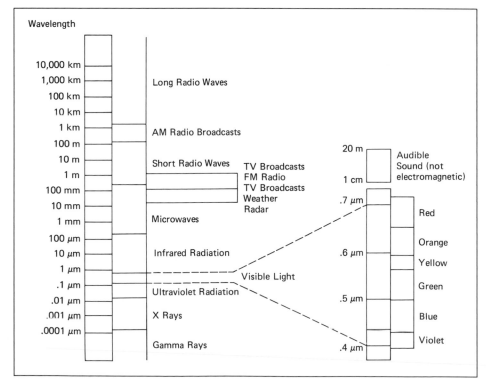

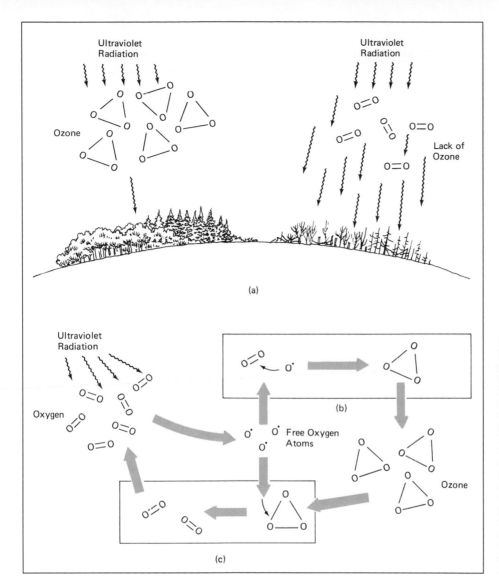

**FIGURE 14-26**
Ultraviolet radiation and the ozone shield. (a) Ozone ($O_3$) in the stratosphere is instrumental in absorbing and screening out ultraviolet radiation. Without this effect ultraviolet radiation would destroy essentially all terrestrial life on earth. (b) Ultraviolet radiation itself produces the ozone by splitting oxygen molecules ($O_2$) to free oxygen atoms which combine with additional oxygen molecules to form ozone. (c) A balance is reached, however, because free oxygen atoms will also react with ozone molecules, breaking them down to re-form oxygen molecules.

light (Fig. 14-25). When UV radiation penetrates living tissues, it is preferentially absorbed by proteins or nucleic acids such as DNA. Its high energy actually causes the breakage of chemical bonds in these molecules. Consequently, UV radiation is extremely destructive to biological tissues and it is capable of causing mutations.

Some UV radiation does penetrate to the surface of earth and this is responsible for sunburns and is involved in some 200,000 to 600,000 cases of skin cancer per year in the United States. However, we are spared the worst effects of UV radiation because about 99 percent is absorbed and hence screened out by the ozone in the stratosphere. Without this ozone "shield," the bi-

ological damage to both plants and animals would be disastrous. Indeed it is doubtful that life could even exist on earth without the protection of ozone in the stratosphere (Fig. 14-26a).

Interestingly, ozone in the stratosphere is a product of UV radiation itself acting on oxygen molecules. The UV radiation causes some oxygen molecules ($O_2$) to split apart into free oxygen atoms, and these in turn may combine with other oxygen molecules to form ozone ($O_3$) as shown in Figure 14-26b. All the oxygen is not converted to ozone, however, because free oxygen atoms may also combine with ozone molecules causing them to break down to oxygen (Fig. 14-26c). Thus, the amount of ozone in the stratosphere is not static;

it represents equilibrium or balance between these two reactions.

This oxygen-ozone balance can be upset by a number of chemicals that catalyze the breakdown of ozone. The most significant ozone catalyst is free chlorine atoms. How they act is shown in Figure 14-27. The presence of a catalyst such as chlorine atoms will not totally destroy all the ozone because the ozone continues to be reformed by the initial reaction. However, it shifts the equilibrium to a new balance point where there is relatively more oxygen and less ozone. Of course, the more catalyst present, the more the equilibrium point is shifted. Furthermore, the shift may be essentially permanent because the chlorine atoms can continue to act indefinitely.

### SOURCES OF FREE CHLORINE ATOMS AND OTHER OZONE CATALYSTS

In the early 1970s scientists pointed out that free chlorine atoms were entering the stratosphere as a result of the wide-scale release of **chlorofluorocarbons** such as freon into the atmosphere. In particular chlorofluorocarbons were being used as the propellants in aerosol cans. They liquefy under modest pressure and are relatively nonreactive. Thus a small amount of liquid freon in an aerosol container can act as an inert ingredient that provides an even pressure over the life of the can. By 1974, the United States alone was spraying chlorofluorocarbons into the air at the rate of about 230 million kg (500 million lb) per year.

Chlorofluorocarbons are highly volatile and relatively insoluble in water. They tend to remain in the gas phase and are not readily flushed out of the atmosphere with precipitation. Thus, given time, some will eventually diffuse into the stratosphere where they may break down under UV radiation and release free chlorine atoms.

When the hazard to the ozone shield was reported, the United States acted quickly and banned the use of chlorofluorocarbons in aerosol cans. Most other countries followed suit. Unfortunately aerosol cans are only one of the places freon is used. A major use of freon is the heat transfer fluid in air conditioners, refrigerators and heat pumps. As these units break down or are ultimately scrapped, the freon is released. Chlorofluorocarbons are also used in industrial solvents

and synthetic foams and they escape into the atmosphere from these sources as well. Therefore, in spite of phasing out chlorofluorocarbons from aerosol cans, worldwide production and release of these compounds are estimated to be still increasing at a net rate of between 5 and 11 percent per year.

Chlorofluorocarbon compounds are not the only threat to the ozone layer. **Carbon tetrachloride** ($CCl_4$) is another substantial and perhaps even more significant source of free chlorine. In addition, **nitric oxide** (NO) can break down ozone in a manner similar to chlorine; high-altitude aircraft, such as the supersonic transport (SSTs), nitrogen fertilizers, and automobile exhaust are all direct or indrect sources of nitric oxide.

### DEGREE AND EFFECTS OF OZONE DEPLETION

All projections of the degree of ozone depletion (shift of the equilibrium) are based on mathematical models and, unfortunately, such models are fraught with uncertainties. In 1979 the National Academy of Science estimated that the continuing release of chlorofluorocarbons at the 1977 rate would lead to an ozone depletion of between 5 and 28 percent (note the range of uncertainty). Taking the average, 16.5 percent, it was estimated that this would cause several hundred thousand additional cases of skin cancer per year in the United States alone. Given a growth in emissions of 7 percent per year to the year 2000, which is likely to occur in the absence of controls, it was estimated that an eventual depletion of 30 percent would occur. This could have catastrophic consequences in UV damage to crops and forests as well as direct effects on humans and animals.

More recent analyses indicate that, given the same assumptions, the degree of ozone depletion will not be as great as initially indicated. Five to 10 percent depletion is now taken as a more accurate estimate. Nevertheless, is it safe to take a "wait-and-see" position before acting?

### COMING TO GRIPS WITH OZONE DEPLETION

The problem with a wait-and-see attitude is that there is a very long lag time between production of chlorofluorocarbons and depletion of the ozone layer. Even if production were halted today, emissions would continue over the next 15

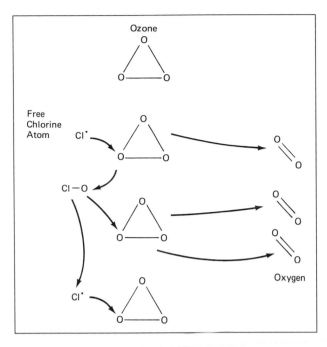

**FIGURE 14-27**
Destruction of the ozone shield. Free chlorine atoms and certain other pollutants entering the stratosphere catalyze the breakdown of ozone as shown. Ozone will not be totally destroyed because it keeps being produced, but the oxygen-ozone balance is shifted to a lesser ozone concentration.

**FIGURE 14-28**
Freon is used for heat transfer in refrigerators, air conditioners, and heat pumps. When these are damaged or eventually scrapped, the freon escapes into the atmosphere.

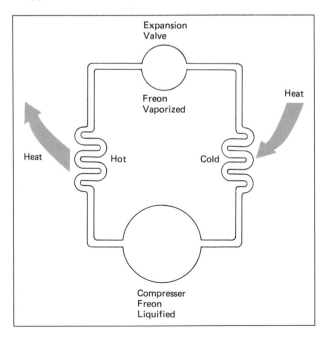

years as failing air conditioners, for example, continue to release freon already in existence. Even without continued emissions it is estimated that chlorofluorocarbons already in the lower atmosphere will continue to diffuse into the stratosphere over the next 10 years. Finally, after chlorine is in the stratosphere, depletion of ozone will occur over the next 50 or so years before a new equilibrium is reached. Then it may be many hundreds of years before various processes eventually remove the chlorine from the stratosphere.

Once chlorine is in the atmosphere we have no known way of removing it. Therefore, if we wait until we detect unacceptable levels of UV radiation, we shall be well past the point of no return. Furthermore, the danger involves the total world, not just one country. Therefore it is only prudent to adopt policies for curtailing the production and use of chlorofluorocarbons in spite of uncertainties.

As noted above, the use of chlorofluorocarbons in aerosol cans was banned by the United States in the 1970s, and most other countries have or are phasing out this use as well. One should note that the banning was a relatively straightforward and acceptable action in this case because alternative propellants, including finger-activated pumps, were readily available. Secondly, chlorofluorocarbon spray cans are hardly a critical aspect of most people's lives. Indeed, it was largely the result of a highly concerned public that enabled their rapid banning in the United States. Interestingly, the French public has clung to the use of chlorofluorocarbon aerososl and, as a result, France has been one of the last holdouts. This is noted to emphasize the importance of public recognition of problems and public willingness to accept some sacrifice. It also points out that solutions to many environmental problems demand international cooperation.

Alternatives also exist for other uses of chlorofluorocarbons. Solvents using these compounds could be more rigorously contained and recycled. Other materials could be used in foams and as heat transfer fluids in refrigerators and air conditioners. However, substitution in these cases is not as straightforward. For example, using an alternative heat transfer fluid in cooling units requires modification of the compressor and other parts as well, and the end product is not as efficient (Fig. 14-28). Lessening of public concern about the ozone shield and a stronger desire for

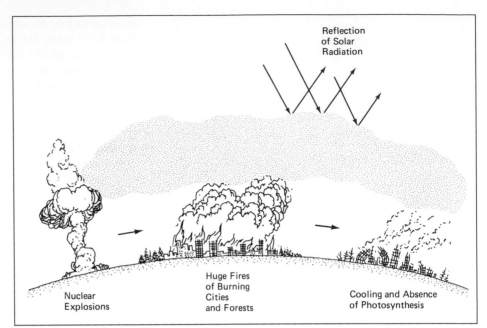

**FIGURE 14-29**
Nuclear winter. In addition to the devastating effects of explosions and radiation, a nuclear conflict would probably usher in a "nuclear winter" as a result of increased smoke and dust particles reflecting solar radiation.

Reflection of Solar Radiation

Nuclear Explosions

Huge Fires of Burning Cities and Forests

Cooling and Absence of Photosynthesis

items such as cooling units have added up to the fact that further restrictions on chlorofluorocarbon production and use have been slow in coming. Whether they come in the near future will depend in large part on the amount of concern expressed.

In conclusion, we again emphasize the failing of humankind's traditional assumption that nature will take care of all wastes. The carbon dioxide effect and the depletion of stratospheric ozone illustrate, above all, that human activities have reached a magnitude such that we are not just polluting our own immediate environment. We are quite capable of upsetting balances that influence the entire globe in ways that are well beyond our ability to control.

## Nuclear Winter

The immediate destruction as well as the radiation sickness that would result from a nuclear conflict is well known. Recently, however, another issue has been added to the nuclear disaster scenario. A precipitous change in climate—a sudden drop in temperature over most of the earth—would follow a nuclear exchange of any appreciable size. This phenomenon, known as **nuclear winter,** would last at least several months.

The sudden drop in temperature would result from the ejection of particulates into the at-mosphere by nuclear explosions and resulting fires. As noted previously, particulates in the atmosphere tend to reflect incoming solar radiation. Without this energy input, temperatures in the atmosphere drop (Fig. 14-29). Some cooling has been observed in the aftermath of severe volcanic eruptions. A single nucelar explosion over a city would have an equivalent effect. Nuclear explosions over many cities would have an effect many times greater. In a larger-scale nuclear conflict, it is estimated that so much particulate matter would be emitted into the atmosphere that the entire hemisphere would darken to a dim, predawn light and summertime temperatures would drop to near or below freezing. Several months would pass before enough material settled and light and termperature conditions began returning to normal. If one survived the immediate conflict, the problems would be much greater than just keeping warm. If the conflict occurred during the spring or summer in the Northern Hemisphere, it would preclude essentially all photosynthesis for that season. This would result in a complete dieoff of nearly all plant life in the Northern Hemisphere. Hence, all animals and humans would starve. If the conflict occurred in the winter, some chance exists that the skies would clear enough by spring to enable some photosynthesis and a greater degree of survival.

However, in any case, light would be reduced and temperatures would remain abnormally low for several years.

Given the effects of nuclear winter, there is little doubt that a nuclear conflict would lead to annihilation of virtually all life on earth. And, the experts point out, an "all-out" conflict is not necessary for such devastation. Estimates show that detonation of only a small fraction of U.S. or Soviet stockpiles of nuclear weapons would be sufficient to bring on a nuclear winter. *Where* the weapons were detonated would not matter. Within days, the entire hemisphere would be covered by the cloud of particulates.

The nuclear arms race between the United States and the Soviet Union is based on the principle that each wishes to be able to assure retaliatory capability after a first strike by the other. The recognition of the nuclear winter phenomenon makes this a meaningless endeavor. Given the nuclear winter effect, a strike by either superpower would be suicidal.

# 15

# RISKS AND ECONOMICS OF POLLUTION

## CONCEPT FRAMEWORK

| Outline | | Study Questions |
|---|---|---|

**Outline**

**I. THE COST-BENEFIT ANALYSIS**      394

**II. PROBLEMS IN PERFORMING COST-BENEFIT ANALYSIS**      396
    **A. Estimating Costs**      396

**Study Questions**

1. What is a *cost-benefit analysis?* How is it used? Give examples of indirect costs and benefits that must be accounted for. Discuss the differences between costs and benefits and pollution control and benefits. Why is 100 percent cleanup not always necessary?

2. What expenses can be objectively estimated and what type of expenses must be projected? Why are pollution control costs higher when initiated than they are later on?

392

3. Give examples of the type of controls used to avoid or reduce risks. Why are these difficult to assess? What factors does risk assessment consider?

4. What types of problems are cost-ineffective in the short term but cost-effective in the long term?

5. Why are those who pay for pollution control and those who benefit from it often two different groups of people?

6. Explain why economic pressures lead businesses to oppose pollution control? What steps have been taken to remedy this situation?

7. Consider factors and examples not mentioned in the text and discuss areas where costs should override benefits and areas where benefits should override costs.

Previous chapters have shown the many serious pollution problems we face. These chapters have also shown that we now have sufficient knowledge and technology to reduce, if not eliminate, many of these problems. The same holds true for environmental problems that will be discussed in subsequent chapters. Why, then, have solutions not been forcefully implemented?

In short, the answer is economics. Scientific information and the availability of control strategies are not enough to solve problems. Numerous competing economic interests often block action. The objective of this chapter is to examine these conflicts and discuss how the economic validity of a proposed project or action may be determined.

## THE COST-BENEFIT ANALYSIS

A cost-benefit analysis is often used as a means of rationally deciding whether to go ahead with a given project. A cost-benefit analysis compares an estimate of the costs of the project to the value of the benefits that will be achieved by the project. A comparison of the costs and benefits is commonly referred to as the **benefit-cost** (or cost benefit) ratio. A favorable benefit-cost ratio (benefit/cost greater than one) means that the benefits outweigh the costs or, the project is **cost-effective** and there is economic justification for proceeding with the project. If costs are projected to outweigh the benefits, the project may be dropped.

In terms of pollution control, the costs include the price of purchasing, installing, operating, and maintaining pollution control equipment and/or implementing a control strategy. Even the banning of an offensive product costs money because jobs are lost, new products must be developed, and machinery may have to be scrapped. In some instances, controlling a pollutant may result in the discovery of a less expensive alternative. However, such money-saving controls are relatively rare. In most cases, any form of pollution control involves additional expenses. Further, costs generally increase exponentially with the percent of control to be achieved (Fig. 15-1a). That is, a small reduction in the level of pollution

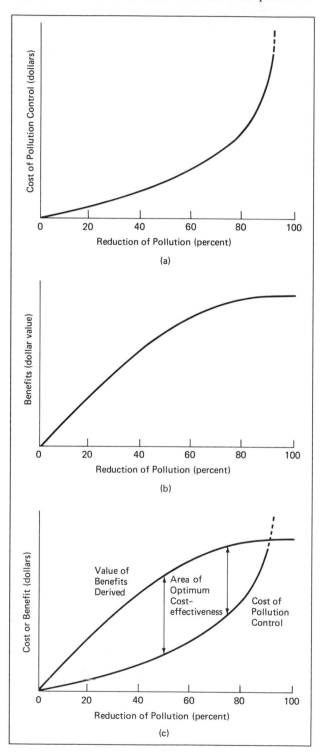

(a)

(b)

(c)

**FIGURE 15-1**
The cost-benefit ratio of pollution reduction. (a) The cost of pollution control increases exponentially with the degree of control to be achieved. (b) However, additional benefits to be derived from pollution control tend to level off and become negligible as pollutants are reduced to near or below threshold levels. (c) When the curves for costs and benefits are compared, we see that the optimum cost-effectiveness is achieved at less than 100 percent control. Expenditures to achieve maximum reduction may yield little if any additional benefit and, hence, may be cost-ineffective.

may be achieved by a few relatively inexpensive measures. However, further reductions generally require increasingly expensive measures and 100 percent control is likely to be impossible at any cost.

Benefits, on the other hand, include such things as improved public health, reduced corrosion and deterioration of materials, and/or reestablishing recreational use of a polluted area. The dollar value of these benefits is derived by estimating, for example, the reduction in health care costs, the reduction in maintenance and/or replacement costs, and the economic value generated by the enhanced recreational activity. Further examples of potential benefits are listed in Table 15-1.

The relationship between the percent reduction of pollution and value of benefits is very different from that for costs. Significant benefits are frequently achieved by modest degrees of cleanup. Yet, as cleanup approaches 100 percent, little if any additional benefits may be realized (Fig. 15-1b). This follows from the fact that organisms can tolerate a certain level of pollution without ill effect. It is only when pollutant levels exceed this "certain level," known as the **threshold level**, that harmful effects are observed and then such effects may increase rapidly. Conversely, reducing a pollutant from its threshold to a still lower level will not yield an observable improvement

When the relationship between costs and benefits are compared graphically, as shown in Figure 15-1c, it is clear that with modest degrees of cleanup, benefits outweigh costs. As cleanup efforts move toward the 100 percent mark, however, the lines cross and costs exceed the value of benefits. Consequently, while it is tempting to argue that we should strive for 100 percent control, demanding upwards of 90 percent control may involve astronomical costs with little or no added benefit. At this point it makes more sense to allocate dollars and effort to other projects where greater benefits may be achieved for the money spent. The optimum cost-effectiveness is achieved at the pollution reduction point where the benefit curve is the greatest distance above the cost curve.

It is important to emphasize that the cost-benefit analysis described here is only a general concept. Costs of control and values of benefits

**Table 15-1**

| **Benefits That May Be Gained by Reduction and Prevention of Pollution** |
| --- |
| 1. Improved human health<br>   a. Reduction and prevention of pollution related illnesses<br>   b. Reduction of worker stress caused by pollution<br>   c. Increased worker productivity<br>2. Improved agriculture and forest production<br>   a. Reduction of pollution-related damage<br>   b. More vigorous growth by removal of pollution stress<br>   c. Higher farm profits benefiting all agriculture-related industries<br>3. Enhanced commercial and/or sport fishing<br>   a. Increased value of fish and shellfish harvests<br>   b. Increased sales of boats, motors, tackle, and bait<br>   c. Enhancement of businesses serving fishermen<br>4. Enhancement of recreational opportunities<br>   a. Direct uses such as swimming and boating<br>   b. Indirect uses such as observing wildlife<br>   c. Enhancement of businesses serving vacationers.<br>5. Extended lifetime of materials and cleaning<br>   a. Reduction of corrosive effects of pollution extending the lifetime of metals, textiles, rubber, paint, and other coatings<br>   b. Cleaning costs reduced<br>6. Enhancement of real estate values<br>   a. Real-estate values depressed in polluted areas<br>   b. Reduction of pollution will enhance them |

achieved will differ with each pollutant in question. Each pollutant or each pollutant category must be analyzed separately and, in each case, there are many uncertainties. Consequently, ongoing questions include: How clean is clean? Is 50 percent clean clean enough? Is 75 percent enough? Do we need 100 percent?

What has been the result of cost-benefit analyses to date? We noted that pollution of air and surface water reached critical levels in many areas and systems in the late 1960s and that huge sums of money were spent on pollution abatement. Cost-benefit analysis shows that, overall, these expenditures have more than paid for themselves in decreased health care costs and enhanced environmental quality. But does this mean that further expenditures on pollution control will prove equally cost-effective? Or, are we at the point where further expenditures will yield little if any benefit and the money will be effectively wasted? At the very least, industry, many economists, and government officials now demand more documentation of presumed benefits before consenting to further expenditures. Some believe that this demand represents real questions about the cost-effectiveness of additional expenditures for environmental protection. Others believe that it represents a more sophisticated method of protecting economic self-interests at the expense of environmental quality and society at large. To understand these views, we must take a more detailed look at the facets and problems of analyzing and comparing the costs and benefits.

## PROBLEMS IN PERFORMING COST-BENEFIT ANALYSIS

The concept behind a cost-benefit analysis is relatively straightforward. Simply estimate costs that may be incurred and estimate the value of the benefits that may be derived from various degrees of cleanup. Then compare the two estimates. The difficulty in implementing this concept lies in obtaining realistic, objective estimates and in making objective comparisons.

### Estimating Costs

In most cases, pollution control technologies and/or strategies are understood and available. Thus, equipment, labor, and maintenance costs can be estimated with a fair degree of objectivity. Unforeseen problems that cause costs to rise may occur, but as technology advances and experience is gained lower-cost alternatives frequently emerge. Note also that pollution control itself provides jobs and hence leads to increased economic prosperity which can be considered a cost reduction or benefit over time. All told, the costs of pollution control are likely to be highest at the time they are initiated and tend to decrease as time passes (Fig. 15-2a). The importance of this will become evident when we consider the time span over which costs and benefits are compared.

### Estimating Benefits and Performing Risk Analysis

The value of many of the benefits to be gained can also be estimated with a fair degree of objectivity. For example, it is well recognized that air pollution episodes cause increases in the num-

ber of people seeking medical attention. Since the medical attention provided has a distinct dollar value, eliminating air pollution episodes provides a health benefit of that value. Similarly, the rate at which materials corrode or deteriorate under polluted and nonpolluted conditions is known and we know how much money is spent to maintain and replace materials. In another example, consider a polluted lake that is upgraded to the point where it will again support water recreation. The benefits of this are estimated by assigning a value of $3 to $5 to each anticipated swimmer-day. This figure is based on the fact that most people will willingly pay this price for admission to a pool.

As we have seen in the preceding chapters, however, prime reasons for controlling present pollutants such as toxic chemicals, ozone, acid rain, and carbon dioxide do not involve benefits to be gained as such. They involve protection against or avoidance of *future risks* of environmental degradation. For example, if toxic wastes are not disposed of properly, we risk widespread groundwater contamination and increased cancer rates. If acid precipitation is not controlled, we risk the widespread dieoff of aquatic life and forests. If fossil fuel consumption is not controlled, we risk climatic change due to the carbon dioxide effect, and so on. In such cases we are really considering reducing certain risks and placing a value on reducing such risks. This is not easy and there are wide differences of opinion regarding the value of reducing any particular risk.

A new field of study has emerged to tackle this subject. It is referred to as **risk analysis** or **risk assessment.** A risk analysis includes consideration of factors such as the following:

1. The number of people that may be affected.
2. The geographic extent or area that may be affected.
3. The nature and/or severity of the effects.
4. The probability of the effects occurring. (Risks may range from "virtual certainty" to "little likelihood of occurring.")
5. The immediacy of the threat.
6. Indirect effects if the threat occurs.
7. The reversibility of the threat.

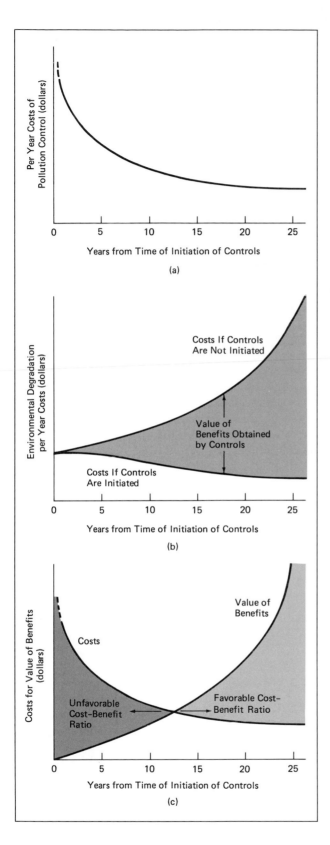

**FIGURE 15-2**
Evaluation of the optimum cost-effectiveness of pollution control expenditures changes with consideration of time. (a) Pollution control strategies generally demand high up-front costs. Costs generally decline as those strategies are absorbed into the overall economy. (b) Benefits may be negligible in the short term, but they increase and continue to accrue as environmental and human health recovers from the impacts of pollution. (c) When these two curves are compared, we see that what may appear as cost-ineffective expenditures in the short term (5–10 years) may, in fact, be very cost-effective expenditures in the long term.

By weighing these factors we can gain a more realistic appraisal of the values that may be attained by embarking on actions that will mitigate the risks. The value to be obtained is the *difference* between the cost of the damage that would probably occur in the absence of controls and the mitigated costs that may occur even with controls.

But here, the time span to be considered again becomes all-important. It is extremely important to recognize that benefits of risk reductions only begin to occur sometime after initial control strategies commence. But they continue to accrue and become greater and greater as time passes (Fig. 15-2b).

## Problems in Comparing Costs and Benefits

Even after valid cost and benefit estimates are obtained, the comparison is often a complicated matter.

### SHORT-TERM VERSUS LONG-TERM VIEW

We have seen that, while costs are high and observable benefits may be few if any during the initial stages of control, as time passes, costs generally moderate, whereas benefits increase and accumulate. Consequently, whether benefits outweigh costs or vice versa depends on whether one takes a short-term or a long-term point of view. A situation that appears to be cost-ineffective in the short term may prove extremely cost-effective in the long term (Fig. 15-2c). This is particularly true in cases involving pollution problems like acid rain or groundwater contamination due to toxic wastes. In these instances, the consequences of delaying control may very seriously affect large geographic areas and many millions of people, and may be irreversible.

## WHO BEARS THE COSTS AND WHO RECEIVES THE BENEFITS?

Those who bear the costs of pollution control and those who receive most of the benefits are frequently different groups of people. For example, industry and its shareholders may bear the costs of curtailing effluents into a river while fishermen gain the benefits. Obviously, the two parties are more than likely to reach different conclusions regarding whether benefits outweigh costs.

This problem is made more complex by the fact that the pollutants produced in one state or country may have their greatest negative impact in another state or country. This is particularly true of acid precipitation. Again, we find little agreement between the parties involved in the conflict over costs and benefits.

## THE NEED FOR REGULATION AND ENFORCEMENT

Many people feel that industry should control its pollutants out of good conscience. However, the economic pressures among competing businesses are often so severe that they preclude meaningful action. For example, consider two competing industries, A and B. Suppose A decides to under-take pollution control measures; it must either pass the costs on to customers in the form of higher prices, whereupon it may lose customers to B who maintains lower prices, or it must accept lower profits, whereupon it begins to lose the financial support of investors. Thus, company A, in its effort to be virtuous, loses competitive advantage to company B regardless of which road it takes. Consequently, it has proven necessary to institute laws and regulations to affect all offending companies equally. In this way, no company gets a competitive edge. Nevertheless, we see many industries, individual companies, and special interest groups attempting to exempt themselves from regulations in order to gain an economic advantage.

## CONCLUSION

It is important to recognize that ultimately all of society will receive the benefits of environmental protection and we will all suffer the costs of environmental degradation. Hence, a broad, long-term perspective of risks and benefits is in order. We can hardly afford to let groups with short-term economic and political interests prevent timely action.

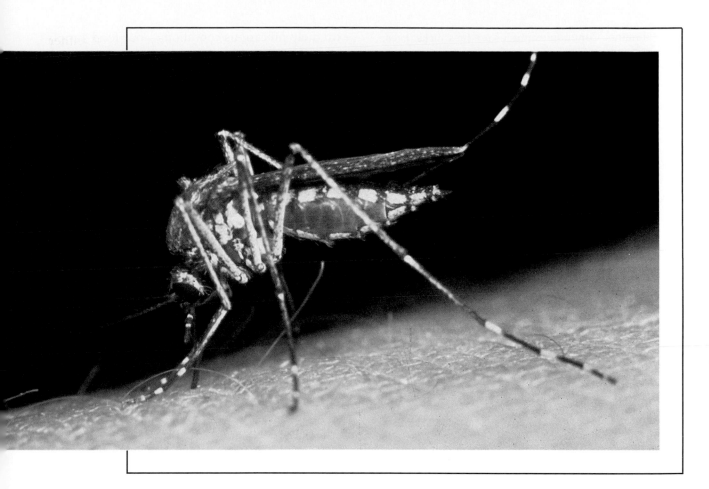

# PESTS
# AND PEST CONTROL

The dictionary defines *pest* as "any organism that is noxious, destructive, or troublesome." The term obviously includes a tremendous variety of organisms that interfere directly or indirectly with humans or their social and economic endeavors. The principal categories of pests are:

1. Organisms that cause disease in humans or domestic plants and animals. These pests include viruses, bacteria, and a wide variety of other parasitic organisms such as intestinal worms and flukes.

2. Organisms that harass people and domestic animals and that may transfer disease by biting or

399

stinging. Common examples are flies, ticks, bees, and mosquitoes.

3. Organisms that feed on ornamental plants or agricultural crops, both before and after they are harvested. The most notorious of these organisms are various insects, but certain worms, snails and slugs, rats, mice, and birds also fit into this category.

4. Animals that attack and kill domestic animals, such as sheep-killing coyotes and chicken-killing foxes and weasels.

5. Organisms that cause wood, leather, and other materials to rot or deteriorate and cause food to spoil. Bacteria and fungi, especially molds, are largely responsible for this, but in warm, moist climates, termites are the primary culprit in the destruction of wood.

6. Plants that compete with agricultural crops, forests, and forage grasses for light and nutrients. Some may poison cattle; others simply detract from the aesthetics of lawns and gardens. A plant in any of these roles is often referred to as a *weed*.

You may ask why humans seem virtually surrounded by pests while in a natural ecosystem the category of "pests" does not seem to exist. The answer lies in one's point of view. When an ecosystem is viewed as a whole, each organism is seen in its particular role as a producer, consumer, or decomposer and all play a role in the transfer of energy and nutrients (Fig. V-1). In considering pests, however, we adopt the viewpoint of a particular species within the system. For example, rabbits compete with woodchucks, insects, and other animals for food. They have their peculiar diseases and parasites and they suffer from biting insects. Other animals compete with them for living space, and they have numerous predators. From the rabbit's point of view, all these organisms would be considered "pests." Ecologically, however, they are considered as factors of natural environmental resistance that serve the important function of keeping the rabbit population in balance with the rest of the ecosystem. This allows the system as a whole to perpetuate itself. Without these "pests," the rabbit population would explode, often to the detriment of the ecosystem.

The prosperity of humans can be largely attributed to pest control. We would still live under extremely precarious conditions—our food supply and physical health at the mercy of such organisms—if it were not for our ability to control pests. Today, there are numerous ways to control pests, but two basic, yet extremely different, philosophies predominate.

One philosophy is based on a purely technological approach. It seeks the development of a "magic bullet," usually in the form of a human-made chemical, which will simply eradicate or exterminate the pest organism. This approach too often gives little if any consideration to the effects of this action on the ecosystem at large.

The second philosophy, which is now called **ecological pest management,** recognizes the importance of sustaining an overall ecological balance. It emphasizes the *protection* of people and domestic plants and animals from pest damage rather than *eradication* of the pest organism. Thus, the benefits of pest control can be obtained while the integrity of the ecosystem is maintained.

Traditionally, humans opted for the purely technological approach. As a result, huge quantities of chemical pesticides are sprayed on agricultural crops, forests, and our own lawns and gardens. Yet, for reasons which will be described in Chapter 16, the success of this approach is reaching its limit. Not only has it failed to eradicate pests, it has actually increased many pest problems. In addition, it has adversely affected wildlife. And, because residues from the chemicals are increasingly tainting water supplies, it poses potentially serious human health hazards. The situation reminds us that anything applied to the surface of the earth may be leached into groundwater or surface water. Some scientists believe that continued widespread use of chemical pesticides is the world's most severe pollution problem.

The futility of the technological approach, along with the pollution hazards it creates, has led more and more people involved in pest control to turn their thoughts and efforts toward ecological pest management. Chapter 17 describes various techniques which this method employs, and addresses some of the economic and social factors that are crucial to shifting from the philosophy of eradication by chemicals to that of ecological pest management.

**FIGURE V-1**
All organisms are natural parts of ecosystems, but when they interfere with
human endeavors we call them pests.

# 16

# THE PESTICIDE TREADMILL

## CONCEPT FRAMEWORK

4. Why are the costs of pesticide use increasing?

5. What led to the ban on DDT in the United States? How do pesticides effect the entire food chain?

6. What is the *pesticide treadmill?*

7. What are the requirements of the Federal Insecticide, Fungicide, and Rodenticide Act.?

8. Why are testing procedures under FIFRA inadequate? What incident brought the problem to the forefront?

9. What do events surrounding kepone and EDB contamination illustrate about the legal control of pesticides?

10. What are the dangers of exporting pesticides?

11. Why does the public have little say over pesticide regulation? How could the situation be changed?

12. What does the term *persistence* mean when associated with halogenated hydrocarbons? Why are nonpersistent chemicals dangerous?

Thousands of chemicals have been developed to eradicate pests. These chemicals are called **pesticides** (from *pest* and *cide*, "to kill"). Pesticides are categorized according to the group of organisms they aim to kill. Thus, there are insecticides (insect killers), rodenticides (mice and rat killers), fungicides (fungi killers), and so on. None of these chemicals, however, is entirely specific for the organisms it is designed to control; they all pose hazards to other organisms and to humans. Therefore, they are also referred to as **biocides,** emphasizing that they may endanger many forms of life.

Our objective in this chapter is to become familiar with the problems inherent in the use of chemical pesticides. We shall focus on **insecticides** and the battle against insect pests, but the principles discussed can be applied to other categories of pesticides and organisms.

## PROMISES AND PROBLEMS OF CHEMICAL PESTICIDES

### Development of Chemical Pesticides and Their Apparent Virtues

That certain chemicals would repel or kill pests was recognized at least 1000 years before the time of Christ. The Greek epic poet, Homer, wrote of "pest-averting sulfur with its properties of divine and purifying fumigation." Sulfur is still used to some extent, but it is hardly deserving of the term *divine*. By modern standards, its effectiveness is very limited.

For thousands of years, people suffered the annoyance and losses from pests as they occurred. For example, the Bible speaks of invasions of locusts that destroyed crops and caused famines. Such devastating losses continued to occur intermittently well into the twentieth century and still occur in some places.

As agricultural practices developed, efforts were made to find effective materials to combat all kinds of pests. The early substances used included toxic compounds like lead, arsenic, and mercury. These inorganic chemicals are frequently referred to as **first-generation pesticides.**

We now recognize that many of these toxic elements may accumulate in soils and can inhibit plant growth. In some places, soils were so poisoned with these elements that now, 40 years later, they are still unproductive. One can speculate as to the number of cases of animal and human poisoning that doubtlessly occurred. Most disconcerting from the grower's point of view, however, was the fact that these chemicals lost their effectiveness—pests became increasingly resistant, or tolerant of them. For example, in the early 1900s citrus growers were able to kill 90 percent of injurious scale insects by placing a tent over the tree and piping in deadly cyanide gas for a short time. By 1930, however, this same technique killed as few as 3 percent of the pests. With agriculture expanding to meet the needs of a rapidly increasing population and first-generation (inorganic) pesticides failing, the farmers of the 1930s were begging for new pesticides.

These new pesticides, or **second-generation pesticides** as they came to be called, were found in synthetic organic chemicals, those human-made chemicals having carbon as a principal part of their structure.

The science of organic chemistry actually began in the early 1800s. Over the next century, chemists synthesized thousands of organic compounds but, for the most part, these compounds sat on shelves with no known use.

It was not until the 1930s that a Swiss chemist, Paul Müller, began systematically testing some of these chemicals for their effect on insects. In 1938, he found that the chemical **DDT** (dichlorodiphenyltrichloroethane) which had actually been synthesized some 50 years before, was extremely toxic to insects, but relatively nontoxic to humans and other mammals.

You are probably familiar with the fact that eventually DDT was found to be extremely dangerous to the environment and in the early 1970s was banned from most uses in the United States. In the early 1940s, however, DDT appeared to be nothing less than the long-sought "magic bullet." First of all, it was very inexpensive. At the height of its use in the early 1960s, it cost no more than about 20¢ a pound to produce. Second, it was a **broad-spectrum pesticide;** that is, it was effective against a multitude of insect pests. This made control simple because the "magic bullet" cured many problems. Third, it was persistent; that is, it did not break down readily in the environment. Consequently, it provided lasting protection. This attribute provided additional economy by eliminating both the material and labor costs of repeated treatments.

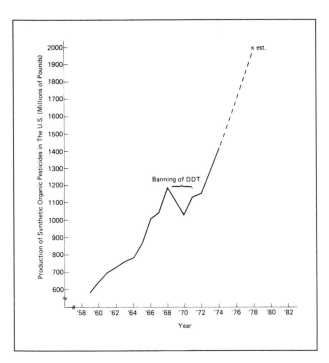

**FIGURE 16-1**
U.S. production of synthetic organic insecticides, herbicides, and fungicides. Use of pesticides is increasing, a trend that has ominous environmental implications. (Data from *Pesticide Review*, USDA.)

throughout the tropical world and greatly reduced the number of deaths caused by malaria. There is little question that DDT saved millions of lives.

In fact, the virtues of DDT seemed so tremendous that Müller was awarded the Nobel Prize in 1948 for his discovery. It is hardly surprising that while DDT fell into disrepute and was banned from use in most industrialized countries in the early 1970s, its initial success ushered in an unending parade of synthetic organic pesticides that continue to be used today in increasing amounts (Fig. 16-1). Yet, while such pesticides continue to provide economic benefits to growers, the problems associated with them are multiplying.

## Problems Stemming from Chemical Pesticide Use

Problems associated with synthetic chemical pesticides can be placed in the three categories discussed below.

### DEVELOPMENT OF PEST RESISTANCE

The most fundamental problem in the use of chemical pesticides is that they gradually lose their effectiveness; as pests become resistant to the substances, larger and larger quantities and/or new and more potent pesticides are required to obtain the same degree of control. This phenomenon was observed with first-generation inorganic pesticides and again with DDT. All synthetic organic pesticides currently in use are suffering the same fate.

Numbers illustrate the situation. In 1946, 1 pound (0.45 kg) of pesticides provided enough protection to produce about 30,000 bushels of corn. In 1971, some 140 pounds (64 kg) were used for the same production, and still, losses due to pests actually increased during this period. Clearly, it is becoming increasingly evident that over the long run, the chemical control strategy has many problems.

The strategy fails because it assumes that pests are a static entity that can be eradicated. In fact, pest populations represent a very dynamic gene pool that is capable of relatively rapid evolution. Pesticide treatments provide selective pressure, which leads to resistance.

Fourth, and most importantly, DDT was extremely effective, at least in the short run. Crop yields dramatically increased in many cases as a result of reduced pest damage. Further, growers could now ignore other more painstaking methods of pest control such as crop rotation and destruction of old crop residues. They could also grow less resistant, but more productive varieties. They could grow certain crops in new locations at different times of year and under climatic conditions that had formerly been prohibitive because of the pest damage that would be incurred. In short, DDT gave growers more options for growing the most economically productive crop.

DDT also successfully controlled important insect disease vectors. During World War II, for example, the military used DDT to control body lice, which spread typhus among the men living in dirty battlefield conditions. As a result, World War II was the first war in which fewer men died of typhus than of battle wounds. The World Health Organization of the United Nations used DDT to control malaria-carrying mosquitoes

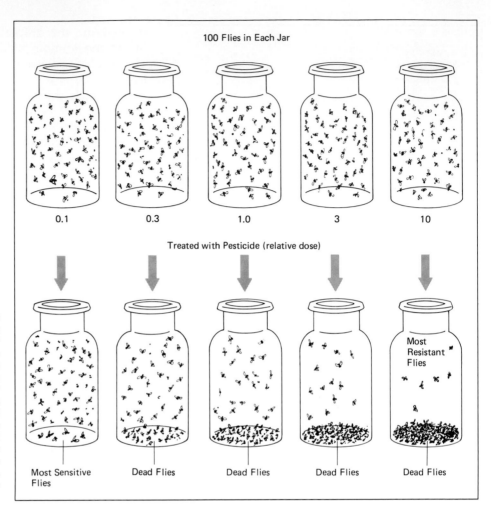

100 Flies in Each Jar

0.1     0.3     1.0     3     10

Treated with Pesticide (relative dose)

Most Resistant Flies

**FIGURE 16-2**
Variation in resistance to pesticides. The kind of experiment shown here reveals that genetic variation in relative resistance to pesticides exists in all populations. Therefore, any pesticide treatment that kills fewer than all the individuals will invariably select those that are most resistant for survival and reproduction (i.e., survival of the fittest will occur). Thus, genes that enhance pesticide resistance are passed on to future generations.

Most Sensitive Flies     Dead Flies     Dead Flies     Dead Flies     Dead Flies

This can be experimentally demonstrated in the following way. If one places 100 flies in each of several jars and treats them with different doses of pesticide, one can observe that the flies show a high degree of variation in their sensitivity to the pesticide. As shown in Figure 16-2, low doses of pesticide will kill a few flies—those that are most sensitive to the pesticide. Higher doses kill higher percentages and eventually all flies may be killed. But it is important to note that a few flies will survive despite a significant dose of pesticide. These are the flies most resistant to the chemical. If these few resistant flies are allowed to breed and the experiment is repeated using the new population, a higher overall degree of resistance is found because the new population carries the genetic traits of resistance from their parents.

Such experiments may be used to interpret field observations. Pesticide treatments in the field destroy the sensitive individuals of the pest population while the more resistant members continue breeding. Incidentally, insects have a phenomenal reproductive capacity. A single pair of houseflies, for example, can produce several hundred offspring which may mature and reproduce again in just two weeks. Consequently, repeated pesticide applications result in the unwitting selection and breeding of genetic lines that are highly, if not totally, resistant to the chemicals that were designed to eliminate them. Cases have been recorded in which resistance has been increased by as much as 25,000-fold.

In 1980, it was documented that more than 400 insect pest species, a substantial portion of the total, showed marked resistance to one or more pesticides, and some 25 major pest species are resistant to all four of the principal classes of pesticides. It is disconcerting to note that when a pest becomes resistant to one pesticide, it may gain cross resistance to other unrelated pesticides though it has not been exposed to the other chemicals.

### RESURGENCES AND SECONDARY PEST OUTBREAKS

Strangely, growers have also observed that, after a pest outbreak has been virtually eliminated with chemicals, the pests not only return; they do so at higher and more severe levels. This is known as a **resurgence.**

To make matters worse, populations of pests, which previously were of no concern because of their low numbers, may suddenly start multiplying after chemical treatment and create new and often times more severe problems. This phenomenon is called a **secondary pest outbreak.** For example, mites, which are related to spiders, have become a serious pest problem only since the use of synthetic organic chemicals. In another example, the number of serious pests on cotton has increased from 6 to 16 with the use of pesticides.

At first, pesticide proponents denied that resurgences and secondary pest outbreaks had anything to do with the use of pesticides. However, careful investigation has shown otherwise. To prove the point, entomologist Paul DeBach has gone so far as to experimentally demonstrate pesticide "recipes" for fostering pests. One example:

California Red Scale on Citrus

> Take four pounds of 50 percent wettable DDT and mix into 100 gallons of water. Using a three-gallon garden sprayer, spray one or two quarts of this mixture lightly over the tree, repeating at monthly intervals until the tree is defoliated or dying as a result of the increase in red scale [which is resistant to DDT]. This may require from about six to twelve applications, depending upon the degree of initial infestation; the higher the numbers initially, the faster the explosion [of red scale population] occurs. Most growers do not like this sort of test and have usually insisted that we stop before trees were killed but it is a wonderful way to raise red scale.*

By contrast, scale populations on untreated trees remain very low (Fig. 16-3).

The reason for resurgences and secondary pest outbreaks is apparent when we recognize that the insect world involves many complex food chains. As in the world of higher animals, populations of plant-eating insects are largely held in check by natural enemies. Frequently, the most effective natural enemies are also insects—species that are parasitic or predatory on other insects

---

*Paul DeBach, *Biological Control by Natural Enemies* (London: Cambridge University Press, 1974), p. 2.

**FIGURE 16-3**
(a) Red scale on lemon. (USDA photo.)

(a)

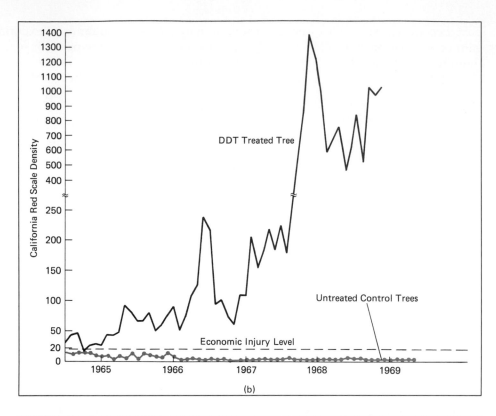

(b)

FIGURE 16-3 (cont.)
(b) Increases in red scale infestation caused by light monthly applications of DDT spray compared to scale populations on untreated trees under biological control in the same grove. Paul Debach, *Biological Control by Natural Enemies*, p 4. London: Cambridge University Press, 1974. (c) Scale insect. (Robert E. Pelham/Photo Researchers.)

(c)

(Fig. 16-4). Pesticide treatments often have a greater impact on the natural enemies than on the herbivorous pest insect. Consequently, populations of the original pest, as well as other plant-eating insects, not only recover; they often attain much higher levels because they are now free from their natural enemies (Fig. 16-5).

There are several reasons for the greater effect of a pesticide on natural enemies than upon

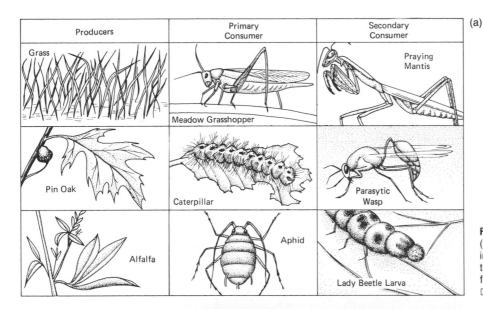

| Producers | Primary Consumer | Secondary Consumer |
|---|---|---|
| Grass | Meadow Grasshopper | Praying Mantis |
| Pin Oak | Caterpillar | Parasytic Wasp |
| Alfalfa | Aphid | Lady Beetle Larva |

**FIGURE 16-4**
(a) Food chains exist among insects just as in other parts of the ecosystem. (b) Aphid lion feeding on aphids. (Courtesy of David Pimentel, Cornell University.)

(a)

(b)

the target pest. First, the herbivore species may be intrinsically more resistant to the pesticide than is its predator. Second, the predator may receive a higher dose because of biomagnification through the food chain. Third, and most important, predatory organisms may be starved out by the temporary lack of prey as well as by the pesticide poisoning. Another factor in resurgences is that the pesticide treatment may subtly alter the chemistry of the plant so that it becomes more susceptible to pest attack.

To illustrate the seriousness of resurgences and secondary pest outbreaks, a recent study in California surveyed a sequence of 25 major pest outbreaks that caused at least $1 million worth of damage. Of the 25, 24 involved resurgences or secondary pest outbreaks. Of course, the species involved in resurgences or secondary outbreaks

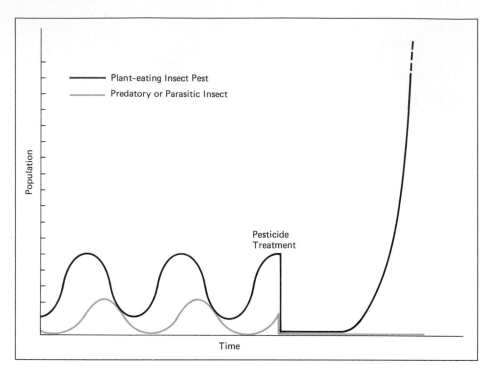

**FIGURE 16-5**
Populations of plant-eating pest insects may be held in check by natural enemies as in other predator-prey relationships. A chemical pesticide may affect the predator more than the pest. Freed of its natural enemy, the pest population increases rapidly.

quickly became resistant to pesticides, thus compounding the problem.

Again, we see that the chemical approach fails because it makes the wrong assumption. It assumes that the ecosystem is a static entity in which one species, the pest, can simply be eliminated. In reality, the ecosystem is a dynamic system of interactions and a chemical assault will inevitably upset the system and produce other undesirable effects.

### INCREASING COSTS

We noted initially that DDT enabled growers to increase their yields and profits dramatically by providing low-cost pest control. It is estimated that $1 worth of pesticide still returns $3 to $4 in increased yields. However, increasing resistance, resurgences, and secondary pest outbreaks are being countered with the more complex and more expensive pesticides that are applied more frequently and in larger quantities. These factors are increasing the cost of pest control so much that the economic advantage they offer is eroding. Indeed, the growing of cotton in some regions is being abandoned because costs of pest control outweigh the value of the cotton.

### ADVERSE ENVIRONMENTAL AND HUMAN HEALTH EFFECTS

Of greatest concern to society, however, is the potential for adverse effects to human and environmental health from pesticides. The now classic story of DDT, which was used increasingly during the 1940s and 1950s, illustrates the hazards.

In the 1950s and 1960s, ornithologists (people who study birds) observed drastic declines in populations of many species of birds that occupied positions at the top of the food chains. Fish-eating birds such as the bald eagle (our national emblem) and the osprey were particularly affected. Investigations showed that the problem was reproductive failure; eggs were breaking in the nest before hatching. In turn, it was shown that the fragile eggs had high concentrations of DDT, and finally that DDT interferes with calcium metabolism, causing birds to lay thin-shelled eggs. It was also shown that birds were acquiring high levels of DDT by bioaccumulation and concentration through the food chain (Fig. 16-6). Indeed, it was these investigations that gave rise to much of our understanding about the propensity of halogenated hydrocarbons such as DDT to

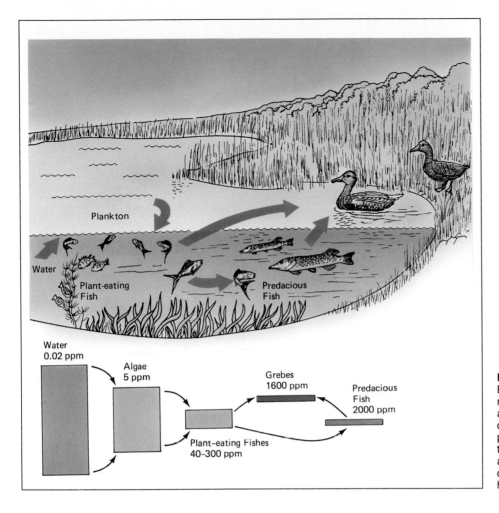

**FIGURE 16-6**
Biomagnification of a DDT-related pesticide. Figures are actual amounts added to or concentrations found in various parts of the ecosystem. Note that the pesticide concentration in animals at the top of the food chain is up to 100,000-fold higher than that of the water.

bioaccumulate. Fish-eating birds were most affected because large amounts of DDT leached into waterways where long aquatic food chains provide more steps for concentration.

In addition, tissue assays showed that DDT was accumulating in the body fat of humans and virtually all other animals, including Arctic seals and Antarctic penguins even though these animals were far removed from the point of application (Fig. 16-7). While harmful effects on humans have not been substantiated, experimental testing has shown that related compounds are carcinogenic, mutagenic, and teratogenic (causing birth defects).

These findings led to the banning of DDT in the United States and most other industrialized countries in the early 1970s. Subsequently, several other related halogenated hydrocarbon pesticides

were banned when their propensity to bioaccumulate and cause adverse human health effects was documented.

In the years since the banning of DDT, observers have noted a marked recovery in the populations of birds that were adversely affected. However, this does not mean that the situation is under control.

Because of increasing resistance, resurgences, and secondary pest outbreaks, the number and type of pesticides in use continue to grow. About 1.7 billion pounds (772 million kg) or nearly 5 pounds (2.72 kg) per person were used in the United States in 1982. The shocking fact is that less than 1 percent of this huge amount ever comes in contact with or is ingested by pest organisms. Because many pesticides are applied by aerial spraying, only about one half of

**FIGURE 16-7**
Bioaccumulation and magnification of DDT in virtually all organisms on earth. Tests revealed that DDT was accumulating in all organisms tested, not only those in or close to areas where DDT was used but also those far removed from the area of use, including Antarctic penguins. It was thus discovered that DDT vaporizes and, given its persistence, it was being carried to and contaminating all food chains on earth. All humans tested were also accumulating DDT in body tissues.

the chemicals even reaches the area treated. The remainder drifts in the air and settles on surrounding ecosystems and water bodies. Then, of the portion reaching the crop, less than 0.1 percent is ingested by the target pest.

The continuing potential for adverse environmental and human health effects is obvious. Both terrestrial and aquatic ecosystems may be upset if pesticides poison nontarget organisms, and the effect of this can be felt throughout the food chain. Further, aquatic ecosystems may receive significant inputs of pesticides long after the application because of gradual leaching of residues from land into waterways.

Human health hazards also remain for agricultural workers, who may be directly exposed to pesticides during application, and to the public, who may be exposed to residues that may linger on treated crops, leach into water supplies from treated soils, or bioaccumulate in the food chain.

Of course, food stuffs are withdrawn from the market and wells are shut off if unsafe levels of contamination are found. Further use of the pesticide may also be prohibited. You may recall that in 1983 and 1984 numerous wells were closed and certain grain products were removed from store shelves because of EDB (ethylene dibromide) contamination, and use of EDB was banned. But do these measures provide adequate protection for human and environmental health? The agrochemical industry says yes. But for a variety of reasons, many experts think not. We shall return to a fuller discussion of this controversy later in the chapter.

## The Pesticide Treadmill

The late entomologist Robert van den Bosch coined the term **pesticide treadmill** to describe attempts to eradicate pests with synthetic organic chemicals. We can readily see how apt this term is. The chemicals do not eradicate the pests. They increase resistance and secondary pest outbreaks, which lead to the use of new chemicals, which in turn lead to more resistance and secondary outbreaks. The process is never-ending. More serious than a treadmill that goes nowhere is the vicious cycle of the pesticide treadmill. Each round involves increasing amounts and kinds of toxic chemicals. This increases the risks to human and environmental health (Fig. 16-8).

## ATTEMPTING TO PROTECT HUMAN AND ENVIRONMENTAL HEALTH

### The Federal Insecticide, Fungicide and Rodenticide Act (FIFRA)

The key legislation to control pesticides in the United States is the Federal Insecticide, Fungicide and Rodenticide Act, commonly known as *FIFRA*. This law requires manufacturers to register pesticides with the government before marketing them. The registration procedure involves testing to determine toxicity to animals (and, by extrapolation to humans). From the test results, standards regarding use are set. For example, highly toxic compounds such as chlordane, which is used to control termites, are not authorized for use on food crops. For pesticides that are approved for food crops, standards are set regarding amounts of residues that may remain at harvest. There have been many cases where foods were withdrawn from the market because residues of certain pesticides were above the established standard.

If health hazards appear after a substance has been registered and marketed, the act provides for "deregistration," whereby the pesticide may be banned from one or more uses.

### Shortcomings of FIFRA

Unfortunately FIFRA has many shortcomings, four of which are discussed below.

#### INADEQUATE TESTING

The perils of bioaccumulation and the potential for long-term exposure that may cause cancer, birth defects, mutations, and other physiological disorders were not fully appreciated until the late 1960s when bitter experience with DDT and other chemicals came to the forefront. This experience made officials and the public aware that many pesticides in use have never been adequately tested. Consequently, Congress amended FIFRA in 1972 to require the Environmental Protection Agency (EPA) to reevaluate and reregister all pesticide products then on the market.

The required testing involves feeding doses of the pesticide to a population of 100 to 200 mice over a one- to two-year period. At the conclusion

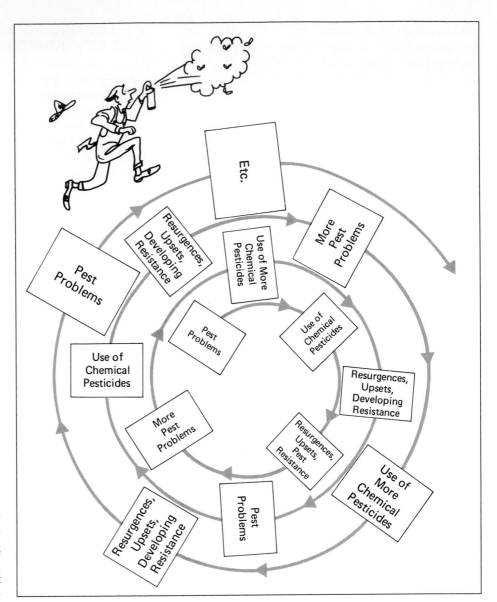

**FIGURE 16-8**
The pesticide treadmill. Use of hard chemical pesticides aggravates many pest problems; this demands the use of increasing amounts of pesticide, which further aggravates pest problems, which . . .

of the testing period, the mice are examined for tumors and other physical disorders, and the findings are compared to a control group.

In 1972, when the amendment was adopted, there were already some 1400 chemicals and 40,000 formulations (mixtures of different chemicals) on store shelves. In addition, there was and still is tremendous pressure from the chemical industry to register numerous new pesticides. Needless to say, the EPA's office of pesticide programs has been overwhelmed. Ongoing budget restrictions have not helped the situation.

Since the law allows existing pesticides to remain in use unless proven hazardous, many products on the market still have not been subjected to adequate tests. To make matters worse, it was discovered that registration of several new chemicals in the 1970s was based on fraudulent tests. Industrial Biotest Laboratory (IBL), the nation's largest supposedly independent laboratory, was contracted by the EPA to study some of the pesticides in question. However, the laboratory had strong ties and support from the chemical industry. Evidence revealed that the firm suppressed information showing the adverse effects of certain chemicals, terminated studies before they were completed, substituted healthy mice for experimental mice that died, and forged data when none existed. In each case, the deception worked toward the manufacturer's benefit. Chemicals were given a "clean bill of health" without proper justification. Many of them remain on the market today.

In 1983, despite the serious implications of this scandal and the negative publicity it received, the EPA closed the only government facility capable of conducting its own tests. Senator Paul S. Sarbanes of Maryland introduced a bill requiring the agency to reopen and maintain the facility, but by 1984, no action had been taken.

## BANS ISSUED ONLY ON A CASE-BY-CASE BASIS WHEN THREATS ARE PROVEN

DDT is just one of many chemically related, persistent, halogenated hydrocarbon pesticides. When the problems involving DDT were discovered, logic and prudence dictated banning the whole group because their chemical similarities made it very likely that they would all bioaccumulate and pose long-term human and wildlife health hazards.

Yet, FIFRA precludes such guilt by association. Each individual compound must be shown to pose a threat before it is subject to a ban. Even then the government must demonstrate that risks to humans are greater than the agricultural benefits derived from using the pesticide. This frequently involves litigation that takes several years. Thus, in the years following the DDT ban, many related pesticides such as chlordane, heptachlor, aldrin, dieldrin, DBCP (dibromachloropane), kepone, mirex, and most recently, EDB (ethylenedibromide) were also banned, but only after serious incidents.

For example, kepone was banned only after a number of workers in a manufacturing plant on the James River in Virginia became seriously ill with a variety of physical and nervous disorders, and after fish swimming in water that received washings from the plant were found to contain so much kepone from bioaccumulation that they were deemed unsafe to eat. This forced a ban on commercial fishing on the James River. As of 1985 the river remained closed because residual kepone persisted in the sediments and continued to bioaccumulate (Fig. 16-9).

The recent story of EDB is perhaps more upsetting. It has been known since the mid-1970s that EDB causes cancer, birth defects, and other illnesses in laboratory animals. Yet its widespread use as a soil fumigant to control root nematodes (small worms that attack roots) continued. By 1982, 20 million pounds of EDB were pumped into the soil each year. It could be predicted that this practice would eventually result in groundwater contamination. In addition, EDB was increasingly used to fumigate and protect grains and other crops in storage. But EDB was not banned until 1984 after traces were found in hundreds of wells in Florida, California, Massachusetts, and other regions, and unacceptable

**FIGURE 16-9**
A fisherman removes his nets from the James River for the last time, out of business because of kepone contamination. (United Press International photo.)

amounts of residues of EDP were found in flour and other food products. Contaminated food products were taken off the market and well-boring was stopped. But how long will the water remain unsafe? Unfortunately, since groundwater is essentially a sealed system with little flushing action, contamination may linger indefinitely. To illustrate the situation, a study conducted in California in the early 1980s showed 50 different pesticides present in the groundwater of 23 separate counties. Some of these chemicals had not been used for more than a decade.

Clearly, waiting until after the consequences to ban such chemicals is hardly a prudent way to protect public health. Yet, this is precisely how the system operates. A few persistent halogenated hydrocarbon pesticides remain in use and the use of some, notably lindane, is increasing. Perhaps most serious is the widespread and increasing use of **herbicides.** They now account for more than half of all pesticides used. Large numbers of people are being exposed to these compounds with unknown effects.

You are probably familiar with the controversy surrounding Agent Orange, the herbicide used in Vietnam to clear jungle vegetation. Veterans who were exposed to Agent Orange have filed hundreds of lawsuits against the federal government, claiming various adverse health effects. The government, on the other hand, holds that many of these claims cannot be substantiated.

### PESTICIDE EXPORTS

Another ominous loophole in FIFRA is the lack of a requirement for registration of pesticides intended for export. The United States currently exports more than 100 million pounds of these chemicals to less-developed countries each year. Some 25 percent of this total consists of products banned in the United States. Chemical companies are free to promote their products in other countries; thus, less-developed countries hear only how pesticides will solve their pest problems. They are not informed about resurgences or secondary pest outbreaks, much less about hazards to human and environmental health. As a result, these chemicals are often overly and carelessly used, leading the World Health Organization to estimate that some 500,000 pesticide poisonings occur each year.

Ironically, a large portion of these pesticides is used by foreign countries on export crops, many of which are imported by the United States.

### LACK OF PUBLIC INPUT

Unlike laws such as the Clean Water Act and the Clean Air Act, FIFRA provides no mechanism for public input. Therefore, the legislation primarily reflects the chemical industry's interests, which are conveyed through intense lobbying efforts as opposed to public forums. The bias is obvious when one considers that the law imposes a $1000 penalty on those who misuse a pesticide and a $10,000 penalty on those who reveal a trade secret about a pesticide's formulation.

## Nonpersistent Pesticides: Are They the Answer?

A key factor underlying the tendency of halogenated hydrocarbons to contaminate the environment and bioaccumulate is their persistence, that is, their slowness to break down. DDT, for example, has a half-life in the order of 20 years—half of the amount applied is still present and active 20 years later; after 20 more years, half of that amount, or one quarter of the amount applied, remains, and so on. Recognizing this fact, the agrochemical industry has, in large measure, substituted nonpersistent pesticides for the banned compounds.

These nonpersistent pesticides are also synthetic organic compounds, but they are compounds that break down in the environment into simple nontoxic products within a few days or weeks after application. Thus, the danger of their migrating long distances through the environment and affecting wildlife or humans long after their application is removed. These nonpersistent pesticides have been touted as "environmentally safe," but are they?

For several reasons, nonpersistent pesticides are not as environmentally sound as their proponents claim.

First of all, total environmental impact is a function of persistence along with three important factors: toxicity, dosage applied, and where applied. Many of the less persistent pesticides are actually more toxic than DDT. This, combined with the frequent applications needed to maintain control, presents a significant hazard to agricultural workers. There have been numerous cases of poisonings due to these compounds.

Second, nonpersistent pesticides may still have far-reaching environmental impacts. For example, to control outbreaks of the spruce budworm in New Brunswick, Canada, forests were sprayed with a nonpersistent organophosphate pesticide that was promoted as environmentally safe (Fig. 16-10). After spraying, however, an estimated 12 million birds died. These birds may have died by direct poisoning or by the loss of their food supply since a bird eats nearly its own weight in insects each day. In either case, visitors commented on the eerie silence and the numerous dead warblers littering the ground after spraying.

Third, nonpersistent pesticides may upset the ecosystems in treated areas. Insects are an integral part of many aquatic food chains. Blooms

(a)

(b)

**FIGURE 16-10**
(a) Damage caused by Spruce budworm. (William J. Jahoda/Photo Researchers.) (b) Spraying for spruce budworm. (U.S. Forest Service.)

of phytoplankton have occurred as insects that normally feed on it were killed. Likewise, soil ecosystems may be upset, affecting decomposition and release of nutrients. Populations of soil-dwelling pests may also increase due to the loss of natural enemies.

Fourth, desirable insects may be just as sensitive as pest insects to these substances. Bees, for example, which play an essential role in pollination, are highly sensitive to nonpersistent pesticides. Thus, use of these compounds creates an economic problem for beekeepers as well as jeopardizing pollination.

Finally, nonpersistent chemicals are just as likely to cause resurgences and secondary pest outbreaks as persistent pesticides and pests become resistant to nonpersistent chemicals just as quickly.

In conclusion, the mechanisms designed to control pesticides and protect human and environmental health leave much to be desired. The National Resources Defense Council, a citizens' organization, has spearheaded lobbying efforts urging Congress to overhaul FIFRA and supply funding for adequate testing. This action is certainly needed and if implemented would help the situation. However, it does not address the underlying pesticide problem. The chemical approach will remain a treadmill generating a growing number of hazards because it does not recognize the dynamics of species or the ecosystem. Our thinking and our approach must shift toward the concept of ecological pest management, techniques of which will be addressed in the following chapter.

# 17

# NATURAL PEST CONTROL METHODS AND INTEGRATED PEST MANAGEMENT

## CONCEPT FRAMEWORK

| Outline | | Study Questions |
|---|---|---|

1. What is *natural control?* Based on the life cycle of a moth as shown in Figure 17-1, speculate on ways in which the population could be controlled.

2. What does *biological control* involve? What is the first element of biological control? What problems plague this approach? Give examples of how this method has been used successfully. What is *Bacillus thuringiensis?* What principle underlies biological methods used to control weeds? Summarize the advantages of biological pest control.

3. What is the essence of *genetic control?*

4. How are resistant varieties bred? Cite examples where this approach has been used. What are its benefits? Why is it not exploited more?

5. What are *chemical barriers?* Give some examples.

6. What are *physical barriers?* What problem hampers them and how may it be overcome?

7. Explain the *sterile male technique.*

8. What are *cultural controls?*

9. How have these controls eliminated certain problems and where are further controls needed?

10. How can paying attention to the ecological preferences of plants help home gardeners? How do farmers use this technique?

11. What simple remedy can help eliminate weed problems in lawns and pastures?

12. How do water and fertilizer management affect pests?

13. What role does time of planting play in pest control?

14. How do crop residues support pest organisms?

15. How have farmers utilized adjacent planting and weed control?

16. What is *crop rotation* and why is it effective?

17. What are the advantages and disadvantages of plowing?

18. Why does polyculture offer better protection against pests than monoculture?

19. Why are quarantines important?

20. What is *natural chemical control?*

21. What hinders the juvenile hormone method?

22. What are *pheromones?* How effective is pest control when sex attractants are manipulated? What are the differences between the use of natural chemicals and the use of synthetic organic chemicals? Summarize the benefits of natural control and the underlying principles upon which it is based.

23. Describe *integrated pest management.*

24. What are the consequences of attempting to

eliminate all "bugs?"

25. What is the *Snow White syndrome?*

26. What are the advantages and disadvantages of insurance spraying?

27. How do chemical companies promote the use of chemicals? What are the results?

28. What is the philosophy behind Migros and how does Migros operate? Do you think this program could be successful here? Why?

29. What role do field scouts play in pest management?

30. What do studies comparing conventional farms with organic farms show? What steps are taken if pests become a problem on organic farms? How has the integrated pest management approach been used to control Medflies?

31. Describe the Medfly and its method of attack. What measures protect against Medfly invasions?

32. Why are synthetic organic pesticides a problem in integrated pest management?

33. Discuss the recommendations for ecological pest management described in the text and consider other ideas.

Numerous ecological and biological factors affect the relationship between a pest and its host (the plant or animal attacked). Ecological pest management seeks to manipulate one or more of these natural factors so that crop protection is achieved without upsetting the rest of the ecosystem or jeopardizing environmental and human health. Since ecological pest management involves working with natural factors instead of synthetic chemicals, the techniques are referred to as *natural control* methods.

Our objective here is to show the major categories of natural control and the opportunities they offer. We shall then focus on some of the economic and sociological factors that need to be overcome in moving from chemical to ecological control methods.

## THE INSECT LIFE CYCLE AND ITS VULNERABLE POINTS

The ecological pest management approach, unlike the chemical biocide approach, depends on an understanding of the pest and its host. The more we know about the organisms involved, the greater our opportunities for natural control.

To illustrate, the life cycle of moths and butterflies is shown in Figure 17-1. Many groups of insects have a similarly complex life cycle. The development of each stage of the life cycle may be influenced by numerous factors such as temperature and humidity. One or another stage may be attacked by viruses, bacteria, or other parasites or predators. Additionally, proper completion of each stage depends on certain internal chemical signals or hormones. The male's ability to locate females for mating and an insect's ability to find suitable food sources depend on external chemical signals between individuals or between pests and hosts. As will be seen in the following pages, all of these findings suggest vulnerable points or

**FIGURE 17-1**
Insect life cycle. Most insects have a complex life cycle that includes a larval and adult stage. Each stage has special food and/or habitat requirements. Affecting any stage of the life cycle can prevent reproduction and affect control.

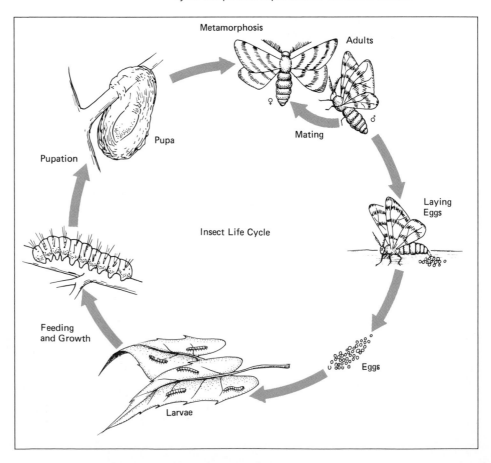

Metamorphosis

Adults

Mating

Pupa

Pupation

Laying Eggs

Insect Life Cycle

Eggs

Feeding and Growth

Larvae

ways in which pest populations may be controlled without resorting to synthetic chemical pesticides.

## METHODS OF NATURAL CONTROL

The five general categories of natural pest control are (1) biological control or use of natural enemies, (2) genetic control, (3) the sterile male technique, (4) cultural control, and (5) use of natural chemicals.

### Biological Control

The term **biological control** is used to denote the control of a pest population by a natural enemy, that is, some kind of parasite or predator. Control by natural enemies is a much more common and widespread phenomenon than is generally recognized. Entomologists (scientists who study insects) estimate that only about 1 percent of some 50,000 known species of plant-eating insects that might be serious pests actually are. The other 99 percent are present in crops and forests, but their populations are not large enough to do significant amounts of damage because they are held in check by one or more natural enemies or other natural factors.

The first element in biological control, then, should be **conservation**—preserving the natural controls that already exist. This means avoiding the use of broad-spectrum chemical pesticides, which may affect natural enemies even more than pests, thus resulting in resurgences and secondary pest outbreaks. Elimination or considerable restriction on the use of broad-spectrum chemical pesticides will, in many cases, allow natural enemies to reestablish themselves.

The major problem with biological control is that effective natural enemies, the ones that keep the pest population at low levels, are not always readily available. In some cases, the lack of natural enemies is the result of accidentally importing the pest without its natural enemy. In other cases, effective natural enemies are not known, but this does not necessarily mean that they do not exist. In either case, natural enemies may be found by searching locations where the pest species or a closely related species is present, but in fairly small numbers. In such locations, entomologists have frequently found one or more natural enemies affecting the pest or related species and controlling its numbers.

When a natural enemy is found, techniques are developed for rearing it in the laboratory to produce suitable numbers for release in the problem area. The introduction of one should not preclude the search for and introduction of even more effective natural enemies. When two or more natural enemies are introduced, a better ecological balance may be established between them and the pest. Often one natural enemy proves more effective in one habitat, while another is more effective in a slightly different habitat. Where two or more control organisms exist together, their effects may be additive. Usually, however, the final size of the pest population is determined by the most effective of its natural enemies.

Hundreds of examples illustrate these principles and demonstrate the potential success of biological control; only a few are cited here.

#### COTTONY-CUSHION SCALE

The cottony-cushion scale insect was first observed in California in 1868 (Fig. 17-2). By 1886 it had spread throughout the state and was devastating citrus trees. Entomologists were sent in search of a natural enemy and two were found in Australia—a small parasitic fly and the predatory ladybird (vedalia) beetle (Fig. 17-3). These were imported and released, and by 1890 the cottony-

**FIGURE 17-2**
Cottony-cushion scale feeding on citrus trees. (Photo by Max E. Badgley.)

**FIGURE 17-3**
Vedalia (ladybird) beetle eating a scale insect. (Photo by Max E. Badgley.)

cushion scale was no longer a significant problem. These two natural enemies maintain control to this day, the parasitic fly being dominant in coastal areas and the ladybird beetle in drier interior regions.

### SUGARCANE LEAFHOPPER

The sugarcane leafhopper was first discovered in the Hawaiian Islands in 1900, but by 1904 it had spread throughout the islands, killing large areas of sugarcane and threatening the survival of the Hawaiian sugar industry, the economic mainstay of the islands at the time. Between 1904 and 1916, several natural enemies that parasitized the eggs of the leafhopper were discovered and introduced: a previously unknown one from Australia, another from Fiji, a third from China, and a fourth from Formosa. All these helped, each adding to the effect of the others, but control was still not complete. In 1919, a parasite was discovered in Australia. Although this species had been known before, entomologists had not observed that it fed on leafhopper eggs. Instead, it was thought to feed on sugarcane itself. Hence, its potential as a biological control had been overlooked. Introduction of this fifth species brought about complete control, which is still maintained.

### WALNUT APHID

The walnut aphid invaded California from Europe around the turn of the century and soon populated all walnut-growing areas of the state. It was controlled originally by increasing doses of pesticides with the problems of secondary pest outbreaks and increasing resistance. In 1959, Robert van den Bosch, of the University of California at Berkeley, found a parasitic wasp in France that proved successful in controlling the aphid, but only in the coastal areas of California. It was not effective in hot, dry areas. However, a genetic variant of the same species from Iran, a hot desert country, was highly effective in controlling the aphid in the hot, dry areas of California. Thus control was and has continued to be complete.

These few examples illustrate that control of the plant-eating insects by predatory or parasitic insects can be highly effective (Fig. 17-4). Once established in its new environment, an effective natural enemy seeks out and destroys the pest, keeping its population at a low level year after year, so long as the situation is not upset by human activity. Rarely, if ever, does an insect become resistant to its natural enemy, presumably because the natural enemy also undergoes constant natural

**FIGURE 17-4**
(a) A parasitic wasp depositing eggs in a gypsy moth pupa. (USDA photo.)

(a)

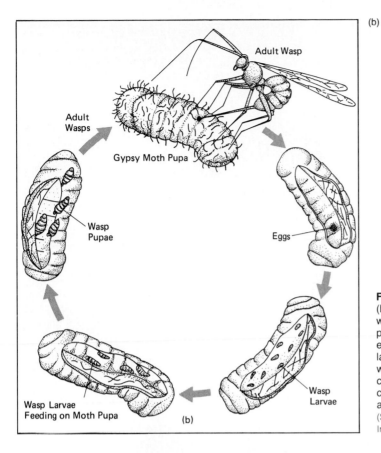

Adult Wasp

Adult Wasps

Gypsy Moth Pupa

Wasp Pupae

Eggs

Wasp Larvae Feeding on Moth Pupa

Wasp Larvae

(b)

**FIGURE 17-4** *(cont.)*
(b) The life cycle of the parasitic wasp. (c) Another insect parasite, a braconid wasp lays its eggs on a tomato hornworm, the larva of the hawk moth. The wasp larvae feed on the caterpillar and shortly before the caterpillar's death they emerge and form cocoons seen here.
(Scott Camazine/Photo Researchers, Inc.)

(c)

selection and hence evolves along with its prey. The total economic benefit of such control is millions of dollars per year, with no undesirable side effects.

The examples emphasize, however, that biological control is not a simple matter. Finding an effective natural enemy, developing techniques for rearing it, and establishing it in problem areas are the result of painstaking efforts by highly trained and dedicated scientists.

Naive or simplistic efforts at biological control are more than likely to be meaningless. For example, many garden and seed catalogues offer the praying mantis as an insect control (Fig. 17-5). The praying mantis is an impressive eater of other insects but, unfortunately, it is totally indiscriminate in its eating habits. It is as likely to eat desired natural enemy insects as pest insects. It is also highly cannibalistic so it quickly reduces its own population to low levels. All in all, it has little if any effect in reducing a specific pest. Likewise, ladybeetles (ladybugs) collected from clusters in the wild are in a resting state analogous to hibernation, and if sold directly they will not feed but will simply fly away to reaggregate. The purchaser must be sure that beetles have been preconditioned so as to be in the feeding state when they arrive.

Biological control may also utilize bacteria or viruses to infect and thus control pest populations. *Bacillus thuringiensis*, for example, is a bacterium that is fatal to a number of species of caterpillars (moth and butterfly larvae). Caterpillars that ingest the bacteria sicken and die within a few days. However, the disease is not highly contagious; the bacteria do not readily spread from one caterpillar to the next. As a result, the disease does not provide effective control under purely natural conditions. The bacteria must be cultured in the laboratory and applied periodically when problems occur. Application of the bacteria is similar to spraying a chemical insecticide. But in this case, it is a nontoxic, biological "insecticide," which is quite specific to a particular pest.

Because of its safety and effectiveness, this bacterium has been used for more than 25 years to control several species of caterpillars in the United States. It is most notably used to control the gypsy moth in urban areas in the Northeast (Fig. 17-6). State agencies responsible for gypsy moth control still prefer to use less expensive

**FIGURE 17-5**
Praying mantis. An impressive and voracious insect predator, but it is indiscriminate in what it eats. (Treat Davidson/Photo Researchers, Inc.)

chemical pesticides in rural areas where the danger of human exposure remains low. This economic argument, however, does not take into account the long-term costs of the ecological damage, resurgences, and outbreaks of secondary pests that may occur.

The milky spore disease of Japanese beetles is an example of a microbial control that establishes itself and maintains lasting control. Japanese beetles feed on many kinds of vegetation, but they are especially fond of the rose family. In addition, their larvae feed on roots in the soil (Fig. 17-7). Larvae, however, are very susceptible to milky spore disease caused by a bacterium that can be cultured in laboratories, packaged, and applied to soil. The feeding larvae ingest the bacte-

(a)

(b)

**FIGURE 17-6**
Gypsy moth and larva. Control is possible by spraying disease-causing bacteria on the leaves which the larvae eat.
(a) Adults laying egg masses on tree trunk. (Nobel Proctor/Photo Researchers, Inc.) (b) Larva feeding. (John M. Barnley/Photo Researchers, Inc.) (c) Larva stricken by virus. (USDA photo.)

(c)

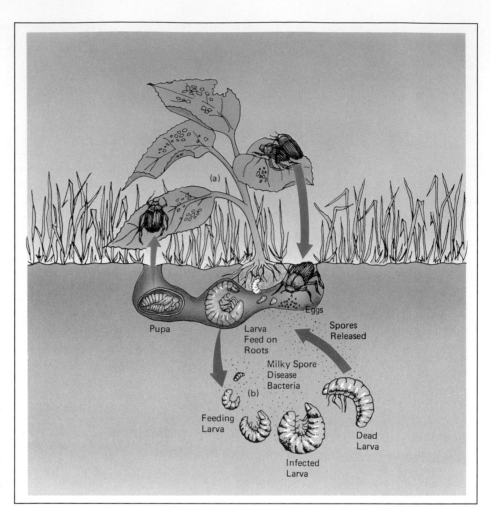

**FIGURE 17-7**
Milky spore disease is an effective control for Japanese beetles. (a) Normal life cycle. (b) Bacteria infect and multiply in the larva. As the larva dies, bacterial spores are released into the soil and the cycle is repeated.

rial spores and the bacteria proliferate in the body of the beetle larvae, leading to their death. As the larvae die, they split open, releasing another generation of spores into the soil to repeat the cycle. The spores can remain viable in the soil for up to 20 years.

Biological methods can also be used to control weeds. But it should be noted that, if the natural enemy thrives on any plant in addition to the weed, its ecological effects could be devastating. Therefore, organisms that will potentially control weeds must be very thoroughly tested to show that they will starve to death before they will eat anything but the target weed. Given this criterion, natural enemies have proven highly effective, particularly in eliminating plants that have been imported and then spread to become weeds. For example, the prickly pear cactus was imported to Australia as an ornamental plant and es-

caped into the wild. By 1925, 30 million acres of what had been food-producing or grazing land were covered so densely that the land was useless and another 30 million acres were heavily infested. Introduction of a cactus-eating insect had the problem overcome seven years later, thus eliminating the need for chemical herbicides (Fig. 17-8).

In addition to insects, other animals are being tested for control of aquatic weeds, particularly the water hyacinth and hydrilla (Fig. 17-9). Parasitic fungi are being extensively tested for possibilities in control of many other weeds.

In conclusion, the advantages of biological control include specificity to the pest, persistence of control, avoidance of secondary pest outbreaks, and few, if any, undesirable ecological side effects. About 70 serious agricultural pests are controlled by biological introductions. But, more than

(a)

(b)

**FIGURE 17-8**
Biological control of the weed prickly pear cactus by a cactus-eating moth larva
in Queensland, Australia. (a) Homestead abandoned because of prickly pear
infestation. (b) Homestead reoccupied following cleanup by the insect.
(Queensland Department of Primary Industries; Australian Information Service.)

**FIGURE 17-9**
Manatees or sea cows might be
used to maintain control of water
hyacinths. Unfortunately
poaching and injury by motor
boats endanger their survival.
(© Douglas Faulkner, Photo
Researchers, Inc.)

2000 serious insect pests are known, and natural enemies have been found for only about 10 percent of them. Why has potential biological control remained largely unexploited? Development of biological controls requires skilled scientists and money to fund their research. Unfortunately, funding so far has been inadequate. Entomologist Paul DeBach states that the lack of funding stems from the fact that biological controls do not generate profits for industry the way that synthetic chemicals do. Without the profit incentive, biological control has little backing. However, we should recognize that biological control may provide tremendous economic and ecological benefits to society at large and we should urge the government to provide the support needed for further development of biological controls.

## Genetic Control

Most plant-eating insects and plant pathogens (bacteria, viruses, and other parasitic organisms) that seriously threaten crops attack only one or a few closely related plant species. This implies a certain genetic compatibility between the pest and its host. By the same token, it implies a genetic incompatibility between a pest and a species that is not attacked. The essence of **genetic control** is to develop genetic characteristics in host species that provide protection or resistance to attack. This has been successfully accomplished in many cases.

### BREEDING RESISTANT PLANTS

Throughout history there have been many instances when microbial pests wiped out entire crops; in some cases, famines followed as a result. The last time this occurred was between 1845 and 1847 in Ireland when a parasitic fungus known as **late blight** destroyed the potato crop. Nearly 1 million people starved to death. Another million escaped starvation only by emigrating to other countries, including the United States.

Some of the first inorganic pesticides were developed to provide protection from such blights; synthetic organic chemicals are still widely used for this purpose, particularly on orchard crops. However, in another approach to the problem, plant scientists worked on breeding resistant varieties. It often goes unappreciated, but it is no overstatement to say that the world owes much of its production of corn, wheat, and other cereal grains to the painstaking work of plant geneticists who selected and bred varieties that are resistant to attack by common parasitic fungi. Look in any seed catalogue and you will see further evidence of this work. Nearly every seed listed is designated as a variety resistant to one or more microbial pests.

The same potential exists for breeding plants that are resistant to insect pests. The relationship between wheat and the Hessian fly provides an example. The Hessian fly was introduced to this country in the straw bedding of Hessian soldiers during the Revolutionary War. The fly spread throughout much of the Midwest, becoming tine most serious pest of wheat. The adult fly lays its eggs on the wheat leaves and, after hatching, the larvae move down the leaves and enter the stems as they feed. This feeding activity weakens the stem so that the wheat is killed or broken in the wind (Fig. 17-10). Scientists at the University of Kansas developed a variety of wheat that causes the larvae to die as they feed on the leaves.

Increasing resistance through breeding may not provide 100 percent protection, but even partial protection can make the difference between profit or loss for the grower. In addition, any degree of resistance lessens the need for chemical pesticides.

It is important to note that the reason resistance has not been exploited more is that breeding each trait is a difficult process that frequently takes seven to ten years or more. Recall that breeding involves crossing carefully selected parents, testing all the progeny for the desired traits, selecting those that show the desired trait to the highest degree, followed by rebreeding and reselection over many generations. However, significant shortcuts in this work may be achieved in the near future. Techniques have been developed that enable scientists to isolate specific genes (segments of DNA) and transfer them from one species to another. These techniques are commonly referred to as **genetic engineering.** Also, techniques have been developed that enable the hybridizing of cells in the test tube. This overcomes many of the problems of sexual incompatibility between plants and allows for the creation of a much wider variety of hybrids.

**FIGURE 17-10**
Genetic control by developing chemical resistance. (a) The Hessian fly, which is a serious pest of wheat. (b) Its life cycle. Control has been gained by developing strains of wheat that cause the flies to die as they feed. (USDA.)

(b)

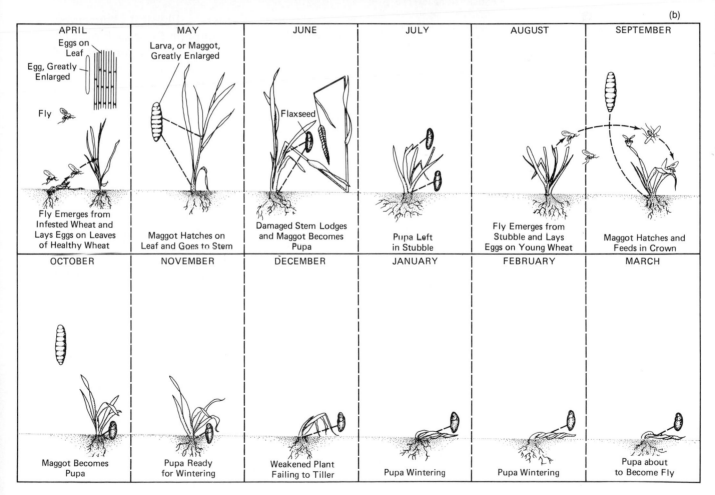

| APRIL | MAY | JUNE | JULY | AUGUST | SEPTEMBER |
| --- | --- | --- | --- | --- | --- |
| Eggs on Leaf / Egg, Greatly Enlarged / Fly — Fly Emerges from Infested Wheat and Lays Eggs on Leaves of Healthy Wheat | Larva, or Maggot, Greatly Enlarged — Maggot Hatches on Leaf and Goes to Stem | Flaxseed — Damaged Stem Lodges and Maggot Becomes Pupa | Pupa Left in Stubble | Fly Emerges from Stubble and Lays Eggs on Young Wheat | Maggot Hatches and Feeds in Crown |

| OCTOBER | NOVEMBER | DECEMBER | JANUARY | FEBRUARY | MARCH |
| --- | --- | --- | --- | --- | --- |
| Maggot Becomes Pupa | Pupa Ready for Wintering | Weakened Plant Failing to Tiller | Pupa Wintering | Pupa Wintering | Pupa about to Become Fly |

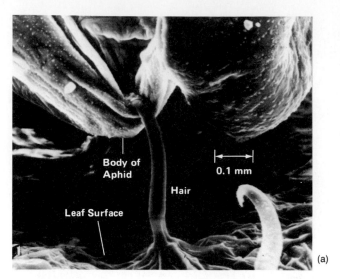

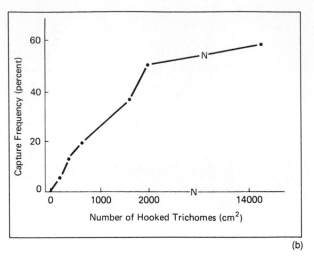

(a)

(b)

**FIGURE 17-11**
Genetic control by developing physical barriers. (a) The leafhopper is controlled when hairs on the leaf, a genetic trait, hook immature leafhoppers. (b) Relationship between the number of hairs and the frequency of capture. (Courtesy of E. A. Pillemer and W. M. Tingey, New York State Agricultural Experiment Station, Ithaca, New York. *Science*, 193 (August 6, 1976), 482–84. © 1976 by the American Association for the Advancement of Science.)

## TRAITS THAT PROVIDE RESISTANCE

What are the actual traits that provide resistance? They may be categorized in two groups: chemical barriers and physical barriers.

**Chemical barriers.**  It has been found that in many cases resistant plants produce some chemical that is lethal or at least repulsive to would-be pests. Can you see how this trait would arise through the process of natural selection? Those individuals without a suitable degree of resistance are eliminated from the population in each generation.

Examples of insecticidal compounds found in plants are pyrethrum, found in chrysanthemums; nicotine, found in tobacco; and rotenone, found in the tropical plants derris and cube. These particular compounds have been isolated and are used as "natural" insecticides, but they are both expensive and not highly effective when used in this unnatural way. The plant's production of these natural pesticides or repellents may be enhanced by breeding. However, a note of caution should be sounded. Some of these chemicals are also highly toxic to humans and some are known to be carcinogenic. In breeding or engineering plants for pest resistance, we must be

careful that we do not also make them unsuitable for human or livestock consumption.

**Physical barriers.**  Physical barriers are structural traits that impede the attack of a pest. For example, leafhoppers are significant worldwide pests of cotton, soybeans, alfalfa, clover, beans, and potatoes. It has been observed that hooked hairs on the leaf surfaces of some plants tend to trap and hold immature leafhoppers until they die (Fig. 17-11). Similarly, alfalfa weevil larvae are fatally entrapped by glandular hairs that exude a sticky substance. Obviously, the greater the number of these hairs, the greater the degree of protection. Traits such as these can be augmented through breeding.

Unfortunately, pests may develop the ability to overcome genetic controls in the same way they develop resistance to pesticides. This does not mean, however, that genetic controls will become obsolete. It means that breeders must continually work for new resistant varieties to substitute for old varieties whose resistance the pest has overcome. This substitution process has occurred some seven times in the case of wheat and the Hessian fly. Such substitution often takes place without the consumer ever knowing that a potential catastrophe is being averted.

## The Sterile Male Technique

A technique that has proven highly effective in controlling or even eliminating certain insect pests involves inundating a natural population with sterile males that have been reared in laboratories. This is known as the **sterile male technique.**

The use of this technique in combating the screwworm fly provides a prime illustration. The screwworm fly is closely related and similar to an ordinary housefly. But this particular species has the obnoxious trait of seeking out and laying its eggs in open wounds of cattle and other animals. The larvae (maggots) feed on blood and lymph, keeping the wound open and festering (Fig. 17-12). Secondary infections frequently occur and often lead to the death of the animal.

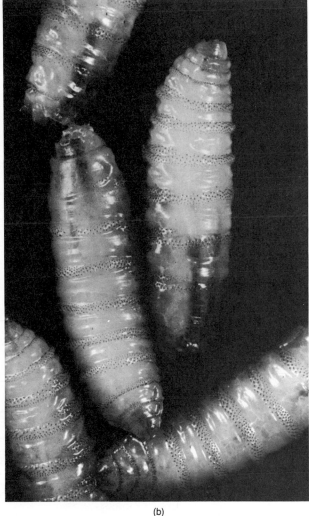

(b)

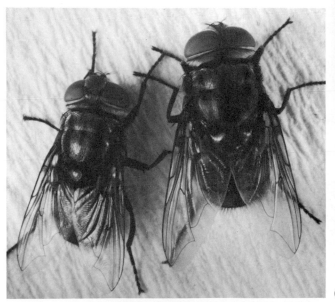

(a)

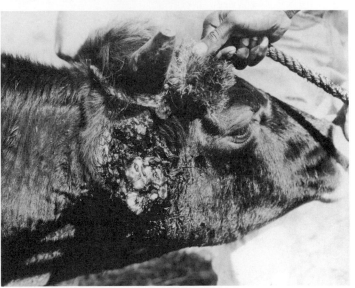

(c)

**FIGURE 17-12**
The screwworm fly, a deadly pest of cattle. (a) Adults; (b) larvae; (c) larvae of the screwworm fly feeding in a wound. Keeping the wound open allows the entry of other infections and frequently results in the death of the animal. (USDA photos.)

Before World War II. this problem became so severe that cattle ranching from Texas to Florida and northward was becoming economically impossible. In studying the situation, Edward Knipling, an entomologist with the U.S. Department of Agriculture, observed two essential features of screwworms: (1) their populations are never very high; and (2) the female fly mates just once, lays her eggs, then dies. Knipling reasoned that if the female mated with a sterile male the problem would be solved; she would lay her eggs but no offspring would be produced. The procedure developed to achieve this begins with the propagation of huge numbers of screwworm larvae on meat in laboratores. As the larvae become pupae, they are collected and subjected to enough high-energy radiation to render them sterile but not enough to preclude other functions. Huge numbers of these sterilized pupae are then air-dropped into the area to be controlled (Fig. 17-13).

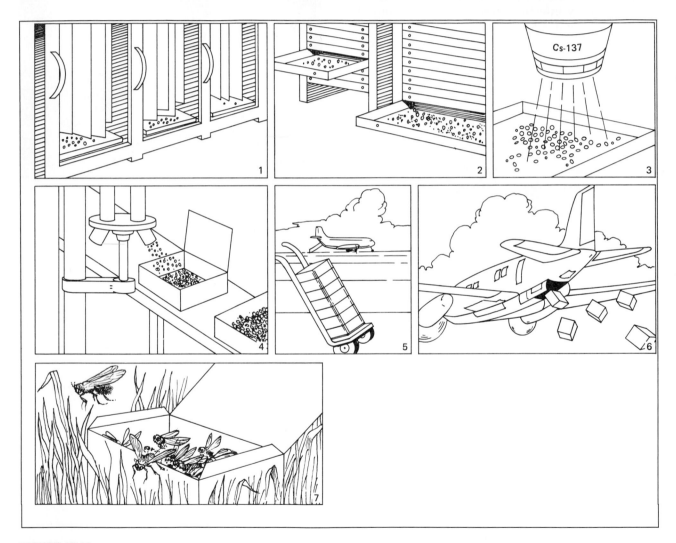

**FIGURE 17-13**
Sterile male technique used in control of screwworm flies. (1) Insects are nurtured through their different life stages in temperature-controlled chambers. (2) Pupae are collected and (3) exposed to radiation from cesium-137, which renders them sterile but not inactive. (4–6) Irradiated pupae are packaged and air-dropped in infected areas. (7) Sterile flies emerge and mate with normal flies in the environment, thus blocking reproduction. (Redrawn from photos from Comision Mexico Americana para la Erradicacion del Gusano Barrenador del Ganado and USDA.)

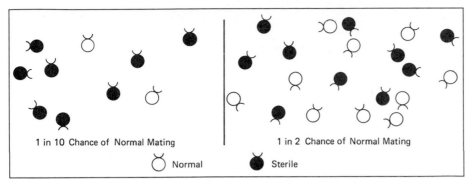

**FIGURE 17-14**
The sterile male technique can only be successful when the pest population is low, permitting one to create a high ratio of sterile to normal males. If the pest population is high, breeding between normal males and females will probably continue despite the presence of sterile males.

Ideally, 100 sterile males are dropped for every normal female in the natural population, making it very likely that the wild females will mate with one of the sterile males. If the population in the wild is so large that it is impossible or impractical to introduce enough sterile males to vastly outnumber normal males, this technique has little effect because breeding between normal individuals is sufficiently common to maintain the population (Fig. 17-14). Under the right circumstances, however, the method is effective. It successfully eliminated the screwworm fly in Florida in 1958–59 and it continues to be used to control it in the Southwest. The savings to the cattle industry is estimated at better than $100 million a year. The technique has also been used with some success to eradicate infestations of imported pests before they become widely established. Stocks of various insects are maintained in facilities around the world so that sterile males may be called up on very short notice if the need arises.

Again, pests may develop the ability to overcome the effects of the sterile male technique. In recent years, for instance, the sterile male technique has been less effective in combating the screwworm fly because a particularly aggressive behavior has evolved in the fertile males in the natural population so that they now fertilize the females despite the numerical odds. We now recognize that if a feature exists by which females distinguish normal from sterile males, a strong selective pressure for the development of that feature will emerge. This means that entomologists must watch for the development of such traits in the wild populations and be sure to breed them into the populations used to produce the sterile flies.

## Cultural Controls

A **cultural control** is any nonchemical technique that alters or controls one or more environmental factors so that it is unsuitable for the pest to live or propagate.

Many cultural controls have been available for centuries, but their use declined with the advent of chemical pesticides because chemicals seemed to offer an easier method. Simply reinstituting known cultural controls could mitigate the use of chemicals in many cases. This point is especially important for urban and suburban residents. Large portions of pesticides are used in homes, gardens, and lawns, where is most instances cultural controls would suffice. Moreover, they are the immediate province of the individual gardener or homeowner. One does not need to wait for future developments as in the case of biological or genetic controls. Cultural controls are also used extensively in agricultural situations. A better understanding of pest-host relationships could increase their use.

The diversity of cultural controls is best illustrated by the examples discussed below.

### CULTURAL CONTROL OF HUMAN DISEASE AND PARASITIC ORGANISMS

We routinely practice many forms of cultural control against diseases and parasitic organisms. Some of these practices are so familiar and well entrenched in our culture that we no longer recognize them as such.

For instance, proper disposal of sewage and the avoidance of drinking water from "unsafe" sources are cultural practices that protect against waterborne disease-causing organisms. Combing

**FIGURE 17-15**
Cultural controls. Many practices of hygiene and sanitation are cultural controls against disease, parasites, and other pests. How many cultural controls can you find in this illustration?

and brushing the hair, bathing, and wearing clean clothing are cultural practices that eliminate head and body lice, fleas, and other ectoparasites (*ecto*, "on the surface"). Regular changing of bed linens protects against bedbugs. Proper and systematic disposal of garbage, sweeping the crumbs from under the table, keeping a clean house, and simply living in a tight house with good screens are all effective cultural controls in keeping down populations of roaches, mice, flies, mosquitoes, and other pests (Fig. 17-15). In turn, control of these pests protects us against numerous diseases that they may carry. Similarly, sanitation requirements in handling and preparing food as well as cleaning cooking and eating utensils are cultural controls designed to prevent the spread of diseases. Refrigeration, freezing, canning, and drying of foods are cultural controls that inhibit the spread of organisms that cause rotting, spoilage, and food poisoning.

Real concern has been expressed that, as practices of personal hygiene and sanitation are dropped, as occurs in some subcultures, we shall again experience outbreaks of parasites and diseases that were virtually eliminated in developed countries for generations. In fact, it is estimated that 80 percent of human illness in less-developed

countries results from contaminated drinking water. The most economical way to improve human health in these areas is through cultural controls such as improved water supplies, sanitation, and personal hygiene.

## CULTURAL CONTROL OF PESTS AFFECTING LAWNS, GARDENS AND CROPS

The following examples illustrate some of the major categories of cultural controls that can be used to manage lawn, garden, and agricultural pests.

**Selection of what to grow and where to grow it.** Plants (including trees and shrubs) grown in conditions to which they are not ecologically well adapted may survive, but they do not thrive. Their consequent lack of vigor frequently renders them more susceptible to pests, and this in turn leads the grower to apply more pesticides. Conversely, plants grown under conditions to which they are best suited thrive and are generally more pest resistant. Thus, one can do much to alleviate pest problems by paying close attention to the ecological preferences of each species and making selections and placements accordingly.

This has particular relevance to the urban and suburban homeowner. Ornamental plantings around homes are frequently misplaced ecologically. Sun-loving plants are put in the shade; shade-loving plants are put in sunny locations; wet-loving plants are put in dry places, and so on. Then when these plants are attacked, home-owners are prone to use prodigious amounts of pesticides to maintain these "mistakes." A more practical alternative may be to let them die or transplant them, and replace them with something that is better suited to the particular location and more pest resistant.

The same principle applies on a broader scale to agricultural crops. Cotton, for example, has a high profit potential so it is grown in many areas where it is particularly vulnerable to pests and requires huge amounts of pesticides. In fact, more pesticides are used on cotton than on any other single crop. Yet, most of the profit potential may be absorbed by the costs of the pesticides. More attention should be given to helping the farmer select and grow crops that might return an equal or greater real profit without the use of pesticides.

### Management of lawns and pastures.
Homeowners are also prone to use excessive amounts of pesticides to maintain a weed-free lawn. Weed problems in lawns are frequently a result of simply cutting the grass too short. If grass is not cut to less than 3 inches it will maintain a dense enough cover to keep out crabgrass and many other noxious weeds. Similarly, weed problems in pastures are frequently a problem of overgrazing, which allows noxious weeds to invade and thrive.

### Water and fertilizer management.
Water, along with temperature, is one of the most important environmental factors for any organism, crop and pest alike. Drought-stricken plants may become more susceptible to certain pests; on the other hand, too much moisture may foster the reproduction of a pest. Therefore, water management needs to be practiced with an eye toward controlling pests as well as simply providing water to the plants.

Management of fertilizer is also important. As might be expected, plants weakened by lack of adequate nutrients are more susceptible to pests. However, very rapid growth promoted by excess fertilizer tends to produce juicy and succulent tissues that are much more prone to attack. Homeowners should take special note of this, because they are particularly prone to over- or underfertilize gardens, lawns, trees, and shrubs.

### Time of planting.
For a pest that emerges early in the spring, the planting of a susceptible crop may be delayed so that most of the pest population starves before the plants are available. On the other hand, if a pest population emerges late, early planting may be the answer. Early planting may allow the crop to get through most of its growth and be harvested before the pest population multiplies to destructive levels.

### Destruction of crop residues (sanitation).
Spores of plant disease organisms and insects may overwinter or complete part of their life cycle in the dead leaves, stems, or other plant residues that remain in the fields after harvest. Plowing or burying the material may be very effective in keeping pest populations to a minimum. In gardens, a clean mulch of material such as grass clipping may be substituted to protect the soil from erosion. Likewise, removal and/or distribution of treetops and branches after logging a forest is particularly important for control of forest pests.

### Adjacent crops and weeds.
It is possible to control some pest problems by controlling the kinds of crops and weeds grown in adjacent areas. Certain plants act as attractants. In seeking the attractant, the pests not only attack the attractant, but other plants that would not attract pests by themselves. The gardener may successfully eliminate the pest by eliminating the attractants.

These principles are also used in agriculture. For example, wheat crops around the turn of the century were devastated by a parasitic fungus known as wheat rust (Fig. 17-16). When it was discovered that rust also requires a second host, barberry, to complete its life cycle barberries, which grew as a weed around wheat fields, were removed. In combination with resistant varieties this control is still practiced in the Wheat Belt of the United States and Canada. Similarly, for certain insect pests that require noncrop plants to complete their life cycle, control may be gained by eliminating these weeds. On the other hand, certain adjacent plantings and habitat situations may be necessary to maintain populations of natural enemies. Adjacent plantings may also be used as a kind of trap. For instance, alfalfa grown next to cotton serves to attract lygus bugs away from the

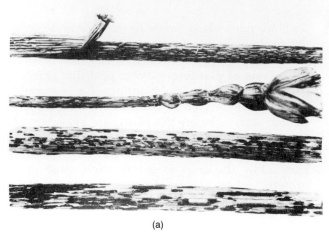

(a)

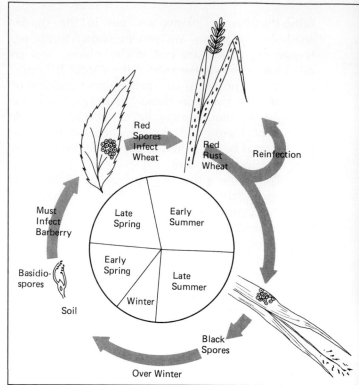

(b)

**FIGURE 17-16**
Wheat rust, a parasitic fungus that is a serious pest on wheat. (a) Injury to wheat caused by the rust. (b) The life cycle of the rust. Since part of the life cycle of wheat rust requires that it infest barberry, eliminating barberry in wheat-growing regions has been an important cultural control, along with developing resistant varieties of wheat. (USDA.)

cotton. If the alfalfa is harvested in rotation so that alfalfa is always present, the lygus bugs will remain on their preferred host, alfalfa.

**Crop rotation.** Crop rotation is the practice of alternating crops from one year to the next in a given field; for example, corn one year, beans the next. As a form of pest control, this practice takes advantage of the fact that most pests are limited to feeding on one kind of plant. If pests emerge in the spring and find no host plants to feed on, the population will starve.

Crop rotation is also effective in terms of controlling root nematodes, roundworms that live in the soil and feed on roots, as well as other pests that do not have the ability to migrate appreciable distances (Fig. 17-17).

**Plowing and cultivation.** Plowing and cultivation have been the traditional cultural controls for weeds. Unfortunately, these practices leave the soil prone to erosion. If conventional techniques will control the erosion, however, plowing and cultivation may provide the most environmentally sound way to control weeds and pests.

**Monoculture versus polyculture.** Agriculture in industrialized countries has progressively moved toward **monoculture,** the growing of a single species over a wide area. Today, the practice often involves not only one species, but a particular cultivated variety of that species.

Monoculture grew with the mechanization of farming because efficient use of machinery requires large uniform fields. Unfortunately, monoculture is ecologically **unstable.** When a pest outbreak occurs, monoculture is conducive to its multiplication and spread. Even if natural enemies are present, they may be overwhelmed by the avalanche effect of spreading pests. On the other hand, the spread of a pest outbreak is impeded if there is a mixture of plant species, some of which are not vulnerable to attack. This, in turn, enables natural enemies to be more effective. It is probable that this principle accounts for the fact that natural ecosystems consist of a mixture of different species. Indeed, much of our prodigious use of pesticides may be seen as a desperate attempt to maintain unstable monocultures by means of synthetic chemicals.

Recognizing this situation, many agriculturalists are experimenting with various systems of **polyculture.** One technique, the growing of two or more species together or in alternate rows, is known as *intercropping*. Polyculture presents few problems if most of the farming is a product of human labor, but is difficult where large machinery is employed.

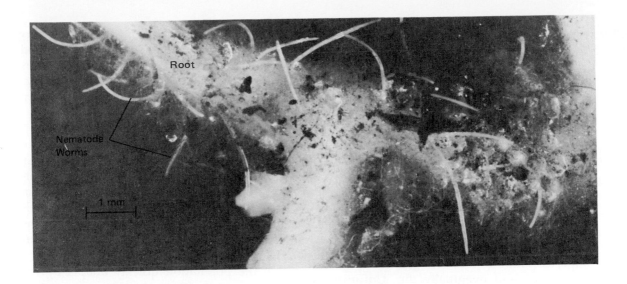

**FIGURE 17-17**
Cultural control of nematodes, parasitic worms that feed on roots and cause serious damage to a number of plants. Populations of these pests may be kept to a minimum by crop rotation with alternate planting of a crop that does not support the nematodes. (USDA.)

The principles of monoculture versus polyculture apply to forestry as well as agriculture. Clearcutting, then replanting a single species creates a uniform environment that persists for many years. This is the ideal condition for a herbivorous insect or pathogen to flourish. Conversely, maintaining a diverse forest of different species at different ages through selective cutting allows natural balances to largely control pest populations.

**Quarantines.** Again, many of our most difficult to control pests are species that have been unwittingly imported from other parts of the world. Numerous other species, if brought into this country, would become serious threats because their natural enemies are not present. Here, the best procedure is to keep the would-be pests out. This is a major function of the customs office. Biological materials that may carry pest insects or pathogens are either prohibited from entering the country or are subjected to quarantines, fumigation, or other treatments that assure they are free of pests. The cost of such procedures is small in comparison to the costs that could be incurred if such pests established themselves. It is extremely important that citizens recognize the need for quarantines and cooperate with procedures. The

system, however, is still far from perfect. Further lines of defense have been developed and will be discussed in the section on integrated pest management.

## Natural Chemical Control

You are probably familiar with the fact that **hormones** control your growth, sexual development, and many other bodily functions. Hormones may be thought of as chemical signals; very tiny amounts of the hormone signal and direct the body to undergo sometimes striking changes. On the other hand, an improper signal, namely, too much or too little hormone, can cause severe abnormalities.

The aim of **natural chemical control** is to isolate and identify hormones that control an insect's life cycle. Once this is accomplished, the potential exists to manipulate the chemical signals in such a way as to prevent the insect from developing and/or reproducing.

While straightforward in concept, this process takes many years of careful research. Scientists must process literally hundreds of kilograms of insects to isolate a few milligrams of active ingredient. The molecular structure of the active chemical must then be determined and a process developed to produce it synthetically. This is the only way that usable amounts of the natural chemical can be obtained at affordable costs. Finally, field experiments must be conducted to find ways of using the compound to achieve the desired results. Despite the difficulties, however, marked progress is being made. It is not unlikely that numerous natural chemical controls will become available within the next five to ten years. The following examples show two ways the method has already been put to some use.

### JUVENILE HORMONE

Scientists have discovered that pupation of insect larvae is triggered by a decrease in the level of a certain chemical called the **juvenile hormone.** If juvenile hormone is artificially given to maintain the level, pupation does not occur, but the larvae simply continue to feed and grow, become grossly oversized, and eventually die. Unfortunately, despite 15 years of research, practical methods for controlling pests in field conditions with juvenile hormone have not been found.

### SEX ATTRACTANTS (PHEROMONES)

Chemicals that provide signals between separate individuals of the species are called **pheromones.** A particular category of pheromones that is being exploited is the **sex attractants.** Adult females secrete a chemical "perfume"; males detect it, fly to the female, and mating occurs. Once identified and synthesized, sex attractants may be used in one of two ways: (1) the **trapping technique** or (2) the confusion technique. In the trapping technique, the pheromone is used to lure males to traps, where they can be disposed of, or to poisoned bait. In the confusion technique, the pheromone is sprayed and dispersed over the field in such large quantities that males are confused, cannot find the females, and thus fail to mate (Fig. 17-18).

You may note that the use of natural chemicals is similar to the use of synthetic organic chemical pesticides in that both are synthesized and applied in much the same way. Three important distinctions exist, however: (1) the quantity of signal chemicals required for pest control is tiny when compared to the amount of pesticide needed; (2) a signal chemical is specific to the target pest and thus will not affect unrelated species to cause resurgences and secondary pest outbreaks; and (3) although signal chemicals are synthesized in the laboratory, they are identical to those that occur naturally and thus can be broken down and metabolized by any number of organisms, including humans. Nearly all of the food we eat contains traces of the hormones present in plants and animals. Therefore, there is every reason to believe that the use of natural chemicals will be environmentally safe.

Until recently, scientists believed that pests could not develop resistance to natural chemicals because it seemed impossible for an organism to become resistant to its own hormones or pheromones. Unfortunately, this is not the case. For example, we now know that insects do not rely on a single chemical sex attractant. They depend on a complex of several pheromones. If only one of these pheromones is used in the confusion technique, the population quickly adapts by relying on other signal chemicals. The opportunity for natural chemical control remains, however, because scientists can shift the pheromones they apply.

In conclusion, it is important to emphasize again that the ecosystem does not contain pests as such. A species only becomes a pest when its population proliferates to the point where it causes significant destruction. It is not necessary—in fact it is ecologically detrimental—to aim at eradication with biocidal chemical "weapons." The ecological balance can be sufficiently maintained if we simply keep the population of the troublesome species within tolerable bounds or prevent it from attacking desired plants and animals. All methods of natural control reflect this distinction. They do not aim for eradication; rather, they attempt to introduce or alter factors that will impede the pest's ability to reproduce and multiply while still maintaining the overall integrity of the ecosystem and the health of the environment.

## INTEGRATED PEST MANAGEMENT

Despite the availability and potential of natural control methods, huge amounts of synthetic organic pesticides continue to be used. Considering the hazards and decreasing effectiveness of synthetic organic pesticides, it is natural to ask why.

Pesticide producers and many agriculturalists and growers insist that unacceptable crop losses would occur if chemical pesticides were abandoned. With respect to some crops, they are

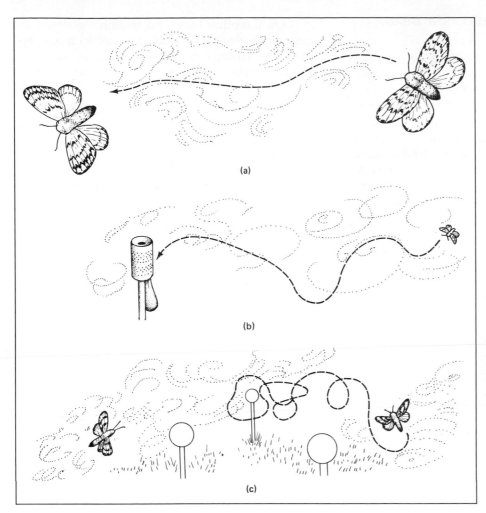

**FIGURE 17-18**
Natural chemical control using a pheromone, which is a sex attractant. (a) The females of many insect species secrete a pheromone that functions to attract males and enable mating. (b) In the trapping technique, synthetic pheromones may be used to lure males into traps. (c) In the confusion technique, synthetic pheromones may be used in excess quantities such that males become confused and are unable to find females. In either case mating is prevented.

right. Natural control methods cannot adequately keep pest populations below damaging levels in all situations. However, entomologists generally agree that the use of synthetic organic pesticides could at least be cut in half without significantly affecting crop production. In other words, while some use of chemical pesticides is required, current use is probably at least twice as high as it has to be.

Why are these chemicals so overused? The reasons are deeply rooted in certain human attitudes and economic interests. If pesticide use is to be reduced, these issues, as well as the ecological problems of natural control, must be addressed.

**Integrated pest management (IPM),** is an approach that combines sociological, economic, and ecological factors in an overall program that aims to minimize use of synthetic organic pesticides without jeopardizing crop protection. Importantly, IPM is not a technique in and of itself. Rather, it is an approach that integrates many techniques. In this section we shall examine the factors that lead to overuse of pesticides and learn how these factors may be addressed through integrated pest management.

## Economic and Sociological Factors Supporting Overuse of Pesticides

The economic and sociological factors that support the overuse of pesticides are closely linked, but they may be broken down into the following four categories.

## "THE ONLY GOOD BUG IS A DEAD BUG" ATTITUDE

The attitude that "The only good bug is a dead bug" is deeply ingrained in many people. They get great satisfaction watching insects literally drop dead as a result of spraying. Seldom do people consider whether the "bug" actually causes damage or whether the insect may be beneficial. As a result, many people tend to apply pesticides more frequently and in quantities far greater than are needed to achieve adequate protection. In fact, an overdose of pesticide often makes people think they are really licking the problem. The fact that they may be aggravating the problem by causing resurgences, secondary pest outbreaks, and increased pest resistance is not commonly recognized (Fig. 17-19).

On the other hand, professional growers are increasingly trying to minimize pesticide use because the rising cost of the chemicals squeezes profit margins. Some pesticides cost as much as $80 per pound. The cost factor, however, is seldom a major consideration of urban and suburban homeowners.

## AESTHETIC QUALITY AND COSMETIC SPRAYING

In any supermarket you can observe people picking over produce to find the best-looking fruits and vegetables. The not-so-good-looking items are frequently left and trashed. This approach to shopping is what I call the "Snow White" syndrome. Like Snow White, we all tend to reach for the perfect unblemished apple, or-

**FIGURE 17-19**

① LITTLE MISS MUFFET SAT ON A TUFFET, EATING HER CURDS AND WHEY.

② ALONG CAME A SPIDER, WHICH IS AN INSECT PREDATOR.

③ BUT MISS MUFFET ZAPPED IT ANYWAY.

④ TO THE DISMAY OF MISS MUFFET THIS CAUSED ECOLOGICAL UPSET FROM WHICH SHE COULDN'T GET AWAY.

⑤ IS THE ONLY GOOD 'BUG' A DEAD 'BUG'?

ange, head of lettuce, and so forth. We seldom consider that pesticide "potions" may have been used to make the products attractive (Fig. 17-20).

Although many of the speckles, spots, and blemishes are caused by pests, in many cases they are only **cosmetic damage;** the defects do not affect the taste, nutrition, or storability of the produce, much less the yield. Yet, growers and re-

tailers find that blemished produce brings in a sharply reduced price—if it can be sold at all. Canneries will not accept blemished produce because it does not reflect their "top quality" image. Thus, the consumer's unquestioning demand for aesthetic perfection effectively forces growers to indulge in **cosmetic spraying**—the use of pesticides to control those pests (certain species of in-

**FIGURE 17-20**
What "potions" have been used to produce the perfect fruit or vegetable?

sects, mites, and fungi) that harm only the item's outward appearance. As a result, millions of kilograms of pesticides, with all their potential hazards to human and environmental health, are applied with absolutely no benefit to yield, nutritional value, or any other practical aspect.

The Snow White syndrome also applies to homeowners in search of the perfectly uniform, unblemished green lawn. Increasingly, residents pour pesticides on their grass or hire commercial chemical lawn services to do the job for them. These services drench the lawn two or three times a year with mixtures of herbicides, insecticides, and fungicides, as well as fertilizers. The short-term results are seen in a beautiful, weed-free lawn. But the long-term costs to the environment may be considerable.

### INSURANCE SPRAYING AND THE NEED OF GROWERS TO PROTECT THEIR INVESTMENT

A farmer's costs are high and profit margins are frequently narrow. It can take many years to recoup the losses incurred from a single crop failure. Two consecutive years of failure can easily cause bankruptcy. Thus, a grower's need for protection against losses cannot be overemphasized. Though pesticide applications are expensive, a grower, understandably, may choose to apply pesticides as a form of insurance against losses.

It makes little difference whether the threat of loss to pests is real or imagined, close at hand or remote; what is important is how the grower perceives the threat. Even if there is no evidence of immediate pest damage, a grower who believes his or her crops are at risk is likely to succumb to **insurance spraying,** the use of pesticides just to be safe. Few such insurance treatments actually prove necessary to protect yields, but they all have negative impacts on the environment.

### CHEMICAL COMPANY PROFITS

The fact that the production and sale of synthetic organic pesticides is a highly profitable business for a huge and very powerful chemical industry is an unfortunate but important reality. More sales mean more profits so chemical companies do all they can to promote and exploit the previously described attitudes. Through advertising and paid-on-commission field representatives, pesticide producers attempt to convince agricul-

turists and growers that the threat of pests is much greater than it actually may be and that chemical "weapons" are the only technique available to counter the threat. It is estimated that at least 90 percent of the pest management information growers received in the 1970s came from chemical companies. Much of this information advocates insurance spraying, recommending regimes such as "spray on May 15 and every two weeks thereafter with X to protect against Y." X can be any number of pesticides and Y can be any number of pests. The companies also promote the Snow White syndrome by emphasizing the enhanced cosmetic quality that can be obtained with pesticides.

Following this advice has intensified growers' problems with resurgences, secondary pest outbreaks, and increasing pest resistance—all of which benefit the chemical companies by creating a market for larger quantities and numbers of pesticides. Hence, one can easily see the profit motive for intense lobbying by pesticide producers who want farmers to continue to approach pest problems with chemicals.

## Aspects of Integrated Pest Management

How can these sociological and economic factors be addressed? Some approaches are discussed in the following pages. These approaches, along with the implementation of natural controls, are important aspects of integrated pest management.

### PRODUCE LABELLING VERSUS COSMETIC SPRAYING

Public concern about the use of synthetic organic pesticides is growing, but the Snow White syndrome persists because growers and marketers do not fully inform the public about the kinds or amounts of pesticides used.

If produce carried labels describing the chemical treatments it was subjected to, consumers would no doubt react. Already, food outlets selling "organically grown" produce (produce that is grown without synthetic chemical pesticides or fertilizers) are thriving, though the produce is not as cosmetically perfect and may be higher priced. As a result, many growers now find it more profitable to rely on natural controls and sell through these specialized markets.

A major supermarket in Switzerland, Migros, is exploiting this situation to the benefit of itself and growers. A special label that stands for health is placed on fruits and vegetables grown with minimum use of synthetic chemical pesticides. These products are marketed side by side with and at the same price as unlabelled produce grown with traditional use of pesticides. The store explains the difference to the consumers on large wall posters proclaiming: We grow our vegetables and fruits with fewer artificial fertilizers and pesticides and that means better soils, better energy use, and less pollution so that we can protect the health of our clients.*

Migros promotes its philosophy by employing farm advisors to instruct growers in the techniques of organic farming. If farmers comply with practices, Migros guarantees to buy their produce. An increasing percentage of Migros produce now bears the special label. As farmers become more experienced and natural controls become more established, standards for allowable synthetic chemicals are made stricter.

A similar scheme is possible in the United States because most supermarket chains contract directly with growers for produce. Legislative action requiring mandatory labelling of produce is another option that could alleviate the situation.

## ADVICE FROM TRAINED FIELD SCOUTS VERSUS INSURANCE SPRAYING

Devastating pest infestations generally occur only when certain weather and crop conditions coincide, making circumstances particularly favorable for the reproduction of pests. At other times, various factors of environmental resistance keep pest populations below the economic threshold. This means that the cost of the damage caused by pests is less than the cost of spraying to eliminate the pests. Clearly, spraying when pest populations are below the economic threshold is economically counterproductive, particularly when spraying may create a real threat by causing a resurgence or secondary pest outbreak.

The solution is to recognize those conditions where pests pose a real threat. But most growers do not have sufficient knowledge in this area and

must rely on help from others. Many universities in the United States are now turning out "field scouts" trained to provide such help. The field scouts, who are usually employed by state or local agricultural extension services or by farm cooperatives, examine fields and orchards, monitor pest populations and other factors, and advise farmers if spraying is necessary.

Field scouts have significantly reduced pesticide use in cotton-growing areas of Texas. Impeding progress is a shortage of trained field scouts and an initial reluctance by growers to adhere to this new advice. As distinct economic advantages are realized, however, this attitude may change.

## ECONOMICS OF ORGANIC FARMING

Making the economic benefits of organic farming known to growers is another important aspect of integrated pest management. Initially, pesticides did enable an increase in yields and profits, hence their widespread adoption. Many farmers still cling to pesticides, believing that they offer the only way to bring in a profitable crop. In some instances, however, the rising costs of pesticides, along with their tendency to aggravate pest problems, eliminates their economic advantage.

The Center for the Biology of Natural Systems of Washington University compared 16 "organic" farms with 16 similar "conventional" farms in the Corn Belt. Organic farms relied on crop rotation and other cultural techniques to control pests; conventional farms used pesticides. Crop production per acre was virtually identical in the two cases. Total value of crops produced was slightly higher on conventional farms because pesticides permitted monocropping and hence a large percentage of the land was in high-value corn. However, this gain was offset by the cost of pesticides and fertilizer so that the economic return to the farmers was not significantly different. Productivity of most crops on Amish farms, where pesticides were never used because of religious reasons, is as high as or higher than that of farms where pesticides are used.

## INTEGRATION OF NATURAL CONTROLS

The U.S. program to protect against the Mediterranian fruit fly, or "Medfly," exemplifies how a number of natural and chemical controls

---

*By permission: "All Things Considered," Dec. 20, 1982, National Public Radio.

may be combined in a program of integrated pest management. The Medfly resembles a common housefly, but it is a foreign species that lays its 200–500 eggs in any of some 200 varieties of fruits and vegetables. The hatching larvae feed as maggots inside the fruit, thus destroying the produce. They then pupate and emerge as adults that repeat and escalate the cycle every four to five weeks. Unlike the common fruit fly, which is attracted only to very ripe fruit, the Medfly attacks unripe fruits and vegetables in the field. The pest can thus cause extensive damage before harvest and during storage and transport. If the Medfly established itself in the United States, huge amounts of pesticides would be needed to protect crops. And still, millions of dollars worth of damage would occur each year.

To prevent this from happening, an integrated pest management program with multiple lines of defense has been adopted. First, since the Medfly is most likely to enter the United States via imported fruits and vegetables, such as Hawaiian pineapple, all incoming produce is checked and fumigated if officials detect any trace of the insect. Some Medflies still manage to slip into the country, so a second line of detection, a network of traps baited with a sex attractant, is maintained. Keeping up enough traps to catch every Medfly that enters is impractical, but monitoring the traps provides an early warning if the pest is in the area.

If Medflies are found, the sterile male technique may be called into play. Stocks of Medflies are maintained in South America and batches of sterile flies can be delivered on short notice and dropped over the infected area. In addition, students of all ages are called out to strip fruit from the trees and thereby deny the pests a place to breed. Officials continue to monitor the pheromone traps to gauge the success of their efforts.

If their efforts have not been successful, they move to a final line of defense: the chemical malathion. Malathion has a relatively low toxicity to humans and animals, but is very effective against Medflies. As an extra safeguard, however, the pesticide is applied as a treated bait, thus increasing its specificity to the pest and decreasing human and animal exposure (Fig. 17-21).

This integrated management program successfully eradicated several Medfly invasions in the United States—in 1956 and 1961 in Florida, in 1966 in Texas, and in 1975, 1980–81, and 1984 in California. Some of these invasions were controlled without the use of malathion. But because of an unfortunate accident, the 1980–81 California episode resulted in widespread spraying and still nearly ended in failure.

The 1980–81 invasion was nearly under control as a result of the sterile male technique when scientists discovered that some of the supposedly sterile flies, identifiable by dye markers, were actually fertile. They immediately realized that a batch of "sterile" flies had been insufficiently irradiated, and, as a result, instead of curtailing the breeding of a few remaining wild Medflies, they had introduced as many as 100,000 additional pairs of fertile Medflies. (Rather than tediously separate male and female insects in the laboratory, both males and females are irradiated and introduced. Normally the presence of the sterile females along with the sterile males does not influence the effectiveness of the technique.)

It is impossible to control an infestation of this magnitude with the sterile male technique because 100 infertile flies are needed for every normal fly. Therefore, immediate spraying was called for. However, public protest against the use of pesticides delayed action and gave the flies another few weeks to establish themselves. Eventually, extensive spraying took place and the invasion was eradicated.

It is interesting to note here that public outcry against the use of synthetic organic pesticides is an important element in promoting the movement toward natural controls. However, it should also be noted that public outcries should be tempered with an understanding of the situation. In this case, public opposition to the use of pesticides resulted in the need for additional spraying.

### PRUDENT USE OF PESTICIDES

As seen above, integrated pest management may reduce the use of synthetic chemical pesticides significantly. But prudent use of these chemicals is still an important issue. Is this approach sufficient to eliminate the negative impacts of these chemicals?

Some entomologists feel that integrated pest management is becoming overly reliant on the use of synthetic organic pesticides and this reliance is aggravating the pesticide treadmill. These entomologists advocate renewed emphasis on the study of pest ecology and the development and implementation of natural controls—in short, a renewed emphasis on the concepts of ecological pest management.

## Recommendations for Promoting Ecological Pest Management

The foregoing discussion suggests a number of issues on which public pressure and support are needed to promote the movement toward ecological pest management. Let us conclude with a summary of these issues and recommended actions.

First, consumers can promote and support legislative efforts requiring that produce be la-belled as to the pesticide treatments used in production and/or storage.

Second, citizens can promote and support public funding for research and development of natural controls. The profit potential in synthetic chemical pesticides has led chemical companies to invest huge sums of money in developing and marketing their products. Natural controls may provide greater benefit to growers, but they generate little if any profit for their developers or promoters. Consequently, the private sector is unlikely to push natural controls; they must be supported and promoted by the public sector.

Third, citizens can also promote and support expenditures of funds for adequate training and employment of field scouts who can advise farmers in ways described earlier in this chapter.

Fourth, the federal government can help

**FIGURE 17-21**
Integrated pest management (IPM). Integrated pest management involves the integration of a number of techniques, none of which would be completely effective by itself. The illustration depicts the techniques or "lines of defense" that are used to keep the Mediterranean fruit fly out of California.

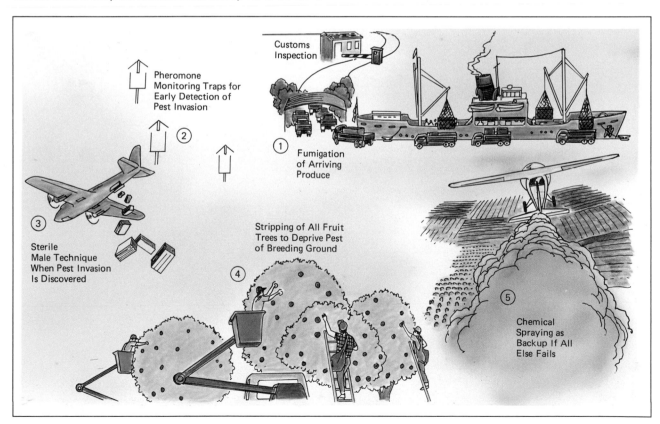

farmers make the transition to natural controls by providing pest loss insurance. Farmers who depend on pesticides to protect their crops are understandably reluctant to stop spraying simply on the advice of newly arrived field scouts. Indeed, the risks of error are real and the chance that pests will destroy a significant portion of a farmer's crop cannot be ignored, especially during the first year pesticides are withheld. At this time, natural controls may not be well established. If pest loss insurance, similar to other disaster insurance provided by the government, were available, farmers may be more willing and financially better able to take the risks they associate with natural controls.

 **RESOURCES: MINERALS, BIOTA, ENERGY, AND LAND**

Human societies depend on many resources besides air, water, and soil. These additional resources may be grouped in four primary categories: minerals, biota, energy, and land. Addition- ally, since the disposition of trash, its disposal or recycling, has some impact on all these areas, trash or refuse is considered another resource category.

**Minerals.**   Minerals are rocklike materials that comprise the earth's crust. Minerals that yield metallic elements such as iron, aluminum, and copper upon smelting are known as ores. Other minerals, such as phosphate rock, are valuable for the fertilizer compounds they contain. Still others, such as asbestos, clays, and mica, are commercially useful in their existing state.

**Biota.**   The term **biota** refers to all living things collectively. Frequently we speak of the biota of a region, referring to all plant, animal, and microbial organisms of that region. Commonly biota is subdivided into flora (plants) and fauna (animals). Our food, wood and wood products, natural fibers, spices, and many drugs are examples of products that come from the biota. Many new or modified species are now propagated through agricultural activities, but we still depend on or desire natural biota for many purposes. Many commercial products such as fish, shellfish, and timber are still harvested largely from natural ecosystems. Natural biota continues to be the reservoir genetic stock used to maintain and improve domestic breeds. Natural biota and the ecosystems they constitute also provide aesthetic and recreational benefits. Further, the nature of the biota is crucial to the maintenance of air and water quality in particular and the overall ecological stability of the earth in general.

**Trash.**   Imprudent disposal of trash has negative impacts on air, water, land, and biological resources and it represents a wanton discard of resources. Learning to utilize trash constructively has important bearing on all other resources.

**Energy.**   The progress of industrial societies can be attributed largely to the harnessing of energy resources such as coal, oil, natural gas, and, in modern times, nuclear and solar energy. Maintaining and extending industrialization require a continuous supply of energy resources.

**Land.**   Any use of the land, be it for mining, agriculture, building, or preserving, is intertwined with and affects all other resources. A growing population and growing demand for resources involve a growing demand for land.

Hence, land itself becomes the most critical of all resources.

We may divide these resources along with air, water, and soil into separate categories, but it must be emphasized that all resources are interdependent. For instance, food production requires land and suitable air, water, and soil. It also draws on mineral and energy resources for the production and/or operation of machinery, fertilizer, and pesticides, and it depends on biota for genetic stock. Conversely, agriculture and its demands for other resources may have serious negative impacts on soil, water, and air quality and on the natural biota. But, we humans play a vital role in the "center" of all resources. While we use all resources for our benefit, we may have positive as well as negative effects on the conservation of resources (Fig. VI-1.).

The demand for all resources is growing dramatically as population and affluence increase. The effect of affluence is demonstrated when we recall that industrialized nations, which have about 25 percent of the world's population, consume about 80 percent of the world's mineral and energy resources. Industrializing less-developed nations would entail a threefold increase in consumption of these resources even if population growth came to a halt.

Can the world's resources sustain increasing demands? This question has been studied and debated extensively in recent years. Unless drastic efforts are made to curtail population growth and resource consumption, some foresee exhaustion of critical resources with cataclysmic consequences within the next 30 to 40 years. People with this point of view are often referred to as **limitists.**

Other people believe, however, that our resources are fully adequate for the foreseeable future and that population and resource availability will ultimately balance each other through natural demographic and economic processes. People with this point of view are often named **cornucopians** after the mythical horn of plenty.

As in most debates, both points of view contain some truth. The "ultimate resource," as Professor Julian Simon calls it, is human wisdom and creativity. The ultimate resource, he holds, will determine the outcome. If humans act without foresight and exploit resources to extinction, soci-

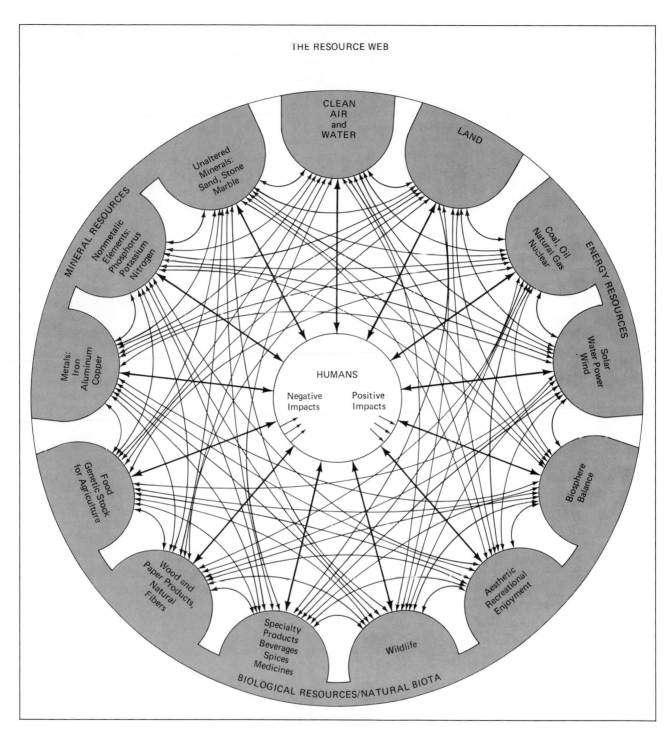

**FIGURE VI-1**

The resource web. Humans play a pivotal role in the center. All resources are necessary for the support of human civilization, but resource availability is not static. Through technology and conservation we may preserve and enhance resource availability, or through overexploitation and carelessness we may deplete or destroy resources. Loss of any one resource may have profound effects on the availability of other resources because, as in an ecological food web, obtaining any one resource requires certain inputs of other resources.

ety will suffer extreme consequences. However, if humans foresee problems and adapt to the situation by conserving and protecting the environment and developing substitutes or alternative resources, the consequences will be much more acceptable.

Our objective in the following chapters is to examine each of the four resource categories and trash to see what problems exist and to learn what is being done or needs to be done to assure a future with adequate resources.

# 18

# MINERAL RESOURCES

5. Why do limitists predict the impending exhaustion of many resources? What do cornucopians predict? Which view has been given more credence in recent years? Why?

6. Discuss the factors that have curtailed use of certain critical resources.

7. Choose a critical resource and discuss how its availability has or may be affected by a free economy.

8. Why is international diplomacy so important in the availability of minerals?

9. Why does energy policy affect the availability of minerals?

10. What type of environmental policy do you believe would enhance the use and availability of critical elements now and in the future?

About 60 elements are critical to our modern technological society and its economy. The best known are metals such as iron, aluminum, copper, lead, zinc, magnesium, nickel, chromium, manganese, tin, mercury, and the fertilizer minerals such as phosphorus and potassium. These elements occur as minerals (rocklike) compounds in the earth's crust. However, the earth's crust is not an even mixture of minerals or elements. Some minerals are very common; others are rare. In both cases, however, geochemical processes that occurred over hundreds of millions of years created relatively rich deposits of particular minerals in particular locations (Fig. 18-1). While it is not economically practical to mine "common rock" to obtain traces of desired elements, it is economically profitable to mine enriched deposits.

Minerals that may be profitably mined to obtain desired elements are called **ores.** Once mined, ore deposits are, for all practical purposes, gone forever. Therefore, mineral deposits are frequently referred to as **nonrenewable resources.**

Factors do exist, however, that lessen the probability of mineral resource exhaustion and/or minimize the economic impact if depletion does occur. The objective of this chapter is to study these factors.

## THE DEBATE: WILL WE RUN OUT OF NECESSARY MINERALS?

As previously noted, some elements are extremely common. In fact, 97 percent of the earth's crust is comprised of only eight elements: oxygen, 46.6 percent; silicon, 27.7 percent; aluminum, 8.3 percent; iron, 5.8 percent; calcium, 3.6 percent; potassium, 2.6 percent; and magnesium, 2.1 per-

**FIGURE 18-1**
Geographic distribution of important mineral deposits in the United States. Distribution of elements in the earth's crust is far from uniform. Deposits of ores rich in particular elements exist at specific locations. (U.S. Department of the Interior, Bureau of Mines.)

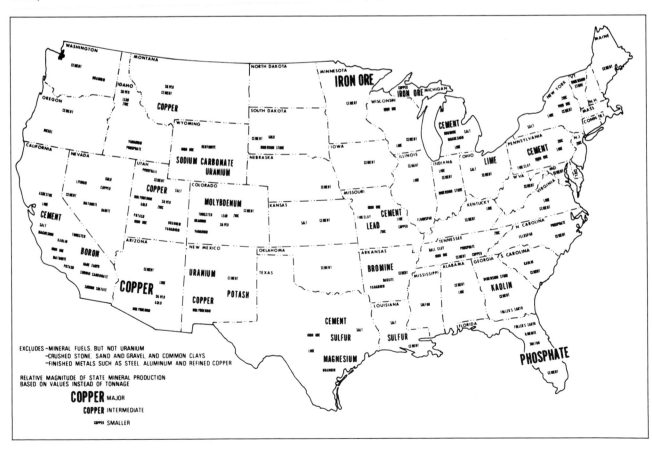

cent. The potential for exhaustion of these minerals is minimal. Using them all up would amount to mining the earth from under our feet and throwing it into space.

For most elements, however, the **average crustal abundance,** that is, the percentage of the earth's crust that the element constitutes, is only a tiny fraction of 1 percent as shown in Table 18-1. Nevertheless, these elements may be found in economically exploitable concentrations in certain deposits. For example, the average crustal abundance of copper is 0.0063 percent, but certain minerals contain as much as 30 percent copper. A mineral with a high concentration of a desired element is called a **high-grade ore;** minerals with minimal, but still exploitable concentrations are called **low-grade ores.** In both cases, areas where these elements are concentrated enough to make them economically exploitable are few. Limitists see the exhaustion of these deposits as the most problematic. How long will supplies of these elements last?

**Table 18-1**

| Average Crustal Abundance for 20 Valuable Elements | | |
|---|---|---|
| | CRUSTAL ABUNDANCE | |
| ELEMENT | (ppm) | (percent) |
| Aluminum | 83,000 | 8.3 |
| Iron | 58,000 | 5.8 |
| Titanium | 6,400 | 0.64 |
| Manganese | 1,300 | 0.13 |
| Phosphorus | 1,200 | 0.12 |
| Carbon | 320 | 0.032 |
| Chromium | 110 | 0.011 |
| Zinc | 94 | 0.0094 |
| Nickel | 89 | 0.0089 |
| Copper | 63 | 0.0063 |
| Cobalt | 25 | 0.0025 |
| Lithium | 21 | 0.0021 |
| Lead | 12 | 0.0012 |
| Tin | 1.7 | 0.00017 |
| Uranium | 1.7 | 0.00017 |
| Molybdenum | 1.3 | 0.00013 |
| Tungsten | 1.1 | 0.00011 |
| Mercury | 0.089 | 0.0000089 |
| Silver | 0.075 | 0.0000075 |
| Gold | 0.0035 | 0.00000035 |

Source: Data from "Limits to Exploitation of Nonrenewable Resources" by Earl Cook, p. 63, in *Materials: Renewable and Nonrenewable Resources*, ed. P. H. Abelson and A. L. Hammond. © 1976 by the American Association for the Advancement of Science.

## Reserves Divided by Rate of Use Equals Years of Availability

The land masses of the earth have been well explored, and fairly reliable estimates exist regarding the quantities of various minerals. Known mineral deposits that can be exploited using current technology at current prices are known as **reserves.** If one divides the reserves for each element by its projected rate of use, one can determine the number of years to its depletion.

Studies conducted in the early 1970s indicated that many critical elements would be exhausted within the first half of the twenty-first century, that is, in 40 to 70 years. This finding caused initial alarm. Subsequent studies, however, concluded that the situation is less dire for many reasons.

## Changing Reserves

### FACTORS INCREASING RESERVES

Contrary to what one might think, reserves are not static quantities that are only depleted by exploitation. The term *reserve* does not refer to the total amount that exists; it includes only known deposits that can be profitably exploited at current prices with current technology. Therefore reserves can be increased in a number of ways.

First, discoveries of new deposits obviously add to reserves. The discovery process has been aided in recent years by Earth Resources Technology Satellites, which allow scientists to observe geological features on a scale not previously possible.

Second and more significant are price increases. A mineral deposit will not be exploited at an economic loss. But, shortages, real or anticipated, cause the price to rise, and this in turn makes it economically practical to mine lower-grade ores. The same can be said for mining ores in remote locations. For instance, much of the ocean floor is covered with mineral nodules formed by crystallization of elements from seawater (Fig. 18-2). When it becomes economical to mine these nodules, there will be a vastly increased supply of the minerals they contain.

Third, advances in technology may substantially increase reserves by reducing the cost of exploiting lower-grade ores. Thus, lower-grade ores, of which there are frequently massive quantities, are added to reserves.

**FIGURE 18-2**
Manganese nodules found on the ocean floor. These nodules, formed by crystallization of various elements from seawater, are rich in manganese, copper, and nickel. Discovery of these nodules and development of efficient mining techniques have resulted in a great increase in the availability of these metals. (a) Nodules on the ocean floor. (b) Closeup. Cut nodule on left shows rings of crystallization. (National Oceanic and Atmospheric Administration photos.)

Fourth, advances in technology turn formerly worthless minerals into valuable resources. To illustrate, aluminum was virtually unknown before 1900 and aluminum ore, for all practical purposes, was just another rock. Then aluminum refining technology was developed and now, aluminum is second only to iron in the quantity used. Indeed, high-grade aluminum ores are being depleted, but aluminum is a common element. As technological advances permit economical exploitation of lower-grade ores, the reserves of aluminum will become much greater.

For the future, research is moving toward the fabrication of ceramics into metallike materials. The fruition of this research will make huge quantities of common clay into a "metal" resource.

Hence, despite increasing exploitation and use of mineral resources over the years, estimates of remaining reserves of a number of resources have actually moved upward (Fig. 18-3). Hence,

more recent analyses have concluded that running short of any critical element before 2050 is unlikely, even when increasing global population and development are considered. This provides time for the development of new technologies that will extend the lifetime of reserves.

These facts seem to support the arguments of the cornucopians but there are still other points to be considered.

### FACTORS LIMITING RESERVES

The mere existence of reserves does not necessarily make them available for exploitation. Certain factors may preclude exploitation or access to known reserves.

**Negative environmental impact.** The only practical way to obtain most minerals, especially low-grade ores, is through open-pit mining (Fig. 18-4). This type of mining devastates the land in-

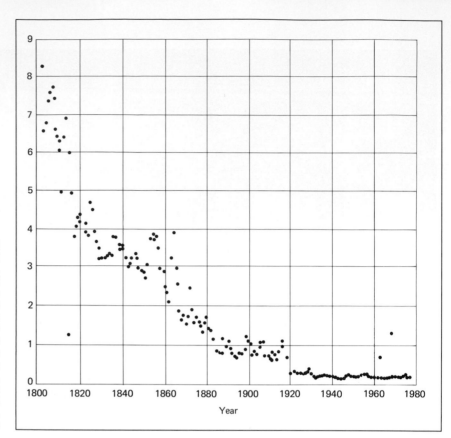

FIGURE 18-3
Scarcity of copper as measured
by its price relative to wages.
Since price is a function of
supply and demand, the
declining price of copper, relative
to wages, indicates that available
supplies of copper have actually
increased over the years despite
exploitation and usage. (From
Julian Simon, "Life on Earth is getting
better, not worse," *The Futurist*,
August 1983, p. 7–14.) Reprinted
with permission from the World
Future Society, 4916 St. Elmo Ave.
Bethesda Md. 20814

FIGURE 18-4
Ecological impacts of mineral
exploitation. The only practical way to
exploit many minerals is by open-pit
mining. This is an open-pit copper mine
near Salt Lake City, Utah. What are the
ecological impacts? (Paolo Koch/Photo
Researchers, Inc.)

volved—it frequently has a negative impact on
groundwater and commonly pollutes surface water with sediment runoff and leaching from mine
spoils (waste rock and soil that remain after the
desired mineral is removed [Fig. 18.5]). Smelting
the ores (extracting the desired element from the

mineral) produces more air pollution than any
other industrial process. Obviously, obtaining desired elements from the earth is fraught with negative environmental impacts. Additionally, reserves may be located in areas we simply do not
wish to sacrifice. Debate is becoming more com-

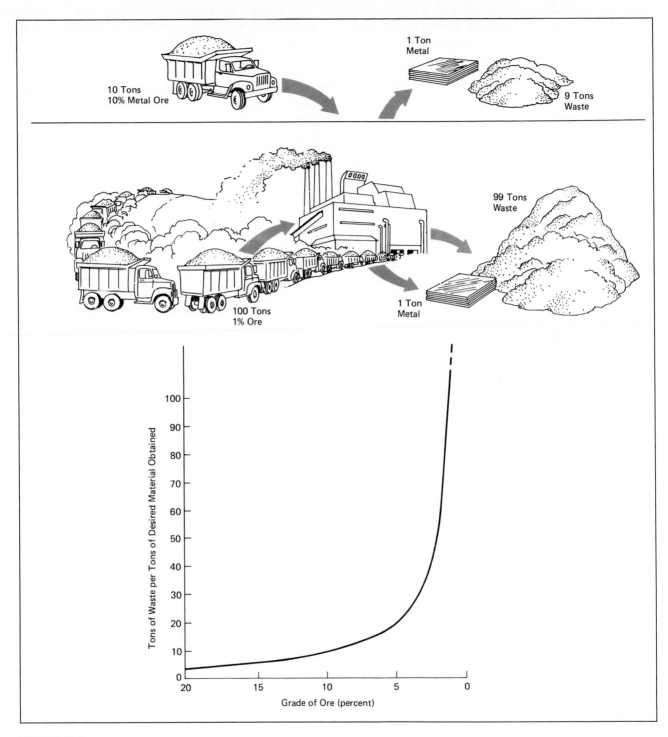

**FIGURE 18-5**
Mine wastes. The percentage of ore that is not the desired
material ends up as waste. Mining lower-grade ores means
increasing tonnages of waste.

mon and heated over questions about whether we should protect an area for its wilderness and wild-life value or sacrifice it for the minerals that lie beneath it. Even existing national parks and monuments, which are legally protected from such exploitation, are increasingly under attack as industries seek exceptions to their protected status to enable exploitation. Unfortunately, the negative environmental impacts will multiply as high-grade ores are depleted and industry begins exploiting lower-grade ores. For example, high-grade ores that contain about 30 percent copper have been exhausted and industry is now getting copper from ores that contain 1 percent copper or less. To obtain a ton of copper from 30 percent ores requires mining only about 3 tons. Extracting the copper generates only about 2 tons of waste. Obtaining a ton of copper from 1 percent ore requires mining 100 tons and produces 99 tons of waste.

**FIGURE 18-6**

Limits of available energy may place limits on mineral resources because the amount of energy required to obtain the ore increases exponentially as the ore grade decreases. This graph shows energy requirements for obtaining copper from different grades of ore. (Redrawn from Earl Cook, "Limits to Exploitation of Nonrenewable Resources," p. 63, in *Materials: Renewable and Nonrenewable Resources*, ed. P. H. Abelson and A. L. Hammond. © 1976 by the American Association for the Advancement of Science.)

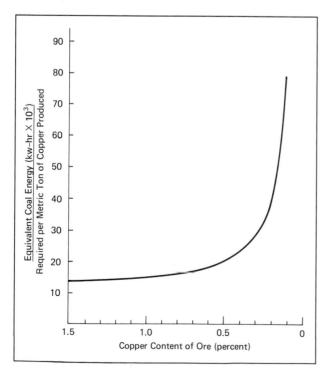

**Limits of energy.** Mining and smelting lower-grade ores also involves an exponential increase in the amount of energy needed (Fig. 18-6). Energy shortages could cause many mineral reserves to become unexploitable. It is tempting to think that, if industry wants an element badly enough, it will expend any effort to get it. In the final analysis, however, an element is only useful if it makes human life more enjoyable, easy, or healthful. If energy shortages arise, society will demand that the energy available be used for essentials and will do without elements that play less important roles.

**Political constraints.** While the United States is the world's most prodigious user of mineral resources, we depend on foreign sources for at least a portion of most of the minerals we desire (Fig. 18-7). Political shifts or revolutions in any of the countries that supply us could severely affect availability. Political events were responsible for the chaotic "energy crisis" in oil supplies of the 1970s.

## Changing Rate of Use

### INCREASING RATE OF USE

Despite studies that allowed for increases in reserves, limitists still predicted impending exhaustion of many resources. They based their prediction on the fact that the rate of use is increasing exponentially. Specifically, in the early 1970s it was observed that, due to increasing population and affluence, the use of many key resources was growing at a rate of 3.5 percent per year, or a doubling of consumption every 20 years (Fig. 18-8). Significantly, each doubling period (20 years in this case) involves the use of more than was used in all the preceding years. This can be seen in any doubling sequence of numbers, such as 1, 2, 4, 8, 16, 32. . . . Each successive number is greater than the sum of all the preceding numbers: 4 is greater than the sum of 1 plus 2, 8 is greater than the sum of 1 plus 2 plus 4, and so on. Thus, assuming that growth in use continues at 3.6 percent per year, more resources will be used in the period from 1980 to 2000 than were used in all the years up to 1980. Then, in the period from 2000 to 2020, the total consumption of all time will be more than doubled again. As illustrated in Figure 18-9, extrapolating **exponential growth** in this

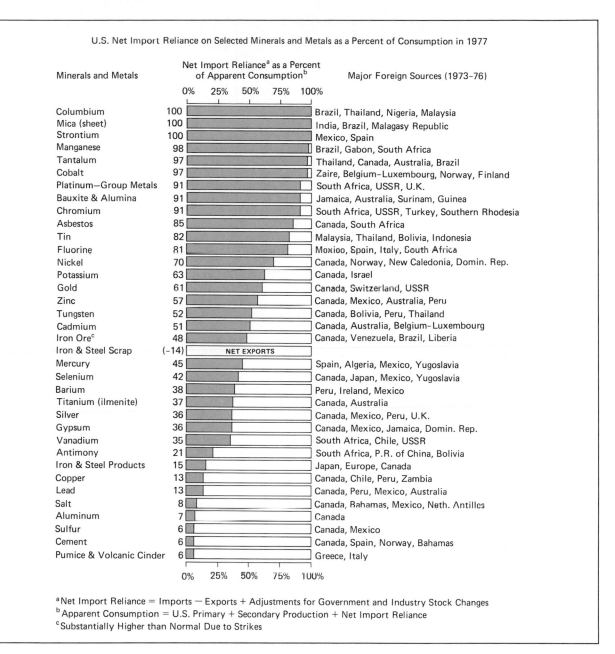

U.S. Net Import Reliance on Selected Minerals and Metals as a Percent of Consumption in 1977

| Minerals and Metals | Net Import Reliance[a] as a Percent of Apparent Consumption[b] | Major Foreign Sources (1973-76) |
|---|---|---|
| Columbium | 100 | Brazil, Thailand, Nigeria, Malaysia |
| Mica (sheet) | 100 | India, Brazil, Malagasy Republic |
| Strontium | 100 | Mexico, Spain |
| Manganese | 98 | Brazil, Gabon, South Africa |
| Tantalum | 97 | Thailand, Canada, Australia, Brazil |
| Cobalt | 97 | Zaire, Belgium–Luxembourg, Norway, Finland |
| Platinum–Group Metals | 91 | South Africa, USSR, U.K. |
| Bauxite & Alumina | 91 | Jamaica, Australia, Surinam, Guinea |
| Chromium | 91 | South Africa, USSR, Turkey, Southern Rhodesia |
| Asbestos | 85 | Canada, South Africa |
| Tin | 82 | Malaysia, Thailand, Bolivia, Indonesia |
| Fluorine | 81 | Mexico, Spain, Italy, South Africa |
| Nickel | 70 | Canada, Norway, New Caledonia, Domin. Rep. |
| Potassium | 63 | Canada, Israel |
| Gold | 61 | Canada, Switzerland, USSR |
| Zinc | 57 | Canada, Mexico, Australia, Peru |
| Tungsten | 52 | Canada, Bolivia, Peru, Thailand |
| Cadmium | 51 | Canada, Australia, Belgium–Luxembourg |
| Iron Ore[c] | 48 | Canada, Venezuela, Brazil, Liberia |
| Iron & Steel Scrap | (–14) NET EXPORTS | |
| Mercury | 45 | Spain, Algeria, Mexico, Yugoslavia |
| Selenium | 42 | Canada, Japan, Mexico, Yugoslavia |
| Barium | 38 | Peru, Ireland, Mexico |
| Titanium (ilmenite) | 37 | Canada, Australia |
| Silver | 36 | Canada, Mexico, Peru, U.K. |
| Gypsum | 36 | Canada, Mexico, Jamaica, Domin. Rep. |
| Vanadium | 35 | South Africa, Chile, USSR |
| Antimony | 21 | South Africa, P.R. of China, Bolivia |
| Iron & Steel Products | 15 | Japan, Europe, Canada |
| Copper | 13 | Canada, Chile, Peru, Zambia |
| Lead | 13 | Canada, Peru, Mexico, Australia |
| Salt | 8 | Canada, Bahamas, Mexico, Neth. Antilles |
| Aluminum | 7 | Canada |
| Sulfur | 6 | Canada, Mexico |
| Cement | 6 | Canada, Spain, Norway, Bahamas |
| Pumice & Volcanic Cinder | 6 | Greece, Italy |

[a] Net Import Reliance = Imports — Exports + Adjustments for Government and Industry Stock Changes
[b] Apparent Consumption = U.S. Primary + Secondary Production + Net Import Reliance
[c] Substantially Higher than Normal Due to Strikes

**FIGURE 18-7**
Political constraints may limit resources. Considerable portions of mineral resources used by the United States are imported from other countries. (U.S. Department of the Interior, Bureau of Mines.)

way creates a scenario in which even vast reserves are rapidly depleted.

Cornucopians basically agree with limitists that exponential growth in resource use cannot be sustained indefinitely. However, cornucopians argue that natural economic forces cause the econ-omy to change and adapt to impending shortages of various elements without undue upsets. Limitists cite the need for what are basically the same economic changes and adapation, but question whether natural economic forces would act sufficiently or in time to avert an economic collapse caused by sudden shortages. Thus, limitists advocate strong regulations to promote the changes.

Experience over the last decade has given more credence to the philosophy of the cornucopians. Even without strong regulatory acts, eco-

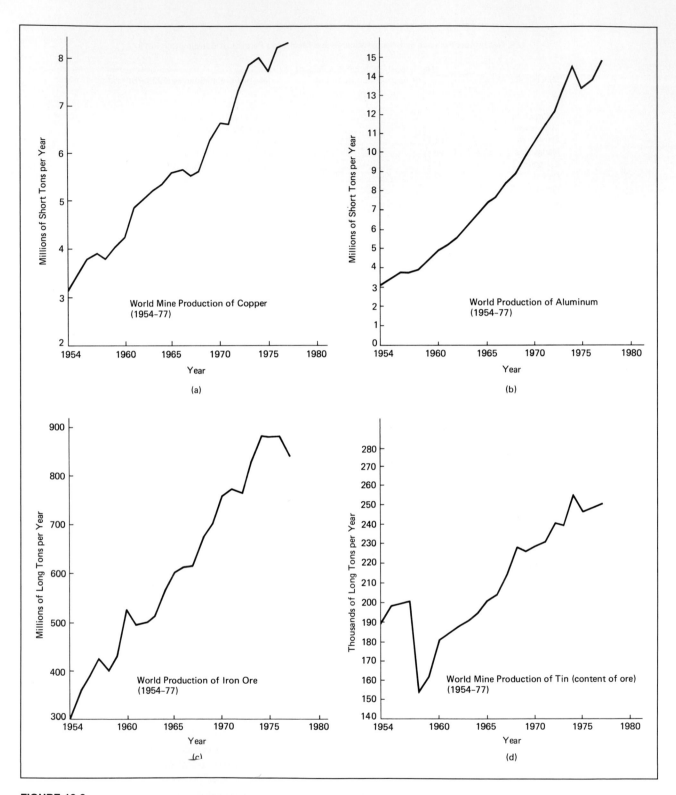

**FIGURE 18-8**
World use of a few important resources. Humans are making
increasing demands on the earth's resources. (U.S. Department
of the Interior, Bureau of Mines.)

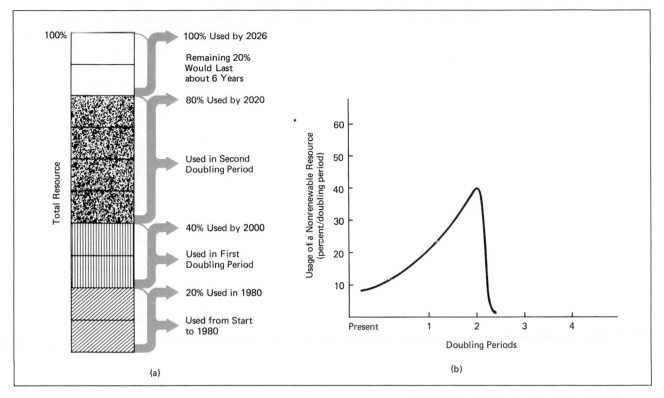

(a)                                              (b)

**FIGURE 18-9**
(a) Consumption of a nonrenewable resource, assuming that in 1980 20 percent of the total had been consumed and growth in use continues at 3.5 percent per year. (b) Continuing growth in use up to the limits will lead to a very sharp decline as the resource is exhausted. What will be the economic, political, and social consequences?

nomic changes and adaptations are occurring; the rate of resource consumption is not growing as rapidly in most cases, and, consequently, the outlook for mineral resource availability is brighter.

## DECREASING RATE OF USE

Growth in consumption of critical mineral resources has been mitigated by the factors discussed below.

**Downsizing and reduction in the amount of critical material used in products.** The most conspicuous example of downsizing is the reduction in the size of cars since the early 1970s. The primary motivation for the size reduction was energy conservation, although a decrease in use of metals also resulted. A more striking downsizing has occurred in electronic instruments and other devices. The amount of mercury, a crucial metal

now used in thermostatic switches, for example, is about 2 percent of what it was in the 1950s. Importantly, the downsizing did not decrease reliability or durability. In fact, these factors increased in many cases.

**Substitutions.** In technology and industry a particular element is desired for its chemical and physical properties. If the same properties can be obtained from other more abundant materials, or if a job can be done in a way that does not utilize a rare element, then the depletion of that element is of little consequence. A good example is the replacement of metal parts with plastic parts in cars, appliances, electronic instruments, and plumbing. Optical fibers are replacing metallic wires used in communications systems. Electronic temperature measuring devices have replaced mercury thermometers in doctor's offices and hospitals. Replacing metals with ceramic materials is yet another example of substitution.

It is likely that science and technology will continue to develop substitutes for materials as well as techniques that will lessen the need for rare elements. These substitutes will allow remaining reserves to be used in areas where substitutes are not yet or never will be available. For instance, the elements used as nutrients in plant fertilizer cannot be replaced with substitutes.

**Increasing durability.** If the useful lifetime of an appliance is, for example, doubled, the demand for material resources for producing that appliance is cut in half. The lifetime of many products seemed to hit a low point in the 1960s and early 1970s. Indeed, "planned obsolescence" all but forced consumers to discard old products and buy new ones. In recent years, however, the trend is toward more durable products, and consumers are keeping major appliances for longer periods of time. The average life expectancy of a car, for example, has increased from about 9 to 15 years.

**Recycling.** Because atoms of elements are never destroyed, the use of an element does not theoretically reduce its quantity. However, as products are used and/or disposed of, it becomes economically impractical to recover the elements in them—that is, the costs of recovery exceed the costs of exploiting ore deposits. But this too is changing. Industries are making increased efforts to recover desired elements from wastes and recycle them. Similarly, recycling of metals from discarded consumer products is on the rise. Inevitably, there will always be some waste; 100 percent recycling seems to be impossible. And recycling eventually becomes a moot issue when all or most of the available supply of an element is tied up in products that are in use. However, recycling will postpone shortages and provide more time for development of substitutes and/or alternative technologies.

## ARE REGULATORY ACTS NECESSARY?

In the 1960s and early 1970s, limitists advocated strong regulatory acts to curtail the growing use of critical mineral resources. Basically, they felt our economic system would not respond appropriately without mandatory steps such as rationing or without crash programs promoting recycling and the development of substitutes. Yet, a transition seems to be occurring without such drastic laws or regulations controlling our system.

### Adaptations in a Free Economy

A free economy seems better able to adapt to impending shortages than many believed possible. The key force that prompts this adaptation is price. Shortages are foreseen and reflected in rising prices. In turn, increasing prices provide the impetus for new discoveries, the exploitation of lower-grade ores, the development of new technologies and substitutes, the use of more abundant and less expensive resources, and an increased reliance on recycling. The increasing cost of metals also leads to downsizing and extended product lifetime.

Of course, raising the price of resources contributes to inflation. It is important to realize, however, that the increased cost of raw materials is a relatively small factor in the price of finished products because raw materials generally represent only 1 to 5 percent of the final cost of a product. Fabrication of components, assembly, distribution, and marketing make up the remaining portion. For instance, there is no more than about $20 worth of raw materials in a $2000 home computer. Even if raw material costs doubled, only about 1 percent would be added to the value of the computer.

Conversely, regulating the price of a resource at a low level is an almost sure way to create a shortage artificially because appropriate adaptations become economically impractical. For example, most economists feel that the natural gas shortages of the late 1970s were caused because prices were regulated at low levels. These controls on the market price prevented the development of new, hard-to-reach fields. In conclusion, the best economic policy for mineral resources appears to be a free market approach.

### Importance of International Diplomacy

When considering economics and the free market, however, it is essential to recognize that significant portions of our mineral resources are imported. International diplomacy aimed at developing and maintaining open trade throughout the world is exceedingly important to the flow of resources.

### Energy Policy

Since mining and processing mineral resources are heavily dependent on energy, any shortage or disruption of energy supplies will effectively cause shortages of numerous mineral resources. Consequently, policies assuring adequate energy supplies in the future are vitally important. This matter will be discussed more fully in Chapters 21, 22 and 23.

## Environmental Policy

Exploitation of mineral resources will always involve some tradeoffs with air and water quality and with natural biota. Some of these tradeoffs may be warranted because certain mineral resources are crucial to the health and benefit of human societies. Unfortunately, however, air, water, and natural biota have traditionally been taken for granted and assumed to be free. Our economic system does not give them value in proportion to their real worth. Water charges, for example, include piping, pumping, and treatment fees. There is no charge for the water itself, despite the fact that no life can survive without it. Hence, while economic forces act to open up new supplies of mineral resources, obtaining these sources often requires the sacrifice of air, water, and natural biota. These sacrifices are being questioned more and more and have become the subject of debate among industry, the government, and environmental organizations. However, the outcome of such debates frequently depends more on who has the financial resources to carry on extended litigation than on the facts and actual benefits to society as a whole. It is therefore necessary to develop policies through which we can project and weigh the various tradeoffs. However, before this can be done, we must understand the value of natural biota and the many negative impacts it faces. This is the subject of Chapter 19.

# 19

# BIOTA: BIOLOGICAL RESOURCES

---

## CONCEPT FRAMEWORK

Outline                                           Study Questions

1. What are *biological resources*? Why are biological resources renewable?

2. If most essential biological products are produced in managed systems, why are natural ecosystems so important? How do natural ecosystems act as a kind of "barometer" for the state of the biosphere?

3. Why is the genetic bank worth preserving?

4. Discuss some of the functions of biological resources. How does society accommodate for these resources when they are unavailable? How do wetlands lost to development illustrate this issue?

5. What limits should be placed on the uses of natural biota? What is the aim of conservation?

6. What is *maximum sustainable yield?* What issues are considered when determining maximum sustainable yield? What factors are difficult to account for?

7. How does overuse affect maximum sustainable yield? How could overuse be curtailed?

8. Discuss ways in which habitat destruction alters the ecosystem beyond the immediate area in question.

9. Recall an incident where pollution was a major cause of a reduction in maximum sustainable yield.

10. Give an example in which introduction of a foreign species altered an ecosystem.

11. Despite knowledge of the threat of extinction, why are some species still exploited?

12. Discuss the controversy over wetlands and waterfront development. How does the law of supply and demand affect desirable species or parcels?

13. How was the term *tragedy of the commons* coined? How is the term used today? Give an example.

14. How do unknown factors complicate and aggravate problems associated with declining populations?

15. In what ways do mineral resources differ from wildlife resources?

16. What three factors must conservation programs address?

17. What animals qualify as *game?* What programs attempt to preserve them? What problems are occurring?

18. What kinds of plants or animals are considered *exotic?* How does the history of the snowy egret illustrate the plight of exotic species? Why are other exotic species not similarly protected? Why are poachers difficult to stop? Discuss the benefits and shortcomings

of the Endangered Species Act. What steps have been taken to protect and propagate endangered species outside of their natural habitats?

19. How do whales illustrate the exploitation of marine species? What accomplishments were made by the United Nations Law of the Sea Conference?

20. What led to most fish declines before 1950? What was the major factor after 1950? What are the economic and ecological consequences?

21. Why is the term *wasteland* applied to some natural areas? Why is this term a misnomer? What steps have been taken to protect these areas? Are they successful? Has the federal government helped or hindered the preservation of natural areas? What changes could be made to better protect natural biota?

22. What two questions must be addressed in a discussion of wildlife preservation?

23. How are yields increased? Why are artificial systems insufficient in and of themselves?

24. Where do artificial systems fail?

25. Summarize the basic problems hindering conservation in today's world and discuss their implications for the future. What areas do you believe need priority attention? Why?

Most of our foods, spices, natural fibers, drugs, wood and paper products come from a variety of plant and animal species. Additionally, many of our recreational and aesthetic pleasures hinge on natural biota and ornamental plantings. All living things that provide necessities and/or enjoyment can be considered biological resources. Biological resources include both domestic species, which are propagated in agricultural or other highly managed systems, and natural biota, the plants, animals, and microbes that constitute natural ecosystems.

Since biological resources may be replenished through reproductive cycles, they are frequently referred to as **renewable resources.** Renewability implies the maintenance of both a breeding population and an environment of suitable size and quality for reproduction to occur. If this situation ceases to exist, the species becomes extinct and, to quote a slogan of the World Wildlife Federation, "extinction is forever." The tragic paradox is that certain biological, renewable resources are more likely to be destroyed by humans than exhausted by them. The following is an overview of the situation.

Primitive humans obtained all of their food and many of the materials they used for clothing, shelter, and implements by hunting and gathering animals and plants from the wild. The natural production of desired species, however, varies widely and, in many cases, quantities are limited. In many cases, overhunting occurred and led to the extinction of valuable species. Humans adapted to this situation by becoming increasingly skilled at developing agricultural and other artificial systems in which desired plants and animals could be propagated. These systems have so greatly improved the quantity and reliability of production that there is now little, if any, need to exploit natural biota as such. Yet direct exploitation of wildlife continues. And, more seriously, massive alteration or outright destruction of natural ecosystems and extinction of biota are accelerating as land is converted from natural ecosystems to various human uses and biota are adversely affected by other human activities such as pollution.

The potential consequences are grave. While mineral resources are necessary for an industrial society, natural biota underpin survival itself. The objective of this chapter is, first, to provide a better understanding of the value and importance of natural biota. Second, we shall examine in detail the factors that influence the exploitation of natural biota. Finally, we shall address the steps that have been taken and need to be taken to protect various biota.

## VALUES OF NATURAL BIOTA AND ECOSYSTEMS

Frequently, the loss of natural biota and ecosystems is simply accepted as a necessary tradeoff for human progress. It is often argued that the loss is small in terms of overall human progress since we can obtain all necessary biological resources from agricultural and other artificial systems. The contention that the entire ecosystem may collapse because of the extinction of a single species is, in large part, false. Indeed, human progress has continued despite the extinction of the great auk, the passenger pigeon, the dodo, and many other species (Fig. 19-1). Nevertheless, very real and important concerns do exist. We must assess what we stand to lose in the sacrifice of natural biota and ecosystems.

### Commercial, Recreational, Aesthetic, and Scientific Values

Most foods and other essential or valuable biological products are produced in agrucultural or otherwise highly managed systems. However, natural ecosystems, with little or no management are still the major source of fish, shellfish, and many forest products. Natural ecosystems also provide the foundation for numerous recreational and/or aesthetic pleasures ranging from sport fishing and hunting to hiking, camping, birdwatching, and the simple enjoyment of scenery. These recreational or aesthetic activities also support commercial interests such as the production and marketing of fishing and camping gear, lodging services, and tour guides. Indeed, as leisure time increases, larger portions of the economy become connected to the natural environment. Conversely, loss or degradation of any portion of the natural environment affects commercial interests and causes an aesthetic and/or recreational loss. Examples abound of businesses folding or going bankrupt when a lake, for example, is polluted or otherwise despoiled (Fig. 19-2).

Society is also indirectly affected by environmental losses and gains. For instance, one may

**FIGURE 19-1**
"Graveyard" of extinct species at Bronx Zoo in New York City. Human impacts have already caused the extinction of numerous species. Only a small portion of the extinct species are depicted on these gravestones. (New York Zoological Society photo.)

never see a whale, but knowing that whales and other exotic species exist provides a certain aesthetic pleasure. Further, knowing that the earth and its biosphere continues to support and maintain such wildlife provides a sense of well-being. When various plants and animals are threatened with or pushed into extinction, we experience a sense of uneasiness about how far the process will go and where it will ultimately end.

In short, wildlife may act as an "early warning system" for humans—a kind of "barometer" for the state of the biosphere. In the early days of coal mining, canaries were taken into mines because they are more sensitive to poisonous methane than humans; thus, when canaries died, it warned miners that dangerous amounts of methane gas were present. Extinction of wildlife species warns us that all is not well in the biosphere and calls into question our management of the planet. If the warning is not heeded, catastrophes of much greater magnitude could lie ahead.

Although scientific study of biota has enabled us to learn much about how organisms develop and what ecosystems need to function and

survive, there is much that we do not know. The destruction and upset of natural biota will ultimately deprive us of the "laboratories" we need to gain a complete understanding of how to operate the planet on a sustainable basis.

Even if we learn to operate without whales, wild tigers, and other forms of exotic wildlife, is this the life we wish for ourselves and future generations? Or is coexistence with these diverse kinds of species in their natural ecosystems part of the quality of life we wish to preserve?

## Genetic Bank

One of the most important reasons for maintaining natural biota is that they provide the foundation for all agricultural plants and animals. The entire range of our domestic plants and animals is derived from wild species that have been modified through the techniques of selective breeding and hybridization. Moreover, only a small por-

**FIGURE 19-2**
Natural biota provide numerous recreational and aesthetic values, a few of which are depicted here.

tion, less than 0.1 percent, of wild species has been thoroughly examined for its potential to produce food, fibers, spices, drugs, and other useful products. Thus, wildlife represents a vast storehouse of plants and animals which, by domestication, could become valuable producers of innumerable products.

Maintaining the vigor of highly inbred do-mestic species requires the periodic infusion of new genes from related wild species, a process carried on by hybridization and reselection. Similarly, remarkable adaptations providing unique resistance to such adverse conditions as drought, high salinity, and attack of pests and disease may be found in certain wild species. Modern techniques of biotechnology (genetic engineering) provide the opportunity to isolate the specific genes for these traits from wild species and introduce them into domestic species. The potential benefits of this process are incalculable. Finally, natural ecosystems undoubtedly hold yet-undiscovered species that may be invaluable in biological pest control.

In short, the plant and animal wildlife present in natural ecosystems represents a reservoir of genetic material that is frequently referred to as a **genetic bank.** Our present and future production of food and other biological resources is and will be utterly dependent on "withdrawals" from this genetic bank. Clearly, destruction of natural ecosystems and extinction of species diminishes this genetic bank and undercuts the potential for production of all biological resources (Fig. 19-3).

The destruction of tropical rainforests is of special concern here. These forests contain a greater species diversity (number of different species) than any other ecosystem. They represent an incredible genetic bank. Yet, in nearly every area where they exist (Central and South America, Africa, Asia, and Indonesia), methods ranging from bulldozing to slash-and-burn techniques are employed to destroy these forests and clear land for development. Undoubtedly, the number and abundance of species is being lost before it is even tallied.

## Nature's Services

In addition to producing various plants and animals that may be valuable biological resources, natural ecosystems perform many other tasks. For example, the vegetative cover of a forest or grass-

**FIGURE 19-3**
Natural biota serve as a "genetic bank." Wild species have many invaluable uses as depicted here. Hence, a healthy, diversified natural biota may be viewed as a genetic bank from which withdrawals for these uses can be made at any time. What will be the effects on the areas illustrated if natural biota are destroyed?

NATURAL BIOTA ARE A GENETIC BANK FOR:

Enhancing Agricultural Production. Traits for:

Disease Resistance

Pest Resistance

Salt Tolerance

Increased Vigor

Increased Productivity

Drought Tolerance

Increased Nutritional Value

Species Useful in Biological Pest Control

Insect Parasites

Weed-eating Insects

Weed-eating Fish

Insect-eating Insects

Species with Medicinal Value. Compounds for Treatment of:

Heart Disease

Cancer

Metabolic Disorders

Nervous Disorders

Other Conditions

Arthritis

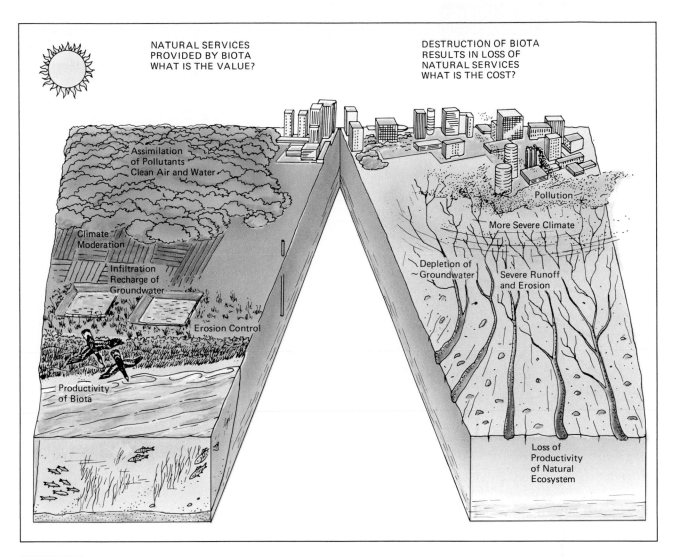

NATURAL SERVICES
PROVIDED BY BIOTA
WHAT IS THE VALUE?

DESTRUCTION OF BIOTA
RESULTS IN LOSS OF
NATURAL SERVICES
WHAT IS THE COST?

Assimilation
of Pollutants
Clean Air and Water

Pollution

More Severe Climate

Climate
Moderation

Infiltration
Recharge of
Groundwater

Depletion of
Groundwater

Severe Runoff
and Erosion

Erosion Control

Productivity
of Biota

Loss of
Productivity
of Natural
Ecosystem

**FIGURE 19-4**
Natural biota perform invaluable natural services, including
water and air purification, control of the water cycle,
propagation of fish, and climate control. Destruction of natural
biota results in the loss of these services. What are the costs
to humans?

land prevents soil erosion. It also helps replenish
groundwater and control flooding by enhancing
infiltraton and reducing runoff of water. Natural
ecosystems assimilate both air and water pollutants, at least those that are biodegradable, and
thus enhance air and water quality. We should
recognize that these are distinct "services" of natural ecosystems (Fig. 19-4).

When we lose these natural services, the
costs are quite enormous. Economically, the burdens may include the need to build additional
treatment plants to provide adequate water quality. More often, however, the costs involve a deg-

radation in the quality of our lives because we are
forced to live with decreased air and water quality.

The development of wetlands exemplifies
this phenomenon. Tidal and nontidal wetlands
exist. Tidal wetlands are the broad expanses of
low-lying grassland surrounding bays and estuaries. Examples are seen along the East and Gulf
coasts of the United States (Fig. 19-5a). Tidal wetlands are immersed in shallow water during high
tides and stand slightly above water at low tides.
Consequently, they are not useful for either water
or land-based human activities and have generally
been considered wastelands. Developers have
turned these "wastelands" into tremendous
money-making projects. By dredging and filling
in the wetlands, they create valuable waterfront
property (Fig. 19-5b).

(a)

**FIGURE 19-5**
Tidal wetlands are supremely important in the biological productivity and, through their capacity to absorb nutrients and pollutants, the protection of water quality of bays and estuaries. (a) Tidal wetlands at Wallops Island, Virginia. Inserts: Mussels; a willet nest. (Photos by author.) (b) Dredging, filling, and bulkheading of wetlands to develop waterfront properties. (EPA-Documerica photo by Flip Schulke; courtesy of U.S. Environmental Protection Agency.)

(b)

Similarly extensive areas of nontidal or freshwater wetlands exist. These areas, which are generally well drained during part of the year and waterlogged or flooded at other times, are known as marshes, swamps, or bogs. Again, these wetlands are considered waste and there is a great deal of pressure to drain them and make the land suitable for agriculture, housing developments, or other seemingly more productive economic uses. But what natural services are lost in the exchange?

First, water entering the estuaries and bays from rivers and streams is normally filtered through the tidal wetlands. This process removes sediments, excess nutrients, and other pollutants, especially during times of flooding. In short, tidal wetlands serve as a water purification system protecting the estuary or bay and its aquatic life from pollution.

Second, the tidal wetlands are actually the source of a large portion of the biological productivity of bays and estuaries. The changing tides flush organic matter produced by the grasses into the water. This material becomes the basis of many food chains. In addition, numerous species of fish and shellfish spawn, or are dependent on those that do spawn, in the shallow, protected water of the many rivulets that permeate the tidal wetlands. Thus, dredging and filling or simply bulkheading shorelines has a pronounced negative effect on both water quality and biological productivity (Fig. 19-6).

Third, freshwater wetlands are very important aquifer recharge areas, that is, areas where water percolates into the ground and replenishes groundwater. In many cases, drainage of wetlands, a process that involves digging channels to drain the water off the surface rather than letting it soak in, has resulted in serious depletion of groundwater.

Finally, both tidal and nontidal wetlands are biologically very productive. Beyond the produc-

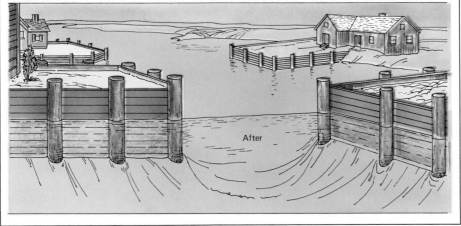

**FIGURE 19-6**
Tidal wetlands provide both the habitat and much of the food production for numerous fish and shellfish. They also serve to filter sediments and nutrients out of water. For each acre of wetlands that is destroyed by development or bulkheading shorelines, society stands to lose up to $100,000 per year in pollution control and wildlife propagation services.

Before

After

**FIGURE 19-7**
Ducks and geese at Klamath, Oregon. Vast flocks of waterfowl are also supported by wetlands. What happens to these when wetlands are developed or drained? (Animals Animals. © 1978, Margat Conte.)

ers (plants) there are huge populations of small fish, clams, mussels, crayfish, crabs, and snails of many kinds. In turn, wetlands are the feeding and breeding grounds for vast flocks of migratory waterfowl, such as geese, ducks, and herons (Fig 19-7).

When wetlands are dredged and filled or drained, these natural forms of pollution control and productivity are lost. How much are they worth? Scientists estimate that it would cost more than $100,000 a year to duplicate the water purification and fish propagation that just 1 acre (0.4 ha) of tidal wetland provides. This sum does not include the aesthetic value of waterfowl and other wildlife the wetlands support and which we are unable to duplicate at any cost. Still, in the United States alone, more than half of the original 215 million acres (86 million ha) of wetlands have been destroyed by development projects and another 458,000 acres (183,000 ha), an area half the size of Rhode Island, are diverted to development each year.

Other examples of natural services include such things as the soil biota, which are instrumental in topsoil formation, nitrogen fixation, and degradation of wastes. These services are lost when soil is destroyed by erosion (Fig. 19-8). Pollination is another tremendously valuable natural service that may be lost when bees are inadvertently killed with pesticides.

## ASSAULT AGAINST NATURAL BIOTA

Despite their values, natural biota and ecosystems are, in general, in a state of rapid and accelerating decline such that countless species are **threatened.** This means their populations are declining precipitously as a direct result of human activity.

**FIGURE 19-8**
Soil biota provide numerous natural services as depicted.

NATURAL SERVICES PROVIDED BY SOIL BIOTA

Breakdown and Assimilation of Wastes

Nitrogen Fixation
Topsoil Formation
Soil Aeration
Water Infiltration

Of these, several hundred species, including both plants and animals, are recognized as **endangered**—they are close to extinction. As a result, we are dangerously close to losing these aesthetic, scientific, and ecological resources forever.

Why has this situation occurred? All too often the issue is cast in simplistic terms of preservation versus greed. However, many factors are actually involved and in order to find meaningful solutions, we must gain a broader perspective.

## Conservation May Permit Use

Natural wildlife can provide biological resources on a renewable basis. Fundamentally, every species has the capacity to overreproduce. We have seen how factors of environmental resistance constantly eliminate the excess population which the ecosystem cannot support (Chapter 3). It is difficult to find fault with hunting that effectively puts some of this excess population to human use. Indeed, removal of excess numbers of herbivores is sometimes an absolute necessity to prevent overgrazing and the destruction of the ecosystem as a whole. However, ethical questions arise when hunters take more than the excess and deplete the breeding populations, threatening or causing extinction of the species. The same can be said about all uses of natural biota.

Conservation of natural biota then, does not, or at least should not imply complete denial of any usage, although it may be temporarily expedient in a management program to allow a certain species to recover its population size. In fact, it is interesting to consider that if we were to deny all use of a resource, we might as well not have the resource in the first place or it might as well be obliterated immediately for it will be of no further use. Consequently, the aim of conservation is to manage the use so that it does not exceed the capacity of the species or system to renew itself.

## Concept of Maximum Sustainable Yield

The concept of sustainable yield is central to conservation. Forests, fish and shellfish, deer, soils, and parks are examples of renewable resources. But how many trees can be cut each year without destroying the forest? How many deer may be killed or fish and shellfish caught each year without depleting the populations beyond the recovery point? How many visitors can a park accommodate without damage? How much cultivation can a soil sustain without degradation? The maximum use a system can sustain without impairing its ability to renew itself is the **maximum sustainable yield.**

Note that "use" can include cutting timber or fishing for personal or commercial interests; it includes visiting a park for aesthetic pleasure; it can even include discharging pollutants into a stream. Natural systems can withstand a certain amount of use (or abuse in terms of pollution) and still remain viable. However, a point exists where increasing use begins to destroy regenerative capacity.

For example, consider a stable population of 50 animals which produce 10 offspring a year. The maximum sustainable yield from this population is 10 animals per year; 10 older animals are replaced by 10 young ones each year. It would be possible to go beyond the maximum sustained yield and kill 10 additional animals per year for a few years, but this would be at the expense of the breeding population and fewer offspring would be produced. Obviously, such a practice could not be sustained; it would lead to extinction of the population in about five years (Fig. 19-9).

This example illustrates the general concept, but real situations involve many other factors. First, replacement of a population depends on more than the reproductive rate. It depends on recruitment—the percentage of young that survive to reach adulthood and the time it takes for the individuals to grow from birth (or hatching) to reproductive age. Some species of fish and shellfish, for example, have tremendous reproductive rates. The females produce thousands of eggs. But recruitment may be so low that actual replacement of harvested stocks is slow. Clearly, if replacement rates are low, a sustainable yield can be no more than a very small portion of the population each year. On the other hand, if replacements rates are high, the population can sustain a substantial portion of the population being harvested each year (Fig. 19-10).

A second important consideration in determining maximum sustainable yield is the **carrying capacity** of which, as noted in Part II, is the maximum population that the ecosystem can support. If a population is well within the carrying capacity of the ecosystem, allowing a population to grow

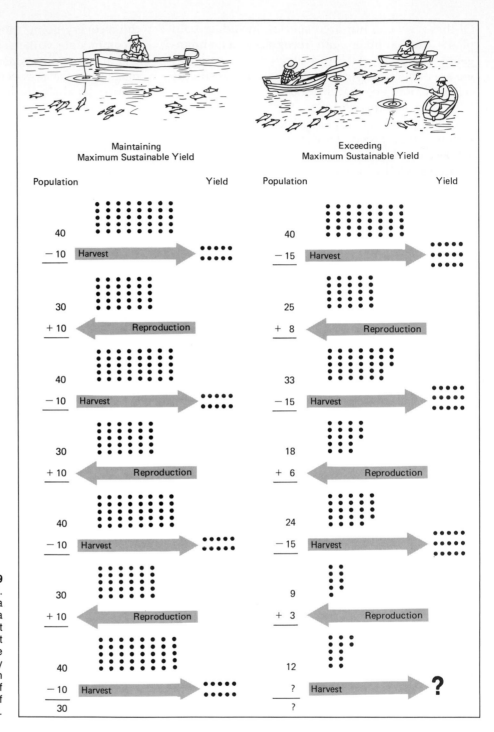

**FIGURE 19-9**
Maximum sustainable yield. Biological resources all have a maximum sustainable yield, a maximum harvest or use that can be sustained without impairing their productivity. The maximum sustainable yield may be exceeded in the short run only at the expense of undercutting the productivity of the system in the long run.

will increase the number of reproductive individuals, and thus, the sustainable yield. However, as the population approaches the carrying capacity of the ecosystem, new individuals must compete with those that are already present for food and living space. As a result, recruitment may fall

drastically. The net production in a mature climax forest, for instance, is essentially zero because new growth is more or less balanced by natural dieoff. If populations are at or near the carrying capacity of the ecosystem, production, and hence sustainable yield, can actually be increased by

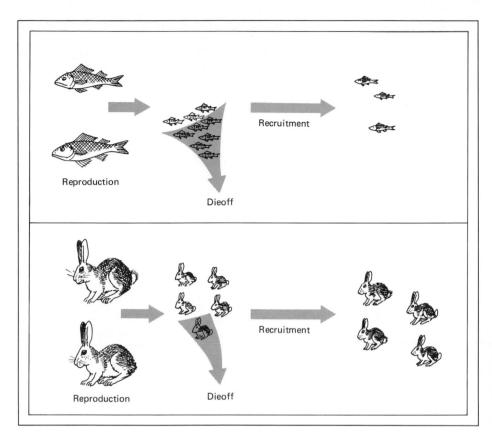

**FIGURE 19-10**
Reproduction, recruitment, and replacement. The portion of adults that may be replaced by young each year depends on both reproduction (birth rate) and recruitment, the percentage of young that survive to become adults. Replacement is often affected more by recruitment level than by reproduction.

thinning the population so that competition is reduced and optimal growth and reproductive rates are achieved. Thus, maximum sustainable yield is obtained with *optimal*, not *maximum* population size (Fig. 19-11).

The matter is further complicated, however, by the fact that neither optimum population size or replacement capacity are constant. For example, reproduction of fish may remain relatively constant. Females may still lay thousands of eggs, but in some years weather may provide conditions that are particularly favorable for recruitment; other years may be unfavorable. Similarly, an optimal population one year may be unsupportable another year if weather or other factors change. For example, drought causes dieoffs of

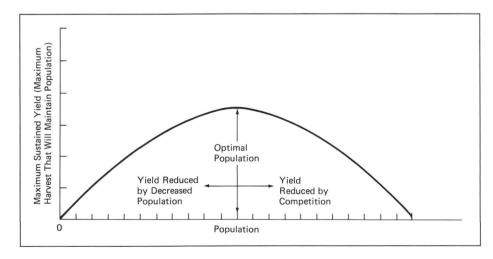

**FIGURE 19-11**
Maximum sustainable yield occurs at an optimum population level. At populations above the optimum, yield is diminished by competition between organisms. At levels below the optimum, yield is diminished by reduced stock.

many animals by reducing vegetation and drying up water holes. Human impacts such as pollution may adversely affect reproductive rates, recruitment, carrying capacity, and, consequently, sustainable yields. With proper management and conservation techniques, however, these negative impacts may be minimized and yields may be increased (Fig. 19-12). We shall examine this further in the following sections.

## Human Factors Undercutting Maximum Sustainable Yield

Despite natural fluctuations in population, clear evidence exists that human factors are reducing sustainable yields and even threatening extinction in countless situations. Four major factors are involved. One is the direct impact of overuse; the remaining three factors involve impacts from other human activities.

### OVERUSE

Despite the shortsightedness evinced by the overhunting, overfishing, and overcutting of forests, these practices are on the rise. Many valuable species are threatened with extinction because they are being hunted, fished or cut faster than they can reproduce.

### HABITAT DESTRUCTION OR ALTERATION

When a habitat is altered it may no longer support certain plants or animals. The change can affect additional plants and animals by altering the food web. An obvious example of destruction is the bulldozing carried out to make way for urban development, highways, and even parks. Habitat destruction in equatorial regions is of particular concern. Millions of acres of tropical rainforests are being destroyed to increase the amount of farmland. Unfortunately, because of soil deterioration under cultivation, this land

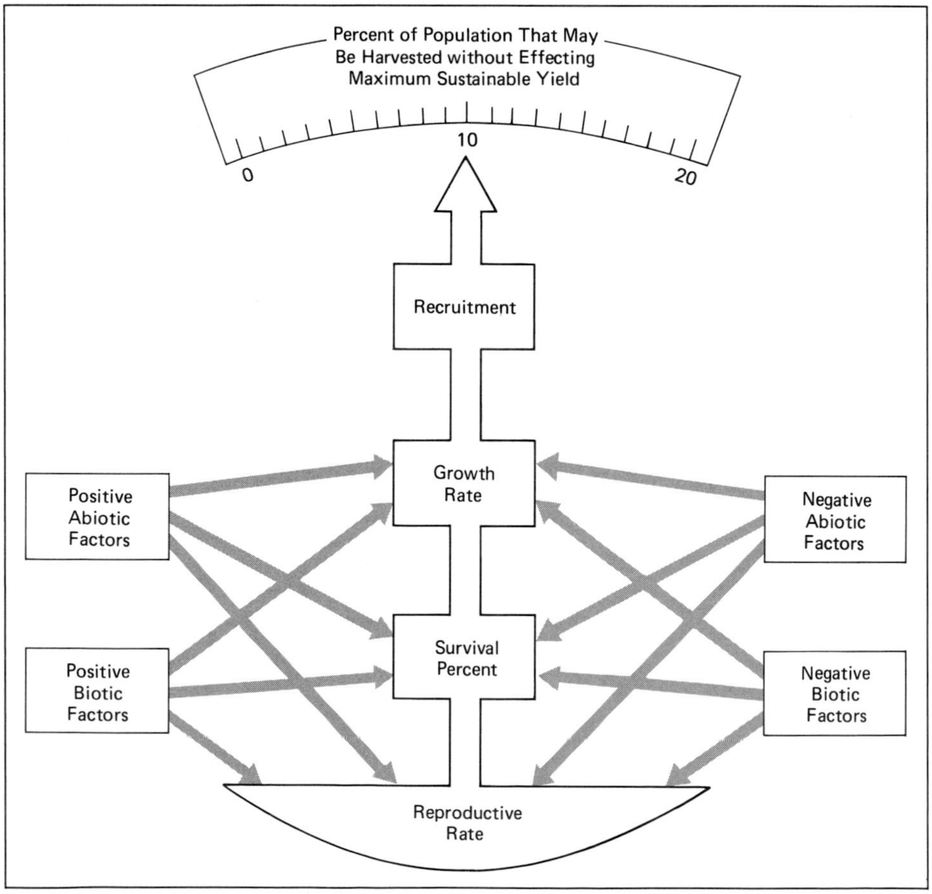

**FIGURE 19-12**
The percentage of the population that can be harvested each year without depleting the stock depends on replacement capacity, which in turn depends on many other interconnected factors.

**FIGURE 19-13**
Burning of tropical rain forest for agriculture in Amazonia, Brazil. Millions of acres of rain forest are being thus destroyed each year, undoubtedly causing the extinction of many species and sharply reducing genetic diversity of innumerable others and perhaps affecting global climate in the long run.
(Earth Scenes. © 1985, Dr. Nigel Smith.)

quickly loses its agricultural capacity and the world loses one of its greatest resources for naught (Fig. 19-13).

When a section of land is bulldozed, all of the larger plants and animals it contains are usually destroyed. The idea that this wildlife will simply move "next door" and continue to live in an undisturbed section is erroneous. Population balances, described in Chapter 2, produce a situation such that each area has all of the wildlife it will support. Any loss of natural habitat can only result in a proportional reduction in all populations that require that habitat. And in some cases, the impact is greater than the proportion of land actually developed. For example, a highway that runs along a river may become a blockade that prevents animals from reaching a water source. Developments can destroy important stopover areas for migrating birds. Loss of tropical forests in Central and South America could have a devastating effect on North American songbirds that winter in these forests. Another point to remember is that many animals can only survive in flocks, packs, or herds. If the natural habitat is reduced to a point where it cannot support the **critical number**—a certain minimum-sized

group—the entire population may perish. Similarly, development that splits a territory and prevents migration between the two halves can cause a population to perish because neither half can adequately support the critical number.

Finally, ecosystems are not isolated from one another. Changes in one ecosystem are more than likely to affect others. A highway along a river, for example, may change the total ecology of the river as runoff, sedimentation, and pollutants are increased, and entry of natural organic matter such as leaf fall is decreased. Temperature may also be increased by reducing natural shading (Fig. 19-13b).

Even unwitting alteration of an ecosystem can have drastic effects. Water diversion projects, as described in Chapter 8, are a prime example. For instance, marsh habitats along the Platte River in Nebraska support hundreds of thousands of cranes, geese, ducks, songbirds, and bald eagles. However, large and increasing portions of the river's flow are being diverted for irrigation. The reduced flow is causing the marshes to dry up and is threatening the survival of all the wildlife they support.

### POLLUTION

Pollutants may kill populations outright or they may act through adversely affecting reproduction and/or recruitment. The many kinds of pollutants and their effects on biota have been described in previous chapters.

### INTRODUCTION OF FOREIGN SPECIES

The destruction of the American chestnut by chestnut blight disease described in Chapter 3 is a prime example of how the introduction of foreign species, in this case a fungus, can alter an ecosystem. You may recall other examples from Chapter 2 as well.

In conclusion, any of these factors may be devastating to biota. What makes matters worse is that frequently two or more of these factors occur simultaneously (Fig. 19-14). For example, a combination of pollution, which decreases recruitment; elimination of wetlands, which destroys breeding habitats; and overfishing are all involved in the decline of many fish and shellfish populations in the Chesapeake Bay in the eastern United States.

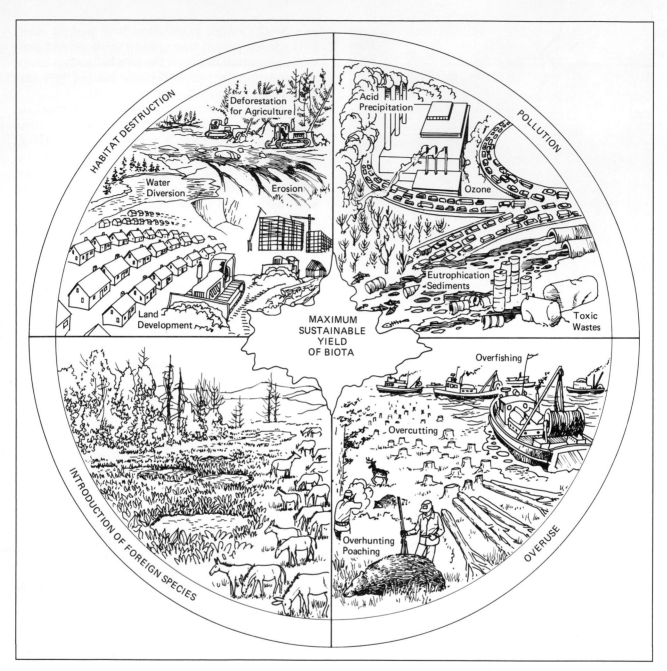

**FIGURE 19-14**
A summary of the ways in which sustainable yields of resources from natural biota are being undercut by human impacts.

## REASONS FOR ASSAULT AGAINST WILDLIFE AND ECOSYSTEMS

We noted that many species are threatened or endangered with extinction due to shortsighted practices of overhunting and/or habitat destruction. Here, we wish to examine closely the reasons for this so that we can understand the kinds of regulations that are necessary to provide protection.

### Economic Pressures (the Profit Motive)

#### *LUST FOR WILDLIFE PRODUCTS*

Polar bear rugs, baskets made from elephant and rhinoceros feet, ivory-handled knives, reptile skin shoes and handbags, leopard skin coats, pets or houseplants of rare or exotic species, and furniture made of exotic tropical hardwoods are examples of items on which society places high value. The list goes on and on, but none of the items on it is essential. In fact, most are trivial (Fig. 19-15). Yet the affluent are willing to pay exorbitant prices for them. Some Indonesian and South American parrots sell for up to $5000. A panda skin rug can bring $25,000.

The costs to the hunters and trappers who kill or capture rare species, however, is relatively low. Huge profits can be made in exploiting wildlife and many eagerly place personal profit ahead of the value of maintaining wildlife. As a result, conservationists warn that about 20 percent of the species now listed as endangered face extinction by the end of this century. In 1983, more than 80,000 African elephants, about 8 percent of the

(a)

**FIGURE 19-15**
Some people will pay exorbitant prices for various items from wild animals. Despite the fact that trade in such products has been made illegal and the animals are supposedly protected under the law, poaching and black market trade to profit from this market are still causing merciless slaughter. Sadly, most of the products for which animals are slaughtered are trivial.
(a) "Products" from endangered species confiscated in a breakup of a black market ring. (U.S. Fish and Wildlife Service photo. Steve Hillebrand.) (b) A giraffe in Kenya killed by a poacher for the tuft from its tail, which some consider to be a prized ornament. (FAO photo by Thane Riney.)

(b)

continent's population, were killed. Most died to satisfy demand by artisans for ivory tusks. The prospect of extinction does not curtail the activities of exploiters of elephants or other endangered species because the opportunity for immediate profit outweighs other considerations. The exploiters assume that when these species are gone, profits can be made in other areas.

Unfortunately, the situation shows no signs of abatement. Increasing numbers of people want and are willing to pay any price for the rare fur or whatever. Particularly severe is the growing fad for unusual animal pets, fish, and houseplants. In virtually every case, these species are trapped or hunted in the wild. Keeping these animals or plants for "pets" may seem more acceptable than killing them, but few of the animals or plants survive the harsh conditions of capture, transportation, and marketing. Fewer yet survive in captivity, and regardless of survival, all are removed from the natural breeding population. Hence, they may as well have been killed. Numerous species of tropical fish, birds, reptiles, and plants are thus joining the ranks of endangered species.

Unfortunately, as certain plants and animals become increasingly rare, the law of supply and demand takes over and people are willing to pay more for them. This, in turn, increases the fortune hunter's incentive to hunt or collect that animal or plant. Laws and regulations do little to curtail the exploitation because the potential profit is so great that many willingly accept the risks of illegal activities. The World Wildlife Federation reports that poaching and black market trade in endangered species has reached epidemic proportions and continues despite efforts to control it.

### LAND USE CONFLICTS AND ECOSYSTEM DESTRUCTION

The profit motives that trigger the destruction of rare wildlife for the purpose of consumer products also trigger the destruction of natural ecosystems for development. As discussed earlier, society places relatively little value on wetlands, despite their natural worth. Much value, however, is placed on waterfront property. Consequently, enormous profit can be made in buying wetlands for a few hundred dollars per hectare, dredging and filling, then selling them as waterfront property. Similarly, mass destruction of tropical rainforests is carried out to make way

for agriculture. But this agriculture is not designed to feed the hungry and malnourished; it is primarily used by already wealthy, multinational industrial concerns that obtain the land cheaply and convert it to grassland on which cattle can feed inexpensively. The company then makes a huge profit selling beef to already overfed markets in North America and Europe. Similarly over one-half of United States wetlands have been drained, destroying their wildlife to expand agriculture. At the same time the United States has been plagued by agricultural surpluses, which are supported by the federal government. Again profits are obviously the motivating factor.

## The Tragedy of the Commons

Overexploitation of natural resources is further aggravated by competition. If only one party is involved in exploiting a resource in excess of its maximum sustainable yield, that party may recognize the long-term disadvantages, limit exploitation, and sacrifice immediate profits for future interest. When two or more independent parties are involved in the exploitation, however, this logic does not prevail. Instead, a phenomenon known as the **tragedy of the commons** occurs.

The American biologist and writer Garrett Hardin, in a now classic essay, described this phenomenon in 1968. Originally, the "commons" referred to areas of pastureland in England that were provided free by the government to anyone who wished to graze cattle. Despite the social virtue of the idea, farmers soon realized that whoever grazed the most cattle stood to benefit most from the commons. If one farmer did not take advantage of the commons, another would. As a result, farmers grazed excessive numbers of cattle on the commons until it was overgrazed and destroyed. Hence, the tragedy (Fig. 19-16).

The concept of the commons applies to any situation where two or more independent people or groups are in competition in the exploitation of a resource. Lobster fishing in New England provides a modern example. The lobster grounds are considered common property, free to be used by anyone with the means to do so. Although lobstermen may recognize that the lobsters are being overharvested, any lobsterman who curtails his own fishing diminishes his own income while his competitors continue fishing and catch what he could have caught. His loss only becomes their

**FIGURE 19-16**
The tragedy of the commons. Whenever a resource is open to be exploited by numerous parties, it will tend to be overexploited because each party benefits maximally from its own exploitive efforts while the loss (destruction of the resource) is shared by all.

gain and the lobsters are depleted in any case. Therefore, the feeling prevails: If I don't, they will. I'd better get them while they last. One can imagine poachers of endangered wildlife assuaging any feelings of guilt by the same argument.

The principle of the tragedy of the commons can also be applied to pollution. Before regulations were issued, air and natural waterways were generally considered a "commons" into which anyone could discharge wastes free of charge. Again, the psychology prevails: If it is to my economic advantage to dump untreated wastes, why not? Everyone else is. Similarly, if regulations do not preclude development, available land effectively becomes a "commons" for developers: If I don't develop it, someone else will.

An interesting note about the tragedy of the commons is that it allows each exploiter to shift the guilt and responsibility to the "other guy." In reality, however, the short-term gains of relatively few are paid for dearly over the long term by society at large, which must deal with the loss of wildlife and suffer the effects of environmental degradation.

## Ignorance or Insensitivity?

It is frequently difficult, if not impossible, to make direct counts of population numbers. More often total populations are estimated from various sampling techniques. Moreover, if populations are declining, it may be even more difficult to determine exact cause or causes. Is it because of overuse? Is it because of pollution and, if so, what particular pollutant is responsible? Or is it because of a disease, predator, or some other environmental factor?

Exploiters commonly find reason to continue their exploitation by pointing to various "unknowns" and, in view of the unknowns, disclaiming any responsibility. For example, fishermen may blame declining catches on pollution while polluters blame it on overfishing. Whaling interests continue their activities by disclaiming that populations are actually being depleted. Utilities continue to receive permission to eject their acid-causing pollutants into the atmosphere by claiming there is insufficient evidence linking them to environmental degradation.

Of course there are valid reasons to delay imposing harsh regulations until a situation has been studied and understood. However, it is significant that those with an economic interest in continuing exploitation often attempt to impede studies and raise every imaginable objection to any study that implicates their activities. In short, it appears that exploiters make extensive efforts to maintain lack of knowledge as a shield for callous insensitivity.

## Contrasting Economics of Renewable and Nonrenewable Resources

In Chapter 18 we observed that economic forces tend to promote adjustments and adaptations that will prevent our running out of essential mineral resources. Unfortunately, wildlife resources present a very different situation. The principles pertaining to mineral resources do not apply to wildlife resources.

First, we know that mineral elements exist in a range of ores from high-grade to low-grade (Chapter 18). As market supplies diminish, prices increase and enable the exploitation of lower-grade ores or more remote deposits. By contrast, species only exist at one "grade," and rare species are restricted to just one or a few regions. Thus, it is as easy to shoot the last herd of elephants for their ivory tusks as it is to shoot the first herd. Cutting the last mahogany tree is as simple as cutting the first. Consequently, as the market supply of ivory, for example, diminishes, the rising price that occurs only serves to increase the profit margin and hence the incentive to continue exploitation into extinction. Figure 19-17 shows these relationships in the case of lobster fishing in New England.

Second, and equally important, elements are not valuable in and of themselves; they are valuable because of their chemical or physical properties. Consequently, society is willing to turn away from scarce, expensive elements to utilize more abundant, less expensive materials with the same properties. Products from wildlife, however, are valued for their own sake. Numerous substitutes exist for polar bear rugs, for example, but, for the person who lusts after the real thing, nothing else will do. Nor are people who desire rare, exotic species for their aquariums satisfied with mere goldfish. As a result, demand and exploitation of dwindling species may continue to grow, despite the number of substitutes or alternatives. Finally, aspects of recycling that stretch the supplies of mineral resources do not apply to products from biota.

One factor that may protect some wildlife is the "farming" of particular species for which there is a stable and suitably priced market. Mink farms providing fur for coats, alligator farms providing skins for handbags and other products, trout farms, and wild turkey farms are examples. Such farms satisfy market demand and hence protect respective wild populations from exploitation. Unfortunately, for the majority of wild species, such farming is impractical, if not impossible. Consequently, no buffer exists between depletion of wild stocks and total extinction.

Maintaining natural biota and ecosystems depends on concrete steps that afford protection. Moreover, protection will require increased understanding, awareness, and sensitivity.

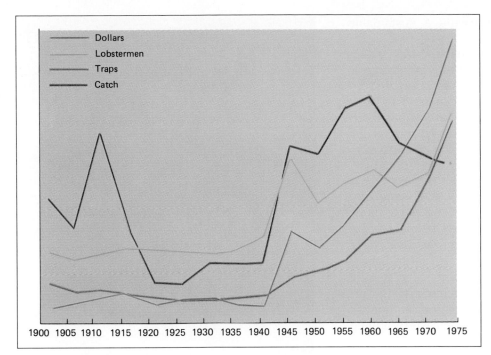

**FIGURE 19-17**
Economics may promote overexploitation, as shown in the case of lobster fishing in New England. As lobster catches have declined, price has increased owing to limited supply. The continued potential for profit fosters continued exploitation of already overharvested lobsters. (Copyright 1977 by the National Wildlife Federation. Reprinted from the April-May issue of *National Wildlife* magazine.)

## CONSERVATION OF NATURAL BIOTA AND ECOSYSTEMS

Though many species are threatened, numerous steps have been taken to provide protection. We shall begin by considering the general principles pertaining to conservation. Then we shall examine where we stand with respect to specific categories of biota.

### General Principles

To protect biota, all of the factors that undercut its support or lead to its overexploitation must be addressed.

1. Tendencies to treat biota as "commons" must be stopped. A single organization or government committed to its long-term survival must take responsibility for regulating use and/or providing protection.

2. Economic incentives for exploitation must be reduced by enforcement of suitable laws and by shifting societal values.

3. Suitable habitats must be preserved and protected from pollution.

### Where We Stand with Regard to Biota

A look at some of the categories of biota will show where these principles have been applied and how successful they have been.

#### GAME ANIMALS

Game animals include deer, rabbits, pheasants, and other species traditionally sought by hunters. A tragedy of the commons occurred with the extinction of the passenger pigeon and near extinction of the American buffalo (bison), wild turkeys, and many others. The threat was largely eliminated by laws authorizing state agencies to control hunting. The controls typically require hunters to obtain licenses, restrict hunting seasons, and limit the number of animals that can be taken. The state determines the length of the hunting season and the number of animals that may be taken by monitoring game species populations. If populations drop too low, a hunting moratorium may be issued. In addition, game animals in general are well adapted to the rural agricultural environment; hence loss of habitat in most cases has not been a critical issue.

Given suitable habitat and regulation of hunting, game animals in the United States are neither endangered nor threatened and the hunting that is permitted represents a sustainable yield.

Nevertheless, some problems are apparent. Significant losses of habitat are now occurring as rural environments are increasingly pocketed by suburban developments and segmented by new and improved roadways. This process is fragmenting the environment into "ecological islands," many of which may be too small to support a breeding population. If crossing a roadway from one "island" to another does occur, there is an added hazard to both wildlife and motorists. Draining of wetlands is diminishing populations of many waterfowl as mentioned before. A few animals such as opossums and raccoons have adapted and are thriving even in highly urbanized areas but this creates hazards of a reverse sort. For example, a rabies epidemic among "urban" raccoons is considered a significant public health risk in several cities in the eastern United States.

### EXOTIC SPECIES

Exotic species are unusual animals or plants with particularly attractive features. Birds with remarkable plumage, finely furred animals such as leopards, ivory-tusked elephants, and unusual plants, such as cacti and wild orchids are examples of prized exotic species.

After severe exploitation, many exotic species are now protected. The story of the snowy egret provides a good example (Fig 19-18). In colonial days, huge flocks of snowy egrets inhabitated coastal wetlands and marshes of the southeastern United States. But in the 1800s, when women's fashion turned to fancy hats adorned with feathers, egrets, along with many other species, were seriously exploited. Without protection, the flocks became a "commons" attacked by countless independent hunters seeking to profit from the lucrative plumage market. By the late 1800s, egrets were almost extinct. The newly formed Audubon Society began an active campaign in 1886 against this "terrible folly." Gradually, attitudes changed and newspapers and society began shaming "feather wearers." The shift in attitudes led Florida and Texas to pass laws protecting plumed birds. However, markets still existed and provided incentive to hunters to ignore the laws. Finally, in 1900, Congress passed the

**FIGURE 19-18**
Snowy egret. Hunted almost to extinction for its plumage in the late 1800s, egrets are now valued most for their living beauty. (Courtesy of Steve Simon, Catonsville Community College.)

Lacey Act, forbidding interstate traffic in illegally killed wildlife. This law effectively cut the supply from the market, which was centered in New York. Without a market, exploitation diminished. Another extremely important step was the establishment of parks, preserves, and refuge areas that protected the birds' breeding habitats. Now these preserves attract millions of visitors each year who seek the aesthetic pleasure of observing the birds in their natural habitat. Attitudes changed so completely that today the thought of hunting these birds would be abhorrent to most, even if official protection were removed.

However, thousands of other exotic species throughout the world are not as well off. A species must be officially recognized as endangered before it can receive full legal protection, including a ban on trade. The U.S. Department of the

Interior maintains an endangered species list, but when the administration is indifferent, the department is extremely slow to add new species. Thus, open and legal exploitation of any rapidly declining populations continues.

If a species does gain legal protection, another problem that frequently occurs is poaching (illegal hunting). Society may frown on feather wearers, but attitudes regarding products like ivory and fur have not changed. Elephants, wanted for their ivory tusks, and some of the animals that produce fine furs are currently threatened with extinction due to two factors: overhunting and/or habitat destruction. Overhunting is perpetuated by the fact that the market for exotic "pets" has grown enormously in recent years. Exotic pets include numerous rare and unusual animals, birds, fish, and plants. Many of the less-developed nations where these species exist do not have laws protecting them. Hence, they are a commons for mercenaries to exploit.

Even in places like the United States where laws exist that provide protection and forbid trade in endangered species, the World Wildlife Foundation reports that poaching and black market trade have reached epidemic proportions. Poaching is stripping the southwestern United States of exotic cacti. In other parts of the world, different plants, birds, fish, and animals are similarly threatened.

Despite stepped up enforcement efforts, only modest changes have been observed because, like the illegal drug market, the market for exotic species is run by highly sophisticated criminals skilled at evading game wardens and custom officials. An intensified effort to shame those who lust for exotic pets and products of endangered species, similar to the humiliation inflicted on feather wearers in the late 1800s, might be more effective.

The second factor, habitat destruction or alteration, is even more important. Even without hunting, hundreds of exotic species and many thousands of less conspicuous species seem doomed to extinction because of loss of their habitats. Unlike most game animals, which can thrive in humanized rural and agricultural environments, exotic species are generally highly adapted and hence restricted to specific natural environments. When these habitats are altered, much less destroyed, these species perish.

Recognizing the plight of many species, Congress passed the **Endangered Species Act** in 1973. This act charges the Department of the Interior with identifying species that are in imminent danger of extinction. Such species are officially designated as endangered and receive the following protection under the act. Hunting and killing is prohibited as is selling, exporting, or importing of birds, pelts, and feathers. Further, the act requires the mapping of habitats of endangered species and forbids any private, state, or federal agency from destroying such habitats in the course of construction projects such as building highways, dams, or airports. Violators may be fined up to $20,000 and sentenced to one year in prison.

This act has several shortcomings, however. First, environmental groups note that the bureaucratic process of getting species listed is painfully slow. More than a decade after enactment, over 1000 candidates for the endangered list still await official designation. Meanwhile, legal killing and trade continues. Many species may be extinct before they receive protection. Second, the United States act cannot prevent hunting or force habitat protection in foreign countries; the only protection it provides for foreign species is through the prohibition of trade, but even this protection may be largely circumvented by black markets.

Still, the Endangered Species Act has protected some species and their habitats in the United States. And, this protection is not without its own controversy. Developers claim that important projects are either foregone or made more expensive because they are required to accommodate species that are of no value to anyone except a few scientists and nature lovers. The controversy is illustrated by problems between the snail darter and the Tellico Dam on the Little Tennessee River in Tennessee. The $116 million dam was nearly complete when a small inconspicuous fish in the river, the snail darter, was discovered to be a new species, and one that would become extinct if the dam were completed (Fig. 19-19a). Completion of the dam was therefore delayed for seven years under the provisions of the Endangered Species Act. Then proponents of the dam lobbied amendments through Congress that provided exceptions and allowed completion of the dam. The snail darter survived because it was transplanted to another river; moreover, it was

(a)

**FIGURE 19-19**
Heroic efforts are being made to save certain species from extinction. (a) The snail darter. Completion of the Tellico Dam in Tennessee was delayed for seven years under provisions of the Endangered Species Act because it was felt that closing the dam would cause the extinction of the darter. (b) A ten-day-old whooping crane chick being reared by sandhill crane foster parents in captivity.
(Courtesy of U.S. Fish and Wildlife Service.)

(b)

discovered that additional populations existed in other rivers.

An interesting sidelight, however, was that economic studies conducted in the interim revealed that Tellico Dam was a "pork barrel" project (i.e., one in which funds were allocated for political patronage reasons, rather than strictly on the merits of the project). Its benefits to the Tennessee economy could have been achieved in much less costly ways while at the same time preserving the natural state of the river and the ecological cultural values of the valley.

In conclusion, controversy surrounding this Endangered Species Act continues. Environmentalists wish to strengthen it as a means to block such "pork barrel" projects as well as to protect valuable species. Other special interest groups wish to weaken it so that they can proceed with their projects without impediment.

It is interesting to note here that along with legal protection, a few exotic species have gained exceptional public attention and heroic efforts have been mounted to save them. Efforts to save the whooping crane, for example, included virtually full-time monitoring and protection of the single remaining flock, which for years numbered only in the teens. To assure a higher recruitment rate, eggs were collected (the female normally lays two but only rarely does more than one chick survive) and artificially incubated. The chicks were then returned to nests of related cranes to be raised by foster parents (Fig. 19-19b). This effort seems to have paid off; the whooping crane is now at least holding its own.

Similar efforts are now being made to save the California condor, which has been reduced to only three to five breeding pairs. Eggs are taken from nests and incubated, and the young, raised in captivity, are returned to the wild. A number of zoos and private organizations are beginning to take an active role in breeding other endangered animals such as rhinoceros and elephants (Fig. 19-20). Likewise, seeds are being collected from wild plants and stored in "seed banks" maintained by international cooperative efforts among universities and governments. Many believe that this is the only way to save some of the endangered species. However, a few animals maintained in captivity or a few seeds propagated in a bank are a poor representation of the total genetic diversity that exists in a wild population. In the end these few representatives may only serve as monuments of our failure to maintain an ecologically balanced world.

### MARINE SPECIES

The hunting of marine species, including whales and many species of fish and shellfish, supports commercial interests in many countries, but the seas, in large part, are still treated as a commons. Although the yields from many prime fishing areas have declined markedly in recent years, indicating that populations have been depleted and another tragedy of the commons is taking place, commercial fleets from many countries continue to exploit these stocks (Fig. 19-21).

Several whale species have been so depleted that many fear their extinction is imminent (Fig. 19-22). However, efforts by an increasing number of concerned people may make the difference in this particular case. Several public interest groups are working diligently to halt exploitation and protect whales as the greatest mammals on earth. The International Whaling Commission works each year to set hunting quotas or moratoriums for various species of whales.

Meanwhile, the United Nations "Law of the Sea Conference" has endeavored since 1974 to draw up and adopt regulations for the use of other marine resources. However, as seems to be true in many international negotiations, success has been limited. Instead of providing leadership in this area, the United States stands out as the single major nation that has refused to sign the agreement, basically because U.S. corporations wish to maintain it as a commons for free exploitation. The most significant move to date occurred in 1977, when a number of nations, including the United States, extended their territorial limits from the former 3 to 12 miles (5.8 to 19.3 km) offshore to 200 miles 322 km) offshore (Fig. 19-23). Since many prime fishing grounds are located within 200 miles of a coastline, this action removes them from the international commons and places them under the jurisdiction of a particular nation. As a result of this action, which also provides for management and control of exploitation, some fishing areas are now recovering. Whale spe-

**FIGURE 19-20**
Exotic animal "ranch" in Florida. Many feel that transferring small herds of exotic animals to ranches where they will be totally protected may be the only way to save them from extinction.
(Gerald Davis Photography ©. All rights reserved.)

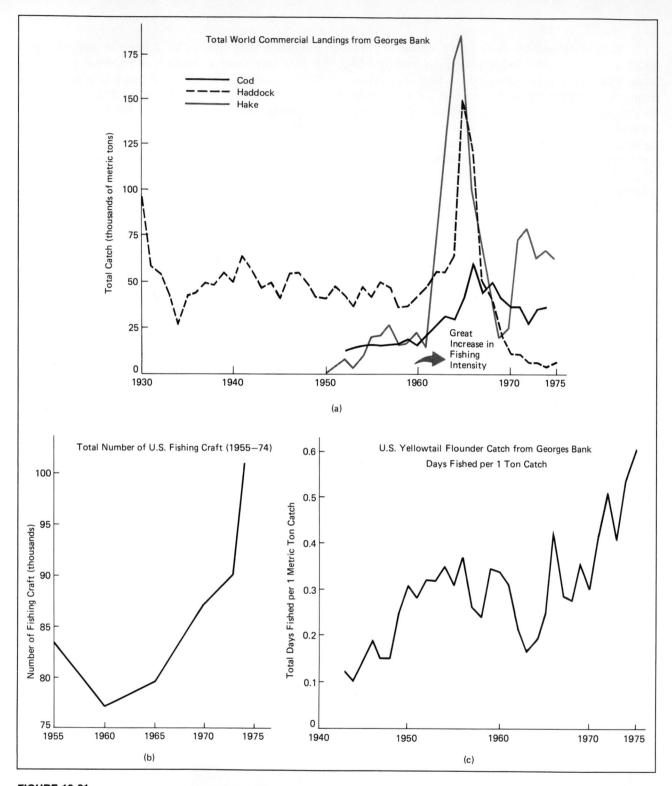

**FIGURE 19-21**
(a) Fish catches for several North Atlantic species. (b) While catches are declining, the number of fishing vessels is increasing, and (c) it takes more and more time fishing for each ton of fish, showing that stocks are being depleted. (Data from Dept. of Commerce, NOAA.)

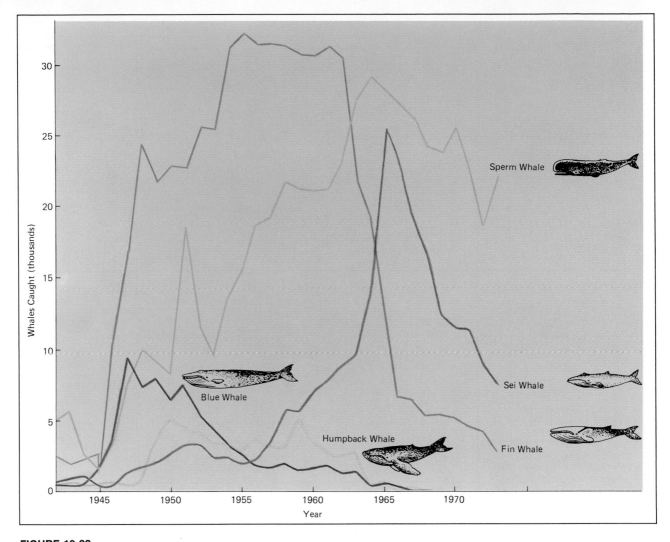

**FIGURE 19-22**
Catches reported for various species of whales from 1942 to 1973. Intensive whaling was resumed after World War II. Yields were temporarily high from overharvesting, but they declined steeply as stocks were depleted. Yet whaling continues. Can it be stopped before extinction occurs? (Data from International Whaling Commission.)

cies that inhabit the commons outside the 200 mile limit are still plagued with protection problems.

Tuna is another species that is seriously overfished today because commercial tuna fleets in the United States managed to exempt themselves from the 200 mile regulation to maintain tuna grounds around the world as a commons.

### FRESHWATER AND ESTUARIAN FISH AND SHELLFISH

Freshwater lakes, rivers, and estuaries, such as the Chesapeake Bay, have traditionally supported both sport and commercial fishing. Until the 1950s, the primary cause of decline in yields was overfishing; then pollution became a major factor. Sediment problems, organic wastes, toxic chemicals, and nutrient overload essentially eliminated aquatic life in many streams and rivers. In other water bodies, such as the Hudson River and the Mississippi River, some game fish continued to live, but they were contaminated with unhealthy levels of toxic chemicals as a result of bioaccumulation and biomagnification. As a result, commercial fishing was halted and sportsmen were advised not to eat their catch.

Some waterways were cleaned up, but others continue to decline. The Willamette River in Oregon and the Detroit River in Michigan were little more than open sewers in the 1960s. Because of pollution control efforts, however, they again support sport fishing. But fish in the Great Lakes continue to be tainted with unhealthy amounts of

**FIGURE 19-23**
Territorial limits. In 1977, countries, including the United States, extended territorial limits from the former 3 to 12 miles to 200 miles offshore. This removed many prime fishing areas from the international commons.

toxic chemicals. Here, the pollution apparently does not result from the discharge of toxics directly into the water, but rather from a fallout of chemicals that are vented or disposed of in the atmosphere.

Chesapeake Bay, the largest and most productive estuary in North America, is still in a state of decline (Fig. 19-24). A massive dieoff of submerged aquatic grasses, which support most of the major food chains of the bay and provide breeding habitat for fish and shellfish, began in the 1970s and today is over 90 percent complete. Intensive study revealed that the prime cause of the dieoff was eutrophication (overloading with nutrients). Phytoplankton shaded the grasses and, in deeper waters, caused a depletion of dissolved oxygen (see Chapter 10). A multimillion dollar cleanup effort aimed at reducing nutrient inputs was begun in the mid-1980s.

As discussed earlier, the dredging and filling of wetlands and bulkheading shorelines are other factors that hurt the Chesapeake Bay and other estuaries. Land practices that lead to erosion and sediment runoff further complicate the situation.

## Where We Stand with Respect to Protecting Ecosystems

Traditionally the use of land for agriculture, urban and suburban development, highways, mining, and other human needs has been deemed of greater value than the existing natural ecosystems. Indeed, the derogatory term "wasteland" is frequently applied to natural areas with biota that do not directly serve human needs. Thus, the millions of years that went into the evolution of the natural ecosystem is ignored and the human changes, which may be exceedingly temporary, are referred to as "development." This euphemism makes it all too easy to sacrifice natural ecosystems to achieve short-term objectives without considering the real situation—loss of na-

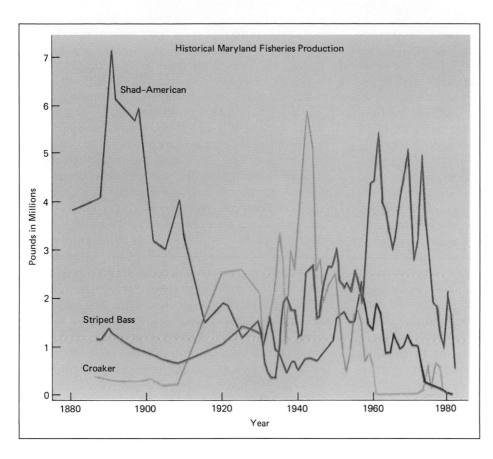

**FIGURE 19-24**
Production of three important fish species from Chesapeake Bay. Declines observed early in the century were due to overfishing. However, further declines in the 70s and failure of recovery in the 80s is attributable to pollution factors and loss of habitat as well. (Maryland Dept. of Natural Resources, Tidewater Administration, Chris Bonzek.)

ture's services, decline of the genetic bank, and disregard of the aesthetic, recreational, and scientific values of natural biota.

As discussed in previous sections, efforts to save biota generally focus on saving certain large conspicuous animals, or in some cases plants, while massive destruction of ecosystems continues in the face of "development." Ironically, the earth's biosphere will continue to function normally even in the absence of such conspicuous forms of wildlife, a fact that unfortunately has already been proven by the extinction of many species of wildlife. However, the earth's biosphere cannot be expected to function normally without the myriad of small organisms that comprise natural ecosystems and perform such basic natural services as degrading and assimilating wastes and controlling the water cycle. For example, many ecologists and climatologists predict that the present destruction of tropical forests may affect the climate of the entire earth. The conversion from tropical forest to agriculture increases the amount of solar energy that is reflected rather than ab-

sorbed; causes a decrease in evapotranspiration, which affects the water cycle of the globe; and involves the oxidation of a tremendous store of biomass in tropical forests and soils, which will exacerbate the carbon dioxide or greenhouse effect (Chapter 14).

It should be clear now that stopping such projects, which have many short- and long-term implications, involves action on many levels. Many states, along with implementing the federal Endangered Species Act, have adopted their own programs to acquire prime natural areas and keep them out of the hands of developers. Such areas may then be converted to parks, wilderness areas, or wildlife refuges. Depending on the designation, various degrees of use are permitted and wildlife protection afforded.

A number of private organizations are also involved in putting land into public trust. One such organization is the Nature Conservancy. With donations from individual members, the Nature Conservancy buys critical areas that the government is interested in acquiring but for which it lacks immediate funds. The Nature Conservancy resells the land to the government when funds are available. Thus, the Nature Conservancy protects critical areas from development during the interim period required for government to respond. Another organization is the Save-the-Redwoods League, a California-based group that gets donations from individuals nationwide. The donations are used to buy prime tracts of redwood forests, which are then donated to the California state government for inclusion in its park system. The World Wildlife Fund is an organization that is internationally active in helping to purchase and set up preserves to protect endangered species.

Another important activity of such environmental organizations is their spearheading of legal efforts to prevent development of certain projects under existing laws. The Tellico Dam episode discussed previously is a notable example. In another case, in the early 1970s, a proposal for a huge jetport in southern Florida was successfully blocked on the grounds that it would have great impact on Everglades National Park through changing the flow of water. Finally, environmental organizations are an important lobbying force in convincing legislators to pass appropriate laws and compelling governmental agencies to take action in protecting certain natural areas.

Still, public support for environmental organizations and their efforts, which are funded mainly by individual membership donations, is relatively meager. On the other side, development interests are commonly represented by powerful, well-entrenched businesses and industries. As a result of development, natural areas and their biota continue to be lost at an alarming pace. Areas that are saved are simply not large enough to maintain the ecological balances that existed in the overall ecosystem. For instance, some African preserves appear very large, but in actuality, they are not large enough and are being overgrazed by elephants. This threatens everything in the preserve (Fig. 19-25). Second, the very small populations of biota saved in preserves may not be large enough to contain the genetic variations necessary for long-term survival. This problem has been observed among zoo animals bred in captivity; after a few generations, the small, inbred populations are not of the same genetic stock as those in the wild. Third, human activity in the areas adjacent to preserves seriously impacts the preserve. For example, redwood parks in California are suffering from adjacent logging activities, particularly as sediments from erosion in logged areas clog streams and cause bank erosion and flooding in the parks.

**FIGURE 19-25**
Preservation of wildlife requires a careful consideration of the habitat area required to support the animals. Photograph shows a preserve in Africa that has been largely destroyed by the overgrazing of elephants. (Mark Boulton/Photo Researchers, Inc.)

Finally, human visitation in protected areas, even without misuse, has some negative impacts. Simply walking on the ground reduces plant growth, compacts the soil, and makes it more vulnerable to erosion. The management of many parks and preserves must now include limitation of visitation for this reason. The impact of "visitation" becomes even greater when people use trail bikes and other off-the-road vehicles (Fig. 19-26) or indulge in illicit hunting, trapping, or collecting of plant or animal wildlife. Controlling such activity in parks is becoming increasingly difficult. Again, as population increases and the amount of open land decreases, the negative impact of such activities is focused on remaining areas.

In our discussion of the need to protect remaining natural ecosystems, it is important and unfortunately discouraging to note that large tracts of federal land—land that has been owned and preserved by the government as a continuing legacy of our country's inception—are also threatened because the government is now selling them. Large areas of wetlands along the East and Gulf coasts are a case in point. Because these extremely productive ecosystems are already owned by the federal government, they can be saved and protected without additional costs. In recent years, however, the federal government has embarked on a program to sell this land to developers, often at almost give-away prices. The government contends that selling these lands makes them economically productive for the country. Certainly, some special interest groups will benefit enormously from the sales. But in the long run, society at large may pay immensely for the deterioration of environmental quality and natural services provided by these areas. The situation in less-developed countries is even more serious. For example, the governments of Latin America turn over huge tracts of tropical rainforest land to corporations at minimal costs for clearance to produce agricultural exports. Along with the loss of genetic resources, conversion of tropical forests to agriculture may lead to unprecedented soil erosion because of the heavy rainfall in these re-

(a)

(b)

**FIGURE 19-26**
Off-the-road vehicles destroy fragile vegetation and leave soil subject to severe erosion. Even a single run can be destructive. Off-the-road vehicles are an inappropriate private use of public lands. (Bureau of Land Management/EPA—Documerica photos; courtesy of U.S. Environmental Protection Agency.)

gions, and that in turn to sedimentation of rivers and flooding, as well as the climatological effects noted earlier.

In conclusion, there must be a much greater awareness at all levels of society concerning the need to protect natural ecosystems. This basically means rethinking our attitudes toward "development" and land use. We shall return to this subject in Chapter 24.

## ARTIFICIAL SYSTEMS

As noted previously, because food and other products can now be produced far more abundantly and with greater reliability through the artificial techniques of agriculture, forestry, and ranching, humans no longer depend on the exploitation of wildlife. However, two important questions must be considered. First, can productivity from existing systems be increased to supply the needs of the growing world population without converting more natural ecosystems to agriculture? Second, are agricultural systems themselves being managed in a manner that is sustainable?

### Increasing Yields

Over the last several decades crop production has risen more by increasing yield per hectare than by increasing the number of hectares cultivated. Yields per hectare are increased by

- Optimizing nutrient inputs through fertilization
- Optimizing water inputs through irrigation
- Improving control of weeds, insects, pathogens, and other pests
- Breeding varieties that produce more abundantly

Productivity of animal products is increased by

- Optimizing feeding by improving nutrition and increasing quantity
- Providing protection from disease, predators, and other pests
- Providing protection from adverse environmental conditions, particularly at times of reproduction
- Giving extra protection and care to young
- Breeding varieties that give improved production of meat, milk, eggs, wool, or other desired products.

Full potential has not been reached in any of these areas. In fact, we are on the threshold of a biotechnology that will enable the transfer of specific genes between species. The process commonly referred to as *genetic engineering* opens up spectacular opportunities to create new varieties of crop plants and animals that are more productive, more tolerant of adverse conditions, and more resistant to pests.

Thus, if techniques for increasing yields are developed and adopted more widely, existing farms, ranches, and plantations can meet growing needs for some time to come without further exploitation of natural ecosystems. Indeed, artificial systems are used and may increasingly be used to support wildlife. Hatcheries, which vastly aid in recruitment of young fish and shellfish, and tree nurseries are prime examples.

The existence and potential of these artificial systems not only makes further exploitation of wildlife and natural ecosystems unjustifiable, it makes their preservation more crucial so that a reservoir of genetic material for future development is available.

### Are Artificial Systems Sustainable?

Whether we learn to save and protect wildlife or exploit it to extinction, humans will eventually become more and more dependent on artificial systems. There is no question that artificial systems can be managed in ways that will sustain production indefinitely and even improve it. Unfortunately, all too often, we find agricultural systems that are operated in ways that are not sustainable. We need only cite the following examples, all of which have been discussed previously in the text:

1. Degradation of soil due to erosion, inadequate addition of organic matter, overgrazing, desertification, and salinization from improper irrigation

2. Depletion of water resources

3. Overreliance on chemical pesticides, which keeps farmers floundering on the pesticide treadmill

4. Depletion of oil-based fuels on which modern agriculture has become dependent

5. Conversion of agricultural land to suburban and exurban development

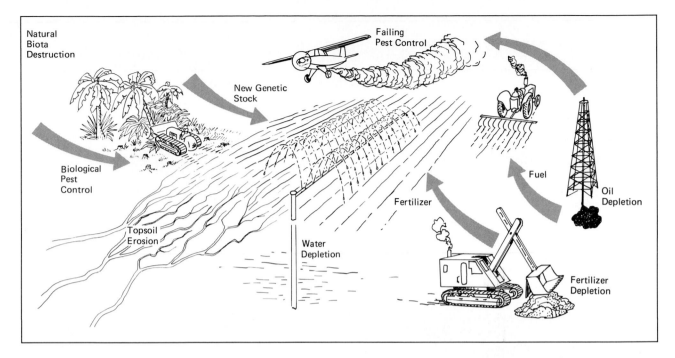

**FIGURE 19-27**
Is agriculture being practiced on a sustainable yield basis?
Agricultural production is dependent on many additional
resources, as depicted, and could be limited by the depletion of
any one or more of these resources.

In short, while fertile soil and water are the most important of renewable resources, we are exploiting them in ways that cannot possibly be sustained.

## CONCLUSION

The outlook for renewable biological resources is not an optimistic one. Many ecologists feel that the next 30 to 50 years will be an age of unprecedented extinctions unless much greater efforts are made to control the development and pollution that are destroying natural biota and ecosystems. At the same time, stricter regulations must be passed to control the overexploitation of natural biota.

Even more importantly, an immense effort to modify agricultural practices and balance them with soil maintenance and water resources is needed. Without agricultural systems that are 100 percent sustainable, saving natural ecosystems and wildlife is a moot point.

Many are inclined to point at the increasing human population as the root of wildlife destruction and environmental degradation. Ultimately, curtailing the growth of the human population is essential in coming to a sustainable ecological balance. However, a stable human population is at least 50 years in the future (Chapter 6). Further, even a much smaller human population would still have the inherent exploitive capacity to destroy all natural biota and ecosystems if that capacity were unchecked. Therefore, population control is neither timely enough nor sufficient to solve the problem of preserving biological resources. On the other hand, the steps to preserve biological resources can be taken despite growing population. Indeed, we emphasize again that a growing population makes preservation of natural biota, ecosystems, soils, water quality, and other renewable resources ever more essential.

# 20

# TRASH: WASTE OR RESOURCE

## CONCEPT FRAMEWORK

5. What products can be produced by recycled paper, glass, metals, food wastes, and textiles? Why are volunteer efforts unable to make significant reductions in the amount of solid waste that must be landfilled?

6. Discuss the factors that make recycling costs so high.

7. Why is the market for reclaimed materials insufficient?

8. What factors must local governments address when considering recycling programs? What formula is used to determine whether recycling is economically feasible? What factors does this formula fail to take into account?

9. What benefits does burning trash as a means of disposal offer? Why is this method not practiced widely?

10. What are "bottle bills"? Why does industry oppose them?

11. Why is solid waste disposal a significant aspect of environmental preservation and ecosystem balance?

In contrast to natural ecosystems where most materials are recycled, modern human societies are characterized by a flow-through process in which materials enter as raw materials and exit as trash. The enormous quantity of trash removed from homes, schools, and commercial establishments each week is a constant reminder of how much waste society generates. Studies show that the average individual generates about five pounds (2 kg) of trash per day. This waste is referred to as **domestic solid waste** to distinguish it from industrial wastes, agricultural wastes, and sewage wastes. It is also referred to as **municipal solid waste** because local governments are generally responsible for its disposal.

Over the years, the amount of domestic solid waste has grown steadily, in part because of increasing population, but more so because of changing lifestyles and the increasing use of disposable materials (Fig. 20-1). This volume of trash represents an inordinate demand on diminishing resources and its disposal poses many pollution and land-use problems. To alleviate these problems, people are looking at ways to reduce the quantity of trash generated and at methods of reclaiming solid wastes as resources. The objective of this chapter is to gain a better understanding of the solid waste problem and to explore the options for both reclamation and volume reduction.

## TRASH: WASTE OR RESOURCE?

### The Nature of Domestic Solid Waste

We assume that everything we no longer want and which no longer has practical value in our homes, schools, and commercial establishments is trash. Studies show that the trash generated by a city is generally composed of:

| | |
|---|---|
| Paper | 41% |
| Food Wastes | 21% |
| Glass | 12% |
| Ferrous Metals | 10% |
| Plastics | 5% |
| Wood | 5% |
| Rubber and Leather | 3% |
| Textiles | 2% |
| Aluminum | 1% |
| Other Metals | 0.3% |

However, the proportions vary greatly depending on the generator (commercial vs. residential), the neighborhood (affluent vs. poor),

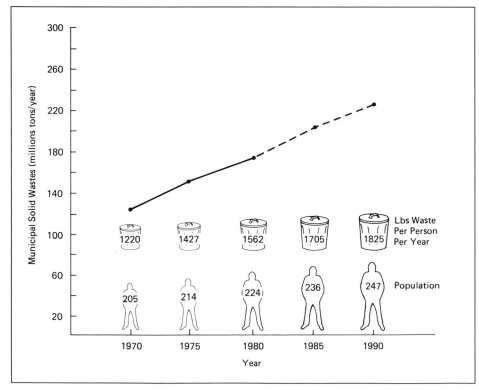

**FIGURE 20-1**
Output of solid wastes in the United States has grown much more rapidly than population and the trend is continuing. Can it be reversed? (Data from S. L. Blum, "Tapping Resources in Municipal Solid Waste," p. 47, in *Materials: Renewable and Nonrenewable Resources*, ed. P. H. Abelson and A. L. Hammond. © 1976 by American Association for the Advancement of Science, Washington, D.C.)

and the time of year. During certain seasons grass clippings, leaves, and other lawn and garden wastes add to the solid waste burden, often equalling all of the categories listed above in weight. This extreme variability of solid waste presents may problems in regard to its use or disposal.

## Trash as Waste: Methods and Problems of Disposal

Traditionally, local governments have assumed the responsibility for collecting and disposing of domestic solid wastes. The local jurisdiction itself may own the equipment and employ workers or it may contract with a private firm to provide the service. In either case, the service is paid for by local tax dollars and the type and quality of the service ultimately depends upon what residents are willing to support and pay for.

### DUMPS

Until the 1960s most municipal solid waste was disposed of by open burning in dumps. Although it reduced the volume of waste, burning was often delayed or incomplete. Consequently, dumps were notorious breeding grounds for rats, flies, and other vermin, and they were always the source of objectionable smoke and odors. Public objection and air pollution laws led to the phasing out of most open-burning dumps during the 1960s and early 1970s.

### LANDFILLS

Sanitary landfills were generally adopted as the substitute for open-burning dumps and some 90 percent of municipal solid waste is now disposed of in landfills. In a sanitary landfill, the wastes are dumped in a hole in the ground and covered with a few inches of dirt. A natural valley or ravine may be used as the initial hole, or a trench may be dug. In either case, further excavations are made to expand the hole and to provide a source of dirt to cover each day's dumping (Fig. 20-2). In addition to the dirt cover, chemicals such as lime are often spread over the waste to absorb and eliminate odors. As long as each day's refuse is treated and well covered with dirt, landfills may be relatively clean and sanitary and air pollution from smoke is eliminated.

When the fill capacity is reached, the top is seeded and the municipal dump commonly becomes a municipal park. Building structures on landfills has not been advisable because paper and other organic material in the refuse gradually decompose, resulting in an inevitable settling of

**FIGURE 20-2**
(a) Sanitary landfill—a dump in which each day's refuse is compacted and covered with a layer of dirt to keep down populations of flies, rats, and other vermin. Finally the landfill may be covered with enough earth to support trees and the area returned to a relatively natural site. (Drawing from *Sanitary Landfill; Design and Operation*, [1972], U.S. Environmental Protection Agency.)

(a)

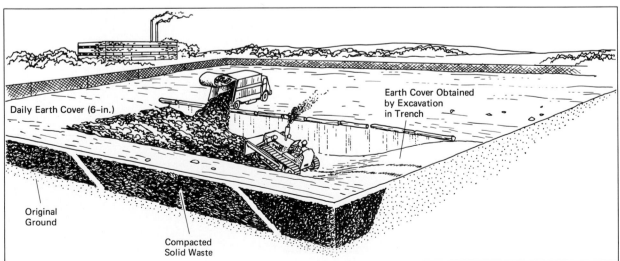

(b)

**FIGURE 20-2** *(cont.)*
(b) landfill in operation. (Photo by author.)

the filled area. However, many pleasant recreational areas exist on old landfills unbeknownst to visitors (Fig. 20-3).

Landfills still have problems, however. The most serious problem is leaching, which can cause groundwater contamination as discussed in Chapter 12. In theory, domestic wastes do not contain toxic materials. Consequently, regulations do not require municipal waste landfills to be as secure as toxic waste landfills. But in fact, recent studies show that homeowners discard sizable

quantities of unused pesticides, cleaning solvents, paints, thinners, and other toxic compounds. It is also not unusual to find toxic industrial wastes illegally dumped in landfills intended only for domestic waste. As a result, groundwater contamination from municipal landfills has occurred in many places.

Another problem, or perhaps benefit, of solid waste landfills is gas production. The compacted paper and other organic material in landfills gradually decompose, but the decomposition

**FIGURE 20-3**
Longview golf course, Baltimore County, Maryland. This area of the golf course is actually the completed top of a landfill. (Photo by author.)

process occurs anaerobically—without air. This results in the production of methane gas (see Chapter 9). The methane gas that escapes from landfills is dangerous to breathe and creates a fire hazard. A few cities resolved the problem by installing "gas wells" in older landfills. The wells trap the methane, which can be used as fuel.

For government officials, however, the most troublesome problem landfills present is finding a place to put them. In the past, burning reduced trash volume by 70 to 80 percent. Since burning is no longer acceptable, the volume of trash that must be buried is enormous and places a tremendous demand on available space. A city of 1 million generates enough trash to fill a hole the size of a major football stadium in just one year.

Hauling expenses represent another problem. Landfills must be relatively close to cities in order to minimize these costs. Yet so many suburban and exurban developments have sprung up that there is literally no place near cities that is not close to a residential community. Residents simply do not want a landfill nearby because the op-

erations at the site seldom fulfill the ideal. It is difficult to prevent trash from being spilled in transit or scattered by wind and it is difficult to completely cover all of the refuse in the landfill. Further, the covering dirt may erode, reexposing waste. As a result, sanitary landfills may produce objectionable odors and sights and breed rats and flies (though not as flagrantly as open-burning dumps). These problems as well as the potential for water pollution from landfills are realistic grounds for objection, and the continuous traffic of trash trucks to and from the landfill is also objectionable. As a result, proposals for new landfills are frequently defeated by vigorous political counterattacks mounted by local residents.

Because of the lack of acceptable, close-in landfill sites, many cities are turning to remote sites. To minimize the additional hauling expense, this practice requires a transfer station where trash is taken from local collection trucks, compacted, bailed, and put on large trucks or even rail cars for final hauling to fills (Fig. 20-4). However, finding locations to build even transfer stations is difficult because they elicit the same objections from nearby residents as do landfills themselves. Frustrated government officials are

**FIGURE 20-4**
It has become necessary to go increasingly far from cities to find suitable landfill sites. (a) This requires a system of transfer stations and long-distance hauling.

(a)

(b)

**FIGURE 20-4** (cont.)
(b) New York City garbage being towed to sea, 1973. (EPA–Documerica photo by Gary E. Miller; courtesy of U.S. Environmental Protection Agency.)

led to say: Everyone wants us to pick the stuff up, but no one wants us to put it down.

The situation is extremely serious. Many cities face the prospect of current landfill sites being filled to capacity within the next 5 to 20 years with no new sites available. Even after landfills are full, trash is still generated, and it must go somewhere. One of the solutions to this dilemma is to look at trash as a resource that may be used rather than as a waste that must be disposed of.

## Trash as a Materials Resource

### RECYCLING POTENTIAL

A possibility for solving at least a part of the municipal solid waste problem is to recycle the usable materials in trash back into products. For example:

1. Paper can be
   a. Repulped and made into paper, cardboard, or other paper products.
   b. Manufactured into cellulose insulation.
   c. Shredded for garden mulch; however, paper is devoid of nutrients.
2. Glass can be
   a. Crushed, remelted, and made into new containers.

b. Crushed and used as a substitute for sand in construction materials or on beaches to replace sand lost to erosion.
3. Metals can be remelted. Making new aluminum from scrap aluminum saves 70 to 80 percent of the energy required to make aluminum from virgin ore.
4. Food wastes can be composted to make humus, which can be used as a soil conditioner.
5. Textiles can be shredded and used in new fiber products.

Recycling, of course, has been practiced for years by volunteer groups who collect paper, glass, and metals and take them to dealers. Such volunteer efforts generally have not managed to recycle more than 1 percent of any solid waste component. Therefore, as currently practiced, volunteer efforts do not have much impact on the solid waste problem. Nevertheless, if recycling were practiced on a municipal scale, as in some parts of Europe, it would alleviate the problem. Also, it might produce rather than consume revenue and would save virgin resources. Why is this not widely done?

### PROBLEMS OF RECYCLING

In spite of the apparent virtue of recycling, one cannot escape economic reality. Even given the value of recovered materials, costs involved in recycling are generally higher than straight disposal. There are two reasons for this: the cost of separation and the paucity of markets for reclaimed materials.

**Cost of separation.** Present patterns of discard and collection mix all components of solid waste into one great mass (or mess). Recycling requires that the components be separated. This can be done either before or after collection, but neither is without problems. In separation before collection, individual residents are asked (or required) to put paper, glass, metal, and garbage into separate containers. Containers are then collected separately, or at least put on different trucks that can deliver the material directly to the dealers. This technique requires exceptional understanding, dedication, and cooperation from residents, to say nothing of a large number of trash cans. Lack of full cooperation can completely defeat this effort.

There are some compromises between no recycling and a multitude of different trash cans and collections. For example, in Greenbelt, Maryland (a suburb of Washington, D.C.), residents put newspapers in a marked can apart from other trash. Newspapers are collected once a week and other trash is collected twice a week. The newspaper goes to a dealer and other trash goes to the landfill. In instituting this program, Greenbelt had several factors in its favor. First, its residents generally have a high degree of environmental awareness. Second, as a well-to-do suburban community, the town already supported three trash collections per week and had the sanitation workers pick up containers from the house, rather than relying on residents to place them by the curbside.

Simply by devoting one of the three weekly collections exclusively to paper, the program did not increase city costs. In fact, there was a net return from the sale of the paper, as well as a savings in keeping it out of the landfill. From the point of view of the residents, there was very little additional thought or effort required to separate newspapers, and the remaining collections were still fully adequate to prevent accumulation of other trash.

Other situations would not be so conducive to this system. For example, if there were only one collection per week, the city would have to double its trucks and personnel to provide a weekly collection of both newspaper and other trash. The value of paper collected would not offset the increased cost. On an every-other-week schedule, trash would accumulate, resulting in unsanitary conditions. Likewise, where residents must put their own trash by the curbside, it has been found that not everyone will remember to put out the right can on the right day.

Alternatively, the pickup of trash can remain unchanged and separation can be conducted after collection. Facilities that separate trash have been built and tested. The general scheme for one such facility is shown in Figure 20-5.

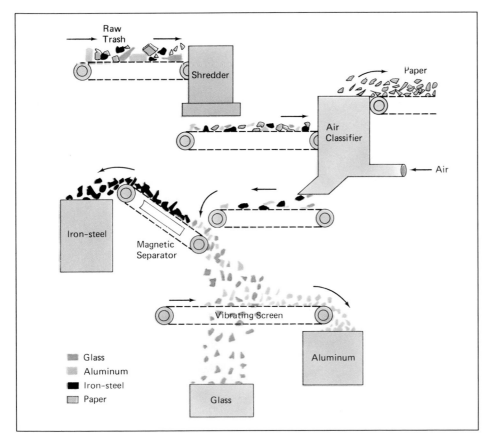

**FIGURE 20-5**
Schematic flow diagram for the separation of municipal solid waste. Separation can be achieved but is it cost-effective? Does the value of the separated materials justify the costs of separation?

Even with this or similar technology, complete separation is hardly possible. This detracts from some of the options for use of recovered material. For example, fragments of plastic bags cannot be readily separated from paper and this detracts from the ability to recycle paper. Cans that combine aluminum and iron lead to inevitable cross-contamination of these metals and reduce scrap value. In addition, the machinery needed to do this job is costly to install, operate, and maintain.

**Insufficient markets for reclaimed materials.** After they are separated, reclaimed materials only have value if there are markets to absorb them. Most of the recycling that currently occurs involves highly uniform "clean waste" that is reused for similar purposes. For instance, when writing paper is produced, the trimmings from cutting pads are a highly uniform clean waste that can be repulped and used for new sheets of paper. Paper from municipal solid waste, on the other hand, is highly variable in quality; it is "dirty" with printing ink and it may contain other debris such as plastic. Consequently, there is little demand for reclaimed municipal solid waste.

The absence of sufficient markets that are willing and able to convert the waste materials into new products or of a consumer market for the recycled products, can readily defeat the best efforts toward recycling (Fig. 20-6). For example, the municipal paper collection effort in Greenbelt, Maryland, had to be temporarily abandoned in the early 1970s when the paper buyers had so much paper that they could not accept, even for free, the municipal wastepaper. There were many other instances in the early 1970s when paper and glass, carefully collected and separated by ardent environmentalists, ended up being taken to landfills along with other wastes because markets were glutted. Obviously, collecting and separating efforts are an additional waste if there are no markets.

Consumer acceptance of recycled products is equally important. No manufacturer can operate without some profit. The virtue of recycling does not alter the fact that income from sales must exceed the cost of materials and production. There-

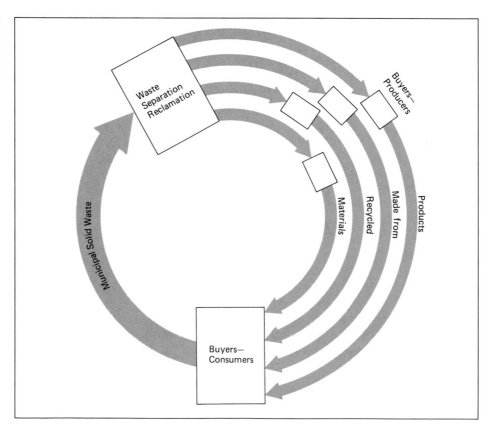

**FIGURE 20-6**
Recycling must involve a complete cycle. Absence of a market for the recycled products can readily defeat the best efforts toward recycling.

fore, products made from recycled materials must compete in the marketplace with those made from virgin materials; they may or may not be able to compete successfully. For instance, in reprocessing wastepaper into paper, there is an inevitable breakage of fibers that provide strength and integrity to the paper. Recycled paper is of lower quality, but the cost of producing it is nearly the same as for virgin paper. Consequently, recycled paper has not found a large market. On the other hand, cellulose insulation (made from paper fibers) has proved very competitive with other types of insulation and now provides a substantial market for wastepaper.

Recycling, then, is as dependent upon supply-and-demand relationships as any other business. Unfortunately, creating a balance between supply and demand with respect to recycled products tends to be impeded by "the chicken or the egg" phenomenon. A city may not invest in a facility to separate its solid wastes because there are too few industries to absorb the reclaimed products (insufficient demand for the supply). On the other hand, a company may not invest in a plant to reprocess waste materials because there is no guarantee of a constant supply of raw materials (insufficient supply for the demand). It is worth noting here that volunteer efforts in collecting materials for recycling tend to be sporadic. Many people become enthusiastic about recycling at one time and produce large amounts of salvageable waste products. Then interest fades and very little is produced. In other words, volunteer efforts generally do not provide the long-term dependability that is necessary for a satisfactory balance between supply and demand.

## ECONOMICS OF RECYCLING

Our previous discussion shows that if a local government considers the pros and cons of switching from landfill disposal to reclamation of solid wastes, the following factors must be addressed:

1. Present and future landfill costs. Reclamation will reduce the amount to be landfilled and this will produce savings on landfill costs. How much will these savings amount to?

2. The value of reclaimed materials. How much revenue will the sale of reclaimed materials provide?

3. Costs of separation and recovery. How much will additional collections and recovery facilities cost to acquire and operate?

Combining these factors will show whether reclamation will yield a net savings:

$$
\begin{array}{r}
\text{Savings on Landfill Costs} \\
+ \text{ Value of Recovered Materials} \\
\underline{- \text{ Costs of Separation and Recovery}} \\
\text{Profit (or Loss) in Recycling}
\end{array}
$$

Unfortunately, experience to date generally shows that, because of the high costs of separation and the relatively low value of reclaimed materials, recycling municipal solid waste involves higher, not lower, costs.

Sometimes environmental enthusiasts are prone to discount the economics and assume that recycling must be good and disposal bad regardless of relative costs. It is well to remember that costs in themselves ultimately reflect the value of resources used. Consequently, a higher cost associated with recycling may be indicative of more precious resources being used than are being recovered. For example, the tungsten used in grinders and shredders is becoming a scarce resource, as is the energy to run them, while iron and silica recovered from cans and bottles are among the most abundant elements on earth. From the city officials' point of view, relative costs must be considered; decisions that lead to higher costs, and hence the need to raise taxes, are unpopular indeed.

Controversy nevertheless continues because factors in the recycling equation are not constant. Landfill disposal costs are escalating with increasing land values and increasing costs of fuel for hauling. Likewise, the value of recovered material may increase or decrease with changes in supply and demand, and costs of separation and recovery may change as technology develops. In addition, the potential pollution costs that arise when, for example, leaching causes water contamination are seldom given sufficient consideration. Finally, if new landfill sites cannot be found at any price, an alternative must be found.

## Trash as an Energy Source

An alternative that is gaining increasing attention and use is the use of municipal solid waste as a source of energy. This concept involves

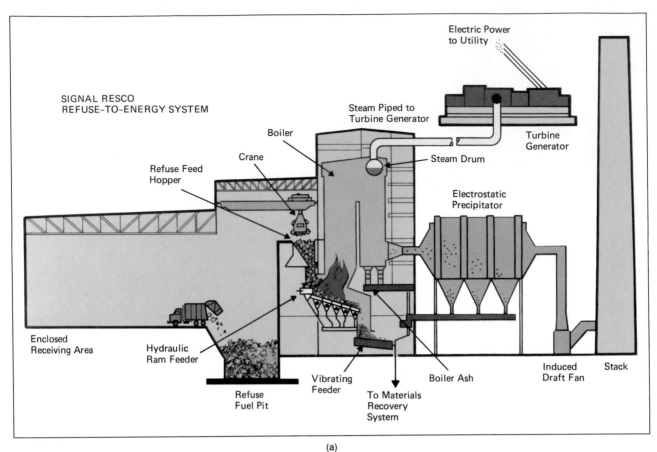

(a)

**FIGURE 20-7**
Conversion of municipal solid waste to energy. It is proving more cost-effective to build special furnaces that will burn the raw (unseparated) refuse efficiently and utilize the heat to produce electrical power for which there is a guaranteed market. This refuse-to-energy plant in Baltimore, Maryland, is capable of consuming 2000 tons of refuse per day to generate 60 megawatts (60 million watts) of on-line power, enough to supply some 60,000 homes. Air pollutants are removed by electrostatic precipitators. (a) Schematic diagram of plant. (b) The outside of the plant. Note the cleanliness and lack of smoke from the stack. This shows what modern "dumps" can be. (Courtesy of Signal Environmental Systems Inc.)

(b)

using the waste as a fuel to generate heat. The heat can be used directly or it can be used to produce steam to drive electric generators. If the waste can be burned directly, the costly problem of separation is avoided, and a ready market for energy always exists. The most valuable materials in the waste—iron and aluminum—can still be recovered from the ash if desired. Other unburnable materials must be landfilled, but because these materials make up only about 10 to 20 percent of the original volume, the life of the landfill will be extended about five- to tenfold. More importantly, since the incinerated material is not subject to further decomposition and settling, the ash may be used as fill dirt in construction sites, road beds, and so forth. In other words, disposal of the incinerated material presents relatively few problems.

Unfortunately, using municipal solid waste as a fuel is seldom this simple. Only about 80 to 90 percent of the material is burnable, and even that portion is highly variable and often wet. Consequently, municipal solid waste makes a very low grade fuel. Designing facilities that will efficiently burn the material and control air pollution continues to challenge engineers. Most of the facilities constructed over the last 15 years have been to some degree experimental and many did not achieve desired results. However, many of the problems have been solved. Some facilities now operate successfully and more are under construction.

A municipal solid waste energy recovery plant in Nashville, Tennessee, for example, supplies steam heat for state and city office buildings in the winter. In the summer, the steam drives compressors that supply air conditioning. In Akron, Ohio, a solid waste energy recovery plant supplies steam heat for city buildings in winter, and in summer sells the steam to the local tire industry for use in processing rubber. A refuse-to-energy plant went into operation in Baltimore, Maryland in 1984. The plant is capable of consuming 2000 tons of raw trash per day (Fig. 20-7). Steam produced by the boilers drive a 60,000 kilowatt generator that produces enough power to service about 60,000 homes. Air pollution from the combustion is controlled by electrostatic precipitators.

It is important to note, however, that despite seemingly huge amounts of trash, the amount of energy produced is small compared to total energy demands. Hence, this method is still considered primarily as a means of trash disposal rather than as an energy alternative.

## REDUCING WASTE VOLUME

We noted earlier that the increased amount of waste produced over the last 30 years is largely a result of changing lifestyles, notably the growing use of disposable products. We cannot ignore this issue when addressing the problem of how to deal with increasing volumes of waste. The question arises: Must we really generate all of this waste or can products be kept in use longer?

### Returnable versus Nonreturnable Bottles

The use of returnable versus nonreturnable beverage containers is a prime case in point. Prior to the 1950s, most soft drinks and beer were marketed by local bottlers and breweries in returnable bottles that required a deposit. Trucks delivered filled bottles and picked up empties to be cleaned and refilled. This procedure is efficient when the distance between the producer and the retailer is relatively short. However, as the distance increases, transportation costs become prohibitive because the consumer pays for hauling the bottles as well as for the beverage. In the 1950s, distributors, bent on expanding markets and growth, observed that transportation costs could be greatly reduced if they used light-weight containers that could be thrown out rather than shipped back. Thus, no-deposit, nonreturnable bottles and cans were introduced. The throwaway container is also an obvious winner for manufacturers who profit by each bottle or can they produce.

Through massive advertising campaigns promoting national brands and the convenience of throwaways, a handful of national distributors gained dominance and countless local breweries and bottlers were driven out of business during the 1950s and 1960s. At the same time, bottle and can manufacturing grew into a multibillion dollar industry.

The average person drinks about a liter of liquid each day. For 237 million Americans, this daily consumption amounts to some 1.3 million barrels of liquid. That a significant portion of this fluid should be packaged in single-serving containers that are used once and then thrown away is bizarre. It is difficult to imagine a more costly, wasteful way to distribute fluids.

Indeed, beverages in nonreturnable containers appear to be priced competitively on the market shelf. But these containers constitute some 6 percent of all solid waste, about 50 percent of the nonburnable portion; they constitute about 90 percent of the nonbiodegradable portion of roadside litter. The broken bottles are responsible for innumerable cuts and other injuries, not to mention flat tires. These containers are also environmentally undesirable because both the mining of the materials they are composed of and the manufacturing process create pollution. All of these are hidden costs that do not appear on the price tag. Consumers not only pay with taxes for litter cleanup, but also suffer the cost of treating injuries, flat tires, environmental degradation, and so on.

In an attempt to reverse the trend, environmental and consumer groups have promoted "bottle bills"—laws that encourage the use of returnable rather than throwaway beverage containers. Such bills generally call for a deposit on all beverage containers—returnable and throwaway. Since the deposit is wasted on nonreturnable containers, consumers gradually change their habits and buy returnables.

Bottle bills have been proposed in virtually every state legislature over the last decade. In every case, however, the proposals have met with fierce opposition from the beverage and container industries and certain other special interest groups. The reason for their opposition is obvious—economic loss. But the arguments they put forth are more subtle. The industry contends that bottle bills will result in loss of jobs and higher beverage costs for the consumer. They also claim that consumers will not return the bottles and litter will not decline. In most cases, the industry's well-financed lobbying efforts have successfully defeated bottle bills.

However, some states—nine as of 1984—have adopted bottle bills despite industry opposition (Table 20-1). Their experience has proven the beverage industry arguments false. More jobs are gained than lost, costs have not risen, a high per-

centage of bottles are returned, and there is a marked reduction in can and bottle litter. In some cases, local breweries and bottlers are making a comeback, thus improving the local economy.

| Table 20-1 | States That Have Bottle Bills | |
|---|---|---|
| | STATE | DATE |
| | Oregon | 1972 |
| | Vermont | 1973 |
| | Maine | 1976 |
| | Michigan | 1976 |
| | Connecticut | 1978 |
| | Iowa | 1978 |
| | Massachusetts | 1978 |
| | Delaware | 1982 |
| | New York | 1983 |

A final measure of the success of bottle bills is the continued public approval of the program after it is initiated. Despite industry efforts to repeal bottle bills, no state that has one has done so. With some additional rise in public concern, a national bottle bill could be enacted in the near future.

## Other Measures

Whenever items are reused rather than thrown away, the effect is a reduction in waste and better conservation of resources. In this respect, it is encouraging to see the growing popularity of yard sales, flea markets, and other "not new" markets. As discussed in Chapter 18, other measures to reduce the amount of material going into the trash include reducing the amount of material in products, downsizing, and increasing durability.

Consideration of our domestic wastes and their disposal emphasizes what a mammoth stream of materials of all kinds flows in one direction from our resource base to disposal sites. Just as natural ecosystems depend upon recycling nutrients, the continuance of a technological society will also ultimately depend upon our learning to recycle or reuse not only nutrients, but virtually all other kinds of materials as well.

# 21

# ENERGY RESOURCES AND THE NATURE OF THE ENERGY PROBLEM

## CONCEPT FRAMEWORK

| Outline | | Study Questions |
|---|---|---|

1. Define *energy*. Why is energy the most critical of all resources?

2. What are the four principal categories of energy use? What are the major sources of fuel in the United States? What factors influence the development of alternative energy sources?

3. What changes have taken place in recent years in regard to energy consumption? What enabled the Industrial Revolution to begin? How did this change energy usage in the United States? What benefits and problems

did the internal combustion engine present? Why have oil-based fuels become the dominant energy source? Compare and contrast oil and natural gas.

4. Discuss how fossil fuels formed and why they are finite resources.

5. How are oil and natural gas fields found and assessed? How can production from these fields be extended? How is the "10 percent" rule used?

6. What role do new discoveries play in the pace of production? What caused the energy problem in the United States in the 1970s?

7. What is *OPEC* and how did it influence energy use and availability in the 1970s? What is an *oil glut*? How did the energy picture change in the 1980s?

8. Discuss the three areas that must be addressed in a discussion of dependency on foreign oil. About how much money does the United States spend annually on the import of foreign oil? What impact does this have? What is the "Easter egg" argument? Discuss the potential economic and political repercussions of a shortfall between production and demand for oil.

9. What two options exist for curtailing shortfalls? Where has progress been made?

# ENERGY RESOURCES

*Energy* is defined as the ability to do work. This definition includes heating, cooling, lighting, transportation, processing, and production of all materials, as well as other conspicuous activities. With adequate supplies of energy, enough work can be conducted to obtain nearly all other resources. Conversely, in the absence of energy supplies, all other resources would be unobtainable. Consequently, energy is the most critical of all resources.

In Chapter 2 we explained that through the process of photosynthesis solar (light) energy is trapped and stored in green plants and then transferred to other organisms in the form of energy-rich organic matter or food. The functioning of our own physical bodies remains dependent on this natural ecosystem process. However, the progress of civilization, especially technological civilization, may be seen as humans learning to harness additional energy sources to perform desired work. This allows humans to increase production and relieve themselves of much physical toil.

Agriculture provides an excellent example. The use of only hand labor to prepare the soil, plant, and harvest provides little more than enough for self-sufficiency. Therefore, before the mechanization of agriculture, more than 90 percent of the population was employed in farming. Society could not have survived otherwise. With advancing mechanization, however, fewer farmers could produce greater quantities. Now only about 2 percent of the population in the United States is directly involved in crop production and these people are able to provide not only enough food to feed this country but enough to supply other countries, permit the energy-wasteful process of conversion to meat, and still have surpluses (Fig. 21-1).

However, as mechanization increases, so does energy consumption. We can look at this in terms of the amount of additional energy used for

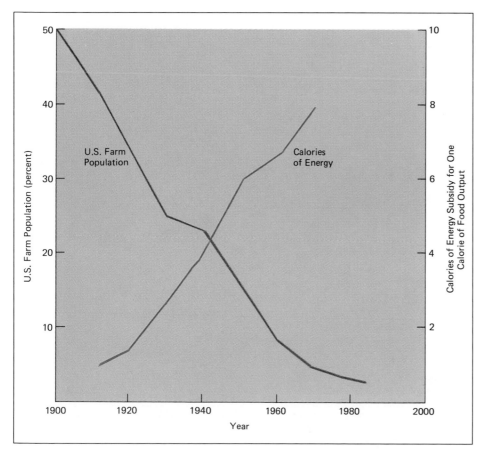

**FIGURE 21-1**
Industrialization has been basically a process of substituting machine labor for human labor. This is illustrated by the fact that a decline in farm population (largely farm labor) has been accompanied by an increase in energy consumed per calorie produced. (Population data from *Statistical Abstracts of the United States*, U.S. Dept. of Commerce, 1985. Calorie data from J. S. and C. E. Steinhart, "Energy Use in the U.S. Food System," *Science*, 184 [1974], p. 312.)

each calorie of food that is produced (Fig. 21-1). In primitive agriculture, where all of the work is performed by hand, almost no energy is used other than human labor and solar energy for photosynthesis. In 1910, the United States food system consumed about 1 calorie of additional energy for every calorie of food produced. By 1970, this figure increased to 8 calories additional energy used for each calorie produced. Clearly, progress in agricultural production has largely involved the substitution of nonhuman energy sources for human labor.

Virtually the same can be said for other industries. Consequently, we find that overall standard of living (as measured by gross national product per capita) is correlated with energy consumption. Note in Figure 21-2 that in less-developed countries such as India, energy consumption per person is very low. It generally increases with development. Also note that the correlation

between standard of living and energy consumption as shown in Figure 21-2 is not perfect. Some countries equal others in standard of living, while consuming markedly less energy. To some extent, this reflects the kinds of industry that prevail. However, it also reflects *efficiency* of energy use. Most machines are relatively inefficient in converting energy into the desired application. For example, only about 10 percent of the energy used by vehicles goes into moving them from one place to another. The other 90 percent is lost in cooling the engine and overcoming friction and wind resistance. Clearly, there is tremendous opportunity here to ''get more for our money'' in terms of energy use. Yet, a basic relationship between energy and work exists. Fundamentally, more work requires more energy (the first law of thermodynamics; see Chapter 2).

The most important point to remember, however, is that energy is not just required to de-

**FIGURE 21-2**
Correlation between energy consumption and gross national product. Gross national product per capita is one measure of standard of living. (Adapted from Earl Cook, ''The Flow of Energy in an Industrial Society,'' *Scientific American,* Sept. 1971, pp. 134–144. Copyright © 1971 by Scientific American, Inc. All rights reserved.)

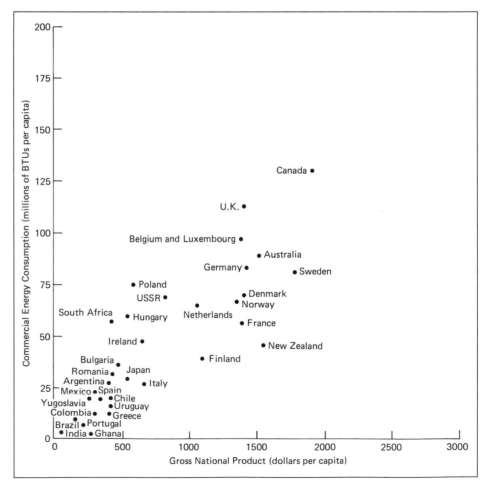

## ELECTRICAL POWER PRODUCTION

Electricity is produced by the phenomenon first described by Michael Faraday in 1831, namely, that when a coil of wire is passed through a magnetic field it induces a flow of electrons in the wire. The flow of electrons is synonymous with an electric current. An electric generator is basically a coil of wire that can be rotated within a magnetic field, or the magnetic field can be rotated within the coil of wire (Fig. 21-3). However, one does not get something for nothing. As an electric current begins to flow through the wire, it creates a magnetic field that is opposite to the existing field, and this resists the movement of the wire through the field. Therefore, an amount of energy proportional to the output must be expended in turning the generator. Since no energy conversion is 100 percent efficient, *more* energy must be put into turning the generator than one gets out in electricity.

Recognizing this principle, the main problem in producing electricity is finding a power source to turn the generator. The most common technique is to boil water to create high-pressure steam. The steam is passed over a turbine (a sophisticated paddle wheel) coupled to the generator (Fig. 21-4a). The combined turbine and generator are called a **turbogenerator.** Any heat source can be used to boil the water; coal, oil, and nuclear energy are most commonly used at present. Wood, wastepaper and garbage, solar energy, and geothermal energy (heat from the earth's interior) may be used more in the future.

In addition to steam turbines, gas and water (hydro) turbines are also used. In a gas turbogenerator, the high-pressure gases produced by the combustion of a fuel, usually natural gas, drive the turbine directly (Fig. 21-4b). In a hydroturbogenerator, the high pressure of water from behind a dam or at the top of a waterfall is used to drive the turbine (Fig. 21-4c). The proportion of our electricity generated by different sources is indicated in each case.

The major point is that despite the apparent cleanness of electricity in its final use, it has distinct environmental impacts related to the fuels or other energy source used to produce it. These impacts may be divided into three categories: (1) land destruction or alteration caused by mining the fuel—or building a dam and reservoir in the case of hydroelectric power, (2) pollution or hazardous waste products resulting from consuming the fuel, and (3) thermal pollution from waste heat.

These problems take on even greater significance when the relatively low efficiency of power production is considered. In order to drive a turbine, steam must pass from high pressure at the boiler side of the turbine to low pressure at the opposite side. In order to create the low pressure, the steam leaving the turbine is passed through a condenser, which removes the heat and condenses it back to water. The heat removed is 60 to 70 percent of the energy that was used to produce the steam. Only 30 to 40 percent of the energy is actually converted into electricity; in terms of efficiency, this is only 30 to 40 percent efficient (Fig. 21-5a). The waste heat removed from condensers may be discharged into the atmosphere through cooling towers (Fig. 21-5b) or it may be discharged into a body of water. Discharge of the waste heat into a natural body of water results in thermal pollution, which may have a severe impact on ecosystems.

**FIGURE 21-3**
Principle of an electric generator. Rotating a coil of wire in a magnetic field induces a flow of electricity in the wire. But more energy must go into turning the generator than is gotten out in electricity.

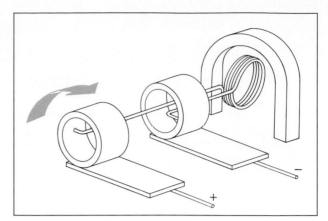

**FIGURE 21-4**
Electricity is produced commercially by driving generators with (a) steam turbines, (b) gas turbines, and (c) water turbines. Percentage of electricity (United States) derived from each source is indicated. (d) Component of steam turbine being assembled. (Photo courtesy of Consulate General of Japan, N.Y.)

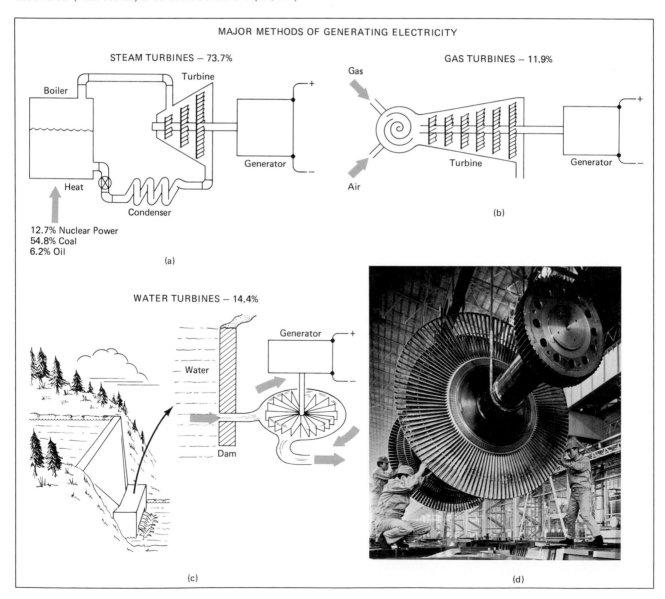

MAJOR METHODS OF GENERATING ELECTRICITY

STEAM TURBINES — 73.7%

Boiler  Turbine  Generator  +  −

Heat  Condenser

12.7% Nuclear Power
54.8% Coal
6.2% Oil

(a)

GAS TURBINES — 11.9%

Gas  Air  Turbine  Generator  +  −

(b)

WATER TURBINES — 14.4%

Water  Dam  Generator  +

(c)

(d)

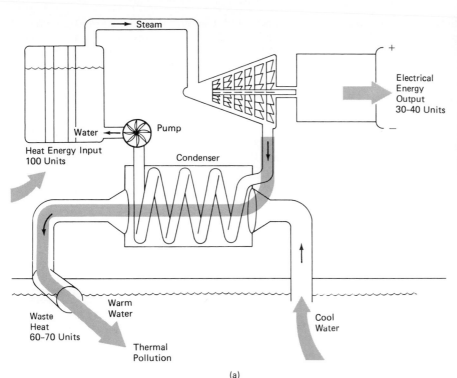

(b)

**FIGURE 21-5**
Waste heat from power generation. Steam leaving the turbine must be recondensed before the water is recycled back to the boiler. Consequently, 60 to 70 percent of the heat used in producing the steam is removed at the condenser as waste heat. The waste heat may be dissipated (a) into waterways by circulating water over the condenser or (b) into the atmosphere by using cooling towers, which are often the most conspicuous feature of a power plant. (TVA photo.)

velop a high standard of living. High inputs are required on a day-to-day basis to maintain life as we know it. If the "energy plug" were pulled, everything would come to a grinding halt and, if the situation continued, the results would be catastrophic. During the "energy crisis" in the 1970s, gas stations ran out of gas. Cars lined up for blocks at the stations still open and people hesitated to drive for fear they might not get another tankful. Shopping malls, motels, and restaurants were nearly vacant and certain industries closed down because they lacked fuel to process materials. This situation awakened the public at large to the reality of the energy problem. It became the talk of the day and launched efforts to develop more fuel-efficient vehicles and alternative sources of energy.

Due at least in part to these measures, fuel supplies again seem abundant. Does this mean the problem has been solved? Unfortunately it does not. The underlying problems remain and may well surface again in the near future. Next time, the results may be more severe and long-lasting unless continuing adjustments are made now. There is much misunderstanding and controversy, however, about the nature of the problem and the adjustments that are appropriate. To

clarify the issue, this chapter presents an overview of energy sources and uses, focusing on crude oil depletion—the crux of the energy problem. The pros and cons of nuclear power and some other alternatives will be presented in Chapter 22, and solar energy and conservation will be discussed in Chapter 23.

## PRIMARY SOURCES AND USES OF ENERGY IN THE UNITED STATES

The energy problem (commonly called the energy crisis) revolves around an impending depletion of crude oil. To achieve timely and appropriate solutions to this problem, we must first understand the role of oil in our overall energy economy and we must understand certain facts about its depletion.

### The Current Situation

Our current uses of energy can be divided into four principal categories:

1. Transportation. This includes trucks, buses, airplanes, trains, and boats as well as private cars, farm tractors, bulldozers, and similar machinery.

2. Industrial processes. Production of all raw materials such as metals and synthetic chemicals requires energy, as do all manufacturing and fabricating processes.

3. Space and water heating and cooling. This includes the heating or cooling of homes and buildings for comfort, and the heating of water in both homes and commercial establishments.

4. Generation of electrical power. Electrical power is used for lighting, operating all electrical equipment such as televisions and computers, and driving electric motors that run all kinds of appliances and machinery in homes and commercial and industrial establishments. Electricity, however, is not a primary source of energy. It is called a secondary source because it is generated from a primary source such as coal, oil, or nuclear power as shown in the box.

The United States consumes roughly equivalent amounts of energy—as measured in units such as BTUs (British Thermal Units) or calories—in each of these four categories. Of the total energy consumed in the United States in 1983, oil

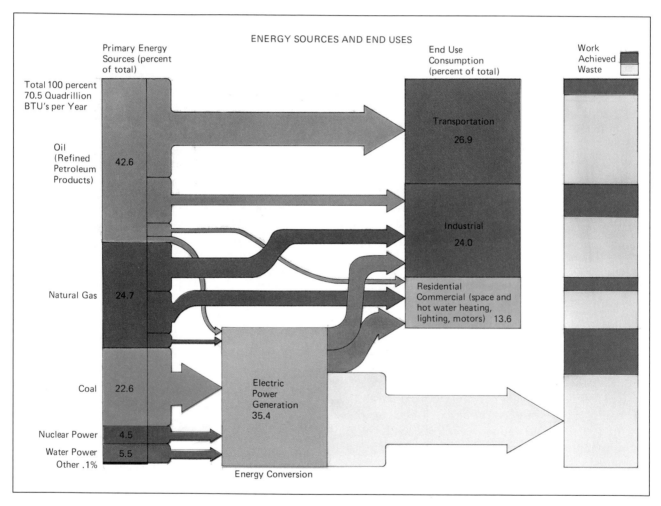

**FIGURE 21-6**
Energy flow from primary sources to end uses. This is a highly simplified diagram showing only major patterns, but it should emphasize that, given current technology, primary sources are connected to end uses in specific ways. Transportation is almost totally dependent on liquid fuels refined from crude oil, while coal and nuclear energy are useful only in producing electricity, almost none of which is used in transportation. Also, a large portion of energy is wasted at each conversion step. (Data from *Statistical Abstracts of the United States*, U.S. Department of Commerce, 1985.)

provided 42.6 percent; natural gas, 24.7 percent; coal, 22.6 percent; water power, 5.5 percent; nuclear power, 4.5 percent; and other sources, 0.1 percent. Figure 21-6 shows how these energy sources are matched with end uses.

Transportation and other vehicles depend by and large on liquid fuels such as gasoline and diesel fuel, which are refined from crude oil. Space and hot water heating uses mostly natural gas and occasionally fuel oil. Nuclear power, water power, and most of the coal are consumed in the generation of electricity. Industrial processes use liquid fuels, natural gas, and coal in roughly equal proportion.

It is significant to note that, as shown in Figure 21-6, the largest portion of total primary energy used in the United States currently comes from crude oil. Additional large portions come from coal and natural gas. Despite the emphasis on developing new sources of energy over the last decade, nuclear power, solar power, and other alternatives still play a relatively small role in the overall energy picture.

It is also important to emphasize the matching of source and use. For example, transportation and other vehicles are matched almost 100 percent with liquid fuels refined from crude oil. Other energy sources are still impractical for this function. Similarly, it is impractical to use nuclear power for anything other than to generate electrical power. This sort of specificity becomes an im-

portant consideration in developing alternative energy sources. A new source of energy is meaningless if there is not a ready means of applying it to perform the desired functions.

## Trends

The energy picture presented above is far from static. First, the total amount of energy consumed has steadily grown with population increases, technological development, and rising standards of living. Second, major sources of energy have shifted with developing technology and the availability of various sources.

Throughout most of human history, the major energy source was human labor. Some people lived in relative luxury by exploiting the labor of others—slaves, indentured servants, and minimally paid workers, for example. Human labor was supplemented to some extent with domestic animals, water power (water wheels), and wind power (windmills). However, animals are extremely inefficient in providing the sustained power required to run machines over long periods. Water wheels are limited to certain locations and by the capacity of flowing streams and rivers; windmills are similarly limited as an energy source.

The breakthrough that enabled the Industrial Revolution to begin was the develoment of the steam engine in the late 1700s (Fig. 21-7). The

**FIGURE 21-7**
Diagram of the first steam engine designed and built in the United States. Steam engines permitting the physical conversions of fuels into useful work fostered the Industrial Revolution. (Library of Congress.)

**FIGURE 21-8**
Colossal steam-driven tractors marked the beginning of modern agriculture. Today tractors of half the size do ten times the work and are much easier to operate. (Courtesy of Carroll County Farm Museum, Westminster, Maryland; photo by Mary J. Wise.)

steam engine provided almost any amount of power at almost any location, and, when mounted on wheels, it provided transportation—the steam locomotive. The first major fuel for steam engines was firewood. Coal was substituted for firewood as demands for energy increased and wood became scarce. Coal became the major fuel by the end of the 1800s. By 1920, the peak of the coal era, coal provided 80 percent of all energy used in the United States.

Coal, however, is messy to handle and transport, and its burning pollutes the air with smoke, soot, and noxious gases, particularly sulfur dioxide. It also requires the disposal of significant quantities of ash, which is also messy and polluting. In addition to the nuisance of starting and maintaining the fire, steam engines are huge and unwieldy to operate (Fig. 21-8). A cleaner, more convenient alternative was highly desirable.

In the late 1800s, simultaneous development of oil-well drilling technology and the internal combustion engine provided the alternative. There is hardly a comparison between stoking up a coal fire and waiting for the boiler to heat, and the almost instant starting of a gasoline engine. The burning of gasoline creates little smoke and no ashes, and gasoline or other oil products are relatively clean to handle and transport. This is not to say that burning liquid fuels does not create pollution. Indeed, serious pollution problems from auto exhaust are now well recognized (see Chapter 14). Still, much less pollution is generated by gasoline than by coal burning.

Further, the internal combustion engine provided a valuable power-to-weight advantage. A 100 horsepower gasoline engine weighs but a small fraction of a 100 horsepower steam engine and its boiler. In addition, a gram of gasoline has a higher energy content than a gram of coal, providing a further power-to-weight advantage. Automobiles and other forms of transportation would be cumbersome, to say the least, without this power-to-weight advantage, and airplanes would be impossible.

Given the advantages, it is not surprising that oil-based fuels became the dominant energy source by 1950 and their use continued to increase. Coal, on the other hand, despite its abundance, was phased out of use for most processes except the generation of electrical power and certain industrial processes.

Natural gas found in association with oil or in the process of looking for oil, burns even more cleanly than oil and is not subject to spillage. Thus, in terms of pollution, it is a most desirable fuel. However, at first there was no means available for transporting natural gas from the well to the consumer. Its energy content per volume is so small that until the technology of liquefaction was developed, one could not put enough into a tank to make transportation practical. Natural gas, where found, was, and in many parts of the world still is, simply flared, that is, vented and burned in the atmosphere, a tremendous waste of valuable fuel. Gradually, however, in the United States a network of pipelines was constructed to carry the gas from wells to consumers. With the completion of these pipeline networks, especially after World War II, use of natural gas for heating and industrial processes escalated rapidly because of its cleanliness, its convenience (no storage bins or tanks are required on the premises), and, perhaps most importantly, its relatively low cost. These trends, as well as total quantities used, are shown graphically in Figure 21-9.

In summary, virtually all of the increases in technology and in standard of living in the United States since 1920 have been supported by the use of oil-based fuels and natural gas. Together, they

**FIGURE 21-9**
Energy consumption in the United States, 1860–1982; total consumption and major primary sources of energy. Note how the mix of primary sources has changed over the years while the total amount of energy consumed has continued to grow. Conspicuous changes in the trends occurred in the 1970s with the recognition of limited oil reserves. (Data from the U.S. Department of Energy.)

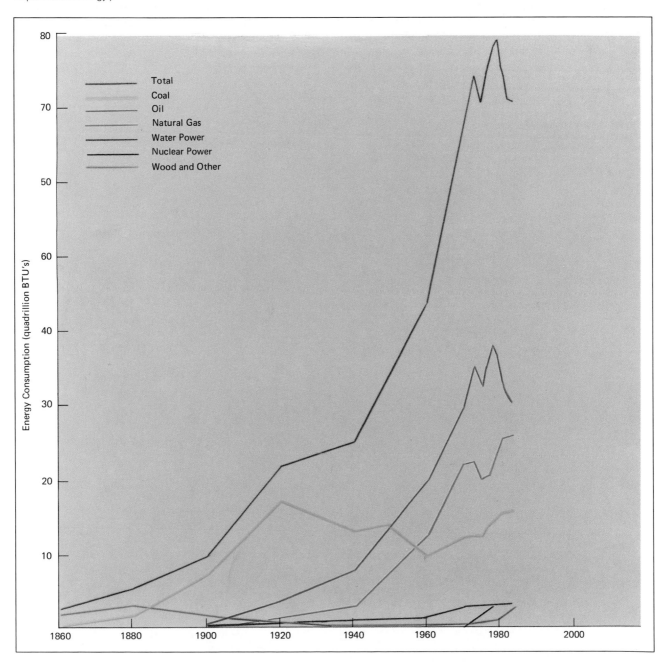

still supply about two thirds of the total energy demand in this country.

Based on existing trends, it was predicted in the early 1970s that the use of oil-based fuels and natural gas would continue to increase at the rate of about 4 percent per year, thus doubling consumption in the next 20 years. But the energy crisis of the 1970s changed the situation.

## THE ENERGY PROBLEM (OR CRISIS): DECLINING AVAILABILITY OF CRUDE OIL

You have probably heard the statement: We are running out of oil. But, in reality, supplies seem more than adequate. The crux of the energy problem, nevertheless, involves a declining availability of crude oil, and to a lesser extent, of natural gas. Understanding how we reach this conclusion is central to understanding the energy problem. First, we must look at the nature of crude oil reserves and their production.

### Formation of Fossil Fuels

Crude oil, natural gas, and coal, which are found in the earth, are actually residues of biological production—hence their group name: **fossil fuels**. Frequently in the history of the earth, photosynthetic production outpaced the activity of consumers and decomposers. Consequently, large

**FIGURE 21-10**
Energy flow through fossil fuels. Coal, oil, and natural gas are derived from photosynthesis of early geological times. Deposits are limited, and as they are used, the energy is gone forever.

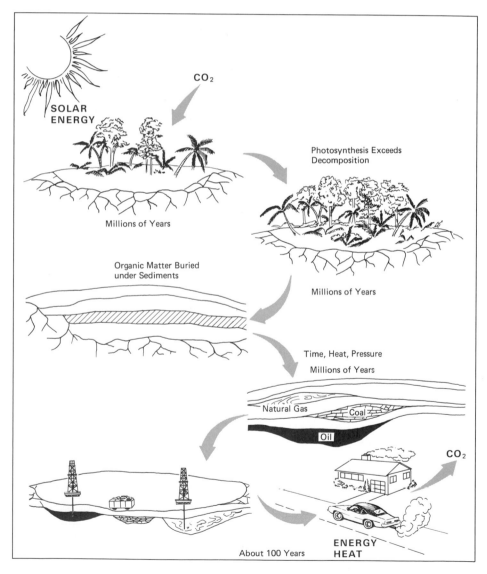

amounts of organic matter accumulated, especially on the bottoms of shallow seas and swamps. Gradually, this material was buried under sediments eroding from the land. Over millions of years of geological history, this material was converted to the materials we now recognize as coal, oil, and natural gas, the particular material being dependent on the specific conditions and time involved (Fig. 21-10).

While fossil fuels are formed by ongoing, natural processes, two factors preclude any practical notion of renewability. First, biological conditions on earth have changed so that significant accumulations of organic matter no longer occur. Second, a striking disparity between the rate of natural formation and the rate of human exploitation of fossil fuels exists. It is estimated that humans currently exploit each day an amount of crude oil that it took nature about 1000 years to produce.

Because supplies are finite, there is no question that sooner or later we shall run out of these fuels. This has been well understood since the early 1900s. However, exhausting supplies is of little practical concern if that event is not going to occur for 100 or more years. The pertinent question is: What size are the reserves and how much remains to be found?

## Exploration, Reserves, and Production

The science of geology provides some knowledge about the location and extent of ancient, shallow seas. Based on this knowledge and past experience, geologists make educated guesses as to where oil and/or natural gas may be located and the amounts that may be found. These are the world's **estimated reserves**.

The next step is exploratory drilling in areas most likely to contain reserves. If oil is found, further exploratory drilling is conducted to determine the size of the area and the thickness of the deposit, known as the oil (or gas) field. When this process is complete, a fairly accurate estimate as to how much oil or gas can be recovered from the field at current prices is made. These are the world's **proven reserves.**.

The final step is production—withdrawing the oil or gas from the field. Production is often a complex process because the oil in an oil field is not like water in an underground reservoir which can easily be withdrawn at a desired rate. Oil is contained in sedimentary rock like water in a sponge. The field may be under pressure such that initial penetration produces a gusher. But gushers are short-lived and the oil then seeps slowly from the rock into a well where it is withdrawn. To further complicate matters, only about 30 percent of the oil present seeps from the rock into a well by itself. Additional oil recovery involves the use of secondary techniques such as injecting steam or other material into adjacent wells to force the oil into the producing well. These techniques can be both time-consuming and costly.

As a result of the slow movement of oil into wells, the maximum amount of oil obtainable from a field in a year averages only about 10 percent of the amount remaining in that field. For example, suppose we have a field with 100 million barrels of recoverable reserves. The maximum production from the field will be about 10 million barrels the first year. Of the remaining 90 million barrels, about 9 million will be recovered the second year. Of the 81 million barrels remaining the third year, about 8.1 million barrels will be recovered, and so on (Fig. 21-11).

You can see from these calculations that production declines each year, but may be extended indefinitely over time. The real cutoff of production from a field, its "running out," is determined by economics. There comes a time when the value of the oil obtained does not justify the costs of further recovery. We should note, however, that the increasing price of oil may make further recovery economically practical (Fig. 21-11). Eventually a limit of zero **net yield** may be reached, a point at which energy expended in production becomes equal or greater than energy produced.

Considering the "10 percent rule" for production, it follows that if one wishes to obtain a certain production level, one must have proven reserves that are at least 10 times greater than that level. For example, a production of 20 million barrels in a given year would demand proven reserves of 200 million barrels (10% × 200 = 20).

Of course, proven reserves are not only decreased by production, they may be increased by new discoveries. To hold oil production at a constant level, new discoveries (reserve additions) must, on the average, at least equal production (reserve subtractions).

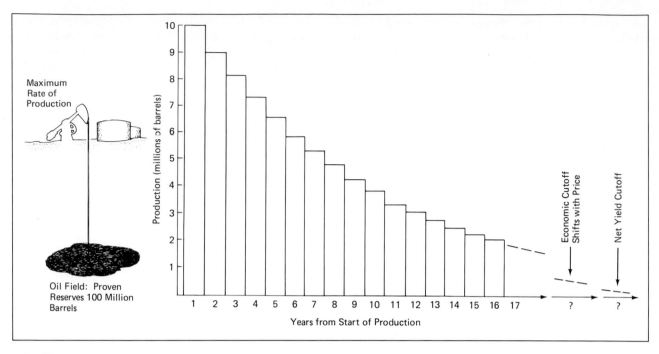

**FIGURE 21-11**
When will we run out of oil? The maximum yearly production from a given oil field is 10 percent of the remaining reserves. Therefore, although the maximum production from given reserves decreases, it may stretch out indefinitely as shown. However, the economic cutoff of production from a field occurs when the value of the oil obtained no longer justifies the costs of producing it. How is this cutoff point affected by the price of oil? Finally a net yield cutoff may occur when energy expended in obtaining oil outweighs the energy value of the oil obtained. The only way that production can be maintained or increased is to discover and develop new reserves at a rate equal to or exceeding consumption.

## Declining Reserves and Production and Increasing Importation of Crude Oil

From the previous discussion, it should be clear that if proven reserves are just ten times larger than annual production, it does not mean that we shall run out after ten years. The important issue is whether new discoveries keep pace with production. If they do not, then proven reserves will be drawn down and production will be diminished accordingly. However, there will not be an abrupt running out as in a car running out of gas.

Here, then, is the energy problem in the United States. Through 1970, new discoveries in the United States, for the most part, kept up with our growing oil-based economy. But the real numbers are staggering. In 1970, United States consumption of crude oil was over 5 billion bar-

rels per year (13.7 million barrels per day) and growing rapidly (Fig. 21-12). The last major discovery in the United States was made in 1968 in Alaska. Since then, new discoveries have not kept up. Consequently, U.S. reserves and production have generally moved downward (Fig. 21-13). Production has increased somewhat in recent years because high prices have allowed for the re-opening and further exploitation of old fields from which easy-to-get oil was already removed. But such increases clearly cannot be sustained.

Despite decreasing production of crude oil in the United States during the 1970s, consumption increased, leading to a widening "energy gap." As shown in Figure 21-12, this gap was filled by importing increasing quantities of crude oil, primarily from the Arab countries of the Mideast. European countries and Japan did likewise. Since imported oil cost only about $2.30 per barrel (42

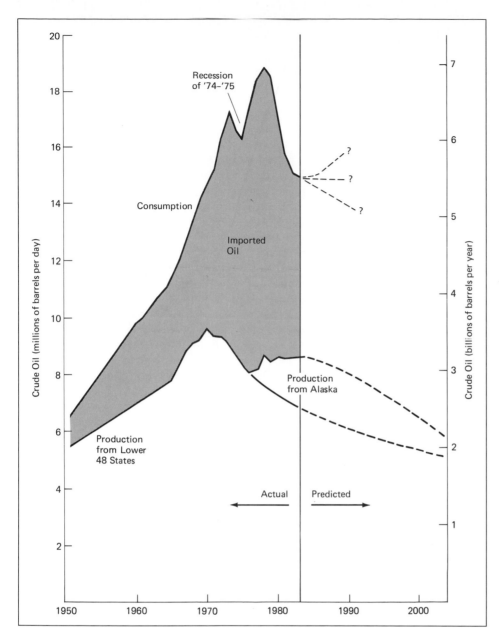

**FIGURE 21-12**
Actual and predicted production of oil from the United States compared to consumption. Since 1968 new discoveries have not kept pace with production. Therefore, production has been on a downward trend that is expected to continue as existing reserves are drawn down. Meanwhile, consumption has continued to rise, making us highly dependent on foreign oil, although this situation has been mitigated somewhat by conservation efforts to reduce oil demand. (Data from the U.S. Department of Energy.)

gallons) in the early 1970s and Mideast reserves were vast, this course seemed to present few problems.

## Crises and Gluts in Crude Oil Supply

By the early 1970s the United States and most other Western nations were dependent on oil imports from a small group of countries known as **OPEC**, the Organization of Petroleum Exporting Countries (Fig. 21-14). In 1973, OPEC decided to take advantage of its unique position and temporarily suspend oil exports. This threw the world into a crisis and allowed OPEC to raise the price of its oil from $2.30 a barrel to about $10.50 a barrel. A second "crisis" occurred in 1979 when political upset in Iran resulted in a cutoff of exports from that country, and the energy turmoil through the period allowed OPEC to move the

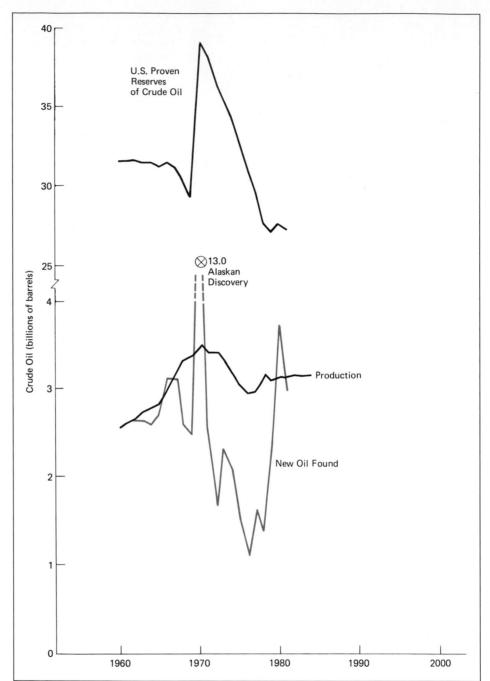

**FIGURE 21-13**
Production, discoveries, and reserves of U.S. crude oil. Since the exceptional Alaskan discovery in 1968, the amount of new oil found has been considerably less than production. Consequently, proven reserves since the Alaskan addition have been drawn down sharply (top curve). This portends a trend of decreasing production in the future. (Data from *Statistical Abstracts of the United States*, Department of Commerce, 1985.)

price of oil further upward. In the early 80s a barrel of oil cost 30 to 35 dollars.

Note that the crisis was not, in fact, a result of the world running out of oil; it was the result of temporary *disruptions in supply*. These cutoffs actually involved only a small portion of total world supply, but the situation was nevertheless perceived as a crisis. To understand why this ov-

erreaction occurred, we must recognize that, compared with the amount of oil used (close to 20 billion barrels per year worldwide), relatively little is maintained in storage. We depend on a continuous flow of oil from the sites of production to the place of consumption.

Consequently, if production is cut by just a few percent, supplies in storage are rapidly de-

**FIGURE 21-14**
The nations in OPEC, the Organization of Petroleum Exporting Countries.

pleted and desired activities are curtailed. This is perceived as a **crisis.** Conversely, if production rises just a few percent and/or consumption drops, available storage tanks soon fill to capacity. There is literally no place to put the excess oil and the situation is referred to as an oil *glut* (Fig. 21-15). As a result of shifting production and consumption, "crises" were interspersed with "gluts". By the mid 1980s, due to both conservation reducing oil demand and increasing production from various sources, a glut situation became severe and the oversupply led to a considerable drop in oil prices. Certainly, these along with changing fuel prices have left many people confused about whether the energy problem was real.

However, that the crises occurred at all should emphasize the underlying problem—in the United States and other industrialized nations, new discoveries are not keeping pace with consumption. Hence reserves are being drawn

down and production has declined accordingly. Since 1980, demand (consumption) has also declined somewhat, but it has remained considerably above domestic production. As long as this situation continues, we shall remain dependent on foreign supplies.

## The Outlook for the Future

Continuing dependency on foreign oil supplies involves three major areas of concern: (1) the threat of political disruptions, (2) the adverse economic impacts of imports, and (3) the problem of diminishing supplies.

### POLITICAL DISRUPTIONS

Political turmoil involving petroleum-exporting countries may again disrupt oil imports and cause a crisis similar to that of the 1970s. The United States is no longer as vulnerable to such disruptions because of our decreased dependence on imports (see Figure 21-12). In addition, we have diversified our sources of supply (the largest

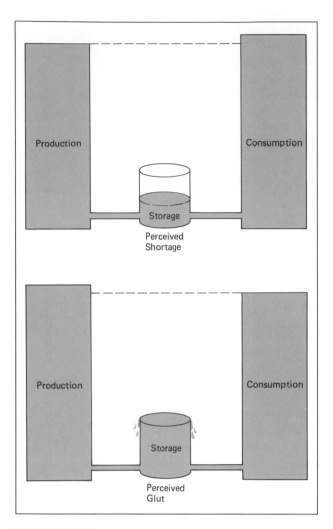

**FIGURE 21-15**
Shortages and gluts. The amount of oil in storage is relatively
small compared to annual consumption. Therefore, production
must be closely balanced with consumption. Any imbalance
can rapidly give the impression of a shortage or glut.

(a)

(b)

ration of oil now comes from Mexico) and we
have "stockpiled" about 300 million barrels of oil
(equivalent to 90 days of imports) in underground
caverns in Louisiana (Fig. 21-16). We are, however, committed to share oil supplies with other
countries if a world situation demands it. And
many countries are even more dependent on oil
imports than we are.

### ECONOMIC IMPACTS

Importing foreign oil is an economic liability.
The outflow of American dollars to pay for imported oil skyrocketed from about $3 billion in
1973 to over $50 billion in 1980 and 1981. The cost

**FIGURE 21-16**
Strategic petroleum reserves. To gain protection from future
cutoffs of foreign oil supplies, the United States has been
buying excess oil and putting it into storage in salt mines in
Louisiana and Texas. The reserve contains about 30 days
supply of oil. (a) Aerial view of Morton Salt surface operation.
(b) Underground at the work site. (U.S. Department of Energy.)

continues presently at about $40 billion (Fig. 21-17). This expense is responsible for a major portion of the U.S. balance of trade deficit, which reached record levels in 1984. Such high deficits cannot be sustained over many years without economic repercussions. Many less-developed countries are on the verge of bankruptcy as a result of the burden of bills for imported oil.

## DIMINISHING SUPPLIES

Finally, and most seriously, is the outlook for the world oil supply itself. World consumption of crude oil is about 20 billion barrels per year. Despite intensive exploration, new discoveries have not kept pace for the last 15 years, some notable finds notwithstanding. In this respect, it is interesting to note that a discovery of a field with 10 billion barrels is now rare and re-

ceives headlines as a "giant discovery" when it occurs. Yet such a discovery represents no more than a six-month supply for the Free World, or less than a two-year supply for the United States alone. Another indication of the problem is that exploratory drilling has shown that areas that were thought to be prime, such as the eastern continental shelf of the United States, are essentially void of oil (Fig. 21-18).

This slowdown in the discovery of new fields despite increasing exploration indicates that the majority of oil present in the earth has already been discovered. This conclusion is based on what is called "the Easter egg argument." When people first begin searching for hidden Easter eggs, the discovery rate is proportional to the searching effort. However, the searchers reach a point where, in spite of their continued effort,

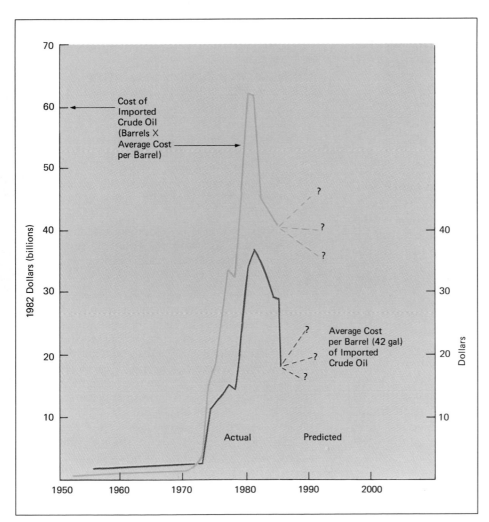

**FIGURE 21-17**
Cost of oil dependency. As the U.S. increased its dependency on foreign oil in the '70s, OPEC was able to increase prices, resulting in skyrocketing costs to the U.S. Conversely, decreasing oil dependency, enabled largely by conservation, brought about "oversupply" and a collapse of oil prices in 1986. Without continuing conservation the episode of the '70s will be repeated even more severely (dotted projections). With conservation and development of alternatives continuing to lessen dependence on oil, price and costs will tend to stabilize (dashed projections). (Data from *Statistical Abstracts of the United States*, U.S. Dept of Commerce, 1985.)

**FIGURE 21-18**
Despite intensive exploration, new finds have been disappointing. Exploratory wells in the Mukluk area of the Bering Sea, a 1.7 billion dollar venture, came up dry. (Lowell Georgia/Photo Researchers.)

fewer and fewer eggs are found. This situation leads to the logical conclusion that most of the eggs have already been found.

In view of the declining rate of discovery, the U.S. Geological Survey recently lowered its estimate of world reserves (estimates of the amount of oil that remains to be discovered). Their conclusion is that, if current consumption patterns continue, there will be shortfalls between world production and world demand by the mid-1990s. When this occurs there will be no alternative to reducing the use of oil-based fuels. At this point, other sources of energy must be available to meet the demand or else energy use and the transportation, industry, and other functions dependent upon it will be curtailed to a level supportable by the energy available. In the words of Earl Hayes, former chief scientist at the U.S. Bureau of Mines, "talk of rising petroleum (and gas) production for long periods is both immoral and nonsensical. Whatever slight gain might be achieved for a very few years will be at the expense of the youth of today."* That is, higher production now can only result in a faster decline in production after the peak is reached. In turn, the faster the decline in production, the greater

will be the economic and social disruptions resulting from it.

The most immediate economic and/or political repercussions of such shortages will take one or a combination of three forms:

1. Fuel prices may escalate rapidly. As at an auction, those with money effectively bid up the price in their competition to gain desired amounts. Those with less money will be gradually squeezed out of the market and forced to get along with less or do without.

2. Various rationing or allocation schemes may be initiated in an attempt to divide the supply equitably so that all suffer some but no one suffers unduly.

3. Prices may be maintained by regulation at artificially low levels without satisfactory allocation or rationing programs being initiated. This will lead to capricious and chaotic shortages such as all the service stations in an area suddenly having no gas to sell. One can speculate as to what may happen at that point; the only thing that cannot happen is that unrestricted supplies remain available at a price we all can afford.

During the shortages of the 1970s we witnessed, to some extent, all of these events—an escalation in fuel prices, discussion of rationing and allocation programs, and some spot shortages.

*Hayes, Earl, "Energy Resources Available to the United States 1985–2000," *Science*, 203 (January 19, 1979), pp. 233–239.

## MEETING THE ENERGY PROBLEM: BASIC OPTIONS

In dealing with the energy problem, it is important to keep in mind that the problem begins with small, but increasing shortfalls in the production of crude oil and natural gas. Two basic options exist: (1) conservation, that is, using the oil-based fuels more efficiently and sparingly; and (2) developing alternative energy sources that may be substituted for oil or gas-based fuels.

When considering the alternatives, more than the energy factor must be considered. It is also important to consider environmental impacts such as potential pollution and land degradation.

In the years since the 1973 Arab oil boycott, much thought, work, and experimentation have gone into both of these areas and progress has been made. New cars, for example, now average about twice the miles per gallon as cars manufactured in 1970. Many alternative energy sources have at least been experimented with. Although development of these ideas has not been exten-

sive, we do have a better understanding of what will and will not work and what their environmental impacts are likely to be. Perhaps most importantly, the last 15 years have shown that much lead time is necessary before new developments can significantly contribute to the energy economy. Consequently, it is vitally important that we do not become complacent during this period of apparent oil surpluses, but that we continue to move ahead with those strategies that appear to have the best potential for providing adequate, long-term energy resources without harming the environment.

The dangerous alternative is that we allow the 1986 drop in oil prices to lull us into relaxing efforts toward further conservation and development of alternatives. Such relaxation will inevitably result in regrowth in our dependence on foreign oil and set the stage for much more severe and intractable energy crises in the 1990s. In the following chapters we shall examine pros and cons of some alternative sources of energy and conservation.

# 22

# NUCLEAR POWER, COAL, AND SYNTHETIC FUELS

## CONCEPT FRAMEWORK

| Outline | | Study Questions |
|---|---|---|
| **I. NUCLEAR POWER: HOW IT WORKS** | 539 | 1. What triggered the United States to promote the use of nuclear energy? What is the role of the Nuclear Regulatory Commission? When did the first nuclear power plants come on line? How has the construction pattern changed? What is *fission*? What is *fusion*? |
| **A. The Fuel for Nuclear Power Plants** | 539 | 2. From what is energy derived in existing nuclear power plants? What makes one isotope different from another? What is the key to a nuclear reaction? Why doesn't this reaction occur in nature? How is nuclear fuel produced? How does the nuclear reaction in a bomb differ from that which occurs in a power plant? |

From the time fossil fuels were first used, geologists recognized that they would not last forever. Sooner or later, other energy sources would be needed. The end of World War II, marked by the awesome power of the atom, was the time of decision. The United States government desperately wanted to show the world that the power of the atom could benefit humankind as well as destroy it and embarked on a course to make nuclear power the next major source of energy.

It was anticipated that nuclear power could produce electricity in such large amounts and so cheaply that we could gradually phase into an economy in which electricity would take over virtually all functions at nominal cost. Thus, the United States supported and promoted nuclear research and development (Fig. 22-1). Utilizing this research, companies such as General Electric and Westinghouse now construct the nuclear power plants that are ordered and paid for by utility companies (Fig. 22-2). The Nuclear Regulatory Commission, an arm of the federal government, was formed to enforce safety standards for the operation and maintenance of the plants.

Utility companies moved ahead with plans for numerous nuclear power plants in the 1960s and early 1970s (Fig. 22-3). By 1975, 53 plants were operating in the United States, producing about 9 percent of the nation's electricity. Another 165 plants were in various stages of planning or construction. Officials estimated that by 1990 several hundred plants would be on line, and by the turn of the century as many as 1000 plants would be operating in the United States.

However, since 1975, the picture has changed dramatically. Another 50 plants, already under construction, have been completed but no new orders have been placed since 1978. Further, most existing orders have been cancelled and construction of some plants was terminated despite the fact that in some cases billions of dollars had already been invested. By 1986, the number of operating plants had increased to 100 but the number of plants headed for completion had dropped

**FIGURE 22-1**
U.S. Government appropriations for research and development of commercial energy sources. The government's commitment to the development and promotion of nuclear power is demonstrated by the fact that nuclear power has consistantly been given the lion's share of government support. Other energy sources and conservation received serious consideration only for a brief period under the Carter administration, 1976–1980.

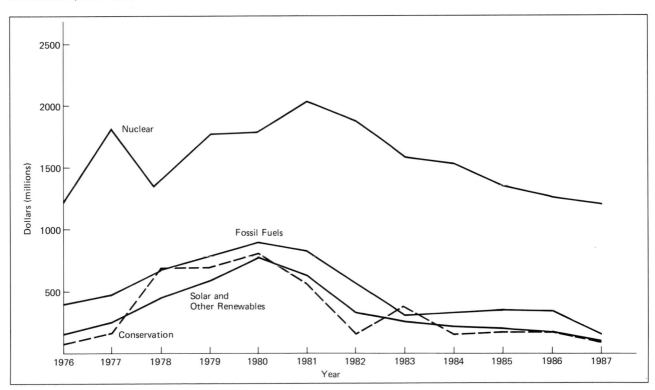

**FIGURE 22-2**
Arkansas Power and Electric's nuclear power plant near Russellville, Arkansas. (U.S. Department of Energy.)

**FIGURE 22-3**
(a) Changing fortunes of nuclear power in the U.S. are evident in graphs of the number of nuclear plants on order and in operation. Since the early 1970s, when orders for plants reached a peak, few utilities have called for new plants and many have canceled earlier orders. (The last plant order not subsequently canceled was placed in 1973.) Nevertheless, the number of plants in service has increased steadily. The figures for the number of plants operating in 1986 and beyond *(circles)* are estimates. (From Richard K. Lester, "Rethinking Nuclear Power," *Scientific American,* March 1986, p. 32. Copyright © 1986 Scientific American, Inc. All rights reserved.)
(b) Locations of operating nuclear power plants (U.S. Department of Energy.)

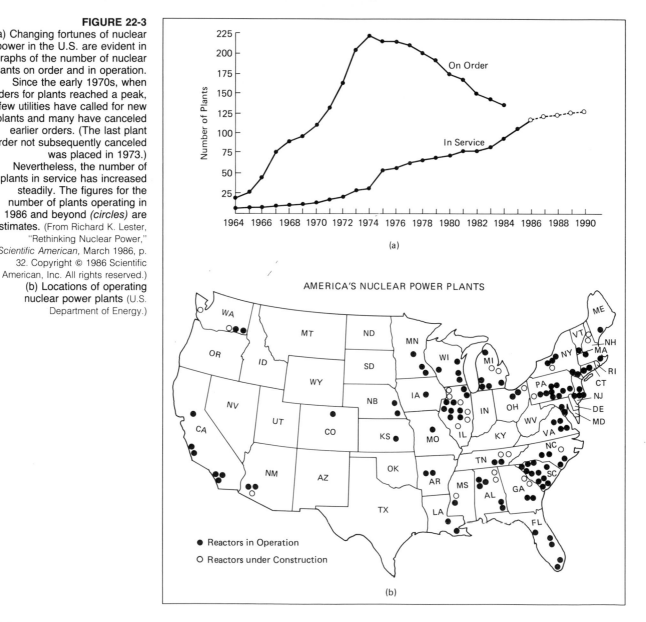

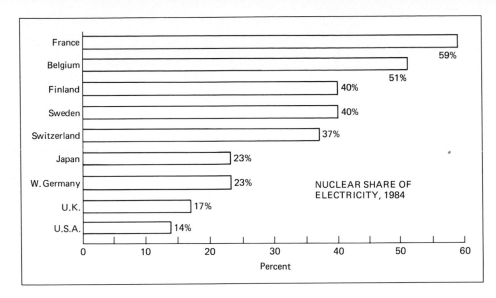

**FIGURE 22-4**
Percentage of total electrical power from nuclear plants in various countries. A number of nations surpass the United States in utilization of nuclear energy for power production. (International Atomic Energy Agency.)

to 25. The proportion of electricity produced by nuclear plants had changed little since 1975—about 14 percent in 1984—and was far lower in the United States than in other industrialized nations (Fig. 22-4).

Clearly, the nuclear scenario that had been anticipated is not materializing. But the underlying philosophy of the government has changed little. A lion's share of the Department of Energy's budget still goes to nuclear power research and development. In an effort to rekindle public interest and growth in nuclear power, the nuclear industry launched a massive media campaign in 1984. The campaign was designed to convince the public that nuclear power does work and is the best choice for a secure energy future. Yet, for a variety of reasons, many people are adamantly opposed to nuclear power. More and more, the average citizen is being brought into this debate and asked to vote for or against nuclear power. Clearly, nuclear power, or the lack of it, will affect our future. The objective of this chapter is to provide an understanding of the nature of nuclear power so that we can respond as informed citizens. Coal and certain other nonsolar alternatives will also be discussed.

## NUCLEAR POWER: HOW IT WORKS

The release of nuclear energy is a completely different phenomenon from the burning of fuels or other chemical reactions we have discussed. In a seeming exception to the usual case of conservation of matter, nuclear energy involves changing atoms through one of two basic processes. In the process known as **fission** a large atom of one element is split into two smaller atoms which constitute two different elements (Fig. 22-5a). In the process known as **fusion** two small atoms may combine to form a larger atom, again constituting a different element (Fig. 22-5b).

In both fission and fusion one finds that the total mass of the products is less than the mass of the starting material. The mass lost is converted into energy, according to the law of mass-energy equivalence ($E = mc^2$) first described by Albert Einstein. The amount of energy released by this conversion is tremendous. The sudden fission or fusion of a kilogram (2.2 lb) or so of material gives rise to the devastatingly explosive force of a nuclear bomb. In nuclear power plants, the fission reaction is controlled so that it progresses gradually and the energy is released as heat. The heat is used to boil water and the resulting steam is used to drive conventional turbogenerators.

### The Fuel for Nuclear Power Plants

In existing nuclear power plants, energy is derived from the fission of uranium atoms, specifically uranium-235. Uranium is an element that occurs naturally in the earth's crust in mineral form like other metallic elements. It exists in two primary forms or **isotopes:** uranium-238 ($^{238}$U) and uranium-235 ($^{235}$U).

The isotope numbers represent the number of subatomic particles present in the nucleus of the atom (e.g., $^{238}$U has three more neutrons than $^{235}$U). Most elements exist in a number of different isotopes. Because different isotopes of a given

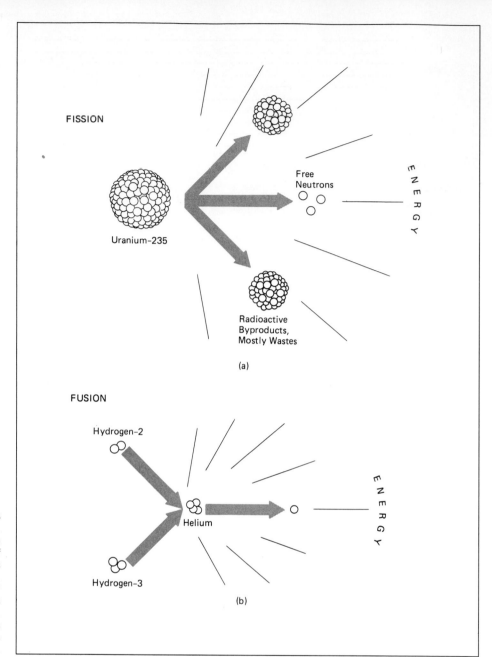

**FIGURE 22-5**
Release of nuclear energy. Nuclear energy is released from either (a) fission, the splitting of certain large atoms such as uranium-235 into smaller atoms, or (b) fusion, the "melting" together of small atoms such as hydrogen to form a larger atom. In both cases some of the mass of the starting atom(s) is converted to energy.

element are chemically identical, the phenomenon has little significance from a strictly chemical point of view. However, other characteristics of isotopes may differ profoundly. In this case, uranium-235 atoms will undergo fission, or split, readily while uranium-238 atoms will not.

Occasionally, a uranium-235 atom undergoes fission spontaneously, but it can be forced to undergo fission if it is struck by a neutron. This is the key to the controlled nuclear reaction. When one uranium-235 atom splits, in addition to releasing energy, it ejects two or three neutrons as well as forming the two smaller atoms. If one of these neutrons strikes another uranium-235 atom, fission occurs again, releasing more energy and more neutrons, which repeat the process. A domino effect, known as a **chain reaction,** occurs (Fig. 22-6a).

A chain reaction does not occur in nature because uranium-235 atoms are too dispersed

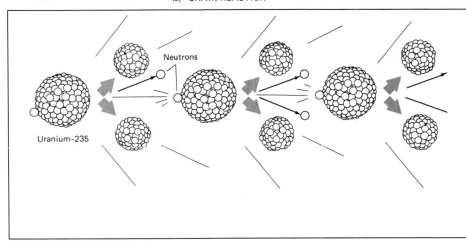

**FIGURE 22-6**
(a) A simple chain reaction. When a uranium atom fissions, it releases two or three high-energy neutrons in addition to the energy and the split "halves." If another uranium-235 atom is struck by a high-energy neutron, it fissions and the process is repeated, causing a chain reaction.

among other elements and stable uranium-238 atoms. Indeed, 99.3 percent of all uranium is $^{238}U$; only 0.7 percent is $^{235}U$. Hence, when a $^{235}U$ atom spontaneously undergoes fission in nature, it seldom triggers another atom and the energy released by a single atom goes unnoticed without the aid of radiation detectors such as geiger counters.

To make nuclear "fuel," uranium ore is mined, purified, and enriched. **Enrichment** involves separating $^{235}U$ from $^{238}U$ to produce a material with a higher concentration of $^{235}U$. Since $^{238}U$ and $^{235}U$ are chemically identical, enrichment is based on the slight difference in mass. The difficulty and sophistication of this enrichment procedure is the major technical hurdle that prevents less developed nations from advancing their own nuclear capability.

When uranium-235 is suitably enriched and placed in an appropriate mass, an atom that undergoes spontaneous fission triggers a chain reaction. In nuclear weapons, small masses of highly enriched uranium are forced together so that the two or three neutrons from a spontaneous fission trigger two or three more reactions and each of these sets off two or three more, and so on. Thus, the whole mass undergoes fission in a fraction of a second, releasing all of the energy in one huge explosion (Fig. 22-6b). A nuclear power plant reactor is designed to sustain a continuous chain reaction but not allow its escalation into an explosion (Fig. 22-6c).

## The Nuclear Reactor

Uranium is enriched to about 3 percent $^{235}U$ to 97 percent $^{238}U$ for nuclear power plant reactors

in the United States. It is then made into pellets that are placed in long steel tubes. These are the **fuel elements.** A single fuel element does not contain enough $^{235}U$ to support a chain reaction. But when many fuel elements are placed close together, neutrons from spontaneous fissions in one may strike $^{235}U$ atoms in another. A chain reaction is thus initiated and sustained.

Rods of neutron-absorbing material, generally carbon, are placed between the fuel elements to keep them isolated from one another. These neutron-absorbing rods are called the **control rods.** When the control rods are removed, a chain reaction is initiated and the fuel elements become intensely hot. The reaction is regulated by moving the control rods in or out as desired. The **nuclear reactor** is simply this system of fuel elements and movable control rods (Fig. 22-7). Water, or some other fluid is circulated through the reactor both to remove the heat for power generation and to prevent the reactor from overheating.

If a reactor gets out of control, it will not explode like a nuclear bomb because the fuel is not sufficiently enriched. However, the reactor has the potential to overheat to the point of melting, an event referred to as a **meltdown,** which will be discussed shortly.

## The Nuclear Power Plant

The general design of United States' nuclear power plants is shown in Figure 22-8. The reactor is mounted in a thick-walled vessel that is filled with water. The water in the reactor vessel is heated by the reactor. The water also keeps the reactor from overheating and melting. The hot water from the reactor vessel is circulated through

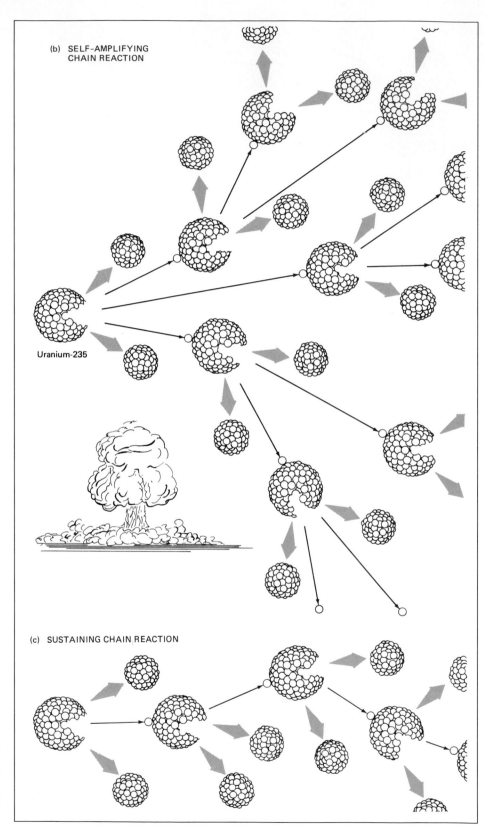

(b) SELF-AMPLIFYING
CHAIN REACTION

Uranium-235

(c) SUSTAINING CHAIN REACTION

**FIGURE 22-6** (cont.)
(b) A self-amplifying chain reaction leading to a nuclear explosion. Since two or three high-energy neutrons are produced by each fission, each fission may cause the fission of two or three additional atoms. Hence the entirety of a suitably concentrated mass of fissionable material may be caused to fission in a tiny fraction of a second, resulting in a nuclear explosion. (c) In a sustaining chain reaction the extra neutrons are absorbed in other materials so that amplification does not occur.

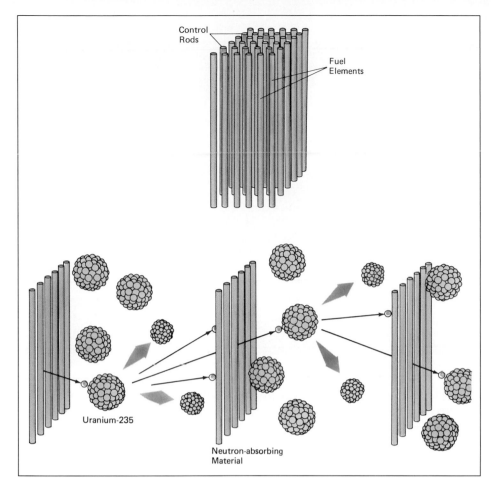

**FIGURE 22-7**
The nuclear reactor. In a nuclear reactor, a large mass of uranium is created by placing uranium in adjacent tubes, the fuel elements. The uranium is *not* sufficiently concentrated to permit a nuclear explosion, but it will sustain a chain reaction that will produce a tremendous amount of heat. The rate of the chain reaction is modulated by inserting or removing rods of neutron-absorbing material (control rods) between the fuel elements. Heat is removed from the fuel elements, and the reactor is kept from overheating, by circulating water through the reactor.

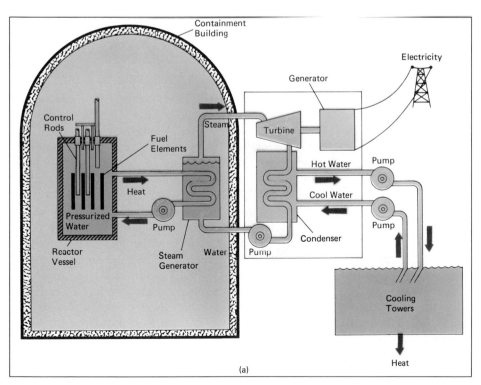

**FIGURE 22-8**
(a) Schematic diagram of a nuclear power plant.

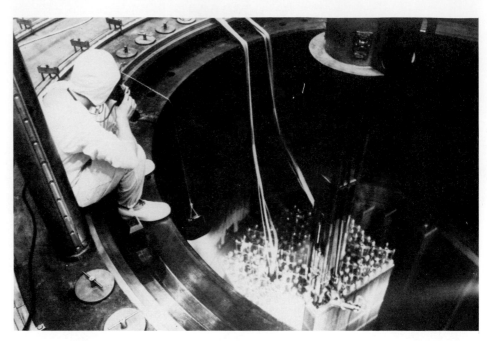

FIGURE 22-8 (cont.)
(b) Technician sitting on edge of reactor vessel is monitoring the fuel-loading operation.
(U.S. Department of Energy photo; courtesy of Combustion Engineering, Inc.)

a heat exchanger where it boils water and produces the steam that is used to drive the turbogenerator. This "double loop" in producing steam aids in isolating hazardous materials in the reactor from the rest of the power plant. There are backup cooling systems to keep the reactor immersed in water should leaks occur, and the entire assembly is housed in a thick concrete **containment building.**

The fissioning of about a pound (0.5 kg) of uranium fuel releases energy equivalent to burning 1000 tons of coal. Thus, one fueling of the reactor (with about 3 tons) is sufficient to run the power plant for one to two years. The spent fuel elements are then removed and replaced with new ones.

## ADVANTAGES AND DISADVANTAGES OF NUCLEAR POWER

For most people, the main concern regarding nuclear power is its safety. Is it, or can it be made, acceptably safe? Assessing the safety of nuclear power necessitates our understanding a little about radioactive substances and their hazards. Therefore, we shall discuss radioactive substances before addressing the safety issue itself. Throughout our discussion of nuclear power, it is important to keep in mind that, if we desire the electrical power, we must produce it. If we do not use

nuclear power plants, we must use an alternative; the major alternative for large-scale power production is currently coal-fired power plants. Therefore, we must contrast the risks of nuclear power with those of using coal.

### The Problem of Radioactive Substances

#### RADIOACTIVE MATERIALS AND THE RADIATION HAZARD

We have noted that when uranium or any other element undergoes fission, the split halves are atoms of lighter elements such as iodine, cesium, strontium, cobalt, or any of some 30 different elements. These newly formed atoms, however, are generally unstable isotopes of their respective elements, which means the number of subatomic particles in the nucleus is unbalanced.

Unstable isotopes gain stability by spontaneously ejecting the extra particles and/or high-energy radiation. The particles and radiations that are thrown off are referred to as **radioactive emissions.** Materials in which some of the atoms are giving off such emissions are known as **radioactive substances.** In addition to the direct fission products, other substances in and around the reactor may become radioactive by absorbing neutrons from the fission process (Fig. 22-9). These direct and indirect fission products are the **radio-**

544    Part VI: Resources: Minerals, Biota, Energy, and Land

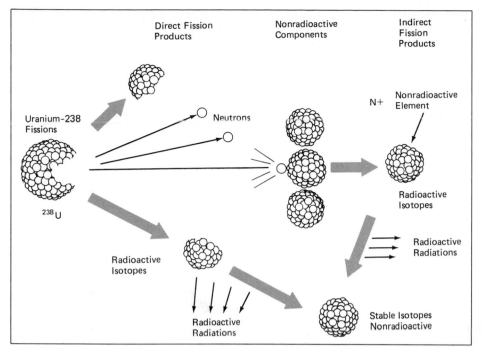

Direct Fission
Products

Nonradioactive
Components

Indirect
Fission
Products

Uranium-238
Fissions

Neutrons

N+  Nonradioactive
Element

$^{238}U$

Radioactive
Isotopes

Radioactive
Isotopes

Radioactive
Radiations

Radioactive
Radiations

Stable Isotopes
Nonradioactive

**FIGURE 22-9**
Radioactive wastes and
radioactive emissions. Nuclear
fission results in the production
of numerous unstable isotopes
as depicted. These unstable
isotopes are the radioactive
wastes. They give off potentially
damaging radiations until they
regain a stable structure.

active wastes of nuclear power. Similarly, radioactive fallout from nuclear explosions consists of these direct and indirect fission products.

Radioactive emissions, particularly radiations, may penetrate through biological tissue in a manner analogous to tiny bullets. They are so small that they do not create a visible mark, nor are they felt, but they are capable of breaking molecules within cells.

In low doses, radiation does not kill cells, but it can damage DNA molecules, the genetic material inside the cell. Cells with damaged (mutated) DNA may then begin dividing and growing out of control, forming malignant tumors or cancer. If an egg or sperm cell is involved, the radiation may cause birth defects in offspring.

In higher doses, radiation may cause enough damage to prevent cell division. Hence radiation can be focused on a cancerous tumor to destroy it. However, if the whole body is exposed to such levels of radiation, a generalized blockage of cell divisions occurs which prevents the normal repair and replacement of blood, skin, and other tissue. This results in what is called radiation sickness which may result in death a few days to months after exposure. Finally, very high levels of radiation may totally destroy cells, causing immediate death. However, outside of nuclear war, there is

no way that large numbers of people would ever be exposed to these high levels of radiation. The major concern is the potential of large numbers of people being exposed to low levels of radiation which might cause a general increase in frequency of cancer and birth defects.

### RELATIVE RISKS

The potential danger of exposure to radiation from nuclear power plants exists throughout the entire course of the nuclear fuel cycle. The hazards of radiation are now well recognized and numerous precautions are taken to protect workers and the public from exposure. For example, highly radioactive substances are handled by remote control; they are placed in casks of lead or other material that shield the radiation when they are transported or stored and they are constantly monitored for leaks and spills. During normal operations at a nuclear power plant, the radioactive fission products remain within the fuel elements and indirect or secondary fission products are maintained within the containment building that houses the reactor. No routine discharge of radioactive materials into the environment occurs.

Due to these precautions, nuclear proponents argue that neither workers nor the public is (nor will be) exposed to "significantly increased"

Chapter 22: Nuclear Power, Coal, and Synthetic Fuels   545

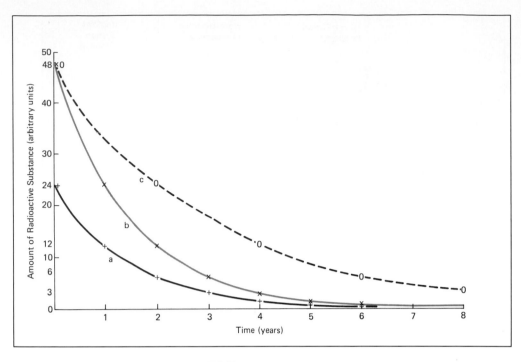

**FIGURE 22-10**
Radioactivity for any isotope declines as shown. Regardless of starting amount, one half decays during each successive half-life. (a) A substance with a half-life of one year starting with 24 units; (b) the same substance starting with 48 units; (c) a substance with a half-life of two years. The half-life for different isotopes may vary from less than one second to many thousands of years.

levels of radiation. They point out that we are exposed to radioactive materials other than those connected with nuclear power. This exposure comes from low **background-level radiation** from natural radioactive materials in the earth's crust and from cosmic particles from outer space. In addition, we subject ourselves to radiation from X-rays. The radiation that the public is exposed to from normally operating nuclear power plants amounts to a tiny fraction (less than 1 percent) of normal background levels. In fact, a radiation detector will pick up more radiation from the earth and concrete on a basement floor than it will when held within 500 feet of a nuclear power plant.

Thus, the argument becomes one of relative risks. The question is not whether low-level radiation will or will not cause cancer, but whether the health risk posed by nuclear power is greater, equal to, or less than the risks posed by other methods of producing power.

Nuclear proponents note that there is no evidence of an increase in the cancer rate as a result of nuclear power. They also note that the nuclear industry has the best health and safety record of any major industry. On the other hand, the health risks associated with the pollution gener-

ated by coal mining and burning are considerable. Still, the future, long-term risks must be examined as well as the current risks.

## DISPOSAL OF RADIOACTIVE WASTES

Critics of nuclear energy point out that as nuclear power progresses, radioactive wastes are produced in ever larger quantities and their accumulation presents increasing risks. What to do with the highly radioactive fission products from spent fuel elements is an extremely controversial subject. It is important to note, however, that the radioactive waste problem is not exclusively a nuclear power plant problem. Ten to 20 times more nuclear waste results from the weapons program than from nuclear energy programs.

To understand the hazards, we must clarify the concept of *radioactive decay*. As unstable isotopes eject radiation, they become stable and cease to be radioactive. This process is known as

**radioactive decay.** As long as radioactive materials are isolated from humans and other organisms, the decay process proceeds harmlessly, and radiation levels decrease accordingly.

How long does radioactive decay take? This is not a simple question to answer. Radioactive decay is measured in terms of **half-life.** This means that half of a particular radioactive isotope decays in a specified period of time. In the next similar period of time, half of the remaining half (or one-fourth of the original amount) decays, and so on. Figure 22-10 shows that radioactive decay seems to proceed rapidly at first, but then slows down and some radioactivity may linger indefinitely. However, it is generally considered that after ten half-lives, radioactivity will have declined to insignificant levels.

The fissioning of uranium results in a very heterogeneous mixture of isotopes, the most common of which are listed in Table 22-1. Note that the half-lives of these radioactive isotopes vary from a few days to many years; note especially plutonium with its half-life of 24,000 years. Thus, much of the radioactivity from fission wastes will dissipate in a period of a few weeks to months as the short-lived isotopes decay. However, to be safe, long-lived isotopes require isolation for up to 240,000 years (ten times the half-life of plutonium).

Is there a repository that will hold wastes this long? Most thought has focused on solidifying wastes, placing them in sealed containers and burying them deep in stable rock formations (Fig. 22-11). But 240,000 years is such a long time that geological changes cannot be predicted. For instance, changes in the earth's crust may occur. It is doubtful that any container or location can hold these wastes and prevent them from leaching into aquifers during this time period. If these radioactive isotopes do escape into the environment, they may bioaccumulate in specific organs. For example, radioactive strontium acts like calcium and accumulates in bones, and radioactive iodine, like nonradioactive iodine, accumulates in the thyroid gland.

Over the last 30 years, various techniques and locations for permanent storage have been and are still being explored. But the issue remains unsolved and some state legislatures have now enacted bills forbidding the disposal of radioactive wastes in their states. In the meantime, nuclear power plants continue to operate and generate more waste. Currently, spent fuel elements containing the radioactive wastes are stored in tanks of water similar to deep pools on the plant site. The water shields the radiation, dissipates the waste heat still produced, and can be monitored for leaks. After a few years of temporary storage, the short-lived isotopes have decayed. The decreased radiation reduces handling and transporting risks. Also, after short-lived isotopes have decayed, it may be practical to reprocess the spent fuel elements and obtain remaining unfissioned uranium, plutonium, and other useful materials. With reprocessing, less radioactive material would require disposal.

Still, the problem of ultimate disposal cannot remain unresolved indefinitely. Temporary storage areas are reaching capacity. Some nuclear

**Table 22-1**

| A Few of the Most Common Radioactive Isotopes Resulting from Uranium Fission and Their Half-lives | |
|---|---|
| *SHORT-LIVED FISSION PRODUCTS* | *HALF-LIFE (DAYS)* |
| Strontium-89 | 54 |
| Yttrium-91 | 59.5 |
| Zirconium-95 | 65 |
| Niobium-95 | 35 |
| Molybdenum-99 | 2.8 |
| Rubidium-103 | 39.8 |
| Iodine-131 | 8.1 |
| Xenon-133 | 15.3 |
| Barium-140 | 12.8 |
| Cerium-141 | 32.5 |
| Praseodymium-143 | 13.9 |
| Neodymium-147 | 11.1 |
| *LONG-LIVED FISSION PRODUCTS* | *HALF-LIFE (YEARS)* |
| Krypton-85 | 10.27 |
| Strontium-90 | 28. |
| Rubidium-106 | 1.0 |
| Cesium-137 | 30 |
| Cerium-144 | .8 |
| Promethium-147 | 2.6 |
| *ADDITIONAL PRODUCTS OF NEUTRON BOMBARDMENT* | |
| Plutonium-239 | 24,000 |

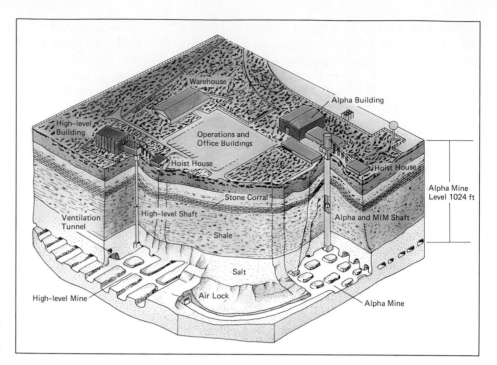

**FIGURE 22-11**
Disposal of radioactive wastes from nuclear power plants. These wastes must be isolated from the environment for thousands of years. Elaborate plans have been made for their disposal, but will this assure that they will stay where they are put? (U.S. Department of Energy.)

power plants will shut down in the near future unless a place to dispose of spent elements is found.

## The Potential for Accidents

While disposal of long-lived radioactive materials will pose a significant problem under even the best of circumstances, most people consider the possibility of an accident at a nuclear power plant a much greater threat. Again, it is not possible for a nuclear power plant to explode like a nuclear bomb because the uranium-235 fuel is not sufficiently enriched. The worst-case event is a meltdown, a situation in which the reactor core overheats and melts. The intense heat of the molten core in the presence of water may generate hydrogen and/or steam, and a resulting hydrogen and/or steam explosion could eject radioactive materials from the core into the atmosphere, and subsequent fallout could contaminate large areas and populations.

The accident which occurred at a nuclear power plant at Chernobyl in eastern Russia in April of 1986 provided the world with a vivid example of such a disaster. The events precipitating the disaster are not clear and may never be known, but on April 25 a meltdown occurred and a resulting hydrogen explosion blew the top off

the reactor and ejected radioactive materials into the atmosphere. While only two people were reportedly killed by the actual explosion, at least 100 others in and near the plant received potentially lethal doses of radiation. Within the next few days, all the people living within an 18 mile radius around the plant were evacuated. This involved not just the evacuation but also the more or less permanent relocation of some 100 thousand people because it may be many years before radiation levels in the evacuation zone decay to acceptable levels. Over a much broader area, efforts were taken to wash down buildings and roadways to flush away radioactive dust. Many tons of food stuffs, caught in fallout areas, were banned from the market. But even with these precautionary measures, many of those in or near the evacuation zone may have received levels of radiation which may lead to cancers and birth defects in future years. Indeed, increased levels of radiation resulting from the disaster were detected around the globe including the United States, but these increases were deemed to have no significance for human health because they were both temporary and only modestly above normal background levels.

Are we in danger of such disasters occurring at U.S. nuclear power plants? Nuclear opponents argue that the answer is yes. However, proponents point out that the *possibility* and the *probability* of such events are very different matters. It is *possible* that two fully loaded 747 airplanes could collide in midair and drop into a crowded football stadium, thus killing 50,000 people. Yet the *probability* of such an event is so remote that we do not consider it seriously.

Therefore, the question is not whether the preceding scenario is possible, but whether it is probable.

Nuclear proponents point out that U.S. reactors are of a fundamentally different design than the Russian reactor at Chernobyl. There are more backup systems to prevent core overheating and, most notably, U.S. reactors are housed in a thick concrete-walled containment building designed to withstand explosions such as the one that occurred at Chernobyl. The Russian reactor had no such containment building.

In order for a disastrous result to occur, it is not only necessary for the initial accident to occur, it is also necessary for all the safety and backup systems to simultaneously fail. The probability of all such systems failing, proponents argue, is remote. Indeed, there have been potentially disastrous accidents at some U.S. nuclear power plants, but systems proved adequate to avert catastrophe.

An example is an accident which occurred at the Brown's Ferry nuclear power station in 1975. A maintenance person was checking for air leaks, using a candle—a flicker of the flame indicates a draft. One air leak created such a draft that the flame was pulled horizontally into the insulating material of electrical wiring, thus starting a fire. Before the fire was brought under control, most of the control systems of the entire plant had malfunctioned and/or been put out of commission. However, the plant operators did perform admirably and, in spite of the confusion of malfunctioning systems, they found an operational mode in which it was possible to insert the control rods and bring the plant to a safe shutdown. No one was injured.

The 1979 episode at Three Mile Island was also largely due to human errors, namely, sloppy maintenance procedures. Here, the operators did not respond properly to the problem. Hence, the accident came precariously close to causing a meltdown. It held the whole nation, to say noth-

ing of the nearly 300,000 residents of Harrisburg, Pennsylvania, who were poised for evacuation, in suspense for several days. In the end, the situation was brought under control and, again, no injuries occurred. The reactor, however, was so badly damaged and so much radioactive contamination occurred inside the containment building, that six years later, the cleanup process was still going on and proving to be as costly as building a new power plant.

Interestingly, these accidents are cited not only by critics, but also by proponents of nuclear power. Proponents point out that in spite of these accidents and mistakes, the plants were still safely shut down. Further, we have learned valuable lessons from these accidents. The Nuclear Regulatory Commission has upgraded safety standards not only in the technical design of nuclear power plants, but also in the maintenance procedures and in the training of the operators. Thus, the proponents contend, nuclear plants were designed to be safe in the beginning and now they are safer than ever. For example, it is estimated that as a result of new procedures instituted after the accident at Three Mile Island, nuclear plants are now six times safer than before.

Another argument against nuclear power plants, and one related to accidents, is that the plants are prime targets for sabotage. Could a saboteur make the remote chance of a nuclear power plant disaster a certainty? Proponents of nuclear power say no. First, nuclear power plants do have security systems and, in the face of criticism, the systems are being strengthened. Therefore, the probability of a successful sabotage attempt is relatively low in itself. Second, an explosive charge, if successfully planted, would only make breakdown in one area a certainty. There is still protection from the various emergency backup systems. Thus, proponents claim that the probability of a nuclear power plant disaster from sabotage is not much greater than from an accident.

Still, as long as the risks of a catastrophe exist, regardless of how remote they are, a lack of public confidence in nuclear power is understandable. However, all risks of a meltdown could be obviated by designing reactors that are "inherently safe." One such concept is to place the reactor vessel in a pressurized pool of water so that, in the event of any break or rupture, all parts of the reactor would be flooded, thus avoiding the potential for a meltdown. Human or mechanical intervention would not be needed for this task. A. H. Winberg, an authority on the subject, states that inherently safe reactors might not be much more expensive than current designs because there would be no need for many of the extensive safety systems now employed.

In summary, the likelihood of an accident is dependent on two factors: equipment failures and human failure. The human factor is by far the more important since suitable institutional controls such as safe designs and suitable inspection and operating procedures can mitigate or even obviate any accident. The question then becomes one of cost. Are we willing to pay the social and economic price demanded to operate nuclear power safely or is there a more economical way to meet our energy needs? The question of economics will be explored further in a later section.

## Environmental Advantages and Disadvantages

### LESS PRODUCTION OF "TRADITIONAL" POLLUTANTS

If the safety and radioactive waste disposal issues can be satisfactorily resolved, nuclear power has some decided environmental advantages over coal. First, most coal is strip-mined (Fig. 22-12). Strip-mining causes destruction of the ecosystem and subsequent erosion, leaching of acid-forming compounds, and water pollution.

**FIGURE 22-12**
(a) Diagram depicts the environmental impacts of nuclear power in contrast to those of coal.

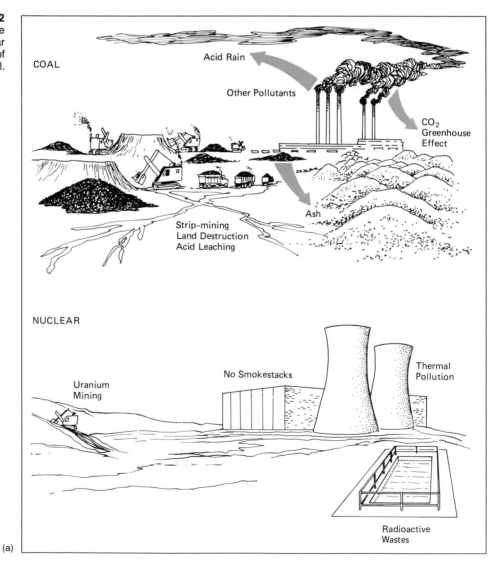

COAL

Acid Rain

Other Pollutants

$CO_2$ Greenhouse Effect

Ash

Strip-mining Land Destruction Acid Leaching

NUCLEAR

No Smokestacks

Thermal Pollution

Uranium Mining

Radioactive Wastes

(a)

**FIGURE 22-12** (*cont.*)
(b) A nuclear power plant. Note the absence of smoke stacks and visible air pollution. (U.S. Department of Energy photo.)

Second, recall that coal-fired power plants are the major source of many air pollutants including sulfur oxides, which result in acid rain, and carbon dioxide, which contributes to the greenhouse effect or warming of the earth's atmosphere. Third, burning coal leaves large amounts of ash which demand large disposal areas and may cause water pollution if they leach.

Compared with these well-known problems of coal burning, nuclear energy is remarkably clean. Since the amount of uranium needed to fuel a power plant is relatively small, the impact of strip-mining uranium is minimal and will continue to be as long as high-grade uranium ore deposits last. In addition, nuclear power plants do not produce any of the "traditional" air pollutants, nor do they produce carbon dioxide. Further, they do not generate large amounts of ash.

## MORE THERMAL POLLUTION

Nuclear power does, however, produce more thermal pollution (waste heat) than coal. Waste heat is characteristic of all steam-generating plants because steam leaving the turbine is still hot and it must be cooled and condensed before completing the cycle (see Fig. 21-5).

Nuclear reactors must be operated at a lower temperature than a coal-fired boiler to avoid damaging the reactor. The lower temperature leads to a lower efficiency, about 30 percent as opposed to 40 percent for coal. As shown in Figure 22-13, this drop in efficiency results in a 55 percent increase in the waste heat. For a large nuclear power plant (1000 megawatt capacity), the waste heat is equivalent to burning 700 tons of coal per day.

In many cases the waste heat is discharged into the atmosphere through huge **cooling towers**

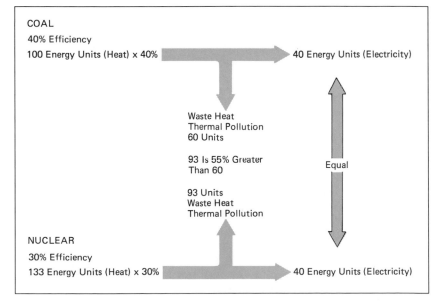

**FIGURE 22-13**
Waste heat or thermal pollution. Nuclear power plants have more potential for thermal pollution than comparable coal-fired power plants because they must be operated at somewhat lower temperatures to avoid damage to the reactor. The lower temperature results in lower efficiency and hence a larger amount of waste heat.

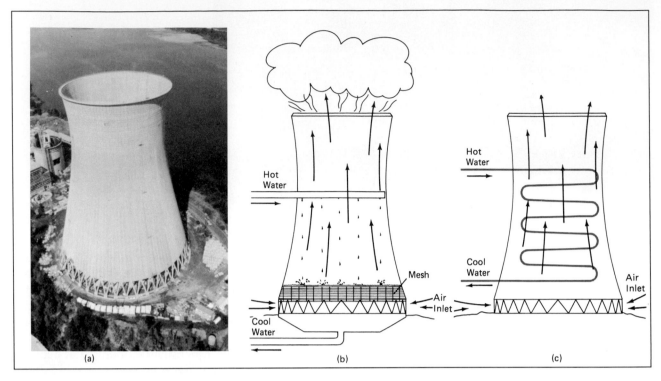

**FIGURE 22-14**
Cooling towers are structures for dissipating waste heat from power plants into the atmosphere.
(a) Huge size is indicated by cars and garages at the lower left. The way the towers work is shown
in the diagrams. (b) Wet type and (c) dry type. (EPA-Documerica photo by Gene Daniels; courtesy of U.S.
Environmental Protection Agency.)

(Fig. 22-14). In other cases, however, water from a natural waterway, a river for example, is passed over the condenser. The waste heat added to the natural waterway is referred to as **thermal pollution**. What are the environmental impacts? Of course this type of disposal is not used unless the total heat load on the natural waterway is determined to be low enough (no more than a few degrees) that aquatic life will presumably not be harmed. (Otherwise cooling towers, which are more expensive, are used instead.) But adverse effects may nevertheless occur.

**Fishkills at intake pipes.** Although intake pipes to condensers are screened to keep fish out, planktonic organisms such as fish and shellfish embryos may be drawn in and killed by the high temperatures inside the condenser. This killoff of planktonic organisms affects the young and significantly reduces the food supply for large organisms. Also, large fish may be killed if they are drawn against the intake screen.

**Eutrophication.** Increasing temperature may promote or intensify the latter phases of eutrophication in which oxygen depletion leads to fishkills. This occurs because warmer water holds less dissolved oxygen than cooler water. At the same time, increased temperature raises the metabolic rate and hence the oxygen consumption of both bacteria and fish. The result may be a large fishkill due to oxygen deprivation.

**Changes in species composition.** Increasing temperature may affect the species composition of the producer level and hence the entire food chain. Many valuable species, namely green algae and diatoms, have lower optimum temperatures for growth than do noxious blue-green algae. Thus, thermal pollution can lead to a replacement of desirable algae by the undesirable blue-greens.

**Disruption of predator-prey relationships.**
Increased temperatures may disrupt critical predator-prey relationships. For example, trout have

a lower optimum temperature than the minnows on which they feed. Consequently, increased temperature enables the minnows to escape from the trout more easily. Hence, the minnow population proliferates while the trout population starves. On the other hand, some desirable fish populations such as bass and catfish may benefit.

**Synergistic effects.** Many synergistic effects come into play as a result of increased temperatures. Fish that are resistant to diseases at lower temperatures may become highly susceptible at increased temperatures. Also, increased temperature may render fish more sensitive to other pollutants such as heavy metals and pesticides.

Because of these potential problems, cooling towers are being constructed on most new power plants despite their higher cost. Dissipating waste heat into the atmosphere does not seem to have any adverse environmental impact.

## The Weapons Connection

Whether nuclear power plants should be built is not just a decision for the United States. Lack of energy resources is a very real handicap for many less-developed nations. Compared with the high cost of importing fossil fuels, a nuclear power plant that relies on uranium fuel is attractive. Hence, the United States and other industrial nations (Canada, West Germany, France, and Great Britain) sell nuclear power plant parts to less-developed countries. Some of these countries already have nuclear plants on line and others can be expected to have them within the next 10 to 20 years.

The major concern here is that spreading use of nuclear energy could increase the number of countries possessing nuclear weapons because power plant technology and weapons technology are linked. In particular, the uranium that fuels the plant, or the plutonium obtained from reprocessing spent fuel, could be diverted to the building of nuclear weapons. The 1968 Non-proliferation of Nuclear Weapons Treaty aims to prevent this occurrence by stating that any nation that buys a nuclear power plant must allow outside inspection and monitoring of the fuel and wastes at the plant. However, there is still concern that some illicit activity may go unnoticed.

Is this possibility one that should curtail the construction of nuclear power plants? The American Nuclear Society (ANS) recently examined the question and concluded that, if a country seriously wants to produce nuclear weapons, there are more direct and less expensive ways to go about it than meddling with the operation of a commercial power plant. Consequently, even in the absence of nuclear power, the nuclear weapons potential would remain.

The key point, according to the ANS, is that a country that sells a nuclear power plant should guarantee a supply of uranium fuel and should take back the spent fuel. Otherwise, the recipient nation is forced to develop its own enrichment and/or reprocessing technologies. It is these technologies (not the power plant, per se) that are used to produce weapon-grade materials. However, in the United States political pressure against accepting such wastes continues because of the disposal problems discussed earlier. It is clearly a no-win situation: accept the waste without knowing how to dispose of it or refuse it and motivate its conversion into weapons.

## ECONOMIC PROBLEMS WITH NUCLEAR POWER

At the beginning of this chapter we noted that no new nuclear power plants were ordered in the United States between 1978 and 1985 and numerous orders were cancelled. This implies that electric utility companies are turning away from nuclear power despite the best efforts of promoters. Safety issues are not the only reason for this course of action. Though widespread public criticism, including numerous demonstrations, acted as a deterrent, the more significant factor is economic.

### Mismatch between Nuclear Power and the Energy Problem

During the period between World War II and the early 1970s, consumption of electricity grew at an average rate of about 7 percent per year, a rate that caused demand to double every ten years. Through the mid-1970s, planners assumed that this trend would continue. Because it takes about ten years to construct a nuclear power plant, the utility companies placed orders and embarked on nuclear construction programs to meet demand anticipated for the 1980s. But nuclear power proved ill suited to meet the energy problems that arose.

Simply stated, there is a basic mismatch, at least in the United States, between nuclear power and the energy problem. As we have emphasized, our energy problem involves a shortage of crude oil for liquid fuels. Nuclear power produces electricity. However, the United States has abundant reserves of relatively low-cost coal that may also readily be used to produce electricity. Consequently, in the United States, nuclear power turned out to be a more costly substitute for abundant coal resources and it did not significantly change our dependence on high-priced foreign oil. The environmental considerations, such as acid rain, may still favor nuclear power, but this factor is not generally included in cost analysis.

Furthermore, conservation efforts stemming from the shortages of the 1970s served to curtail demand for electricity as well as other fuels. Demand for electricity has grown by only about 2 percent per year since 1978.

The financial dilemma should be clear. Even if a utility company cannot sell the nuclear power it generates, it must still pay the billions of dollars it cost to construct the plant. To make up the difference, many utilities have sharply increased (some two to threefold) the prices they charge

**FIGURE 22-15**
Over the past decade, the cost of electricity has risen even more sharply than inflation at large. This is, in part, attributable to the huge costs of nuclear power plants which consumers are obliged to assume as the plants are brought on line. Even so, there are additional costs hidden in tax moneys supporting nuclear research and development, regulation, and disposal of nuclear wastes. (Statistical Abstracts of the United States, U.S. Department of Commerce, 1985.)

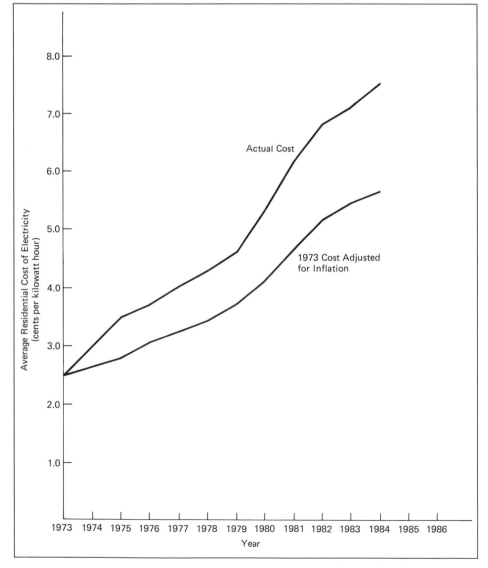

consumers for the electricity they do use (Fig. 22-15).

The situation might be resolved if we were moving towards a totally electric economy. For example, if we had acceptable electric cars, nuclear-generated electricity could be substituted for oil-based fuels. Unfortunately, electric cars have not proved practical; despite a few commercial efforts and much experimentation, the outlook for them in the near future remains dim. Similarly, the use of electricity for space and hot water heating has proved more expensive than various alternatives.

Eventually, growth in electrical demand will probably restore the balance. In the meantime, utilities hardly need to embark on further nuclear power plant construction. Even when a balance is achieved, the need for more nuclear power plants may be carefully reexamined because of the development of other technologies that may be less expensive and pose less risk to human health and the environment (see Chapter 23).

## Short Lifetime of Nuclear Power Plants and Decommissioning

Another factor that promises to increase the cost of nuclear-generated electricity is a shorter than expected lifetime for nuclear power plants. Originally, it was thought that nuclear plants would have a lifetime of about 40 years. It now appears that their lifetime will be considerably less, perhaps only 30 years. This difference substantially increases the cost of the power produced because the cost of the plant must be repaid in the shorter period.

The shorter than expected lifetime is due to a problem known as **embrittlement.** As well as maintaining a chain reaction, some of the neutrons from fission bombard the reactor vessel and other hardware. Gradually, this neutron bombardment causes the metals to become brittle such that they may crack under stress. When the reactor vessel becomes too brittle to be considered safe, the plant must be shut down, or in technological jargon, **decommissioned.**

Decommissioning itself may be an extremely costly process. By this time, the power plant components will have accumulated so much radioactivity from neutron bombardment that the only safe course of action will be to seal off the entire containment building for an indefinite period of time and construct a new plant. Of the plants operating in the United States, it is estimated that at least ten will be decommissioned in the 1990s and most of the rest in the ten years thereafter.

On the positive side of nuclear economics, however, proponents point out that nuclear power plants in the United States are more expensive than necessary because nearly every plant is the result of a new, custom design. Further, most plants have been retrofitted with additional safety features after potential problems were discovered. Some plants, estimated to cost about $1 billion initially ended up costing between $5 and $10 billion.

Proponents contend, however, that we are now in a position to design and construct standardized nuclear plants that are safer and less complex. Also, new building materials and techniques could extend the lifetime of new plants. Together, these elements would make nuclear power less costly. Both France and Switzerland now utilize a standardized approach and they are producing about 30 percent of their electricity via nuclear power with fewer problems than United States plants.

## MORE ADVANCED REACTORS

### Breeder Reactors

Uranium, especially uranium-235, is not a highly abundant mineral on earth. At the height of nuclear optimism in the 1960s, when as many as 1000 plants were envisioned by the turn of the century, it was foreseen that shortages of $^{235}U$ would develop. Conversion to the **breeder reactor** was seen as the solution.

Breeder reactors are similar in principle to the uranium fission reactors already discussed. Their distinctive feature is that they are designed to maximize the bombardment of nonfissionable uranium-238 with neutrons, which causes its conversion to plutonium-239. Plutonium-239 is fissionable and can be used as a nuclear fuel. Since the fission of a single uranium-235 atom may produce three neutrons, it may create two plutonium atoms and maintain the chain reaction (Fig. 22-16). Thus, a breeder reactor creates more fuel (plutonium-239) than it consumes (uranium-235). With suitable breeder reactors, one can envision

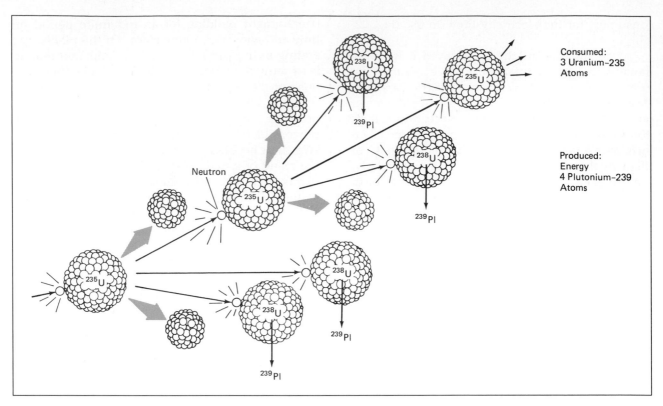

Consumed:
3 Uranium-235
Atoms

Produced:
Energy
4 Plutonium-239
Atoms

Neutron

**FIGURE 22-16**
Breeder reactor produces fuel by converting nonfissionable uranium-238 to
plutonium-239, which in turn fissions to release energy. Note that the amount of
plutonium-239 produced may be greater than the uranium-235 used.

$^{238}$U becoming as useful as $^{235}$U. Since $^{235}$U is 99.3 percent of the total, this would increase the world's nuclear fuel resource at least 100-fold.

Breeder reactors, however, heighten the risks of nuclear war. They present all of the problems and hazards of standard fission reactors plus additional problems. If a meltdown occurred, the dangers would be much more serious because of the presence of large amounts of plutonium-239, which has the exceedingly long half-life of 24,000 years. In addition, because plutonium can be purified and fabricated into nuclear weapons more easily than uranium-235, the potential for the diversion of breeder fuel into weapons production is greater. Hence, the safety and security precautions needed for breeder reactors are greater. Can they be great enough?

With its scaled-down nuclear program, the United States currently has enough uranium stockpiled to fuel all reactors that are operating or under construction through their lifetimes. Thus, there is no urgency for the United States to develop breeders, and construction of the prototype breeder at Clinch River, Tennessee, has been terminated for the time being. France, however, is proceeding with a breeder reactor development program.

## Fusion Reactors

Fusion is another form of nuclear power. Rather than splitting a large atom to release energy, fusion "melts" together smaller atoms into a larger atom. As in fission, fusion results in a loss of mass, which is released as energy.

The vast energy emitted by the sun and other stars comes from this fusion process. The sun, as well as other stars, is composed mostly of hydrogen. Solar energy is the result of fusion of this hydrogen into helium (Fig. 22-17). Scientists have duplicated this process in the hydrogen bomb; but hydrogen bombs hardly constitute a useful release of energy. The aim of fusion technology is to carry out this fusion in a controlled manner so as to provide a practical heat source for boiling water to power a steam turbogenerator.

Since hydrogen is an abundant element on earth (two atoms in every molecule of water), and helium is an inert, nonpolluting, and nonradioactive gas, hydrogen fusion is dreamed of and promoted as the ultimate solution to all our energy problems—that is, pollution-free energy from a virtually inexhaustible resource, water. However, the dream is still a long way from reality. Indeed a fusion plant has not yet been proven possible much less a practical alternative.

At the present state of the art, fusion power is still an energy consumer rather than a producer. The problem is that it takes extremely high temperatures (some 100 million degrees Celsius) and pressure such as exist in the sun to get hydrogen atoms to undergo fusion. In the hydrogen bomb, the temperature and pressure are achieved by using a fission bomb as an igniter—hardly a practical alternative to sustained, controlled fusion.

A major problem is the container to hold the hydrogen while it is being heated to the tremendously high temperatures required for fusion. No material known can withstand these temperatures without vaporizing; however, two concepts are undergoing experimental testing. One is the Tokamak design, in which ionized hydrogen is contained within a magnetic field while it is heated to the necessary temperature (Fig. 22-18). The second concept is laser fusion, in which a tiny pellet of frozen hydrogen is dropped into a "bull's-eye" where it is hit simultaneously from all sides by

FIGURE 22-17
Hydrogen fusion is the source of solar energy. In fact, the sun has been referred to as a fusion reactor properly located in space.

powerful laser beams (Fig. 22-19). The laser beams simultaneously heat and pressurize the pellet to the point of fusion.

Some fusion has been achieved in these devices, but as yet, the breakeven point has not been reached. More energy is required to run the magnets or lasers than is obtained by the fusion. The most optimistic workers in the field feel that with sufficient money for research—some $10 billion—the breakeven point might be reached in

(a)

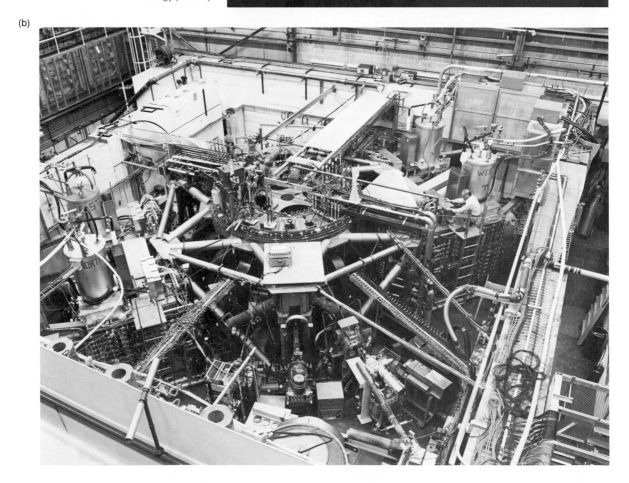

**FIGURE 22-18**
TOKAMAK. A magnetic containment vessel for hydrogen fusion. (a) Scale model. (b) The real thing. (U.S. Department of Energy photos.)

(b)

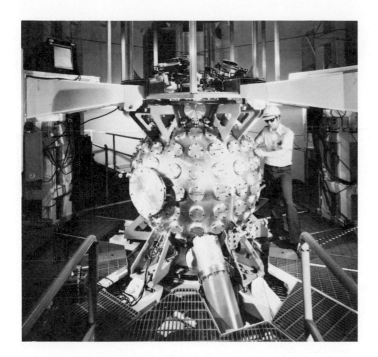

**FIGURE 22-19**
Laser fusion. In this experimental instrument at Lawrence Livermore Laboratory, 30 trillion watts of optical power are focused onto a tiny pellet of hydrogen smaller than a grain of sand located in the center of this vacuum chamber. For less than a billionth of a second the fusion fuel is heated and compressed to temperatures and densities like those found on the sun.
(U.S. Department of Energy photo.)

the 1990s. Even if this goal is achieved, however, it is still a long way from a practical commercial fusion reactor power plant. Developing, building, and testing such a plant would require at least another 20 to 30 years and many more billions of dollars. Additional plants would require additional years. Thus, fusion is, at best, a very long-term option. It cannot solve the energy problems that will probably grip the 1990s when world oil production begins to decline.

Still, many believe that fusion power will continue to be the elusive pot of gold at the end of the rainbow. As disturbing as the fact that the breakeven point may never be reached is the fact that fusion energy, if it does become a reality, promises to be neither clean nor resource unlimited. Current designs use not regular hydrogen but isotopes of hydrogen, namely, **deuterium** ($_2$H) and **tritium** ($_3$H). Fusion of regular hydrogen ($_1$H) would demand even greater temperatures and pressures. Deuterium is a naturally occurring isotope that can be isolated in almost any desired amounts from the normal hydrogen in seawater. Tritium, however, is an unstable radioactive isotope that must be produced. Current plans call for the production of tritium by bombarding the element lithium with neutrons. The neutrons will be produced by the fusion of deuterium and tritium. The overall reaction is

$$
\begin{array}{ccccccccc}
^{2}\text{H} & + & ^{3}\text{H} & \longrightarrow & ^{4}\text{He} & + & n & + & \text{Energy} \\
\text{Deuterium} & & \text{Tritium} & & \text{Helium} & & \text{Neutron} & & \\
\text{Isolated} & & & \uparrow & & & & \downarrow & \\
\text{from Water} & & ^{3}\text{H} & + & ^{4}\text{He} & \longleftarrow & n & + & ^{6}\text{Li} \\
& & \text{Tritium} & & \text{Helium} & & \text{Neutron} & & \text{Lithium}
\end{array}
$$

Lithium is not an abundant element and it could easily become the limiting factor in the wide-scale use of fusion reactors. Also, tritium is radioactive, hence very hazardous and difficult to contain. As a result, fusion reactors could easily become a source of radioactive tritium leaking into the environment. Finally, the hardware of the reactor itself will be embrittled and made highly radioactive by the constant bombardment from neutrons. Thus, there will be the cost of constantly replacing reactor components and the problem of disposing of components that have been made radioactive. Finally, fusion reactors promise to be the source of unprecedented thermal pollution. A steam turbogenerator itself is only 30 to 40 percent efficient and, if half the power produced must be fed back into the reactor to sustain the fusion process, the overall reactor is only 15 to 20 percent efficient. That is to say, 80 to 85 units of heat energy will be dissipated into the environment for every 15 to 20 units of electrical energy produced.

In conclusion, fusion research may continue

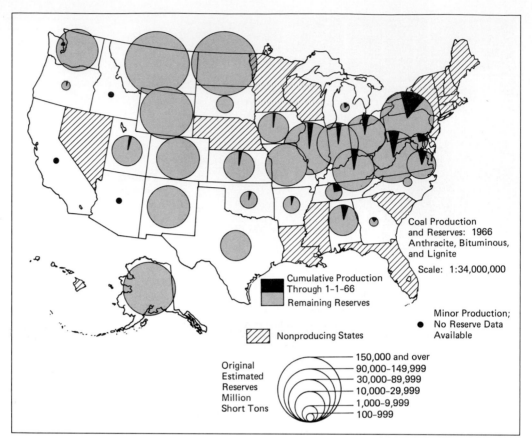

Coal Production
and Reserves: 1966
Anthracite, Bituminous,
and Lignite

Scale: 1:34,000,000

■ Cumulative Production
Through 1-1-66
▨ Remaining Reserves

▨ Nonproducing States

● Minor Production;
No Reserve Data
Available

Original
Estimated
Reserves
Million
Short Tons
150,000 and over
90,000–149,999
30,000–89,999
10,000–29,999
1,000–9,999
100–999

**FIGURE 22-20**
Major coal deposits in the United States: a solution or a
potential ecological disaster? (U.S. Geological Survey.)

in an effort to increase human understanding of physical processes. However, it seems likely that solar technologies will be producing power less expensively and with fewer risks before fusion reactors become a reality.

## COAL

In contrast to the rather limited reserves of crude oil, the United States is exceptionally well endowed with coal reserves. As noted in Chapter 21, coal followed wood as the major energy source in the late 1800s and early 1900s, becoming the major source of energy for heating and driving steam engines. The later transition to oil and natural gas occurred because they polluted less, could be used in less bulky engines, and were more convenient. Aside from its industrial use, coal is used today primarily as boiler fuel in the generation of electrical power. Still, huge reserves of coal remain unexploited in the United States (Fig. 22-20). Certain reserves could supply this country's needs for energy for over 100 years.

Can this be the answer to the energy problem?

When the energy problem came to the forefront in the 1970s, the United States generated about 30 percent of its electrical power with oil and natural gas. Coal has since been substituted in many plants, and, to that extent, it has alleviated the problem of shortages. However, coal use is fraught with the environmental problems already discussed.

Most coal is in deposits that can be exploited practially only by strip-mining. In underground mining, at least 50 percent of the coal must stay in place to support the mine roof. In strip-mining, gigantic power shovels turn aside the rock and soil above the coal seam and then remove the coal (Fig. 22-21). It is evident that this procedure results in total destruction of the ecosystem. Although such areas may be reclaimed—that is, regraded and replanted—it takes many years before an ecosystem like the original is reestablished. In

**FIGURE 22-21**
(a) Coal being strip-mined near Zanesville, Ohio. Forty-five feet of overlying rock and soil are being removed to get at a 3-foot seam of coal. (Fred McConnaughey/Photo Researchers, Inc.) (b) Giant strip-mine shovel located near Marissa, Illinois. (U.S. Department of Energy photo by Schneider.)

arid areas of the West, limits of water are such that it is questionable whether the ecosystem could ever be reestablished. Consequently, such areas, if strip-mined, may be turned into permanent deserts. Furthermore, erosion and acid leaching from the disturbed earth may have numerous adverse effects on waterways and groundwater that drains from the site.

Also, coal-burning power plants have been identified as the major contributors to acid rain because sulfur in the coal converts to sulfur dioxide and then to acids in the atmosphere (Chapter 14). Acid-forming nitrogen oxides are also produced, along with particulates. In addition, burning coal generates many other wastes including trace elements such as arsenic, lead, and mercury, as well as small amounts of a number of radioactive substances. Finally, coal is responsible for the long-term, but not insignificant, carbon dioxide or greenhouse effect.

Even with tighter regulations and control regarding the mining and burning of coal, which seem extremely slow in coming, we will pay a high environmental and human health price for increasing coal consumption. Even then, will it really solve the energy problem?

Coal is practical for little more than the generation of electrical power. Now that coal is already substituting for oil in this area further increasing its use only serves to substitute for nuclear power and other means of generated power. It will not help the oil situation.

## SYNTHETIC FUELS

A way in which coal might be used to help the oil shortage is through its chemical conversion to liquid compounds, which can then be substituted for oil-based fuels. Such coal-derived fuels are referred to as **synthetic fuels** or **synfuels.** With both government and corporate support, a great deal of research and development went into synfuel production in the late 1970s and early 1980s. While some government support remains, the major oil companies have now abandoned these projects because they proved too expensive, at least for the present. Profitable production of synthetic fuels would involve prices that are about twice the current cost of oil-based fuels. In addition, if synfuel production becomes economically competitive, it promises to be an exceedingly "dirty" industry, which will generate numerous pollutants. Furthermore, it can only aggravate the negative impacts of strip-mining.

The second major potential source of synfuels is oil shale, a sedimentary rock that contains a tarlike organic material called **kerogen.** Vast formations of oil shale exist in several areas of the United States (Fig. 22-22). When oil shale is heated to about 600° C, the kerogen releases hydrocarbon vapors that may be recondensed to form a black, viscous, crude oil, which can in turn be refined into gasoline and other petroleum products. It is estimated that ultimately 600 billion barrels, about 100 years' supply of oil at current rates of use, might be recovered from oil shale; at least 50 billion barrels, 10 years' supply, could be recovered by current technology.

Nevertheless, oil shale is not the panacea for our energy shortages. In order to extract shale oil, the rock must be mined, crushed, and heated to distill off the kerogen, and then one must dispose of the rock waste. A ton of high-grade oil shale will yield little more than half a barrel of oil. The mining, transportation, and disposal of wastes necessitated by an operation producing, for example, a million barrels a day would be a herculean task to say nothing of its environmental impact. Perhaps for our environmental good fortune, oil shale, like oil from coal projects, has proven economically impractical for the present.

But in the long term, the most compelling

**FIGURE 22-22**
(a) Map shows locations of major oil shale deposits in the United States. (U.S. Geological Survey.)

Oil Shale and
Tar Sands: 1965

OIL SHALE

▨ Principal deposits

--- Area concealed or
uncertain

TAR SANDS AND
RELATED DEPOSITS

● Major commercial
deposits, 10,000,000 or
more barrels.

○ Minor deposits, possibly
commercial, 1,000,000
or more barrels.

(a)

(b)

(c)

**FIGURE 22-22** (cont.)
(b) Self sustaining combustion in high grade "paper shale." Most shale will not burn but must be heated to remove oil-like material. (c) Anvil Points oil shale processing facility near Rifle, Colorado. (U.S. Department of Energy.)

reason to avoid all-out exploitation of every source of fossil carbon fuel is the existence of the greenhouse or carbon dioxide effect. As was described in Chapter 14, carbon dioxide in the atmosphere has a warming effect on the earth's climate, and there is no question that burning fossil fuels is significantly increasing the carbon dioxide concentration of the atmosphere. If humans persist in burning more and more fossil carbon fuel, there is little question that they may precipitate climatic changes that could be extremely disruptive to agriculture all over the world.

In conclusion, we find that two of the most touted energy options, nuclear power and coal, are fraught with long term risks to both human and environmental health. Also, although the United States has relatively abundant reserves of coal and uranium, neither is a renewable resource. Opting to meet the energy problem by all-out exploitation of these resources would assuredly lead to another energy resources crisis within 100 years. This may seem like a long time, but over the course of history, it is a very short period.

Recognizing these facts, a great deal of research and development has gone into solar and other "renewable" energy options. Progress in these areas will be the subject of Chapter 23.

# 23

# SOLAR ENERGY, OTHER "RENEWABLE" ENERGY SOURCES, AND CONSERVATION

---

## CONCEPT FRAMEWORK

| Outline | | Study Questions |
|---|---|---|

1. Why is low density a problem with solar energy?

2. Into what forms must solar energy be converted? Why?

3. Why is storage needed for solar power?

4. What primarily hinders a solution to the above problems and how might they be overcome?

5. Why is electricity wasted when used to heat homes and water? What is a *flat-plate col-*

21. Describe the two energy pictures for the future. Discuss steps that have been taken to help us improve our energy situation and consider steps that would put us further down the road.

Energy always flows "downhill." This concept is basic to the production of useful energy. Energy can be converted from one form to another and used to perform useful work as this flow proceeds. But, in the end, low-temperature heat cannot be recaptured and reused. (This fact is encompassed in the second law of thermodynamics described in Chapter 2). Consequently, any and every energy source can ultimately be exhausted. A number of natural energy sources, however, such as sunlight, wind, falling water, tides, and heat from the earth's interior, will continue at a more or less constant rate for hundreds of millions of years and human use will not fundamentally alter them. Since, for all practical purposes, these energy sources are everlasting, they are commonly referred to as **renewable** energy resources (Fig. 23-1).

The advantages of developing renewable energy resources are obvious. The problems lie in how to harness them so that they will perform useful work in an industrial society at an affordable cost. Much research and developement has been devoted to this problem in recent years and progress has been made. Some avenues have proved practical and promising for the future. Indeed, nearly 10 percent of the primary energy used in the United States already comes from renewable resources and further growth seems assured. Some avenues, however, while economically practical, may be detrimental to the environment. Others are still in the research stage. Our first objective in this chapter is to gain an understanding of what the various renewable energy resources are, their environmental impacts, and their potential for the future.

We shall also look at a closely related area: conservation. Prior to the 1970s, during an era of apparently abundant energy resources, relatively little attention was paid to how efficiently energy was used. Many engines, furnaces, lighting systems, and other uses allowed as much as 90 percent of the energy to "flow by" without performing any function. In the last decade, significant strides have been made to increase the efficiency of energy use, but much improvement can still be made. Our second objective is to examine where conservation has been achieved and where opportunities still lie.

# DIRECT SOLAR ENERGY

Sunlight trapped through photosynthesis is the energy source for all major natural ecosystems. It is both nonpolluting and infinitely renewable (astronomers estimate that the sun will "burn out" in several billion years, but for all practical purposes, this is an infinite amount of time). Also, there is a tremendous excess of solar energy. Natural ecosystems utilize only a fraction of 1 percent of the sunlight that falls on the earth, and all human needs could theoretically be supplied by a similarly tiny fraction. Whether or not we constructively use this energy before it is radiated back into space has no negative impact on the environment or on the earth.

Considering the virtues of solar energy, it is hardly surprising that it has many ardent supporters and that development of solar energy has made marked progress. The Center for Renewable Resources estimates that over 1 million homes in the United States obtain at least part of their heating via direct sunlight and a number of other applications are on the verge of wide-scale use. What is surprising is that progress has not been greater and that, in recent years, public interest in solar energy has diminished because consumers believe that it is too costly or impractical or that the technology is insufficiently developed. In addition, further government support for research and development has been all but terminated.

To understand this situation, we shall first examine the basic problems in the direct use of solar energy and then we shall look at how these problems are being or may be overcome in various applications.

## Problems to Overcome

To adopt solar energy, four general problems must be considered: it must be (1) concentrated, (2) converted, (3) stored, and (4) matched to suitable applications.

### NEED FOR CONCENTRATION

Solar energy is a *low-density* energy; at no single point is there high intensity. Its abundance derives from the fact that it falls over the entire earth. But most of our energy-using equipment,

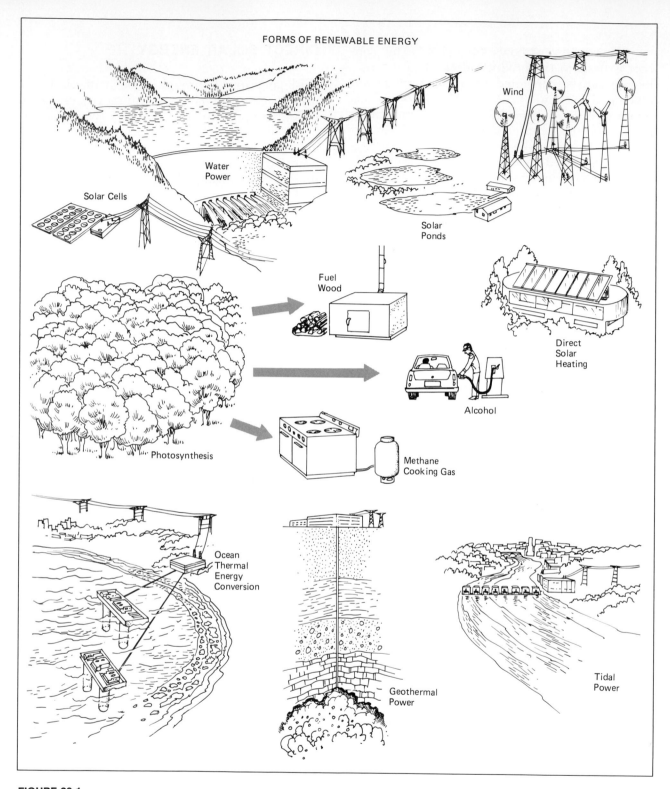

FORMS OF RENEWABLE ENERGY

Wind

Water Power

Solar Cells

Solar Ponds

Fuel Wood

Direct Solar Heating

Photosynthesis

Alcohol

Methane Cooking Gas

Ocean Thermal Energy Conversion

Geothermal Power

Tidal Power

**FIGURE 23-1**
Renewable energy resources. The natural energy sources shown will continue for hundreds of millions of years with little change, and they will not be altered appreciably whether or not humans utilize them. Therefore, these energy resources are referred to as *renewable.*

such as furnaces and engines, requires high-density energy—a lot of energy concentrated in a relatively small space. To make solar energy more useful, there is a need to collect it over a broad area and concentrate it into a small area.

### NEED FOR CONVERSION

Solar energy is largely in the form of light while most of our needs require energy in the form of heat, fuels for engines and furnaces, or electrical power. Thus, solar energy must be converted into these forms. The conversion is also necessary if solar energy is to be collected in one area and used in another because it cannot be otherwise transported. The same is true for storage purposes.

### NEED FOR STORAGE

Obviously, solar energy is not available during the night or in periods of cloudy weather. It is also greatly reduced during winter months. This exacerbates the need for some kind of energy storage.

### GENERAL CONSIDERATIONS

There are numerous ways to solve the above problems. However, the technology and/or hardware involved is frequently expensive, driving the costs beyond the point of practicality. The key to utilizing solar energy now involves finding applications where these costs are minimal or where they may be bypassed. If a suitable match is made between existing solar energy and the desired end use, concentration and storage may be unnecessary and costs of conversion may be minimal. Then, as technological development lowers costs, solar energy may find additional applications.

## Ways of Using Direct Solar Energy

Solar energy is well suited to provide the modest temperatures needed to heat buildings and hot water. Thus, solar energy is already cost-effectively used for this purpose. Technology that will convert sunlight to electrical power at costs suitable for widespread use is not far off. In the more distant future, technology may enable the solar production of hydrogen, which can be substituted for natural gas and liquid fuels. We shall consider each of these areas in more detail.

### SPACE AND WATER HEATING

About 25 percent of the total U.S. energy budget is used to produce *low-temperature heat*—heat that warms homes, buildings, and hot water. The fact that we generally use a high energy density flame with a temperature of over 1000° C to heat a room to 20° C (68° F) or to heat water to 60° C (140° F) is exceedingly wasteful. Using electrical power for this purpose is even more wasteful. It is like using a bomb to kill a fly. The job can be done using less force and with fewer undesirable side effects. Low energy density sunlight is ideally suited for this task.

As seen in Figure 23-2, the peak output of solar radiation occurs in the visible light portion of the spectrum. However, when light is absorbed by any black surface, it is readily converted to heat and raises temperatures at the surface up to 50°C (120°F). Anyone who walks barefoot on an asphalt pavement in the summer can testify to this. Hence, concentration of sunlight is not required for such heating. A **flat-plate collector** will suffice. There are countless variations on the flat-plate collector, but basically it consists of a black surface that absorbs the sunlight and thus converts it to heat. A glass or plastic "window" over the black surface traps the heat by preventing it from radiating out (Fig. 23-3). Air is heated by passing it between the window and the black surface. Water may be heated by passing it through tubes in the surface. Thus, there is minimal cost in this light-to-heat conversion.

Beyond this, heating systems may be *active* or *passive* and may or may not include a means of heat storage. An active system is defined as one that uses pumps or blowers to "actively" circulate the air or water through the system to the desired location. A passive system relies on convection currents (the fact that hot air or water rises) to move the air or water passively. A schematic diagram of an active system with storage is shown in Figure 23-4. Note the number of pumps, blowers, valves, and plumbing features. Active systems may work, but they are costly and tend to be economically impractical. In addition, they suffer from maintenance problems.

Passive systems, on the other hand, can be relatively inexpensive and maintenance free. Innumerable plans are available for passive solar homes, but the basic concept is that shown in Fig-

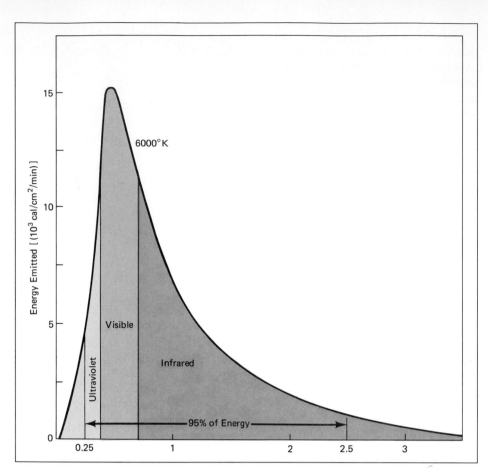

**FIGURE 23-2**
The solar energy spectrum. The greatest output of solar energy is in the visible light part of the spectrum. From: Joe R. Eagleman, Meteorology: *The Atmosphere in Action,* Copyright © 1980 by Litton Educational Publishing, Inc. Reprinted by permission of Wadsworth Publishing Co.

**FIGURE 23-3**
The principle of a flat-plate solar collector. Sunlight is converted to heat as it is absorbed by a black surface. A glass or clear plastic window over the surface allows the sunlight to enter but traps the heat. Air or water is heated by passing it over or through tubes in the surface.

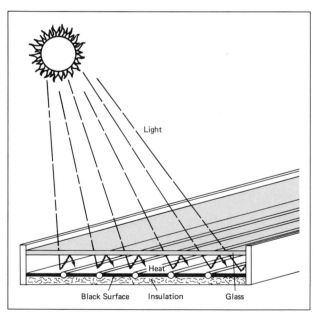

ure 23-5. Relying on large south-facing windows, the building itself acts as the collector. In winter, sunlight beams through the window, heating the interior; at night, insulated drapes or shades are pulled to trap the heat inside. To avoid excessive heat load in the summer months, an awning or an overhang can shield the window from the hot summer sun. Landscaping may also be a significant part of passive solar designs. For example, deciduous trees will shade the home in summer, but let the solar energy through in the winter.

A common criticism of solar heating, active or passive, is that a backup heating system is still required for periods of especially inclement weather. The question that arises is whether the need for backup heating offsets the advantage of solar systems? The answer is no. First, we must keep in mind that the energy problem involves the diminishing production of oil; it does not involve completely running out of oil. Thus, even if

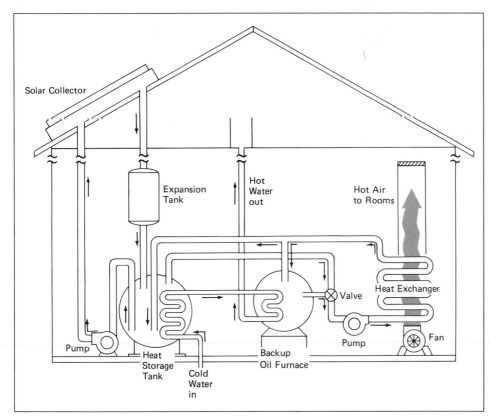

**FIGURE 23-4**
An active solar heating system with heat storage. As shown, a solar system that will provide all heating needs and function automatically like current central heating systems can be designed. In practice, however, such systems have generally proven to be cost-ineffective and frought with maintenance problems. Note the amount of piping and number of pumps and blowers. (From Bruce Anderson, with Michael Riordan, *The Solar Home Book*, pp. 118, 32. Harrisville, N.H.: Brick House Publishing Co., 1976.)

**FIGURE 23-5**
Passive solar heating. In contrast to expensive active solar systems, solar heating may be achieved by suitable architecture and orientation of the home at little or no additional cost. (a) The fundamental feature is large sun-facing windows that permit sunlight to enter during winter months. Insulating drapes are pulled to hold in the heat when the sun is not shining. (b) Suitable overhangs, awnings, or deciduous plantings will prevent excessive heating in the summer. (From Bruce Anderson, with Michael Riordan, *The Solar Home Book*, p. 87. Harrisville, N.H.: Brick House Publishing Co., 1976.)

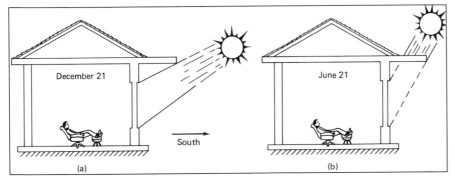

solar systems only lessen the need for heating oil by a few percent, they still help us adapt to the diminishing supply. Then as solar technologies become more advanced and more widely used, we can expect the backup needs to decrease. Also, in a passive solar home, a small wood stove is often an adequate backup system.

Second, from the consumer's point of view, the major cost of heating over a period of years is the cost of the fuel (or electricity), not the cost of the furnace. Consequently, even if the solar system only reduces the need for conventional heat by 50 percent, that 50 percent will show up as a 50 percent savings in each year's fuel bill.

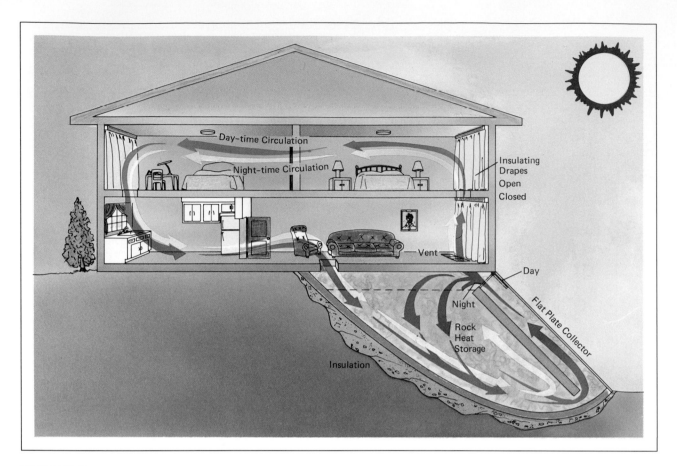

**FIGURE 23-6**
Passive solar heating may include additional heat storage.
Rocks are the preferred heat-storage material because they
readily absorb and give off heat, and they are inexpensive. The
heat storage area may be separate, or it may be incorporated
into the architecture of the home in the form of interior stone or
brick walls.

In regard to heat storage, many of the early solar designs involved circulating air through a "reservoir" of rocks that readily absorb excess heat and then release it as the temperature drops (Fig. 23-6). The need for conventional backup heating is decreased by increasing storage capacity. However, experience has shown that such storage is not cost-effective; it is the most expensive component of the solar installation in comparison to the amount of energy that is stored and later used. Further, the storage problem can be bypassed by improving the insulation of the building. Heat storage can also be incorporated into the design of the building in the form of interior brick or stone walls. In either case, the building interior itself becomes the storage unit. It is ironic to note that, in terms of solar energy, we traditionally construct buildings inside out. For solar efficiency, the insulation should be on the outside and the brick on the inside.

The Center for Renewable Resources holds that a well designed passive solar home can reduce energy bills by 75 percent with an added construction cost of only 5 to 10 percent in almost any climate. In other words, the cost of the solar features will pay for themselves in about seven years. Also, in innumerable situations, passive solar features can be retrofitted into existing homes (Fig. 23-7). Such additions, especially if homemade, may pay for themselves and provide an overall savings in less than seven years. More than a decade after recognizing the energy crisis and the virtues of solar energy, why are we still building and heating most homes in traditional ways? In their publication, "Renewable Energy at the Crossroads," The Center for Renewable Resources concludes, "the chief barrier to more widespread use of passible solar design is ignorance. Many builders and consumers are unaware

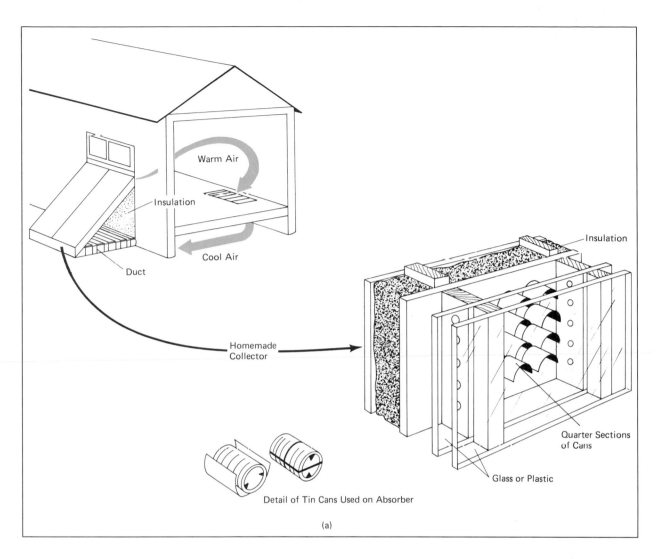

Insulation

Warm Air

Insulation

Cool Air

Duct

Homemade
Collector

Insulation

Quarter Sections
of Cans

Glass or Plastic

Detail of Tin Cans Used on Absorber

(a)

(b)

**FIGURE 23-7**
Many homeowners could gain heating
economy by adding solar collectors to
south-facing, exterior walls. a) Do-it-
yourself pannels can be made from
inexpensive materials as shown. (Bruce
Anderson with Michael Rirodan, The Solar
Hove Book, p. 118, Brick House Publishing
Co. 1976.) b) Home with retrofitted solar
pannels. (Photo by author)

of the potential benefits of passive solar design. . . . [Further] they are often ignored by policymakers and have received meager government support." At least one factor in helping to maintain this "state of ignorance" has been intensive advertising campaigns purporting that solar energy is not practical or cost-effective at present. It should be noted that these advertising campaigns are conducted by utility and oil companies which profit with each unit of energy they sell regardless of its cost.

Much the same holds true for solar hot water heating. There are an estimated 800,000 solar hot water systems operating in the United States. Most people do not realize it, but about one-third to one-half of the total energy consumed in an average household goes into heating water. Solar hot water systems work all year around, not just in winter. This adds to cost-effectiveness. In nonfreezing climates, passive systems suffice. In freezing climates, circulating antifreeze systems are preferred (Fig. 23-8).

With current high costs and relatively stable energy prices, solar hot water heating is currently of marginal economic benefit in the United States except in specialized applications such as pools. However in Israel, over 60 percent of the homes are equipped with solar hot water heaters.

Again, the solar hot water system frequently does not fill total needs; additional or backup heating is required. But, it is important to recognize that it takes the same amount of energy to raise water one degree whether that degree is from 10° to 11° C or from 49° to 50° C. Thus, just using the solar system as a preheater to warm water halfway saves 50 percent of the energy cost.

## SOLAR PRODUCTION OF ELECTRICITY

Sunlight can be used to produce electrical power by direct conversion through **photovoltaic cells** or by concentrating heat to drive turbogenerators.

**Photovoltaic cells.** The most promising technology for solar production of electricity is the photovoltaic cell, or, as it is more commonly called, the **solar cell.** Solar cells look absurdly simple, just a thin layer of material imbedded in a plastic wafer (Fig. 23-9). But this appearance is deceptive; highly sophisticated technology is involved. The thin layer of material is actually a "sandwich" of two layers of ultrapure silicon, a common nonmetallic element. One layer has been treated so that some of the electrons surrounding the atoms do not have a stable place to reside— they are "loose." The other layer has been treated so that there are "electron holes"—places where electrons might reside if they were present. When light radiation strikes this sandwich, the loose electrons are jarred from their positions and they fall across the sandwich into the holes, thus creating a surplus of electrons on the hole side and

**FIGURE 23-8**
Solar water heater. Since needs for hot water exist all year, solar water heaters may be very cost-effective. In nonfreezing climates, simple water convection systems may suffice. Where freezing occurs, an antifreeze fluid is circulated.

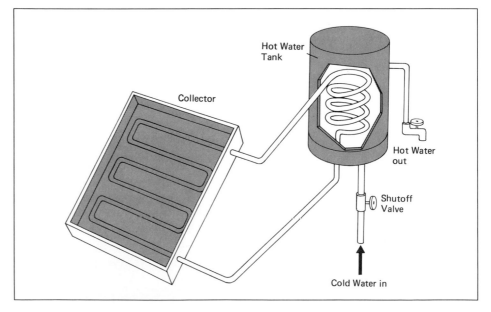

Hot Water Tank

Collector

Hot Water out

Shutoff Valve

Cold Water in

**FIGURE 23-9**
The thin "wafer" of material with wires attached is a photovoltaic cell. Converting light to electrical energy, this cell provides enough energy to run a small electric motor. (Photo by author.)

a shortage on the other. This flow of electrons provides the basis for an electrical current (Fig. 23-10). The circuit is completed by connecting a wire to the two sides so that the "surplus" electrons continue their flow through a motor or other device to do useful work and back to the other side. Thus, without any moving parts, a solar cell collects, converts, and concentrates light energy into an electrical current.

Under direct sunlight, a single 5 cm (2 in.)

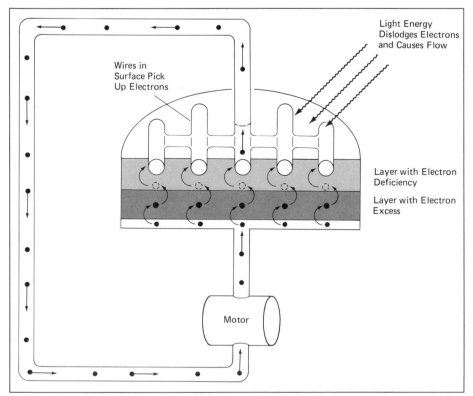

Light Energy Dislodges Electrons and Causes Flow

Wires in Surface Pick Up Electrons

Layer with Electron Deficiency

Layer with Electron Excess

Motor

**FIGURE 23-10**
The photovoltaic cell is actually a "sandwich" of two very thin layers of silicon treated in such a way that one layer tends to have a surplus of electrons while the other layer tends to have a deficiency. When light energy strikes the cell, electrons are dislodged and tend to pass from one layer to the other, producing the electric current.

solar cell, which is about the largest size that is practical to manufacture, provides about the same power output as a standard flashlight battery. However, solar cells can be connected together to obtain any amount of power desired (Fig. 23-11).

While the science of photovoltaic cells has been understood for some 50 years, high production costs have restricted their use to such things as space satellites, signal devices, and radio transmitters in remote locations where other sources of power are exceedingly expensive or impossible to provide. However, this situation is changing rapidly as more efficient manufacturing techniques are developed. The cost of cells has been reduced 100-fold since they were introduced in the 1950s; between 1977 and 1984, costs dropped from $20 to $8 per watt. This is five to ten times the cost of conventional generating plants, but photovoltaic cells are nevertheless widely used in pocket calculators, toys, and novelty items. A number of large-scale demonstration projects have been built or are under construction, and utility companies are beginning to study the feasibility of building photovoltaic power plants.

The concept of photovoltaic power plants is attractive to utilities because small units can be constructed and put into operation quickly and enlarged as the need develops. This eliminates problems associated with projecting needs many years in advance and making the huge upfront investments that are required for construction of nuclear power plants. In the near future (ten years) it may also be cost-effective for consumers to mount photovoltaic panels on their rooftops to generate a portion of their own electrical needs and thus offset utility bills.

Again, questions involving storage arise. Electrical demands are actually much higher during daytime hours when industries and offices are in operation. Also, the heavy demand for air conditioning is concentrated in daylight hours. If solar electricity provided power for the increased daytime load, existing facilities could carry the remaining load for many years to come. Thus, no urgent need for storage exists as long as present generating facilities are retained.

**Power towers.** Everyone has probably used a magnifying glass to focus sunlight onto a tiny spot and burn a hole through a piece of paper. A "power tower" is the ultimate expression of this concept. An array of sun-tracking mirrors are used to focus several hectares of sunlight on a boiler mounted on a tower (Fig. 23-12). The in-

(b)

(a)

**FIGURE 23-11**
Solar cells as a power source. (a) A sufficient number of photovoltaic cells may be wired together to produce any desired amount of power. (b) An array of solar cells being used to power an irrigation project near Mead, Nebraska. (U.S. Department of Energy photos.)

(a)

**FIGURE 23-12**
The "power tower" method of producing electrical power from sunlight. Sun-tracking mirrors are used to focus a broad area of sunlight onto a boiler mounted on a tower in the center. The steam produced is used to drive a conventional turbogenerator. (a) A facility in southern California under construction. (U.S. Department of Energy.) (b) The completed facility. (Courtesy of Southern California Edison Co.)

(b)

tense heat produced generates steam to drive a conventional turbogenerator that produces electricity. The facility shown in Figure 23-12 was constructed to test this concept and began operating in 1983. At present, power towers seem to be more cost-effective than photovoltaic cells, but because of construction and maintenance costs, power towers are not inexpensive. However, projections for the future suggest that the photovoltaic approach will probably be more cost-effective in the long run.

**Solar ponds.** Solar ponds are another innovative but low-cost method for collecting and storing solar energy. The pond is partially filled with brine (very salty water). Fresh water is placed over the concentrated brine. Since brine is much denser than fresh water, the brine remains on the bottom and little or no mixing occurs. Sunlight passes through the transparent fresh water, but is absorbed and converted to heat in the brine. The fresh water on top then acts as an insulating "blanket," which holds the heat in (Fig. 23-13). In short, solar ponds produce a type of greenhouse effect with fluids. The hot brine solution can be used for direct heating purposes or it can be converted to electrical power by using it to vaporize fluids with low boiling points; the vapors in turn can be used to drive low-pressure turbogenerators. Note here that the pond also acts as a very effective heat storage unit. Israel is pioneering the development of solar ponds. They have much potential under suitable environmental conditions.

However, all of these methods of producing

power from sunlight require considerable land area for collection. Further, the area must be located no more than a few miles from the place where the power is used; otherwise, an unacceptable amount of power is lost in transmission. Finding suitable areas for ponds or power towers near cities could pose a problem in some regions. Solar cells, on the other hand, can be dispersed on rooftops of existing buildings (Fig. 23-14).

## PRODUCTION OF HYDROGEN

With crude oil and natural gas supplies diminishing, hydrogen has frequently been advocated as the "fuel of the future." Hydrogen gas ($H_2$) is highly flammable. It could, in theory, be used in place of natural gas with little change in distribution networks or furnaces. Automobiles have been run on hydrogen gas with only minor modification in the carburetor. Further, hydrogen is clean burning; the only waste product is water vapor. Thus, pollution problems would be greatly reduced.

However, a significant problem often goes unmentioned. Essentially no free hydrogen exists

**FIGURE 23-13**
Solar ponds. Heat may be trapped and stored in brine (very salty water) covered with a layer of fresh water. Effectively the pond acts as a large flat-plate collector. The hot brine may be used for direct heating or to vaporize low-boiling-point fluids to drive low-pressure turbines.

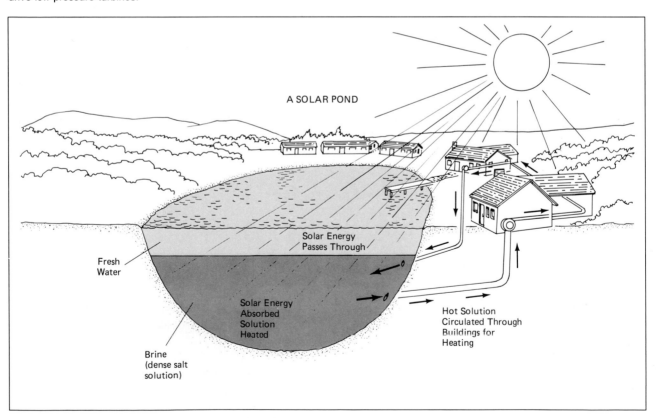

A SOLAR POND

Solar Energy
Passes Through

Fresh
Water

Solar Energy
Absorbed
Solution
Heated

Brine
(dense salt
solution)

Hot Solution
Circulated Through
Buildings for
Heating

**FIGURE 23-14**
Georgetown University's Intercultural Center (Washington, D.C.) supports an array of 4,400 photovoltaic modules providing a power output of 300 kilowatts under full sun. Such rooftop arrays may become commonplace in the future as production costs of photovoltaic cells are reduced. (U.S. Department of Energy.)

**FIGURE 23-15**
Electrolysis of water—one way to produce hydrogen gas. But the energy content of the hydrogen obtained only amounts to about 20 percent of the electrical energy used.

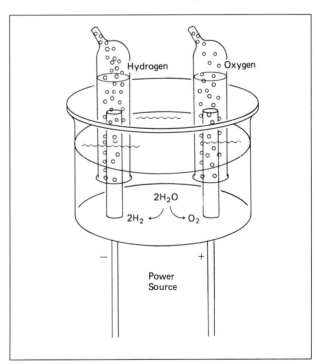

on earth. It has all been oxidized to water. Free hydrogen must be produced and this requires energy. One common method involves the electrolysis of water: electrical energy passed through water causes water molecules to disassociate into hydrogen and oxygen (Fig. 23-15). But this process is only about 20 percent efficient—100 calories of electrical energy are used for every 20 calories worth of hydrogen produced. Consequently, if hydrogen were to be the fuel of the future, and we were to produce it with nuclear and/or coal power, the environmental impacts and costs would be unspeakable. In short, hydrogen can only become a widely used fuel if there is an inexhaustible, nonpolluting energy source to produce it. Solar energy fits the bill.

During the initial reactions of photosynthesis, hydrogen atoms are disassociated from water molecules (these go on to become attached to carbon atoms to form organic compounds of the plant body while the free oxygen is released). Thus, the problem that must be solved is developing a stable, artificial system that will duplicate the initial step of photosynthesis. Some success has been achieved in laboratory experiments (Fig. 23-16). But the system must be scaled up to commercial size at affordable costs. Unfortunately, the amount of money allocated to this research is extremely limited, despite the fact that the concept holds much greater promise than fusion nuclear power, for example, which received nearly $500 million in 1984.

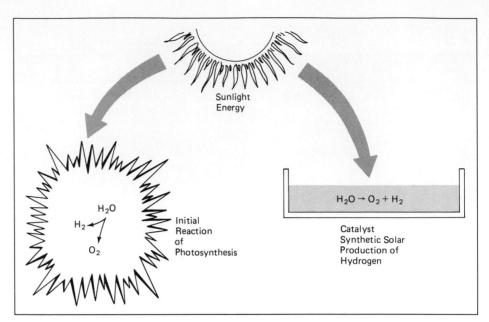

**FIGURE 23-16**
Solar production of hydrogen gas. (a) In photosynthesis, light energy causes the splitting of water into hydrogen and oxygen. (b) Research is being conducted to accomplish this in a stable artificial system. Some experiments have proven partially successful. If success is achieved, we may have solar production of a fuel (hydrogen) that can readily be substituted for gasoline.

In the figure: Sunlight Energy; $H_2O$, $H_2$, $O_2$, Initial Reaction of Photosynthesis; $H_2O \rightarrow O_2 + H_2$, Catalyst Synthetic Solar Production of Hydrogen

## INDIRECT SOLAR ENERGY

As solar radiation interacts with air, water, and biota, much of it is absorbed and converted to other forms of energy, principally, wind, hydropower (energy in falling water), and biomass (energy in the material of organisms). In effect, air, water, and biota collect sunlight and convert it into forms that are easier to harness and use.

### Biomass Energy or Bioconversion

The term **biomass** refers to all materials derived from living organisms. As we learned in Chapter 2, biomass originates from the process of photosynthesis. Therefore energy content of biomass or **biomass energy** represents solar energy trapped in biological materials. As well as being a food source for humans and other consumers and decomposers, biomass can be converted in a number of ways for use as **biomass fuels** (Fig. 23-17). The use of organisms to collect, convert, and store sunlight as chemical energy is referred to as bioconversion. The following are most significant means of **bioconversion:**

1. Heat for space heating, cooking, or power generation can be produced by burning wood or wood and paper wastes.

2. Methane (natural gas) can be produced by anaerobic digestion of sewage sludge, manure, and other agricultural wastes.

3. Alcohol, which can be substituted for most liquid fuels, can be produced by fermentation of sugars and grains and subsequent distillation.

All of our desired forms of energy can be produced from biomass and, in recent years, all of these methods have been developed to some extent. However, the key question is, can the techniques of bioconversion be expanded to fill the gap between declining production of crude oil and desired supplies? Although these methods utilize certain wastes, consumption of biomass for energy poses a number of severe environmental and/or economic problems. Also, the renewability of biomass is constrained by the concept of maximum sustainable yield discussed in Chapter 19. We shall note these problems as we consider each method in more detail.

### DIRECT BURNING

The use of wood stoves has enjoyed a tremendous resurgence in recent years. It is estimated that about 5 million homes in the United States rely entirely on wood for heating and another 20 million homes use it to partially heat. Hence, air pollution from wood stoves has be-

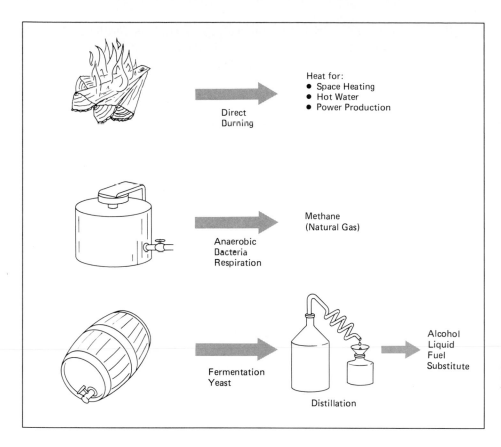

Direct
Durning

Heat for:
• Space Heating
• Hot Water
• Power Production

Anaerobic
Bacteria
Respiration

Methane
(Natural Gas)

Fermentation
Yeast

Distillation

Alcohol
Liquid
Fuel
Substitute

**FIGURE 23-17**
Bioconversion. As depicted, there are a number of ways of using biomass directly as a useful fuel or converting it to useful fuels.

come a problem in some communities, and restrictions are being applied in some areas.

In addition, many sawmills and woodworking companies burn wood wastes to supply all or most of their power. The burning of municipal trash, largely wastepaper, for power production is another example (see Chapter 20). In these cases, air pollution is controlled by precipitators.

Some studies conclude that direct burning of wood and wood wastes could supply as much as 20 percent of U.S. energy needs by the turn of the century, but environmental considerations hinder this goal. As long as wastes are utilized and air pollution is controlled, adverse impacts are not a threat and alternative disposal problems may be alleviated. However, wastes can supply no more than about 5 percent of our energy needs. Forests would have to be harvested for firewood to meet such demands. Individuals who cut appropriate amounts of firewood from their own land will not generate immense problems, but the amount of cutting necessary to supply more than a small percent of the nation's energy needs could have

severe impacts. Access roads and ruts caused by vehicles used in harvesting firewood would greatly aggravate erosion problems. Managing forests for wood production would seriously affect the diversity of natural biota. Pest control problems may also increase as a result of forest uniformity. Illicit cutting on both private and public lands is already a problem in some areas and would become a much greater problem. Crops such as sunflowers, which some advocate growing for the express purpose of burning, would compete with food crops.

These problems, as well as air pollution from wood smoke, would persist even if firewood use were kept within the limits of maximum sustainable yield. But what if we go beyond the maximum sustainable yield? This situation is occurring in many developing nations, especially in Africa. Nine-tenths of the people in less-developed countries, or about a third of the world population, are still dependent upon firewood as a source of fuel for cooking—they have never used more sophisticated sources of energy. In his paper, "The

Other Energy Crisis: Firewood,'' Erik Eckholm, a senior researcher with Worldwatch Institute, notes that the area for several miles around communities in many poor countries has been picked bare—overgrazed as it were—in the quest for fuel. Severe soil degradation and erosion are setting in as a result. Indeed the famine in Africa had its origins in erosion and soil degradation resulting from deforestation in the search for firewood.

### METHANE (NATURAL GAS) PRODUCTION

Bacteria feeding on organic matter under anaerobic conditions results in the production of **biogas,** which is about two-thirds methane gas. Recall the production of methane from sewage sludge described in Chapter 12 and the production of methane from landfills described in Chapter 20. This concept has been exploited by, for example, the Peoples Gas Company of Oklahoma, which constructed digesters to use manure from cattle feedlots. The methane that is produced supplements supplies of natural gas and the nutrient-rich humus residue is sold as fertilizer (Fig. 23-18). A number of dairy operations are, to some extent, self-powered by biogas from manure (Fig. 23-19). In China, millions of small farmers maintain a simple digester in the form of a sealed pit into which they put agricultural wastes. The biogas produced is used as a source of gas for cooking.

However, producing methane from biomass presents a number of problems. First, the most suitable materials for methane production are sewage sludge, manure, and other agricultural wastes, careless handling of which presents significant disease hazards. Second, these materials contain significant levels of sulfur; therefore, the biogas produced is contaminated with sulfur compounds. Since these compounds are converted to acids, burning the gas directly may damage the combustion equipment and cause air pollution. In large operations, it is economically practical to purify the methane.

Third, methane production is economically practical only when the raw material is already

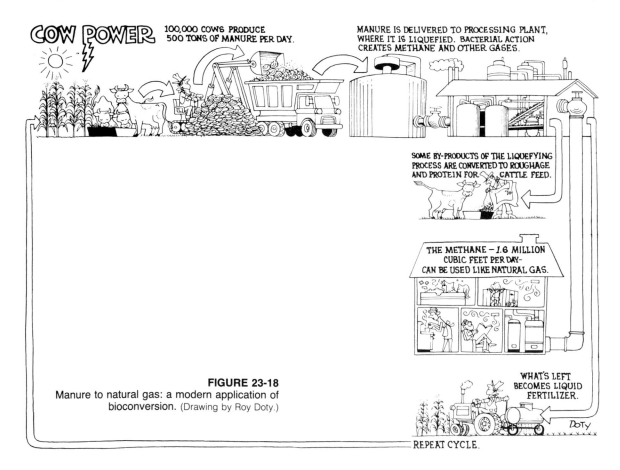

**FIGURE 23-18**
Manure to natural gas: a modern application of bioconversion. (Drawing by Roy Doty.)

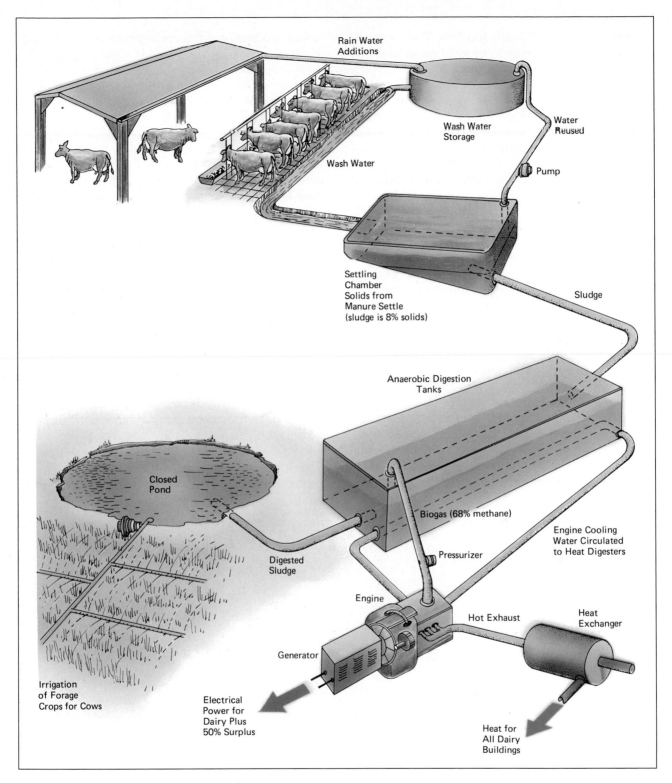

**FIGURE 23-19**
Dairy farm operated on cow manure. The total power needs for the Mason
Dixon Dairy located near Gettysburg, Pennsylvania, are obtained as a
byproduct of cow manure as shown. Excess power, nearly half of what is
produced, is sold to the local utility.

"concentrated," as in cattle feedlots. The economy lies in the fact that biogas production offsets the need for other means of disposal. Even if all such raw materials were utilized in methane production, it would contribute, at most, about 3 percent to the U.S. energy budget. Gathering additionl agricultural wastes for methane production would be environmentally, as well as economically, unsound because the return of such material to the ground is necessary for maintenance of soil quality.

## ALCOHOL PRODUCTION

Yeast cells feeding on sugars and/or high-starch grains under anaerobic conditions produce alcohol as a metabolic waste product; the process is known as **fermentation.** The alcohol can be further concentrated by **distillation,** a process involving boiling and recondensing the alcohol. Fermentation and distillation have been used for millennia in the production of drinking alcohol. Here we are simply considering producing a sufficient quantity for fuel use (Fig. 23-20).

Brazil has pioneered the development of large-scale fermentation plants that utilize sugarcane. A large percentage of Brazil's cars now run on the alcohol or a mixture of alcohol and gasoline known as **gasohol.** In the United States, alcohol production from corn for use as gasohol increased rapidly in the early 1980s (Table 23-1). In 1984, the United States used about 160 million bushels of corn, about 2 percent of the total, to produce 420 million gallons of alcohol (about 0.5 percent of our liquid fuel).

How much alcohol can be produced? The important point is that alcohol is produced from traditional food crops such as sugarcane, corn, and potatoes. Thus, a conflict between food and fuel is inevitable, especially if alcohol producers offer better prices than food producers. Currently, because of large grain surpluses in the United States, alcohol production poses little problem. The social consequences, however, can be diabolical. It is reported that in Brazil, sugarcane is cultivated at record levels for alcohol production while food crops are down by 10 to 15 percent. This is occurring despite widespread malnutrition and a rapidly growing population.

Another problem is pollution. While alcohol is promoted as clean burning, its production generates much pollution because inexpensive, dirty-burning fuels such as soft coal are used for distillation and an amount equivalent to at least one-half gallon of fuel is used for every gallon of alcohol produced. This makes alcohol expensive—roughly double the cost of gasoline in 1985. The apparent competitive pricing of gasohol is due to

**FIGURE 23-20**
Sweet sorghum being grown for alcohol production. Such "energy crops" will inevitably compete with food crops for land and other agricultural inputs. (U.S. Department of Energy.)

Table 23-1

| Year | Corn Used (million bushels) | Area in Corn (thousand hectares) | Alcohol Fuel Produced (million gallons) | Alcohol Fuel Consumed (million gallons) | Share of Fuel (percent) |
|---|---|---|---|---|---|
| **Production of Alcohol Fuel from Corn in the United States 1980–1984** | | | | | |
| 1980 | 15 | 57 | 37.8 | 79.8 | 0.08 |
| 1981 | 35 | 134 | 75.6 | 84.0 | 0.09 |
| 1982 | 80 | 308 | 210.0 | 231.0 | 0.24 |
| 1983 | 135 | 518 | 373.8 | 432.6 | 0.45 |
| 1984[a] | 160 | 615 | 420.0 | 504.0 | 0.52 |

[a]preliminary estimate

Source: Worldwatch Institute estimates based on unpublished data from U.S. Department of Agriculture. Reprinted from *Renewable Energy at the Crossroads,* Washington, D.C.: Information Resources Inc., 1983.

tax subsidies amounting to 60¢ to 90¢ per gallon of alcohol.

In summary, a small, but significant energy contribution with positive environmental impact would be obtained by developing bioconversion of municipal and other organic wastes. Bioconversion on a more massive scale, however, runs the risk of unacceptable environmental problems and competition with food production.

## Hydropower

Because sunlight drives the water cycle, hydro or water power is another indirect form of solar energy. Of course, water power has been used for millennia by diverting water from natural falls over various kinds of paddle wheels or turbines (Fig. 23-21). But relatively few natural falls with significant volume exist in the United States. Therefore, over the last century or so, the trend has been to build huge dams to create artificial falls that will generate substantial **hydroelectric** power (Fig. 23-22). About 13.5 percent of the overall electrical power, 5.5 percent of total energy, in the United States currently comes from **hydroelectric dams,** most of it from some 300 major dams concentrated in the Northwest and Southeast. This is slightly more power than currently comes from nuclear power plants. Can the use of hydroelectric power increase?

**FIGURE 23-21**
Utilizing the force of falling water was one of the earliest sources of power. Mill near Pigeon Forge, Tenn., built in 1831. (Dick Foster/Stockpile.)

**FIGURE 23-22**
Hoover Dam. About 13.5 percent of the electrical power used in the United States comes from large hydroelectric dams such as this. Water flowing through the base of the dam drives turbines.
(U.S. Department of Energy.)

While hydroelectric power is basically a non-polluting, renewable energy source, it still involves tremendous tradeoffs.

Dams have drowned out some of the most beautiful stretches of river in North America as well as wildlife habitats; productive farmlands; forests; and areas of historic, archeological and geological value. Glen Canyon Dam [on the border of Arizona and Utah] drowned one of this world's most spectacular canyons. Tellico Dam in Tennessee eliminated an ancient Cherokee village and the site of the oldest continuous habitation on the North American continent.*

The reservoir behind the Aswan High Dam in Egypt has caused the spread of a parasitic worm, shistosoma, which causes a serious debilitating disease. The dam has also increased humidity, which is now causing rapid deterioration of ancient monuments and artifacts that stood virtually unchanged for many centuries.

Since water flow is regulated according to the need for power, dams also play havoc downstream because water levels may go from near flood levels to virtual dryness and back in a single day. Other ecological factors are also affected as different amounts of nutrients and sediments reach the rivers' mouth.

Consequently, in the United States and many other countries, proposals for more dams are creating increasing controversy over whether the projected benefits justify the ecological and sociological tradeoffs. In any case, few sites conducive to large dams remain in the United States. It is notable that a proposal to construct a dam on the New River in Virginia was defeated in 1976 after years of litigation on the part of environmentalists. The New River was the *last* large undammed river on the East Coast. Similarly, dam proposals on the Buffalo River were defeated, preserving it as the *only* large undammed stream in the Arkansas Ozarks.

However, the potential for increasing hydroelectric power with small dams continues to exist. In the early days of hydroelectric power, numerous small dams were constructed on small streams and rivers to generate local power. As power production went to large, centralized facilities, these small dams were abandoned (Fig. 23-23). Nevertheless, they continue to exist and groups of independent investors are now renovating these small projects and selling the power to local utility companies. Also, many dams that were built for purposes such as flood control and water supply could be retrofitted for power gen-

*Blackwelder, Brent. "Dams: A Change of Course." *National Parks*, 58 (July/Aug. 1984), pp. 8–13.

**FIGURE 23-23**
This dam on the Patapsco River near Baltimore, Maryland, was used for power generation in the early 1900s but was abandoned as large centralized facilities came into use. If the thousands of such dams were renovated, as some have been, production of hydropower could be nearly doubled. (Photo by author.)

eration. If all of these alternatives were undertaken, it is estimated that production of hydroelectric power in the United States could be doubled. The amount of energy thus generated would equal that of 50 to 60 new nuclear or coal-fired power plants.

## Wind Power

Winds result from the solar heating of the atmosphere; therefore, wind power is also a form of indirect solar energy. Using wind as a source of energy is far from a new idea. In fact, along with water power and domestic animals, it is one of the oldest sources of energy and has been used to a greater or lesser extent throughout history. Until the 1930s, most farms in the United States used windmills for pumping water and/or generating small amounts of electricity (Fig. 23-24). However, these small windmills do not provide the amounts of power desired for modern living. Therefore, in the 1930s and 1940s, as transmission

**FIGURE 23-24**
Windmills like this were widely used on American farms for pumping water until the 1940s. (USDA photo.)

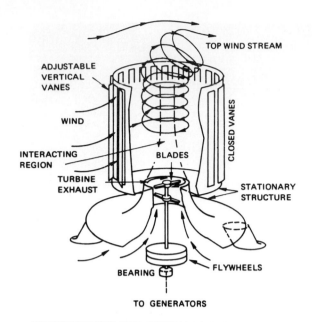

**VORTEX TOWER FOR OMNIDIRECTIONAL WINDS**

**FIGURE 23-25**
Various types of wind turbines. The simple airplane-type propeller on a
horizontal shaft still appears the most practical. (U.S. Department of Energy, Sandia
Laboratories, and NASA photos.)

lines brought low-cost power from central generating plants, the windmills fell into disuse. However, wind is again being looked at as an energy source with tremendous potential and windmills are taking on new roles as wind generators or wind turbines.

The use of windmills for large-scale power generation is new technology, and problems involving stress and vibration have arisen. But over the past decade, many experimental wind turbines have been built and tested to determine the most efficient and trouble-free designs (Fig. 23-25). Testing is still underway, but using an airplane-type propeller on a generator shaft seems to be as efficient as any method. What size will prove most economical, however, is still to be seen.

Currently, development is proceeding along two lines. Several utility companies are considering the feasibility of buying and operating huge wind generators with blades about 30 m (100 ft) long mounted on 50 m (150 ft) towers (Fig. 23-26). Such machines will generate between 1 and 10 megawatts of power, each megawatt being enough to supply about 1000 homes. For comparison, nuclear power plants produce about 1000 megawatts. After several experimental failures, the first commercial giant windmills are now in operation.

The second line of development is "wind farms," which are arrays of between 50 and several hundred moderately sized windmills (Fig. 23-27). Each machine, with blades 20 to 30 feet long, generates 10 to 50 kilowatts (1 kilowatt = 0.001 megawatt). While the smaller machines are less efficient, they do not have many of the problems of the huge machines and they are now being mass-produced by several companies. Thus, the use of wind power is growing rapidly. Over 100 wind farms have now been established, mostly in California, by groups of independent investors. Of course, individuals can put up their own machines if zoning permits, and some farmers have done so. However, these single installations are not generally cost-effective at current prices.

One might be concerned about the amount of land devoted to wind farms. However, the land beneath the towers can still be used for farming and ranching. Windmills do have a negative visual impact, which prevents development in

**FIGURE 23-26**
MOD 2 wind turbine in Washington State; blades, 300 ft from tip to tip; tower, 200 ft tall; capacity, 2.5 megawatts in winds 14–45 mph. Giant wind generators such as these may produce power at a cost competitive with fossil-fuel power plants. (U.S. Department of Energy photo.)

some areas. However, the sight of windmills may also be refreshing in that they signify that power is being generated with far less pollution than coal and with none of the hazards of nuclear power. A final advantage of wind power is that, like photovoltaic cells, it can be added in small increments with very little lead time required. Again, this circumvents the need for utilities to make huge investments for power plants that may not be needed.

Cost-effective wind power generation does require relatively consistent winds of about 12 miles per hour (20 km per hour) or more, but large areas of the country do meet this criterion. Wind generators are tied into regular electric power grids and, thus, are backed up by conventional power sources. Increasing wind power would reduce the need for conventional fuels, and with this savings, there would be no need to consider storage of power from wind for some time.

**FIGURE 23-27**
Wind farm located 40 miles east of San Francisco. Large arrays of modestly sized wind machines are proving more practical than a few very large wind machines.
(Photo by author.)

In conclusion, generation of power from wind, after just ten years of development, is already becoming economically competitive with power from nuclear or coal-fired plants. If the costs of the negative environmental impacts of coal and/or nuclear power were fully considered, wind power would have a definite cost advantage. In 1985, just five years after installation began, about 1000 megawatts of wind power were on line. This is equivalent to the power produced by one nuclear power plant. Thus, wind power is likely to see rapid growth.

## GEOTHERMAL, TIDAL, AND WAVE POWER

### Geothermal Energy

The interior of the earth is very hot, as evidenced by the eruption of molten lava from volcanoes. **Geothermal energy** involves using this heat source for the production of useful energy. It may be done in two ways.

First, natural groundwater may come in contact with hot rock. Such heated water may come to the surface in natural steam vents as are observed in Yellowstone National Park, or the steam may be obtained by drilling into such superheated aquifers (Fig. 23-28). Second, the "wells" may be drilled into hot, dry rock. Water injected into such wells is heated by the rock and comes back up as steam (Fig. 23-29). In both cases, the steam is piped through turbogenerators to produce electricity, or, if the geothermal well is near a city, the steam may be used directly to heat buildings and homes.

Operating geothermal facilities presently exist in many parts of the world. In the United States, the Pacific Gas and Electric Company uses natural steam vents in northern California for the production of electricity (Fig. 23-30). However, large-scale development of geothermal power presents many problems. Hot steam and water brought to the surface are frequently heavily laden with salt and other contaminants, particularly sulfur compounds. These contaminating compounds are highly corrosive to turbines and other equipment, and they result in both air and water pollution as they are finally released into the environment. Sulfur pollution from a geothermal plant may be equivalent to that from a fossil-fuel plant burning high-sulfur coal, and the hot brines released into streams or rivers could be ecologically disastrous. There may be less pollution from "dry" wells but, to date, no dry well is operating so their practicality remains theoretical.

Production of geothermal power from "wet" wells has grown from about 500 to 1500 megawatts since 1978. However, this can be attributed to further exploitation of known "clean" geothermal facilities. The difficulty and cost of exploratory drilling in search of additional clean formations make the potential of geothermal power highly speculative.

**FIGURE 23-28**
Geothermal well on the island of Hawaii. (U.S. Department of Energy photo.)

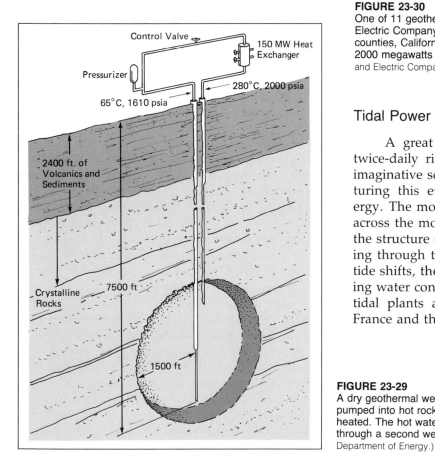

**FIGURE 23-29**
A dry geothermal well. Water is pumped into hot rock, where it is heated. The hot water is returned through a second well. (U.S. Department of Energy.)

**FIGURE 23-30**
One of 11 geothermal units operated by the Pacific Gas and Electric Company at The Geysers in Sonoma and Lake counties, California. The field may be capable of supporting 2000 megawatts by the late 1980s. (Photo courtesy of Pacific Gas and Electric Company.)

## Tidal Power

A great deal of energy is inherent in the twice-daily rise and fall of the tides and many imaginative schemes have been proposed for capturing this eternal, pollution-free source of energy. The most straightforward is to build a dam across the mouth of a bay and mount turbines in the structure (Fig. 23-31). The incoming tide flowing through the tubines generates power. As the tide shifts, the blades are reversed so the outflowing water continues to generate power. Two such tidal plants are presently in operation—one in France and the other in the Soviet Union.

The problem with developing more tidal power is that it requires a fluctuation between high and low tide of about 6 meters (20 feet) or more in order to produce enough "head" of water pressure to make the tidal dam worthwhile. Otherwise, the net energy is zero or less; that is, more energy is required in construction and operation of the facility than is ever produced in power. There are only some 15 locations in the entire world that have tides of this magnitude, only one of which is in North America: the Bay of Fundy in Nova Scotia where a major tidal power plant is under consideration.

But tidal power, while nonpolluting, is not without environmental impact. In addition to the loss of unique aesthetic and recreational pleasures in these areas, there would be far-reaching environmental effects due to the dam's trapping of sediments, impeding of the migration of marine organisms, and most of all from changing circulation and mixing fresh and salt water. Shipping and boating might also be blocked. In conclusion, tidal power does not have the potential to contrib-

**FIGURE 23-31**
Tidal power. (a) Tidal power dam located on the Rance estuary in northern France. (b) One of 24 turbines mounted in the dam.
(U.S. Department of Energy photos; Michel Brigaud, French Embassy.)

ute more than a minute amount to overall energy use in the near or long term, and this only at the sacrifice of unique bays and estuaries. Indeed, potential impacts may extend over much broader areas; some scientists project that the Bay of Fundy project may affect tides as far south as Boston.

## Wave Power

Waves are generated by wind, but they are mentioned here because they have much in common with tidal power. The energy inherent in waves holds the same general allure as the energy inherent in tides. It also presents the same difficulties with respect to harnessing it. Along coasts of the United States, consistent wave action is not great enough to develop a head of water pressure high enough to generate a significant amount of power with any device of practical size. Again, the net energy output of the wave generators conceived today appears to be zero or less. Therefore, wave power also has little, if any, potential for contributing to the solution of our energy problem in the United States. England and Ireland, however, have access to consistently higher waves and are experimenting with various wave generators.

Again, all such schemes must be economically competitive to be practical. It appears that these alternatives may soon be outpaced by one or more of the solar technologies already discussed.

## CONSERVATION

After looking at all of the major options for alternative energy sources, we should be impressed (or depressed) by the fact that none provides a simple solution to the critical problem of the declining availability of crude oil, which is the base for all liquid fuels currently used in transportation. Most of the alternatives involve production of heat and/or electrical power. Alternative sources of liquid fuels, namely, synthetic fuels from coal and alcohol from biomass, are both expensive and involve a severe environmental impacts. Methods of producing hydrogen from solar energy are not sufficiently developed to have any bearing on current problems.

Consider then how we might greet news that We have just discovered a new oil field twice the size of the Alaskan field. It has a potential production of 4 million barrels a day, perhaps much more. It is inexhaustible and its exploitation will not adversely affect the environment. Of course, such an oil field is only a dream, but this is effectively what can be achieved through **conservation.**

## What Is Conservation?

Generally, energy conservation brings to mind such things as turning off lights, turning down thermostats, and car pooling. These activities can, and do, produce immediate fuel savings and they have helped get us through periods of limited supplies, such as those that occurred in the 1970s. However, in view of the long-term energy problem, such activities have distinct limitations. First, the actual savings that can be achieved by such efforts is only 2 to 4 percent of overall energy use. Second, such efforts do entail some inconvenience and/or discomfort. Consequently, conservation has gained a reputation that has some people believing that it means "freezing in the dark." This is wrong.

The real objective of conservation and its lasting potential lies in being able to perform the same functions—maintaining physical comfort, lighting, transportation, and so on—but with less energy. To be sure, natural laws dictate that a certain minimum amount of energy will be required for each function. However, as we have pointed out, in many if not most situations, we use five- to tenfold more energy than the theoretical minimum and the excess is simply wasted. Thus, conservation involves making machinery more efficient so that less energy is consumed, and/or utilizing the waste energy (heat) for further useful functions before it is dissipated. Much progress has been made in this direction over the past decade, but the full potential of conservation has hardly been tapped.

## Examples of Conservation

New cars averaged about 13 miles per gallon in the early 1970s. By the early 1980s, new cars averaged about 25 miles per gallon. This represents a savings of nearly 2 billion barrels of crude oil per year (5 million barrels per day). Experts

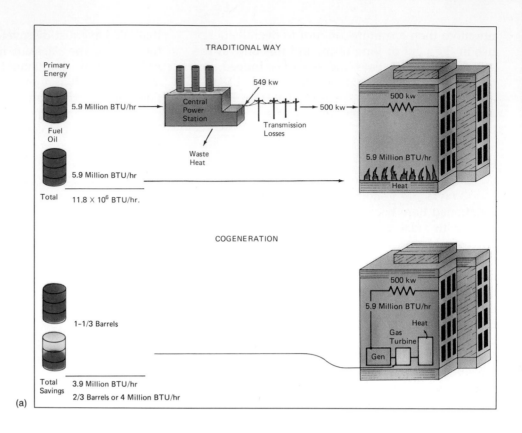

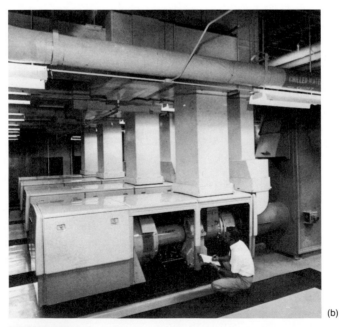

**FIGURE 23-32**
Cogeneration: (a) Traditional means is to provide electricity from a central power plant and then burn additional fuel for heat. A 30 percent fuel savings may be realized by generating electricity in the building where waste heat is utilized. (b) Installed, cogeneration units provide both heat and electricity for an office building. (Photo courtesy of the Garrett Corporation.)

estimate that by 1995 cars could average about 90 miles per gallon if current technologies for increasing efficiency were fully exploited.

Another example of conservation progress is illustrated by the increased insulation that now goes into buildings and homes, and the reinsulation of older homes and buildings. Again, full potential has hardly been tapped; it is estimated that well over half of the existing homes still have substandard insulation and, as noted, little attention has been paid to the potential benefits of passive solar heating. At least another billion barrels of oil per year could be saved in this area.

The industrial sector of the economy is conscious of energy costs and has made tremendous strides in conservation. Yet, more can be done. A particular area that is open to further exploitation is **cooperative energy use.** The custom has been to supply every energy-utilizing process with its own primary source of energy and to discard the waste heat in each case. Improvements in efficiency can be achieved by using the waste heat from one process to supply the needs of another process. In the auto industry, for example, waste heat from a high-temperature process such as heating metals for forging can be used to supply the low-temperature heat required for paint drying. Similar transfers might take place between two industries or between industry and the private or the government sector. For instance, the waste heat from power plants that presently becomes thermal pollution can be used for heating buildings if they are located close enough to make transfer possible. This practice, known as **district heating,** is already used in many European cities. Certainly, heat from waste incineration could be used in district heating.

A related concept with great potential is **cogeneration,** which is the production of electrical power in the process of obtaining heat. For example, virtually every building and factory requires both electricity and heat. (In the summer, heat can be used to drive air conditioning units.) The traditional means is to obtain electricity from a central power station where 60 to 70 percent of the heat from the fuel is wasted; then additional fuel is burned to provide heat. With cogeneration, the building "furnace" is an appropriately sized turbogenerator that produces electrical power for the building while at the same time the hot exhaust provides the required heat. The result is a 30 percent saving in the overall amount of fuel required (Fig. 23-32). Cogeneration has the added advantage of providing protection against the wide-scale blackouts that may occur with highly centralized systems.

New technologies may also allow the same jobs to be done with less energy. For example, solid state electronics has greatly reduced the power requirements as well as the size of computers and other electronic instruments. Fluorescent light bulbs are about 95 percent efficient in converting electrical power into light whereas traditional incandescent bulbs are only 10 percent efficient. Microwave ovens reduce the energy needed in certain kinds of cooking.

Progress in all of these areas has been so significant in recent years that, for the first time in history, we have seen the economy expand without a parallel rise in energy consumption. If the relationship between energy and gross national product that existed in the early 1970s existed in 1983, we would have used about 100 quadrillion BTUs in 1983. In fact, in 1983, total energy consumption was about 77 quadrillion BTUs (Fig. 23-33). The difference of 23 quadrillion BTUs can be attributed to conservation and improved efficiency and the savings amounted to about $150 billion.

This improved efficiency has cost money. The automobile industry, for instance, spent many billions of dollars redesigning and retooling to produce more efficient cars. The savings, however, far outweigh the costs. Hence, conservation has proven to be the most cost-effective means of mitigating the energy problem. Importantly, this was achieved without significant changes in lifestyle. Further gains can still be made without lifestyle changes because we are nowhere near the limits of energy efficiency. Once we reach that point, some lifestyle changes may further increase savings.

For example, people who move to a location where the commuting distance is cut by half will effectively double the efficiency of their gasoline usage because the trip is being made with half the amount of gas. Moving from a detached house to a cluster or townhouse may effectively double heating efficiency. Great energy savings can also come through recycling bottles, keeping items in use longer, and other resource-saving measures discussed in Chapter 20.

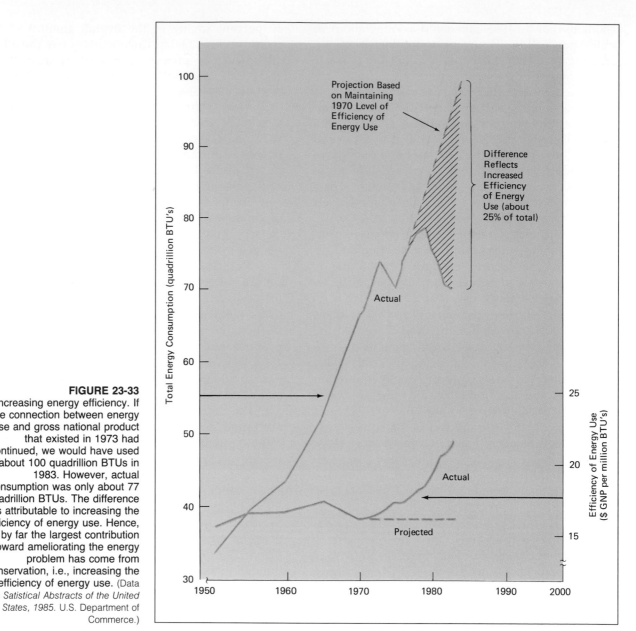

**FIGURE 23-33**
Increasing energy efficiency. If the connection between energy use and gross national product that existed in 1973 had continued, we would have used about 100 quadrillion BTUs in 1983. However, actual consumption was only about 77 quadrillion BTUs. The difference is attributable to increasing the efficiency of energy use. Hence, by far the largest contribution toward ameliorating the energy problem has come from conservation, i.e., increasing the efficiency of energy use. (Data from *Satistical Abstracts of the United States, 1985.* U.S. Department of Commerce.)

In conclusion, regardless of which alternative sources of energy are pursued in the future, it is difficult to see how we can get through the transition period of declining crude oil supplies without conservation. But, just as new sources of energy need time to develop, experience has shown that the changes involved in conservation and improved efficiency require substantial investments and long lead times. Making the changes will require continuous effort; conservation must not be viewed as something to be done just when a shortage occurs.

## ENERGY POLICY

We have seen that considerable progress has been made toward coping with the energy problem. In addition to exploiting the existing potential of coal and nuclear power, marked progress has been made in conservation, and a number of renewable energy technologies have been developed to the point where rapid expansion is possible. However, the fact remains that we are still dependent on crude oil for about 40 percent of our total energy. Thus, we remain extremely vulnerable to

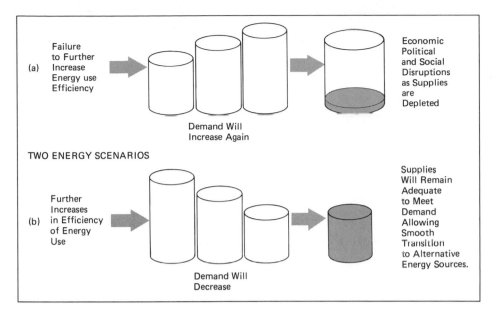

**FIGURE 23-34**
Two energy scenarios for the future. (a) Abandon further efforts at conservation and increasing energy efficiency, and allow demand to increase again. This will lead to crippling consequences as demands will inevitably exceed supplies. (b) Continue efforts toward increasing energy efficiency and conservation to gradually reduce demand in accordance with supply. This will allow a smooth economic transition into energy sources of the future.

the reality of declining reserves and the possibility of import restrictions.

Scenarios for the future involve two possibilities (Fig. 23-34). With continuing efforts at conservation and development of appropriate alternative energy sources, we can make a smooth economic transition to an economy that is less and less crude oil dependent. Or, we can ignore conservation and development of alternative sources of power and allow our consumption of crude oil to increase again until it is suddenly cut short by falling production, which is inevitable. Developing new technologies at that time would be very difficult because development itself is an energy-consuming process. In essence, waiting until shortages actually occur is a positive way to precipitate disaster.

Which of the many alternative technologies to pursue becomes another important question. It is not practically necessary, environmentally sound, nor economically possible to engage in unlimited crash programs in all areas simultaneously. Therefore, an energy policy must be adopted that will allocate economic and technological resources toward development of those methods that hold the most promise of meeting our future needs in an environmentally, economically and socially acceptable way.

A glaring aspect of the energy dilemma remains. Use of various motor vehicles that are dependent on oil-based fuels continues to grow and no suitable substitute for liquid fuels has yet de-veloped. Deriving synthetic fuels from coal and alcohol from biomass involves environmental and/or social impacts that are unacceptable and expensive. Development of electric cars or solar production of hydrogen are still too far in the future. What has enabled us to reduce our dependence on crude oil in large measure has been the increased fuel efficiency of cars. Further conservation in this respect is definitely in order. Progress to date in this area was motivated *both* by sharp increases in the cost of fuel, which led consumers to seek cars that got the most miles per gallon, *and* by the Energy Power and Conservation Act of 1975, which mandated that new cars should achieve an *average* of 27.5 miles per gallon by 1985. Under the law manufacturers can still produce cars that get fewer miles per gallon if these are balanced by the sale of cars getting more than 27.5 miles per gallon. The goal of this legislation was not quite met, but it is remarkable that it was even nearly met.

However, as oil supplies appeared to be more abundant in the mid-1980s, introduction of further energy efficiency technology lagged and consumers were again buying cars that get fewer miles per gallon. Many energy experts feel that, to meet the projected crude oil shortages ahead, more of the same ''medicine'' is required—a law requiring an average of 60 miles per gallon for cars by 1995, for example.

Another area where opportunities exist for crude oil conservation is in home heating through

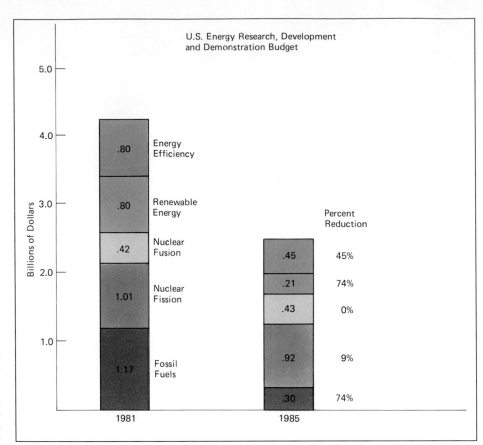

**FIGURE 23-35**
Energy research, development, and demonstration budget in the United States. Despite the remarkable contributions from the areas of energy efficiency and renewable energy sources, funding for these areas has been slashed in recent years while that for nuclear power has remained constant or increased.
(Source: U.S. Department of Energy. Redrawn with permission from *Renewable Energy at the Crossroads*, Center for Renewable Resources, Washington D.C., 1985.)

improved insulation and/or installation of solar heating. Some progress in this area has been spurred by legislation providing tax writeoffs for weatherizing and requiring utility companies to provide consumers with information on weatherization and methods of reducing energy consumption. Higher fuel costs have also provided homeowners with a greater incentive to take advantage of the provisions. However, the opportunity for further improvement is vast.

The Public Utilities Regulatory and Policies Act of 1978 has been a key factor in the development of renewable energy alternatives that produce electricity. This act requires publicly owned utilities to buy power from independent producers at "full cost," a price defined as the same price it would cost the utility to produce the power. This means that a group of private investors who wish to set up a wind farm or renovate a small hydroelectric dam, for example, are guaranteed a market for the power they generate. Without this guarantee, such investments would be impossible

**Table 23-2** | **United States Government Spending on Renewable Energy (millions)**

| ENERGY SOURCE | 1981 | 1985 |
|---|---|---|
| Alcohol Fuels | 21 | 31 |
| Other Biomass | 46 | — |
| Hydropower | 22 | — |
| Solar Thermal | 219 | 46 |
| Photovoltaics | 160 | 57 |
| Geothermal Energy | 199 | 32 |
| Windpower | 86 | 29 |
| Other | 44 | 17 |
| Total | 797 | 212 |

Source: U.S. Department of Energy. Reprinted with permission from *Renewable Energy at the Crossroads*, Center for Renewable Resources, Washington, D.C. 1985.

for private parties. The investors also receive tax writeoffs for part of their investment. This, too, has helped make such projects profitable and has further encouraged private investment and development of alternative power sources.

Finally, development of alternative energy technologies has been promoted by federal money allocated to research, development, and demonstration projects. Funding for renewable energy alternatives, as well as traditional sources of energy, increased steadily through the late 1970s. This illustrates that a policy including legislation that mandates conservation can be instrumental in achieving our energy goals, and suggests that stronger policies may be in order.

In their report "Renewable Energy at the Crossroads," the Center for Renewable Resources stated that, "With modest, but effective support, the country can move well along the path to a sustainable energy system and a renewable energy industry that is second to none." Conversely, "Sudden withdrawal of federal incentives would be an enormous mistake, crippling an industry that the government has carefully nurtured up to the point of economic viability."

It is ironic that during the early 1980s, despite the progress that had been made under previous funding, the government slashed budgets for conservation and development of renewable resources while maintaining budgets for nuclear power (both fission and fusion) (Fig. 23-35; Table 23-2). In addition, several bills that would provide incentives are now in jeopardy. Dennis Hayes, chairman of the Solar Lobby, notes that the largest recipients of tax breaks are energy industries. However, most such breaks are received by oil, coal, and nuclear industries due to their well established, powerful lobbies. Informed and active citizens can do much to promote and support a policy that is more equitable and that will provide a more sustainable future.

# 24

# LAND USE

---

## CONCEPT FRAMEWORK

Outline                                                          Study Questions

1. Describe current land use trends. What is *urban sprawl* and when did it begin? What factors contributed to its emergence? Why was urban sprawl not regulated from the beginning? How does urban sprawl affect farmers?

2. Why is 90 percent of the world's population situated on about 2 percent of the land? Why is the maintenance of agriculture in the United States of international importance?

3. What effects does development have on water resources? Why is depletion of water re-

sources one of the most critical problems facing future agriculture and development?

4. How does highway construction affect natural ecosystems?

5. Why does the suburban lifestyle translate into energy consumption?

6. What sources of pollution may be associated with urban sprawl?

7. What is *exurban migration?* How does exurban migration contribute to discrimination and segregation? How does urban sprawl represent a gentrification of society?

8. How does exurban migration affect the tax base? What effect does this have on a city? What city services are most affected?

9. What are the social and environmental costs of highways?

10. Describe the vicious cycle of exurban migration and urban decay. Why do businesses leave the city? How does this affect inner-city residents? Why hasn't the "rebirth" of some cities solved the problem?

11. Why have cities failed to become the pleasant places Gruen describes? What are the disadvantages of public housing projects?

12. What are the benefits of cluster housing?

13. Why does urban sprawl continue despite efforts at control? Discuss the importance of environmental impact statements. How does population growth affect land use? Why are population growth and land use patterns so important in relation to all other resources?

In previous chapters we discussed resources that are essential to humans: soil, water, air, minerals, biota, and energy. These resources are interconnected in many ways, but all are fundamentally dependent on land. Consider the following:

1. Water. Land contains the largest reservoir of fresh water, both surface water and groundwater.

2. Biota. Land supports all terrestrial biota, both natural and domesticated. This biota provides over 90 percent of our food, and all of our timber and natural fibers. Aquatic biota are also dependent on land because they are affected by runoff containing sediments, nutrients, and pollutants.

3. Natural services. By supporting natural biota and ecosystems, land provides natural services affecting water and air quality.

4. Minerals. Nearly all mineral resources are mined from the land.

5. Energy. Coal, oil, natural gas, and uranium are mined from the land. In addition, land provides space for hydroelectric reservoirs, power plants, and solar installations.

6. Human activities. Land provides space for all human activities, homes, stores, schools, offices, factories, roadways, parking lots, airports, landfills, and other facilities that make up cities and towns and connections between them. Also, land is used for recreational and aesthetic pleasure. Even water-based activities require land facilities such as beaches and marinas.

As population and affluence grow, we need and desire more land for all of the above. But, as Mark Twain said, "they don't make it anymore." Land is a strictly limited resource. A few situations exist where land has multiple uses. For instance, a national forest can produce timber and also support natural biota and various recreational activities. However, in most cases, a decision to use land for a specific purpose precludes all or nearly all other uses. Hence, land use decisions often have effects far beyond the immediate location of the land in question. A highway does a lot more than provide for a human activity. It affects natural biota, water, and air quality. When land use decisions are made, they frequently determine the future of all other resources.

In this chapter we shall examine current patterns of land use and see how these patterns affect other resources. We shall also discuss how strategies aimed at resource conservation and at sustaining society must include a consideration of current and future lifestyles.

## TRENDS IN CURRENT LAND USE

Rampant development is the term that best characterizes current land use trends in the United States and Canada. More and more, farms and natural areas surrounding cities are giving way to new housing developments, shopping and industrial centers, highways, and other facilities that

**FIGURE 24-1**
Somehow escaping the initial phase of land clearing, a jack-in-the-pulpit gives its last bloom. In the United States, some 2 million acres of land per year goes into development. (Photo by author.)

**FIGURE 24-2**
Urban sprawl. Complexes of housing developments, highways, and shopping centers spread over the countryside in a largely unplanned, helter-skelter fashion. (NASA photo.)

have become a normal part of everyday life (Fig. 24-1). These projects are accepted and promoted as indicators of a strong economy. For young people who grew up in this landscape of spreading development, it is difficult to believe that it is a phenomenon that did not begin until after World War II and that it was largely unplanned, uncontrolled, and unintended. It is commonly referred to as **urban sprawl** (Fig. 24-2).

Because urban sprawl is the root of many resource and environmental problems, we shall take a brief look at the primary factors responsible for it.

The automobile is a key element in the origin and continuing growth of urban sprawl. Prior to the development of the car, cities were compact with homes, stores, workplaces, and schools within walking distance because walking was the only practical means of transport for everyday business. Residences were typically situated above the small stores, offices, and shops scattered throughout the city. Often, proprietors simply walked downstairs in the morning to open their businesses (Fig. 24-3). Cities were provided with parks which were heavily used (Fig. 24-4). Public transportation systems such as horse-drawn trolleys and, later, buses eventually found their way into cities, but they did not change the compact structure because people still needed to walk to the trolley or bus line. The small towns and villages surrounding cities—the suburbs—were compact for the same reasons. Between the cities and towns were farms and open country.

Automobiles entered the picture around the turn of the century, but widespread ownership was held in check initially by high costs, then by the Great Depression, and later by World War II. However, the abundance of jobs and the shortage of consumer goods during the war made it possible for people to accumulate considerable amounts of money. After the war, many of these people, as well as veterans, entered the housing and car market with a tremendous amount of purchasing power.

But cities were not necessarily where these people wanted to settle Despite today's ideals, walking and public transit did not make cities desirable places to live. Poor housing, inadequate sewage systems, inadequate refuse collection, pollution from home furnaces and industry, and generally congested conditions combined to make cities trying places to live. A decrease in services during the war aggravated these problems. Hence, the desire to escape the city and move to a home in the suburbs prevailed. The problem of

**FIGURE 24-3**
Before the advent of widespread use of cars, cities had an integrated structure. A wide variety of small stores and offices on ground floors with residences above placed everyday needs within walking distance. This structure still exists in certain areas of some cities. This is a location in North Baltimore. (Photo by author.)

**FIGURE 24-4**
Central Park, New York City, about 1900. Before the advent of the automobile; city parks were heavily used because there was not ready access to other open areas. (From the collection of the Library of Congress.)

lack of convenient transportation for commuting to work was eliminated by the availability of cars. Consequently, a move to the suburbs was now not only desirable, it was plausible.

Developers responded quickly to the new demand. They bought farms wherever they could, and put up houses, each with its own septic system. The government provided incentives by offering low-interest mortgages through the Veterans Administration and the Federal Housing

**FIGURE 24-5**
Scattered groups of homes shown here reflect which farms developers were able to buy. An overall plan which would include schools, shopping centers, parks, offices, and other facilities does not exist. (USDA–Soil Conservation Service photo.)

Administration. Interest payments on mortgages were made tax deductible. In addition, property taxes in the suburbs were much lower than in the city. These financial factors meant that, for the first time in history, making monthly payments on one's own home in the suburbs was actually cheaper than paying rent for equivalent or less living space in the city.

Motivated by these factors, people bought the new homes as fast as they could be put up. With cars, location was of little consequence as long as the commute to jobs and other facilities was reasonable.

It is important to note here that urban sprawl began without any overall plan. New developments were situated wherever builders could buy land (Fig. 24-5). Overall planning was not merely neglected, it was actually prevented by the fact that large cities were surrounded by a maze of more or less autonomous local jurisdictions (towns, townships, counties, municipalities). No governing body existed to come up with an overall plan, much less enforce it. But, like today's new developments, the swelling suburban populations of the 1940s and 1950s required new schools, roads, sewers, water systems, and other public facilities. Due to the lack of prior planning, local governments were thrown into a "catch-up" situation. To a large extent, such reactive planning still occurs today and in many cases it does not produce a logically consistent plan; it produces urban sprawl (Fig. 24-6).

Farmers were and still are caught in the middle of the process. A farmer may abhor the sprawling development, but when the time comes for him to sell his farm, it is logical for him to seek the highest price. And developers usually offer the best prices. Farmers who do not initially plan to sell are often forced to do so because development boosts land values and increases property taxes. These increased taxes often cut farm profits to untenable levels. The problems of the farmer were further aggravated by harassment from suburbanites who walked across their fields and molested their livestock.

Some farmers choose to hold on to their land, and in many instances, speculators have purchased farms and left them undeveloped, waiting for market prices to further increase. Again, you can see how this contributes to a "crazy quilt" pattern of urban sprawl.

One of the consequences of this growing suburban-exurban trend and the increasingly car-dependent population that goes with it is intolerably congested roadways and the need for new and improved highways. Note that these new highways not only serve urban sprawl, they perpetuate it. Time, not distance, is usually the limiting factor for commuters. Studies show that the average person is willing to spend about 20 to 40 minutes commuting to a job. If one has to walk, this time limit means living a maximum of 1 to 2 miles from work. But, if one travels on a high-speed expressway, this time limit means that the commuter can live 20 to 40 miles from work and

**FIGURE 24-6**

A large number of environmental problems and conflicts arising from lack of sufficient planning are evident in this area. Indicated on the photograph: (1) Storm runoff from parking lots causes severe flooding problems along the stream. (2) The stream is badly polluted by sewage overflows from the office complex. (3) Park versus highway conflict has kept this interstate a dead end for more than ten years. Highways block convenient routes between many areas. (Baltimore County, Department of Planning)

still get there in the same time. Consequently, new highways, which solve problems of local congestion in the short run, permit suburban-exurban growth at more distant locations in the long run. It is interesting to note that highway construction is largely financed by a tax on gasoline. The vicious cycle created by urban sprawl should be obvious (Fig. 24-7).

As mentioned earlier, urban sprawl originally began with housing developments. These developments were followed by shopping centers, then huge shopping malls, and more recently industrial parks, office complexes, and entertainment centers. At this point, you may ask why urban sprawl is a problem. It may not have been planned, but it does provide a better and more enjoyable lifestyle for many. Furthermore, the continuing construction of new homes and highways and the activity of automotive-related industry and business has come to represent a major portion of the country's economic activity. In the following sections, we shall examine the aspects of urban sprawl that are not as beneficial.

**FIGURE 24-7**
The gasoline tax, which is specifically designated to build new highways, creates a self-perpetuating cycle of increasing urban sprawl.

# ENVIRONMENTAL AND SOCIAL COSTS OF URBAN SPRAWL

The problems inherent in urban sprawl fall into two areas. One involves the environmental problems related to profligate development of land and consumption of resources. The second involves sociological problems stemming from the fact that suburban growth is largely the result of the migration of the affluent out of the city, while the poor and disadvantaged minorities remain in the city.

## Environmental Costs

### LOSS OF AGRICULTURAL LAND

It is frequently pointed out that about 90 percent of the world's population lives on only about 2 percent of the land surface. Therefore, the argument is made that any apparent shortage of land is simply a matter of human distribution rather than a shortage of land itself. While true in theory, this argument is highly misleading in that it implies that all land is of equal quality and that the skewed distribution of people is a matter of happenstance that could easily be changed. In

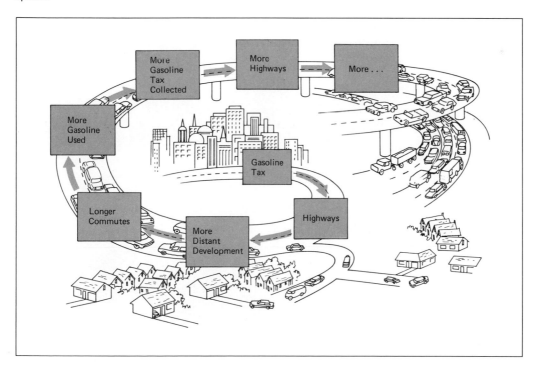

fact, all land is not equal in quality. Some 70 percent of the world's land is tundra, rugged mountains, severe deserts, or wetlands, all of which are extremely unsuitable for human use. Another 19 percent of the land is considered marginal; it is excessively dry, wet, hilly, or cold, but given suitable inputs, it can be used. Only 11 percent has the climate, relatively level terrain, and good soil that make it naturally suitable for cultivation. Not surprisingly, most humans find the same 11 percent the most desirable place to live.

More importantly, food has always been the most basic resource so it was natural to locate cities in agricultural regions near the heart of supply. Seaport cities are usually at the mouths of rivers where delta deposits of the deepest, richest soils frequently occur. Highways generally follow the "easy" terrain of river basins, which also generally contain the best agricultural soils. Then, urbanization tends to follow highways. Thus, the 1 million hectares (2.5 million acres) that undergo development in the United States each year are not just "any old land"; they are largely the best agricultural lands (Fig. 24-8). Consequently, there is a serious and growing conflict between urbanization and the loss of good agricultural land.

The loss of 1 million hectares of agricultural land per year in the United States is partially offset by the annual development of some 0.5 mil-lion hectares of new agricultural land. But, as mentioned previously, this takes place at the expense of natural ecosystems, the biota they support, and the natural services they perform. Then, the land is frequently of only marginal agricultural quality. Development may require extensive water diversion projects for irrigation or drainage, or forests on hillsides may have to be cleared. Such projects have a further impact upon the natural environment. Also, this land is more difficult to maintain because it is prone to erosion, salinization, or waterlogging. Finally, the amount of new land that is potentially arable is limited, about 30 million hectares in the United States. It is important to note that one of the compensations for the loss of agricultural land is that, due to better technology and farming methods, productivity on remaining parcels has increased. Still, increasing production per hectare cannot compensate for the loss of land beyond a certain point.

Yet, the United States is abundantly endowed with agricultural land in proportion to its population. We are not about to squeeze ourselves into a famine. However, many other coun-

**FIGURE 24-8**
Land that goes into urban development is generally prime agricultural land. The same is true around all major American cities. (USDA–Soil Conservation Service photo.)

tries are not so fortunate and these nations are likely to become increasingly dependent on agricultural exports from the United States. Thus, agricultural land must be seen as a world resource. Squandering it for any purpose is unconscionable.

## DEGRADATION AND DEPLETION OF WATER RESOURCES

Although the effects of development on the water cycle were discussed in detail in Chapter 8, a brief summary here is pertinent.

By more or less covering the ground with hard surfaces, development increases surface runoff and decreases infiltration. The increased runoff carries surface pollutants and debris into waterways, increases the frequency of flooding, and increases stream bank erosion, which undercuts trees. This makes channels wider and shallower and gradually destroys the stream valleys.

Decreased infiltration plus the exorbitant quantities of water used by suburbanites for lawns, gardens, washing cars, and so on, may seriously deplete groundwater supplies. In turn, wells and springs go dry and in coastal regions there is danger of saltwater encroachment and land subsidence. Dry springs mean dry streams and riverbeds between rains and, hence, destruction of aquatic life. In short, there is a serious conflict between spreading urbanization and maintenance of the quantity and quality of water resources, including stream and river valleys that provide aesthetic and recreational pleasure along with practical benefits.

Also, as agriculture is pushed to drier regions, the demand for water for irrigation becomes a significant factor in the depletion of water resources. As discussed in Chapter 8, depletion of water resources is likely to be the most critical factor in limiting both agriculture and development in years ahead.

## SACRIFICE OF RECREATIONAL AND SCENIC AREAS FOR NEW HIGHWAYS

The increasing commuting demands of suburban-exurban growth create a perpetual demand for new and improved highways. The government must buy the land for the highway and, in a populated metropolitan area, this is often the most expensive aspect of highway construction. Therefore, new highways are frequently directed through parks or along stream valleys where the government already owns land or where relatively few people will be displaced. This results in the sacrifice of aesthetic, recreational, and wildlife values in the metropolitan areas where they are so important. Highway planners argue that a highway through a relatively large park will take a relatively small portion of the total area. If a park is to provide humans with a measure of peace and tranquility, it is dubious that two halves split by a pollution- and noise-generating highway still add up to the whole. Certainly wildlife must find their two halves incomplete if they are blocked from their source of drinking water, as frequently happens when highways are placed along rivers or stream valleys.

In addition, highway construction is responsible for more erosion and sedimentation of waterways than any other form of building. After completion, highways are responsible for more runoff, decreased infiltration, and other water resource problems than all other forms of development.

## CONSUMPTION OF ENERGY (CRUDE OIL) RESOURCES

The suburban lifestyle directly translates into an increasing demand for energy resources, especially crude oil, which is needed to fuel motor vehicles. Likewise, individual suburban homes require 1.5 to 2 times more energy for heating and cooling than comparable attached city dwellings. The tremendous energy appetite of the suburban-exurban lifestyle largely precipitated the energy problem (if not crisis) and all of its related environmental, social, and economic problems. A more energy conservative lifestyle could mitigate the depletion of crude oil reserves and reduce the environmental impacts of power production via coal and nuclear energy.

## THE AGGRAVATION OF POLLUTION

In Chapter 14 we noted that most air pollution results from the burning of fuels. Thus, the energy appetite of the suburban-exurban lifestyle is also responsible for a large portion of air pollution problems. Pollution from commuting cars is obvious. Another major factor is the electrical demands of modern suburban homes, which have increased the need for coal-burning power plants that not only generate local air pollution but are largely responsible for acid rain.

Pesticides, weed killers, and fertilizers, which are often overused or misused on suburban lawns and gardens, contribute significantly to water pollution problems.

## Social Costs—Decline of Central Cities

Urban sprawl exacts social as well as environmental costs. The social costs stem from the fact that urban sprawl is largely a phenomenon of **exurban migration.** Exurban migration begins with the movement of people from the central city to the suburbs. As this occurs, people in older suburban neighborhood move outward to exurbs. This leapfrog migration creates an expanding "donut" of urban sprawl around the central city and profoundly affects the central city.

### EXURBAN MIGRATION AND ECONOMIC AND RACIAL SEGREGATION

Historically, cities included a wide economic and cultural variety of people; however, with exurban migration, significant economic and racial

segregation occurred. Only the relatively affluent could afford cars and down payments on homes. Therefore, movement to the suburbs primarily involved middle- and upper-income people. For the most part the poor, the elderly, and the handicapped who could not drive remained in the cities.

Discrimination added to the problem. Generations of discrimination in education and jobs made blacks, Hispanics, and other minorities and ethnic groups disproportionately poor. In addition, banks frequently practiced discriminatory lending practices. When money was not a factor, discriminatory real estate practices often kept minorities out of new developments. Civil rights laws enacted in the 1960s made such overt discrimination illegal, yet underlying prejudices persist. As a result, urban sprawl continues to reflect a **gentrification** of society—a segmentation in which communities are composed of people with similar economic, social, and cultural backgrounds (Fig. 24-9a). By and large, new developments continue to be occupied by the economi-

**FIGURE 24-9**
(a) Major migrations from rural areas to metropolitan areas and from cities to suburbs have also tended to segregate people along economic and racial lines. Such separation is called gentrification.

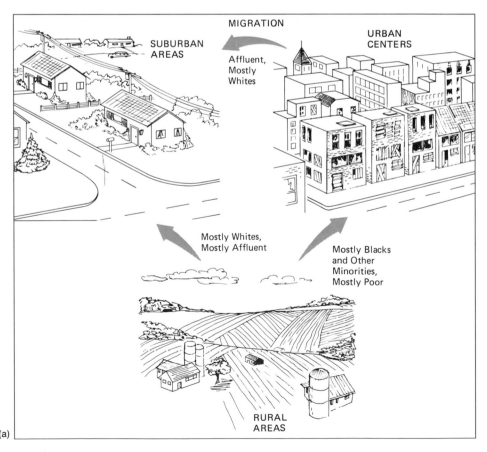

SUBURBAN AREAS

MIGRATION

URBAN CENTERS

Affluent, Mostly Whites

Mostly Whites, Mostly Affluent

Mostly Blacks and Other Minorities, Mostly Poor

RURAL AREAS

(a)

(b)

(c)

**FIGURE 24-9** (cont.)
(b) an area of suburbia and (c) an area of inner-city Baltimore contrast the extremes. (Photos by author.)

cally advantaged while older areas are left to the poor (Fig. 24-9b and c). Thus, the areas that comprised the original central cities now have majority populations of poor and economically de-pressed blacks, Hispanics, and other minorities.

A considerable migration of people from rural areas to metropolitan areas has also occurred. More recently, there has been a tremendous migration from the Northeast to the southern "sunbelt" states. The same gentrification pattern emerges when these people settle in new areas.

## DECLINING TAX BASE AND DETERIORATION OF PUBLIC SERVICES

The migration of people to the suburbs seemed harmless enough at first, but it created a downward spiral of urban decay, or blight, which remains today as one of society's most vexing problems. To understand this downward spiral, we must understand several points concerning local government. First, the local government (usually city or county, though the particular entities vary from state to state) is responsible for providing

- Schools
- Local roads
- Police and fire protection
- Public water and sewer systems
- Refuse collection
- Delivery of welfare services

Close to 90 percent of the revenue for these services is raised from property taxes. Each year the local government levies a tax on all real estate which is proportional to the value of the property. Thus, if real estate values increase or decrease, the property tax is adjusted accordingly. The final, and perhaps most important, point is that,

in most cases, the central city is a governmental jurisdiction separate from the surrounding suburbs.

As affluent people move to the suburbs, suburban property values escalate and suburban jurisdictions enjoy increasing amounts of tax revenue. However, in the central city, demand for housing drops and prices of homes decrease until the less affluent can afford them. The decreased values mean less tax revenue for the city, a situation which is often referred to as an eroding or declining **tax base.** In extreme cases, properties do not sell at all; they are abandoned and the taxes left unpaid. In these situations, the government takes ownership of the property. In the end, these properties, which were once a source of tax revenue, become a government liability (Fig. 24-10).

As a result of an eroding tax base, city governments are forced to cut the quality and quantity of services provided and/or to increase the tax rate. Hence, the property taxes on a home in the city are often two to three times greater than on a comparably priced home in the suburbs, while schools, trash collection, and other services are often not as good.

The services that tend to suffer most are maintenance of roads and bridges and water and

**FIGURE 24-10**
Growing effects of exurban migration. While suburban areas receive the economic benefits of an active home construction industry, cities receive the burden of dealing with the problems of abandoned buildings. (EPA-Documerica photo; courtesy of the U.S. Environmental Protection Agency.)

sewer systems because the money available goes to more politically sensitive areas. These systems, which are referred to as the city's infrastructure, are approaching a state of chronic breakdown in many older cities.

### SEGMENTATION AND DESOCIALIZATION CAUSED BY NEW HIGHWAYS

Many people who move to the suburbs maintain jobs in the city so each day the city is swamped in a tide of commuter traffic. To accommodate this traffic, new highways are constructed. A city's parks and its poorer areas are usually seen as the least expensive rights of ways. Untold numbers of parks have been destroyed and millions of city residents have been crudely displaced to make way for new highways. Even when highways do not displace people, they frequently act as a barrier between older neighborhoods and the locations where people work and shop. Many cities have become more devoted to moving and parking commuter cars than to serving their own residents. In some cities as much as two-thirds of the land area is deveoted to highways, parking lots, service stations, and other car-related facilities (Fig. 24-11). Using land for road-

(a)

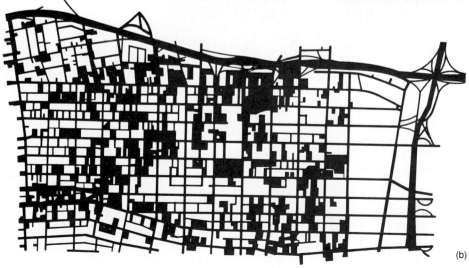

(b)

**FIGURE 24-11**
A large portion of the land area of many American cities has been used to accommodate automobiles rather than people. (a) Dallas, Texas. (EPA-Documerica, Bob W. Smith; courtesy of the U.S. Environmental Protection Agency.) (b) Map is an area of downtown Los Angeles in which about two-thirds of the land, indicated in black, is used for automobiles. (By permission from the Victor Gruen Center for Environmental Planning.)

ways further subtracts from the tax base, but it also segments and depersonalizes the area and further erodes the environment with noise and pollution. Recall that a high percentage of inner-city children have elevated levels of lead in their blood as a result of their exposure to the exhausts of leaded gasoline.

### VICIOUS CYCLE OF EXURBAN MIGRATION AND URBAN DECAY

It should be apparent now that exurban migration and urban decay are tied together in a vicious cycle. Eroding tax bases lead to declining schools and other public services and/or sharply increased tax rates aimed at maintaining the services. In either case, more and more people decide to leave the city for better conditions in the suburbs. This leads to further erosion of the tax base.

Importantly, it is not just people who leave. As the affluent consumers move, stores, restaurants, theaters, and other commercial establishments move with them or go out of business. Recently, manufacturing facilities and offices that were typically located in the cities have also moved to new suburban and exurban **industrial parks** (Fig 24-12). The migration of business not only represents a further loss of tax revenues for the city, it represents a serious loss of employment opportunities for city residents. The poor do not generally own reliable cars, and public transit systems are often inadequate. Hence, inner-city residents can seldom take advantage of job opportunities in the suburbs. As a result, unemployment in cities has become extremely serious. The unemployment rate among young urban blacks has risen and remained near 50 percent in recent years.

Such high levels of unemployment are not only connected with financial problems; they breed contempt for schools and education and tend to promote drug use, crime, and other forms of asocial and antisocial behavior. Welfare and other social services aimed at combating these problems place greater demands on tax revenues and further aggravate the declining tax base. These factors can force even the most fervent city dwellers to leave if they have the means to do so. Thus, the downward spiral continues (Fig. 24-13).

In recent years a conspicuous rebirth of the core areas of many cities has occurred. This rebirth includes new shops, office buildings, and residences and some affluent people are moving back into the city (Fig. 24-14). As encouraging as this may sound, however, it does not mean the problem is solved. As the affluent move back into redeveloped core areas, the poor are displaced into areas between the new core and the suburbs. Statistics show a significant number of blacks are moving from the city into older suburbs while

**FIGURE 24-12**
One of many plants in an industrial park being developed in the rural countryside north of Baltimore, Maryland. Such industrial parks are the latest but accelerating trend in exurban migration. By transferring employment opportunities from urban to exurban areas, they will cause further hardships for city dwellers. They will also promote the sacrifice of huge amounts of additional surrounding land for residential and commercial uses.
(Photo by author.)

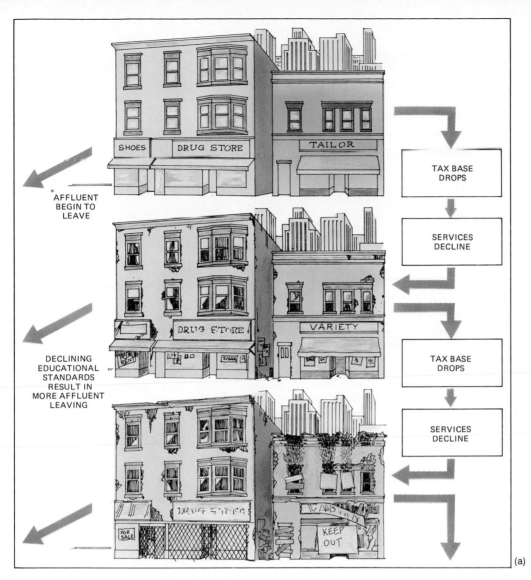

AFFLUENT
BEGIN TO
LEAVE

DECLINING
EDUCATIONAL
STANDARDS
RESULT IN
MORE AFFLUENT
LEAVING

SHOES    DRUG STORE    TAILOR

DRUG STORE    VARIETY

FOR RENT    SPECIAL

DRUG STORE

FOR SALE    KEEP OUT

TAX BASE
DROPS

SERVICES
DECLINE

TAX BASE
DROPS

SERVICES
DECLINE

(a)

(b)

**FIGURE 24-13**
(a) Migration of affluent people
from cities to suburbs creates a
vicious cycle leading to further
urban decay. (b) Boarded-up
shops in an inner-city area, the
result of exurban migration.
(Photo by author.)

**FIGURE 24-14**
In recent years there has been a conspicuous rehabilitation of the core areas of many cities. Shown here is the new inner harbor area of Baltimore, Maryland. Left foreground: harbor. Green-roofed building: a complex of small variety shops and eating places surrounded by a brick promenade, small amphitheater, park, and fountain. White building at left: McCormick Spice factory, (not new) with new and renovated housing to the back and left. Center: new hotels and convention center. Right: new office buildings. Note elevated pedestrian walkway from shopping center to hotels. But the basic problems remain. Huge areas affected by urban blight are present in the background.

whites move still further out. Hence, erosion of the tax base and urban decay continue despite the rebirth and infusion of revenues from the new "downtowns."

## REVERSING THE TREND OF SUBURBAN SPRAWL AND URBAN DECAY

Since both the sprawl of the suburbs and the decay of cities result largely from the urban exodus, it follows that the problems of both may be solved if the migratory trend could be reversed and growth and development could be redirected into cities rather than further into suburbs and exurbs.

### Cities Can Be Beautiful

As a first step, we need to remember that, while most visions of the city conjure up images of decay, cities can in fact be beautiful aesthetically, socially, and culturally. Many examples can be seen in Europe and in some places in the United States (Fig. 24-15). Famous architect-planner Victor Gruen provides us with an eloquent description of the positive aspects of the city.

The city is the sum total of countless features and places, of nooks and crannies, of vast spaces and intimate spots, an admixture of the public and private domain, of rooms for work and rooms for living, of rooms for trade, where

(a)

**FIGURE 24-15**
Various aspects of a "real" city.

(b)

**FIGURE 24-15** *(cont.)*

(c)

(d)

STREET SCENE IN PARIS

STREET SCENE IN NEW YORK

**FIGURE 24-16**
"A Humiliating Contrast." Historically, the cities of the United States, growing rapidly, neglected the creation of urban quality in their early stages. *Harper's Weekly*, in 1881, criticized New York by printing two contrasting drawings with this original caption.

money and wares change hands, and rooms where music and drama lift the soul, of churches and night spots, of landmarks expressing the spirit of the community and homes for the comfort of the individual.

It is the fountains and flower beds, the trees shading streets and boulevards, the sculptures and monuments, the rest benches placed in thousands of spots. The city is the little merchants who make their living on the streets, the vendors of balloons and pretzels, of newspapers, chestnuts, ice cream, flowers, lottery tickets and souvenirs. And the city, of course, is also the buildings. . . . the relationship of these buildings to each other and, most important of all, to the spaces created between them.

The city is the countless cafes . . . where a person with little money may spend hours over a cup of coffee and a newspaper. . . .

The city is the crowded sidewalks, the covered galleries in Italy, the arcades and colonnades and the people on them and in them, some bustling, some walking for pleasure. . . . The city is the

parks: the tiny green spots with benches, the middle-sized ones that interrupt rhythmically the sea of stone and brick, the big ones that act as lungs and recreational places. . . .

The city is the community of soul and spirit rising from an audience in a theater or at the opera or from those attending services in a church. The city is the daily chat (with the butcher, the newsman, the baker, and numerous other merchants with whom one shops).

A real city is full of life, with ever-changing moods and patterns: the morning mood, the bustling day, the softness of evening and the mysteries of night, the city on workdays so different from the city on Sundays and holidays.*

---

*The Heart of Our Cities; The Urban Crisis: Diagnosis and Cure.* Copyright © 1964 by Victor Gruen. Reprinted by permission of Simon & Schuster, a Division of Gulf & Western Corporation.

Unfortunately, most American cities have not achieved the status of "real cities" in the sense described by Gruen. Driven by rapidly expanding commerce and industrialization, they have failed to provide the social amenities that make cities pleasant places to live (Figs. 24-16 and 24-17).

Public housing projects as shown in Figure 24-18 and urban renewal projects as shown in Figure 24-19 were constructed during the 1960s and 1970s in an attempt to alleviate growing urban blight, but these projects did not recognize or provide for the complex social interactions that de-

(a)

**FIGURE 24-17**
The inhumanity of many urban areas persists. (Photos by author.)

(b)

**FIGURE 24-18**
A public housing development. Clean, efficient, but impersonal. Lacking are the small stores and places that form the interconnections of urban ecosystems. (Photo by author.)

**FIGURE 24-19**
Urban renewal has replaced the core of many cities with marvelous architecture, expansive stone plazas, fountains, trees in great planters—but where are the people? (Photo by Tommy Noonan, Washington, D.C.)

velopment must encourage to be successful. Hence, decline continued despite such projects.

In recent years, however, the significance of architecture in fostering social interactions has been recognized and efforts have been made to redevelop downtown areas with a mixture of small shops, restaurants, and entertainment and living places, and this has attracted some people back to the city; but whether this trend expands depends on continuing interest and allocation of resources (Fig. 24-20). Also, whether this redevelopment benefits the poor or further displaces them depends on the establishment of businesses that will provide jobs.

## Clustered versus Detached Housing

The negative environmental effects of urban sprawl could be reduced not only by focusing more attention on redevelopment of cities, but by greater use of **clustered development.**

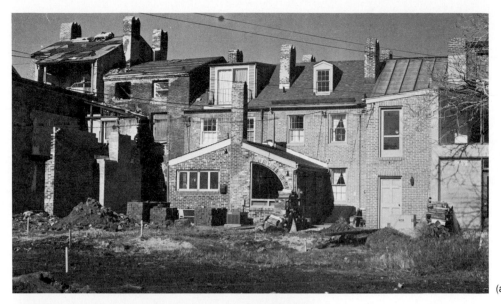
(a)

(b)

**FIGURE 24-20**
"Dollar homes" being renovated in Baltimore, Maryland.
(a) Note the still-abandoned structures on the left which are typical of the "before" condition. In the center are homes nearing completion of renovation. (b) Front side of homes.
(Photos by author.)

Two distinct approaches to housing may be used on a given tract of land. The first and most common approach is to divide the tract into lots and construct individual homes or detached housing. The second approach involves the same number of homes, but constructs them as a group of townhomes on a smaller tract of land (Fig. 24-21). The clustered arrangement provides distinctive, attractive homes, but leaves much of the land in its natural state or available for recreational and other uses. Compared to individual housing, clustered developments greatly reduce negative environmental impacts both during and after construction. Consider the following:

1. Clustered units are more efficient in their use of energy for heating and cooling.
2. Clustered developments can be efficiently serviced by public transportation because each cluster makes a single convenient stop.
3. Services such as trash collection are similarly more efficient.
4. Greater efficiency of transportation lessens air pollution.
5. Provision of water, sewer, and other utilities is more efficient.
6. Greater open space preserves water resources in terms of quality and quantity of runoff.

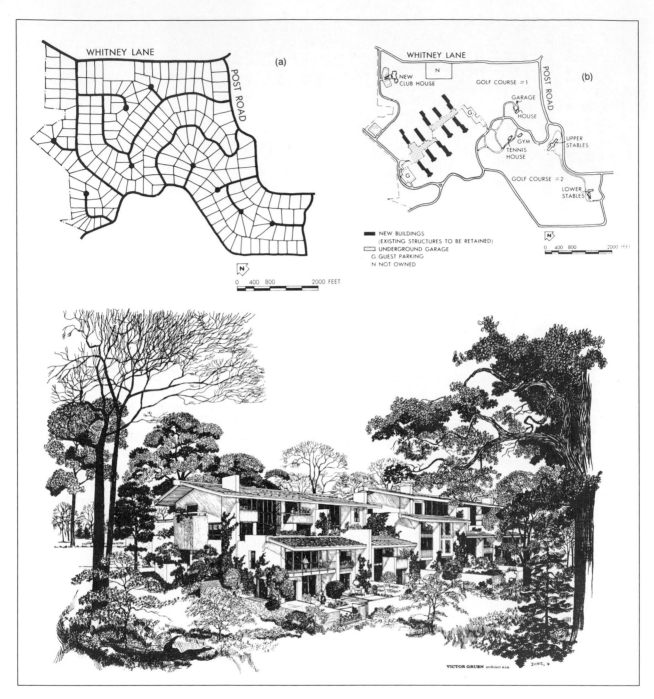

**FIGURE 24-21**
Alternative plans for developing the same tract of land. (a) Typical subdivision
into half-acre lots. (b) The same number of units are clustered into groups of
attached dwellings. The rest of the land is retained for open space and other
amenities, and is cooperatively owned, used, and maintained by the residents.
(c) Attached dwellings planned for clustered development. (By permission from the
Victor Gruen Center for Environmental Planning.)

7. Greater open space can provide for recreation and garden plots, or can be put into forests.

8. Woods left growing near clusters could provide wood to supplement energy demands.

A study of the Baltimore metropolitan region determined that a more centralized pattern of housing in the area, including increased clustering of houses in the suburbs as opposed to the past pattern of separate houses spread out on lots, would

1. Reduce the burden of stormwater pollution from development by about one-half.

2. Reduce septic tank failures by 98 percent.

3. Reduce space heating requirements by 25 percent.

4. Reduce motor vehicle fuel consumption by 17 percent.

5. Reduce air pollution from space heating by 24 percent.

6. Reduce air pollution from motor vehicles by 10 to 18 percent.

7. Reduce farmland losses by 76 percent.

8. Reduce forest losses by 65 percent.

9. Reduce transportation costs for solid waste collection and haulings.

10. Increase public transit patronage.

11. Reduce highway congestion for most locations in the region.

12. Reduce school pupil transportation costs by 38 percent.

13. Provide more affordable housing.

## Land Use Regulations and Environmental Impacts

The negative environmental impacts of urban sprawl have become critical in many regions, and localities throughout the nation are studying, enacting, and enforcing land use regulations. A summary of the types of restrictions is shown in Table 24-1.

Despite the merits and absolute necessity of such regulations, opposition to them is often tremendous. Many times, these regulations are blocked or amended to the point where they are essentially meaningless. This is why urban sprawl generally still occurs in an uncontrolled fashion. In most cases, the highest bidder decides what use land will be put to. Clearly, society would benefit from a better awareness and understanding of how environmental problems are largely the cumulative product of our individual choices about where to live and how to live.

The environmental impacts of development came to national attention in the National Environmental Policy Act of 1969 (NEPA). An important provision of this act is the requirement that **environmental impact statements** be submitted for proposed projects like dams and highways that receive federal funding. The environmental impact statement addresses the following:

1. The need for the project

2. The probable environmental impact (positive, negative, direct, and indirect)

3. The effects of alternatives including cancellation of the project

4. The short-term and long-term impacts

NEPA does not contain the "teeth" to prevent construction of a project regardless of its negative environmental effects. However, the act requires that the statements be open to public scrutiny, and many citizens' groups have been spurred to oppose plans. Actions by citizens have resulted in the modification or cancellation of numerous dams, highways, and airports.

Though NEPA applies only to large projects receiving federal funding, the concept of NEPA can be broadly applied. By 1983, about 32 states had adopted laws or executive orders requiring environmental impact statements for state projects. Unfortunately, this does not affect a significant amount of the development associated with urban sprawl. Again, a greater understanding by society of the problems tied to urban sprawl may be the only way to improve the situation. More than anything else, land use will determine our future quality of life. As individuals or as groups, we must see to it that land is used in ways that will protect the future of our natural resources. We are beyond the point where we can continue to treat land as a commodity to be sliced up and passed out to the highest bidder for whatever use the buyer chooses. We cannot afford to let short-

| Table 24-1 | Methods Which Local or State Governments May Use to Control Land Use | | |
|---|---|---|---|
| KIND OF LEGISLATION | ACTIONS ENABLED UNDER LEGISLATION | EFFECTIVENESS IN CONTROLLING LAND USE |
| Zoning ordinances | Allow local governments to specify the kind of development-residential, commercial, industrial-and density-units per acre-that may occur in given areas. | Used to prevent "undesirable" kinds of development from occurring in given areas. Not used to prevent development as such. Areas remain subject to rezoning and granting of exceptions. |
| Adequate facilities ordinances | Allow local governments to withhold building permits until adequate water, sewer, and other services are in place. | Enables control of timing and direction of development. Cannot be used to prohibit development since it is deemed that governments are obligated to provide services. |
| Districting or easements | Governments may define agricultural or other such districts. Farms in agricultural districts cannot be subdivided. | Used to maintain the viability of agriculture or other such activity of an area. Districting frequently opposed because it limits free choice and blocks what is deemed as natural change. |
| Preferential taxation | Governments may levy taxes on land at different rates depending on its use. Undeveloped land is taxed at a lower rate than residential or industrial land. | Provides some incentive to owners not to develop land, but this is often outweighed by potential profits derived from development. Does not prevent scattered development. |
| Right of purchase to protect critical resources | Governments are considered to have the right to purchase land to protect critical resources as well as for parks, highways, and other uses but owners must receive just compensation. | Most effective but amount of land that governments can save by purchase is very limited by lack of funds. |
| Purchase of development rights | Governments may purchase landowner's right to develop. | Less costly than purchase of land but still limited by lack of funds. Landowner cannot be forced to sell rights. |

term narrow interests determine land use and hence the fate of the natural resources that support the biosphere.

In our efforts to preserve the land, we cannot forget that population growth is integrally connected to resource consumption. Pressures on land for living space and resources will mount and eventually reach a breaking point if population growth is allowed to continue. Thus, in the end, our ability to control population and to properly steward the land will be the ultimate factor that determines the sustainability of the earth.

# EPILOGUE

In the introduction, I stated that humankind might have a bright or dismal future depending upon which of the many trends and counter-trends prevail. The specific nature of these trends, such as soil degradation versus conservation, energy depletion versus conservation and development of new sources, overpopulation versus population control, has undoubtedly become more clear to you. As the seriousness and gravity of the situation have come into sharper focus, the ultimate question of whether humankind will succeed in the long run has probably become more acute. In preparing this book, it has become clearer to me.

I am often asked whether I am an optimist or a pessimist about environmental matters. My answer is, neither. Both pessimism and optimism can be self-defeating. Pessimism can be self-defeating because believing that the case is hopeless leads one to make no effort to save the situation. Likewise, undue optimism can be self-defeating because believing that everything will be all right says that one's efforts are not necessary and again leads to inaction.

A particularly dangerous form of optimism has been called "millennium fever." It is expressed by statements that we have plenty of resources, water, energy, or whatever, to last until the year 2000. Millennium fever seems to assume that in the year 2000 we will step off into a golden age where all problems will be solved—the millennium will have arrived and everything will be beautiful. But the year 2000 is now less than 15 years away. We are no longer talking about distant generations and as-yet-undreamed-of technology. Most of us who are alive today will enter the year 2000. In reality, that day will be just like today that followed yesterday. All the unsolved problems and consequences of shortcomings will still be with us, only much more acute, if we fail to take corrective actions between now and then.

What is necessary, then, is neither optimism nor pessimism, but a forthright willingness to face problems boldly and make every possible effort to bring about the necessary changes.

Besides the question concerning optimism and pessimism, I am also asked, "How much time do we have left?" or "Isn't it already too late?" Such questions seem to carry the misimpression that the biosphere is like a soap bubble that will keep its beautiful form until, without warning, it will suddenly burst and disappear all at once. In fact, environmental degradation occurs relatively slowly, bit by bit, first one place and then another. Yes, it is too late to save the passenger pigeon and other things that have already been lost, and, as I have tried to show throughout this book, if we continue on the same course of environmental exploitation we shall, step by step, lose more and more. But at any time it is not too late to change the course of our actions and save what does remain. Furthermore, nature has great regenerative capacity. By understanding natural processes and making suitable corrections in our actions, much of the damage done by past abuse may be repaired. Again we must face problems boldly and make every effort to bring about necessary change.

But then, I am asked, what can one individual do when the problems are so enormous and the forces with vested interest in the *status quo* seem so entrenched? There are many things that you as an individual can do. First, it is important to recognize that you are by no means alone. There are hundreds of thousands of individuals, both professional and nonprofessional, who recognize the problems and who are working toward solutions. Further, hundreds of environmental organizations exist. Through organizations, members give each other mutual support. Even more important, organizations provide the structure by which talents and financial resources can be focused effectively on particular issues. For example, many national organizations (see Appendix A) use membership dues to support professional lobbyists and lawyers who work to promote the passage, strengthening, and enforcing of environmental legislation. Such organizations have been largely responsible for the environmental laws and regulations that we now have. With additional members the force of such organizations could be much greater. Without members, such organizations would fold, the legislative pressure would be lost, and environmental gains would be rapidly reversed by counterforces. Thus joining one or more such organizations can be a very significant step.

Organizations are also crucial in the dissemination of information. Publications and/or direct contacts keep members informed on the status of particular issues and enable efforts to be concentrated at the needed place at the proper time. For example, when particular environmental legislation is pending, it is important for individuals to express their opinions to their elected representatives.

The same process holds true on state and local levels. Most cities have environmental organizations through which the individual may join with others concerned about local and regional issues. Here, opportunities to interact directly with government officials through public hearings, public advisory committees, and so on are even more abundant than at the national level.

Candidates for public office frequently have contrasting views regarding environmental issues. Also, environmental issues may come to a direct vote in the form of propositions or a referendum. Consequently, the influence of the individual as a voter should not be underestimated.

Finally, the ways of altering one's habits toward an environmentally conscious lifestyle are more numerous and more significant than generally recognized. In addition to commonly mentioned steps toward conservation such as driving smaller cars, turning down heat, and turning off lights, one can apply principles of biological pest control, soil conservation, and fertilization to plantings, lawns, and gardens. The number of children a couple has is a private decision. In the use of resources, it is only individual buying and use habits that can influence the balance between cheap disposable items and durable ones that can be serviced or recycled. In the final analysis, individual decisions determine whether an item will be discarded or serviced and used again. Individual choices on where to live have great impact on land use and consequent transportation and energy demands. Individual investments in solar energy will largely determine the rate at which solar energy comes into use. These avenues of action are summarized in Figure E-1.

I recognize that one's own efforts and actions in these areas seem virtually useless when the behavior of most of one's neighbors is unchanged. However, one is not nearly as alone as it sometimes appears. There are many people in the nation and the world who feel and do likewise. Then too, large, rapid changes are as disruptive to the human social system as they are to ecosystems. Gradual change leaves time for making adjustments all through the system. Gradual changes must start with relatively few individuals and then spread. Ultimately it comes down to looking at oneself and asking, "Have I been part of the problem or part of the solution?"

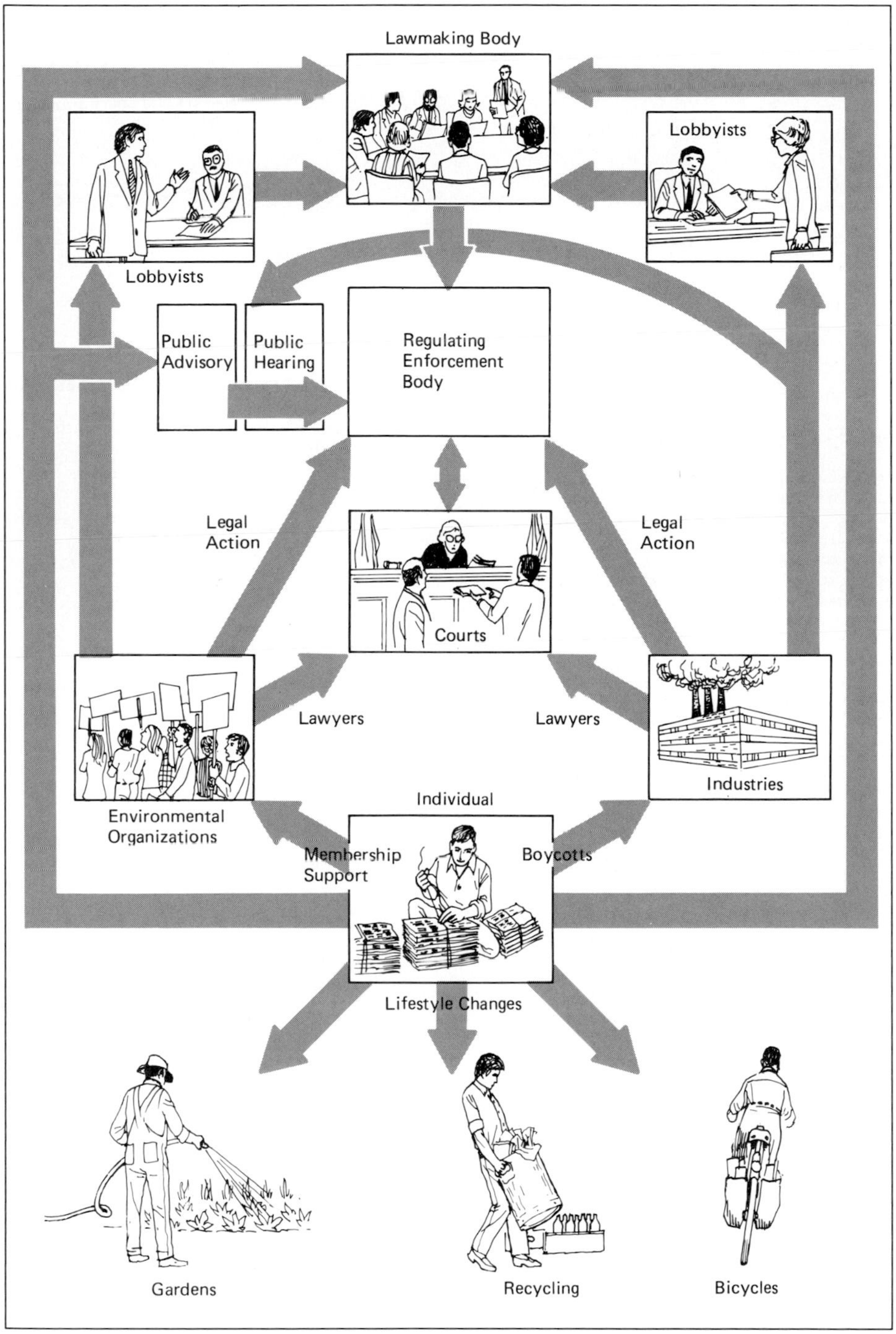

Avenues of environmental action.

Lawmaking Body

Lobbyists

Lobbyists

Public Advisory

Public Hearing

Regulating Enforcement Body

Legal Action

Legal Action

Courts

Lawyers

Lawyers

Environmental Organizations

Industries

Individual

Membership Support

Boycotts

Lifestyle Changes

Gardens

Recycling

Bicycles

# BIBLIOGRAPHY

## GENERAL REFERENCES

**Allen, John, ed.** *Annual Editions: Environment 84/85.* Guilford, Conn.: Dushkin Publishing Group, Inc., 1984. An anthology of key papers in environmental science; updated yearly.

**Billings, W.D.** *Plants, Man and the Ecosystem* 2nd ed. Belmont, Cal.: Wadsworth Publishing Co., 1970.

**Brown, Lester R., et al.** *State of the World.* New York: W.W. Norton and Company, 1985. A yearly report from Worldwatch Institute.

**Brown, Lester R., Patricia L. McGrath, and Bruce Stokes.** "Twenty-Two Dimensions of the Population Problem," *Worldwatch Paper 5.* Washington, D.C.: Worldwatch Institute, March 1976.

**Conservation Foundation.** *State of the Environment,* Annual, Washington D.C.: Conservation Foundation. An annual publication.

**Eckholm, Erik P.** *Down to Earth.* New York: W.W. Norton and Co., 1982.

**Goudie, Andrew.** *The Human Impact.* Cambridge, Mass.: The M.I.T. Press, 1982.

**Kormondy, E.J.** *Concepts of Ecology* 3rd ed. Englewood Cliffs, N.J.: Prentice-Hall, Inc., 1984.

**Luoma, Samuel N.** *Introduction to Environmental Issues.* New York: Macmillan Publishing Co., 1984.

**Miller, G. Tyler.** *Living in the Environment,* 4th ed. Belmont, Cal.: Wadsworth Publishing Co., 1985.

**Moran, Joseph M., M.D. Morgan, and J.H. Wiersma.** *Introduction to Environmental Science,* 2nd ed. New York: W.H. Freeman and Company, 1986.

**National Wildlife Federation.** "The Annual Environmental Quality Index," *National Wildlife* (February issue of each year). An annual assessment of various environmental quality indicators.

**Odum, E.P.** *Fundamentals of Ecology* 3rd ed. Philadelphia: W.B. Saunders Company, 1971.

**ReVelle, Penelope, and C. ReVelle.** *The Environment: Issues and Choices for Society* 2nd ed. Boston: PWS Publishers, 1984.

**Smith, R.L.** *Ecology and Field Biology* 3rd ed. New York: Harper and Row, 1980.

**Southwick, Charles H., ed.** *Global Ecology.* Sunderland, Mass.: Sinaver Associates, Inc., 1985.

**U.S. Executive Office, Council on Environmental Quality.** *Annual Environmental Quality Report.* Washington, D.C.: U.S. Government Printing Office.

**World Resources Institute and International Institute for Environment and Development.** *World Resources 1986.* New York: Basic Books, Inc., 1986.

## CHAPTER REFERENCES

### Introduction

**Baker, Jeffrey J.W., and Garland E. Allen.** *Hypothesis, Prediction, and Implication in Biology.* Reading, Mass.: Addison-Wesley Publishing Co., 1968.

**Brady, Donald.** *Logic of the Scientific Method.* New York: Irvington Publishers, 1973.

**Gieve, Ronald N.** *Understanding Scientific Reasoning* 2nd ed. New York: Holt, Rinehart & Winston, Inc., 1984.

**McCain, Garvin, and Erwin M. Segal.** *The Game of Science.* 4th ed. Florence, Ky.: Brooks-Coal Publishing Co., 1981.

**Lastrucci, Carlo.** *Scientific Approach.* Cambridge, Mass.: Schenkman Bks., Inc., 1967.

**Walker, M.** *The Nature of Scientific Thought.* Englewood Cliffs, N.J.: Spectrum Book/Prentice-Hall, Inc., 1963.

### Chapter 1

**Cloud, Preston.** "The Biosphere," *Scientific American,* 249 (September 1983), pp. 176–189.

**Cloudsley-Thompson, J.L.** *Terrestrial Environments.* New York: Halsted Press, 1975.

**Curry-Lindahl, K.** *Wildlife of the Prairies and Plains.* New York: Harry N. Abrams, Inc., 1981.

**Hughes, Carol and David.** "Teeming Life of a Rain Forest," *National Geographic,* 163 (January 1983), pp. 49–64.

**Hutchinson, G. Evelyn.** "The Biosphere" In *The Biosphere,* A Scientific American Book, pp. 1–11. San Francisco: W.H. Freeman and Company, 1970.

**Margulis, Lynn, et al.** "Microbial Communities," *BioScience,* 36 (March 1986), pp. 160–170.

**Pimm, S.L.** "Properties of Food Webs," *Ecology,* 61 (1980), pp. 219–225.

**Sutton, A., and M. Sutton.** *Wildlife of the Forests.* New York: Harry N. Abrams, Inc., 1979.

**Wagner, F.H.** *Wildlife of the Deserts.* New York: Harry N. Abrams, Inc., 1980.

**Whittaker, R.H.** *Communities and Ecosystems* 2nd ed. Toronto, Ontario: The Macmillan Co., 1974.

**Whittaker, R.H., and G.E. Likens, eds.** "The Primary Production of the Biosphere," *Human Ecology,* 1, 4, (1973), pp. 299–369.

### Chapter 2

**Bolin, B., and R.B. Cook.** *The Major Biogeochemical Cycles and Their Interactions.* New York: John Wiley and Sons, 1983.

**Brill, W.J.** "Biological Nitrogen Fixation," *Scientific American,* 236 (March 1977), pp. 68–74, 79–81.

**Frieden, E.** "The Chemical Elements of Life," *Scientific American,* 227 (July 1972), pp. 52–60.

**Gosz, J.R., R.T. Holmes, G.E. Likens, and F.H. Bormann.** "The Flow of Energy in a Forest Ecosystem," *Scientific American,* 238 (March 1978), pp. 92–102.

**Janick, Jules.** "Cycles of Plant and Animal Nutrition," *Scientific American,* 235 (September 1976), pp. 74–84. Excellent overview of nutrient cycles and energy flow.

**Jordan, Carl F.** *Nutrient Cycling in Tropical Forest Ecosystems.* Somerset, N.J.: John Wiley & Sons, Inc., 1985.

**Oort, A.H.** "The Energy Cycle of the Earth," *Scientific American,* 223 (September 1970), pp. 54–63.

**Payne, W.J.** "Bacterial Denitrification: Asset or Defect?" *BioScience,* 33 (May 1983), pp. 319–325.

**Pimentel, David, et al.** "Environmental Quality and Natural Biota," *BioScience,* 30 (November 1980), pp. 750–755.

Rounick, J.S., and M.J. Winterbourn. "Stable Carbon Isotopes and Carbon Flow in Ecosystems," *BioScience,* 36 (March 1986), pp. 171–177.

*Scientific American,* 223 (September 1970). Issue devoted to articles describing nutrient cycles and energy flow in ecosystems.

Wood, Tim, et al. "Phosphorus Cycling in a Northern Hardwood Forest: Biological and Chemical Control," *Science,* 223 (27 January 1984), pp. 391–393.

## Chapter 3

Barrett, Gary W., and Rutger Rosenberg, eds. *Stress Effects on Natural Systems.* Somerset, N.J.: John Wiley and Sons, Inc., 1981.

Boerner, Ralph E.J. "Fire and Nutrient Cycling in Temperate Ecosystems," *BioScience,* 32 (March 1982), pp. 187–192.

Detwyler, Thomas R., ed. *Man's Impact on Environment.* New York: McGraw-Hill Book Company, 1971. Part Six, "The Spread of Organisms by Man," pp. 443–501.

Golgel, Monica, and Susan Bratton. "Exotics in the Parks," *National Parks,* 57 (January/February 1983), pp. 25–29.

Hartline, Beverly Karplus. "Fighting the Spreading Chestnut Blight," *Science,* 209 (22 August 1980), pp. 892–893.

Horn, Henry S. "Forest Succession," *Scientific American,* 232 (May 1975), pp. 90–98.

May, R.M. "Parasitic Infections as Regulators of Animal Populations," *American Scientist,* 71 (1983), pp. 36–45.

Odum, Eugene P. "Trends Expected in Stressed Ecosystems," *BioScience,* 35 (July/August 1985), pp. 419–422.

Peterson, R.O., et al. "Wolves, Moose, and the Allometry of Population Cycles," *Science,* 224 (22 June 1984), pp. 1350–1352.

Smuts, G.L. "Interrelationships between Predators, Prey, and their Environment," *BioScience,* 28 (1978), pp. 316–320.

Stevens, Larry. "King Kong Kudzu, Menace to the South," *Smithsonian,* 7 (December 1976), pp. 93–99.

Wright, Henry A., and Arthur W. Bailey. *Fire Ecology—United States and Southern Canada.* Somerset, N.J.: John Wiley and Sons, Inc., 1982.

## Chapter 4

Adovasio J.M., and R.C. Carlisle. "An Indian Hunters' Camp for 20,000 Years." *Scientific American,* 250 (May 1984), pp. 130–136.

Bishop, J.A., and L.M. Cook. "Moths, Melanism, and Clean Air," *Scientific American,* 232 (January 1975), pp. 90–99.

Bradshaw, A.D., and T. McNeilley. *Evolution and Pollution.* Baltimore, Md.: Edward Arnold Publishers, 1981.

Cairns-Smith, A.G. "The First Organisms," *Scientific American,* 252 (June 1985), pp. 90–100.

Clarke, B. "The Causes of Biological Diversity," *Scientific American,* 233 (August 1975), pp. 50–60.

Ghiglieri, Michael P. "The Social Ecology of Chimpanzees," *Scientific American,* 252 (June 1985), pp. 102–113.

Howells, William W. "Homo Erectus," *Scientific American,* 215 (November 1966), pp. 46–53.

Isaac, Glynn. "The Food-Sharing Behavior of Protohuman Hominids," *Scientific American,* 238 (April 1978), pp. 90–108.

Lewin, Roger. "How Did Vertebrates Take to the Air?" *Science,* 221 (1 July 1983), pp. 38–39.

Naylor, Bruce G., and Paul Handford. "In Defense of Darwin's Theory," *BioScience,* 35 (September 1985), pp. 478–484.

Pilbeam, David. "The Descent of Hominoids and Hominids," *Scientific American,* 250 (March 1984), pp. 84–96.

Richerson, Peter J., and Robert Boyd. "Natural Selection and Culture," *BioScience,* 34 (July/August 1984), pp. 430–434.

*Scientific American.* 239 (September 1978). This issue is devoted to various aspects of evolution.

*Scientific American,* 249 (September 1983). This issue is devoted to various aspects of change in and on the planet Earth over time.

Vidal, Gonzalo. "The Oldest Eukaryotic Cells," *Scientific American,* 250 (February 1984), pp. 48–57.

Wilson, Allan C. "The Molecular Basis of Evolution," *Scientific American,* 253 (October 1985), pp. 164–173.

## Chapter 5

Anderson, Dennis, and Robert Fishwick. *Fuelwood Consumption and Deforestation in African Countries.* Washington, D.C.: The World Bank, 1985.

Berardi, Gigi, M., ed. *World Food, Population and Development.* Totowa, N.J.: Rowman and Littlefield Publishers, 1986.

Birdsall, Nancy. "Population Growth and Poverty in the Developing World," *Population Bulletin* 35. Washington, D.C.: Population Reference Bureau, 1980.

Bouvier, Leon F. "Planet Earth 1984–2034: A Demographic Vision," *Population Bulletin* 39. Washington, D.C.: Population Reference Bureau, 1984.

Brown, Lester, R., et al. "Twenty-two Aspects of the Population Problem," *The Futurist,* 10 (October 1976), pp. 238–245.

Crosson, Pierre. "Agricultural Land: Will There Be Enough?" *Environment,* 26 (September 1984), pp. 16–20, 40–45.

McDowell, Bart. "Mexico City: An Alarming Giant," *National Geographic,* 166 (August 1984), pp. 138–172.

Odell, Rice, ed. "Cairo: A Third World City in Growing Trouble," *Conservation Foundation Letter,* (September 1983), pp. 1–8.

Odell, Rice, ed. "Egypt: A Case of Ecological Vulnerability," *Conservation Foundation Letter,* (August 1983) pp. 1–8.

Population Reference Bureau, Washington, D.C. A source of current data and information concerning human population, both U.S. and world. Numerous publications are available on an individual or subscription basis.

U.S. Congress Council on Environmental Quality and the U.S. Department of State. *The Global 2000 Report to the President: Entering the Twenty-First Century,* vol. 1, *The Summary Report.* Washington, D.C.: U.S. Government Printing Office, 1981.

Vining, Jr., Daniel R. "The Growth of Core Regions in the Third World," *Scientific American,* 252 (April 1985), pp. 42–49.

World Bank. *World Development Report 1984.* New York: Oxford University Press, 1984. See also World Development Reports for other years.

## Chapter 6

Brown, Lester R. "The Changing Global Economic Context," *Environment,* 25 (July/August 1983), pp. 28–34.

Brown, Lester R., and Edward C. Wolf. "Reversing Africa's Decline," *Worldwatch Paper,* 65. Washington, D.C.: Worldwatch Institute, June 1985.

Chandler, William V. "Improving World Health: A Least Cost Strategy," *Worldwatch Paper* 59. Washington, D.C.: Worldwatch Institute, July 1984.

Eckholm, Erik. "Human Wants and Misused Lands," *Natural History,* 91 (June 1982), pp. 33–48.

Jacobsen, Judith. "Promoting Population Stabilization: Incentives for Small Families," *Worldwatch Paper* 54. Washington, D.C.: Worldwatch Institute, June 1983.

Keyfitz, Nathan. "The Population of China," *Scientific American,* 250 (February 1984), pp. 38–47.

Mott, Frank L., and Susan H. Mott. "Kenya's Record Population Growth: A Dilemma of Development," *Population Bulletin* 35. Washington, D.C.: Population Reference Bureau, Oct 1980.

Odell, Rice, ed. "Critics Fault World Bank For Ecological Neglect," *Conservation Foundation Letter,* (November/December 1984), pp. 1–7.

Palmer, Jay. "The Debt-Bomb Threat," *Time,* 121 (10 January 1983), pp. 42–51.

Plucknett, Donald L., and Nigel J.H. Smith. "Sustaining Agricultural Yields," *BioScience,* 36 (January 1986), pp. 40–45.

Stokes, Bruce. "Global Housing Prospects: The Resource Constraints," *Worldwatch Paper* 46. Washington, D.C.: Worldwatch Institute, Sept 1981.

Wortman, Sterling, and Ralph W. Cummings, Jr. *To Feed This World.* Baltimore, Md.: Johns Hopkins University Press, 1978.

Yinger, Nancy, et al. "Third World Family Planning Programs: Measuring the Costs," *Population Bulletin* 38. Washington, D.C.: Population Reference Bureau, Feb. 1983.

## Chapter 7

Batie, Sandra S. *Soil Erosion: Crisis in America's Cropland?* Washington, D.C.: The Conservation Foundation, 1981.

Batie, Sandra S., and Robert G. Healy. "The Future of American Agriculture," *Scientific American,* 248 (February 1983), pp. 45–53.

Breman, H., and C.T. de Wit. "Rangeland Productivity and Exploitation in the Sahel," *Science,* 221 (30 September 1983), pp. 1341–1347.

Brown, Lester R. "The Worldwide Loss of Cropland," *Worldwatch Paper* 24. Washington, D.C.: Worldwatch Institute, October 1978.

Brown, Lester R., and Edward C. Wolf. "Soil Erosion: Quiet Crisis in the World Economy," *Worldwatch Paper* 60. Washington, D.C.: Worldwatch Institute, Sept. 1984.

Carter, Vernon G., and T. Dale. *Topsoil and Civilization*. Norman, Oklahoma: University of Oklahoma Press, 1974.

Cole, John N. "We're Losing Our Precious Farmland!" *National Wildlife*, 20(September 1982), pp. 12–15.

Cohen, Ellen. "This Fall Grow Soil," *Rodale's Organic Gardening*, 32 (September 1985), pp. 55–61.

Crosson, Pierre R., ed. *The Cropland Crisis—Myth or Reality?* Baltimore: Published for Resources for the Future by Johns Hopkins University Press, 1982.

Eckholm, Erik, and Lester R. Brown. "Spreading Deserts—the Hand of Man," *Worldwatch Paper No. 13*. Washington, D.C.: Worldwatch Institute, August, 1977.

Gibbons, Boyd. "Do We Treat Our Soil Like Dirt?" *National Geographic*, 166 (September 1984), pp. 350–363, 375–388.

Steiner, Frederick. *Ecological Planning For Farmlands Preservation*. Chicago: A.P.A. Planners Press, 1981.

Triplett, Jr., Glover B., and David M. Van Doren, Jr. "Agriculture Without Tillage." *Scientific American*, 236 (January 1977), pp. 28–33.

## Chapter 8

Eshman, Robert. "The Jonglei Canal: A Ditch Too Big?" *Environment*, 25 (June 1983), pp. 15–20, 32.

Ferguson, ·Bruce K. "Whither Water? The Fragile Future of the World's Most Important Resource," *The Futurist*, 17 (April 1983), pp. 29–36.

Francko, David A., and Robert G. Wetzel. *To Quench Our Thirst: The Present and Future Status of Freshwater Resources of the United States*. Ann Arbor: University of Michigan Press, 1983.

Hansen, Kevin. "South Florida's Water Dilemma: A Trickle of Hope for the Everglades," *Environment*, 26 (June 1984), pp. 14–20, 40–42.

Johnsgard, Paul A. "The Platte:.A River of Birds," *The Nature Conservancy News*, 33 (September/October 1983), pp. 6–10.

Jubak, Jim. "The Drying of America," *Environmental Action*, 13 (June 1982), pp. 20–23.

Karr, James R., and Isaac J. Schlosser. "Water Resources and the Land Water Interface," *Science*, 201 (July 21, 1978), pp. 227–234.

Micklin, Philip P. "The Vast Diversion of Soviet Rivers," *Environment*, 27 (March 1985), pp. 12–20, 40–45.

National Wildlife. "Water: A Special Report," *National Wildlife*, 22 (February/March 1984), pp. 6–21. Several authors/articles.

Pimentel, David. "Water Resources in Food and Energy Production," *BioScience*, 32 (December 1982), pp. 861–867.

Postel, Sandra. "Conserving Water: The Untapped Alternative," *Worldwatch Paper 67*. Washington, D.C.: Worldwatch Institute, Sept. 1985.

Sinclair, T.R., et al. "Water-Use Efficiency in Crop Production," *BioScience*, 34 (January 1984), pp. 36–40.

Splinter, William E. "Center-Pivot Irrigation," *Scientific American*, 234 (June 1976), pp. 90–99.

Stokes, Bruce. "Water Shortages: The Next Energy Crisis," *The Futurist*, 17 (April 1983), pp. 37–41, 45–47.

## Chapter 9

Clark II, Edwin H., et al. *Eroding Soils—The Off-Farm Impacts*. Washington, D.C.: The Conservation Foundation, 1985.

DeVore, R. William, ed. *Proceedings 1983 International Symposium on Urban Hydrology, Hydraulics and Sediment Control*. Lexington, Ky.: O.E.S. Publications, 1983.

Flick, Art. "Must We Rape Our Streams?" *Trout*, 15 (Autumn 1974), pp. 18–20, 41–42.

Towsend, Colin R. *The Ecology of Streams and Rivers*. Baltimore, Md.: Edward Arnold Publishers, 1980.

## Chapter 10

Aberley, Richard C., and Susan Berg. "Finding Uses For Sludge," *American City and County*, 101 (June 1986), pp. 38–46.

Alth, Max and Charlotte. *Constructing and Maintaining Your Well and Septic System*. Blue Ridge Summit, Pa.: Tab Books, Inc., 1984.

Goldstein, Jerome. *Sensible Sludge, A New Look at a Wasted Natural Resource*. Emmaus, Pa.: Rodale Press, 1977.

Payton, Beverly M. "Ocean Dumping in the New York Bight," *Environment*, 27 (November 1985), pp. 26–32.

Tucker, Jonathan B. "Schistosomiasis and Water Projects: Breaking the Link," *Environment*, 25 (September 1983), pp. 17–20.

Valdes-Cogliano, Sally J. "International Drinking Water Decade," *Environment*, 27 (October 1985), pp. 41–42.

## Chapter 11

American Society of Limnology and Oceanography. *Nutrients and Eutrophication: The Nutrient Limiting Controversy*, Lawrence, Kans.: American Society of Limnology and Oceanography, 1972.

Chapra, S.C., and A. Robertson. "Great Lakes Eutrophication: The Effect of Point Source Control of Total Phosphorus," *Science*, 196 (June 24, 1977), pp. 1448–1449.

Lowrance, Richard, et al. "Riparian Forests as Nutrient Filters in Agricultural Watersheds," *BioScience*, 34 (June 1984), pp. 374–377.

McCloskey, William. "Hard Times Hit The Bay," *National Wildlife*, 22 (April/May 1984), pp. 6–14.

National Academy of Sciences. *Eutrophication: Cause, Consequence, Corrections*. Washington, D.C.: National Academy of Sciences, 1969.

Officer, Charles B., et al. "Chesapeake Bay Anoxia: Origin, Development, and Significance," *Science*, 223 (6 January 1984), pp. 22–27.

Organization For Economic Cooperation and Development (O.E.C.D.). *Eutrophication of Waters: Monitoring, Assessment, and Control*. Washington, D.C.: O.E.C.D. Publications, 1982.

Smith, Val H. "Low Nitrogen to Phosphorus Ratios Favor Dominance by Blue-Green Algae in Lake Phytoplankton," *Science*, 221 (12 August 1983), pp. 669–671.

Wetzel, R.G. *Limnology*. Philadelphia: Saunders, 1983. a textbook that covers the physical, chemical, and biologic aspects of aquatic ecosystem function.

## Chapter 12

Culver, Alicia, and Rose Marie Audette. "Danger's in the Well," *Environmental Action*, 16 (March/April 1985), pp. 15–19.

Dowling, Michael. "Defining and Classifying Hazardous Wastes," *Environment*, 27 (April 1985), pp. 18–20, 36–41.

Epstein, S., L. Brown, and C. Pope. *Hazardous Waste in America*. San Francisco, Cal.: Sierra Club Books, 1982.

Ghosal, D., et al. "Microbial Degradation of Halogenated Compounds," *Science*, 228 (12 April 1985), pp. 135–142.

Gordon, Wendy. *A Citizen's Handbook on Groundwater*. New York: Natural Resources Defense Council, 1984.

Magnuson, Ed. "A Problem That Cannot Be Buried (The Poisoning of America Continues)," *Time*, 126 (14 October 1984), pp. 76–84.

Morell, Virginia. "Fishing For Trouble—A Cancer Epidemic in Fish Is Warning Us: You May Be Next," *International Wildlife*, 14 (July/August 1984), pp. 40–43.

National Research Council. *Groundwater Contamination*. Washington, D.C.: National Research Council, 1984.

Organization for Economic Cooperation and Development. *Hazardous Waste Problem Sites*. Wasington, D.C.: O.E.C.D. Publications, 1983.

Piasecki, Bruce. "Unfouling the Nest," *Science 83*, (September 1983), pp. 76–81.

Segel, Edward et al. *The Toxic Substances Dilemma A Plan For Citizen Action*, Washington, D.C.: The National Wildlife Federation, 1985.

Shute, Nancy. "The Selling of Waste Management," *The Amicus Journal*, 7 (Summer 1985), pp. 8–17.

Tangley, Laura. "Groundwater Contamination: Local Problems Becomes National Issue," *BioScience*, 34 (March 1984), pp. 142–146, 148.

Taylor, Ronald A. "$1 Billion Later, Toxic Cleanup Barely Begun," *U.S. News & World Report*, 99 (22 April, 1985), pp. 57–58.

U.S. Congress Office of Technology Assessment. *Superfund Strategy*. Washington, D.C.: U.S. Government, 1985.

## Chapter 13

American Medical Association. *Journal of the American Medical Association*, 225 (28 February 1986). Entire issue devoted to effects of smoking tobacco.

Bormann, F.H. "The Effects of Air Pollution on the New England Landscape," *Ambio*, 11, (5) 1982, pp. 338–346.

Bormann, F.H. "The New England Landscape: Air Pollution Stress and Energy Policy," *Ambio*, 11 (4) 1982, pp. 188–194.

Hughes, Kathleen, and Phil Simon. "Murder on the Job," *Environmental Action*, 15 (November 1983), pp. 18–21.

Kirsch, Laurence S. "Behind Closed Doors: II. Indoor Air Pollution and Government Policy," *Environment*, 25 (April 1983), pp. 26–39.

Lang, Robert D. "Asbestos in Schools: Low Marks For Government Action," *Environment*, 26 (November 1984), pp. 14–20, 38–40.

National Clean Air Coalition. *The Clean Air Act*. Washington, D.C.: National Clean Air Coalition, 1983.

Parbery, D.G. "Pollutants and Plant Health." In *Air Pollution Control*. Part 4. Eds. Gordon M. Bragg and Werner Strauss. New York: Wiley-Interscience, 1981. pp. 81–119.

Strauss, W., and S.J. Mainwaring. *Air Pollution.* Baltimore, Md.: Edward Arnold Publisher, 1983.

Uzych, Leo. "Law: Chemical Disclosure Laws," *Environment,* 26 (June 1984), pp. 4–5.

Wark, Kenneth, and Cecil F. Warner. *Air Pollution: Its Origin and Control* 2nd ed. New York: Harper and Row Pubs., Inc., 1981.

Woodwell, G.M. et al. *Ecological and Biological Effects of Air Pollution.* New York: Irvington Publishers, 1973.

## Chapter 14

Allar, Bruce "Atmospheric Fluidized Bed Combustion: No More Coal-Smoked Skies?" *Environment,* 26 (March 1984), pp. 25–30.

Boyle, Robert H. "An American Tragedy," *Sports Illustrated,* 55 (21 September, 1981), pp. 68–82. Acid precipitation killing lakes in U.S. & Canada.

Bryson, Reid A., and Thomas J. Murray. *Climates of Hunger: Mankind and the World's Changing Weather.* Madison, Wisconsin: The University of Wisconsin Press, 1977.

Hawkins, David G. "Controlling Acid Rain—A Modest Proposal," *The Amicus Journal,* 4 (Winter 1983), pp. 12–13.

Kellogg, William W., and Robert Schware. *Climate Change and Society: Consequences of Increasing Atmospheric Carbon Dioxide.* Boulder, Colorado: Westview Press, 1981.

Kiester, Edwin Jr. "A Deadly Spell is Hovering above the Black Forest," *Smithsonian,* (November 1985), pp. 211–230.

MacCracken, Michael C., and Harry Moses. "The First Detection of Carbon Dioxide Effects: Workshop Summary 8–10 June 1981, Harpers Ferry, W. Va.," *Bulletin of the American Meteorological Society,* 63 (October 1982), pp. 1164–1177.

National Academy of Sciences. *Acid Deposition: Long-Term Trends.* Washington, D.C.: National Academy Press, 1986.

National Academy of Sciences. *Causes and Effects of Changes in Stratospheric Ozone: Update 1983.* Washington, D.C.: National Academy Press, 1984.

Parry, Martin, et al. "Climatic Change: How Vulnerable Is Agriculture?" *Environment,* 27 (January/February 1985), pp. 4–5, 43.

Postel, Sandra. "Air Pollution, Acid Rain, and the Future of Forests," *Worldwatch Paper* 58. Washington, D.C.: Worldwatch Institute, March 1984.

Revelle, R. "Carbon Dioxide and World Climate," *Scientific American,* 247 (August 1982), pp. 35–43.

Roberts, Leslie. "California's Fog Is Far More Polluted Than Acid Rain," *BioScience,* 32 (November 1982), pp. 778–779.

Roberts, Walter Orr. "It Is Time to Prepare for Global Climate Changes," *Conservation Foundation Letter,* (April 1983), pp. 1–7.

Turco, Richard P., et al. "The Climatic Effects of Nuclear War," *Scientific American,* 251(August 1984), pp. 33–43.

U.S. Congress Office of Technology Assessment. *The Regional Implications of Transported Air Pollutants: An Assessment of Acidic Deposition and Ozone.* Washington, D.C.: U.S. Government, 1982.

Wetstone, Gregory, and Sarah Foster. "Acid Precipitation: What Is It Doing to Our Forests?" *Environment,* 25 (May 1983), pp. 10–12, 38–40.

## Chapter 15

Ames, Bruce N. "Dietary Carcinogens and Anticarcinogens," *Science,* 221 (23 September 1983), pp. 1256–1264.

Carpenter, Richard A., and John A. Dixon. "Ecology Meets Economics: A Guide to Sustainable Development," *Environment,* 27 (June 1985), pp. 6–11, 27–32.

Davies, J. Clarence, et al. *Determining Unreasonable Risk under the Toxic Substances Control Act.* Washington, D.C.: The Conservation Foundation, 1979.

Goldfarb, Theodore D. *Taking Sides: Clashing Views on Controversial Environmental Issues.* Guilford, Conn.: The Dushkin Publishing Group, 1983.

Hohenemser, C., et al. "The Nature of Technological Hazard," *Science,* 220 (22 April, 1983), pp. 378–384.

Hueper, M.D., W.C. "Environmental Cancer Risks in an Industrialized Economy," in *Industrial Pollution.* Ed. N. Irving Sax. New York: Van Nostrand Reinhold Co., 1974, pp. 118–149.

Kneese, Allen V. *Measuring the Benefits of Clean Air and Water.* Baltimore, Md.: Resources for the Future, 1984.

Long, F.A., and Glenn E. Schweitzer. *Risk Assessment at Hazardous Waste Sites.* Washington, D.C.: American Chemical Society, 1982.

Maugh II, Thomas H. "Just How Hazardous Are Dumps?" *Science,* 215 (29 January, 1982), pp. 490–493.

Odell, Rice, ed. "In a Deluge of Problems, Where Are Worst Threats?" *Conservation Foundation Letter,* (December 1983), pp. 1–8.

U.S. Executive Office, Council on Environmental Quality. *Environmental Quality—The Sixth Annual Report of the Council on Environmental Quality,* 1975. Chapter 4, "Environmental Economics," pp. 494–570, is a particularly good discussion of the costs of pollution.

Weinstein, Milton C. "Cost-Effective Priorities for Cancer Prevention," *Science* 221 (1 July, 1983), pp. 17–23.

## Chapter 16

Boralko, Allen A. "The Pesticide Dilemma," *National Geographic,* 157 (February 1980), pp. 144–183.

Carson, Rachel. *Silent Spring.* Boston: Houghton Mifflin Company, 1962.

Dover, Michael J., and Brian A. Croft. "Pesticide Resistance and Public Policy," *BioScience,* 36 (February 1986), pp. 78–85.

Marshall, Eliot. "The Murky World of Toxicity Testing," *Science,* 220 (10 June 1983), pp. 1130–1132.

May, Elizabeth E. "Canada's Moth War," *Environment,* 19 (August/September 1977) pp. 16–23.

Michel, Nancy. "The Fire That Won't Die Out," *National Wildlife,* 22 (April/May 1984), pp. 47–49.

Pimentel, David. *Ecological Effects of Pesticides on Non-Target Species.* Washington, D.C.: U.S. Government Printing Office, 1971.

Pimentel, David, and Lois Levitan. "Pesticides: Amounts Applied and Amounts Reaching Pests," *BioScience,* 36 (February 1986), pp. 86–91.

Schneider, Keith. "Faking It: The Case Against Industrial Bio-Test Laboratories." *The Amicus Journal,* 4 (Spring 1983), pp. 14–26.

Van den Bosch, Robert. *The Pesticide Conspiracy.* Garden City, N.Y.: Doubleday & Co., Inc., 1978.

Vietmeyer, Noel D. "We Haven't Zapped the Boll Weevil Yet," *Smithsonian,* 13 (August 1982), pp. 60–68.

Weir, David, and Mark Schapiro. "The Circle of Poison," *The Nation,* 231 (15 November 1980), pp. cover, 514–516.

## Chapter 17

Barnett, Mary. "A Better Way to Fight Bugs: Integrated Pest Management," *The Amicus Journal,* 4 (Spring 1983), pp. 27–31.

*BioScience,* 30 (October 1980). Special issue on integrated pest management.

Dahlsten, Donald L. "Pesticides in an Era of Integrated Pest Management," *Environment,* 25 (December 1983), pp. 45–54.

Debach, Paul. *Biological Control by Natural Enemies.* London: Cambridge University Press, 1974.

Dethier, Vincent. "Smart Strategies for Insect Control," *Horticulture,* 61 (January 1983), pp. 48–59.

Edwards, Peter J., and Stephen D. Wratten, Ph.D. *Ecology of Insect-Plant Interactions.* Baltimore, Md.: Edward Arnold Publishers, 1980.

Horwith, Bruce. "A Role for Intercropping in Modern Agriculture," *BioScience,* 35 (May 1985), pp. 286–291.

Maxwell, Fowden G., and Peter R. Jennings, Eds. *Breeding Plants Resistant to Insects.* New York: John Wiley and Sons, Inc., 1980.

Metcalf, Robert L., and William H. Luckmann, eds. *Introduction to Insect Pest Management.* Somerset, N.J.: John Wiley & Sons, Inc., 1982.

Miller, Lois K., et al. "Bacterial, Viral, and Fungal Insecticides," *Science,* 219 (11 February 1983), pp. 715–721.

Pillemer, Eric A., and Ward M. Tingey. "Hooked Trichomes: A Physical Plant Barrier to a Major Agricultural Pest," *Science,* 193 (August 6, 1976), pp. 482–484.

Rosenthal, Gerald A. "The Chemical Defenses of Higher Plants," *Scientific American,* 254 (January 1986), pp. 94–99.

Walsh, J. "Cosmetic Standards: Are Pesticides Overused for Appearance's Sake?" *Science,* 193 (August 27, 1976), pp. 744–747.

## Chapter 18

Abelson, Philip H., and A.L. Hammond, eds. *Materials: Renewable and Nonrenewable Resources.* Washington D.C.: American Association for the Advancement of Science, 1976.

Birchall, J.D., and Anthony Kelly. "New Inorganic Materials," *Scientific American,* 248 (May 1983), pp. 104–115.

Chandler, William V. "Materials Recycling: The Virtue or Necessity," *Worldwatch Paper* 56. Washington, D.C.: Worldwatch Institute, October 1983.

Goeller, H.E., and A. Zucker. "Infinite Resources: The Ultimate Strategy," *Science,* 223 (3 February 1984), pp. 456–462.

Goetz, Alexander F.H., et al. "Mineral Identification from Orbit: Initial Results from the Shuttle Multispectral Infrared Radiometer," *Science*, 218 (3 December 1982), pp. 1020–1024.

Kozlovsky, Ye. A. "The World's Deepest Well," *Scientific American*, 251 (December 1984), pp. 98–104.

Larson, Eric, et al. "Beyond the Era of Materials," *Scientific American*, 254 (June 1986), pp. 34–41.

Matthews, Olen Paul, et al. "Mining and Wilderness: Incompatible Uses or Justifiable Compromise?" *Environment*, 27 (April 1985), pp. 12–17, 30–36.

Meadows, Donella, H.D.L. Meadows, J. Randers, and W.W. Behrens III. *The Limits to Growth, A Report for the Club of Rome's Project on the Predicament of Mankind*. New York: Universe Books, A Potomac Associates Book, 1972.

Simon, Julian L. "Life on Earth Is Getting Better not Worse," *The Futurist*, 17 (August 1983), pp. 6–12.

## Chapter 19

Altieri, Miguel A., et al. "Developing Sustainable Agroecosystems," *BioScience*, 33 (January 1983), pp. 45–49.

Bean, Michael, et al. "The Endangered Species Program Needs Rejuvenation," *Conservation Foundation Letter*, (January/February 1984), pp. 1–6.

Borlaug, Norman E. "Contributions of Conventional Plant Breeding to Food Production," *Science*, 219 (11 February 1983), pp. 689–693.

Cohn, Jeffrey P. "A New Plan to Conserve the Earth's Biota," *BioScience*, 35 (June 1985), pp. 334–339.

Crawford, Mark. "The Last Days of the Wild Condor?" *Science*, 229 (30 August 1985), pp. 844–845.

D'Aulaire, Emily, and Per Old. "Pangolins Are All the Rage: These Artichokes on Legs Are Victims of the Latest Fad in Footwear," *International Wildlife*, 13 (January/February 1983), pp. 14–16.

Doherty, Jim. "Hail, Lobsterman . . . and Farewell," *National Wildlife*, 15 (April-May 1977), pp. 42–49.

Hardin, Garrett. "The Tragedy of the Commons," *Science*, 162 (December 1968), pp. 1243–1248.

Karr, James R., et al. "Fish Communities of Midwestern Rivers: A History of Degradation," *BioScience*, 35 (February 1985), pp. 90–95.

Nations, James D., and Daniel I. Komer. "Rainforests and the Hamburger Society," *Environment*, 25 (April 1983), pp. 12–20.

Nature Conservancy News, 33 (May/June 1983). Entire issue devoted to importance and preservation of wetlands.

Odell, Rice. "Can We Count on Fisheries to Fight Hunger?" *Conservation Foundation Letter*, (March/April 1985), pp. 1–7.

Peters, Robert. "Wildlife Reserves: Can They Do the Job?" *Conservation Foundation Letter*, (January/February 1984), pp. 1–7.

Raven, Peter H. "Disappearing Species: A Global Tragedy," *The Futurist*, 19 (October 1985), pp. 8–14.

Russell, Dick, and Brian Keating. "Sushi Today, Gone Tomorrow: The Plight of the Bluefin Tuna," *The Amicus Journal*, 6 (Winter 1984), pp. 38–46.

Timberlake, Lloyd, et al. "Saving the World's Genetic Bank: The Benefits of the Earth's 'Wild' Resources," *World Press Review*, (December 1982), pp. 32–35.

Wellborn, Stanley N. "Exotic Animals Find New Home on the Range," *U.S. News & World Report*, 99 (April 29, 1985), pp. 77–80.

White, Peter T. "Nature's Dwindling Treasures—Rain Forests," *National Geographic*, 163 (January 1983), pp. 2–9, 20–46.

Wolkomir, Richard. "Draining the Gene Pool," *National Wildlife*. 21 (October/November 1983), pp. 24–28.

## Chapter 20

Abert, J.G., H. Alter, and J.F. Bernheisel. "The Economics of Resource Recovery from Municipal Solid Waste." In *Materials: Renewable and Nonrenewable Resources*, pp. 54–60. Eds. Philip H. Abelson and Allen L. Hammond. Washington D.C.: American Association for the Advancement of Science, 1976.

Blum, S.L. "Tapping Resources in Municipal Solid Waste," In *Materials: Renewable and Nonrenewale Sources*, pp. 47–54. Eds. Philip H. Abelson and Allen L. Hammond. Washington, D.C.: American Association for the Advancement of Science, 1976.

Hasselriis, Floyd. *Refuse-Derived Fuel Processing*. Stoneham, Mass.: Butterworth, 1984.

Henstock, M.E. *Recycling & Disposal of Solid Waste*. Elmsford, N.Y.: Pergamon Press, Inc. 1983.

Noll, Kenneth E., et al. *Recovery, Recycle, and Reuse of Industrial Waste*. Chelsea, Mich.: Lewis Pubs, Inc., 1985.

Organization for Economic Cooperations and Development. *Product Durability and Product Life Extension: Their Contribution to Solid Waste Management*. Washington, D.C.: O.E.C.D. Publications, 1982.

Parker, C., and T. Roberts, eds. *Energy from Waste: An Evaluation of Conversion Technologies*. New York: Elsevier Science Publishing Co., Inc., 1985.

Pawley, Martin. *Building for Tomorrow: Putting Waste to Work*. San Francisco: Sierra Club Books, 1982.

Purcell, Arthur H. "The World's Trashiest People," *The Futurist*, 15 (February 1981), pp. 51–59.

Sobetzer, John G., and Lynn Carson. *Solid Waste Management in Michigan: A Guide for Local Government and Citizens*. East Lansing, Mich.: Michigan State Univ. Community Development Programs, 1982.

White, Peter T. "The Fascinating World of Trash," *National Geographic*, 163 (April 1983), pp. 424–440, 447–456.

## Chapter 21

Cole, Lamont C. "Thermal Pollution," *BioScience*, 19 (November 1969), pp. 989–992.

Ebinger, Charles, interviewed. "A 'Complacent' U.S. Courts New Oil Crisis," *U.S. News & World Report*, 98 (27 May 1985), pp. 37–38.

Flower, Andrew. "World Oil Production," *Scientific American*, 238 (March 1978), pp. 42–49.

Gever, John, et al. *Beyond Oil*. Cambridge, Mass.: Published for Carrying Capacity, Inc. by Ballinger Publishers, 1986.

Landsberg, Hans H. "Relaxed Energy Outlook Masks Continuing Uncertainties," *Science*, 218 (3 December 1982), pp. 973–974.

Marshall, Eliot. "Fill the Oil Reserve, Academy Report Says," *Science*, 232 (25 April 1986), pp. 441–442.

Norman, Colin. "Interior Slashes Offshore Oil Estimates," *Science*, 228 (24 May 1985), p. 974.

Ourisson, Guy, et al. "The Microbial Origin of Fossil Fuels," *Scientific American*, 251 (August 1984), pp. 44–51.

Pugash, James Z., et al. "The Geopolitics of Oil," *Science*, 210 (19 December 1980), pp. 1324–1327.

Ruedisili, Lon C., and Morris W. Firebaugh, Eds. *Perspectives On Energy: Issues, Ideas, and Environmental Dilemmas*. 3rd ed. New york: Oxford University Press, Inc., 1982.

Sheets, Kenneth R., et al. "If U.S. Faced Another Oil Embargo Today—," *U.S. News & World Report*, 95 (24 October 1983), pp. 27–29.

U.S. Department of Energy/Energy Information Administration. *1982 Annual Energy Outlook: With Projections to 1990*. Washington, D.C.: U.S. Government Printing Office, 1983.

U.S. Department of Energy/Energy Information Administration. *1982 Annual Energy Review*. Washington, D.C.: U.S. Government Printing Office, 1983.

Weeks, W.F., and G. Weller. "Offshore Oil in the Alaskan Arctic," *Science*, 225 (27 July 1984), pp. 371–378.

## Chapter 22

Conn, Robert W. "The Engineering of Magnetic Fusion Reactors," *Scientific American*, 249 (October 1983), pp. 60–71.

Douglis, Carole. "The Economics of Contamination, " *The Atlantic*, 254 (October 1984) pp. 29–36.

Eisenbud, Merril. "Sources of Ionizing Radiation Exposure," *Environment*, 26 (December 1984), pp. 6–11, 30–33.

Gould, Stanhope, and Brian McTigue. "Radioactive Wastes: A Barrell of Trouble," *National Wildlife*, 21 (April/May 1983), pp. 20–23.

Hohenemser, C., et al. "Chernobyl: An Early Report," *Environment*, 28 (June 1986), pp. 6–13; 30–43.

Kaku, Michio, and Jennifer Trainer, Eds. *Nuclear Power: Both Sides—The Best Arguments For and Against the Most Controversial Technology*. New York: W.W. Norton and Co., 1983.

Lester, Richard K. "Rethinking Nuclear Power," *Scientific American*, 254, (March 1986), pp. 31–39.

Manning, Russ. "The Future of Nuclear Power—Where Do We Go From Here?" *Environment*, 27 (May 1985), pp. 12–17, 31–37.

Marshall, Eliot. "NRC Reviews Brittle Reactor Hazard," *Science*, 215 (26 March 1982), pp. 1596–1597.

Marshall, Eliot. "The Synfuels Shopping List," *Science*, 223 (6 January 1984), pp. 31–32.

Mynatt, F.R. "Nuclear Reactor Safety Research Since Three Mile Island," *Science*, 216 (9 April 1982), pp. 131–135.

Environmental Action Foundation. *Rate Shock: Confronting the Cost of Nuclear Power*. Washington, D.C.: Environmental Action Foundation, 1984.

Perry, Harry. "Coal in the United States: A Status Report," *Science*, 222 (28 October 1983), pp. 377–384.

Pollock, Cynthia. *Decommissioning: Nuclear Power's Missing Link,"* Worldwatch Paper 64. Washington, D.C.: Worldwatch Institute, 1986.

Roberts, Leslie. "Ocean Dumping of Radioactive Waste," *BioScience,* 32 (November 1982), pp. 773–776.

Starr, Chauncey. "Uranium Power and Horizontal Proliferation of Nuclear Weapons," *Science,* 224 (1 June 1984), pp. 952–957.

Weinberg, Alvin M., and Irving Spiewak. "Inherently Safe Reactors and a Second Nuclear Era," *Science,* 224 (29 June 1984), pp. 1398–1402.

**Chapter 23**

Anderson, Bruce, with Michael Riordan. *The Solar Home Book: Heating, Cooling and Designing with the Sun.* Harrisville, N.H.: Brick House Publishing Co., 1976.

Bruce, James T. "A Plan Not to Plan Is the Plan," *Environment,* 25 (March 1983), pp. 6–9, 28–35.

Flavin, Christopher. "Photovoltaics: A Solar Technology for Powering Tomorrow," *The Futurist,* 17 (June 1983), pp. 41–50.

Flavin, Christopher. *Renewable Energy at the Crossroads.* Washington, D.C.: Center for Renewable Resources, 1985.

Galligan, Mary. "In Solar Village, Sunshine Is Put In Harness," *U.S. News & World Report,* 98 (11 February 1985), pp. 72–73.

Greenbaum, E., et al. "Biological Solar Energy Production With Marine Algae," *BioScience,* 33 (October 1983), pp. 584–585.

Heller, Adam. "Hydrogen-Evolving Solar Cells," *Science,* 223 (16 March 1984), pp. 1141–1148.

Holden, Constance. "Hawaiian Rainforest Being Felled," *Science,* 228 (31 May 1985), pp. 1073–1074.

Kakela, Peter, et al. "Low-Head Hydropower For Local Use," *Environment,* 27 (January/February 1985), pp. 31–38.

Kerr, Richard A. "Extracting Geothermal Energy Can Be Hard," *Science,* 218 (12 November 1982), pp. 668–669.

Marshall, Eliot. "The Procrastinator's Power Source," *Science,* 224 (20 April 1984), pp. 268–270.

Mears, Leon G. "Energy from Alcohol," *Environment,* 20 (December 1978), pp. 17–20.

Metz, William D., and Allen L. Hammond. *Solar Energy in America.* Washington, D.C.: American Association for the Advancement of Science, 1978.

Moretti, Peter M. and Louis V. Divone. "Modern Windmills," *Scientific American,* 254 (June 1986), pp. 110–118.

Munson, Richard. "International: Israel's Solar Ponds Grow Larger," *Environment,* 26 (January/February 1984), pp. 41–42.

Pimentel, D., et al. "Environmental and Social Costs of Biomass Energy," *BioScience,* 34 (February 1984), pp. 89–94.

**Chapter 24**

Carter, Luther J. *The Florida Experience.* Baltimore, Md: Johns Hopkins University Press, 1975.

Castells, Manuel. *The City and the Grassroots.* Berkeley: University of California Press, 1983.

Gruen, Victor. *The Heart of Our Cities; the Urban Crisis; Diagnosis and Cure.* New York: Simon and Schuster, 1964.

Hendler, Bruce. *Caring for the Land: Environmental Principles for Site Design and Review.* Chicago: APA Planners Press, 1977.

Jacobs, Allan B. *Making City Planning Work.* Chicago: APA Planners Press, 1978.

McHarg, Ian L. *Design with Nature.* Garden City, N.Y.: Doublday & Company Inc., 1971.

Pimentel, David. "The Vanishing Land," in *The National Research Council Issues and Studies 1981–1982.* Washington, D.C.: National Academy Press, 1982, pp. 104–115.

Pimental, D., et al. "Land Degradation: Effects on Food and Energy Resources," *Science,* 194 (October 8, 1976), pp. 149–155.

Richardson, David B. "To Rebuild America—$2.5 Trillion Job," *U.S. News & World Report,* 93 (27 September 1982), pp. 57–61.

Stokes, Bruce. "Recycled Housing," *Environment,* 21 (January/February 1979), pp. 6–14.

von Weizsacker, Carl Friedrich. "Designing the Next Century," *World Press Review,* 32 (February 1985), pp. 27–29.

White, Peter T. "This Land of Ours—How We Are Using It," *National Geographic,* 150 (July 1976), pp. 20–67.

**Epilogue**

Brown, Lester R., and Pamela Shaw. "Putting Society on a New Path," *Environment,* 24 (September 1982), pp. 29–33.

Brown, Lester R., and Pamela Shaw. "Six Steps to a Sustainable Society," *Worldwatch Paper,* 48 (March 1982), Washington D.C.: Worldwatch Inst, 1982.

Cobb, R.W., and C.D. Elder. *Participation in American Politics.* 2nd ed. Baltimore: Johns Hopkins University Press, 1983.

Norman, Colin. "Soft Technologies, Hard Choices," *Worldwatch Paper* 21. Washington, D.C.: Worldwatch Institute, June, 1978.

Schumacher, E.F. *Small Is Beautiful: Economics As If People Mattered.* New York: Perennial Library, Harper and Row, 1973.

# APPENDIX A
## Environmental Organizations

This is a list of organizations active in environmental matters. Included here are national organizations as well as some small, specialized ones. Many of these have internship positions available for those wishing to do work for an environmental group. A more complete listing can be found in the Conservation Directory put out by National Wildlife Federation (address below). This directory includes local, regional and national organizations. The cost is $15.00 plus $2.00 for postage.

**American Lung Association,** 1740 Broadway, New York, N.Y. 10019. Research, education: air pollution effects and means of control.

**American Rivers Conservation Council,** 322 4th Street, N.E. Washington, D.C. 20002. Lobbying: wild and scenic rivers.

**Center for Renewable Resources,** 1001 Connecticut Avenue, N.W., Suite 638, Washington, D.C. 20036. Research and education: energy policy community organizing on issues concerning renewable resources.

**Center for Science in the Public Interest,** 1501 16th Street, N.W., Washington, D.C. 20036. Research and education: food, nutrition, health.

**Chesapeake Bay Foundation, Inc.,** "Church," 162 Prince George St., Annapolis, Md. 21401. Research, education, and litigation: environmental defense and management of Chesapeake Bay and surrounding land.

**Clean Water Action Project,** 733 15th Street, N.W., Suite 1110, Washington, D.C. 20005. Lobbying: water quality.

**Common Cause,** 2030 M Street, N.W., Washington, D.C. 20036. Lobbying: government reform, energy reorganization, clean air.

**Concern Inc.,** 1794 Columbia Road, N.W., Washington, D.C. 20009. Research and education: environmental education.

**Congress Watch,** 215 Pennsylvania Avenue, S.E., Washington, D.C., 20003. Lobbying: energy.

**Conservation Foundation,** 255 23rd Street, N.W., Washington, D.C., 20037. Research and education: land use, energy conservation, air and water quality.

**Consumer Federation of America,** 1424 16th Street, N.W., Suite 6604, Washington, D.C. 20036. Lobbying: energy policy.

**Critical Mass Energy Project,** 215 Pennsylvania Avenue, S.E., Washington, D.C. 20003. Research: nuclear power, alternative energy.

**Defenders of Wildlife,** 1244 19th Street, N.W., Washington, D.C. 20036. Research, education and lobbying: endangered species.

**Environmental Action, Inc.,** 1525 New Hampshire Avenue, N.W., Washington, D.C. 20036. Lobbying: transportation, solid waste, water quality, solar energy, energy conservation, toxic substances, deposit legislation.

**Environmental Action Foundation,** 1525 New Hampshire Avenue, N.W., Washington, D.C. 20036. Research and education: electric utility rate reform, alternative energy, solid waste, transportation, water quality planning, deposit legislation.

**Environmental Defense Fund, Inc.,** 444 Park Avenue South, New York, N.Y. 10016. Research, litigation, and lobbying: cosmetic safety, drinking water, energy, transportation, pesticides, wildlife, air pollution, cancer prevention, radiation.

**Environmental Law Institute,** 1616 P Street, N.W., Suite 200, Washington, D.C. 20036. Research and education: institutional and legal issues affecting the environment.

**Environmental Policy Institute,** 218 D Street, S.E., Washington, D.C. 20003. Lobbying: all aspects of energy development.

**Environmentalists for Full Employment,** 1000 Wisconsin Avenue, N.W., Washington, D.C. 20007. Research and education: job potential of environmental protection.

**Friends of the Earth,** 1045 Sansome St., San Francisco, Calif. 94111. Lobbying and Education: preservation, restoration and rational use of the earth.

**Greenpeace, USA, Inc.,** 1611 Connecticut Ave., N.W., Washington, D.C. 20009. Non-violent direct action: whales, ocean waste disposal, acid rain, nuclear weapons testing, seal pups, and Antarctica.

**Institute for Local Self-Reliance,** 2425 18th Street, N.W., Washington, D.C. 20009. Research and education: appropriate technology for community development.

**Izaak Walton League of America, Inc.,** 1701 North Fort Myer Drive, Suite 1100, Arlington, VA 22209. Research and education: conservation, air and water quality, streams.

**League of Conservation Voters,** 320 4th Street, N.E., Washington, D.C. 20002. Political action: evaluation of environmental records of public officials.

**League of Women Voters of the U.S.,** 1730 M Street, N.W., Washington, D.C. 20036. Education and lobbying: general environmental issues.

**Monitor, Inc.,** 1506 19th Street, N.W., Washington, D.C. 20036. Lobbying: endangered marine species.

**National Audubon Society,** 950 Third Avenue, New York, N.Y. 10012. Research, lobbying, education: wildlife, wilderness, public lands, endangered species, water resource management.

**National Park Foundation,** P.O. Box 57473, Washington, D.C. 20037. Lobbying, education, and acquisition of land: National Parks.

**National Park and Conservation Association,** 1701 18th St., N.W., Washington, D.C. 20009. Research and education: parks, wildlife, forestry, general environmental quality.

**National Resources Defense Council,** 122 E. 42nd Street, New York, N.Y. 10019. Research and litigation: water and air quality, land use, energy.

**National Wildlife Federation,** 1412 16th Street, N.W., Washington, D.C. 20036. Research, education, lobbying: general environmental quality, wilderness and wildlife.

**Nature Conservancy,** 1800 N. Kent Street, Suite 800, Arlington, VA 22209. Research and education: identification, protection and management of natural areas.

**New Directions,** 2700 Q Street, N.W., Washington, D.C. 20007. Lobbying: international energy policy and self-reliant economic development.

**Planned Parenthood Federation of America,** 810 Seventh Avenue, New York, N.Y. 10019. Education, services, and research: fertility control, family planning.

**Public Interest Research Group,** 1346 Connecticut Avenue, N.W. Suite 413, Washington, D.C. 20036. Research and education: alternative energy, utilities regulation.

**Rachel Carson Council,** Inc., 8940 Jones Mill Road, Chevy Chase, MD 20815. Publication distribution and educational conferences: pesticides, toxic substances.

**Resources for the Future,** 1616 P Street, N.W., Washington, D.C. 20036. Research and education: conservation of natural resources, environmental quality.

**Rural America,** 1312 18th Street, N.W., Washington, D.C. 20036. Research and education: health, community development, housing, environmental quality.

**Scientists Institute for Public Information,** 355 Lexington Avenue, New York, N.Y. 10017. Education: scientists of full range of disciplines provide information for public.

**Sierra Club,** 730 Polk Street, San Francisco, CA, 94109. Education, lobbying: water quality, energy, offshore energy development, wilderness areas, urban recreation, wilderness trips.

**Solar Lobby,** 1001 Connecticut Avenue, N.W., 5th Floor, Washington, D.C. 20036. Lobbying: solar energy legislation and legislative oversight.

**Wilderness Society,** 1400 I Street, N.W., 10th Floor, Washington, D.C. 20005. Research, education and lobbying: wilderness, public lands.

**World Wildlife Fund,** 1255 23rd Street, N.W., Washington, D.C. 20037. Research and education: endangered species.

**Worldwatch Institute,** 1776 Massachusetts Avenue, N.W., Washington, D.C. 20036. Research and education: energy, food, population, health, women's issues, technology, the environment.

**Zero Population Growth, Inc.,** 1346 Connecticut Avenue, N.WA., Suite 603, Washington, D.C. 20036. Research, education, lobbying: population.

# APPENDIX B
## Units of Measure

| The Metric System and Equivalent English Units | | | |
|---|---|---|---|
| **LENGTH** | 1 centimeter (cm) × 10 =<br>1 cm = 0.39 inches<br>1 inch = 2.54 cm | 1 decimeter (dm) × 10 =<br>1 dm = 3.94 inches<br>1 foot = 3.05 dm | 1 meter (m) × 1000 =<br>1 m = 1.09 yards<br>1 yard = .91 m | 1 kilometer (km)<br>1 km = 0.62 miles<br>1 mile = 1.61 km |
| **AREA** | a square 1 cm on each<br>side is 1 square<br>centimeter (cm²)<br><br>1 cm² = 0.155 square<br>inches<br>1 square inch = 6.45 cm² | a square 1 m on each<br>side is 1 square meter<br>(m²)<br><br>1 m² = 10.8 square feet<br>1 m² = 1.20 square<br>yards<br>1 square yard = .836 m² | a square 100 m on<br>each side is 1<br>hectare (ha)<br>1 ha × 100 =<br>1 km²<br>1 ha = 2.47 acres<br>1 acre = 0.405 ha | a square 1 km on<br>each side is 1<br>square kilometer<br>(km²)<br><br>1 km² = 0.39<br>square miles<br>1 square mile =<br>2.59 km² |
| **VOLUME** | a cube 1 cm on each side<br>is 1 cubic centimeter<br>(cm³) or 1 milliliter (ml)<br>1 ml × 1000 = 1 L<br><br>1 ml = .203 teaspoons<br>1 teaspoon = 4.9 ml | a cube 1 dm on each side<br>is 1 cubic decimeter<br>(dm³) or 1 liter (L)<br>1 L × 1000 = 1 m³<br><br>1 L = 1.06 quarts<br>1 quart = .95 L | a cube 1 m on each<br>side is 1 cubic<br>meter (m³)<br><br>1 m³<br>1 m³ = 264.2 gallons<br>1 m³ = 36.5 cubic feet<br>1 m³ = 28.4 bushels<br>(dry)<br>1 m³ = 1.31 cubic<br>yards<br>1 cubic yard = .76 m³ | |
| **MASS (WEIGHT)** | 1 ml of water at 4°C<br>weighs 1 gram (g)<br>1 g × 1000 = 1 kg<br><br>1 g = .035 ounces<br>1 ounce = 28.4 g | 1 liter of water at 4°C<br>weighs 1 kilogram (kg)<br>1 kg × 1000 = 1<br>metric ton<br>1 kg = 2.2 pounds<br>1 pound = 0.45 kg | 1 cubic meter of water<br>at 4°C weighs 1<br>metric ton (t), also<br>called a long ton<br>1 t = 2200 pounds<br>1 t = 1.1 short tons<br>1 short ton (2000<br>pounds) = .91 t | |

# ENERGY UNITS AND EQUIVALENTS

1 Calorie, food calorie, or kilocalorie—The amount of heat required to raise the temperature of one kilogram of water one degree Celsius (1.8°F).

1 BTU (British Thermal Unit)—The amount of heat required to raise the temperature of one pound of water one degree Fahrenheit.

    1 Calorie = 3.968 BTU's
    1 BTU = 0.252 calories

1 therm = 100,000 BTU's
1 quad = 1 quadrillion BTU's

1 watt standard unit of electrical power

    1 watt-hour (wh) = 1 watt for 1 hr. = 3.413 BTU's

1 kilowatt (kw) = 1000 watts

    1 kilowatt-hour (kwh) = 1 kilowatt for 1 hr. = 3413 BTU's

1 megawatt (Mw) = 1,000,000 watts

    1 megawatt-hour (Mwh) = 1 Mw for 1 hr. = 34.13 therms

1 gigawatt (Gw) = 1,000,000,000 watts or 1,000 megawatts

    1 gigawatt-hour (Gwh) = 1 Gw for 1 hr. = 34,130 therms

1 horsepower = .7457 kilowatts; 1 horsepower-hour = 2545 BTU's

1 cubic foot of natural gas (methane) at atmospheric pressure = 1031 BTU's

1 gallon gasoline = 125,000 BTU's

1 gallon No. 2 fuel oil = 140,000 BTU's

1 short ton coal = 25,000,000 BTU's

1 barrel (oil) = 42 gallons

# APPENDIX C
## Some Basic Chemical Concepts

### ATOMS, ELEMENTS, AND COMPOUNDS

All matter, whether gas, liquid, or solid, living or nonliving, organic or inorganic, is comprised of fundamental units called **atoms.** Atoms are extremely tiny. If all the world's people, about 5,000,000,000 of us, were reduced to the size of atoms, there would be room for all of us to dance on the head of a pin. In fact, we would only occupy a tiny fraction (about 1/10,000) of the pin's head. Given the incredibly tiny size of atoms, even the smallest particle which can be seen with the naked eye consists of billions of atoms.

The atoms comprising a substance may be all of one kind; or they may be of two or more different kinds. If the atoms are all of one kind, the substance is called an **element.** If the atoms are of two or more different kinds bonded together, the substance is called a **compound.**

Through countless experiments, chemists have ascertained that there are only 96 distinct kinds of atoms which occur in nature. They are listed in Table C-1 with their chemical symbols. By scanning Table C-1, you can see that a number of familiar substances such as aluminum, calcium, carbon, oxygen, and iron are elements; that is, they are a single distinct kind of atom. However, most of the substances with which we interact in every-day life, such as water, stone, wood, protein, and sugar, are not on the list. Their absence from the list is indicative that they are not elements; rather, they are compounds which means they are actually comprised of two or more different kinds of atoms bonded together.

### ATOMS, BONDS, AND CHEMICAL REACTIONS

In chemical reactions, atoms are neither created, nor destroyed, nor is one kind of atom changed into another. What occurs in chemical reactions, whether mild or explosive, is simply a rearrangement of the ways in which the atoms involved are bonded together. An oxygen atom, for example, may be combined and recombined with different atoms to form any number of different compounds but a given oxygen atom always has been, and always will be, an oxygen atom. The same can be said for all the other kinds of atoms. In order to understand how atoms may bond and rearrange to form different compounds, it is necessary to first have some concepts concerning the structure of atoms.

#### Structure of Atoms

In every case, an atom consists of a central core called the nucleus (not to be confused with the cell nucleus). The nucleus of the atom contains one or more **protons** and, except for hydrogen, one or more **neutrons** as well. Surrounding the nucleus are particles called **electrons.** Each proton has a positive (+) electric charge and each electron has an equal but opposite negative (−) electric charge. Thus, the charge of the protons may be balanced by an equal number of electrons making the whole atom neutral. Neutrons have no charge.

Atoms of all elements have this same basic structure consisting of protons, electrons, and neutrons. The distiction among atoms of different elements is in the number of protons. The atoms of each element have a characteristic number of protons which is known as the **atomic number** of the element (see Table C-1). The number of electrons characteristic of the atoms of each element

| Table C-1 | The Elements | | | | | |
|---|---|---|---|---|---|---|

| ELEMENT | SYMBOL | ATOMIC NUMBER | ELEMENT | SYMBOL | ATOMIC NUMBER |
|---|---|---|---|---|---|
| Actinium | Ac | 89 | Mercury | Hg | 80 |
| Aluminum | Al | 13 | Molybdenum | Mo | 42 |
| Americium | Am | 95 | Neodymium | Nd | 60 |
| Antimony | Sb | 51 | Neon | Ne | 10 |
| Argon | Ar | 18 | Neptunium | Np | 93 |
| Arsenic | As | 33 | Nickel | Ni | 28 |
| Astatine | At | 85 | Niobium | Nb | 41 |
| Barium | Ba | 56 | Nitrogen | N | 7 |
| Berkelium | Bk | 97 | Nobelium | No | 102 |
| Beryllium | Be | 4 | Osmium | Os | 76 |
| Bismuth | Bi | 83 | Oxygen | O | 8 |
| Boron | B | 5 | Palladium | Pd | 46 |
| Bromine | Br | 35 | Phosphorus | P | 15 |
| Cadmium | Cd | 48 | Platinum | Pt | 78 |
| Calcium | Ca | 20 | Plutonium | Pu | 94 |
| Californium | Cf | 98 | Polonium | Po | 84 |
| Carbon | C | 6 | Potassium | K | 19 |
| Cerium | Ce | 58 | Praseodymium | Pr | 59 |
| Cesium | Cs | 55 | Promethium | Pm | 61 |
| Chlorine | Cl | 17 | Protoactinium | Pa | 91 |
| Chromium | Cr | 24 | Radium | Ra | 88 |
| Cobalt | Co | 27 | Radon | Rn | 86 |
| Copper | Cu | 29 | Rhenium | Re | 75 |
| Curium | Cm | 96 | Rhodium | Rh | 45 |
| Dysprosium | Dy | 66 | Rubidium | Rb | 37 |
| Einsteinium | Es | 99 | Ruthenium | Ru | 44 |
| Erbium | Er | 68 | Samarium | Sm | 62 |
| Europium | Eu | 63 | Scandium | Sc | 21 |
| Fermium | Fm | 100 | Selenium | Se | 34 |
| Fluorine | F | 9 | Silicon | Si | 14 |
| Francium | Fr | 87 | Silver | Ag | 47 |
| Gadolinium | Gd | 64 | Sodium | Na | 11 |
| Gallium | Ga | 31 | Strontium | Sr | 38 |
| Germanium | Ge | 32 | Sulfur | S | 16 |
| Gold | Au | 79 | Tantalum | Ta | 73 |
| Hafnium | Hf | 72 | Technetium | Tc | 43 |
| Helium | He | 2 | Tellurium | Te | 52 |
| Holmium | Ho | 67 | Terbium | Tb | 65 |
| Hydrogen | H | 1 | Thallium | Tl | 81 |
| Indium | In | 49 | Thorium | Th | 90 |
| Iodine | I | 53 | Thulium | Tm | 69 |
| Iridium | Ir | 77 | Tin | Sn | 50 |
| Iron | Fe | 26 | Titanium | Ti | 22 |
| Krypton | Kr | 36 | Tungsten | W | 74 |
| Lanthanum | La | 57 | Uranium | U | 92 |
| Lawrencium | Lr | 103 | Vanadium | V | 23 |
| Lead | Pb | 82 | Xenon | Xe | 54 |
| Lithium | Li | 3 | Ytterbium | Yb | 70 |
| Lutetium | Lu | 71 | Yttrium | Y | 39 |
| Magnesium | Mg | 12 | Zinc | Zn | 30 |
| Manganese | Mn | 25 | Zirconium | Zr | 40 |
| Mendelevium | Md | 101 | | | |

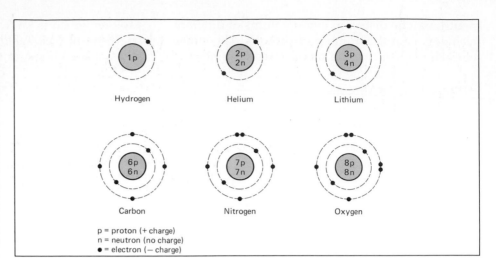

**FIGURE C-1**
Structure of atoms. All atoms consist of fundamental particles: protons (P), which have a positive electric charge, neutrons (n), which have no charge, and electrons, which have a negative charge. Protons and neutrons are located in a central core, the nucleus. The positive charge of the protons is balanced by an equal number of electrons, which occupy various levels or orbitals around the nucleus. The uniqueness of each element is given by its atoms having a distinct number of protons, its atomic number.

p = proton (+ charge)
n = neutron (no charge)
● = electron (− charge)

Hydrogen  Helium  Lithium

Carbon  Nitrogen  Oxygen

also differs corresponding to the number of protons. The general structure of the atoms of several elements is shown in Figure C-1.

The number of protons and electrons, i.e., the atomic number of the element, determines the chemical properties of the element. However, the number of neutrons may also vary. For example, most carbon atoms have six neutrons in addition to the six protons as indicated in figure C-1. But some carbon atoms have eight neutrons. Atoms of the same element which have different numbers of neutrons are known as **isotopes** of the element. The total number of protons plus neutrons is used to define different isotopes. For example, the usual isotope of carbon is referred to as carbon-12 while the isotope noted above is referred to as carbon-14. The chemical reactivity of different isotopes of the same element is identical. However certain other properties may differ. Many isotopes of various elements prove to be radioactive as is carbon-14.

## Bonding of Atoms

The chemical properties of an element are defined by the ways in which its atoms will react and form bonds with other atoms. By examining how atoms form bonds, we shall see how the number of electrons and protons determines these properties. There are two basic kinds of bonding: (1) **covalent bonding,** and (2) **ionic bonding.**

In both kinds of bonding, it is first important to recognize that electrons are not randomly dis-

tributed around the atom's nucleus. Rather, there are, in effect, specific spaces in a series of layers, or **orbitals,** around the nucleus. If an orbital is occupied by one or more electrons but not filled, the atom is unstable; it will tend to react and form bonds with other atoms to achieve greater stability. A stable state is achieved by having all the spaces in the orbital filled with electrons. But, it is also important to keep the charge neutral, i.e., the total number of electrons equal to that of the protons.

### COVALENT BONDING.

These two requirements, filling all the spaces and keeping the charge neutral, may be satisfied by adjacent atoms sharing one or more pairs of electrons as shown in Figure C-2. The sharing of a pair of electrons holds the atoms together in what is called a **covalent bond.**

Covalent bonding, by satisfying the charge-orbital requirements, leads to descrete units of two or more atoms bonded together. Such units of two or more covalently bonded atoms are called **molecules.** A few simple but important examples are shown in Figure C-2.

A chemical formula is simply a shorthand description of the number of each kind of atom in a given molecule. The element is given by the chemical symbol and a subscript following the symbol gives the number present, no subscript being understood as one. A molecule with two or more different kinds of atoms may also be called a compound, but a molecule comprised of a single kind of atom, oxygen ($O_2$) for example, is still defined as an element.

**FIGURE C-2**

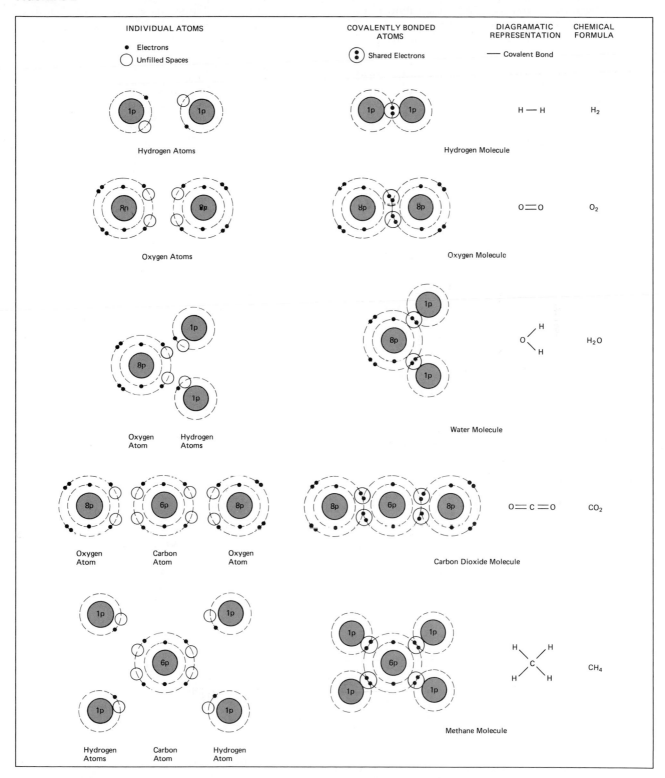

Only a few elements, namely carbon, hydrogen, oxygen, nitrogen, phosphorus and sulfur, have configurations of electrons which lend readily to the formation of covalent bonds. But, carbon specifically, with its ability to form four covalent bonds, can produce long, straight or branched chains, or rings (Fig. C-3). Thus, an infinite array of molecules can be formed by using covalently bonded carbon atoms as a "backbone" and filling in the sides with atoms of hydrogen or other elements. Thus, it is covalent bonding among atoms of carbon and these few other elements that produces all natural organic molecules, those molecules that comprise all the tissues of living things, and also synthetic organic compounds such as plastics.

**FIGURE C-3**
Covalent bonding and organic molecules. The ability of carbon and a few other elements to readily form covalent bonds leads to an infinite array of complex molecules, organic molecules, which constitute all living things. A few major kinds of groupings are shown here. Note that each element forms a characteristic number of bonds: carbon, 4; nitrogen, 3; oxygen, 2; hydrogen, 1; sulfur, 2; phosphorus, 5. Bonds (dashed lines) left "hanging" indicate attachments to other atoms or groups of atoms.

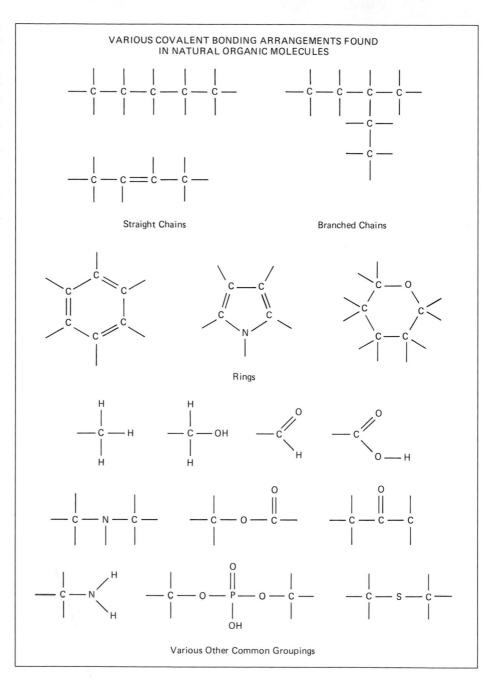

VARIOUS COVALENT BONDING ARRANGEMENTS FOUND IN NATURAL ORGANIC MOLECULES

Straight Chains

Branched Chains

Rings

Various Other Common Groupings

## IONIC BONDING.

Another way in which atoms may achieve a stable electron configuration is to gain additional electrons to complete the filling of an orbital, or lose electrons which are over a completed orbital. In general, the maximum number of electrons that can be gained or lost by an atom is three. Therefore, an element's atomic number determines whether one or more electrons will be lost or gained. If an atom's outer orbital is one to three electrons short of being filled, it will always tend to gain additional electrons. Conversely, if an atom has one to three electrons over its last complete orbital it will always tend to give them away.

Of course gaining or losing electrons results in the number of electrons being greater or less than the number of protons, and the atom consequently having an electric charge. The charge will be one negative for each electron gained or one positive for each electron lost (Fig. C-4). A cova-

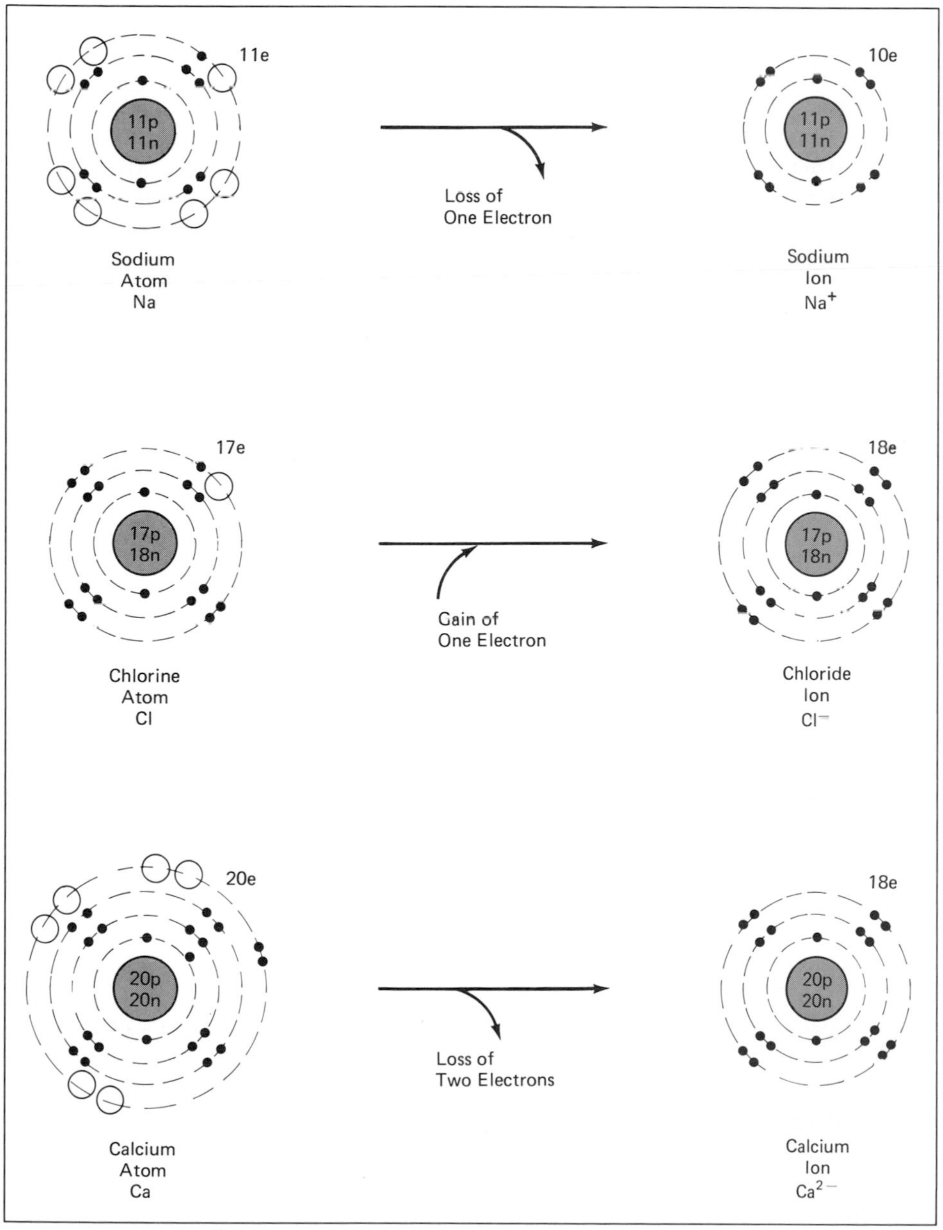

**FIGURE C-4**
Formation of ions. Many atoms will tend to gain or lose one or more electrons in order to achieve a state of complete (electron-filled) orbitals. In doing so they become positively or negatively charged ions as indicated.

**Table C-2**

| Ions of Particular Importance to Biological Systems | | | |
|---|---|---|---|
| *NEGATIVE (−) IONS* | | *POSITIVE (+) IONS* | |
| Phosphate | $PO_4^{3-}$ | Potassium | $K^+$ |
| Sulfate | $SO_4^{2-}$ | Calcium | $Ca^{2+}$ |
| Nitrate | $NO_3^-$ | Magnesium | $Mg^{2+}$ |
| Hydroxyl | $OH^-$ | Iron | $Fe^{2+}$, $Fe^{3+}$ |
| Chloride | $Cl^-$ | Hydrogen | $H^+$ |
| | | Ammonium | $NH_4^+$ |
| | | Sodium | $Na^+$ |

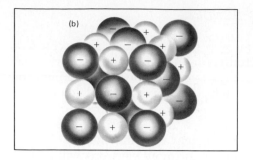

**FIGURE C-5.**
Positive and negative ions bond together by their mutual attraction.

lently bonded group of atoms may acquire an electric charge in the same way. An atom or group of atoms which has acquired an electric charge in this way is called an **ion,** positive or negative. Ions are designated by a superscript following the chemical symbol giving the number of positive or negative charges. Absence of superscripts indicates that the atom or molecule is neutral. Some important ions are listed in Table C-2.

Since unlike charges attract, positive and negative ions tend to join and pack together in dense clusters in such a way as to neutralize the overall electric charge. This joining together of ions through the attraction of their opposite charges is called **ionic bonding.** The result is the formation of hard, brittle, more or less crystalline substances of which all rocks and minerals are examples (Fig. C-5).

It is significant to note that whereas covalent bonding leads to descrete molecules, ionic bonding does not. Any number and combination of positive and negative ions may enter into an ionically bonded cluster to produce crystals of almost any size. The only restriction is that the overall charge of positive ions is balanced by that of negative ions. Thus, ionicly bonded substances are properly called compounds but not molecules. When chemical formulas are used to describe such compounds, they define the ratio of various elements involved, not specific molecules.

## Chemical Reactions and Energy

While atoms themselves do not change, the bonds between atoms may be broken and reformed with different atoms producing different compounds and/or molecules. This is essentially what occurs in all chemical reactions. What deter-

mines whether a given chemical reaction will occur or not? We noted above that atoms form bonds because they achieve a greater stability by doing so. But some bonding arrangements may provide greater overall stability than others. Consequently substances with relatively unstable bonding arrangements will tend to react to form one or more different compounds which have more stable bonding arrangements. Common ex-

**FIGURE C-6.**
Some bonding arrangements are more stable than others. Chemical reactions will go spontaneously toward more stable arrangements, releasing energy in the process. But, reactions may be driven in the opposite direction with suitable energy inputs.

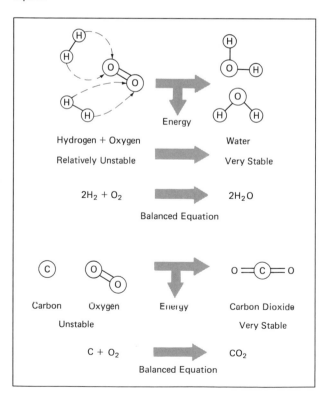

Hydrogen + Oxygen
Relatively Unstable

Energy

Water
Very Stable

$2H_2 + O_2$

Balanced Equation

$2H_2O$

Carbon    Oxygen
Unstable

Energy

Carbon Dioxide
Very Stable

$C + O_2$

Balanced Equation

$CO_2$

amples are the reaction between hydrogen and oxygen to produce water, and the reaction between carbon and oxygen to produce carbon dioxide (Fig. C-6).

Additionally, energy is always released in the process of gaining greater overall stability as indicated in Figure C-6. Thus, energy being released from a chemical reaction is synonomous with the atoms achieving more stable bonding arrangements. Thus, it may be said that chemical reactions always tend to go in a direction that releases energy as well as one which gives greater stability.

However, chemical reactions can be made to go in a reverse direction. With suitable energy inputs and under suitable conditions, stable bonding arrangements may be broken and less stable arrangements formed. As described in Chapter 2, this is the basis of photosynthesis occurring in green plants. Light energy is brought to bear on splitting the highly stable hydrogen-oxygen bonds of water and forming less stable carbon-hydrogen bonds thus creating high-energy organic compounds.

# GLOSSARY

**abiotic.** Pertaining to factors or things that are separate and independent from living things; nonliving.

**acid.** Any compound that releases hydrogen ions when dissolved in water. Also, a water solution that contains a surplus of hydrogen ions.

**acid deposition.** Any form of acid precipitation and also fallout of dry acid particles. (See **acid precipitation**)

**acid precipitation.** Includes acid rain, acid fog, acid snow and any other form of precipitation that is more acidic than normal, i.e., less than pH 5.6. Excess acidity is derived from certain air pollutants, namely sulfur dioxide and oxides of nitrogen.

**activated charcoal.** A form of carbon that readily adsorbs organic material. Therefore it is frequently used in air and/or water filters to remove organic contaminants. It does not remove ions such as those of the heavy metals.

**activated sludge system.** A system for removing organic wates from water. The system uses microorganisms and active aeration to decompose such wastes. The system is most used as a means of secondary sewage treatment following the primary settling of materials.

**adaption** (ecological or evolutionary). A change in structure or function that produces better adjustment of an organism to its environment, and hence enhances its ability to survive and reproduce.

**adsorption.** The process of chemicals (ions or molecules) sticking to the surface of other materials.

**advanced treatment** (sewage treatment). Any of a variety of systems that follow secondary treatment and that are designed to remove one or more nutrients, such as phosphate, from solution.

**aeration.** *Soil:* The property of a soil relating to its ability to allow the exchange of oxygen and carbon dioxide, which is necessary for the respiration of roots. *Water:* The bubbling of air or oxygen through water to increase the dissolved oxygen.

**alga,** pl. **algae.** Any of numerous kinds of photosynthetic plants that live and reproduce entirely immersed in water. Many species, the planktonic forms, exist as single or small groups of cells that float freely in the water. Other species, the "seaweeds," may be large and attached.

**algal bloom.** A relatively sudden development of a heavy growth of algae, especially planktonic forms. Algal blooms generally result from additions of nutrients, which are normally limiting.

**anaerobic digestion.** The breakdown of organic material by microorganisms in the absence of oxygen. The process results in the release of methane gas as a waste product.

**anaerobic respiration.** Respiration carried on by certain bacteria in the absence of oxygen. Methane, which can be used as fuel gas (it is the same as natural gas), may be a byproduct of the process.

**appropriate technology.** Technology which seeks to increase the efficiency and productivity of hand labor without displacing workers. That is, it seeks to enable people to improve their well-being without disrupting the existing social and economic system.

**aquaculture.** A propagation and/or rearing of any aquatic (water) organism in a more or less artificial system.

**aquifer.** An underground layer of porous rock, sand, or other material that allows the movement of water between layers of nonporous rock or clay. Aquifers are frequently tapped for wells.

**artificial selection.** Plant and animal breeders' practice of selecting individuals with the greatest expression of desired traits to be the parents of the next generation.

**asbestos fibers.** Crystals of asbestos, a natural mineral, which have the form of minute fibers.

**assimilate.** To incorporate into the natural working or functioning of the system as, for example, natural organic wastes are assimilated (broken down and incorporated) into the nutrient cycles of the ecosystem.

**atom.** The fundamental unit of all elements.

**autotroph.** Any organism that can synthesize all its organic substances from inorganic nutrients, using light or certain inorganic chemicals as a source of energy. Green plants are the principal autotrophs.

**average crustal abundance.** The percentage of the earth's crust that is composed of a given element.

**bacteria.** Any of numerous kinds of microscopic organisms which exist as simple, single cells that multiply by simple division. Along with fungi, they comprise the decomposer component of ecosystems. A few species cause disease.

**base.** Any compound that releases hydroxyl ($OH^-$) ions when dissolved in water. A solution that contains a surplus of $OH^-$ ions.

**bed load.** The load of coarse sediment, mostly coarse silt and sand, that is gradually moved along the bottom of a river bed by flowing water rather than being carried in suspension.

**benthic plants.** Plants which grow underwater attached to or rooted in the bottom. For photosynthesis, they depend on light penetrating the water.

**bioaccumulation.** The accumulation of higher and higher concentrations of potentially toxic chemicals in organisms. It occurs in the case of chemicals such as heavy metals and chlorinated hydrocarbons that may be absorbed or ingested with food but can neither be broken down nor excreted. Consequently, organisms act as strainers, accumulating increasing amounts. Through a food chain, organisms at higher tropic levels may accumulate concentrations as much as a millionfold higher than those present in the environment. Also called **biomagnification.**

**biocide.** Applies to any pesticide or other chemical which is more or less toxic to many, if not all, kinds of living organisms.

**bioconversion** (energy). The use of biomass as fuel. Burning materials such as wood, paper, and plant wastes directly to produce energy, or converting such materials into fuels such as alcohol and methane.

**biodegradable.** Able to be consumed and broken down to natural substances such as carbon dioxide and water by biological organisms, particularly decomposers. Opposite: **nonbiodegradable.**

**biological control** (pest control). Control of a pest population by introduction of predatory, parasitic, or disease-causing organisms.

**biological oxygen demand (BOD).** A measure of water quality; the amount of dissolved oxygen that will be consumed by biological organisms in the process of decomposing organic material present. The greater the BOD, the lower the water quality.

**biological treatment** (sewage treatment). See **secondary treatment.**

**biomagnification.** See **bioaccumulation.**

**biomass.** Mass of biological material. Usually the total mass of a particular group or category; for example, biomass of producers.

**biomass energy** or **biomass fuels.** Energy, or fuels such as alcohol and methane, produced from current photosynthetic production of biological material. See **bioconversion.**

**biomass pyramid.** Refers to the structure that is obtained when the respective biomasses of producers, herbivores and carnivores in an ecosystem are compared, i.e. producers are found to have the largest biomass followed by herbivores and then carnivores.

**biome.** A group of ecosystems that are related by having a similar type of vegetation governed by similar climatic conditions. Examples include prairies, deciduous forests, arctic tundra, deserts, and tropical rain forests.

**biosphere.** The overall ecosystem of the earth. It is the sum total of all the biomes and smaller ecosystems, which are ultimately all interconnected and interdependent through global processes such as water and atmospheric cycles.

**biota.** Refers to any and all living organisms and the ecosystems in which they exist.

**biotic.** Living or dervied from living things.

**biotic potential.** The potential of a species for increasing its population and/or distribution. The biotic potential of every species is such that, given optimum conditions, its population will increase. Contrast environmental resistance.

**birth control.** Any means, natural or artificial, which may be used to reduce the number of live births.

**BOD.** See **biological oxygen demand.**

**breeder reactor.** A nuclear reactor that in the course of producing energy also converts nonfissionable uranium-238 into fissionable plutonium-239, which can be used as fuel. Hence, a reactor that produces as much nuclear fuel as it consumes, or more.

**broad-spectrum pesticides.** Chemical pesticides that kill a wide range of pests. They also kill a wide range of nonpest and beneficial species; therefore, they may lead to environmental upsets and resurgences. Contrast narrow spectrum pesticides and biorational pesticides.

**BTU (British Thermal Unit).** A fundamental unit of energy in the English system. The amount of heat required to raise the temperature of one pound of water one degree Fahrenheit.

**buffer.** A substance that will maintain the pH of a solution by reacting with the excess acid. Limestone is a natural buffer which helps to maintain water and soil at a pH near neutral.

**buffering capacity.** Refers to the amount of acid that may be neutralized by a given amount of buffer.

**calorie.** A fundamental unit of energy. The amount of heat required to raise the temperature of 1 g of water 1 degree Celsius. All forms of energy can be converted to heat and measured in calories. Calories used in connection with food are kilocalories or "big" calories, the amount of heat required to raise the temperature of 1 liter of water 1 degree Celsius.

**capillary water.** Water that clings in small pores, cracks, and spaces against the pull of gravity, such as water held in a sponge.

**carbon dioxide effect.** See **greenhouse effect.**

**carcinogenic.** Having the property of causing cancer, at least in animals and by implication in humans.

**carnivore.** An animal that feeds more or less exclusively on other animals.

**carrying capacity.** The maximum population of a given animal or humans that an ecosystem can support without being degraded or destroyed in the long run. The carrying capacity may be exceeded, but not without lessening the system's ability to support life in the long run.

**catalyst.** A substance that promotes a given chemical reaction without itself being consumed or changed by the reaction. Enzymes are catalysts for biological reactions. Also catalysts are used in some pollution control devices, e.g., the **catalytic converter.**

**catalytic converter.** The device used by American automobile manufacturers to reduce the amount of carbon monoxide and hydrocarbons in the exhaust. The converter contains a catalyst that oxidizes these compounds to carbon dioxide and water as the exhaust passes through.

**cell** (biological). The basic unit of life, the smallest unit that still maintains all the attributes of life. Many microscopic organisms consist of a single cell. Large organisms consist of trillions of specialized cells functioning together.

**cell respiration.** The chemical process that occurs in all living cells wherein organic compounds are broken down to release energy required for life processes. Higher plants and animals require oxygen for the process as well and release carbon dioxide and water as waste products, but certain microorganisms do not require oxygen (see **anaerobic respiration**).

**cell wall.** A more or less rigid wall, composed mainly of cellulose, which surrounds plant cells and provides the supporting structure of plant tissues.

**cellulose.** The organic macromolecule that is the prime constituent of plant cell walls and hence the major molecule in wood, wood products, and cotton. It is composed of glucose molecules, but since it cannot be digested by humans its dietary value is only as fiber, bulk, or roughage.

**chain reaction** (nuclear). Reaction wherein each atom that fissions (splits) causes one or more additional atoms to fission.

**channelization.** The straightening and deepening of stream or river channels to speed water flow and reduce flooding.

**chemical energy.** The potential energy that is contained in certain chemicals; most importantly, the energy contained in organic compounds such as food and fuels, which may be released through respiration or burning.

**chemosynthesis.** The ability of some microorganisms to utilize the chemical energy contained in certain inorganic chemicals such as hydrogen sulfide for the production of organic material. Such organisms are producers.

**chlorinated hydrocarbons.** Synthetic organic molecules in which one or more hydrogen atoms have been replaced by chlorine atoms. They are extremely hazardous compounds, because they tend to be nonbiodegradable, they tend to bioaccumulate, and many have been shown to be carcinogenic. Also called organochlorides.

**chlorination.** The processes of adding chlorine to drinking water or sewage water in order to kill microorganisms that may cause disease.

**chlorofluorocarbons.** Synthetic organic molecules that contain one or more of both chlorine and fluorine atoms.

**chlorophyll.** The green pigment in plants, responsible for absorbing the light energy required for photosynthesis.

**clearcutting.** Cutting every tree, leaving the area completely clear.

**climax ecosystem.** The last stage in ecological succession. An ecosystem in which populations of all organisms are in balance with each other and existing abiotic factors.

**clustered development.** The development pattern in which homes and other facilities are arranged in dense clusters on a relatively small portion of the land considered for development, allowing the rest of the land to remain open.

**cogeneration.** The joint production of useful heat and electricity. For example, furnaces may be replaced with gas turbogenerators which produce electricity while the hot exhaust still serves as a heat source. An important avenue of conservation, it effectively avoids the waste of heat that normally occurs at centralized power plants.

**compaction.** Packing down. *Soil:* Packing and pressing out air spaces present in the soil. Reduces soil aeration and infiltration and thus reduces the capacity of the soil to support plants. *Trash:* packing down trash to reduce the space that it requires.

**composting.** The process of letting organic wastes decompose in the presence of air. A nutrient-rich humus or compost results.

**composting toilet.** A toilet that does not flush wastes away with water but deposits them in a chamber where they will compost (See **composting**).

**compound.** Any substance (gas, liquid, or solid) that is made up of two or more different kinds of atoms bonded together. Contrast **element.**

**condensation.** The collecting together of molecules from the vapor state to form the liquid state, as, for example, water vapor condenses on a cold surface and forms droplets. Opposite: **evaporation.**

**confusion technique** (pest control). Applying a quantity of sex attractant to an area so that males become confused and are unable to locate females. The actual quantities of pheromes

applied are still very small because of their extreme potency.

**conservation.** The management of a resource in such a way as to assure that it will continue to provide maximum benefit to humans over the long run. Conservation may include various degrees of use or protection, depending on what is necessary to maintain the resource over the long run. *Energy:* Saving energy. It not only entails cutting back on use of heating, air conditioning, lighting, transportation, and so on, but also entails increasing the efficiency of energy use. That is, developing and instigating means of doing the same jobs, e.g., transporting people, with less energy.

**consumers.** In an ecosystem, those organisms that derive their energy from feeding on other organisms or their products.

**contour farming.** The practice of cultivating land along the contours across rather than up and down slopes. In combination with strip cropping it reduces water erosion.

**contraceptive.** Any device or drug that is designed to allow normal sexual intercourse but prevent unwanted pregnancies from occurring.

**control group.** The group in an experiment that is the same as and is treated like the experimental group in every way except for the particular factor being tested. Only by comparison to a control group can one gain specific information concerning the effect of any test factor.

**controlled experiment.** An experiment with adequate control groups (see **control group**).

**control rods** (nuclear power). Part of the core of the reactor, the rods of neutron-absorbing material that are inserted or removed as necessary to control the rate of nuclear fissioning.

**convection currents.** Wind or water currents promoted by the fact that warming causes expansion, decreases density, and thus causes the warmer air or water to rise. Conversely, the sinking of cooler air or water.

**cooling tower.** A massive tower designed to dissipate waste heat from a power plant (or other industrial process) into the atmosphere.

**cooperative energy use.** Utilizing the waste heat from one process as the source of heat for another process which requires a lesser temperature.

**cornucopians.** Those who believe that new discoveries, advances in technology, and finding substitutes will continue to make resources effectively unlimited and allow our present economic-social system to continue more or less indefinitely. Contrast **limitists.**

**cosmetic damage** (of fruits and vegetables). Damage to the surface that affects appearance but does not otherwise affect taste, nutritional quality, or storability.

**cosmetic spraying.** Spraying of pesticides which is done to control pests which only damage the surface appearance.

**cost-benefit ratio** or **benefit-cost ratio.** The value of the benefits to be gained from a project divided by the costs of the project. If the ratio is greater than 1, the project is economically justified; if less than 1, it is not economically justified.

**cost-effective.** Pertaining to a project or procedure that produces economic returns or benefits that are significantly greater than the costs.

**covalent bond.** A chemical bond between two atoms, formed by sharing a pair of electrons between the two atoms. Atoms of all organic compounds are joined by covalent bonds.

**criteria pollutants.** Certain pollutants the level of which is used as a gage for the determination of air (or water) quality.

**critical level.** The level of one or more pollutants above which severe damage begins to occur and below which few if any ill effects are noted.

**critical number.** Refers to the minimum number of individuals of a given species that is required to maintain a healthy, viable population of the species. If a population falls below its critical number its extinction will almost certainly occur.

**crop rotation.** The practice of alternating the crops grown on a piece of land. For example, corn one year, hay for two years, then back to corn.

**crude birth rate.** Number of births per 1000 individuals per year. It is found by taking the total number of births occurring in a population during a year, dividing by the total population at midyear, and then multiplying the result by 1000.

**crude death rate.** Number of deaths per 1000 individuals per year. It is found by taking the total number of births occurring in a population during a year, dividing by the total population at midyear, and then multiplying the result by 1000.

**crystallization.** The joining together of molecules or ions from a liquid (or sometimes gaseous) state to form a solid state.

**cultural control** (pest control). A change in the practice of growing, harvesting, storing, handling, or disposing of wastes that reduces the susceptibility or exposure to pests. For example, spraying the house with insecticides to kill flies is a chemical control; putting screens on the windows to keep flies out is a cultural control.

**DDT (dichlorodiphenyltrichloroethane).** The first and most widely used of the synthetic organic pesticides belonging to the chlorinated hydrocarbon class.

**debt crisis.** Refers to the fact that many less-developed nations are so heavily in debt that they may not be able to meet their financial obligations, e.g. interest payments. Failing to meet such obligations could have severe economic impacts on the entire world.

**declining tax base.** The loss of tax revenues that occurs as a result of affluent taxpayers and businesses leaving an area and a subsequent decline of property values. It has been especially severe in inner cities as a result of migration to suburbs and exurbs.

**decommissioning** (of nuclear power plants). Refers to the inevitable need to take nuclear power plants out of service after 25-35 years because the effects of radiation will gradually make them inoperable.

**decomposers.** Organisms the feeding action of which results in decay, rotting, or decomposition. The primary decomposers are fungi and bacteria.

**degrade.** To lower the quality or usefulness of.

**demographic transition.** The transition from a condition of high birth rate and high death rate through a period of declining death rate but continuing high birth rate finally to low birth rate and low death rate. This transition may result from economic development.

**deoxyribonucleic acid.** See **DNA**.

**desertification.** The process of semiarid grasslands being degraded to pure desert. Overgrazing by livestock is the major causative factor.

**detritus.** The dead organic matter, such as fallen leaves, twigs, and other plant and animal wastes, that exists in any ecosystem.

**detritus feeders.** Organisms such as termites, fungi, and bacteria that obtain their nutrients and energy mainly by feeding on dead organic matter.

**deuterium ($^2$H).** A stable, naturally occurring isotope of hydrogen. It contains one neutron in addition to the single proton normally in the nucleus.

**development rights.** Legal documents that grant permission to develop a given piece of property. They must be owned by the developer before development can occur. They can be bought and sold apart from the property itself.

**dioxin.** A synthetic organic chemical of the chlorinated hydrocarbon class. It is one of the most toxic compounds known to humans, having many harmful effects, including induction of cancer and birth defects, even in extremely minute concentrations. It has become a widespread environmental pollutant because of the use of certain herbicides which contain dioxine as a contaminant.

**disinfection.** The killing (as opposed to removal) of microorganisms in water or other media where they might otherwise pose a health threat. For example, chlorine is commonly used to disinfect water supplies.

**dissolved oxygen (DO).** Oxygen gas molecules ($O_2$) dissolved in water. Fish and other aquatic organisms are dependent upon dissolved oxygen for respiration. Therefore concentration of dissolved oxygen is a measure of water quality.

**distillation.** A process of purifying water or other liquids by boiling the liquid and recondensing the vapor. Contaminants remain behind in the boiler.

**district heating.** The heating of an entire community or city area through circulating heat (e.g. steam) from a central source; particularly, utilizing waste heat from a power plant or from incineration of refuse.

**diversion** (of water). Taking some or all of the flow of a natural waterway and carrying it to other places for uses such as municipal water supplies or irrigation.

**DNA (deoxyribonucleic acid).** The natural organic macromolecule that carries the genetic or hereditary information for virtually all organisms.

**DO.** See **dissolved oxygen**.

**domestic solid wastes.** Wastes that come from homes, offices, schools, and stores, as opposed to wastes that are generated from agricultural or industrial processes.

**doubling time.** The time it will take a population to double in size assuming the continuation of current fertility rate and no change in survival rate.

**ecological pest management.** Control of pest populations through understanding the various ecological factors that provide natural control and so far as possible utilizing these factors as opposed to using synthetic chemicals.

**ecologists.** Scientists who study ecology, i.e., the ways in which organisms interact with each other and their environment.

**ecology.** The study of any and all aspects of the structure, function, and perpetuation of ecosystems.

**ecosystem.** A grouping of plants, animals, and other organisms interacting with each other and their environment in such a way as to perpetuate the grouping more or less indefinitely. Ecosystems have characteristic forms such as deserts, grasslands, tundra, deciduous forests, and tropical rain forests.

**ectoparasite.** See **parasites**.

**electrons.** Fundamental atomic particles that have a negative electrical charge but virtually no mass. They surround the nuclei of atoms and thus balance the positive charge of protons in the nucleus. A flow of electrons in a wire is synonymous with an electric current.

**element.** A substance that is comprised of one and only one distinct kind of atom. Contrast **compound.**

**embrittlement.**   Becoming brittle. Pertains especially to the reactor vessel of nuclear power plants gradually becoming brittle as a result of continuous bombardment by radiation. It is prime factor forcing the decommissioning of nuclear power plants.

**endangered species.**   A species the total population of which is declining to relatively low levels, a trend which if continued will result in extinction.

**endoparasite.**   See **parasites.**

**energy.**   The ability to do work. Common forms of energy are light, heat, electricity, motion, and chemical bond energy inherent in compounds such as sugar, gasoline, and other fuels.

**enrichment.**   With reference to nuclear power, refers to the separation and concentration of uranium-235 so that, in suitable quantities, it will sustain a chain reaction.

**entomologist.**   A scientist who studies insects, their life cycles, physiology, behavior, and so on.

**environment.**   The combination of all things and factors external to' the individual or population of organisms in question.

**environmental impact statement.**   A study of the probable environment impacts of a development project. The National Environmental Policy Act of 1968 (NEPA) requires such studies prior to proceding with any project receiving federal funding.

**environmental resistance.**   The totality of factors such as adverse weather conditions, shortage of food or water, predators, and diseases which tend to cut back populations and keep them from growing or spreading. Contrast **biotic potential.**

**EPA.**   United States Environmental Protection Agency. The federal agency responsible for control of all forms of pollution and other kinds of environmental degradation.

**erosion.**   The process of soil particles being carried away by wind or water. Erosion moves the smaller soil particles first and hence degrades the soil to a more coarse, sandy, stony texture.

**estimated reserves.**   See **reserves.**

**euphotic zone.**   The layer or depth of water through which there is adequate light penetration to support photosynthesis.

**eutrophication.**   Nutrient enrichment of a body of water which leads, in turn, to excessive growth of algae and then depletion of dissolved oxygen as dead algae is consumed by decomposers.

**evaporation.**   Molecules leaving the liquid state and entering the vapor or gaseous state as, for example, water evaporates to form water vapor.

**evolution.**   The theory that all species now on earth are descended from ancestral species through a process of gradual change brought about by natural selection.

**evolutionary succession.**   The succession of different species that have inhabited the earth at different geological periods, as revealed through the fossil record. The process of new species coming in through the process of speciation while other species pass into extinction.

**experimental group.**   The group in an experiment which receives the experimental treatment in contrast to the control group, used for comparison, which does not receive the treatment.

**exponential growth.**   The growth produced when the base increases by a given *percentage* (as opposed to a given amount) each year. It is characterized by doubling again and again,

each doubling occurring in the same period of time. It produces a J-shaped curve.

**extinction.**   The death of all individuals of a particular species. When this occurs, all the genes of that particular line are lost forever.

**exurban migration.**   Refers to the pronounced trend since World War II· of peope relocating homes and businesses from the central city and older suburbs to more outlying suburbs.

**FAO.**   Food and Agriculture Organization of the United Nations.

**fecal coliform test.**   A test for the presence of E. coli, the bacterium that normally inhabits the gut of humans and other mammals. A positive test indicates sewage contamination and the potential presence of disease-causing microorganisms carried by sewage.

**fermentation.**   A form of respiration carried on by yeast cells in the absence of oxygen. It involves a partial breakdown of glucose (sugar) which yields energy for the yeast and the release of alcohol as a byproduct.

**fertility rate.**   The number of live births occurring per 1000 women aged 15-44. Specifically referred to as the general fertility rate. See also **total fertility rate.**

**fertilizer.**   See **inorganic fertilizer** and **organic fertilizer.**

**field capacity.**   A measure of the maximum volume of water that a soil can hold by capillary action, i.e., against the pull of gravity.

**FIFRA** (Federal Insecticide, Fungicide and Rodenticide Act)

**filtration.**   The passing of water (or other fluid) through a filter to remove certain impurities.

**fission.**   The splitting of a large atom into two atoms of lighter elements. When large atoms such as uranium or plutonium fission, tremendous amounts of energy are released.

**flat-plate collector** (solar energy).   A solar collector that consists of a stationary, flat, black surface oriented perpendicular to the average sun angle. Heat absorbed by the surface is removed and transported by air or water (or other liquid) flowing over or through the surface.

**food chain.**   The transfer of energy and material through a series of organisms as each one is fed upon by the next.

**food web.**   The combination of all the feeding relationships that exist in an ecosystem.

**fossil fuels.**   Mainly crude oil, coal, and natural gas, which are derived from prehistoric photosynthetic production of organic matter on earth.

**fuel elements** (nuclear power).   The pellets of uranium or other fissionable material that are placed in tubes, which, with the control rods, form the core of the reactor.

**fungus, pl. fungi.**   Any of numerous species of molds, mushrooms, brackets, and other forms of nonphotosynthetic plants. They derive energy and nutrients by consuming other organic material. Along with bacteria they form the decomposer component of ecosystems.

**fusion.**   The joining together of two atoms to form a single atom of a heavier element. When light atoms such as hydrogen are fused, tremendous amounts of energy are released.

**gasohol.**   A blend of 90 percent gasoline and 10 percent alcohol, which can be substituted for straight gasoline. It serves to stretch gasoline supplies.

**genes.**   The chemical units, passed from parents to offspring through the egg and sperm, that determine hereditary features. There include physical, physiological, and, to some extent, behavioral characteristics. Genes may be changed by mutations, and new combinations of genes in the offspring may lead to characteristics not observed in parents.

**genetic bank.** The concept that natural ecosystems with all their species serve as a tremendous repository of genes that is frequently drawn upon to improve domestic plants and animals; further, that large segments of all natural ecosystems should be perceived to maintain this repository of genes.

**genetic control** (pest control). Selective breeding of the desired plant or animal to make it resistant to attack by pests. Also, attempting to introduce harmful genes—for example, those that cause sterility—into the pest populations.

**genetic engineering.** The artificial transfer of specific genes from one organism to another.

**genetics.** The study of heredity and the processes by which inherited characteristics are passed from one generation to the next.

**gentrification.** The trend seen in modern society of people moving into more or less isolated communities with others of similar economic, ethnic, and social backgrounds.

**geothermal energy.** Useful energy derived from the naturally hot interior of the earth.

**glucose.** A simple sugar, the major product of photosynthesis. Serves as the basic building block for cellulose and starches and as the major "fuel" for the release of energy through respiration in both plants and animals.

**gravitational water.** Water that is not held by capillary action in soil but percolates downward by the force of gravity.

**gray water.** Wastewater, as from sinks and tubs, that does not contain human excrements. Such water can be reused for many purposes since it does not pose a health problem.

**greenhouse effect.** A warming of the atmospheric temperature, resulting from solar radiation (light) coming in but outgoing heat radiation being absorbed. Carbon dioxide in the atmosphere is a potent absorber of heat radiation. Therefore increasing the carbon dioxide content of the atmosphere may increase global temperatures because of this effect.

**groundwater.** Water that has accumulated in the ground, completely filling and saturating all pores and spaces in rock and/or soil. Groundwater is free to move more or less readily. It is the reservoir for springs and wells, and is replenished by infiltration of surface water.

**gully erosion.** Gullies, large or small, resulting from water erosion.

**half-life.** The length of time it takes for half of an unstable isotope to decay. The length of time is the same regardless of the starting amount. Also refers to the amount of time it takes compounds to break down in the environment.

**halogenated hydrocarbon.** Synthetic organic compound containing one or more atoms of the halogen group, which includes chlorine, fluorine, and bromine.

**hard water.** Water that contains relatively large amounts of calcium and/or certain other minerals that cause soap to precipitate.

**heavy metals.** Any of the high atomic weight metals such as lead, mercury, cadmium, and zinc. All may be serious pollutants in water or soil because they are toxic in relatively low concentrations and they tend to bioaccumulate.

**herbicide.** A chemical used to kill or inhibit the growth of undesired plants.

**herbivore,** adj., **herbivorous.** An organism such as rabbit or deer that feeds primarily on green plants, or plant products such as seeds or nuts. Synonym: **primary consumer.**

**heterotroph,** adj. **heterotrophic.** Any organism that consumes organic matter as a source of energy.

**homologous structures.** Structures of different animals which, while outwardly appearing different, have many anatomical features in common, the wing of bat and the arm of human for example.

**hormones,** and **pheromones.** Natural chemical substances that control development, physiology, and/or behavior of an organism. Hormones are produced internally and affect only that individual. Pheromones are secreted externally and affect the behavior of other individuals of the same species, as, for example, attracting mates. Both hormones and pheromones are coming into use in pest control. See **natural chemical control.**

**host.** In feeding relationships, particularly parasitism, refers to the organism which is being fed upon, i.e., supporting the feeder.

**human ecosystem.** The system involving humans and their agricultural plants and animals.

**humidity.** The amount of water vapor in the air. See also **relative humidity.**

**humus.** A dark brown or black, soft, spongy residue of organic matter that remains after the bulk of dead leaves, wood, or other organic matter has decomposed. Humus does oxidize, but relatively slowly. It is extremely valuable in enhancing physical and chemical properties of soil.

**hybridization.** Cross-mating between two more or less closely related species.

**hydrocarbons.** *Chemistry:* Natural or synthetic organic substances that are composed mainly of carbon and hydrogen. Crude oil, fuels from crude oil, coal, animal fats, and vegetable oils are examples. *Pollution:* A wide variety of relatively small carbon-hydrogen molecules that result from incomplete burning of fuel, especially in internal combustion engines and that are emitted into the atmosphere through the exhaust. They are a prime factor in the formation of photochemical smog.

**hydroelectric dam.** A dam and associated reservoir used to produce electrical power by letting the high-pressure water behind the dam flow through and drive a turbogenerator.

**hydroelectric power.** Electrical power that is produced from hydroelectric dams or, in some cases, natural waterfalls.

**hydrogen bonding.** A weak attractive force which occurs between a hydrogen atom of one molecule and, usually, an oxygen atom of another molecule. It is responsible for holding water molecules together to produce the liquid and solid states.

**hydrological cycle.** See **water cycle.**

**hydroponics.** The culture of plants without soil. The method uses water with the required nutrients in solution.

**hypothesis.** An educated guess concerning the cause behind an observed phenomenon which is then subjected to experimental tests to prove its accuracy or inaccuracy.

**infiltration.** The process of water soaking into soil as opposed to its running off the surface.

**infiltration-runoff ratio.** The ratio between the amount of water soaking into the soil and that running off the surface. (The ratio is given by dividing the first amount by the second.)

**inorganic.** *Classical definition:* All things such as air, water, minerals, and metals that are neither living organisms nor products uniquely produced by living things. *Chemical definition:* All chemical compounds that do not contain carbon atoms as an integral part of their molecular structure. Contrast **organic.**

**inorganic fertilizer.** Also called chemical fertilizer; any inorganic compound or mixture of such compounds that will supply plant nutrients. Especially inorganic compounds of nitrogen, phosphorous, and potassium.

**insecticide.** Any chemical used to kill insects.

**integrated pest management (IPM).** Two or more methods of pest control carefully integrated into an overall program designed to avoid economic loss from pests. The objective is to minimize the use of environmentally hazardous, synthetic chemicals. Such chemicals may be used in IPM, but only as a last resort to prevent significant economic losses.

**inversion.** See **temperature inversion.**

**ion.** An atom or group of atoms that has lost or gained one or more electrons and consequently acquired a positive or negative charge. Ions are designated by + or − superscripts following the chemical symbol.

**ion exchange capacity** (soil). The property relating to the ability of a soil to bind, hold, and release plant nutrients.

**ionic bond.** The bond formed by the attraction between a positive and a negative ion.

**IPM.** See **integrated pest management.**

**irrigation.** Any method of artificially adding water to crops.

**isotope.** A form of an element in which the atoms have more (or less) than the usual number of neutrons. Isotopes of a given element have identical chemical properties, but they differ in mass (weight) as a result of the additional (or lesser) neutrons. Many isotopes are unstable and give off radioactive radiation. See **radioactive decay, radioactive emissions,** and **radioactive substances.**

**juvenile hormone.** The insect hormone sufficient levels of which preserve the larval state. Pupation required deminished levels; hence artificial applications of the hormone may block development.

**kerogen.** A hydrocarbon material contained in oil shale that vaporizes when heated and can be recondensed into a material similar to crude oil.

**kinetic energy.** The energy inherent in motion or movement including molecular movement (heat) and movement of waves, hence radiation including light.

**Lamarckianism.** After Lamarck. The false concept of inheritance of acquired characteristics being responsible for evolution.

**landfill.** A site where wastes (municipal, industrial, or chemical) are disposed of by burying them in the ground or placing them on the ground and covering them with earth.

**land subsidence.** The phenomenon of land gradually sinking in elevation. It may result from removing groundwater or oil, which is frequently instrumental in supporting the overlying rock and soil.

**larva,** pl. **larvae,** adj. **larval.** A free-living immature form that occurs in the life cycle of many organisms and that is structurally distinct from the adult. For example, caterpillars are the larval stage of moths and butterflies.

**law of conservation of matter.** In chemical reactions, atoms are neither created, changed nor destroyed. They are only rearranged.

**law of limiting factors.** Also known as Liebig's law of minimums. A system may be limited by the absence or minimum amount (in terms of that needed) of any required factor (see limiting factor).

**leaching.** The process of materials in or on the soil gradually dissolving and being carried by water seeping through the soil. It may result in valuable nutrients being removed from the soil, or it may result in wastes buried in the soil being carried into and contaminating groundwater.

**legumes.** The group of land plants that is virtually alone in its ability to fix nitrogen (see **nitrogen fixation**). The legume group includes such common plants as peas, beans, clovers, alfalfa, and locust trees but no major cereal grains.

**Liebig's law of minimums.** See **law of limiting factors.**

**life cycle.** The various stages of life, progressing from the adult of one generation to the adult of the next.

**limiting factor.** A factor primarily responsible for limiting the growth and/or reproduction of an organism or a population. The limiting factor may be a physical factor such as temperature or light, a chemical factor such as shortage of a particular nutrient, or a biological factor such as a competing species. The limiting factor may differ at different times and places.

**limitists.** Those who believe that our present economic-social system may be sharply curtailed in the future by resource shortages. Consequently, those who believe that changes in our system must be instigated now in order to accommodate future resource shortages. Contrast **cornucopians.**

**limits of tolerance.** The extremes of any factor, e.g., temperature, which an organism or a population can tolerate and still survive and reproduce.

**lipids.** A class of natural organic molecules that includes animal fats, vegetable oils, and phospholipids, the latter being an integral part of cellular membranes.

**litter.** In an ecosystem, the natural cover of dead leaves, twigs, and other dead plant material. This natural litter is subject to rapid decomposition and recycling in the ecosystem, whereas human litter, such as bottles, cans, and plastics is not.

**loam.** A soil consisting of a mixture of about 40 percent sand, 40 percent silt, and 20 percent clay.

**macromolecules.** Very large, organic molecules such as proteins and nucleic acids which comprise the structural and functional parts of cells.

**malnutrition.** Improper nutrition. It may consist of too much as well as too little of particular nutrients or calories, or it may be the lack of a proper balance among nutrients and calories.

**mariculture.** The propagation and/or rearing of any marine (saltwater) organism in more or less artificial systems.

**matter.** Anything that occupies space and has mass. Refers to any gas, liquid, or solid. Contrast with **energy.**

**maximum sustainable yield** (renewable resources). The maximum amount that can be taken year after year without depleting the resource. It is the maximum rate of use or harvest that will be balanced by the regenerative capacity of the system—as, for example, the maximum rate of tree cutting that can be balanced by tree regrowth.

**meltdown** (nuclear power). The event of a nuclear reactor getting out of control or losing its cooling water so that it melts from its own production of heat. The melted reactor would continue to produce heat and could melt its way out of the reactor vessel and eventually down into groundwater, where it would cause a violent eruption of steam which could spread radioactive materials over a wide area.

**metabolism.** The sum of all the chemical reactions that occur in an organism.

**methane.** A gas, $CH_4$. It is the primary constituent of natural gas. It is also produced as a byproduct of anaerobic respiration carried on by certain bacteria. Hence, it may be commercially produced by decomposing organic wastes.

**microbe.** A term used to refer to any microscopic organism. primarily bacteria, viruses, and protozoa.

**microclimate.** The actual conditions percieved by an organism in its particular location. Due to numerous factors such as shading, drainage and sheltering, the microclimate may be quite distinct from the overall climate.

**microfiltration** or **reverse osmosis.** A process for purifying water, in which water is forced under very high pressure through a membrane that filters out ions and molecules in solution.

**microorganism.** Any microscopic organism particularly bacteria, viruses, and protozoans.

**midnight dumping.** The wanton illicit dumping of materials, particularly hazardous wastes, frequently under the cover of darkness.

**Minamata disease.** A "disease" named for a fishing village in Japan where an "epidemic" was first observed. Symptoms, which included spastic movements, mental retardation, coma, death, and crippling birth defects in the next generation, were found to be the result of mercury poisoning.

**mineral.** Any hard, brittle, stonelike material that occurs naturally in the earth's crust. All consist of various combinations of positive and negative ions held together by ionic bonds. Pure minerals or crystals are one specific combination of elements. Common rocks are comprised of mixtures of two or more minerals. See also **ore.**

**mineralization** (soil science). The process of gradual oxidation of the organic matter (humus) present in soil which leaves just the gritty mineral component of the soil.

**mobilization.** In soil science, the bringing into solution of normally insoluble minerals. Presents a particular problem when the elements of such minerals have toxic effects.

**molecule.** A specific union of two or more atoms held together by covalent bonds. The smallest unit of a compound that still has the characteristics of that compound.

**monocropping.** The practice of growing the same crop year after year on the same land. Contrast **crop rotation.**

**monoculture.** The practice of growing a single crop over very wide areas as for example thousands of square kilometers of wheat, and only wheat, grown in the Midwest.

**municipal solid waste.** The entirety of refuse or trash generated by a residential and business community. The refuse that a municipality is responsible for collecting and disposing of, distinct from agricultural and industrial wastes.

**mutation.** A random change in one or more genes of an organism. Mutations may occur spontaneously in nature, but their number and degree are vastly increased by exposure to radiation and/or certain chemicals. Mutations generally result in a physical deformity and/or metabolic malfunction.

**mutualism.** Refers to a close relationship between two organisms in which both organisms benefit from the relationship.

**NASA.** National Aeronautics and Space Administration.

**natural** (adjective to describe a substance or factor). Occurring or produced as a normal part of nature apart from any activity or intervention of humans. Opposite of artificial, synthetic, human-made, or caused by humans.

**natural chemical control** (pest control). The use of one or more natural chemicals such as hormones or pheromones to control a pest.

**natural control methods** (pest control). Any of many techniques of controlling a pest population without resorting to the use of synthetic organic or inorganic chemicals. See **biological control, genetic control, cultural control, hormones and pheromones.**

**natural increase** (in populations). The number of births minus the number of deaths. It does not include immigration and emigration.

**natural laws.** Derivations from our observations that matter, energy and certain other phenomena apparently always act (or react) according to certain 'rules.'

**natural rate of change** (for a population). The percent of growth (or decline) of a population during a year. It is found by subtracting the crude death rate from the crude birth rate and changing the result to a percent. It does not consider immigration or emigration.

**natural selection.** The process whereby the natural factors of environmental resistance tend to eliminate those members of a population that are least well adapted to cope and thus, in effect, select those best adapted for survival and reproduction.

**NEPA.** National Environmental Policy Act.

**net yield** (resources). The amount of a resource obtained minus the amount that must be expended in mining, refining, and delivery of the resource to consumers.

**neutron.** A fundamental atomic particle found in the nuclei of atoms (except hydrogen) and having one unit of atomic mass but no electrical charge.

**niche** (ecological). The total of all the relationships that bear on how an organism copes with both biotic and abiotic factors it faces.

**nitrogen fixation** or **nitrogen fixing.** The process of chemically converting nitrogen gas ($N_2$) from the air into compounds such as nitrates ($NO_3^-$) or ammonia ($NH_3$) that can be used by plants in building amino acids and other nitrogen-containing organic molecules.

**nitrogen oxides.** A group of molecules all composed of various combinations of nitrogen and oxygen atoms. They are chemically symbolized as $NO_x$. Produced by internal combustion engines along with hydrocarbons, they are a primary factor in the formation of photochemical smog.

**NOAA.** National Oceanic and Atmospheric Administration.

**nonbiodegradable.** Not able to be consumed and/or broken down by biological organisms. Nonbiodegradable substances include plastics, aluminum, and many chemicals used in industry and agriculture. Particularly dangerous are nonbiodegradable chemicals that are also toxic and tend to accumulate in organisms. See **biodegradable, bioaccumulation.**

**nonpersistent** (with respect to pesticides and other chemicals). Chemicals that break down readily to harmless compounds, as, for example, natural organic compounds break down to carbon dioxide and water.

**nonrenewable resources.** Resources such as ores of various metals, oil, and coal that exist as finite deposits in the earth's crust and that are not replenished by natural processes as they are mined.

**no-till agriculture.** The farming practice in which weeds are killed with chemicals (or other means) and seeds are planted and grown without resorting to plowing or cultivation. The practice is very effective in reducing soil erosion.

**nuclear power.** Electrical power that is produced by using a nuclear reactor to boil water and produce steam which, in turn, drives a turbogenerator.

**Nuclear Regulatory Commission.** Independent governmental body charged with assuring and upholding safety standards for nuclear power plants.

**nuclear winter.** A pronounced global cooling that would occur as a result of a large-scale nuclear conflict. It is based on theoretical projections concerning the amount of dust and smoke that would be ejected into the atmosphere and the resulting decrease in solar radiation.

**nucleic acids.** The class of natural organic macromolecules that function in the storage and transfer of genetic information.

**nucleus.** *Physics:* The central core of atoms, which is made up of neutrons and protons. Electrons surround the nucleus. *Biology:* The large body contained in most living cells that contains the genes or hereditary material, DNA.

**nutrient.** *Plant:* An essential element in a particular ion or molecule that can be absorbed and used by the plant. For example, carbon, hydrogen, nitrogen, and phosphorus are essential elements; carbon dioxide, water, nitrate ($NO_3^-$), and phosphate ($PO_4^{-3}$) are respective nutrients. *Animal:* Materials such as protein, vitamins, and minerals that are required for growth, maintenance, and repair of the body and also materials such as carbohydrates that are required for energy.

**nutrient cycles.** The repeated pathway of particular nutrients or elements from the environment through one or more organisms back to the environment. Nutrient cycles include the carbon cycle, the nitrogen cycle, the phosphorus cycle, and so on.

**nutrient-holding capacity.** The capacity of a soil to bind and hold nutrients (fertilizer) against their tendency to be leached from the soil.

**observations.** Things or phenomena which are perceived through one or more of the basic five senses in their normal state. In addition, to be accepted as factual, the observations must be verifiable by others.

**oil shale.** A natural sedimentary rock that contains a material, kerogen, which can be extracted and refined into oil and oil products.

**omnivore.** An animal that feeds more or less equally on both plant material and other animals.

**OPEC.** Organization of Petroleum Exporting Countries.

**optimum.** The condition or amount of any factor or combination of factors that will produce the best result. For example, the amount of heat, light, moisture, nutrients, and so on that will produce the best growth. Either more or less than the optimum is not as good.

**optimum population** (resources). The population that will provide the maximum sustainable yield. The yield is reduced at higher or lower populations.

**ore.** A mineral rich in a particular element such as iron, aluminum, or copper which can be economically mined and refined to produce the desired metal. High-grade ore contains a relatively high percentage and low-grade ores contain a relatively low percentage of the desired element.

**organic.** *Classical definition:* All living things and products that are uniquely produced by living things, such as wood, leather, and sugar. *Chemical definition:* All chemical compounds, natural or synthetic, that contain carbon atoms as an integral part of their structure. Contrast **inorganic.**

**organic fertilizer.** Any natural organic material such as manure and other agricultural crop wastes that may be applied to soils to increase their fertility.

**organic gardening or farming.** Gardening or farming without the use of inorganic fertilizers, synthetic pesticides, or other human-made materials.

**organism.** Any living thing—plant, animal, or microbe.

**osmosis.** The phenomenon of water diffusing through a semipermeable membrane toward where there is more material in solution. (Where there is more material in solution there is a relatively lower concentration of water.) Has particular application regarding salinization of soils where plants are unable to grow because of osmotic water loss.

**overgrazing.** The phenomenon of animals grazing in greater numbers than the land can support in the long run. There may be a temporary economic gain in the short run, but the grassland (or other ecosystem) is destroyed and its ability to support life in the long run is vastly diminished.

**overreproduction.** Refers to the observation that all species produce far more offspring, eggs, or seeds than would seem to be necessary to sustain their population.

**oxidation.** A chemical reaction process that generally involves breakdown through combining with oxygen. Both burning and cellular respiration are examples of oxidation. In both cases, organic matter is combined with oxygen and broken down to carbon dioxide and water.

**ozone.** A gas, $O_3$, which is a pollutant in the lower atmosphere, but necessary to screen out ultraviolet radiation in the upper atmosphere. May also be used for disinfecting water.

**ozone shield.** The layer of ozone gas ($O_3$) in the upper atmosphere that screens out harmful ultraviolet radiation from the sun.

**PANs (peroxyacetylnitrates).** A group of compounds present in photochemical smog which are extremely toxic to plants and irritating to eyes, nose, and throat membranes of humans.

**parasites.** Organisms (plant, animal, or microbial) that attach themselves to another organism, the host, and feed on it over a period of time without killing it immediately but usually doing harm to it. Commonly divided into *ectoparasites,* those that attach to the outside and *endoparasites,* those that live inside their hosts.

**parent material.** The rock material, the weathering and gradual breakdown of which is the source of the mineral portion of soil.

**parts per million (ppm).** A frequently used expression of concentration. It is the number of units of one substance present in a million units of another. For example, one gram of phosphate dissolved in one million grams (= 1 ton) of water would be a concentration of 1 ppm.

**pathogen,** adj. **pathogenic.** An organism, usually a microbe, that is capable of causing disease.

**PCBs (polychlorinated biphenyls).** A group of very widely used industrial chemicals of the chlorinated hydrocarbon class. They have become very serious and widespread pollutants, contaminating most food chains on earth, because they are extremely resistant to breakdown and are subject to bioaccumulation. They are known to be carcinogenic.

**percolation.** The process of water seeping through cracks and pores in soil or rock.

**persistent** (with respect to pesticides or other chemicals). Nonbiodegradable and very resistant to breakdown by other means. Such chemicals therefore remain present in the environment more or less indefinitely.

**pesticide.** A chemical used to kill pests. Pesticides are further categorized according to the pests they are designed to kill—for example, herbicides to kill plants, insecticides to kill insects, fungicides to kill fungi, and so on.

**petrochemical.** A chemical made from petroleum (crude oil) as a basic raw material. Petrochemicals include plastics, synthetic fibers, synthetic rubber, and most other synthetic organic chemicals.

**pH.** Scale used to designate the acidity or basicity (alkalinity) of solutions or soil. pH 7 is neutral; values decreasing from 7 indicate increasing acidity; values increasing from 7 indicate increasing basicity. Each unit from 7 indicates a tenfold increase over the preceding unit.

**pheromone.** A chemical substance secreted by certain members which affect the behavior of other members of the population. The most common examples are sex attractants which female insects secrete to attract males.

**phosphate.** An ion composed of a phosphorus atom with four oxygen atoms attached. $PO_4^{3-}$. It is an important plant nutrient. In natural waters it is frequently the limiting factor. Therefore, additions of phosphate to natural water are frequently responsible for algal blooms.

**photochemical smog.** The brownish haze that frequently

forms on otherwise clear sunny days over large cities with significant amounts of automobile traffic. It results largely from sunlight-driven chemical reactions among nitrogen oxides and hydrocarbons, both of which come primarily from auto exhausts.

**photosynthesis.** The chemical process carried on by green plants through which light energy is used to produce glucose from carbon dioxide and water, and oxygen is released as a byproduct.

**photosynthetic organism.** An organism capable of carrying on photosynthesis. Opposite: nonphotosynthetic.

**photovoltaic cells.** (See **solar cells**)

**phytoplankton.** Any of the many species of algae that consist of single cells or small groups of cells that live and grow freely suspended in the water near the surface. Given abundant nutrients, they may become so numerous as to give the water a green "pea soup" appearance and/or form a thick green scum over the surface.

**plankton,** adj. **planktonic.** Any and all living things that are found freely suspended in the water and that are carried by currents as opposed to being able to swim against currents. It includes both plant **(phytoplankton)** and animal **(zooplankton)** forms.

**plant community.** The array of plant species, including numbers, ages, distribution, etc., that occupies a given area.

**pollutant.** A substance the presence of which is causing pollution.

**pollution.** Contamination of air, water, or soil with undesirable amounts of material or heat. The material may be a natural substance, such as phosphate, in excessive quantities, or it may be very small quantities of a synthetic compound such as dioxin which is exceedingly toxic.

**polyculture.** The growing of two or more species together in contrast to monoculture, the usual practice of growing only one species in a field.

**population.** A group within a single species, the individuals of which can and do freely interbreed. Breeding between populations of the same species is less common because of differences in location, culture, nationality, and so on.

**population explosion.** The exponential increase observed to occur in a population when or if conditions are such that a large percentage of the offspring are able to survive and reproduce in turn. Frequently leads, in turn, to over-exploitation, upset, and eventually collapse of the ecosystem.

**population profile.** A bar graph that shows the number of individuals at each age or in each five-year age group.

**population structure.** Refers to the proportion of individuals in each group. For example, is the population predominantly made up of young people, old people, or a more or less even distribution of young and old?

**potential energy.** The ability to do work that is stored in some chemical or physical state. For example, gasoline is a form of potential energy; the ability to do work is stored in the chemical state and is released as the fuel is burned in an engine.

**ppm.** See **parts per million.**

**practical availability** (nonrenewable resources). The fraction of a material such as copper that can be obtained from the earth's crust given restraints of unacceptable pollution, expenditure of energy, and other costs associated with production from low-grade ores or remote sources.

**precipitation.** Any form of moisture condensing in the air and depositing on the ground.

**predator.** An animal that feeds upon another.

**predator-prey relationship.** A feeding relationship existing between two kinds of animals. The predator is the animal feeding upon the prey. Such relationships are frequently instrumental in controlling populations of herbivores.

**pretreatment** (of sewage). Passing sewage water through a coarse screen to remove large pieces of debris and slowing the water flow enough to let coarse grit and sand settle out.

**prey.** In a feeding relationship, the prey is the animal that is killed and eaten by the other.

**primary consumer.** An organism such as a rabbit or deer that feeds more or less exclusively on green plants or their products such as seeds and nuts. Synonym: **herbivore.**

**primary standard.**

**primary succession.** See **succession.**

**primary treatment** (of sewage). The process that follows pretreatment. It consists of passing the water very slowly through a large tank, which permits 30–50 percent of the organic material in the water to settle out. The settled material is raw sludge.

**producers.** In an ecosystem, those organisms, mostly green plants, that use light energy to construct their organic constituents from inorganic compounds.

**profligate growth.** Growth characterized by extravagant and wasteful use of resources.

**protein.** The class of organic macromolecules that is the major structural component of all animal tissues and that function as enzymes in both plants and animals.

**proton.** Fundamental atomic particle with a positive charge, found in the nuclei of atoms. The number of protons present equals the atomic number and is distinct for each element.

**protozoan,** pl. **protozoa.** Any of a large group of microscopic organisms that consist of a single, relatively large complex cell or in some cases small groups of cells. All have some means of movement. Amoebae and paramecia are examples.

**proven reserves.** See **reserves.**

**puddling** (soil science). The phenomenon of bare soil being beaten down and compacted by falling rain. It reduces soil aeration and infiltration, and this reduces the ability of the soil to support plants.

**radioactive decay.** The reduction of radioactivity that occurs as an unstable isotope (radioactive substance) gives off radiation and becomes stable.

**radioactive emissions.** Any of various forms of radiation and/or particles that may be given off by unstable isotopes. Many such emissions have very high energy and may destroy biological tissues or cause mutations leading to cancer or birth defects.

**radioactive substances.** Substances that are or that contain unstable isotopes and that consequently give off radioactive emissions. See **isotope, radioactive emissions.**

**radioactive wastes.** Waste materials which are or contain or are contaminated with radioactive substances. Many materials used in the nuclear industry become wastes because of their contamination with radioactive substances.

**rain shadow.** The low-rainfall region that exists on the leeward (downwind) side of mountain ranges. It is the result of the mountain range causing the precipitation of moisture on the windward side.

**range of tolerance.** The range of conditions over which an organism or population can survive and reproduce, as, for example, the range from the highest to lowest temperature which can be tolerated. Within the range of tolerance is the optimum or best condition.

**raw sludge.** The untreated organic matter that is removed from sewage water by letting it settle. It consists of organic particles from feces, garbage, paper, and bacteria.

**recharge area.** With reference to groundwater, the area over which infiltration and resupply of a given aquifer occurs.

**recruitment.** With reference to populations, the maturation and entry of young into the adult breeding population.

**recycling.** The practice of processing wastes and using them as raw material for new products as, for example, scrap iron is remelted and made into new iron products. Contrast **reuse.**

**relative humidity.** The measure in percent of how much moisture is in the air compared to how much the air can hold at the given temperature.

**replacement capacity** (biological resources). The capacity of a system to recovery to its original state after a harvest or other form of use.

**replacement fertility.** The fertility rate that will just replace the reproducing population, given the average survivorship from birth through the preproductive period. Replacement fertility equals 2 divided by the survivorship from birth through reproductive years.

**reproductive rate.** Refers to the rate at which offspring, eggs, or seed are produced.

**reserves.** With reference to a mineral resource. The amount remaining in the earth that can be exploited using current technologies and at current prices. Usually given as *proven reserves,* those which have been positively identified, and *estimated reserves,* those which have not yet been discovered but which are estimated to exist.

**respiration, cellular respiration.** The chemical process through which organic molecules are broken down in cells to release energy required for the functioning of the cell. In most organisms, respiration involves the breakdown of glucose in the presence of oxygen and the release of carbon dioxide and water as waste products. Certain microbes are able to derive sufficient energy from the partial breakdown of organic molecules which can occur in the absence of oxygen, and in this case other waste products result. See **fermentation** and **anaerobic respiration.**

**resurgence** (with respect to populations, especially of pests). The rapid comeback of a population after a severe dieoff, especially populations of pest insects, after being largely killed off by pesticides, returning to even higher levels than before the treatment.

**reuse.** The practice of reusing items as opposed to throwing them away and producing new items, as, for example, bottles can be collected and refilled. Compare **recycling.**

**risk analysis or risk assessment.** A study seeking to identify the exact nature and probability of risks involved in pursuing any particular project, or policy.

**rivulet erosion.** A treelike pattern of numerous tiny (less than 15 cm) gullies caused by water erosion.

**runoff.** That portion of precipitation which runs off the surface as opposed to soaking in.

**salinization.** The process of soil becoming more and more salty until finally the salt prevents the growth of plants. It is caused by irrigation because salts brought in with the water remain in the soil as the water evaporates.

**saltwater intrusion, saltwater encroachment.** The phenomenon of seawater moving back into aquifers or estuaries. It occurs when the normal outflow of fresh water is diverted or removed for use.

**sand** Mineral particles 0.2-2.0 mm in diameter.

**scientific method.** The methodology by which scientific information is generated. Involves observations, formulating specific questions and hypotheses regarding the question's answer, then testing the hypotheses through experimentation.

**SCS.** U.S. Department of Agriculture—Soil Conservation Service.

**secondary consumer.** An organism such as a fox or coyote that feeds more of less exclusively on other animals that feed on plants.

**secondary pest outbreak.** The phenomenon of a small, and therefore harmless, population of a plant-eating insect suddenly exploding to become a serious pest problem.

**secondary succession.** See **succession.**

**secondary treatment** (of sewage). Also called biological treatment. A process that follows primary treatment. Any of a variety of systems that remove most of the remaining organic matter by enabling organisms to feed on it and oxidize it through their respiration. Trickling filters and activated sludge systems are the most commonly used methods.

**secured landfill.** A landfill with suitable barriers, leachate drainage and monitoring systems such that there is deemed adequate security against hazardous wastes in the landfill contaminating groundwater.

**sediment.** Soil particles, namely sand, silt, and clay, being carried by flowing water. The same material after it has been deposited. Because of different rates of settling, deposits are generally pure sand, silt, or clay.

**sediment trap.** A device for trapping sediment and holding it on a development or mining site.

**sedimentation.** The filling-in of lakes, reservoirs, stream channels, and so on with soil particles, mainly sand and silt. The soil particles come from erosion, which generally results from poor or inadequate soil conservation practices in connection with agriculture, mining, and/or development. Also called **siltation.**

**selection pressure.** With reference of evolution, an environmental factor which results in individuals with certain traits, which are not the norm for the population, to survive and reproduce more than the rest of the population and this results in a shift in the genetic makeup of the population. For example, the presence of insecticides provides a selection pressure toward increasing pesticide resistance in the pest population.

**selective cutting.** The cutting only of particular trees in a forest. Contrast **clearcutting.**

**sex attractant.** A natural chemical substance (pheromone) secreted by the female of many insect species which serves to attract males for the function of mating. Sex attractants may be used in traps or by the **confusion technique** to aid in the control of insect pests.

**sexual reproduction.** Reproduction involving segregation and recombination of chromosomes such that the offspring bear some combination of genetic traits from the parents. Contrast with asexual reproduction where all the offspring are exact genetic copies of the parent.

**sheet erosion.** The loss of a more or less even layer of soil from the surface due to the impact and runoff from a rainstorm.

**silt.** Soil particles between the size of sand particles and clay particles; namely, particles 0.002 to 0.2 mm in diameter.

**siltation.** See **sedimentation.**

**slash-and-burn agriculture.** The practice, commonly exercised throughout tropical regions, of cutting and burning jungle vegetation to make room for agriculture. The process is highly destructive of soil humus and may lead to rapid degradation of soil.

**smog.** See **photochemical smog.**

**Snow White syndrome.** The human attribute that expects or demands cosmetic perfection in produce even though this is often obtained only by using huge quantities of environmentally hazardous pesticides.

**soft water.** Water with little or no calcium, magnesium, or other ions in solution that will cause soap to precipitate (form a curd that makes a "ring around the bathtub").

**soil erosion.** The loss of soil due to particles being carried away by wind and/or water.

**soil structure.** The phenomenon of soil particles (sand, silt, and clay) being loosely stuck together to form larger clumps and aggregates, generally with considerable air spaces in between. Structure enhances infiltration and aeration. It develops as a result of organisms feeding on organic matter in and on the soil.

**solar cells.** Technically known as photovoltaic cells, devices which enable a direct conversion of light energy to electrical energy. Generally constructed from a "sandwich" of two thin layers of silica which are treated such that light striking one causes electrons to fall to the other.

**solid waste.** The total of materials that are discarded as "trash" and handled as solids, as opposed to those that are flushed down sewers and handled as liquids.

**solubility.** The degree to which a substance will dissolve and enter into solution.

**solution.** A mixture of molecules (or ions) of one material in another. Most commonly, molecules of air and/or ions of various minerals in water. For example, seawater contains salt in solution.

**specialization.** With reference to evolution, the phenomenon of species becoming increasingly adapted to exploit one particular niche but, thereby, becoming less able to exploit other niches.

**speciation.** The evolutionary process whereby populations of a single species separate and, through being exposed to differing forces of natural selection, gradually develop into distinct species.

**species.** All the organisms (plant, animal, or microbe) of a single kind. The "single kind" is determined by similarity of appearance and/or by the fact that members do or potentially can mate together and produce fertile offspring. Physical, chemical, or behavioral differences block breeding between species.

**splash erosion.** The destruction and compaction of soil structure that results from rainfall impacting bare soil.

**springs.** Natural exits of groundwater.

**standards** (air or water quality). Set by the federal or state governments, they are the maximum levels of various pollutants that are to be legally tolerated. If levels go above the standards, various actions may be taken.

**standing biomass.** That portion of the biomass (population) which is not available for consumption but which must be conserved to maintain the productive potential of the population.

**starvation.** The failure to get enough calories to meet energy needs over a prolonged period of time. It results in a wasting away of body tissues until death occurs.

**sterile male technique** (pest control). Saturating the area of infestation with sterile males of the pest species that have been artifically reared and sterilized by radiation. Matings between normal females and sterile males render the eggs infertile.

**stomas.** Microscopic pores in leaves, mostly in the undersurface, that allow the passage of carbon dioxide and oxygen into and out of the leaf and that also permit the loss of water vapor from the leaf.

**stormwater.** In cities, the water that results directly from rainfall as opposed to municipal water and sewage water piped to and from homes, offices, and so on. The extensive hard surfacing in cities creates a vast amount of stormwater runoff, which presents a significant management problem.

**stormwater management.** Policies and procedures for handling stormwater in acceptable ways to reduce the problems of flooding and erosion of stream banks.

**stormwater retention reservoirs.** Reservoirs designed to hold stormwater temporarily and let it drain away slowly in order to reduce problems of flooding and stream bank erosion.

**strip cropping.** The practice of growing crops in strips alternating with grass (hay) at right angles to prevailing winds or slopes in order to reduce erosion.

**strip mining.** The mining procedure in which all the earth covering a desired material such as coal is stripped away with huge power shovels in order to facilitate removal of the desired material.

**subsoil.** In a natural situation, the soil beneath topsoil. In contrast to topsoil, subsoil is compacted and has little or no humus or other organic material, living or dead. In many cases, topsoil has been lost or destroyed as a result of erosion or development and subsoil is at the surface.

**succession** (ecological). The gradual, or sometimes rapid, change in species that occupy a given area, some species invading and becoming more numerous while others decline in population and disappear. Succession is caused by a change in one or more abiotic or biotic factors benefitting some species but at the expense of others. *primary succession*. The gradual establishment, through a series of stages, of a climax ecosystem in an area that has not been occupied before, e.g., a rock face. *secondary succession*. The reestablishment, through a series of stages, of a climax ecosystem in an area from which it was previously cleared.

**superfund.** The popular name for the Comprehensive Environmental Response, Compensation, and Liability Act of 1980. This Act provides the mechanism and funding for the cleanup of potentially dangerous hazardous waste sites.

**surface water.** Includes all bodies of water, lakes, rivers, ponds, etc., that are on the surface of the earth in contrast to groundwater which lies below the surface.

**survival of the fittest.** The concept that individuals best adapted to cope with both biotic and abiotic factors in their environment are the "fittest" and most likely to survive and reproduce.

**survival rate.** See **survivorship**.

**survivorship.** The proportion of individuals in a specified group alive at the beginning of an interval, e.g., a five-year period, who survive to the end of the period.

**survivorship graph.** A graph that shows the probability of an individual's surviving from birth to any particular age. Survivorship graphs are constructed (and may differ) for specific groups, such as sex, race, nationality, economic status, and so on.

**suspension.** With reference to materials contained in or being carried by water, materials kept 'afloat' only by the water's agitation and which settle as the water becomes quiet.

**symbiosis.** The intimate living together or association of two kinds of organisms, especially in a way that provides a mutual benefit to both organisms.

**synergism, synergistic effect, synergistic interactions.** The phenomenon in which two factors acting together have a very much greater effect than would be indicated by the sum of their effects separately—as, for example, modest doses of certain drugs in combination with modest doses of alcohol may be fatal.

**synfuels, synthetic fuels.** Fuels similar or identical to those that come from crude oil and/or natural gas, produced from coal, oil shale, or tar sands.

**synthetic.** Human-made as opposed to being derived from a natural source. For example, synthetic organic compounds are those produced in chemical laboratories, whereas natural organic molecules are those produced by organisms.

**synthetic organic chemicals.** Human-made chemical compounds with a structure based on carbon atoms but different from those found to occur in living organisms.

**taxonomy.** The science of identification and classification of organisms according to evolutionary relationships.

**technological assessment.** A study aimed at projecting the environmental and social impacts of introducing a new technology into a less-developed nation. Choices between alternative development projects may be made accordingly.

**technology.** The application of scientific information to solve practical problems or achieve desired goals.

**temperature inversion.** The weather phenomenon in which a layer of warm air overlies cooler air near the ground and prevents the rising and dispersion of air pollutants.

**terracing.** The practice of grading sloping farmland into a series of steps and cultivating only the level portions in order to reduce erosion.

**territoriality.** The behavioral characteristic seen in many animals, especially birds and mammalian carnivores, to mark and defend a given territory against other members of the same species.

**tertiary treatment.** Third stage of sewage treatment, following primary and secondary, designed to remove one or more of the nutrients, usually nitrogen and/or phosphate, from the wastewater. Necessary to reduce the problem of eutrophication.

**test group.** Synonym for experimental group. See **experimental group.**

**texture.** With reference to soils, the sizes of the particles, sand, silt, and/or clay, which make up the mineral portion.

**theory.** A conceptual formulation which provides a rational explanation or framework for numerous related observations.

**thermal pollution.** The addition of abnormal and undesirable amounts of heat to air or water. It is most significant with respect to discharging waste heat from electric generating plants, especially nuclear power plants, into bodies of water.

**threatened species.** A species the population of which is declining precipitously because of direct or indirect human impacts.

**threshold level.** The maximum degree of exposure to a pollutant, drug, or other factors that can be tolerated with no ill effect. The threshold level will vary depending on the species, the sensitivity of the individual, the time of exposure, and the presence of other factors that may produce synergistic effects.

**topsoil.** The surface layer of soil, which is rich in humus and other organic material, both living and dead. As a result of the activity of organisms living in the topsoil, it generally has a loose, crumbly structure as opposed to being a compost mass. In many cases, because of erosion, development, or mining activity, the topsoil layer may be absent.

**total fertility rate.** The average number of children that would be born alive to each woman during her total reproductive years, assuming she follows the average fertility at each age.

**total watershed planning.** A consideration of the entire watershed and planning development and other activities so as to maintain the overall water flow characteristics of the area.

**trace elements.** Those essential elements that are needed in only very small or trace amounts.

**tragedy of the commons.** The overuse or overharvesting and consequent depletion and/or destruction of a renewable resource that tends to occur when the resource is treated as a commons, that is, when it is open to be used or harvested by any and all with the means to do so.

**transpiration.** The loss of water vapor from plants. Evaporation of water from cells within the leaves and exiting through stomas.

**trickling filters.** Systems where wastewater trickles over rocks or a framework coated with actively feeding microorganisms. The feeding action of the organisms in a well-aerated environment results in the decomposition of organic matter. Used in secondary or biological treatment of sewage.

**tritium ($^3$H).** An unstable isotope of hydrogen that contains two neutrons in addition to the usual single proton in the nucleus. It does not occur in significant amounts naturally but is human-made.

**trophic level.** Feeding level with respect to the primary source of energy. Green plants are at the first trophic level; primary consumers at the second, secondary consumers at the third, and so on.

**turbine.** A sophisticated "paddle wheel" driven at a very high speed by steam, water, or exhaust gases from combustion.

**turbogenerator.** A turbine coupled to and driving an electric generator. Virtually all commercial electricity is produced by such devices. The turbine is driven by gas, steam, or water.

**ultraviolet radiation.** Radiation similar to light but with wavelengths slightly shorter than violet light and with more energy. The greater energy causes it to severely burn and otherwise damage biological tissues.

**upset** (ecological). A vast shift in the relative size of one or more populations within an ecosystem. It may result in tremendous damage or even total destruction of the original ecosystem. It may be caused by any factor that changes the normal balances between species, such as introduction of a foreign species or eliminating a predator that was instrumental in controlling the population of an herbivore. Especially the phenomenon of an economically insignificant insect becoming a serious threat when its predators are killed off by pesticide treatment.

**urban decay.** General deterioration of structures and facilities such as buildings and roadways, and also the decline in quality of services such as education, that has occurred in inner city areas as growth has been focused on suburbs and exurbs.

**urban sprawl.** The rapid expansion of metropolitan areas through building housing developments and shopping centers farther and farther from urban centers and lacing them together with more and more major highways. Widespread development that has occurred without any overall land-use plan.

**vestigial organs.** A rudimentary anatomical part or organ which has no apparent function for the organism in which it is found but which is highly developed and has a distinct function in related species.

**USDA.** U.S. Department of Agriculture.

**vitamin.** A specific organic molecule that is required by the body in small amounts but that cannot be made by the body and therefore must be present in the diet.

**water balance.** Refers to the capacity and necessity of all organisms to control the relative volume of water inside vs. outside their cells.

**water cycle.** The movement of water from points of evaporation through the atmosphere, through precipitation, and through or over the ground, returning to points of evaporation.

**water-holding capacity** (soil). The property of a soil relating to its ability to hold water so that it will be available to plants.

**watershed.** The total land area that drains directly or indirectly into a particular stream or river. The watershed is generally named from the stream or river into which it drains.

**water table.** The upper surface of groundwater. It rises and falls with the amount of groundwater.

**water vapor.** Water molecules in the gaseous state.

**weathering.** The gradual breakdown of rock into smaller and smaller particles, caused by natural, chemical, physical, and biological factors.

**wetlands.** Areas that are constantly wet and are flooded at more or less regular intervals. Especially marshy areas along coasts that are regularly flooded by tide.

**work** (physics). Any change in motion or state of matter. Any such change requires the expenditure of energy.

**workability.** With reference to soils, the relative ease with which a soil can be cultivated.

# INDEX